# ADVANCES IN HYPERSPECTRAL IMAGE PROCESSING TECHNIQUES

**Edited by**

**Chein-I Chang**
*Remote Sensing Signal and Image Processing Laboratory,*
*Department of Computer Science and Electrical Engineering,*
*University of Maryland Baltimore County, Baltimore, MD, USA*

WILEY

IEEE PRESS

This edition first published 2023


*Registered Office*
John Wiley & Sons, Inc., 111 River Street, Hoboken, NJ 07030, USA

*Editorial Office*
111 River Street, Hoboken, NJ 07030, USA

For details of our global editorial offices, customer services, and more information about Wiley products visit us at www.wiley.com.

Wiley also publishes its books in a variety of electronic formats and by print-on-demand. Some content that appears in standard print versions of this book may not be available in other formats.



*Library of Congress Cataloging-in-Publication Data*
Name: Chang, Chein-I, editor.
Title: Advances in hyperspectral image processing techniques / edited by Chein-I Chang.
Description: Hoboken, NJ : Wiley, 2023. | Includes bibliographical references and index.
Identifiers: LCCN 2022022693 (print) | LCCN 2022022694 (ebook) | ISBN 9781119687764 (cloth) | ISBN 9781119687757 (adobe pdf) | ISBN 9781119687771 (epub)
Subjects: LCSH: Hyperspectral imaging.
Classification: LCC TR267.73 .A38 2023 (print) | LCC TR267.73 (ebook) | DDC 771–dc23/eng/20220829
LC record available at https://lccn.loc.gov/2022022693
LC ebook record available at https://lccn.loc.gov/2022022694

Cover Design: Wiley
Cover Image: © Shutterstock/ASVMAGZ

Set in 10/12pt Times by Straive, Pondicherry, India.

# Contents

# Editor Biography

**Chien-I Chang** received his Ph.D. degree in electrical engineering from the University of Maryland, College Park, and has been with the University of Maryland Baltimore County (UMBC), since 1987 where he is a professor in the Department of Computer Science and Electrical Engineering. Dr. Chang authored four books: *Hyperspectral Imaging: Techniques for Spectral Detection and Classification* published by Kluwer Academic Publishers in 2003; *Hyperspectral Data Processing: Algorithm Design and Analysis*, John Wiley & Sons, 2013; *Real Time Progressive Hyperspectral Image Processing: Endmember Finding and Anomaly Detection* in 2016 by Springer; and *Recursive Hyperspectral Sample and Band Processing: Algorithm Architecture and Implementation*, Springer 2017. In addition, he has edited two books, *Recent Advances in Hyperspectral Signal and Image Processing*, 2006, and *Hyperspectral Data Exploitation: Theory and Applications*, and co-edited with A. Plaza a book on *High Performance Computing in Remote Sensing*, CRC Press, 2007. Dr. Chang is a life fellow of IEEE and a fellow of SPIE.

# List of Contributors

**Adam Bekit**, Remote Sensing Signal and Image Processing Laboratory, Department of Computer Science and Electrical Engineering, University of Maryland Baltimore County, Baltimore, MD, USA

**Chein-I Chang**, Center for Hyperspectral Imaging in Remote Sensing (CHIRS), Information and Technology College, Dalian Maritime University, Dalian, China
Remote Sensing Signal and Image Processing Laboratory, Department of Computer Science and Electrical Engineering, University of Maryland Baltimore County, Baltimore, MD, USA

**Shuhan Chen**, Department of Electrical Engineering, Zhejiang University, Hangzhou, China

**Charles J. Della-Porta**, Remote Sensing Signal and Image Processing Laboratory, Department of Computer Science and Electrical Engineering, University of Maryland Baltimore County, Baltimore, MD, USA

**Qian Du**, Department of Electrical and Computer Engineering, Mississippi State University, Starkville, MS, USA

**Lianru Gao**, Key Laboratory of Digital Earth Science, Aerospace Information Research Institute, Chinese Academy of Sciences, Beijing, China

**Chiru Ge**, School of Information Science and Engineering, Shandong Normal University, Jinan, China

**Zhu Han**, Key Laboratory of Digital Earth Science, Aerospace Information Research Institute, Chinese Academy of Sciences, Beijing, China
and
College of Resources and Environment, University of Chinese Academy of Sciences, Beijing, China

**Risheng Huang**, College of Electrical Engineering, Zhejiang University, Hangzhou, Zhejiang, China

**Bernard Lampe**, Remote Sensing Signal and Image Processing Laboratory, Department of Computer Science and Electrical Engineering, University of Maryland Baltimore County, Baltimore, MD, USA

**Jie Lei**, State Key Laboratory of Integrated Services Networks, Xidian University, Xi'an, China

**Hsiao-Chi Li**, Department of Electrical Engineering, National Taipei University of Technology, Taipei, Taiwan, Republic of China

**Jiaojiao Li**, State Key Laboratory of Integrated Services Networks, School of Telecommunications Engineering, Xidian University, Xian, China

**Jun Li**, Guangdong Provincial Key Laboratory of Urbanization and Geo-Simulation, School of Geography and Planning, Sun Yat-sen University, Guangzhou, China

**Tong Li**, Satellite Application Division, Shanghai Institute of Satellite Engineering, Shanghai, China

**Wei Li**, College of Information and Electronics, Beijing Institute of Technology, Beijing, China
and
School of Information and Electronics, Beijing Institute of Technology, Beijing, China

**Xiaorun Li**, College of Electrical Engineering, Zhejiang University, Hangzhou, Zhejiang, China

**Yunsong Li**, State Key Laboratory of Integrated Services Networks, Xidian University, Xi'an, China

**Kai Liu**, State Key Laboratory of Integrated Services Networks, Xidian University, Xi'an, China

**Xiaochen Lu**, School of Information Science and Technology, Donghua University, Shanghai, China

**Wenfei Luo**, School of Geographical Science, South China Normal University, Guangzhou, China

**Mingyuan Peng**, Land Satellite Remote Sensing Application Center, MNR, Beijing, China

**Shen-En Qian**, Canadian Space Agency, Saint-Hubert, Quebec, Canada

**Yuntao Qian**, College of Computer Science, Zhejiang University, Hangzhou, China

**Xiaodi Shang**, Center for Hyperspectral Imaging in Remote Sensing (CHIRS), School of Information and Technology, Dalian Maritime University, Dalian, China

**Meiping Song**, Center for Hyperspectral Imaging in Remote Sensing (CHIRS), Information and Technology College, Dalian Maritime University, Dalian, China

**Xu Sun**, Key Laboratory of Digital Earth Science, Aerospace Information Research Institute, Chinese Academy of Sciences, Beijing, China

**Xuejian Sun**, Hyperspectral Remote Sensing Application Division, Aerospace Information Research Institute, Chinese Academy of Sciences, Beijing, China

**Keyan Wang**, State Key Laboratory of Integrated Services Networks, Xidian University, Xi'an, China

**Lin Wang**, School of Physics and Optoelectronic Engineering, Xidian University, Xi'an, China

**Yulei Wang**, Center for Hyperspectral Imaging in Remote Sensing (CHIRS), School of Information and Technology, Dalian Maritime University, Dalian, China

**Weiying Xie**, State Key Laboratory of Integrated Services Networks, Xidian University, Xi'an, China

**Fengchao Xiong**, School of Computer Science and Engineering, Nanjing University of Science and Technology, Nanjing, China

**Bai Xue**, Department of Computer Science and Electrical Engineering, University of Maryland Baltimore County, Baltimore, MD, USA

**Minchao Ye**, College of Information Engineering, China Jiliang University, Hangzhou, China

**Chunyan Yu**, Center for Hyperspectral Imaging in Remote Sensing (CHIRS), Information and Technology College, Dalian Maritime University, Dalian, China

**Haoyang Yu**, Center for Hyperspectral Imaging in Remote Sensing (CHIRS), Information Science and Technology College, Dalian Maritime University, Dalian, China

**Bing Zhang**, Key Laboratory of Digital Earth Science, Aerospace Information Research Institute, Chinese Academy of Sciences, Beijing, China
and
College of Resources and Environment, University of Chinese Academy of Sciences, Beijing, China

**Junping Zhang**, School of Electronics and Information Engineering, Harbin Institute of Technology, Harbin, China

**Lifu Zhang**, Hyperspectral Remote Sensing Application Division, Aerospace Information Research Institute, Chinese Academy of Sciences, Beijing, China

**Xia Zhang**, Hyperspectral Remote Sensing Application Division, Aerospace Information Research Institute, Chinese Academy of Sciences, Beijing, China

**Liaoying Zhao**, School of Computer Science, Hangzhou Dianzi University, Hangzhou, Zhejiang, China

**Xiaoyang Zhao**, Hyperspectral Remote Sensing Application Division, Aerospace Information Research Institute, Chinese Academy of Sciences, Beijing, China
and
College of Natural Resources and Environment, University of Chinese Academy of Sciences, Beijing, China

**Shengwei Zhong**, School of Computer Science and Engineering, Nanjing University of Science and Technology, Nanjing, China

**Jun Zhou**, School of Information and Communication Technology, Griffith University, Nathan, QLD, Australia

**Lina Zhuang**, Department of Mathematics, Hong Kong Baptist University, Hong Kong, China

# Preface

Hyperspectral imaging (HSI) has been around more than three decades and also become a mature advanced remote sensing technology in many applications ranging from defense and intelligence, agriculture, environmental monitoring to food inspection and safety, and medical imaging. Since my first book, *Hyperspectral Imaging: Techniques for Spectral Detection and Classification*, was published by Kluwer Academic/Plenum Publishers in 2003, nearly 20 years have been passed. It has also been my 30 years of research devoted to HSI that has evidenced a rapid growth in remote sensing over the past years with tens of thousands of reports published every year in this fascinating area.

This book entitled "Advances in Hyperspectral Image Processing Techniques" does not intend to cover all the areas in hyperspectral imaging but rather addresses several key areas of interest in HSI based on my most recent research carried out in the Remote Sensing Signal and Image Processing Laboratory (RSSIPL) at the University of Maryland Baltimore County (UMBC) and the Center of Hyperspectral Imaging in Remote Sensing (CHIRS), Dalian Maritime University, Dalian, China. As a result, the topics of many chapters in this book reflect this nature and are new results that have not been discussed and covered in any of my previous books. To make this book a valuable and useful reference, all contributors are experts in their own areas and were invited to write book chapters sharing their research experiences and findings. In particular, all the contributors are either my long-time close friends, my colleagues in CHIRS, my former PhD students, post-doctors, visiting scholars, or exchanging PhD students. To this end, this book simply represents a very small portion of research in HSI and never tries to cover a full range of HSI areas.

Twenty chapters are included in this book and written by many renowned scholars and researchers in HSI who were cordially invited to make contributions. Many of them have established reputation in their own fields and have been kind to agree to write their own book chapters. All the book chapters can be categorized in six parts – Part I: General Theory, Part II: Band Selection for Hyperspectral Imaging, Part III: Compressive Sensing for Hyperspectral Imaging, Part IV: Fusion of Hyperspectral Imaging, Part V: Hyperspectral Data Unmixing, and Part VI: Hyperspectral Image Classification – each of which is summarized as follows.

## Part I: General Theory

Chapter 1: Introduction: Two Fundamental Principles Behind Hyperspectral Imaging
Chein-I Chang

This chapter explores two fundamental principles: pigeon-hole principle and orthogonality principle behind design and development of hyperspectral image processing algorithms in various applications such as pigeon-hole principle that finds its applications in determining the number of endmembers needed for spectral unmixing, spectrally distinct signatures by virtual dimensionality, orders of low rank and sparse representation/low rank and sparse matrix decomposition, and the number of bands to be selected for band selection and orthogonality principle which can be used as a criterion to derive the orthogonal subspace projection (OSP) technique in developing algorithms for finding endmembers, detecting subpixel targets, anomalies as well as OSP-based low rank and sparse matrix decomposition in target/anomaly detection, and OSP-based detection theory applied to classification.

Chapter 2: Overview of Hyperspectral Imaging Remote Sensing from Satellites
Shen-En Qian

This chapter offers a valuable overview of design and development hyperspectral imaging sensors currently operated in space and their future developments for space exploration.

Chapter 3: Efficient Hardware Implementation for Hyperspectral Anomaly and Target Detection
Jie Lei, Weiying Xie, Jiaojiao Li, Keyan Wang, Kai Liu, and Yunsong Li

This chapter introduces some real-time detectors including deep pipelined background statistics (DPBS)-constrained energy minimization (CEM), deep brief network (DBN)-Reed-Xiaoli anomaly detector (RXD), Fast-automatic target-generation process (ATGP), and fast-morphological reconstruction and the simplified guided filtering detector (MGD), implemented on FPGA considering the concept of model-based design, which overcomes the drawbacks of high difficulty and long period in the development of traditional hardware systems.

## Part II: Band Selection for Hyperspectral Imaging

Chapter 4: Constrained Band Selection for Hyperspectral Imaging
Chein-I Chang

This chapter reviews a particular type of band selection methods, constrained band selection that has two versions, band-constrained band selection (BCBS) and target-constrained band selection (TCBS), both of which are derived from the well-known subpixel target detection, constrained energy minimization (CEM), and linearly constrained minimum variance (LCMV) filter.

Chapter 5: Band Subset Selection for Hyperspectral Imaging
Chein-I Chang

This chapter extends band selection (BS) in Chapter 4 to band subset selection (BSS) in the sense that multiple bands are selected as a band subset rather than band by band selected sequentially as band selection does, particularly, single band-constrained BS (BCBS) to multiple band-constrained BSS (MBC-BSS) and single-target-constrained BS (TCBS) to multiple-target-constrained BSS (MTC-BSS).

Chapter 6: Progressive Band Selection Processing for Hyperspectral Image Classification
Chunyan Yu, Meiping Song, and Chein-I Chang

This chapter develops progressive band selection processing (PBSP) of hyperspectral image classification, which performs image classification stage-by-stage progressively in the sense that each stage performs HSIC by band subsets according to a $p$-ary Huffman coding tree constructed by class classification priority (CCP).

**Part III: Compressive Sensing for Hyperspectral Imaging**

Chapter 7: Restricted Entropy and Spectrum Properties for Hyperspectral Imaging
Chein-I Chang and Bernard Lampe

This chapter derives two new compressive sensing properties for hyperspectral imaging, restricted entropy property (REP) and restricted spectrum property (RSP), which can be shown to preserve the entropy of a hyperspectral signature vector and the spectral similarity between two hyperspectral signatures in both the original data space and compressively sensed band domain so that REP and RSP preserve exploitation algorithm performance without the need for decompression so as to avoid specifying a sparse basis.

Chapter 8: Endmember Finding in Compressively Sensed Band Domain
Chein-I Chang and Adam Bekit

This chapter presents a compressive sensing approach to N-FINDR that can find a $p$-vertex simplex with the maximal simplex volume found by Sequential/Successive N-FINDR with the restricted simplex volume property being proved to be preserved in compressively sensed band domain (CSBD).

Chapter 9: Hyperspectral Image Classification in Compressively Sensed Band Domain
Charles J. Della-Porta and Chein-I Chang

This chapter explores the viability of performing classification for hyperspectral data via compressive sensing where a mathematical analysis is derived to show that the classification error can be expressed in terms of the restricted isometry constant so that the hyperspectral image classification in the original data space can be achieved in compressively sensed band domain (CSBD).

**Part IV: Fusion for Hyperspectral Imaging**

Chapter 10: Hyperspectral and LiDAR Data Fusion
Qian Du, Wei Li, and Chiru Ge

This chapter reviews three deep learning based approaches to fusion of HSI and LiDAR data, Two-Branch CNN for Joint Classification, Hierarchical Random Walk Network (HRWN), and Residual Network-Based Probability Reconstruction Fusion (RNPRF), and conducts experiments to evaluate their comparative performance in land cover classification.

Chapter 11: Hyperspectral Data Fusion Using Multi-Dimensional Information
Lifu Zhang, Xia Zhang, Mingyuan Peng, Xuejian Sun, and Xiaoyang Zhao

This chapter selects six simple and easy-to-promote hyperspectral fusion methods to fuse multiple domestic satellite multispectral data with GF-5 hyperspectral data and conducts a comprehensive study on their advantages and disadvantages based on five classic evaluation indicators, classification application accuracy, and running time cost.

Chapter 12: Fusion of Band Selection Methods for Hyperspectral Imaging
Yulei Wang, Lin Wang, and Chein-I Chang

This chapter develops an approach for fusing different band selection methods where two versions are derived, called progressive band selection fusion (PBSF) and simultaneous band selection fusion (SBSF), to fuse multiple band subsets selected by different band selection methods.

**Part V: Hyperspectral Data Unmixing**

Chapter 13: Model-Inspired Deep Neural Networks for Hyperspectral Unmixing
Yuntao Qian, Fengchao Xiong, Minchao Ye, and Jun Zhou

This chapter introduces several model-inspired learning-based methods derived from different spectral unmixing models and their corresponding optimization algorithms with the prior knowledge of the unmixing model and optimization algorithm incorporated into the network architectures.

Chapter 14: Analytical Fully Constrained Least Squares Linear Spectral Mixture Analysis
Chein-I Chang and Hsiao-Chi Li

This chapter derives an analytical spectral unmixing technique, to be called analytical FCLS (AFCLS) that can find fully constrained least squares (FCLS) solutions in closed forms by Cramer's rule analytically so that the AFCLS-unmixed results using analytical solutions are more accurate than FCLS-unmixed resulting from numerical solutions.

Chapter 15: Swarm Intelligence Optimization-Based Spectral Unmixing
Lianru Gao, Xu Sun, Zhu Han, Lina Zhuang, Wenfei Luo, and Bing Zhang

This chapter develops swarm intelligence algorithms that can be implemented in three spectral mixing models, linear mixing model (LMM), normal compositional model (NCM), and non-linear mixing model (NLMM).

Chapter 16: Spectral-Spatial Robust Nonnegative Matrix Factorization for Hyperspectral Unmixing
Risheng Huang, Xiaorun Li, and Liaoying Zhao

This chapter develops a robust NMF using $l_{1,2}$ norm ($l_{1,2}$-RNMF) and further proposes a spectral-spatial robust NMF model (SSRNMF) by incorporating $l_{2,1}$ norm and $l_{1,2}$ norm to take advantage of both spatial dimension and spectral dimension into consideration and achieves robustness to band noise and pixel noise simultaneously.

**Part VI: Hyperspectral Image Classification**

Chapter 17: Sparse Representation-Based Hyperspectral Image Classification
Haoyang Yu, Jun Li, Wei Li, and Bing Zhang

This chapter conducts studies and proposes four new classification methods, locality-preserving sparse representation-based classification (LPSRC), locality-sensitive discriminant analysis-group sparse representation-based classification (LSDA-GSRC), activity degree-driven representation-based classification (ADRC), and neighborhood ADRC (NADRC) for sparse representation-based hyperspectral image classification.

Chapter 18: Collaborative Classification Based on Hyperspectral Images
Junping Zhang, Xiaochen Lu, and Tong Li

This chapter focuses on the hyperspectral-centered multi-source image collaborative classification methods and analyzes in depth the probable problems and challenges in multi-source image synergetic processing by fully exploring the complementary information of multi-source images.

Chapter 19: Class Feature-Weighted Hyperspectral Image Classification
Shengwei Zhong, Jiaojiao Li, Xiaodi Shang, Shuhan Chen, and Chein-I Chang

This chapter extends the commonly used hyperspectral image classification (HSIC) to class feature weighted HSIC where traditional classification measures, overall accuracy, and average accuracy are considered as special cases of several new introduced class feature measures to deal with imbalanced class and background issues.

Chapter 20: Target Detection Approaches to Hyperspectral Image Classification
Chein-I Chang, Bai Xue, and Chunyan Yu

This chapter shows that hyperspectral image classification can be in fact indeed interpreted by statistical detection theory to address background and imbalanced class issues where a 3D receiver operating characteristics (3D ROC) analysis is further used as evaluation tool for performance evaluation.

The ultimate goal of this book is to offer readers with a peek of the cutting-edge research in HSI. In particular, it is my belief that this book can provide a useful guide and assistance for practitioners and engineers who are interested in HSI. Hopefully, the chapters presented in this book have just achieved what this book would like to accomplish.

This book was originally planned to invite my close friends and colleagues, former PhD students, post-doctors, visiting scholars, and visiting exchanging PhD students to contribute to the book chapters to celebrate my seventieth birthday in 2020. It was scheduled to be published in 2020. Unfortunately, due to unexpected pandemic of COVID-19, the original schedule has been pushed back to 2022.

Last but not least, I would like to express my sincere gratitude to all the contributors for their efforts in providing their book chapters to complete this book project. There is no doubt that this book cannot be completed without their dedicated involvement.

Chein-I Chang
Center of Hyperspectral Imaging in Remote Sensing (CHIRS)
Information and Technology College
Dalian Maritime University
Dalian, China

Remote Sensing Signal and Image Processing Laboratory
Department of Computer Science and Electrical Engineering
University of Maryland Baltimore County
Baltimore, MD, USA

December 2021

# Part I

## General Theory

# 1

# Introduction: Two Fundamental Principles Behind Hyperspectral Imaging

Chein-I Chang
*Center for Hyperspectral Imaging in Remote Sensing (CHIRS), Information and Technology College, Dalian Maritime University, Dalian, China*
*Remote Sensing Signal and Image Processing Laboratory, Department of Computer Science and Electrical Engineering, University of Maryland Baltimore County, Baltimore, MD, USA*

## 1.1 Introduction

Hyperspectral imaging (HSI) has been around more than 30 years since the first maiden flight made by airborne visible/infrared imaging spectrometer (AVIRIS) in 1987 [1]. Due to its use of hundreds of contiguous spectral bands, HSI is capable of solving many issues that cannot be resolved by multispectral sensors which use only tens of discrete spectral bands. In particular, with very high spectral resolution provided by HSI, many subtle objects and material substances can be now uncovered and extracted by very narrow diagnostic bandwidths for detection, discrimination, classification, identification, recognition, and quantification including new applications still yet to be explored. Intuitively, many think that HSI is a natural extension to multispectral imaging (MSI). In fact, this is not true as discussed in [2, chapter 1] and will be also discussed in Sections 1.3 and 1.4 of this chapter. The early development of HSI arises from military applications where spatial domain information does not provide needed spectral information, generally referred to as nonliteral analysis, i.e. spectral analysis in [3] to distinguish it from literal analysis, referred to as commonly used spatial domain analysis. As a result, HSI and MSI have their own different applications. In early days, MSI has been used in remote sensing mainly for land cover/use classification, agriculture applications, disaster assessment and management, ecology, environmental monitoring, geology, geographical information system (GIS), etc. In these cases, low spectral resolution multispectral imagery may provide sufficient information for data analysis and the techniques developed for MSI are primarily derived from traditional two-dimensional spatial domain-based image processing which takes advantage of spatial correlation to perform various tasks. On the other hand, HSI has been developed primarily for problems which generally cannot be addressed by MSI such as subpixel target

*Advances in Hyperspectral Image Processing Techniques*, First Edition. Edited by Chein-I Chang.

detection, anomaly detection, endmember finding, and spectral unmixing Therefore, HSI has fundamental differences from MSI, specifically, two principles, pigeon-hole principle and orthogonality principle, which are new concepts found in HSI but not MSI. This chapter explores these two principles which lead to development of various HSI algorithms.

## 1.2 Why Is Hyperspectral Imaging?

In order to answer the question of "why is hyperspectral imaging?," we need to know what HSI can offer, while MSI cannot. The following example available on the website Max Larin, XIMEA provides the answer [4].

There are five different types of leaves, fake paper leaves, fake shop leaves, real leaves, dead leaves, and plastic leaves, with their respective spectra shown in Figure 1.1(a), multispectral RGB and hyperspectral images in Figure 1.1(b), and their classification results in Figure 1.1(c).

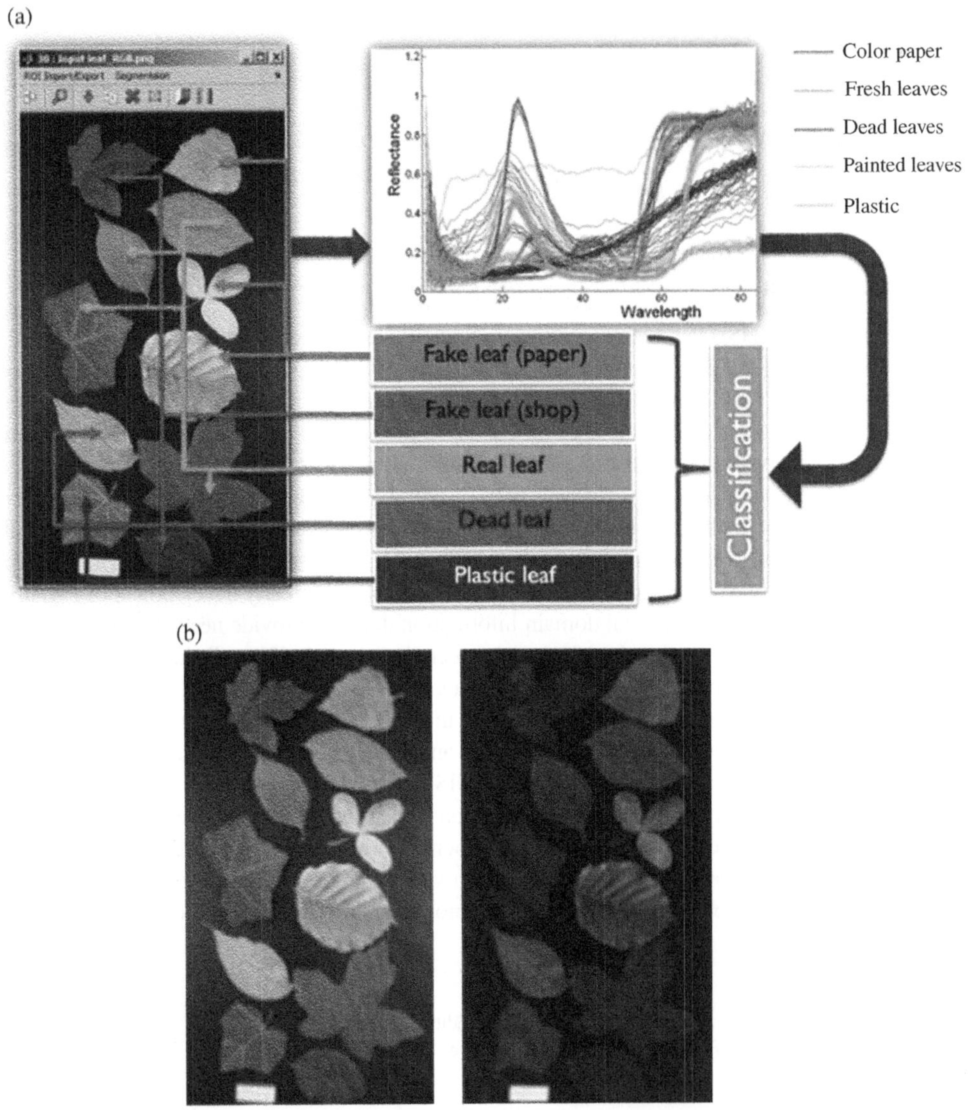


**Figure 1.1** Multispectral RGB and hyperspectral images. (a) Five different types of leaves along with their respective spectra, (b) multispectral RGB image and hyperspectral image, and (c) classification results.

(c)

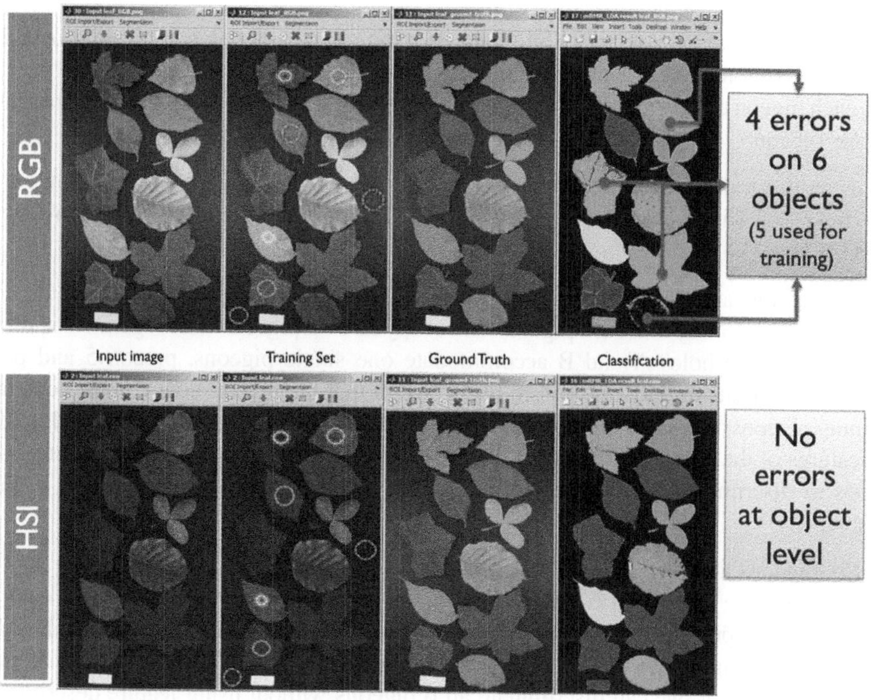


**Figure 1.1** (*Continued*).

According to Figure 1.1(b), it seems that the multispectral RGB image has a better visual classification than the hyperspectral image. However, it is indeed not true. As shown in Figure 1.1(c), using the same set of training samples using the hyperspectral image to perform classification makes no errors, while the multispectral RGB image committing 4 errors. This simple example demonstrates how unreliable human visual assessment is.

## 1.3 Two Principles for Hyperspectral Imaging

Since HSI is quite different from MSI in many aspects, we need to reinvent the wheel to find new concepts to explore the features of HSI that cannot be provided by MSI. Two principles are crucial and keys to understanding HSI, i.e., pigeon-hole principle and orthogonality principle.

### *1.3.1 Pigeon-Hole Principle*

Pigeon-hole principle is a fundamental concept of discrete mathematics [5]. It simply says that if there are $p$ pigeons to fly in $L$ different custom-built pigeon-holes with $p > L$, then there must be at least one pigeon-hole to accommodate two or more pigeons.

Let $L$ and $p$ be the number of pigeon-hole and number of pigeons, respectively. There are three scenarios resulting from the pigeon-hole principle.

1. $L<p$, at least one pigeon-hole must accommodate two or more pigeons and cannot discriminate its accommodated pigeons.
2. $L>p$, all pigeons are allowed to fly into their own individual and separate pigeon-holes. In other words there is more than one pigeon-hole that can be used to accommodate a pigeon.
3. $L=p$, each pigeon can fly into its own pigeon-hole so that no pigeon-hole can accommodate more than one pigeon.

Now, as an illustrative example, assume that there are $p = 6$ pigeons. Then, there are two cases, (1.1) $p = 6 > L = 3$ with 3 pigeon-holes, A,B,C and (1.2) $p = 6 < L = 9$ with 9 pigeon-holes, A,B, C,D,E,F,G,H,I.

Case 1: Six pigeons and three pigeon-holes, A,B,C.
In this case, there is no way that each pigeon can occupy its own pigeon-hole. Figure 1.2 shows one s-cenario where pigeon-holes A and B accommodate one single pigeons, pigeon 5 and pigeon 3, respectively. This leads to only one possibility which is that the pigeon-hole C must accommodate all remaining pigeons, 1,2,4,6. In the latter case, we need to use feature extraction methods to describe distinct features of the 4 pigeons (pigeon 1, pigeon 4, pigeon 2, and pigeon 6) such as size, weights, and shapes to discriminate one pigeon from another, while they are accommodated in the same pigeon-hole, C.

Case 2: Six pigeons and nine pigeon-holes, A,B,C,D,E,F,G,H,I.
In this case, we have more than 6 pigeon-holes to accommodate six pigeons so that each pigeon can occupy its own pigeon-hole. Figure 1.3 shows one scenario where pigeon-holes B,D,F,G,H,I accommodate one single pigeons, i.e. pigeon 1, pigeon 6, pigeon 5, pigeon 2, pigeon 3, pigeon 4, respectively. As a result, we can discriminate all 6 pigeons without using feature descriptors.

Using Figures 1.2 and 1.3 as an illustrative example, we can interpret pigeon-hole and pigeon as spectral band and target to be discriminated. Then case 1 in Figure 1.2 and case 2 in Figure 1.3 correspond to MSI and HSI, respectively. Case 1 also explains why MSI requires feature extraction methods to discriminate pigeons 1,4,2,6 in the pigeon-hole C. Therefore, to tell these four pigeons apart in the pigeon-hole C, feature descriptors are needed. On the other hand, case 2 also explains a fact that

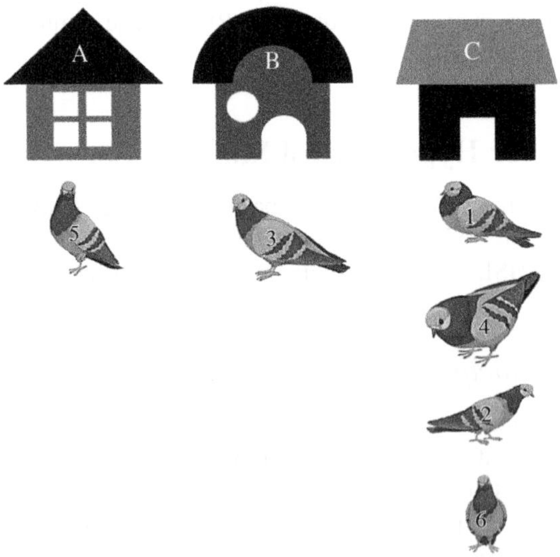


**Figure 1.2** Six pigeons to be accommodated by three pigeon-holes.

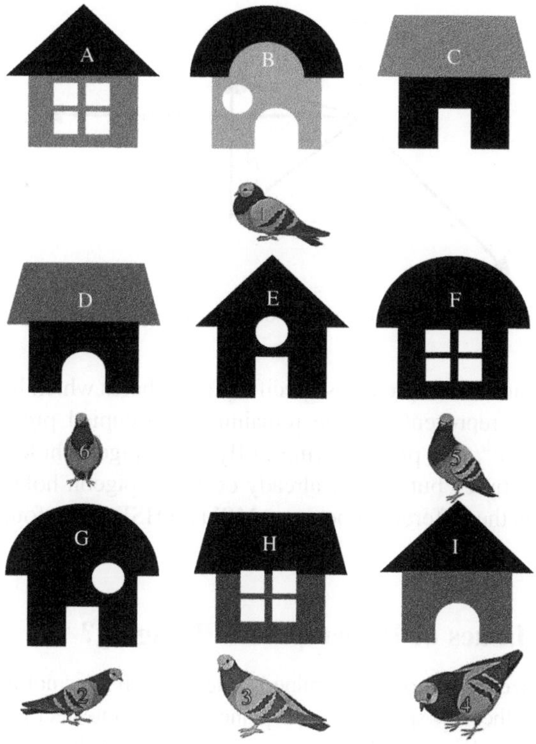


**Figure 1.3** Six pigeons to be accommodated by nine pigeon-holes.

when there are nine pigeon-holes which are sufficient to accommodate six pigeons; each pigeon-hole can be used to discriminate and separate individual pigeons. Therefore, there is no need of feature descriptors to discriminate the six pigeons one from another. This suggests that with sufficient spectral information, no feature extraction methods are required for HSI. As a result, HSI has strengths in taking advantage of rich spectral information to perform subpixel and mixed pixel analysis.

### *1.3.2 Orthogonality Principle*

According to the pigeon-hole principle with $L > p$, if a pigeon flies into a pigeon-hole, this pigeon-hole is then occupied and cannot be further used to accommodate another pigeon. So, the occupied pigeon-hole must be removed to prevent any pigeon from being flying into this pigeon-hole. The orthogonal principle [6] is used for this particular purpose.

For any data sample vector $\mathbf{r}$, it can be decomposed by orthogonal projection into an orthogonal sum as $\mathbf{r} = \mathbf{r}^{\|} + \mathbf{r}^{\perp}$, where $\mathbf{r}^{\|} = \mathbf{r}\cos\theta$, $\mathbf{r}^{\perp} = \mathbf{r}\sin\theta$, and $\theta = \cos^{-1}\langle \mathbf{r}, \mathbf{r}^{\|}\rangle$ [7]. In particular, $\mathbf{r}^{\|}$ is perpendicular to $\mathbf{r}^{\perp}$, i.e. $\mathbf{r}^{\|} \perp \mathbf{r}^{\perp}$. Alternatively, we can define a projection matrix [8, chapter 9] as

$$P_{\mathbf{r}} = \mathbf{r}\left(\mathbf{r}^T\mathbf{r}\right)^{-1}\mathbf{r}^T \text{ and } P_{\mathbf{r}}^{\perp} = \mathbf{I} - P_{\mathbf{r}}. \tag{1.1}$$

Then, $\mathbf{r}^{\|} = P_{\mathbf{r}}\mathbf{r}$ and $\mathbf{r}^{\perp} = P_{\mathbf{r}}^{\perp}\mathbf{r}$. Figure 1.4 shows the relationship between $\mathbf{r}^{\|}$ and $\mathbf{r}^{\perp}$ via the orthogonality principle.

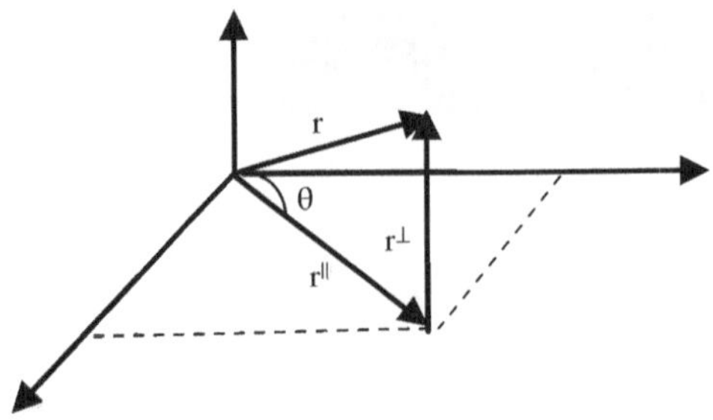


**Figure 1.4** Relationship between $\mathbf{r}^{||}$ and $\mathbf{r}^{\perp}$ via OP.

So, in terms of OP, $\mathbf{r}^{||}$ can be interpreted as specific pigeon-holes which have been used to accommodate pigeons. Then, $\mathbf{r}^{\perp}$ represents all the remaining unoccupied pigeon-holes which are not specified by $\mathbf{r}^{||}$. Since $\mathbf{r}^{||} \perp \mathbf{r}^{\perp}$, any pigeon trying to fly into a pigeon-hole can only fly into one of the pigeon-holes specified by $\mathbf{r}^{\perp}$ but not the already occupied pigeon-holes, $\mathbf{r}^{||}$.

A detailed discussion on the differences between MSI and HSI can be found in [2, chapter 31] and [8, Section 9.2].

## 1.4 What Are the Issues of Hyperspectral Imaging?

A HSI system is characterized by an overdetermined system which does not have solution as shown in Figure 1.5. This is because the HSI-specified hyperplane does not intersect the data sphere. In order to find solutions, the high dimensions of HSI must be reduced to lower dimensions until it has a solution marked by a dark gray filled dot, which is a hyperplane tangent to the data sphere. This also explains why data dimensionality reduction (DR) is required for HSI. Without DR, the HSI system does not have a solution in which case the best approximate solution is the one obtained by the orthogonality principle and marked by a dark filled dot in Figure 1.5 which has the minimum orthogonal projection to the unique solution at the tangent point. On the other hand, an MSI system can be characterized by an under-determined system which has an infinite number of solutions specified by the dot-dashed circle shown in Figure 1.5. To find a unique solution, a dimensionality expansion is required to increase the low-dimensional multispectral image until it reaches a unique solution which is the tangent point, the dark gray filled dot.

According to Figure 1.5, it is obvious that HSI is not an extension to MSI. There are three distinct scenarios associated with Figure 1.5.

1. $L < p$, two or more material substances discriminated by a single distinct spectral band (異物同譜). It is a multispectral image corresponding to an infinite number of solutions. In this case, feature extraction must be performed to discriminate material substances accommodated by one single dimension. Or alternatively, data dimensionality must be expanded from $p$ to $L$ so that each substance will have its own unique dimension. Such dimensionality expansion has been proposed in the past [9, 10].
2. $L > p$, a single material substance discriminated by two or more distinct spectral bands (同物異譜). It is a hyperspectral image corresponding to no solutions. In this case, the best solution is one which yields the minimum orthogonal projection according to the orthogonality principle. This leads to the linear mixing model,

$$\mathbf{r} = \mathbf{M}\boldsymbol{\alpha} + \mathbf{n} \tag{1.2}$$

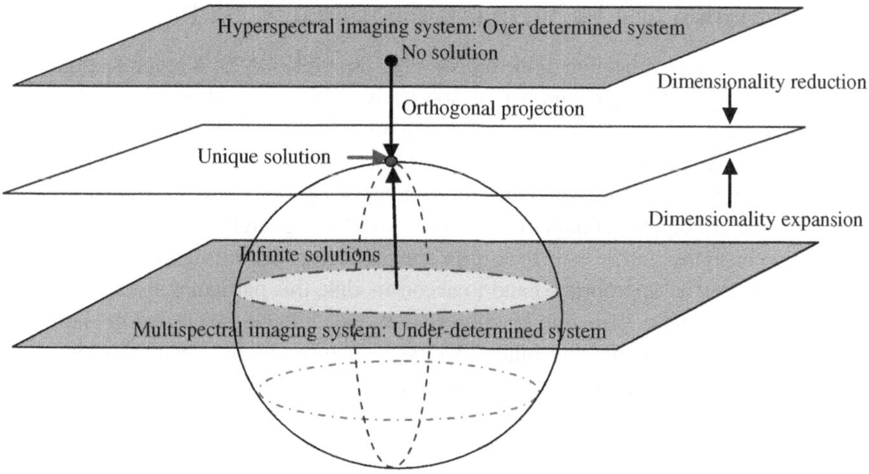


**Figure 1.5** Illustration of relationship between hyperspectral imaging and multispectral imaging.

which has been widely used in hyperspectral unmixing where **r**, **M**, **α**, and **n** are data sample vector, spectral signature matrix, and abundance fractional vector associated with **r**, and noise vector, respectively. The presence of **n** in (1.2) can be interpreted as noise or measurement error or model error. It is this **n** that causes no solution. In order to find the best possible approximate solution to (1.2), the OP-based least squares approach has been widely used, referred to as hyperspectral unmixing. As an alternative to spectral unmixing for (1.2), an approach, called non-negative matrix factorization (NMF) was proposed in [11] to decompose the data matrix **X** into

$$\mathbf{X} = \mathbf{W}\mathbf{H} \tag{1.3}$$

where $\mathbf{W} \geq 0$ is an $L \times p$ matrix and $\mathbf{H} \geq 0$ is a $p \times N$ matrix with $N$ being the total number of data samples, both of which have all non-negative elements as their entries. Despite the fact that both (1.2) and (1.3) can be used for spectral unmixing, the difference between (1.2) and (1.3) is that the noise term **n** in (1.2) is absent in (1.3). Interestingly, the idea of what (1.3) is to (1.2) is very similar to what the LRaSR by robust principal component analysis (RPCA) [12] is to the LRaSMD [13]. More details of discussions on these representations by the orthogonality principle can be found in Section 1.13. Unlike spectral unmixing which reduces data dimensionality from $L$ to $p$ via (1.2) and (1.3), BS can be also used to reduce $L$ to $p$ by selecting appropriate $p$ bands to accommodate $p$ particular substances so that each substance can be discriminated by its own unique band where these bands must be sufficiently effective to be able to diagnose these $p$ substances. As an example of pigeon-hole principle, a large pigeon should have a large pigeon-hole to accommodate it. This explains why spectral unmixing and BS are two major trends in HSI. Discussions on BS will be detailed in Sections 1.7 and 1.8. More details on BS are also provided in Chapters 4 and 5 of this book.

3. $L = p$, a substance discriminated by its unique distinct spectrum (同物同譜). It corresponds to a single unique solution where each substance finds its own unique band for discrimination.

## 1.5 Determination of *p* by Virtual Dimensionality via Pigeon-Hole Principle

The first and foremost best utilization of the pigeon-hole principle is VD. A natural assumption made on hyperspectral imagery is that the number of spectrally distinct signatures present in hyperspectral imagery, denoted by $n_s$, is generally unknown and assumed to be much smaller than the total number of spectral bands, $L$ spectral bands, i.e. $n_s \ll L$. In terms of the pigeon-hole principle, the number of pigeons, $p$, is unknown but much smaller than the number of pigeon-holes, $L$, which is known and fixed. So, the pigeon-hole principle guarantees that for each spectral signature in a hyperspectral image, there exists at least one band that can be used to specify one particular spectral signature. This is equivalent to saying that we can find an appropriate band to accommodate this particular spectral signature. On the other hand, the orthogonality principle guarantees that once a band is selected or sampled, this band cannot be used again to specify any other spectral signature. More specifically, the orthogonality principle makes sure that no single band can be used to accommodate more than one spectral signature. As a result of these two principles, the VD developed in [2, 14–19] can be used to determine $n_s$, i.e. $p$.

Based on the pigeon-hole principle and the orthogonality principle, an effective method for estimating VD was developed by Harsanyi-Farrand-Chang (HFC) [20], commonly referred to as HFC method along with its MATLAB codes also available in [2, pp. 997–998], where one pigeon-hole is specified by one particular spectral band and each pigeon is specified by one signature. It then formulates the VD estimation problem as a binary composite hypothesis testing problem where the null hypothesis $H_0$ and alternative hypothesis $H_1$ are specified by covariance matrix-obtained eigenvalues, $\{\lambda_l\}_{l=1}^{L}$ and correlation matrix-obtained eigenvalues, $\{\tilde{\lambda}_l\}_{l=1}^{L}$ to represent the absence and presence of a signature, respectively. More precisely, for each spectral band $l$, a binary composite hypothesis testing problem can be cast as follows:

$$
\begin{aligned}
&H_0(\text{signature absence}) : p_0(z_l) \approx \tilde{\lambda}_l - \lambda_l = 0 \\
&\text{versus} \\
&H_1(\text{signature presence}) : p_1(z_l) \approx \tilde{\lambda}_l - \lambda_l > 0.
\end{aligned}
\tag{1.4}
$$

In terms of the pigeon-hole principle, the signal in (1.4) can be interpreted as a pigeon specified by its weight. If we interpret the weight of the $l$th pigeon-hole as $\lambda_l$, then $\tilde{\lambda}_l$ is the weight of the $l$th pigeon-hole occupied by a pigeon where $\tilde{\lambda}_l - \lambda_l$ is the weight of the pigeon. With this interpretation, when $H_0$ is true, i.e. $\tilde{\lambda}_l - \lambda_l = 0$, there is no pigeon in the $l$th pigeon-hole, that is, the pigeon weight is zero. When $H_1$ is true, i.e. $\tilde{\lambda}_l - \lambda_l > 0$, it indicates that a pigeon with its weight $\tilde{\lambda}_l - \lambda_l$ is accommodated in the $l$th pigeon-hole.

A Neyman–Pearson detector of $z_l = \hat{\lambda}_l - \lambda_l$, denoted by $\delta^{\text{NP}}(z_l)$ for (1.4), can be obtained by maximizing the detection power $P_D$ subject to the false alarm probability $P_F$ being fixed at a given value, $\beta$ [6]. The level of significance for the detector is specified by $\beta$ which determines the threshold value $\tau_l$ in the following randomized decision rule:

$$
\delta^{\text{NP}}(z_l) = \begin{cases} 1; & \text{if } \Lambda(z_l) > \tau_l \\ 1 \text{ with probability } \kappa; & \text{if } \Lambda(z_l) = \tau_l \\ 0; & \text{if } \Lambda(z_l) < \tau_l \end{cases}
\tag{1.5}
$$

where the likelihood ratio test $\Lambda(z_l)$ is given by $\Lambda(z_l) = p_1(z_l)/p_0(z_l)$ with $p_0(z_l)$ and $p_1(z_l)$ being probability distributions under $H_0$ and $H_1$ for (1.5), respectively. So, a case of $z_l = \hat{\lambda}_l - \lambda_l > \tau_l$ indicates that $\delta^{\text{NP}}(z_l)$ in (1.5) fails the test, which means that there is signal energy assumed to contribute to the eigenvalue, $\hat{\lambda}_l$, in the $l$th spectral band. It should be noted that the test (1.5) must be performed for each of $L$ spectral bands. Therefore, for each pair of $\hat{\lambda}_l - \lambda_l$, the threshold $\tau_l$ varies with the fixed value of $\beta$ and is spectral band-dependent. Using (1.5), VD can be determined by calculating

$$\mathrm{VD}_{\mathrm{HFC}}^{\mathrm{NP}}(\mathrm{P_F} \le \beta) = \sum_{l=1}^{L} \lfloor \delta^{\mathrm{NP}}(z_l) \rfloor \tag{1.6}$$

with a predetermined false alarm probability $\mathrm{P_F} = \beta$ where $\lfloor \delta(z_l) \rfloor = 1$ only if $\delta^{\mathrm{NP}}(z_l) = 1$ and $\lfloor \delta(z_l) \rfloor = 0$ if $\delta^{\mathrm{NP}}(z_l) < 1$. By virtue of (1.6), we can estimate $n_s$ by

$$n_s(\beta) = \mathrm{VD}_{\mathrm{HFC}}^{\mathrm{NP}}(\mathrm{P_F} \le \beta). \tag{1.7}$$

It is the constraint value $\beta$ in (1.7) that determines $n_s$ and varies with different applications.

It is interesting to look into insight that how the pigeon-hole principle and orthogonality principle play their roles in the HFC method. The pigeon-hole principle guarantees that the number of spectral bands is greater than or equal to the number of signatures, i.e. $L \ge n_s$. In this case, if there is a signature in a hyperspectral image, there exists at least one spectral band that can be used to accommodate this signature. On the other hand, the orthogonality principle guarantees that once one spectral band is used to specify a signature, this particular spectral band will be designated to this particular signature and cannot be used again to accommodate any other signature. Furthermore, the HFC method was later extended in [2, 14–19], referred to as noise-whitened HFC (NWHFC) method, which whitens the noise prior to the HFC method. The resulting VD is defined as $\mathrm{VD}_{\mathrm{NWHFC}}^{\mathrm{NP}}(\mathrm{P_F} \le \beta)$. More details about VD can be found in [2, 14–19].

There are three intriguing facts resulting from the above pigeon-hole principle and orthogonality principle which can be summarized as follows:

- The pigeon-hole principle guarantees that there are a sufficient number of bands greater than or equal to $n_s$ used to accommodate $n_s$ signatures.
- The orthogonality principle guarantees that if a band is used to specify one spectral signature, it cannot be used again to specify any other signature.
- Since $n_s < L$, there may be one spectral signature that can be specified by more than one band, viz. two or more bands. In this case, BS is crucial to select the most effective and representative band for each spectral signature.

## 1.6 Order Determination of Low Rank and Sparse Matrices by Virtual Dimensionality via Pigeon-Hole Principle

Another best utilization of the pigeon-hole principle is determination of the rank of low-rank matrix, $m$ and the rank of sparse matrix, $j$ which are crucial parameters commonly used in LRaSR [12, 13].

Interestingly, a recent approach developed by Kuybeda et al. in [21], called maximum orthogonal complement algorithm (MOCA), can be used to determine $p$ followed by the Minimax-SVD (MX-SVD) to determine $m$. Its detailed process is described as follows.

Let $\mathbf{X}$ be a data matrix formed by data sample vectors, $\{\mathbf{r}_i\}_{i=1}^{N}$, i.e. $\mathbf{X} = [\mathbf{r}_1\mathbf{r}_2\cdots\mathbf{r}_N]$. Define the norm of the data matrix $\mathbf{X}$ by

$$\|\mathbf{X}\| = \max_{1 \le i \le N} \|\mathbf{r}_i\| \tag{1.8}$$

where $\|\mathbf{r}_i\|$ is the length of the vector $\mathbf{r}_i = (r_{i1}, r_{i2}, \ldots, r_{iL})^T$ defined by $\|\mathbf{r}_i\|^2 = \sum_{l=1}^{L} r_{il}^2$. Assume that

$$i^* = \arg\{\max_{1 \le i \le N} \|\mathbf{r}_i\|\}. \tag{1.9}$$

The norm of the matrix $\mathbf{X}$ in (1.8) can be further expressed by

$$\|\mathbf{X}\| = \|\mathbf{r}_{i^*}\| \tag{1.10}$$

which is exactly the brightest pixel $\mathbf{r}_{i^*}$ whose norm has the maximum vector length.

We first assume that $p$ is given and then determine $j$. Initially, let $j = 1$, i.e. $\mathbf{T}_j = \emptyset$. We then use SVD to find the first $p$ principal left singular vectors of the data matrix $\mathbf{X} = [\mathbf{r}_1\mathbf{r}_2\cdots\mathbf{r}_N]$, denoted by $\mathbf{s}_1^{(0)}, \mathbf{s}_2^{(0)}, \ldots, \mathbf{s}_p^{(0)}$ in which case the $p$-dimensional signal subspace is given by $\mathbf{S}_p = \left[\mathbf{s}_1^{(0)}\mathbf{s}_2^{(0)}\cdots\mathbf{s}_p^{(0)}\right]$. Let $\mathbf{t}_1 = \arg\left\{\max_{\mathbf{r}\in\{\mathbf{r}_i\}_{i=1}^N}\left[\|P_{\mathbf{S}_p}^{\perp}\mathbf{r}\|\right]\right\}$. Next, we increase $j$ by one, i.e. $j = 1$ and let $\mathbf{T}_1 = [\mathbf{t}_1]$. We once again use SVD to find the first $p - 1$ principal left singular vectors, $\mathbf{s}_1^{(1)}, \mathbf{s}_2^{(1)}, \ldots, \mathbf{s}_{p-1}^{(1)}$ of the matrix $P_{\mathbf{T}_1}^{\perp}\mathbf{X}$ obtained by operating an orthogonal subspace projector $P_{\mathbf{T}_1}^{\perp}$ defined by

$$P_{\mathbf{T}_1}^{\perp}\mathbf{X} = \mathbf{I} - \mathbf{T}_1\left(\mathbf{T}_1^T\mathbf{T}\right)^{-1}\mathbf{T}_1^T \tag{1.11}$$

on the original data matrix $\mathbf{X}$ to form $\mathbf{S}_p = [\mathbf{B}_{p-1} \mid \mathbf{T}_1]$, where $\mathbf{B}_{p-1} = \left[\mathbf{s}_1^{(1)}\mathbf{s}_2^{(1)}\cdots\mathbf{s}_{p-1}^{(1)}\right]$. Then, we further find $\mathbf{t}_2 = \arg\left\{\max_{\mathbf{r}\in\{\mathbf{r}_i\}_{i=1}^N}\left[\left\|\mathbf{P}_{[\mathbf{B}_{p-1}|\mathbf{T}_1]}^{\perp}\mathbf{r}\right\|\right]\right\}$ and add it to the current target subspace $\mathbf{T}_1$ to form a new target subspace $\mathbf{T}_2 = [\mathbf{t}_1\mathbf{t}_2]$. We repeat the same procedure until $j = p$ and find $\min_{1\le j\le p}\left[\left\|\mathbf{P}_{[\mathbf{B}_{p-j}|\mathbf{T}_j]}^{\perp}\mathbf{t}_j\right\|\right]$, which is the minimum of the maximum orthogonal projections $\{\mathbf{t}_j\}_{j=1}^{p}$ found in each of previous stages. The $j^* = \arg\{\min_{1\le j\le p}[\|\mathbf{t}_j\|]\}$, which yields the minimum, turns out to be the rank of the desired target subspace and $m = p - j^*$ is the rank of the background subspace.

The key idea of the above process is to work backward to find target pixels by a sequence of orthogonal complement subspace projections $\left\{\mathbf{P}_{\mathbf{T}_j}^{\perp}\mathbf{r}_i\right\}_{j=1}^{p}$ within the data space until it reaches $p$. Then, $j^*$ is determined by $j^* = \arg\{\min_{1\le j\le p}[\|\mathbf{t}_j\|]\}$, referred to as $j^{\text{MX-SVD}}$. The set of target signal sources $\mathbf{t}_1, \mathbf{t}_2, \ldots, \mathbf{t}_{j^*}$ form a target matrix $\mathbf{T}_{j^*} = \left[\mathbf{t}_1\mathbf{t}_2\cdots\mathbf{t}_{j^*}\right]$ as the desired target signal subspace and the complement subspace within $\mathbf{S}_p$ orthogonal to $\mathbf{T}_{j^*}$ is the background subspace $\mathbf{B}_{m^*}$ with $m^* = p - j^*$.

The process described above is called MX-SVD named by Kuybeda et al. [21]. The details of MX-SVD implementation can be summarized as follows.

---

***MX-SVD***

1. Initialization: Let $p$ be given and $j = 1$, i.e. $\mathbf{T}_1 = \emptyset$.
2. Use SVD to find the first $p$ principal left singular vectors of the data matrix $\mathbf{X} = [\mathbf{r}_1\mathbf{r}_2\cdots\mathbf{r}_N]$, denoted by $\mathbf{s}_1^{(0)}, \mathbf{s}_2^{(0)}, \ldots, \mathbf{s}_p^{(0)}$ and let $\mathbf{t}_1 = \arg\left\{\max_{\mathbf{r}\in\{\mathbf{r}_i\}_{i=1}^N}\left[\|\mathbf{P}_{\mathbf{S}_p}^{\perp}\mathbf{r}\|\right]\right\}$.
3. Let $j \leftarrow j + 1$. Form $\mathbf{T}_j = \left[\mathbf{T}_{j-1}\mathbf{r}_j^*\right]$. Use SVD to find the first $p - j$ principal left singular vectors, $\mathbf{s}_1^{(j)}, \mathbf{s}_2^{(j)}, \ldots, \mathbf{s}_{p-j}^{(j)}$ of the matrix $\mathbf{P}_{\mathbf{T}_j}^{\perp}\mathbf{X}$ and find

$$\mathbf{t}_j = \arg\left\{\max_{\mathbf{r}\in\{\mathbf{r}_i\}_{i=1}^N}\left[\left\|P_{[\mathbf{B}_{p-j}|\mathbf{T}_j]}^{\perp}\mathbf{r}\right\|\right]\right\} \tag{1.12}$$

4. Repeat step 3 until $j = p$. Continue
5. At this stage, calculate $j^{\text{MX-SVD}}$ by

$$j^{\text{MX-SVD}} \equiv \arg\left\{ \min_{1 \le j \le p} \left[ \left\| \mathbf{P}^{\perp}_{[\mathbf{B}_{p-j}|\mathbf{T}_j]} \mathbf{t}_j \right\| \right] \right\}. \tag{1.13}$$

---

It should be noted that the $j^{\text{MX-SVD}}$ in (1.13) is obtained by assuming that the rank of a signal subspace, $p$ is given *a priori*. Unfortunately, finding an appropriate $p$ used for (1.13) is a challenging issue. There are two ways to determine the value of $p$. One is the maximum likelihood detector used by MOCA. The other is VD discussed in Section 1.5. Nevertheless, itwill be nice to implement MX-SVD without actually determining $p$ in (1.13). That is, we can assume that there are $L$ dimensions of a hyperspectral image cube (i.e. band dimensionality) and then carry out MX-SVD by making the rank of a signal subspace, $l$ ranging from 1 to $L$, in which case, the rank of a signal subspace can be examined by all possible subspaces in data space of dimensionality $L$. Since the sequence $\left\{ \left\| \mathbf{P}^{\perp}_{[\mathbf{B}_{l-j}|\mathbf{T}_j]} \mathbf{t}_j \right\| \right\}_{j=1}^{l}$ is monotonically decreasing at $l$, i.e.

$$\left\| \mathbf{P}^{\perp}_{[\mathbf{B}_{l-j}|\mathbf{T}_j]} \mathbf{t}_j \right\| \ge \left\| \mathbf{P}^{\perp}_{[\mathbf{B}_{(l+1)-j}|\mathbf{T}_j]} \mathbf{t}_j \right\| \text{ for } l = 1, \ldots, L-1 \tag{1.14}$$

we can plot the values of $\left\{ \left\| \mathbf{P}^{\perp}_{[\mathbf{B}_{l-j}|\mathbf{T}_j]} \mathbf{t}_j \right\| \right\}_{l=1}^{L}$ against dimensionality $l$ from 1 to $L$ in a similar manner that an eigenvalue distribution is plotted for eigen-energy-based methods so that the dimension $p$ that causes a sudden drop is the desired rank of the signal subspace, $\mathbf{S}_p$. We then use such obtained $p$ in MX-SVD to find the rank of the desired target subspace as $j^{\text{MX-SVD}}$ and $p - j^{\text{MX-SVD}}$ as the rank of the background signal subspace by letting $m^{\text{MX-SVD}} = p - j^{\text{MX-SVD}}$.

## 1.7 Band Selection by Pigeon-Hole Principle

A third and probably most important utilization of the pigeon-hole principle is BS. Since the number of hyperspectral bands, $L$, and the number of material substances, $p$, can be interpreted as the number of pigeon-holes and the number of pigeons, respectively, with $L > p$, there are sufficient number of $L$ pigeon-holes to accommodate all the $p$ pigeons. However, like spectral bands which can be used to characterize different material substances, each pigeon-hole can be also custom-built to accommodate particular pigeons (see Figure 1.3). So, it is necessary to design and develop pigeon-hole selection criteria according to features of various pigeons, such as their size, shape, and origin to select an appropriate pigeon-hole for each individual pigeon. By virtue of the pigeon-hole principle, selection of desired pigeon-holes for pigeons can be translated to BS which selects appropriate bands to specify particular material substances.

BS has been widely used to reduce data dimensionality by removing redundant or unwanted bands, while maintaining a certain desired level of performance [22]. Many various approaches have been developed and reported in the literature. Basically, BS can be categorized according to strategies of how to select bands. A simple and also effective approach is band prioritization (BP) which ranks all bands according to certain criteria, such as variance, signal-to-noise ratio entropy, information divergence criteria [23], constrained BS [24] with details discussed in Chapter 4 of this book. A second approach is band clustering based on grouping bands with similar spectral properties into clusters and then one representative band is selected from each cluster such as [25–27]. A third approach is to adopt certain band search strategies to find optimal bands in a feed forward or backward manner such as minimum estimated abundance covariance (MEAC) [28], orthogonal projection-based BS [29], constrained-target BS [30] and those also discussed in Chapter 4 of this book. A fourth approach is band subset selection (BSS) which is also discussed in Chapter 5 of this book. It is also referred to as band combination, which selects an optimal band subset of multiple bands as a whole instead of finding bands one at a time, for example particle swarm optimization [31], firefly algorithm [32], dominant set extraction band selection (DSEBS) in [33], multigraph determinantal point process (MDPP) in [34], channel capacity-based BSS [35], band subset for anomaly detection [36], linearly constrained minimum variance-based BSS for classification [37], and optimal neighborhood reconstruction (ONR)-based BSS based on correlated neighborhood property (CNP) to exploit strong correlation between neighboring bands [38].

In terms of the pigeon-hole principle, BP corresponds to pigeon-hole prioritization which prioritizes pigeon-holes according to their features. Band clustering can be interpreted as pigeon-hole clustering which groups pigeon-holes characterized by similar features into individual clusters and selects a representative pigeon-hole for each cluster. BS search strategy is to find pigeon-holes according to their prioritized features sequentially in a forward or backward manner for a given set of $p$ pigeons without using prioritization. Finally, BSS is to consider multiple-band subsets as multiple-pigeon-holes as a whole and attempt to find an optimal set of pigeon-holes to accommodate all pigeons together simultaneously.

Two major key issues arising in BS are requirement of prior knowledge of how many bands needed to be selected, i.e. number of bands to be selected, $n_{\mathrm{BS}}$ and how to take care of inter-band correlation among selected bands. Regarding the first issue, finding an appropriate $n_{\mathrm{BS}}$ still remains a challenging task and can be estimated by VD in Section 1.5. As for the second issue, BSS has been developed for this purpose and will be discussed in great details in Chapter 5 of this book.

## 1.8 Band Selection by a Hyperspectral Band Channel via Pigeon-Hole Principle

Interestingly, the pigeon-hole principle also finds its role in spectral inter-band discrimination (SIBD) in [39].

First of all, we introduce a hyperspectral band channel by interpreting a band image, $\mathbf{B}_j$, as a channel input, a band image to be discriminated as a channel output, $\mathbf{B}_k$, and the band channel transition probability between these two band images, $\mathbf{B}_j$ and $\mathbf{B}_k$, $p(\mathbf{B}_k \mid \mathbf{B}_j)$, characterized by their band discrimination quantified by a spectral measure such as spectral angle map (SAM) [14] or spectral information divergence (SID) [14, 40, 41]. So, the total number of band images, denoted by $\Omega = \{\mathbf{B}_l\}_{l=1}^{L}$, will be used as both channel input space, $\boldsymbol{\Omega}_X = \boldsymbol{\Omega}_L$ and channel output space $\boldsymbol{\Omega}_Y = \boldsymbol{\Omega}_L$, and the band channel transition probability matrix, $\mathbf{Q} = \left\{Q_{k|j}\right\}_{j=1,k=1}^{L,L}$ with $Q_{k|j} = p(\mathbf{B}_k \mid \mathbf{B}_j)$ being specified by a conditional band discrimination probability of band image $\mathbf{B}_k$ against band image $\mathbf{B}_j$. With this formulation we can define a hyperspectral band channel by a triplet, $\left(\Omega_X = \{\mathbf{B}_l\}_{l=1}^{L}, \Omega_Y = \{\mathbf{B}_l\}_{l=1}^{L}, \mathbf{Q} = \left\{p\left(\mathbf{B}_k \mid \mathbf{B}_j\right)\right\}\right)$ described in Figure 1.6.

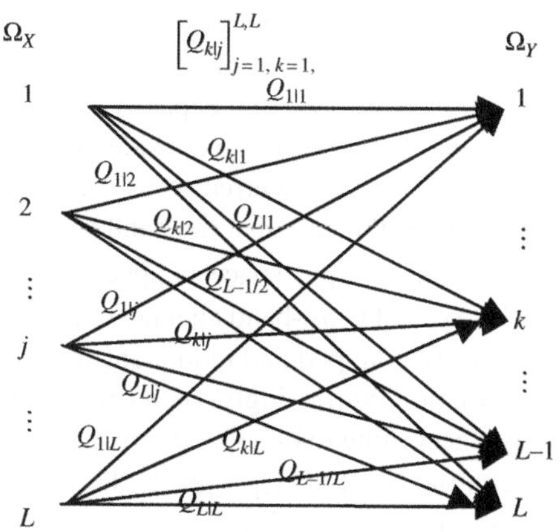


**Figure 1.6** A hyperspectral band channel derived from a hyperspectral image specified by $\Omega_X = \Omega_Y = \{l_j\}_{j=1}^L = \{l_k\}_{k=1}^L = \{l\}_{l=1}^L$.

The hyperspectral band channel mutual information between $\Omega_X = \Omega_L = \{\mathbf{B}_l\}_{l=1}^L$ and $\Omega_Y = \Omega_L = \{\mathbf{B}_l\}_{l=1}^L$ is then defined as band discrimination mutual information (BDMI) which can be used to quantify inter-band spectral discrimination information.

In information theory, the channel capacity is defined as the maximal mutual information of a given channel specified by input space $\mathbf{\Omega}_X$, output space $\mathbf{\Omega}_Y$, and channel transition probabilities given by $\{Q(y \mid x)\}_{x\in\Omega_X, y\in\Omega_Y}$. Following a similar treatment, the maximal BDMI can be further used to define $L$-band capacity ($L$-BC) of a hyperspectral image specified by its entire bands, $\Omega_L = \{\mathbf{B}_l\}_{l=1}^L$, denoted by BC($\mathbf{\Omega}_L$), which measures the maximal capacity of SIBD for a hyperspectral image. Since channel capacity is measured by bit rate, the defined BC($\mathbf{\Omega}_L$) can be also interpreted as spectral discrimination per band measured by bits. Accordingly, BC($\mathbf{\Omega}_L$) can be considered as maximal SIBD of a hyperspectral image where the full discrimination of a band from other bands is encoded by 1 and 0 for two identical bands. Furthermore, the discriminatory capability of each band can be also prioritized by its probability assigned by the found input probability distribution that yields BC($\mathbf{\Omega}_L$). The higher the prior probability, the better the band discriminatory capability. As a result, these priority probabilities can be further used for BS. Such BC approach is quite different from existing BS methods since BC is not the same as band de-correlation but both are indeed closely related to each other. In other words, two-band spectral discrimination measures the power of discriminating one band from another band, while two-band de-correlation measures the degree of one band uncorrelated with another. Accordingly, the less the two bands are correlated, the less similarity the two bands have, thus the better the two-band discrimination is. To the contrary, more spectral bands indicate less SIBD. Thus, the more inter-band correlation means the less inter-band de-correlation and also the more inter-band similarity as well as less discrimination. It may first look odd. But let us consider the following example. Suppose that there is a fixed-size box containing two identical balls with the box corresponding to a fixed range of wavelengths and a ball corresponding to a band. The probability of discriminating one ball from the other is ½. Now, if we add one more identical ball into the box, the probability of discriminating one ball from the other two is reduced to 1/3. Consequently, more balls are added to the box. Then, the probability of discriminating one ball from another is reduced. This example explains that for a given fixed range of wavelengths, adding more bands does not increase but rather decrease inter-band discriminatory power.

Similar to the development for the full band channel, we can also extend it to a $p$-band channel with $\Omega_X = \Omega_p = \{\mathbf{B}_l\}_{l=1}^{p}$ and $\Omega_Y = \Omega_p = \{\mathbf{B}_l\}_{l=1}^{p}$, where $\{\mathbf{B}_l\}_{l=1}^{p}$ is a set of arbitrary selected $p$ bands. The resulting maximal BDMI is referred to as $p$-band capacity ($p$-BC), denoted by BC($\boldsymbol{\Omega}_p$).

One immediate extension to BC($\boldsymbol{\Omega}$) and BC($\boldsymbol{\Omega}_p$) is to consider a band channel with different input and output spaces, i.e. $\boldsymbol{\Omega}_X = \boldsymbol{\Omega}_q$ and $\boldsymbol{\Omega}_Y = \boldsymbol{\Omega}_p$, referred to as ($q$,$p$)-band channel. Two interesting scenarios are of particular interest. One is the case that the band channel input space is the full band set, $\Omega_X = \Omega_L = \{\mathbf{B}_l\}_{l=1}^{L}$ but the channel output space is specified by a particular $p$-band subset $\Omega_Y = \Omega_p = \{\mathbf{B}_i\}_{i=1}^{p}$ with $p < L$. This type of ($L$,$p$)-band channel occurs in many practical applications when hyperspectral imagery contains overlapped and redundant spectral information and there is no need of using all the full band information. Surprisingly, such a band channel formulation can be also used to address a BS problem. Also, according to [42, Corollary 3, p. 96] and [43], we can find at most $p$ bands, $\tilde{\Omega}_p$ in the input space from full band set $\boldsymbol{\Omega}_L$ with BC between $\tilde{\Omega}_p$ and $\Omega_p$ that achieves the exactly same BC between $\boldsymbol{\Omega}_L$ and $\Omega_p$. In this case, $\tilde{\Omega}_p$ is a desired $p$-band subset needed to be selected. As a complete opposite case to ($L$,$p$)-band channel, the other scenario is the ($p$,$L$)-band channel where the channel input space is specified by a particular $p$-band subset $\Omega_X = \Omega_p = \{\mathbf{B}_i\}_{i=1}^{p}$ with $p < L$ and the band channel output space is the full band set, $\Omega_Y = \Omega_L = \{\mathbf{B}_l\}_{l=1}^{L}$. In this case, BC is increased as $p$ is increased. These two scenarios can be better illustrated by the pigeon-hole principle. Specifically, we interpret the band channel input space as the number of pigeon-holes and the band channel output space as the number of pigeons as discussed in Section 1.3.1. Accordingly, BC can be interpreted as the ability in assigning pigeon-holes to accommodate pigeons. So, the first scenario represents the case that the number of pigeon-holes, $L$, as channel inputs is greater than the number of pigeons, $p$, as channel outputs in which we only need $p$ pigeon-holes to accommodate $p$ pigeons. As a result, BC becomes stable and constant when $L \geq p$ since there have sufficient pigeon-holes to accommodate all pigeons in individual and separate pigeon-holes. As a second scenario, it represents the case that there are only $p$ pigeon-holes as channel inputs which are fewer than $L$ pigeons as channel outputs in which BC is increased as $p$ is increased. This is because more pigeon-holes are used to accommodate $L$ pigeons. So, it can be also used to explain the concept of VD using bands as pigeon-holes to discriminate material substances as pigeons.

Alternatively, we can also interpret a hyperspectral band channelby specifying channel input space, $X$ as pigeons and channel output space, $Y$ as pigeon-holes. Then, a hyperspectral image and a multispectral image can be characterized in Figure 1.7 by a lossless channel where each pigeon can be accommodated by more than one pigeon-hole and a noiseless channel where more than one pigeon can be accommodated by the same pigeon-hole described [42].

## 1.9 Band Sampling via Pigeon-Hole Principle

This section looks into an innovational notion called band sampling (BSam) as opposed to BS discussed in Sections 1.7 and 1.8. Its concept was first envisioned in [44] and has never been explored in the HSI literature. It deviates the conventional wisdom using BS to reduce data dimensionality where the $n_{\mathrm{BS}}$ can be estimated by VD. However, it should be noted that VD only provides an estimate of $n_{\mathrm{BS}}$ which is not necessarily the true value of $n_{\mathrm{BS}}$. So, it may occur that the VD-determined $n_{\mathrm{BS}}$ might actually be smaller than the true value of $n_{\mathrm{BS}}$ in which case BS may not work effectively.

A recent breakthrough in communications and signal processing is the compressive sensing (CS) theory [45–47] which is developed to cope with signal under-sampling problems. According to the classic Shannon sampling theorem, the Nyquist rate provides the greatest lower bound on lossless signal reconstruction. In other words, if a signal is reconstructed with the signal sampling below

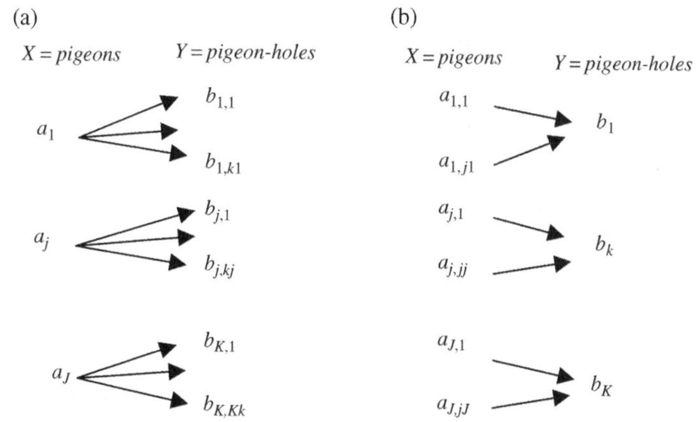


**Figure 1.7** Hyperspectral and multispectral channels. (a) Lossless (hyperspectral) channel and (b) deterministic (multispectral) channel.

the Nyquist rate, it will result in signal aliasing which causes imperfect signal recovery. Interestingly, the CS theory shows otherwise. That is, CS proves that a perfect signal reconstruction under the Nyquist rate is still possible, provided that two properties, sparsity and incoherent sampling, are satisfied [45, 46]. Specifically, for CS to accomplish this task, it must satisfy two constraints, signal representations to be sparse and sampled signals to be incoherent. CS makes use of a random sensing matrix to sample sparse signals to solve the under-sampling problem. Accordingly, CS is often referred to as compressive sampling or random signal sampling.

The idea of BSam originated from the nature of the random signal sampling used by CS. Instead of selecting bands as BS does, BSam randomly samples bands without using specific sampling strategies. In correspondence to signal sampling, we can also define band sampling rate (BSamR) by the ratio of number of bands to be sampled, $n_{\text{BSam}}$, to the total number of spectral bands, $L$ as

$$\text{BSamR} = \frac{n_{\text{BSam}}}{L} \tag{1.15}$$

which is similar to the Nyquist rate. The definition of (1.15) can be also referred to as the information rate described by Shannon's first noiseless source coding theorem in information theory [42].

According to [47], if there is no concern on signal recovery, the signal sparse representation by a sparse matrix $\mathbf{\Psi}$ can be implicitly included in so-called universality model which only involves the sensing matrix $\mathbf{\Phi}$ with no need of referring to the sparse matrix $\mathbf{\Psi}$. This universality model greatly reduces complexity of BSam by only focusing on how to sample bands to achieve the maximal signal incoherence.

In order to make BSam work, two strategies are explored. It is known from information theory [42] that maximal entropy is achieved by uniform distribution. In translation to the BSam language, it means that for a given $n_{\text{BSam}}$, the maximal incoherence can be achieved by sampling bands uniformly. This leads to an approach to BSam, called uniform band sampling (UBSam). However, if $n_{\text{BSam}}$ is under-sampled (viz. $n_{\text{BSam}}$ is smaller than it should be in reality) like signal sampling rate lower than the Nyquist rate, UBSam may not work effectively. In this case, we need to appeal for CS, which uses random signal sampling to deal with the under-signal sampling rate. The concept of random signal sampling in CS gives rise to a second approach, called random band sampling (RBSam). Unfortunately, a direct use of sensing matrix $\mathbf{\Phi}$ in CS via the universality model cannot be applicable to RBSam. We must reinvent the wheel by coming up with a new idea to tackle this problem. Interestingly, a recent work in [48] developed iterative training sampling (ITS) that indeed provides a possible solution. In other words, the idea of using iterative sampling of training samples can be readily applied to random sampling of bands if we interpret training samples as bands. The idea of ITS is to use a set of spatial filters to capture spatial classification information produced by a classifier and then feed the spatial filtered classification maps back to be added to the current data cube to create a new expanded data cube. ITS is then once again implemented on this new data cube to randomly reselect training samples which contain more spatial classification information via feedback of spatial filtered classification maps to reduce misclassification error. The same process is repeated iteratively. Since BSam is a band process and does not take the spatial information into account, the way that ITS introduces spatial information may be in fact applicable to BSam if training samples are replaced by sampled bands. By taking advantage of ITS, BSam becomes realizable to correspond to its counterpart, random sampling in CS. More details can be found in [44].

Finally, several salient differences of BSam from BS discussed in Sections 1.7 and 1.8 can be summarized as follows:

- The first and the foremost difference between BS and BSam is conceptual. For given $n_{\mathrm{BS}}$ and $n_{\mathrm{BSam}}$ in (1.15), both can be estimated by VD. Nevertheless, the key difference lies on how BS and BSam are carried out. From an estimation point of view [6], BS can be considered as a deterministic process of finding $n_{\mathrm{BS}}$ unknown constants, each of which corresponding to a band to be selected. By contrast, BSam is a process of estimating $n_{\mathrm{BSam}}$ random parameters, each of which corresponding to a band to be randomly sampled. More specifically, the bands to be selected by BS are unknown parameters of their occurrence at certain wavelengths with probability one as opposed to bands sampled by BSam which occur at everywhere in the wavelength range with their sampled probabilities equally likely, i.e. uniform distribution. By virtue of the pigeon-hole principle, BS finds specific $n_{\mathrm{BS}}$ pigeon-holes, i.e. $n_{\mathrm{BS}}$ bands, to accommodate $n_{\mathrm{BS}}$ pigeons, while BSam considers all the pigeon-holes equally likely to be used to accommodate $n_{\mathrm{BSam}}$ pigeons, i.e. $n_{\mathrm{BSsam}}$ uniformly sampled bands.
- In BS, if $n_{\mathrm{BS}}$ is too small and less than what it should be, BS methods may not work effectively no matter how effective BS methods are designed. This inherent issue has been overlooked and was not investigated in the past. However, unlike BS, BSam can still work even if $n_{\mathrm{BSam}}$ is lower than its true value.
- BSam does not require any specific means of sampling bands compared to BS, which requires a specific rule such as BP criteria or BS/search strategies.
- One key difference of BSam from BS is that bands sampled by BSam are random and cannot be reproduced as opposed to bands selected by BS which are deterministic and can be reproduced. The beauty of random signal sampling is derived from CS which allows signals to be sampled signals under the Nyquist rate with two constraints, sparsity and incoherence, but does not affect the end results. So, including ITS in BSam also allows BSam to perform effectively in the same way that random signal sampling does in CS.
- BSam samples each band equally likely to maximize band incoherence (i.e. inter-band decorrelation), whereas BS considers certain bands to be more significant than other bands. Consequently, BSam can be also used in the same manner that the uniform signal sampling is used by the Nyquist rate. This also explains why the uniform band selection (UBS) generally performs well compared to many BS methods as long as $n_{\mathrm{BS}}$ is appropriately determined and greater than the true value of $n_{\mathrm{BS}}$.

## 1.10 Spectral Unmixing via Orthogonality Principle

As pointed out in (1.2), since the number of signatures, $p$, used to unmix the data sample vector $\mathbf{r}$ in (1.2) is smaller than the number of spectral bands, $L$, according to the pigeon-hole principle, there is no solution to (1.2). The best possible solution that we can have is to minimize the least squares error of the unmixing result given by

$$\min_{\boldsymbol{\alpha}}(\mathbf{r}-\mathbf{M}\boldsymbol{\alpha})^T(\mathbf{r}-\mathbf{M}\boldsymbol{\alpha}) \tag{1.16}$$

where $\mathbf{M}$ is an $L \times p$ substance spectral signature matrix, denoted by $[\mathbf{m}_1\ \mathbf{m}_2\ \cdots\ \mathbf{m}_p]$. That is, the best approximate solution obtained HSI system is the filled dark dot which has the minimum orthogonal projection to the real solution is the filled dark gray dot of the data sphere as shown in Figure 1.5.

A classical approach to solving (1.16) is the least squares estimation given by

$$\hat{\boldsymbol{\alpha}}^{\mathrm{LS}}(\mathbf{r}) = \left(\mathbf{M}^T\mathbf{M}\right)^{-1}\mathbf{M}^T\mathbf{r} \tag{1.17}$$

where $\hat{\boldsymbol{\alpha}}^{\mathrm{LS}}(\mathbf{r}) = \left(\hat{\alpha}_1^{\mathrm{LS}}(\mathbf{r}), \hat{\alpha}_2^{\mathrm{LS}}(\mathbf{r}), \ldots, \hat{\alpha}_p^{\mathrm{LS}}(\mathbf{r})\right)$ and $\hat{\alpha}_j^{\mathrm{LS}}(\mathbf{r})$ is the abundance fraction of the $j$th substance spectral signature $\mathbf{m}_j$ estimated from the data sample vector $\mathbf{r}$. Here, the data sample vector is included to emphasize that the abundance estimate is determined by $\mathbf{r}$. However, in fact, the solution (1.17) can be obtained by the orthogonality principle as follows.

Using (1.1) and (1.2), we construct the orthogonal projection matrix $P_{\mathbf{M}}$ given by

$$P_{\mathbf{M}} = \mathbf{M}\left(\mathbf{M}^T\mathbf{M}\right)^{-1}\mathbf{M}^T\mathbf{r}. \tag{1.18}$$

Then, applying (1.18) to (1.2) yields

$$P_{\mathbf{M}}(\mathbf{r}) = P_{\mathbf{M}}(\mathbf{M}\boldsymbol{\alpha}+\mathbf{n}) = \hat{\boldsymbol{\alpha}}^{\mathrm{LS}}(\mathbf{r}) = \left(\hat{\alpha}_1^{\mathrm{LS}}(\mathbf{r}), \hat{\alpha}_2^{\mathrm{LS}}(\mathbf{r}), \ldots, \hat{\alpha}_p^{\mathrm{LS}}(\mathbf{r})\right)^T. \tag{1.19}$$

Now, we assume that $\mathbf{m}_p = \mathbf{d}$ is the desired target signature to be detected without loss of generality. Since $\mathbf{M}\boldsymbol{\alpha}$ is a mixed signal vector and we are only interested in detecting the desired spectral signature $\mathbf{d}$, a direct standard detection technique cannot be applied to (1.2). To resolve this dilemma, we divide the set of the $p$ target signatures, $\mathbf{m}_1, \mathbf{m}_2, \ldots, \mathbf{m}_p$ into two types of signature classes. One class is made up of the single desired signature, $\mathbf{m}_p = \mathbf{d}$, with its proportional abundance fraction $\alpha_p$ and the other class consists of the remaining signatures, $\mathbf{m}_1, \mathbf{m}_2, \ldots, \mathbf{m}_{p-1}$, as undesired spectral signatures to form an undesired spectral signature matrix denoted by $\mathbf{U}_{p-1} = [\mathbf{m}_1\mathbf{m}_2\cdots\mathbf{m}_{p-1}]$ with $\boldsymbol{\gamma} = (\alpha_1, \alpha_2, \ldots, \alpha_{p-1})^T$ being its corresponding abundance fraction vector. By virtue of such $p$-signatures decomposition (i.e. desired signature specified by $\mathbf{d}$, undesired signatures specified by $\mathbf{U}_{p-1}$) we can further re-express the mixed signal detection model in (1.2) as a $(\mathbf{d},\mathbf{U}_{p-1})$-signal decomposition detection model represented by

$$\mathbf{r} = \mathbf{d}\alpha_p + \mathbf{U}_{p-1}\boldsymbol{\gamma} + \mathbf{n}. \tag{1.20}$$

Once the undesired signatures $\mathbf{m}_1, \mathbf{m}_2, \ldots, \mathbf{m}_{p-1}$ in $\mathbf{U}$ are separated from the desired signature $\mathbf{d}$ in (1.20), we can eliminate their interfering effects before the detection of $\mathbf{d}$ takes place so as to enhance detectability of $\mathbf{d}$. In doing so, we design an orthogonal subspace projector

$$P_{\mathbf{U}_{p-1}}^{\perp} = \mathbf{I} - \mathbf{U}_{p-1}\mathbf{U}_{p-1}^{\#} \tag{1.21}$$

to annihilate **U** from the pixel vector **r** prior to detection of **d** where $\mathbf{U}_{p-1}^{\#} = \left(\mathbf{U}_{p-1}^{T}\mathbf{U}_{p-1}\right)^{-1}\mathbf{U}_{p-1}^{T}$ is the pseudo-inverse of **U**. The notation $^{\perp}_{\mathbf{U}_{p-1}}$ in $P_{\mathbf{U}_{p-1}}^{\perp}$ indicates that the projector $P_{\mathbf{U}_{p-1}}^{\perp}$ maps the observed pixel vector **r** into the orthogonal complement of $\langle \mathbf{U}_{p-1} \rangle$, denoted by $\langle \mathbf{U}_{p-1} \rangle^{\perp}$. Applying $P_{\mathbf{U}_{p-1}}^{\perp}$ in (1.21) to (1.20) results in a $P_{\mathbf{U}_{p-1}}^{\perp}$-transformed data space, that is, for each data sample vector **r**

$$P_{\mathbf{U}_{p-1}}^{\perp}\mathbf{r} = P_{\mathbf{U}_{p-1}}^{\perp}\mathbf{d}\alpha_p + P_{\mathbf{U}_{p-1}}^{\perp}\mathbf{n} \tag{1.22}$$

where the original data sample vector **r** and the noise **n** become $P_{\mathbf{U}_{p-1}}^{\perp}\mathbf{r}$ and $P_{\mathbf{U}_{p-1}}^{\perp}\mathbf{n}$, and the effects of all the undesired signatures $\mathbf{m}_1, \mathbf{m}_2, \ldots, \mathbf{m}_{p-1}$ on **d** and **n** have been eliminated by $P_{\mathbf{U}_{p-1}}^{\perp}$. By further applying a matched filter using the matching signal **d**, we can obtain the OSP detector, $\delta^{\text{OSP}}$, derived by Harsanyi and Chang in [49]

$$\delta^{\text{OSP}}(\mathbf{r}) = \mathbf{d}^T P_{\mathbf{U}_{p-1}}^{\perp}\mathbf{r} = \hat{\alpha}_p(\mathbf{r}). \tag{1.23}$$

Interestingly, it has been shown in [50–55] that the abundance fraction of the $p$th signature $\mathbf{m}_p$ of $\hat{\boldsymbol{\alpha}}^{\text{LS}}(\mathbf{r})$ in (1.17), $\hat{\alpha}_p^{\text{LS}}(\mathbf{r})$, turns out to be the OSP detector, $\delta^{\text{OSP}}$ in (1.23) scaled by a constant $\left(\mathbf{d}^T P_{\mathbf{U}_{p-1}}^{\perp}\mathbf{d}\right)^{-1}$ as

$$\hat{\alpha}_p^{\text{LS}}(\mathbf{r}) = \frac{\mathbf{d}^T P_{\mathbf{U}_{p-1}}^{\perp}\mathbf{r}}{\mathbf{d}^T P_{\mathbf{U}_{p-1}}^{\perp}\mathbf{d}} = \frac{\delta^{\text{OSP}}(\mathbf{r})}{\mathbf{d}^T P_{\mathbf{U}_{p-1}}^{\perp}\mathbf{d}}. \tag{1.24}$$

In order to faithfully restore the spectral information contained in data sample vectors via LSMA, two physical constraints, abundance sum-to-one constraint (ASC), $\sum_{j=1}^{p}\alpha_j = 1$ and abundance non-negativity constraint (ANC), $\alpha_j \geq 0$ for all $1 \leq j \leq p$ are generally imposed on (1.2) to yield the following fully abundance constrained optimization problem:

$$\min_{\boldsymbol{\alpha}\in\Delta}\left\{(\mathbf{r}-\mathbf{M}\boldsymbol{\alpha})^T(\mathbf{r}-\mathbf{M}\boldsymbol{\alpha})\right\} \text{ subject to } \Delta = \left\{\alpha \mid \alpha_j \geq 0 \text{ for} \forall j, \sum_{j=1}^{p}\alpha_j = 1\right\}. \tag{1.25}$$

Due to the use of the inequality constraint, ANC in (1.25), there is no analytical solution available. To solve (1.25), it must rely on numerical algorithms to find approximate solutions. One of such solutions is called fully constrained least squares (FCLS) developed by Heinz and Chang in [56]. However, there are also some other solutions that can be found for (1.25). These solutions are discussed in great detail in Chapter 14 of this book.

## 1.11 Target Detection by Orthogonality Principle

OSP in (1.23) was originally developed for DR and classification via the orthogonality principle. It was further extended to an unsupervised OSP detector, called automatic target generation process (ATGP) [57].

## *1.11.1 ATGP*

OSP implemented in (1.23) requires the prior knowledge of the target signature matrix **M**. In many practical applications, such **M** may be difficult to obtain. To alleviate this problem, it is highly desirable if we can acquire **M** in an unsupervised manner. The ATGP is a result of this need and its implementation can be described as follows.

#### 1.11.1.1 *Automatic Target Generation Process* (ATGP)

1. Initial condition:
   Select an initial target pixel vector $\mathbf{t}_0 = \arg\{\max_{\mathbf{r}}\mathbf{r}^T\mathbf{r}\}$ and $n_{\text{ATGP}}$ is the number of endmember required to generate ATGP. Set $p = 1$ and $\mathbf{U}_0 = [\mathbf{t}_0]$.
2. At the $p$th iteration, apply $P_{\mathbf{t}_0}^{\perp}$ via (1.1) to all image pixels $\mathbf{r}$ in the image and find the $n$th target $\mathbf{t}_n$ satisfying

$$\mathbf{t}_p = \arg\left\{\max{}_{\mathbf{r}}\left[\left(P_{\mathbf{U}_{p-1}}^{\perp}\mathbf{r}\right)^T\left(P_{\mathbf{U}_{p-1}}^{\perp}\mathbf{r}\right)\right]\right\} \tag{1.26}$$

   where $\mathbf{U}_{p-1} = [\mathbf{t}_1\mathbf{t}_2\cdots\mathbf{t}_{p-1}]$ is the target matrix generated at the $(n-1)$st stage.
3. If $p < n_{\text{ATGP}} - 1$, let $\mathbf{U}_p = [\mathbf{U}_{p-1}\mathbf{t}_p] = [\mathbf{t}_1\mathbf{t}_2\cdots\mathbf{t}_p]$ be the $p$th target matrix. Let $p \leftarrow p+1$ and go to step 2. Otherwise, continue.
   At this stage, ATGP is terminated and the final set of produced target pixel vectors comprises $p$ target pixel vectors, $\{\mathbf{t}_0, \mathbf{t}_1, \mathbf{t}_2, \ldots, \mathbf{t}_p\} = \{\mathbf{t}_0\} \cup \{\mathbf{t}_1, \mathbf{t}_2, \ldots, \mathbf{t}_p\}$ that are assumed to be a desired set of $p$ endmembers.

## *1.11.2 Constrained Energy Minimization (CEM)*

Suppose that a hyperspectral image is represented by a collection of image pixel vectors, denoted by $\{\mathbf{r}_i\}_{i=1}^N$ with $\mathbf{r}_i = (r_{i1}, r_{i2}, \ldots, r_{iL})^T$ being an $L$-dimensional pixel vector and $L$ being the total number of spectral bands. More specifically, let original HSI sample space be $\Omega_{\text{data}}$, where $\Omega_{\text{data}} = \{\mathbf{r}_n\}_{n=1}^N$ and $\mathbf{r}_n \in \Re^{L\times 1}$ be a $n$th data sample with $L$ bands. Let $\delta(\mathbf{r}_n)$ be a linear filter specified by a weighting vector $\mathbf{w} = (w_1, w_2, \ldots, w_L)^T$ as

$$\delta(\mathbf{r}_n) = \mathbf{w}^T\mathbf{r}_n. \tag{1.27}$$

Then, its averaged output energy over the entire image scene is represented by

$$\frac{1}{N}\sum_{n=1}^{N}\delta^2(\mathbf{r}_n) = \mathbf{w}^T\mathbf{R}\mathbf{w}. \tag{1.28}$$

where $\mathbf{R} = \frac{1}{N}\sum_{n=1}^{N}\mathbf{r}_n\mathbf{r}_n^T \in \mathbb{R}^{L\times L}$ is the sample autocorrelation matrix, $N$ is the total number of data samples, and $L$ is the number of spectral bands. Now, we can formulate the following constrained optimization problem as follows:

$$\min{}_{\mathbf{w}}\left\{\mathbf{w}^T\mathbf{R}\mathbf{w}\right\} \text{ subject to } \mathbf{d}^T\mathbf{w} = 1. \tag{1.29}$$

The optimal solution to (1.29) is given by

$$\mathbf{w}^{\text{CEM}} = \mathbf{R}^{-1}\mathbf{d}\left(\mathbf{d}^T\mathbf{R}^{-1}\mathbf{d}\right)^{-1}. \tag{1.30}$$

Substituting (1.30) into (1.27) yields the constrained energy minimization (CEM) detector, $\delta^{\text{CEM}}(\mathbf{r})$, developed in [14, 58, 59] and given by

$$\delta^{\text{CEM}}(\mathbf{r}) = \left(\mathbf{w}^{\text{CEM}}\right)^T \mathbf{r} = \frac{\mathbf{d}^T \mathbf{R}^{-1} \mathbf{r}}{\mathbf{d}^T \mathbf{R}^{-1} \mathbf{d}}. \tag{1.31}$$

Now, we consider the following equations:

$$\begin{aligned}
\frac{\left(d^T R^{-1} r\right)^2}{d^T R^{-1} d} &= \frac{\left(d^T R^{-1} r\right)^T \left(d^T R^{-1} r\right)}{d^T R^{-1} d} = \frac{r^T R^{-1} d d^T R^{-1} r}{d^T R^{-1} d} \\
&= r^T R^{-1} d \left(d^T R^{-1} d\right)^{-1} d^T R^{-1} r \\
&= r^T R^{-1/2} R^{-1/2} d \left(d^T R^{-1/2} R^{-1/2} d\right)^{-1} d^T R^{-1/2} R^{-1/2} r \\
&= \left(R^{-1/2} r\right)^T \left[\left(R^{-1/2} d\right)\left(\left(R^{-1/2} d\right)^T R^{-1/2} d\right)^{-1} \left(R^{-1/2} d\right)^T\right] R^{-1/2} r \\
&= \left(R^{-1/2} r\right)^T P_{R^{-1/2} d}\left(R^{-1/2} r\right) = \left(R^{-1/2} r\right)^T P_{R^{-1/2} d}^T P_{R^{-1/2} d}\left(R^{-1/2} r\right) \\
&= \left\| P_{R^{-1/2} d}\left(R^{-1/2} r\right)\right\|^2 = \left\| P_{\bar{d}}(\bar{r})\right\|^2
\end{aligned} \tag{1.32}$$

where $\bar{\mathbf{d}} = \mathbf{R}^{-1/2}\mathbf{d}$, $\bar{\mathbf{r}} = \mathbf{R}^{-1/2}\mathbf{r}$, $P_{\mathbf{R}^{-1/2}\mathbf{d}}^T = P_{\mathbf{R}^{-1/2}\mathbf{d}}$ and $P_{\mathbf{R}^{-1/2}\mathbf{d}}^2 = P_{\mathbf{R}^{-1/2}\mathbf{d}}$. Using (1.32), we can derive

$$\left(\delta^{\text{CEM}}(\mathbf{r})\right)^2 = \left(\frac{\mathbf{d}^T \mathbf{R}^{-1} \mathbf{r}}{\mathbf{d}^T \mathbf{R}^{-1} \mathbf{d}}\right)^2 = \frac{\left\| P_{\mathbf{R}^{-1/2}\mathbf{d}}\left(\mathbf{R}^{-1/2}\mathbf{r}\right)\right\|^2}{\mathbf{d}^T \mathbf{R}^{-1} \mathbf{d}} = \frac{\left\| P_{\bar{\mathbf{d}}}(\bar{\mathbf{r}})\right\|^2}{\left\|\bar{\mathbf{d}}\right\|^2}. \tag{1.33}$$

By virtue of (1.33), $\delta^{\text{CEM}}(\mathbf{r})$ in (1.31) can be further represented by

$$\left|\delta^{\text{CEM}}(\mathbf{r})\right| = \frac{\left\| P_{\bar{\mathbf{d}}}(\bar{\mathbf{r}})\right\|}{\left\|\bar{\mathbf{d}}\right\|} \tag{1.34}$$

which shows that CEM can be also indeed derived by the orthogonality principle.

Furthermore, we can combine CEM with OSP to derive an extended version to OSP and CEM, called target-constrained interference-minimized filter (TCIMF) in [60] which takes advantages of benefits of $\delta^{\text{OSP}}(\mathbf{r})$ in (1.23) and $\delta^{\text{CEM}}(\mathbf{r})$ in (1.31).

## 1.12 Anomaly Detection by Orthogonality Principle

According to [50, 51], OSP detector in (1.23) has been shown to be the Mahalanobis distance. Coincidentally, the detector developed by Reed-Xiaoli detector (RXD) in [61] is also the Mahalanobis distance given by

$$\delta^{\text{RXD}}(\mathbf{r}) = (\mathbf{r} - \boldsymbol{\mu})^T \mathbf{K}^{-1} (\mathbf{r} - \boldsymbol{\mu}) \tag{1.35}$$

where $\mathbf{K}$ and $\boldsymbol{\mu}$ are the global sample covariance matrix and global sample mean, respectively.

Multiplying the right-hand side of (1.35) with $1 = ((\mathbf{r}-\boldsymbol{\mu})^T\mathbf{K}^{-1}(\mathbf{r}-\boldsymbol{\mu})/(\mathbf{r}-\boldsymbol{\mu})^T\mathbf{K}^{-1}(\mathbf{r}-\boldsymbol{\mu}))$, we can re-express (1.35) as

$$\begin{aligned}
\delta^{\mathrm{RXD}}(r) &= (r-\mu)^T K^{-1}(r-\mu) \\
&= \frac{(r-\mu)^T K^{-1}(r-\mu)(r-\mu)^T K^{-1}(r-\mu)}{(r-\mu)^T K^{-1}(r-\mu)} \\
&= (r-\mu)^T K^{-1}(r-\mu)\left((r-\mu)^T K^{-1}(r-\mu)\right)^{-1}(r-\mu)^T K^{-1}(r-\mu) \\
&= (r-\mu)^T\left(K^{-1}(r-\mu)\left((r-\mu)^T K^{-1}(r-\mu)\right)^{-1}(r-\mu)^T\right)K^{-1}(r-\mu) \\
&= (r-\mu)^T\left(K^{-1/2}K^{-1/2}(r-\mu)\left((r-\mu)^T K^{-1/2}K^{-1/2}(r-\mu)\right)^{-1}(r-\mu)^T\right)K^{-1/2}K^{-1/2}(r-\mu) \\
&= \left(K^{-1/2}(r-\mu)\right)^T\left(K^{-1/2}(r-\mu)\left(\left(K^{-1/2}(r-\mu)\right)^T K^{-1/2}(r-\mu)\right)^{-1}\left(K^{-1/2}(r-\mu)\right)^T\right)\left(K^{-1/2}(r-\mu)\right) \\
&= \left(K^{-1/2}(r-\mu)\right)^T P_{K^{-1/2}(r-\mu)}\left(K^{-1/2}(r-\mu)\right) = \left\|P_{K^{-1/2}(r-\mu)}\left(K^{-1/2}(r-\mu)\right)\right\|^2
\end{aligned} \tag{1.36}$$

which implies that RXD can be also derived from the orthogonality principle by the orthogonal projection matrix, $P_{\mathbf{K}^{-1/2}(\mathbf{r}-\boldsymbol{\mu})}$, given by

$$P_{\mathbf{K}^{-1/2}(\mathbf{r}-\boldsymbol{\mu})} = \mathbf{K}^{-1/2}(\mathbf{r}-\boldsymbol{\mu})\left(\left(\mathbf{K}^{-1/2}(\mathbf{r}-\boldsymbol{\mu})\right)^T\mathbf{K}^{-1/2}(\mathbf{r}-\boldsymbol{\mu})\right)^{-1}\left(\mathbf{K}^{-1/2}(\mathbf{r}-\boldsymbol{\mu})\right)^T. \tag{1.37}$$

Interestingly, if we let $\tilde{\mathbf{r}} = \mathbf{K}^{-1/2}\mathbf{r}$ and $\tilde{\boldsymbol{\mu}} = \mathbf{K}^{-1/2}\boldsymbol{\mu}$ in which case $\mathbf{K}^{-1/2}$ is the whitening matrix in the signal detection theory [6], then

$$\begin{aligned}
(\mathbf{r}-\boldsymbol{\mu})^T\mathbf{K}^{-1}(\mathbf{r}-\boldsymbol{\mu}) &= \left(\mathbf{K}^{-1/2}(\mathbf{r}-\boldsymbol{\mu})\right)^T\left(\mathbf{K}^{-1/2}(\mathbf{r}-\boldsymbol{\mu})\right). \\
&= (\tilde{\mathbf{r}}-\tilde{\boldsymbol{\mu}})^T(\tilde{\mathbf{r}}-\tilde{\boldsymbol{\mu}}) = \|\tilde{\mathbf{r}}-\tilde{\boldsymbol{\mu}}\|^2
\end{aligned} \tag{1.38}$$

By combining (1.37) and (1.38), we can have

$$P_{\mathbf{K}^{-1/2}(\mathbf{r}-\boldsymbol{\mu})}\left(\mathbf{K}^{-1/2}(\mathbf{r}-\boldsymbol{\mu})\right) = P_{\tilde{\mathbf{r}}-\tilde{\boldsymbol{\mu}}}(\tilde{\mathbf{r}}-\tilde{\boldsymbol{\mu}}) = \|\tilde{\mathbf{r}}-\tilde{\boldsymbol{\mu}}\|^2. \tag{1.39}$$

Substituting (1.39) in (1.37) yields

$$\delta^{\mathrm{RXD}}(\mathbf{r}) = \|\tilde{\mathbf{r}}-\tilde{\boldsymbol{\mu}}\|^2. \tag{1.40}$$

Alternatively, from the derivations in [62], we can also define

$$\delta^{\mathrm{RXD}}(\mathbf{r}) = \max_{\mathbf{t}}\delta_{\mathbf{t}}^{\mathrm{GLRT}}(\mathbf{r}) \tag{1.41}$$

where $\mathbf{t}$ is an unknown target signature and $\delta_{\mathbf{t}}^{\mathrm{GLRT}}(\mathbf{r})$ can be found by generalized likelihood ratio test (GLRT) [62] as

$$\delta_{\mathbf{t}}^{\mathrm{GLRT}}(\mathbf{r}) = \frac{(\mathbf{t}^T\mathbf{K}^{-1}(\mathbf{r}-\boldsymbol{\mu}))^T(\mathbf{t}^T\mathbf{K}^{-1}(\mathbf{r}-\boldsymbol{\mu}))}{\mathbf{t}^T\mathbf{K}^{-1}\mathbf{t}} = \frac{(\mathbf{r}-\boldsymbol{\mu})^T\mathbf{K}^{-1}\mathbf{t}\mathbf{t}^T\mathbf{K}^{-1}(\mathbf{r}-\boldsymbol{\mu})}{\mathbf{t}^T\mathbf{K}^{-1}\mathbf{t}}. \tag{1.42}$$

As a result, RXD can be also expressed as the orthogonal projection of $\mathbf{K}^{-1/2}(\mathbf{r}-\boldsymbol{\mu})$ on itself via the same orthogonal projection matrix (1.37) as follows:

$$\begin{aligned}\delta^{\text{RXD}}(\mathbf{r}) &= \max_{\mathbf{t}}\delta_{\mathbf{t}}^{\text{GLRT}}(\mathbf{r}) = \max_{\mathbf{t}}\left\|P_{\mathbf{K}^{-1/2}\mathbf{t}}\left(\mathbf{K}^{-1/2}(\mathbf{r}-\boldsymbol{\mu})\right)\right\|^2 \\ &= \left\|P_{\mathbf{K}^{-1/2}(\mathbf{r}-\boldsymbol{\mu})}\left(\mathbf{K}^{-1/2}(\mathbf{r}-\boldsymbol{\mu})\right)\right\|^2 = \|\tilde{\mathbf{r}}-\tilde{\boldsymbol{\mu}}\|^2\end{aligned} \tag{1.43}$$

which is identical to (1.36) and (1.40) and also derived in Theorems 1.1 and 1.2 in [62].

As an alternative, [14, 63] have also proposed a sample correlation matrix **R**-based anomaly detector given by

$$\delta^{\text{R-RXD}}(\mathbf{r}) = \mathbf{r}^T\mathbf{R}^{-1}\mathbf{r}. \tag{1.44}$$

Interestingly, most recently, CEM in (1.31) can be further extended to constrained energy minimization anomaly detector (CEMAD) which includes (1.44) as a special case [64].

## 1.13 Endmember Finding by Orthogonality Principle

The following three widely used endmember finding algorithms, pixel purity index (PPI) [65], vertex component analysis (VCA) [66], and simplex growing algorithm (SGA) [67], are indeed derived from the concept of the orthogonality principle.

### *1.13.1 Pixel Purity Index (PPI)*

The best-known orthogonality principle-based technique for endmember finding was PPI developed by Boardman [65]. It is a simplest approach to materializing convexity via the orthogonality principle where the data sample vectors will be considered as potential endmember candidates if their orthogonal projections fall at one of the ends of particularly generated random unit vectors, referred to as skewers. In other words, any data sample vector whose orthogonal projection lies between these two end points will be considered their mixed sample vector.

The PPI has been widely used for finding endmembers due to availability of the popular software ENVI (Environment for Visualizing Images) which is developed by L3Harris Geospatial and has been widely used for remote sensing image analysis and has found applications in different areas. Unfortunately, several drawbacks have been identified in [2] as obstacles in its implementation. One is the prior knowledge of number of skewers, $K$, required for PPI to generate. Another is inconsistent results due to the use of randomly generated initial endmembers, which are skewers. A third one is a need of knowing a value $t$ used to threshold the PPI counts generated for data sample vectors. Most importantly, it requires a visualization tool to manually select endmembers by human intervention. In this case, it requires skilled, experienced, and well-trained analysts to produce best possible sets of endmembers. Additionally, because PPI is a one-shot operation method, it does not guarantee that this method is always effective. As a result, it needs a large number of skewers, $K$. Due to limited details of how PPI is implemented in the ENVI, in what follows, we describe our own version of interpreting the PPI in terms of its design rationale.

Let $\{\mathbf{r}_i\}_{i=1}^N$ be a given set of data sample vectors. For a given value of $K$, we now use a random generator to produce a set of $K$ random unit vectors, referred to as skewers, $\{\mathbf{skewer}_k\}_{k=1}^K$, which cover $K$ different random directions. All the data sample vectors $\{\mathbf{r}_i\}_{i=1}^N$ are then orthogonally projected on this randomly generated skewer set, $\{\mathbf{skewer}_k\}_{k=1}^K$. According to convexity geometry, an endmember which is considered as a pure signature should occur at end points of some of these skewers with either maximum projection or minimum projection. For each sample vector $\mathbf{r}_i$, we further calculate the number of skewers, denoted by $N_{\text{PPI}}(\mathbf{r}_i)$ at which this particular sample vector occurs as an end point to tally the PPI count for $\mathbf{r}_i$.

Figure 1.8 illustrates how the concept of the PPI works where three skewers, $\mathbf{skewer}_1$, $\mathbf{skewer}_2$, and $\mathbf{skewer}_3$, are indicated by three random unit vectors, the sample vectors are shown by open circles, three endmembers $\mathbf{e}_1$, $\mathbf{e}_2$, and $\mathbf{e}_3$ by solid circles located at three vertices of the triangles, and a cross "x" used to indicate a maximum or minimum projection of an endmember on a skewer. Because of convexity, all the sample vectors inside the triangle should have their PPI counts = 0 in the sense that they are mixtures of the three endmembers at the vertices of the triangle indicated by dashed lines. It should be noted that a maximum or minimum occurred at a skewer is one that yields the maximum or minimum value among all the sample vectors. Also, a projection can be positive or negative depending upon whether it occurs in the same or opposite direction of a skewer.

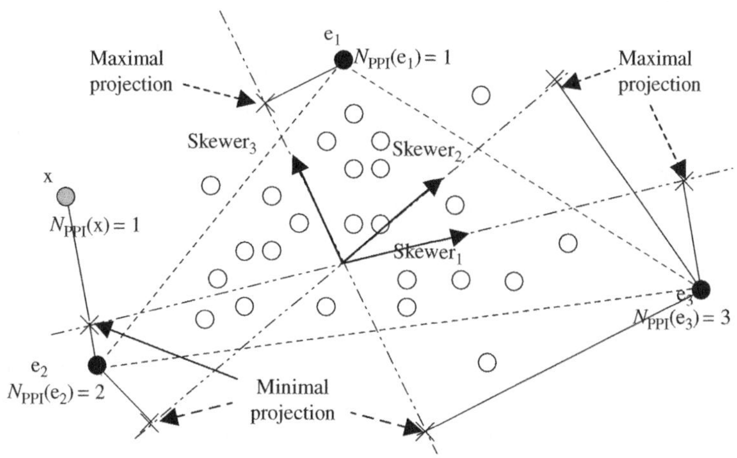


**Figure 1.8** An illustration of PPI with three endmembers $\mathbf{e}_1$, $\mathbf{e}_2$, and $\mathbf{e}_3$.

---

***PPI***

Let $\{\mathbf{r}_i\}_{i=1}^N$ be a given set of data sample vectors.

1. *Initialization:*
   (i) Use the VD to determine the number of dimensions, $q$, required to be preserved after DR.
   (ii) Perform DR to reduce data dimensionality to $q$.
   (iii) Randomly generate a set of $K$ unit vectors called "skewers," $\{\mathbf{skewer}_k\}_{k=1}^K$, where $K$ is a pre-assumed sufficiently large positive integer.
2. *PPI count calculation:*
   Project all the data sample vectors $\{\mathbf{r}_i\}_{i=1}^N$ orthogonally onto all the skewers, $\{\mathbf{skewer}_k\}_{k=1}^K$. For each data sample vector $\mathbf{r}_i$, compute inner products of $\mathbf{r}_i$ with all skewers, $\{\mathbf{skewer}_k\}_{k=1}^K$, i.e. $< \mathbf{r}_i, \mathbf{skewer}_k > \; = \mathbf{r}_i^T\mathbf{skwer}_k$. Let the maximal and minimal projections of each skewer, $\mathbf{skewer}_k$, produced by the data sample vector, $\{\mathbf{r}_i\}_{i=1}^N$ be $\max_k = \max_{1 \le i \le N}\{\mathbf{r}_i^T\mathbf{skewer}_k\}$ and $\min_k = \min_{1 \le i \le N}\{\mathbf{r}_i^T\mathbf{skewer}_k\}$. The PPI candidate set of $\mathbf{r}_i$ is calculated by including skewers with which $\mathbf{r}_i$ produces either maximal or minimal inner products, i.e.

$$\text{PPI}(\mathbf{r}_i) = \left\{\mathbf{skewer}_k, 1 \le k \le K \mid \mathbf{r}_i^T\mathbf{skewer}_k = \max{}_k \text{or} \min{}_k\right\} \quad (1.45)$$

The PPI count of the $\mathbf{r}$ is defined as the number of skewers in PPI($\mathbf{r}_i$) found by (1.45), i.e. $N_{\text{PPI}}(\mathbf{r}_i) = |\text{PPI}(\mathbf{r}_i)|$ where $|A|$ is defined as the number of elements in a set $A$.

3. *Endmember extraction:*
   A data sample $\mathbf{r}_i$ is selected as an endmember if its PPI count $N_{\text{PPI}}(\mathbf{r}_i) \ge t$ with a predetermined value $t$.

---

However, in order for PPI to be effective, a large number of skewers, $K$, are generally required to guarantee that these skewers can cover as many random directions as possible. For example, in Figure 1.8, the shade sample vector labeled by $\mathbf{x}$ outside the triangle has the same $N_{\text{PPI}} = 1$ as $\mathbf{e}_1$ does, but it is obviously not an endmember. The occurrence of such an incident is the result of insufficient numbers of skewers used for the PPI. Unfortunately, by far there is no guideline suggested for users to determine how many skewers should be used for PPI. Empirically, this number is generally hundreds if not thousands for PPI to perform well. As a result of doing so, PPI also suffers from three setbacks. One is that the range of PPI counts is significantly increased. Consequently, it becomes more difficult to find an appropriate value $t$ for thresholding PPI counts to determine a final set of endmembers. A second drawback is un-reproducibility of final selected endmembers due to the random nature arising in skewers. Accordingly, the results are inconsistent. A third drawback is requirement of DR to reduce high-dimensional hyperspectral data. However, it also comes with an issue, how many dimensions, $q$, must be retained to include all the endmembers.

### *1.13.2 Vertex Component Analysis (VCA)*

Since an endmember should have its components to be non-negative and PPI is an abundance unconstrained method, PPI may find its endmembers in negative coordinates. The VCA developed in [66] was developed to address this issue by limiting its search area in the first quadrant with all coordinates to be non-negative. As a result, all VCA-found endmembers should satisfy the constraint specified by ASNC. Technically speaking, VCA can be considered a special case of ATGP with its search region constrained by ANC.

---

*VCA*

---

1. Let the number of endmembers to be generated be $p$. Set a counter to $k = 0$.
2. Perform a DR transform to reduce the original space $\mathbf{X}$ with dimensionality $L$ to a reduced data space $\hat{\mathbf{X}}$ with dimensionality of $p$.
3. Set the initial vector, $\mathbf{e}^{(0)} = \left(\underbrace{0, 0, \ldots, 1}_{p}\right)$ and let a $p \times p$ auxiliary matrix $\mathbf{A}^{(0)}$ be $\mathbf{A}^{(0)} = [\mathbf{e}^{(0)}\mathbf{0}\cdots\mathbf{0}]$.
4. At iteration $k \ge 1$, generate a Gaussian random vector, $\mathbf{w}^k$, to be used to produce $\mathbf{f}^k$.

$$\mathbf{f}^{(k)} = ((\mathbf{I} - \mathbf{A}^{(k-1)}(\mathbf{A}^{(k-1)})^{\#})\mathbf{w}^k)/(\|(\mathbf{I} - \mathbf{A}^{(k-1)}(\mathbf{A}^{(k-1)})^{\#})\mathbf{w}^k\|). \quad (1.46)$$

5. Find $\mathbf{e}^{(k)}$ that maximizes $\mathbf{f}^{(k)T}\hat{\mathbf{x}}$ over $\hat{\mathbf{x}} \in \hat{\mathbf{X}}$, that is,

$$\mathbf{e}^{(k)} = \arg\left\{\max{}_{\hat{\mathbf{x}}\in\hat{\mathbf{X}}}\left[|\mathbf{f}^{(k)T}\hat{\mathbf{x}}|\right]\right\}. \quad (1.47)$$

6. Use $\mathbf{e}^{(k)}$ to replace the $k$th column of $\mathbf{A}^{(k)}$ to let $\mathbf{A}^{(k)} = [\mathbf{e}^{(1)}\cdots\mathbf{e}^{(k)}\mathbf{0}\cdots\mathbf{0}]$.
   If $k = L$, the algorithm is terminated. Otherwise, let $k \leftarrow k + 1$ and go to step 4.

---

### 1.13.3 Simplex Growing Algorithm (SGA)

VCA is not a fully abundance constrained endmember finding method because it does not impose ASC as N-finder (N-FINDR) developed by Winter in [68] which satisfies both ASC and ANC. However, N-FINDR has practical issues to be implemented since it requires exhausting all possible combinations of endmembers. To mitigate this problem, the SGA was developed in [67] as a sequential version of N-FINDR which finds one endmember at a time instead of all endmembers altogether at the same time as N-FNDR does.

---

*SGA*

---

1. Initialization:
   (a) Estimate the number of endmembers to be generated, $p$. A recent concept, VD, developed in [14, 15] can be used for this purpose.
   (b) Find the two sample vectors with maximal distance as the desired initial endmember pixels and set $n = 2$.
2. At $n \geq 2$ and for each sample vector r, we calculate, simplex volume (SV) of the simplex specified by vertices $e_1, e_2, \ldots, e_n$, r, $SV(e_1, \ldots, e_n, r)$ defined by

$$SV(\mathbf{e}_1, \ldots, \mathbf{e}_n, \mathbf{r}) = \frac{\left| \det \begin{bmatrix} 1 & 1 & \ldots & 1 & 1 \\ \mathbf{e}_1 & \mathbf{e}_2 & \ldots & \mathbf{e}_n & \mathbf{r} \end{bmatrix} \right|}{n!}. \tag{1.48}$$

   Since the matrix $\det \begin{bmatrix} 1 & 1 & \ldots & 1 & 1 \\ \mathbf{e}_1 & \mathbf{e}_2 & \ldots & \mathbf{e}_n & \mathbf{r} \end{bmatrix}$ in (1.48) is not necessarily a square matrix, a DR technique such as PCA or MNF is required to reduce the original data dimensionality $L$ to the dimension $n$.
3. Find $\mathbf{e}_{n+1}$ that yields the maximum of (48), that is,

$$\mathbf{e}_{n+1} = \arg\{\max_{\mathbf{r}}[V(\mathbf{e}_1, \ldots, \mathbf{e}_n, \mathbf{r})]\}. \tag{1.49}$$

4. Stopping rule:
   If $n < p - 1$, then $n \leftarrow n + 1$ and go step 2. Otherwise, the final set of $\{\mathbf{e}_1, \mathbf{e}_2, \ldots, \mathbf{e}_p\}$ is the desired $p$ endmembers.

   Interestingly, as shown in [69], SGA is indeed equivalent to ATGP. In fact, SGA can be also shown to be an OSP-based endmember finding algorithms in [70, 71].

---

## 1.14 Low Rank and Sparse Representation by OSP via Orthogonality Principle

In Section 1.4, the pigeon-hole principle has been used to determine the orders of low rank space and sparse space via VD in Section 1.6. In order for these representations to be effective, the matrix decomposition must be orthogonal.

Most recently, an approach to focus on the decomposition of low rank and sparse spaces is to model a data matrix **X** as

$$\mathbf{X} = \mathbf{L} + \mathbf{S} \tag{1.50}$$

by a linear decomposition of a low-rank matrix, **L**, and a sparse matrix, **S**. In this case, the randomized approximate matrix decomposition [72] and the RPCA [11] can be used to solve the decomposition problem. However, the model in (1.50) does not separate the noise from the sparse space **S**. To address this issue, a general model is to represent a data matrix **X** as

$$\mathbf{X} = \mathbf{L} + \mathbf{S} + \mathbf{n} \tag{1.51}$$

by a linear decomposition of a low-rank matrix, **L**, and a sparse matrix, **S**, plus a noise matrix, **n**, referred to as LRaSMD [12]. Of particular interest is an approach called go decomposition (GoDec) developed in [12] can be used to find the low-rank matrix **L** and the sparse matrix **S**. Recently, Chang et al. has derived OSP-based GoDec (OSP-GoDec) which shows that GoDec can be also derived from OSP [73]. In addition to OSP-GoDec, OSP has been recently applied to background (BKG) removal and suppression in target detection [74] and anomaly detection [75].

It is known that the OSP performance is sensitive to its used prior target knowledge, which is significantly affected by target BKG. To resolve this issue, three approaches were developed to extend OSP in improving its performance in [74]. One is data sphering (DS) which can suppress BKG via removing the first- and second-order data statistics. Another takes advantage of a recently developed LRaSMD to separate BKG and target signal sources in two subspaces characterized by the low-rank matrix and sparse matrix, respectively, and then annihilates BKG via the low-rank matrix, referred to as BKG-annihilated OSP (BA-OSP). A third approach combines DS and LRaSMD to further improve OSP over target detectability and BKG suppressibility. Experiments show that implementing OSP in conjunction with DS and LRaSMD significantly improves OSP in target detection and BKG suppression, and also performs as well as a widely used CEM-based subpixel target detection.

Due to its inherent requirement of prior target knowledge, OSP has not been explored in anomaly detection. Chang et al. takes advantage of an unsupervised OSP-based algorithm, ATGP, developed in Section 1.11.1 and OSP-GoDec along with DS in [74] to make OSP applicable to anomaly detection. Its idea is to implement ATGP on the BKG and target subspaces constructed from the low-rank matrix **L** and sparse matrix **S** in (1.50) generated by OSP-GoDec to derive an OSP-based anomaly detector (OSP-AD). In particular, OSP-AD also includes DS to remove BKG interference from the target subspace so as to enhance anomaly detection. Surprisingly, operating data samples on different constructions of the BKG subspace and the target subspace yield various versions of OSP-AD. Experiments show that given an appropriate construction of the BKG subspace and the target subspace, OSP-AD can be shown to outperform existing anomaly detectors including Reed-Xiaoli anomaly detector, RXD [63].

It should be noted that several recent approaches developed in [76, 77] have used a mixture of Gaussian distributions (MoG) to model BKG. However, there are two issues arising in these approaches. One is derived from the fact that a linear mixture of Gaussian distribution is also a Gaussian distribution. This implies that BKG described by a single Gaussian distribution is also heavily determined by the number of Gaussian distribution to be mixed for a Gaussian mixture. Unfortunately, such finite Gaussian mixture model (MoG) issue was not discussed and investigated. The other issue is that there may be distributions such as Rician, Raleigh distributions, other than Gaussian distributions that can be used to characterize BKG. In other words, BKG cannot be completely characterized by Gaussian distributions. In this case, using MoG to estimate and reconstruct BKG is ineffective. So, to address this issue, Chen et al. [78] developed a component decomposition analysis (CDA) model which represented the data space **X** as a three-component orthogonal decomposition by

$$\mathbf{X} = \mathbf{PC}^{m} + \mathbf{IC}^{j} + \mathbf{n}_{\mathrm{CDA}} \tag{1.52}$$

which indeed extended LRaSMD in (1.51) by replacing the low-rank matrix **L** and sparse matrix **S** with principal component (**PC**) space and independent component (**IC**) space, respectively. It further showed that the CDA coupled with the sparsity cardinality (SC) proposed in [12] in fact outperformed LRaSMD model-based anomaly detection approaches in [78]. As a matter of fact, OSP-GoDec in [73] and CDA in [78] are very good examples to take advantage of the pigeon-hole and orthogonality principles to represent a data space **X** in (1.52) in a very effective means. It first makes use of OSP in (1.21) to orthogonally decompose into three component spaces and then appeals for VD in (1.6) and (1.7) to estimate the value of $p = m + j$ and MX-SVD to determine the value of $j$ so that $m = p - j$ and $j$ can be used to specify the order of low-rank space **L** in (1.51) or the number of PCs in (1.52) and the sparsity order of the sparse space **S** (1.51) and the number of ICs in (1.52), respectively.

## 1.15 Hyperspectral Classification

Up to now, we have discussed applications of the pigeon-hole and orthogonality principles to HSI in spectral unmixing, subpixel detection, endmember finding, etc., and have not discussed their applications in classification because of several reasons. Classification has been a fundamental task for land cover/use applications in remote sensing where the maximum likelihood classification has been widely used for MSI classification [79, 80]. Since land cover/use classification usually does not require subpixel and mixed pixel information but rather spatial information to do well, the spatial information provided by MSI is generally sufficient. So, when HSI sensors were developed with abundant spectral information acquired by hundreds of contiguous spectral bands, the spatial domain-based maximum likelihood classification is no longer effective. To resolve this dilemma, spectral-spatial classification (SSC) has been introduced for HSIC to take both spectral and spatial information into account [81–83] and can be considered a natural extension to MSI classification. However, there are several serious issues which are encountered in HSI but not in MSI need to be addressed.

One is the requirement of ground truth to exclude BKG from classification. In HSIC, BKG pixels are considered as uninteresting data samples. In this case, commonly used classification measures, overall accuracy (OA), average accuracy (AA), or kappa coefficient, have been used to evaluate classification performance. Unfortunately, BKG pixels in classification are technically considered as unlabeled data samples which may be due to the fact that these samples cannot be identified by prior knowledge or difficult to be labeled because of their complexity and mixing structures, some of which are indeed significant samples. Discarding BKG pixels seems not appropriate and classification results with exclusion of BKG may also be misleading.

A second issue is that if BKG issue must be factored in, how to deal with BKG pixels is challenging; specifically, OA, AA, and kappa coefficient cannot be used for this purpose. A third issue is how to take care of small classes since the OA, AA, and kappa coefficient of a small class can be very easily dominated and compromised by the accuracies of large classes. On the other hand, in many practical applications, small classes may provide critical and vital information for data analysis. Ignoring small classes may result in incorrect conclusions. The last and most crucial issue is its used pure pixel classification. Recall that the major strength of HSI is its capability of uncovering and revealing unknown material substances at the subpixel level and mixed pixel scale by its wealthy spectral information. The pure pixel classification does not take this important advantage.

### 1.15.1 Hyperspectral Mixed Pixel Classification (HMPC)

One of the major issues arising in hyperspectral pure pixel classification is its classification performed by hard decisions. This is particularly troublesome for those data samples nearby class boundaries which exhibit mixing properties by boundary samples. To address this issue, Xue et al. first developed an iterative CEM (ICEM)-based SSC in [84] which took advantage of CEM as a subpixel target detection to perform hyperspectral mixed pixel classification (HMPC). In the meantime, since CEM is a target detector, it can deal with the BKG issue. Specifically, [84] introduced precision rate (PR) to deal with classification accuracy when the BKG pixels are included for consideration. In particular, Chang [85] developed a statistical detection theory approach by including signal detection theory in classification where a 3-dimensional receiver operating characteristics (3D ROC) analysis introduced in [86] to take care of the BKG issue by model BKG+noise as the null hypothesis. In this case, the originality principle can be applied to operate a detector on the space orthogonal to the BKG+noise space. The CEM in Section 1.11.2 takes exactly such advantage of the orthogonality principle to remove or suppress the background effect. Furthermore, as also shown in [87], the originality principle can be further used to determine VD.

When CEM is implemented by (1.31), the used target signature **d** is assumed to be known. In HSIC, this **d** is calculated by the complete ground truth. Since HSIC generally uses training samples to perform classification, it is highly desirable that ICEM can also use training samples to calculate **d**. This leads to iterative random training sampling CEM (IRTS-CEM) developed in [88]. While IRTS-CEM improved ICEM by using IRTS, there is one disadvantage of CEM resulting from its inherent limitation on linearity implemented in (1.27), which cannot resolve linear non-separability issue. This is particularly severe to those data samples mixed with class boundary samples. As a result, ICEM cannot outperform SSC to some extent because SSC uses nonlinear classifiers such as support vector machine (SVM) [89]. To make CEM a nonlinear classifier, a kernel version of IRTS-CEM (IRTS-KCEM) was developed in [90] so that IRTS-KCEM improved IRTS-CEM significantly and even performed better than any existing SSC method. It is known that classification difficulty often occurs on edge data samples which may be nonlinearly mixed with class boundaries. The impressive performance of IRTS-KCEM results from the capability of KCEM in detecting subpixel targets that allows KCEM to classify mixed samples for HPMC [90]. It is known that classification difficulty often occurs on edge data samples which may be nonlinearly mixed with class boundaries.

To further address the issue in classification of small classes, Zhong et al. recently introduced the concept of class features to account for significance of small classes so that OA was further extended to class feature-based OA (CF-OA) [91]. In this case, each class has its own class features to be recognized and the issue of small classes can be addressed by the pigeon-hole principle where each pigeon-hole can be specified by the class feature of a particular pigeon without compromise. In addition, if we consider each class can be characterized by its class unique feature, then the pigeon-hole principle also tells us that for each class, we should be able to generate its effective class feature to select an appropriate spectral band in the sense that all the data samples in this particular class should be accommodated by the selected band. In terms of the pigeon-hole principle, the same type of pigeons will be accommodated by the same spectral band which will be distinct from other spectral bands. So, the number of spectral bands needed to be selected for HSIC must be at least equal to and greater than the number of classes of interest, $n_{\text{class}}$ [92–94]. Of course, the true number of bands to be selected is determined by spectral variability of each class. With this in mind, one class may be specified by more than one spectral band. This issue provides a further direction of BS in HSIC.

A final comment on the BKG issue is noteworthy. The BKG is not an exclusive issue for classification. As a matter of fact, the BKG issue is more serious and severe in anomaly detection than classification. This is mainly due to the fact that anomaly detection has no prior knowledge about the targets it detects as opposed to classification which has at least prior knowledge of classes of interest. Accordingly, the BKG issue has received considerable interest in anomaly detection where the BKG must be suppressed or removed before anomaly detection takes place. This is similar to HSIC which removes BK pixels by ground truth prior to classification. The difficulty with anomaly detection for removing/suppressing BKG is that the BKG is completely unknown and must be estimated or reconstructed so that it can be separated from anomalies. Detailed discussion on anomaly detection can be found in [62–64]. If the BKG pixels cannot be identified by ground truth in HSIC, then HSIC will become more difficult and challenging than anomaly detection since HSIC has many classes of interest compared to the only targets of interest which are anomalies. To mitigate this BKG issue, one-class classification approach has been investigated [95, 96].

## *1.15.2 Number of Sampled Bands Lower Than Number of Classes*

In HSIC, it generally assumes that each class requires its own particular band to perform its classification. If we interpret the number of classes, $n_{\text{class}} = p$ as the number of pigeons and the number of bands, $L$, as the number of pigeon-holes, then according to the pigeon-hole principle, it will need $L \geq p$. To be precise, for HSIC to perform effectively, the number of bands to be selected, $n_{\text{BS}}$, should be at least equal to or greater than $p$, $n_{\text{BS}} \geq p$. This section is of particular interest and is included to show that when the pigeon-hole principle is violated by $n_{\text{BS}} < p$, BS-based HSIC methods indeed perform very poorly. Interestingly, BSam-based HMPC methods show otherwise when $n_{\text{BSam}} < p$. In other words, HMPC can perform very well even if $n_{\text{BSam}}$ in (1.15) is well below $n_{\text{class}} = p$.

The data to be used for experiments is the Purdue Indiana Indian Pines test site with an aerial view shown in Figure 1.9(a) along with its ground truth of 17 class maps in Figure 1.9(b) and their class labels in Figure 1.9(c). It is an AVIRIS image scene and has size of $145 \times 145 \times 220$ pixel vectors with water absorption bands (bands 104–108 and 150–163, 220). It should be noted that in our experiments, 200 bands were used by excluding 20 of water absorption bands.

Table 1.1 tabulates the number of data samples in parentheses collected for each class where there are three small classes, 1, 7, and 9 with less than 50 data samples and three large classes, 2, 11, and 14 with more than 1000 data samples. So, this data set has an imbalanced class issue in classification. Table 1.2 tabulates the number of training samples and test samples used for experiments.

This section used the Purdue data scene in Figure 1.9 to conduct experiments that particularly show the great advantage of using BSam over BS for HMPC. That is, when $n_{\text{BS}} < p$, BS performs poorly but BSam still performs well even when $n_{\text{BSam}} < p$.

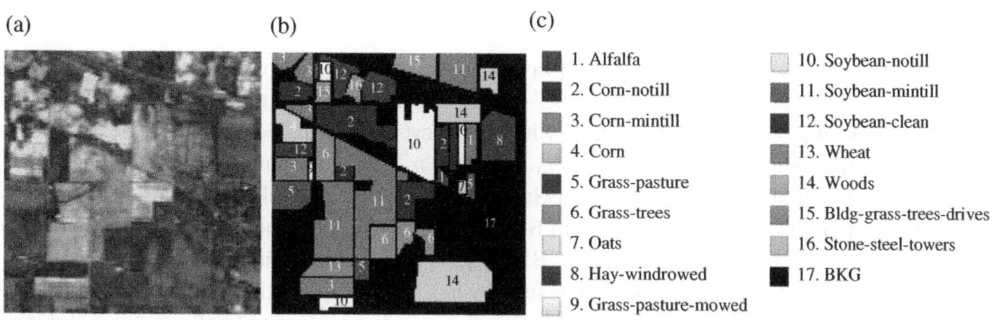


**Figure 1.9** Purdue Indiana Indian Pines. (a) false-color composite image, (b) reference map, and (c) classes by colors, where the color maps of the Purdue data scene can be found in [2], figure 1.13/with permission of John Wiley & Sons.

**Table 1.1** Class labels of purdue's data

| | | | | | |
|---|---|---|---|---|---|
| Class 1 (46) | Alfalfa | Class 7 (28) | Grass/pasture-mowed | Class 13 (205) | Wheat |
| Class 2 (1428) | Corn-notill | Class 8 (478) | Hay-windrowed | Class 14 (1265) | Woods |
| Class 3 (830) | Corn-min | Class 9 (20) | Oats | Class 15 (386) | Bldg-grass green-drives |
| Class 4 (237) | Corn | Class 10 (972) | Soybeans-notill | Class 16 (93) | Stone-steel towers |
| Class 5 (483) | Grass/pasture | Class 11 (2455) | Soybeans-min | Class 17 (10249) | BKG |
| Class 6 (730) | Grass/trees | Class 12 (593) | Soybeans-clean | | |

**Table 1.2** Number of training samples allocated into each class for Purdue data

| | Indian Pines | |
|---|---|---|
| Class | $n_{\text{training}}$ | $n_{\text{test}}$ |
| 1 | 23 | 23 |
| 2 | 36 | 1391 |
| 3 | 35 | 792 |
| 4 | 32 | 203 |
| 5 | 32 | 448 |
| 6 | 35 | 696 |
| 7 | 14 | 14 |
| 8 | 32 | 444 |
| 9 | 10 | 10 |
| 10 | 36 | 937 |
| 11 | 43 | 2416 |
| 12 | 34 | 560 |
| 13 | 33 | 172 |
| 14 | 40 | 1227 |
| 15 | 34 | 354 |
| 16 | 31 | 62 |
| Total | 500 | 9749 |

Table 1.3 tabulates the oval accuracy ($P_{OA}$)/average accuracy ($P_{AA}$) produced by BSam-NKCEM, BSam-IRTS-NKCEM, BS-NKCEM, and BS-IRTS-NKCEM in [90] using the Purdue data for experiments with $n_{BS} = n_{BSam} = 15, 10, 5$, where the best results are boldfaced. For comparison, we also include the results obtained by $n_{BS} = n_{BSam} = 18 > n_{class} = 16$, where $n_{BS} = 18$ was determined by VD in [92, 93]. It should be noted that BSam in Table 1.3 was implemented by two versions, random BSam (RBSam) and random uniform BSam (RUBSam), developed in [44] and BS was also implemented by UBS and constrained BS (CBS) in [24] and also in Chapter 4.

According to Table 1.3, the classification results by $n_{BSam}$ were nearly the same when low BSamR was used. Conversely, BS performed very poorly when $n_{BS}$ is small, specifically $n_{BS} = 5$. Interestingly, if IRTS was implemented in conjunction with BS, their results were not affected much by small values of $n_{BS}$. This is because IRTS allowed new spatial classification information to be added to iteration after iteration during its iterative process so as to improve classification rates.

**Table 1.3** Averaged $P_{OA}/P_{AA}$ resulting from running 10 times NKCEM using different $n_{BSam}$ for Purdue Indiana Indian Pines data

| Method | | $n_{BSam}$ | | | |
|---|---|---|---|---|---|
| | | 18 (VD) | 15 | 10 | 5 |
| BSam | RBSam-NKCEM | **99.98**/99.97 | **99.95/99.94** | 99.91/**99.92** | **99.90**/99.87 |
| | RUBSam-NKCEM | 99.96/**99.98** | **99.95/99.93** | **99.92**/99.91 | **99.91/99.92** |
| | RBSam-IRTS-NKCEM | 99.16/99.63 | 99.12/99.57 | 99.10/99.60 | 98.83/99.36 |
| | RUBSam-IRTS-NKCEM | 99.19/99.62 | 99.18/99.59 | 98.86/99.51 | 98.86/99.52 |
| BS | UBS-NKCEM | 80.05/82.90 | 79.02/81.75 | 72.12/75.69 | 58.40/60.48 |
| | CBS-NKCEM | 66.60/69.82 | 63.92/67.59 | 56.14/58.42 | 39.69/42.08 |
| | UBS-IRTS-NKCEM | 98.96/99.52 | 98.97/99.47 | 99.82/99.46 | 98.81/99.39 |
| | CBS-IRTS-NKCEM | 99.05/99.47 | 99.82/99.38 | 98.77/99.38 | 98.75/99.31 |

As noted above, we have explained why IRTS did well in classification even when $n_{BS} = n_{BSam} = 15$, 10, 5 was lower than the $n_{class} = 16$ in Table 1.3. In order to see why it is this case, we also performed experiments to show how data samples in a class can be compacted through an iterative process and eventually clustered very tightly in one class. For the illustrative purpose of visual inspection, only three classes, 3, 10, and 12 were implemented using a lower BSam rate with $n_{BSam} = 5$ which is much smaller than the total number of classes, 16 for the Purdue data. The reason of selecting these three classes was simply because their classification rates have been shown to be low in [90–94]. To save space, only final classification results are presented in Figure 1.10(a–d) which show 3D visual representations of final classification of the three classes, 3, 10, and 12 using the first three principal components produced by (a) UBS, (b) CBS, (c) RBSam-NKCEM, and (d) RUBSam-NKCEM. As we can see from Figure 1.10, these three classes were still mixed heavily and very poorly classified by BS. By contrast, BSam still performed very well with these three classes well separated. This indicated that BSam could do well even though the $n_{BSam}$ was well below $n_{class} = 16$.

Figure 1.11 further showed that the results in Figure 1.10 could be extended to six classes which included three small classes, classes 1, 7, and 9 in addition to classes 3, 10, and 12 in Figure 1.10, particularly, $n_{class} = 6$ is greater than the $n_{BSam} = n_{BS} = 5$. Under this circumstance, Figure 1.11 showed that BS did even poorly compared to Figure 1.10. On the other hand, Figure 1.11 also demonstrated that BSam not only remained performing well for three classes difficult to be classified but also could do well for small classes in dealing with imbalance class issue. However, BS even did poorly compared to Figure 1.10.

According to Figures 1.10 and 1.11, BSam worked significantly well to separate different classes even when $n_{BSam}$ was well below $n_{class}$. But this is not true for BS which performed very poorly with all the classes heavily mixed. It is our belief that no BS methods can do what BSam does when $n_{BS}$ is less than $n_{class}$. These further showed that BSam inherited the advantage of random signal sampling from CS that can go below the Nyquist rate. More details on BSam of KCEM for HMPC can be found in [97].

### *1.15.3 Potential and Promise of Band Sampling in HMPC*

The most significant novelty resulting from BSam is its use of iterative feedback loops to randomly sample bands to improve classification. Such a unique advantage has several benefits which cannot be provided by BS and CS as demonstrated in Section 1.15.2. One is to include a set of spatial filters

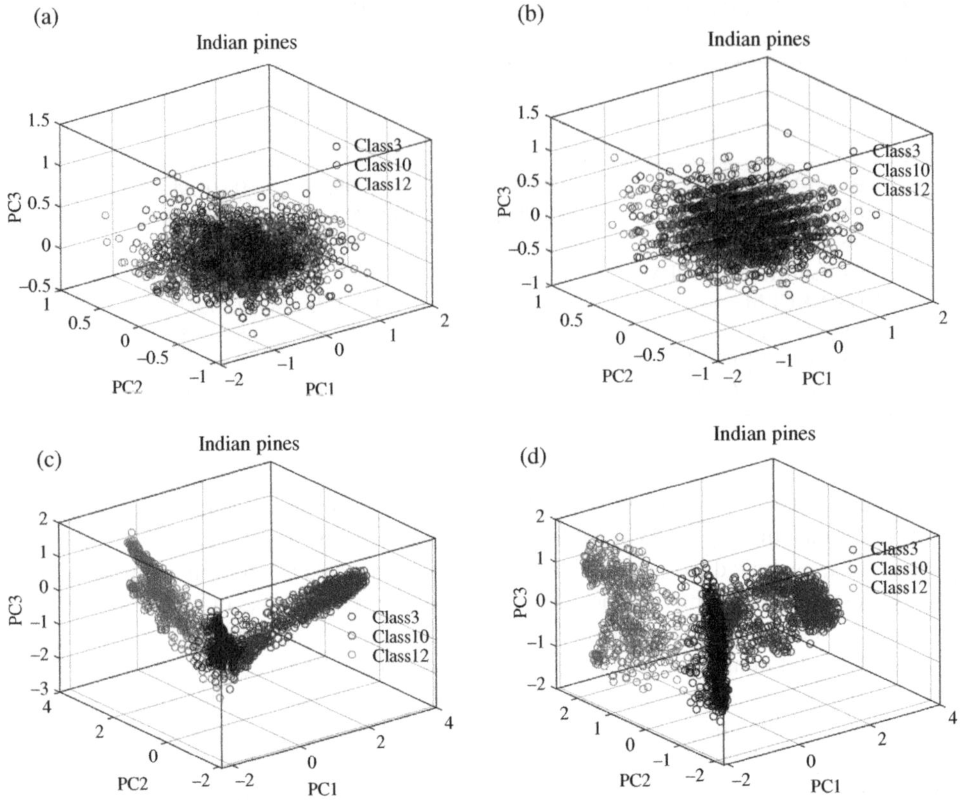


**Figure 1.10** 3D visual representations of classes, 3, 10, and 12 using the first three principal components with $n_{\mathrm{BSam}} = 5$ for Purdue data by (a) UBS, (b) CBS, (c) RBSam-NKCEM, and (d) RUBSam-NKCEM.

to capture spatial information from classified data samples using only sampled bands. This obtained spatial-filtered information is then added to expand the current data cube for the next round BSam. Another is that the new expanded data cube can reduce uncertainty caused by RBSam. In addition, the RBSam allows all bands to be resampled equally likely iteration by iteration. So, the worst scenario is that the bands are randomly sampled at the first iteration and the same bands are then fixed without further BSam throughout iterations. As a result, those bands which were not sampled in the first iteration would not be sampled afterward. This scenario corresponds to BS and CS, both of which select and sense bands and then stay with these same bands during entire data processing. As opposed to iterative fixed BSam, BSam randomly samples bands iteration after iteration, in which case all bands will be resampled equally likely. So, the best scenario would be the case that all bands are sampled throughout the iterative process. If this occurs, it corresponds to the case that all full bands have been sampled. Consequently, the results will certainly be better than that using full bands.

BSam has shown to be a very promising technique. Specifically, it has great potential in HMPC. Its superiority also comes from its ability in mixed pixel classification, an advantage that many HSIC techniques do not have. However, this is also traded for its applications because the stopping rule of the iterative process implemented by random sampling is determined by a specific application. For our case of classification, MAP is used to determine if the iterative should be terminated. Interestingly, we can turn this disadvantage to an advantage in the sense that the stopping rule can be

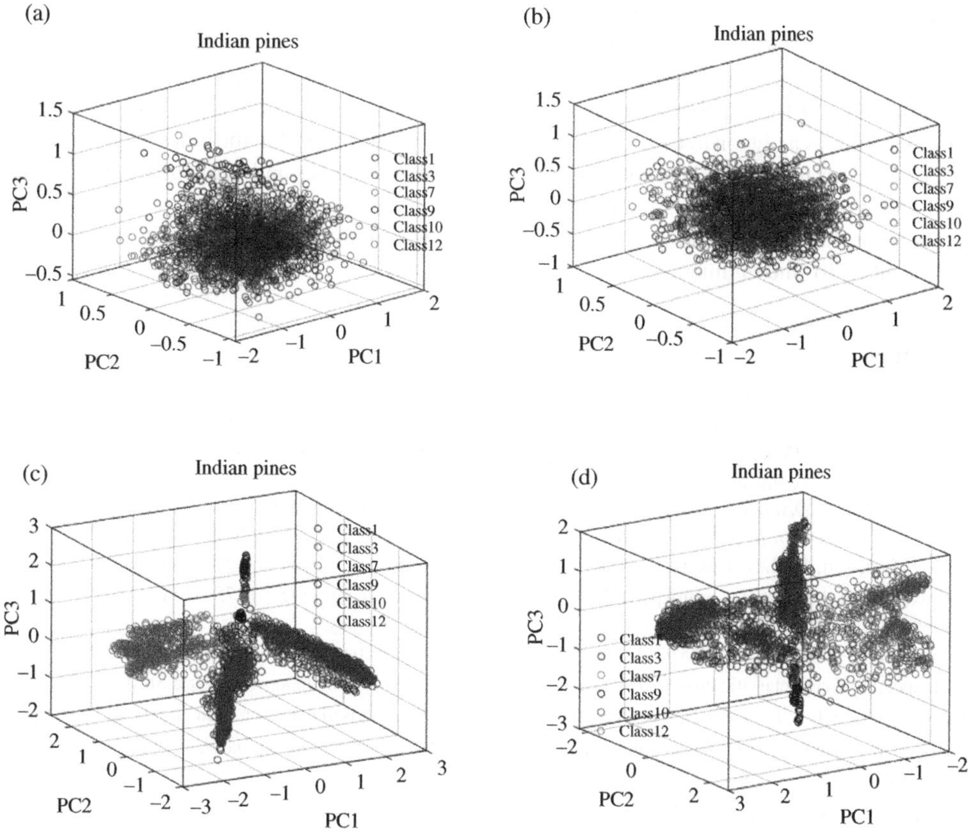


**Figure 1.11** 3D visual representations of classes, 1, 3, 7, 9, 10, and 12 using the first three principal components with $n_{\mathrm{BSam}} = 5$ for Purdue data by (a) UBS-NKCEM, (b) CBS-NKCEM, (c) RBSam-NKCEM, and (d) RUBSam-NKCEM.

custom-designed by a particular application, for example, in target detection, BSam can further improve KCEM performance if the area of an ROC curve, AUC, can be used to replace MAP. As demonstrated in Table 1.3 and Figures 1.10 and 1.11, BSam could do well below BSamR compared to BS. It is anticipated that in the future, BSam-KCEM may eventually replace BS-based classification.

## 1.16 Conclusion

This chapter presents two fundamental principles behind HSI, pigeon-hole principle and orthogonality principle, which lead to design and development of many HSI algorithms. While the pigeon-hole principle arises from discrete mathematics, orthogonality principle is a fundamental concept in statistical signal processing. It is the pigeon-hole principle that can be used to explain why HSI can be interpreted as an overdetermined system as opposed to MSI that can be considered as underdetermined system. With no solution for an overdetermined system, the pigeon-hole principle is used to form linear mixing model in (1.2) and NMF model in (1.3). In addition, the pigeon-hole principle also leads to BS and BSam. Most importantly, the pigeon-hole principle also plays a key role in the development of VD and construction of hyperspectral band channel. In particular, a most recent work [44] shows that the pigeon-hole principle can be used to determine the number of bands to be selected and BSamR by VD for hyperspectral imagery where each spectral band can be used to specify a spectral signature as a pigeon to be accommodated by one pigeon-hole. On the other hand, the orthogonality principle is used to find endmembers for hyperspectral images and least squares solutions for spectral unmixing. In particular, the orthogonality principle can be further used to derive the target detection algorithms, CEM, and commonly used RXD as well as to design algorithms such as PPI, VCA, and SGA to find endmembers for hyperspectral images.

## References


1. https://aviris.jpl.nasa.gov/aviris/index.html.
2. Chang, C.-I (2013). *Hyperspectral Data Processing: Algorithm Design and Analysis*. NJ: Wiley.
3. HYMSMO 1998. HYperspectral MASINT Support to Military Operations Program.
4. https://www.ximea.com/support/attachments/download/7294/All%20over%20IP%20-%20HSI.pdf.
5. Epp, S.S. (1995). *Discrete Mathematics with Applications*, 2e. Pacific Grove, CA: Brooks/Cole.
6. Poor, H.V. (1991). *An Introduction to Signal Detection and Estimation*, 2e. New York: Springer-Verlag.
7. Chang, C.-I, Liu, J.-M., Chieu, B.-C. et al. (2000). A generalized constrained energy minimization approach to subpixel target detection for multispectral imagery. *Optical Engineering* 39 (5): 1275–1281.
8. Scharf, L.L. (1991). *Statistical Signal Processing*. MA: Addison-Wesley.
9. Ren, H. and Chang, C.-I (2000). A generalized orthogonal subspace projection approach to unsupervised multispectral image classification. *IEEE Transactions on Geoscience and Remote Sensing* 38 (6): 2515–2528.
10. Ouyang, Y.C., Chen, H.M., Chai, J.W. et al. (2008). Band expansion-based over-complete independent component analysis for magnetic resonance image analysis. *IEEE Transactions on Biomedical Engineering* 55 (6): 1666–1677.
11. Lee, D.D. and Seung, H.S. (1999). Learning the parts of objects by non-negative matrix factorization. *Nature* 401 (6755): 788–791.
12. Candes, E.J., Li, X., Ma, Y., and Wright, J. (2009). Robust principal component analysis? *Journal of the ACM* 58 (3): 1027–1063.
13. Zhou, T. and Tao, D. 2011. GoDec: randomized low-rank & sparsity matrix decomposition in noisy case. In *Proceedings of the 28th International Conference on Machine Learning*, Bellevue, WA, USA.
14. Chang, C.-I (2003). *Hyperspectral Imaging: Techniques for Spectral Detection and Classification*. Kluwer Academic/Plenum Publishers.
15. Chang, C.-I and Du, Q. (2004). Estimation of number of spectrally distinct signal sources in hyperspectral imagery. *IEEE Transactions on Geoscience and Remote Sensing* 42 (3): 608–619.
16. Chang, C.-I (2006). Exploration of virtual dimensionality in hyperspectral image analysis. *Algorithms and Technologies for Multispectral, Hyperspectral, and Ultraspectral Imagery XII*, SPIE Defense and Security Symposium, Orlando, Florida.
17. Chang, C.-I (2006). Utility of virtual dimensionality in hyperspectral signal/image processing, Chapter 1. In: *Recent Advances in Hyperspectral Signal and Image Processing* (ed. C.-I Chang). Trivandrum, Kerala: Research Signpost, India.
18. Chang, C.-I (2012). A unified theory for virtual dimensionality of hyperspectral imagery. *Proceedings of Conference High-Performance Computing in Remote Sensing*, SPIE 8539, Edinburgh United Kingdom.
19. Chang, C.-I (2018). A review of virtual dimensionality for hyperspectral imagery. *IEEE Journal of Selected Topics in Applied Earth Observation and Remote Sensing* 11 (4): 1285–1305.
20. Harsanyi, J.C., Farrand W., and Chang, C.-I (1994). Detection of subpixel spectral signatures in hyperspectral image sequences. *Proceedings of American Congress on Surveying & Mapping (ACSM)/American Society of Photogrammetry & Remote Sensing (ASPRS) Annual Convention and Exposition*, Baltimore, vol. 1, pp. 236–247.

21. Kuybeda, O., Malah, D., and Barzohar, M. (2007). Rank estimation and redundancy reduction of high-dimensional noisy signals with preservation of rare vectors. *IEEE Transactions on Signal Processing* 55 (12): 5579–5592.
22. Sun, W. and Du, Q. (2019). Hyperspectral band selection: a review. *IEEE Geoscience and Remote Sensing Magazine* 7 (2): 118–139.
23. Chang, C.-I, Du, Q., Sun, T.S., and Althouse, M.L.G. (1999). A joint band prioritization and band decorrelation approach to band selection for hyperspectral image classification. *IEEE Transactions on Geoscience and Remote Sensing* 37 (6): 2631–2641.
24. Chang, C.-I and Wang, S. (2006). Constrained band selection for hyperspectral imagery. *IEEE Transactions on Geoscience and Remote Sensing* 44 (6): 1575–1585.
25. Wang, Q., Zhang, F., and Li, X. (2018). Optimal clustering framework for hyperspectral band selection. *IEEE Transactions on Geoscience and Remote Sensing* 56 (10): 5910–5922.
26. Wang, Q., Li, Q., and Li, X. (2018). Hyperspectral band selection via adaptive subspace partition strategy. *IEEE Journal of Selected Topics in Applied Earth Observation and Remote Sensing* 12 (12): 4940–4950-4333.
27. Li, Q., Wang, Q., and Li, X. (2019). An efficient clustering method for hyperspectral optimal band selection via shared nearest neighbor. *Remote Sensing* 11: 350. https://doi.org/10.3390/rs11030350.
28. Yang, H. and Du, Q. (2011). An efficient method for supervised hyperspectral band selection. *IEEE Geoscience and Remote Sensing Letters* 8 (1): 138–142.
29. Zhang, W., Li, X., Dou, Y., and Zhao, L. (2018). A geometry-based band selection approach for hyperspectral image analysis. *IEEE Transactions on Geoscience and Remote Sensing* 58 (8): 4318–4333.
30. Wang, Y., Wang, L., Yu, C. et al. (2019). Constrained-target band selection for multiple-target detection. *IEEE Transactions on Geoscience and Remote Sensing* 57 (8): 6079–6103.
31. Xu, Y., Du, Q., and Younan, N.H. (2017). Particle swarm optimization-based band selection for hyperspectral target detection. *IEEE Geoscience and Remote Sensing Letters* 14 (4): 554–558.
32. Su, H., Yong, B., and Du, Q. (2016). Hyperspectral band selection using improved firefly algorithm. *IEEE Geoscience and Remote Sensing Letters* 13 (1): 68–72.
33. Zhu, G., Huang, Y., Lei, J. et al. (2016). Unsupervised hyperspectral band selection by dominant set extraction. *IEEE Transactions on Geoscience and Remote Sensing* 54 (1): 227–239.
34. Yuan, Y., Zheng, X., and Lu, X. (2017). Discovering diverse subset for unsupervised hyperspectral band selection. *IEEE Transactions on Image Processing* 26 (1): 51–64.
35. Chang, C.-I, Lee, L.C., Xue, B. et al. (2017). Channel capacity approach to band subset selection for hyperspectral imagery. *IEEE Journal of Selected Topics in Applied Earth Observation and Remote Sensing* 10 (10): 4630–4644.
36. Wang, L., Chang, C.-I, Lee, L.-C. et al. (2017). Band subset selection for anomaly detection in hyperspectral imagery. *IEEE Transactions on Geoscience and Remote Sensing* 55 (9): 4887–4898.
37. Yu, C., Song, M., and Chang, C.-I (2018). Band subset selection for hyperspectral image classification. *Remote Sensing* 10: 113. https://doi.org/10.3390/rs10010113.
38. Wang, Q., Zhang, F., and Li, X. (2020). Hyperspectral band selection via optimal neighborhood reconstruction. *IEEE Transactions on Geoscience and Remote Sensing* 58 (12): 8465–8476.
39. Chang, C.-I (2018). Spectral inter-band discrimination capacity of hyperspectral imagery. *IEEE Transactions on Geoscience and Remote Sensing* 56 (3): 1746–1766.
40. Chang, C.-I (2000). An information theoretic-based approach to spectral variability, similarity and discriminability for hyperspectral image analysis. *IEEE Transactions on Information Theory* 46 (5): 1927–1932.
41. Du, Y., Chang, C.-I, Chang, C.-C. et al. (2004). A new hyperspectral measure for material discrimination and identification. *Optical Engineering* 43 (8): 1777–1786.
42. Gallager, R.G. (1968). *Information Theory and Reliable Communication*. Wiley.
43. Chang, C.-I and Davisson, L.D. (1988). On calculating the capacity of an infinite-input finite (infinite)-output channel. *IEEE Transactions on Information Theory* 34 (5): 1004–1010.
44. Chang, C.-I (2022). Band sampling for hyperspectral imagery. *IEEE Transactions on Geoscience and Remote Sensing* 60: 1–24.

45. Candes, E.J. and Wakin, M.B. (2008). An introduction to compressive sampling. *IEEE Signal Processing Magazine* 21–30.
46. Baraniuk, R.G. (2007). Compressive sensing. *IEEE Signal Processing Magazine* 116–124.
47. Baraniuk, R., Davenport, M., De Vore, R., and Wakin, M.B. (2008). A simple proof of the restricted isometry property for random matrices. *Constructive Approximation* 28 (3): 253–263.
48. Chang, C.-I, Ma, K.Y., Liang, C.-C. et al. (2020). Iterative random training sampling spectral-spatial classification for hyperspectral images. *IEEE Journal of Selected Topics in Applied Earth Observation and Remote Sensing* 13: 3986–4007.
49. Harsanyi, J.C. and Chang, C.-I (1994). Hyperspectral image classification and dimensionality reduction: an orthogonal subspace projection approach. *IEEE Transactions on Geoscience and Remote Sensing* 32 (4): 779–785.
50. Settle, J.J. (1996). On the relationship between spectral unmixing and subspace projection. *IEEE Transactions on Geoscience and Remote Sensing* 34 (4): 1045–1046.
51. Chang, C.-I (1998). Further results on relationship between spectral unmixing and subspace projection. *IEEE Transactions on Geoscience and Remote Sensing* 36 (3): 1030–1032.
52. Tu, T.M., Chen, C.-H., and Chang, C.-I (1997). A posteriori least squares orthogonal subspace projection approach to weak signature extraction and detection. *IEEE Transactions on Geoscience and Remote Sensing* 35 (1): 127–139.
53. Chang, C.-I, Zhao, X., Althouse, M.L.G., and Pan, J.-J. (1998). Least squares subspace projection approach to mixed pixel classification in hyperspectral images. *IEEE Transactions on Geoscience and Remote Sensing* 36 (3): 898–912.
54. Chang, C.-I (2007). Information-processed matched filters for hyperspectral target detection and classification, Chapter 3, *Hyperspectral Data Exploitation: Theory and Applications*, (ed. C.-I Chang), pp. 47–74, Wiley.
55. Chang, C.-I (2005). Orthogonal subspace projection revisited: a comprehensive study and analysis. *IEEE Transactions on Geoscience and Remote Sensing* 43 (3): 502–518.
56. Heinz, D. and Chang, C.-I (2001). Fully constrained least squares linear mixture analysis for material quantification in hyperspectral imagery. *IEEE Transactions on Geoscience and Remote Sensing* 39 (3): 529–545.
57. Ren, H. and Chang, C.-I (2003). Automatic spectral target recognition in hyperspectral imagery. *IEEE Transactions on Aerospace and Electronic Systems* 39 (4): 1232–1249.
58. Harsanyi, J.C. (1993). *Detection and Classification of Subpixel Spectral Signatures in Hyperspectral Image Sequences*. PhD dissertation. Department of Electrical Engineering, University of Maryland, Baltimore County, MD.
59. Farrand, W. and Harsanyi, J.C. (1997). Mapping the distribution of mine tailing in the coeur d'Alene river valley, Idaho, through the use of constrained energy minimization technique. *Remote Sensing of Environment* 59: 64–76.
60. Ren, H. and Chang, C.-I (2000). Target-constrained interference-minimized approach to subpixel target detection for hyperspectral imagery. *Optical Engineering* 39 (12): 3138–3145.
61. Reed, I.S. and Yu, X. (1990). Adaptive multiple-band CFAR detection of an optical pattern with unknown spectral distribution. *IEEE Transactions on Acoustics, Speech, and Signal Processing* 38 (10): 1760–1770.
62. Chang, C.-I (2022). Hyperspectral anomaly detection theory: a dual theory of hyperspectral target detection. *IEEE Transactions on Geoscience and Remote Sensing* 60, 5511720: 1–20.
63. Chang, C.-I and Chiang, S.-S. (2002). Anomaly detection and classification for hyperspectral imagery. *IEEE Transactions on Geoscience and Remote Sensing* 40 (6): 1314–1325.
64. Chang, C.-I (2022). Constrained energy minimization anomaly detection for hyperspectral imagery. *IEEE Transactions on Geoscience and Remote Sensing* 60: 1–19.
65. Boardman, J.W. (1994). Geometric mixture analysis of imaging spectrometry data. *International Geoscience and Remote Sensing Symposium* 4: 2369–2371.
66. Nascimento, J.M.P. and Bioucas-Dias, J.M. (2005). Vertex component analysis: a fast algorithm to unmix hyperspectral data. *IEEE Transactions on Geoscience and Remote Sensing* 43 (4): 898–910.
67. Chang, C.-I, Wu, C., Liu, W., and Ouyang, Y.C. (2006). A growing method for simplex-based endmember extraction algorithms. *IEEE Transactions on Geoscience and Remote Sensing* 44 (10): 2804–2819.

68. Winter, M.E. (1999). N-finder: an algorithm for fast autonomous spectral endmember determination in hyperspectral data. *Image Spectrometry V, Proceedings SPIE* 3753: 266–277.
69. Chang, C.-I, Chen, S.Y., Li, H.C., and Wen, C.-H. (2016). A comparative analysis among ATGP, VCA and SGA for finding endmembers in hyperspectral imagery. *IEEE Journal of Selected Topics in Applied Earth Observation and Remote Sensing* 9 (9): 4280–4306.
70. Li, H.-C. and Chang, C.-I (2016). Recursive orthogonal projection-based simplex growing algorithm. *IEEE Transactions on Geoscience and Remote Sensing* 54 (7): 3780–3793. https://doi.org/10.1109/TGRS.2016.2527737.
71. Chang, C.-I, Li, H.C., Wu, C.C., and Song, M. (2017). Recursive geometric simplex growing analysis for finding endmembers in hyperspectral imagery. *IEEE Journal of Selected Topics in Applied Earth Observation and Remote Sensing* 10 (1): 296–308.
72. Halko, N., Martinsson, P.G., and Tropp, J.A. (2009). Finding structure with randomness: stochastic algorithms for constructing approximate matrix decompositions, *arXiv*: 0909.4061.
73. Chang, C.-I, Cao, H., Chen, S. et al. (2021). Orthogonal subspace projection-based GoDec for low rank and sparsity matrix decomposition for hyperspectral anomaly detection. *IEEE Transactions on Geoscience and Remote Sensing* 59 (3): 2403–2429.
74. Chang, C.-I and Chen, J. (2021). OSP using data sphering and low-rank and sparse matrix decomposition for hyperspectral target detection. *IEEE Transactions on Geoscience and Remote Sensing* 59 (10): 8704–8722.
75. Chang, C.-I, Cao, H., and Song, M. (2021). Orthogonal subspace projection target detector for hyperspectral anomaly detection. *IEEE Journal of Selected Topics in Applied Earth Observation and Remote Sensing* 14: 4915–4932.
76. Qu, J., Du, Q., Li, Y. et al. (2020). Anomaly detection in hyperspectral imagery based on Gaussian mixture model. *IEEE Transactions on Geoscience and Remote Sensing* 59 (11): 9504–9517.
77. Li, L., Li, W., Du, Q., and Tao, R. (2021). Low-rank and sparse decomposition with mixture of Gaussian for hyperspectral anomaly detection. *IEEE Transactions on Cybernetics* 51 (9): 4363–4372.
78. Chen, S., Chang, C.-I, and Li, X. (2022). Component decomposition analysis for hyperspectral target detection. *IEEE Transactions on Geoscience and Remote Sensing* 60: 1–22.
79. Richards, J.A. and Jia, X. (2013). *Remote Sensing Digital Image Analysis*, 4e. Springer Verlag.
80. Landgrebe, D.A. (2003). *Signal Theory Methods in Multispectral Remote Sensing*. New York: Wiley.
81. Fauvel, M., Tarabalka, Y., Benediktsson, J.A. et al. (2013). Advances in spectral-spatial classification of hyperspectral images. *Proceedings of IEEE* 101 (3): 652–675.
82. Ghamisi, P., Mura, M.D., and Benediktsson, J.A. (2015). A survey on spectral–spatial classification techniques based on attribute profiles. *IEEE Transactions on Geoscience and Remote Sensing* 53 (5): 2335–2353.
83. Ghamisi, P., Maggiori, E., Li, S. et al. (2018). New frontiers in spectral-spatial hyperspectral image classification. *IEEE Geoscience and Remote Sensing Magazine* 10–43.
84. Xue, B., Yu, C., Wang, Y. et al. (2017). A subpixel target approach to hyperpsectral image classification. *IEEE Transactions on Geoscience and Remote Sensing* 55 (9): 5093–5114.
85. Chang, C.-I (2019). A statistical detection theory approach to hyperspectral image classification: performance analysis. *IEEE Transactions on Geoscience and Remote Sensing* 57 (4): 2057–2074.
86. Song, M., Shang, X., and Chang, C.-I (2020). 3D receiver operating characteristic analysis for hyperspectral image classification. *IEEE Transactions on Geoscience and Remote Sensing* 58 (11): 8093–8115.
87. Yu, C., Lee, L.C., Chang, C.-I et al. (2018). Band-specified virtual dimensionality: an orthogonal subspace approach. *IEEE Transactions on Geoscience and Remote Sensing* 56 (5): 2822–2832.
88. Shang, X., Song, M., and Chang, C.-I (2021). Iterative random training sample constrained energy minimization for hyperspectral image classification. *IEEE Geoscience and Remote Sensing Letters* 18 (9): 1625–1629.
89. Kang, X., Li, S., and Benediktsson, J.A. (2014). Spectral-spatial hyperspectral image classification with edge-preserving filtering. *IEEE Transactions on Geoscience and Remote Sensing* 52: 2666–2677.
90. Ma, K.Y. and Chang, C.-I (2022). Kernel-based constrained energy minimization approaches to hyperspectral mixed pixel cassification. *IEEE Transactions on Geoscience and Remote Sensing* 60: 1–23.

91. Zhong, S., Chang, C.-I, Li, J.J. et al. (2019). Class feature weighted hyperspectral image classification. *IEEE Journal of Selected Topics in Applied Earth Observation and Remote Sensing* 12 (12): 4728–4745.
92. Yu, C., Song, M., and Chang, C.-I (2018). Band subset selection for hyperspectral image classification, *Remote Sensing*, 10, 113; doi: 10.3390/rs10010113.
93. Yu, C., Song, M., and Chang, C.-I (2019). Class signature-constrained background suppressed approach to band selection for hyperspectral image classification. *IEEE Transactions on Geoscience and Remote Sensing* 57 (1): 14–31.
94. Song, M., Shang, X., Wang, Y. et al. (2019). Class-information based band selection for hyperspectral image classification. *IEEE Transactions on Geoscience and Remote Sensing* 57 (11): 8394–8416. https://doi.org/10.1109/TGRS.2019.2920891.
95. Moya, M. and Hush, D. (1996). Network constraints and multi- objective optimization for one-class classification. *Neural Networks* 9 (3): 463–474.
96. Khan, S.S. and Madden, M.G. (2014). One-class classification: taxonomy of study and review of techniques, Knowledge Engineering Review, Published online by Cambridge University Press. https://www.cambridge.org/core/journals/knowledge-engineering-review/article/oneclass-classification-taxonomy-of-study-and-review-of-techniques/EEDBC97CFD54B2B65BEB5FCD71593782.
97. Chang, C.-I and Ma, K.Y.K. (2022). Band sampling for kernel constrained energy minimization in hyperspectral mixed pixel classification. *IEEE Transactions on Geoscience and Remote Sensing* 60, 5514024: 1–21.

# 2

# Overview of Hyperspectral Imaging Remote Sensing from Satellites

Shen-En Qian
*Canadian Space Agency, Saint-Hubert, Quebec, Canada*

## 2.1 Hyperspectral Imaging Remote Sensing from Airplanes to Satellites

Hyperspectral satellites have emerged as a new generation of remote sensing satellites since the beginning of this millennium. Compared to multispectral satellites, such as Landsat and SPOT satellites, the primary advantage of hyperspectral satellites is that, because an entire spectrum is acquired for each pixel in the scene of the acquired imagery, an operator needs no prior knowledge of the pixels, and post-processing allows all available information from the dataset to be exploited [1]. Hyperspectral imagery can provide direct identification and analysis of the surface materials. Hyperspectral imaging can also take advantage of the spatial relationships among the different pixel spectra in a neighborhood, allowing more elaborate spectral-spatial models for a more accurate segmentation and classification of the image [2]. Hyperspectral imagery has been used in a wide range of remote sensing applications, including geology, agriculture, forestry, environment, ocean, atmosphere, climate change, defense and security, and law enforcement. The fascinating detailed spectral information provided by hyperspectral imagery often provides results not possible with multispectral or other types of imagery [3].

Goetz et al. [4] originally defined *hyperspectral imaging* as "the acquisition of images in hundreds of contiguous, registered spectral bands such that for each pixel a radiance spectrum can be derived." It sets this type of spectral remote sensing apart from *multispectral imaging* by requiring the spectral bands to be contiguous, so that no gaps occur through which precious spectral information of the pixels might slip undetected. This original definition, however, did not explicitly mention the bandwidth of the spectral bands with respect to that of multispectral imaging. The bandwidth of hyperspectral sensors is typically 10 nm or less, which is much narrower than that of multispectral sensors whose bandwidth is about 100 nm or so. The narrowness of the spectral bands in hyperspectral imaging is more important than the number of spectral bands being hundreds. An imaging sensor that produces tens of spectral images with a bandwidth of 10 nm or less should be categorized as a hyperspectral sensor, even if its total number of bands is less than a hundred. Qian [5] updated the original definition of hyperspectral imaging as "the acquisition of <u>many</u> images of contiguous,

*Advances in Hyperspectral Image Processing Techniques*, First Edition. Edited by Chein-I Chang.

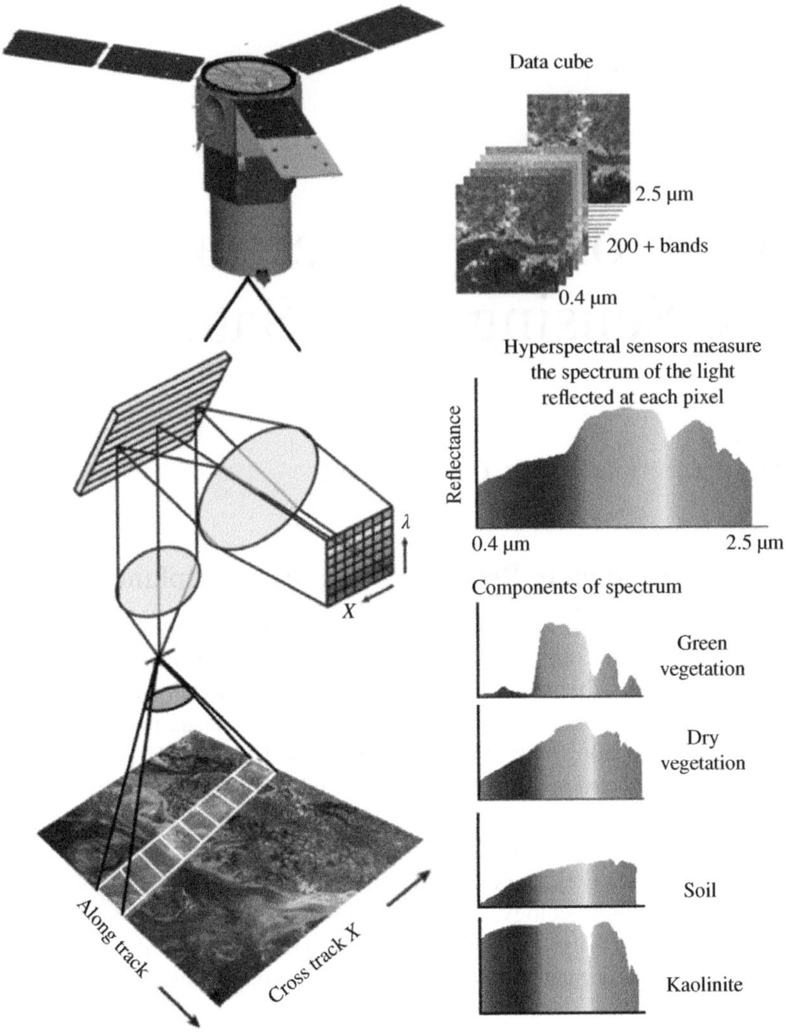


**Figure 2.1** Principle of a hyperspectral satellite. A data cube and spectrum of each pixel are generated by a hyperspectral satellite.

narrow, registered spectral bands such that for each pixel a radiance spectrum can be derived" to explicitly mention the bandwidth of the spectral bands.

Figure 2.1 shows the principle of a hyperspectral satellite. In the figure, the hyperspectral imager is a grating-based imaging spectrometer using a two-dimensional (2D) detector array operating in pushbroom mode by scanning a cross-track line on ground one after another in the satellite flight (along track) direction. It acquires images of a given scene in hundreds (or tens sometimes) of continuous, narrow, and registered spectral bands over wavelengths that range from the visible, near-infrared to the short-wave infrared (SWIR). Each image, often referred to as spectral image or band image, corresponds to a particular wavelength or spectral band number (e.g. 0.4 and 2.5 μm). The collected "data cube" contains both spatial and spectral information of the materials of the pixels within the scene. Each pixel in the scene is sampled across hundreds (or tens) of narrow band images at a particular spatial location in the data cube, resulting in a one-dimensional (1D) spectrum. It is a plot of wavelength versus radiance or reflectance. The spectrum for a single pixel acquired by a hyperspectral satellite appears similar like a spectrum collected by a spectrometer in a spectroscopy laboratory. The spectra can be used to identify and characterize the particular feature of the pixels within the scene based on the unique spectral signatures or "fingerprints."

### 2.1.1 *History of Development of Airborne Hyperspectral Imagers*

The development of hyperspectral imaging technology for remote sensing began in the 1970s from field spectral measurements in support of NASA's Landsat-1 data analysis. The need for hyperspectral imaging came from the recognition based on laboratory and field spectral measurements, primarily of minerals and soils, that multispectral imaging in four broad spectral bands with the Landsat's Multispectral Scanner (MSS) was not adequate to discriminate among minerals on the earth's surface that were important in resource exploration and environmental assessment.

The history of airborne hyperspectral began with the development of the first hyperspectral imager for remote sensing – the airborne imaging spectrometer (AIS) in the early 1980s. At that time neither the technologies for detectors, optics, and electronics nor computers for rapid data analysis were readily available. The AIS was built as an engineering test bed using the first commercially available hybrid infrared 2D detector arrays, as well as a tool for research in hyperspectral remote sensing [4, 6].

The AIS was a pushbroom imaging spectrometer in the SWIR region with a spectral sampling interval of 9.3 nm. It used a hybrid 2D-detector array consisting of a matrix of mercury cadmium telluride (HgCdTe) with $32 \times 32$ elements covering a wavelength range from 1.2 to 2.4 μm after rotating the grating in four positions. Figure 2.2 shows exploded view of the AIS instrument. AIS had several shortcomings, among them, lack of full spectral coverage in the visible and near-infrared (VNIR) and a very narrow field-of-view (FOV) (3.7°), analogous to observing the earth from an aircraft through a soda straw. A later version of the instrument, AIS-2, incorporated a second-generation $64 \times 64$ element HgCdTe detector array, and covered a wavelength range from 0.8 to 2.4 μm region.

The promising results of hyperspectral imaging remote sensing produced by AIS and the demonstrated need led to the development of an imaging spectrometer program within NASA. In 1984, the initial program contained the next-generation aircraft system, the Airborne Visible/Infrared

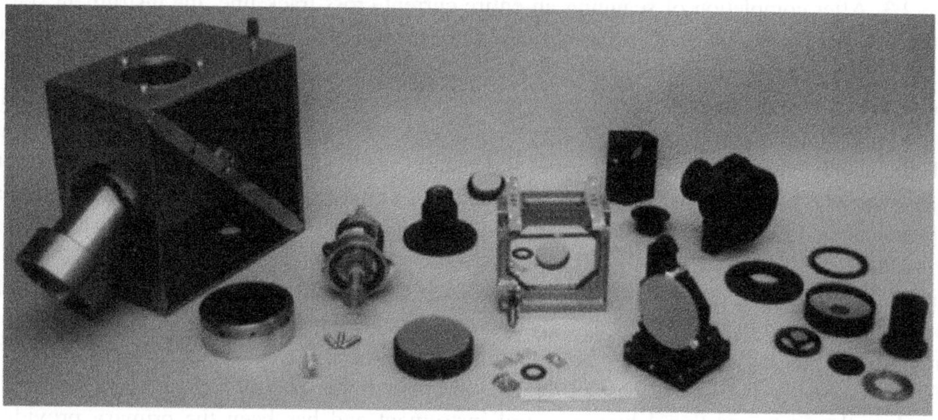

**Figure 2.2** Exploded view of the AIS instrument. (Courtesy of NASA / JPL).

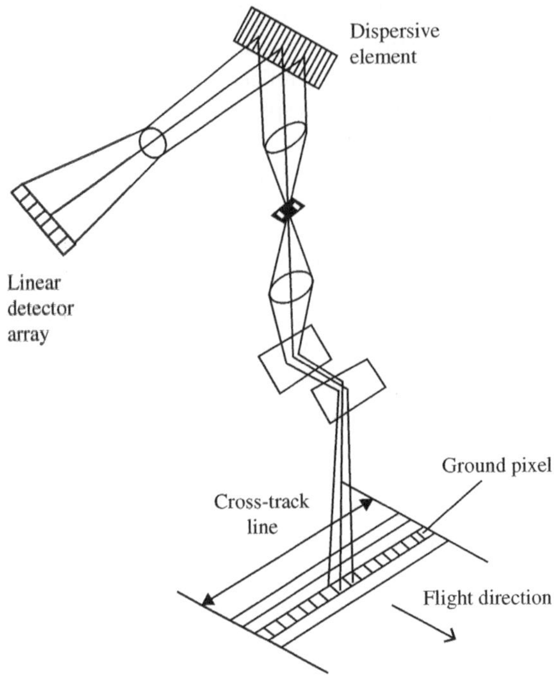


**Figure 2.3** Illustration of an imaging spectrometer operating in whiskbroom mode.

Imaging Spectrometer (AVIRIS). Alexander F. H. Goetz and his colleagues at Jet Propulsion Laboratory (JPL) proposed to design and develop the AVIRIS. The goal was to introduce the newly available solid-state detector technology into remote sensing applications and to enable the acquisition of hyperspectral images from visible to SWIR in hundreds of contiguous spectral bands, such that a complete radiance spectrum is available for each pixel in the FOV of the instrument [7].

The development of the AVIRIS began in 1984. The AVIRIS is a whiskbroom imaging spectrometer using 1D linear detector arrays. Since linear detector arrays were used that could record the monochromatic components of the spectrum of only one ground pixel at one moment, the AVIRIS needs to scan the ground pixels one after another in a cross-track line in the whiskbroom mode as shown in Figure 2.3. After completion of scanning an entire current cross-track line, the instrument starts to scan the ground pixels of the next cross-track line when the aircraft flies forward in the flight direction (also referred to as along-track direction).

The AVIRIS covers a wavelength range from 0.4 to 2.45 μm with 224 spectral bands using four separate spectrometers (referred to as S1–S4), each of which covers a portion of the wavelength range [5]. The instantaneous field-of-view (IFOV) is 1.0 mrad, equivalent to a 20 m footprint on ground when it is on an airplane at a 20 km altitude. The width of a cross-track line (also referred to as swath) is approximately 12 km containing 614 pixels [8, 9]. Figure 2.4 shows the AVIRIS instrument in a rack in laboratory.

The AVIRIS instrument first flew aboard a NASA ER-2 aircraft at a 20 km altitude in 1986, first science data delivered in 1987, and was fully operational since 1989. Since then it had gone through several major upgrades as technology advanced in detector arrays, electronics, and computing. The AVIRIS is arguably the finest, best-calibrated, airborne hyperspectral imager ever flown. It is regarded as the first operational hyperspectral instrument and has been the primary provider of airborne hyperspectral data to the remote sensing research community.

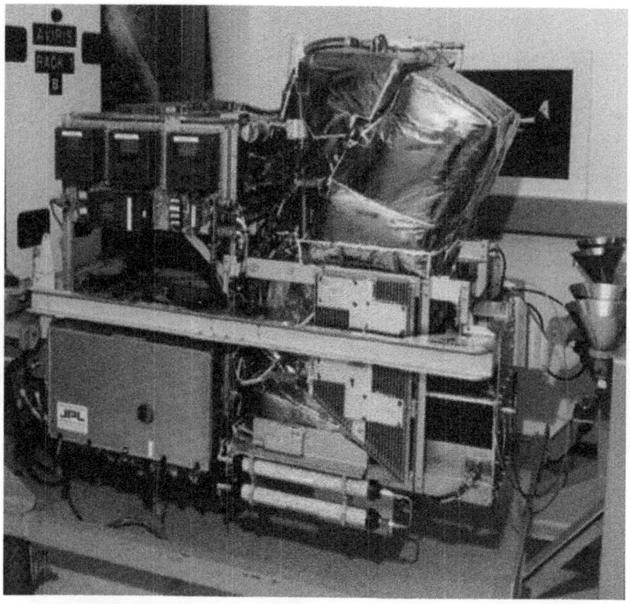

**Figure 2.4** AVIRIS instrument. (Courtesy of NASA / JPL).

The development of hyperspectral imaging technology in Canada began in the early 1980s when the Fluorescence Line Imager (FLI), an airborne imaging spectrometer, was designed and constructed for Canadian government – the Department of Fisheries and Oceans (DFO) in 1981 as the first step toward the development of an advanced satellite imager based on CCD detector arrays for remote sensing of the Earth's surface [10].

Canadian researchers and engineers pioneered this field, together with their American colleagues at NASA's JPL, and built the FLI, one of the first imaging spectrometry imagers in the world. FLI used five 2D CCD detector arrays of format 385 × 288 pixels to cover a wide swath (i.e. FOV). Because the CCD frame rate was limited by the speed of the readout, the FLI operated in either spatial mode or spectral mode. In spatial mode, it operated in pushbroom mode with a FOV of 70° covered by five cameras for a total of 1925 pixels in a cross-track line, each pixel has a IFOV of 1.3 mrad, equivalent to 2.5 m ground sampling distance (GSD) at altitude 2 km of the airplane. In spectral mode, it operated in rake mode with 288 spectral bands covering a wavelength range from 430 to 800 nm. The optics consisted of a transmission grating with five lenses and effective f-number 1.4. The key concept was the charge summation on chip to form bands, thus reducing data bandwidth and increasing SNR. The SNR of 1900 : 1 was achieved for a band of 16 elements. Figure 2.5 shows the photograph of the FLI instrument.

Starting in 1983, the FLI flew over a wide variety of land and sea targets in North America and Europe to acquire hyperspectral imagery with higher spectral resolution and sensitivity than other comparable imagers [11, 12]. In many of these flights, it was the same target molecule, the photosynthetic pigment chlorophyll *a*, whose fluorescence or absorption effects were being studied on land and in water.

In the late 1980s, several commercial airborne hyperspectral imagers entered the market. In 1989, ITRES Research Limited based in Calgary, Canada, marketed the Compact Airborne Spectrographic Imager (CASI), a 2D CCD array-based pushbroom imaging spectrometer, which covers a wavelength range from 400 nm to 900 nm with a bandwidth of 1.8 nm and 512 active pixels in the spatial imaging

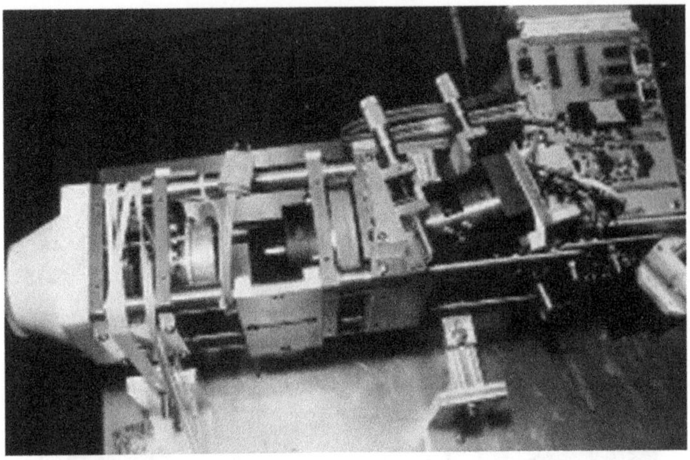

**Figure 2.5** Photograph of Canadian first airborne hyperspectral imager – Fluorescence Line Imager (FLI).

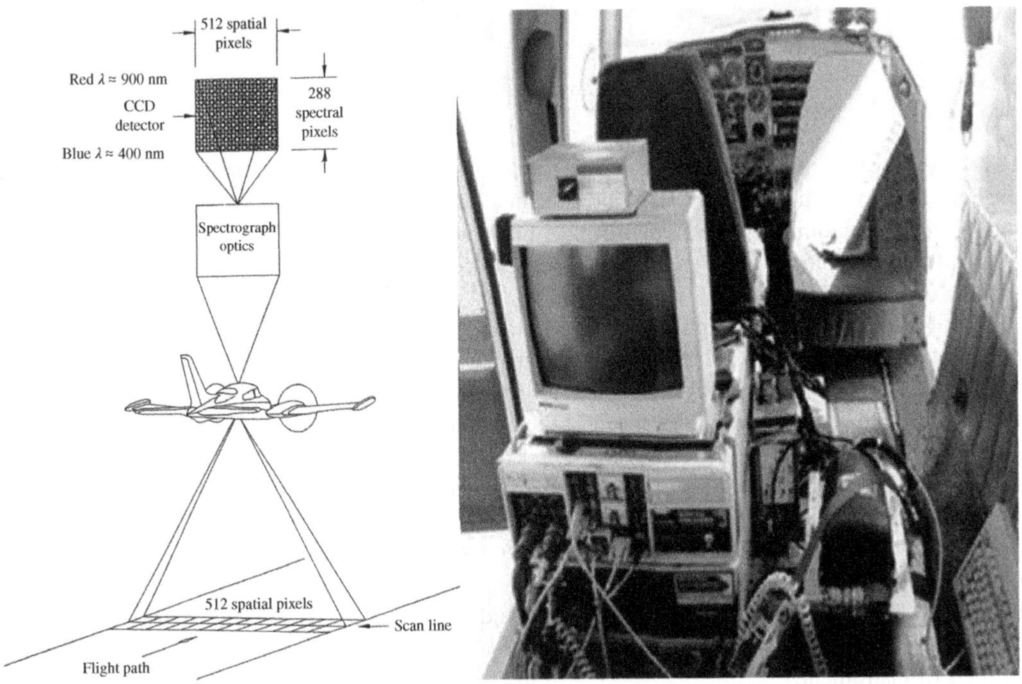


**Figure 2.6** Canadian commercial airborne hyperspectral imager CASI. (Babey et al. [13], Weale et al. [14]).

direction as shown in Figure 2.6 [13, 14]. ITRES has been in commercial airborne hyperspectral imagers since then and extended their product line to cover the SWIR and thermal regions of the spectrum [5].

The Canada Centre for Remote Sensing developed the first Canadian SWIR imaging spectrometer – SWIR Full Spectrographic Imager (SFSI) in 1992 and flew it in the next year [15]. It is an airborne imaging spectrometer covering a wavelength range from 1220 to 2420 nm in the SWIR region with a

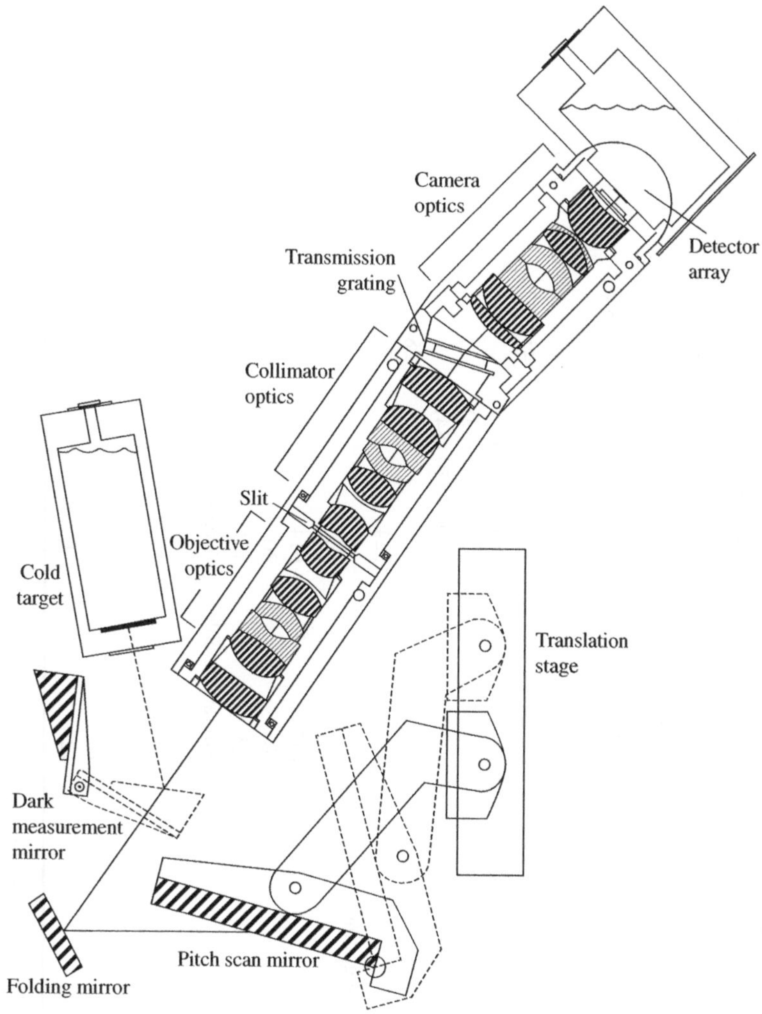


**Figure 2.7** Optics and instrument assembly of the Canadian first airborne SWIR hyperspectral imager SFSI.

spectral sampling interval of 10.4 nm. It utilizes a 2D detector array, refractive optics, and a transmission grating. The fore-optics and spectrograph are f/1.8, and the FOV is 9.4°. The detector is a 488 × 512 pixel PtSi Schottky barrier photodiode array. Figure 2.7 shows the optics and instrument assembly of the SFSI instrument.

Canadian design and applications experience from the FLI and CASI were used to define the specifications for ESA's MEdium Resolution Imaging Spectrometer (MERIS) on board ESA's ENVISAT, which was launched in 2002. Table 2.1 compares the specifications of the AIS, FLI, AVIRIS, CASI, and SFSI in early years of hyperspectral technology development between 1980s and early 1990s.

The Canadian FLI, CASI, and SFSI were developed in the middle 1980s to early 1990s, a time of rapid development in the field of electro-optics. Several other instruments were also being constructed in the same timeframe.

**Table 2.1** Comparison of specifications of selected airborne imaging spectrometers in early years between 1980s and early 1990s

| Parameters | AIS | FLI | AVIRIS | CASI | SFSI |
|---|---|---|---|---|---|
| Period of operation | 1983–85 | 1984–90 | Since 1987 | Since 1989 | Since 1992 |
| Country | USA | Canada | USA | Canada | Canada |
| Operation mode | Pushbroom | Pushbroom | Whiskbroom | Pushbroom | Pushbroom |
| Detector array format | 32 × 32 | 385 × 288 | 1 × 32 (S1)<br>1 × 64 (S2–S4) | 612 × 288 | 512 × 488 |
| Number of detector arrays | 1 | 5[a] | 4[b] | 1 | 1 |
| Spectral range (nm) | 1200–2400[c] | 430–805 | 410–700 (S1)<br>680–1270 (S2)<br>1250–1860 (S3)<br>1840–2450 (S4) | 418–926 | 1200–2400 |
| Number of spectral elements | 32 | 288 | 224 | 288 | 122 |
| Number of spectral bands formed | 4 x 32 = 128<br>(after rotating the grating 4 angles) | 8 | 224 | 72 | 122 |
| Spectral sampling interval (nm) | 9.3 | 1.3 | 10 | 1.8 | 10 |
| Number of spatial elements | 32 | 1925 | 614 | 512 | 512 |
| Field-of-view (°) | 3.7 | 70 | 30 | 35 | 11.7 |
| IFOV (mrad) | 1.9 | 1.3 | 1.0 | 1.2 | 0.4 |
| Platform | AircraftC-130 | Aircraft flexible | Aircraft ER-2 | Aircraft flexible | Aircraft flexible |
| Nominal platform altitude (km) | 6 | 2 | 20 | 2 | 2 |
| Swath (km) | 0.37 | 4.8 | 11 | 1–5 | 0.26 |
| Ground sampling distance (m) | 11.4 | 2.5 | 20 | 2–5 | 0.5 |
| Data rate (Mbps) | 0.394 | 7 | 17 | 0.875 | |
| Digitization | 8 | 12 | 10 | 12 | 12 |

[a] Five detector arrays were used to cover the full swath of 1925-pixel wide.

[b] Four detector arrays associated with the four spectrometers were used to cover the full spectral range (S1–S4).

[c] After rotating the grating four positions. One position covers ¼ spectral range (i.e. 300 nm).

**Table 2.2** Parameters of Multispectral Infrared and Visible Imaging Spectrometer (MIVIS)

| Spectrometer | Wavelength range (μm) | Bandwidth (nm) | Number of bands |
|---|---|---|---|
| Visible | 0.43–0.83 | 20 | 20 |
| Near-IR | 1.15–1.55 | 50 | 8 |
| Mid-IR | 1.983–2.478 | 8 | 64 |
| Thermal-IR | 8.18–12.70 | 340–540 | 10 |

Daedalus Inc. manufactured the Multispectral Infrared and Visible Imaging Spectrometer (MIVIS) in 1993. It was a modular instrument with four imaging spectrometers designed to collect radiation from the earth's surface in the visible, near-IR, mid-IR, and thermal-IR regions. It had a total of 102 spectral bands. Table 2.2 lists the parameters of the MIVIS. It was a second-generation imaging spectrometer developed for its use in environmental remote sensing studies across a broad spectrum of scientific disciplines [16].

An airborne imaging spectrometer called Hyperspectral Data and Information Collection Experiment (HYDICE) was developed and flew in 1994 under sponsorship of the Naval Research Laboratory (NRL) [17]. This imaging spectrometer was designed around a prism dispersion concept and a single, hybrid HgCdTe 2D array to cover a wavelength range of 0.4–2.5 μm. It had an IFOV of 0.5 mrad, equivalent to a 3 m footprint on the ground from a 6 km altitude. The instrument was mounted on a stabilized platform. Although the results were preliminary, its signal-to-noise ratios considerably exceeded those of the early version AVIRIS that had been achieved. Based on the results from data analysis, HYDICE data could make possible the identification of materials that cover only 1–2% of a pixel, provided that their reflectance spectra are unique. The HYDICE began as a dual-use program; it soon reverted to an all Department of Defense (DoD) program.

Geophysical Environmental Research of Millbrook, NY, developed the Digital Airborne Imaging Spectrometer (DAIS), which started operation in 1994. DAIS is an airborne imaging spectrometer with 79 spectral bands covering the visible and SWIR regions as well as the thermal infrared [18].

Other commercial imagers in wavelength range from 0.4 to 2.5 μm were also developed. The one most like AVIRIS is the Australian Hyperspectral Mapper (HyMap) built by the HyVista Corporation, which marketed a full-range (0.4–2.5 μm) hyperspectral imaging service on a global basis. The design of HyMap was modular and this led to an ability to configure the spectral and spatial characteristics to suit the customer's specific requirements. The visible (VIS) module includes a silicon CCD detector array operated at ambient temperature, and the near-infrared (NIR) and short-wave infrared-1 (SWIR1) and 2 (SWIR2) modules incorporate InSb arrays cooled by liquid nitrogen to 77K. Each spectrographic module provides 32 spectral bands. Table 2.3 lists parameters of the HyMap imager. Its FOV is 60° (512 pixels) corresponding to a swath of 2.3 km with 5 m GSD at 1.3 km altitude or 4.6 km with 10 m GSD at 2.6 km altitude [19].

**Table 2.3** Parameters of HyMap imaging spectrometer

| Spectral region | Wavelength range (μm) | Bandwidth (nm) | Number of bands |
|---|---|---|---|
| Visible (VIS) | 0.45–0.89 | 15 | 32 |
| Near-infrared (NIR) | 0.89–1.35 | 15 | 32 |
| Short-wave infrared-1 (SWIR1) | 1.40–1.80 | 13 | 32 |
| Short-wave infrared-2 (SWIR2) | 1.95–2.48 | 17 | 32 |

## 2.1.2 *Early Development of Spaceborne Hyperspectral Imagers*

The development of spaceborne hyperspectral imagers began in the early 1980s. In the late 1970s, JPL developed the Shuttle Multispectral Infrared Radiometer (SMIRR), which was scheduled to fly on the second flight of Space Shuttle (STS-2) in 1979 but was delayed until November 1981. The development of SMIRR could be considered as the beginning of the development of spaceborne imaging spectrometers. SMIRR was installed in the payload bay of STS-2 alongside the Shuttle Imaging Radar A (SIR-A). The purpose of the experiment was to test the feasibility of direct mineral identification using hyperspectral imaging from Earth orbit. The instrument was designed as a profiler to acquire radiance data in 10 spectral bands covering the VNIR and SWIR regions along a 100 m-wide track beneath the spacecraft. The instrument consisted of a spare Mariner Venus–Mercury mission telescope with a rotating filter wheel in front of an HgCdTe detector array in the focal plane. Two 16 mm fighter-aircraft gun cameras recorded the ground track. The experiment was successful and resulted in the first direct identification of soil minerals from space through the use of the five spectral bands in the 2.2–2.5 μm region [20]. The minerals kaolinite and calcite, in the form of limestone, were identified in an Egyptian dry lake and the surrounding hills and confirmed based on returned samples.

The promising results coming from airborne imaging spectrometers and SMIRR and the demonstrated needs led to the creation of an imaging spectrometer program within NASA. In 1984, the initial program included the next-generation airborne hyperspectral imager AVIRIS, the Shuttle Imaging Spectrometer Experiment (SISEX), and the free-flyer High Resolution Imaging Spectrometer (HIRIS) as shown in Figure 2.8. As described in Section 2.1.1, the AVIRIS successfully reached operational status while SISEX was canceled after the space shuttle Challenger disaster in 1986 when the Shuttle was destroyed by a booster rocket malfunction shortly after liftoff.

The HIRIS concept was developed further in 1988 in response to the Earth Observing System (EOS) flight proposal opportunity [21]. The mission goal was to deploy on the EOS platform a hyperspectral imager covering the wavelength range 0.4–2.5 μm with 10 nm bandwidth for 192 contiguous spectral

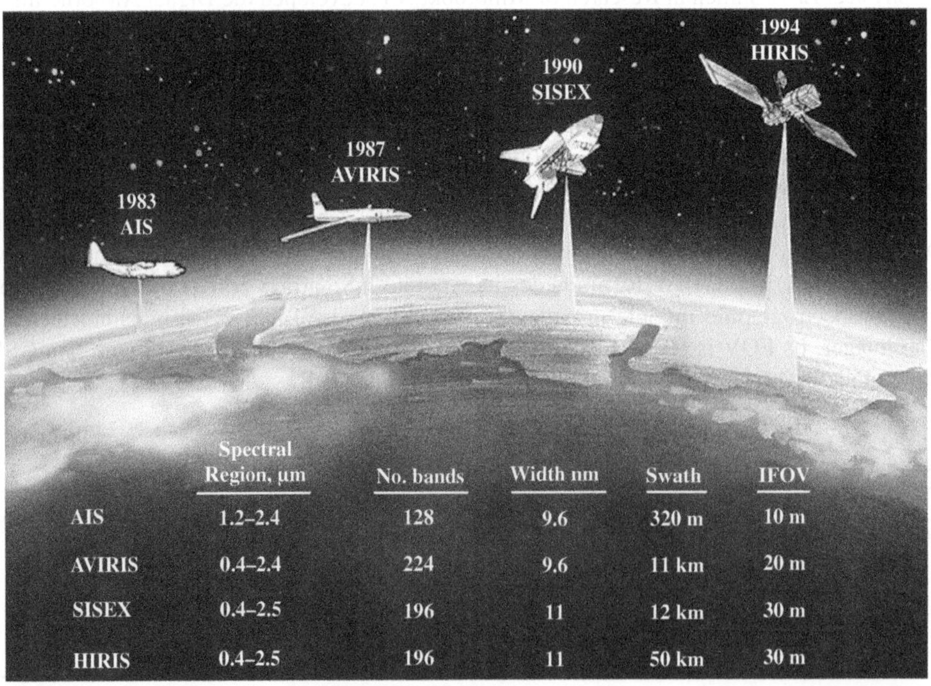


**Figure 2.8** NASA's plan of early time imaging spectrometer program. (Courtesy of NASA / JPL).

bands and spanning a 50 km swath from the 705 km orbital altitude [22]. The designed IFOV (i.e. GSD, or footprint size) was 30 × 30 m. The HIRIS proposal was accepted for EOS; however, HIRIS did not fly as a result of the overall program was scaled back to approximately one-fourth of its original size.

At the time of the demise of HIRIS, only a handful of investigators had actually worked with hyperspectral data mainly because there existed only a few hyperspectral datasets, and they were of poor quality compared to what are available today. Additionally, there were not any readily available software tools to process the datasets that were orders of magnitude larger than the multispectral datasets from Landsat or SPOT satellites, and desktop and powerful computing systems were not yet up to the task. Therefore, it was not surprising that HIRIS was not included on EOS.

In the beginning of the 1990s, NASA also worked with TRW to develop a spaceborne spectrometric system called HyperSpectral Imager (HSI) for the LEWIS mission as a technology demonstration as part of its Small Spacecraft Technology Initiative (SSTI) program [23]. HSI was designed to have 128 bands in the spectral range 0.4–1.0 μm and another 256 bands in 0.9–2.5 μm, for a total of 384 bands. The spectral sampling intervals in both spectral ranges were 5.0 nm and 6.5 nm, respectively, which are still very competitive to the present-day spaceborne hyperspectral imagers. The swath width was 7.7 km and GSD was 30 m. Unfortunately, HSI was unsuccessful. Three days after the launch on 23 August 1997, the control of the satellite was lost and subsequently entered the Earth atmosphere in September 1997 [24].

In November 2000, NASA launched Hyperion, a spaceborne hyperspectral imager built by TRW, aboard the Earth Observing-1 (EO-1) satellite [25]. If one compares the successful Hyperion with HIRIS, it can be easily seen that Hyperion bears some of the HIRIS instrument characteristics. EO-1 was created to develop and validate a number of instrument and spacecraft bus breakthrough technologies. It was intended to enable the development of future Earth imaging observatories that would have a significant increase in performance while also having reduced cost and mass. EO-1 was decommissioned in March 2017 after over 16 years of successful operation.

## 2.2 Development of Spaceborne Hyperspectral Imagers in the Last Two Decades

Airborne hyperspectral imaging has been widely utilized because of its easy adaption to variation of altitude, flight schedule to avoid weather problems, such as clouds, and flexibility of adaption to the task requirements in the course of flight. However, a low-flying aircraft imaging has a narrow coverage swath; a number of flybys are needed to cover a large ground area due to the small swath. Aircrafts are less stable, thus the image quality of the instrument onboard them suffers.

Since the beginning of this millennium, the research and development of spaceborne hyperspectral imagers has been grown exponentially. With the advancement of technologies for focal plane arrays, optical materials manufacture and diamond tuning, data acquisition, data storage, computation, and telemetry, it is made possible to build high-performance hyperspectral imagers and reduce the cost of development of such space systems. Satellite hyperspectral imagers are now more readily available not only for scientific applications but also for mandate operations. Spaceborne hyperspectral imagers proved their capability to provide critical information in numerous application areas such as civilian origin and military.

### *2.2.1 Survey of Spaceborne Hyperspectral Imagers*

Qian [5] made a survey on spaceborne hyperspectral imagers to date and reported that there exist at least 26 spaceborne hyperspectral imagers that have been deployed to orbits of Earth, moon, Mars, Venus, and comet for surface hyperspectral imaging remote sensing. Table 2.4 tabulates a list of

**Table 2.4** List of spaceborne hyperspectral imagers deployed since the beginning

| Number | Hyperspectral imager | Satellites/Platform | Orbit | Launch year | Demission year | Number of bands | Spectral range (μm) | Spectral sampling interval (nm) | Ground sampling distance (m) | Swath width (km) |
|---|---|---|---|---|---|---|---|---|---|---|
| 1 | SPIMs 1-5 | MSX | LEO | 1996 | 2008 | 272 | 0.11–0.9 | 0.5–4.3 | 770 | 15 |
| 2 | HSI | LEWIS | LEO | 1997 | 1997 | 128<br>256 | 0.4–1.0<br>0.9–2.5 | 5.0<br>6.5 | 30 | 7.7 |
| 3 | MODIS | Terra<br>Aqua | LEO | 1999<br>2002 | Active | 36 | 0.41–14.4 | 10–50 | 250: b1–b2<br>500: b3–b7<br>1000: b8–b36 | 2330 |
| 4 | Hyperion | EO-1 | LEO | 2000 | 2017 | 220 | 0.4–2.5 | 10 | 30 | 7.7 |
| 5 | CHRIS | PROBA | LEO | 2001 | Active | 19–62 | 0.4–1.0 | 1.25–11 | 25–50 | 13 |
| 6 | MERIS | ENVISAT | LEO | 2002 | 2012 | 520 (transmit 15) | 0.39–1.04 | 1.25 | 300 | 1150 |
| 7 | VIRTIS | Rosetta | Comet 67P | 2004 | 2016 | 432<br>432 | 0.28–1.10<br>1.05–5.13 | 1.89<br>9.47 | 0.014°× 0.014° (IFOV) | 3.67°× 3.67° (FOV) |
| 8 | VIRTIS | Venus express | Venus | 2005 | 2015 | 432<br>432 | 0.28–1.10<br>1.05–5.13 | 1.89<br>9.47 | 0.014°× 0.014° (IFOV) | 3.67°× 3.67° (FOV) |
| 9 | CRISM | MRO | Mars orbit (300 km) | 2005 | Active | 455 | 0.37–3.92 | 6.55 | 18 | 10.8 |
| 10 | VIRTIS | NASA-Dawn | Vesta, Ceres | 2007 | 2018 | 432<br>432 | 0.28–1.10<br>1.05–5.13 | 1.89<br>9.47 | 0.014°× 0.014° (IFOV) | 3.67°× 3.67° (FOV) |
| 11 | M3 | Chandrayaan-1 | Lunar orbit | 2008 | 2009 | 260 | 0.43–3.0 | 10 | 70 | 40 |
| 12 | FTHSI | HJ-1A | LEO | 2008 | Active | 115 | 0.45–0.95 | 4 | 100 | 50 |
| 13 | HySI | IMS-1 | LEO | 2008 | 2012 | 64 | 0.40–0.95 | 8 | 500 | 130 |
| 14 | ARTEMIS | TacSat-3 | LEO | 2009 | 2012 | 400 | 0.4–2.5 | 5 | 4 | 4 |
| 15 | HICO | ISS | ISS orbit | 2009 | 2014 | 128 | 0.35–1.08 | 5.7 | 90 | 51 |
| 16 | VNIS | Change'E | Lunar rover | 2013 | 2015 | 100 | 0.45–0.95<br>0.9–2.4 | 5 | — | 8.5°× 8.5° (FOV) |
| 17 | OLCI | Sentinel-3A | LEO | 2016 | Active | 520 (transmit 21) | 0.39–1.04 | 1.25 | 300 | 1270 |

| | | | | | | | | | | |
|---|---|---|---|---|---|---|---|---|---|---|
| 18 | MHRIS | GHGSat-D | LEO | 2016 | Active | 512 | 1.6–1.7 | 0.2 | 50 | 15 |
| 19 | AaSI | Aalto-1 | LEO | 2017 | Active | 6–20 | 0.50–0.90 | 7–10 | 192 | 97 |
| 20 | DESIS | ISS | ISS orbit | 2018 | Active | 235 | 0.40–1.0 | 2.55 | 30 | 30 |
| 21 | HyperScout | GomX-4B | LEO | 2018 | Active | 45 | 0.4–1.0 | 15 | 50 | 200 |
| 22 | AHSI | Gaofen-5 | LEO | 2018 | Active | 330 | 0.40–2.50 | 5 (VNIR)<br>10 (SWIR) | 30 | 60 |
| 23 | HysIS | IMS-2 | LEO | 2018 | Active | 256 | 0.40–2.4 | 10 | 30 | 30 |
| 24 | PRISMA | PRISMA | LEO | 2019 | Active | 237 | 0.40–2.51 | 12 | 30 | 30 |
| 25 | HISUI | ISS | LEO | 2019 | Active | 185 | 0.40–2.50 | 10 (VNIR)<br>12.5 (SWIR) | 30 | 30 |
| 26 | EnMAP | German HS | LEO | 2022 | N/A | 244 | 0.42–2.50 | 6.5 (VNIR)<br>10 (SWIR) | 30 | 30 |
| 27 | MAJIS | JUICE | Ganymede orbit (500 km) | Planned 2023 | N/A | 508<br>508 | 0.50–2.35<br>2.55–5.54 | 3.6<br>6.4 | 75 | 30 |

spaceborne hyperspectral imagers since the beginning in the order of years chronologically. This list may miss some spaceborne hyperspectral imagers. It is worth noting that during a short period from 2016 to 2019, nine (9) spaceborne hyperspectral imagers have been launched to space. There is a leap in the number of spaceborne hyperspectral imagers launched in 2018. Many new spaceborne hyperspectral imagers have been planned and will come up.

In terms of platforms and orbits of these spaceborne hyperspectral imagers, majority of them are aboard satellites on low Earth orbits (LEOs). Three of them are deployed on the International Space Station (ISS). There are also six hyperspectral imagers out of Earth orbits, one (CRISM) on a Mars orbit, one (M3) on a lunar orbit, and one (VNIS) on a lunar rover for in-situ observation. The Visible and Infrared Thermal Imaging Spectrometer (VIRTIS) and its two slight variants were deployed onboard the space probes of three planetary missions on orbits of a comet, Venus, and two proto-planets.

With respect to spectral dispersion means, twenty (20) spaceborne hyperspectral imagers use dispersive elements, either gratings or prisms. One (MODIS) uses band-pass filters. Two (2) of them use linear variable filters (LVFs) to disperse the spectrum. These two imagers are Hyperspectral Imager (HySI) onboard the Indian Mini Satellite-1 (IMS-1), and HyperScout on ESA's GomX-4B nanosatellite. Three (3) hyperspectral imagers utilize electronically tunable filters (ETFs), of which two imagers use tunable Fabry–Perot filters and one imager uses acousto-optic tunable filter. There is also one spaceborne hyperspectral imager that uses Fourier transform (FT) interferometer to disperse the spectrum.

**Acronyms List**

| | |
|---|---|
| Active: | when the manuscript was written in March 2020 |
| AaSI: | Aalto-1 Spectral Imager |
| AHSI: | Advanced Hyperspectral Imager on GaoFen-5 satellite |
| ARTEMIS: | Advanced Responsive Tactically Effective Military Imaging Spectrometer |
| CHRIS: | Compact High-Resolution Imaging Spectrometer |
| CRISM: | Compact Reconnaissance Imaging Spectrometer for Mars; MRO: Mars Reconnaissance Orbiter |
| DESIS: | DLR Earth Sensing Imaging Spectrometer |
| EnMAP: | Environmental Mapping and Analysis Program |
| FTHSI: | HyperSpectral Imager on HJ-1A satellite |
| HICO: | Hyperspectral Imager for Coastal Ocean |
| HISUI: | Hyperspectral Image Suite |
| HSI: | HyperSpectral Imager; LEWIS mission |
| Hyperion: | Hyperspectral imager; EO-1: Earth Observing-1 Mission |
| HyperScout: | Hyperspectral camera of size 1U onboard ESA's nanosatellite GomX-4B |
| HySI: | Hyperspectral Imager (HySI) onboard Indian Mini Satellite-1 (IMS-1) |
| M3: | Moon Mineralogy Mapper |
| MAJIS: | Moons And Jupiter Imaging Spectrometer onboard spacecraft of JUpiter ICy moons Explorer |
| MERIS: | (MEdium Resolution Imaging Spectrometer); ENVISAT: ESA's Environmental Satellite |
| MHRIS: | Miniature high-resolution imaging spectrometer |
| MODIS: | Moderate Resolution Imaging Spectroradiometer |
| OLCI: | Ocean and Land Color Imager |

| | |
|---|---|
| PRISMA: | PRecursore IperSpettrale della Missione Applicativa |
| SPIMs 1-5: | Spectrographic Imagers1-5; MSX: Midcourse Space Experiment satellite |
| VIRTIS: | Visible and Infrared Thermal Imaging Spectrometer |
| VNIS: | Visible and Near-infrared Imaging Spectrometer aboard Chang'E 3 Spacecraft |

Regarding the operation mode, among the 27 spaceborne hyperspectral imagers listed in Table 2.4, all of them use 2D detector arrays and operated or are operating in pushbroom mode, except MODIS instruments that use 1D linear detector arrays and operate in whiskbroom mode.

In Table 2.4, one upcoming spaceborne hyperspectral imager is also listed. It is Moons And Jupiter Imaging Spectrometer (MAJIS).

## 2.2.2 *Brief Description of Spaceborne Hyperspectral Imagers*

This section provides a brief description of the spaceborne hyperspectral imagers listed in Table 2.4 from the perspective of instrument. The applications of hyperspectral satellites are not the scope of this chapter and have been reported by Qian [5].

### 2.2.2.1 Visible Imagers and Spectrographic Imagers (UVISI) Onboard the MSX Satellite

The earliest spaceborne hyperspectral imager was the Ultraviolet and Visible Imagers and Spectrographic Imagers (UVISI) onboard the Midcourse Space Experiment (MSX) mission of the US DoD, which was launched in April 1996. The UVISI was composed of five ultraviolet and visible hyperspectral imagers (called spectrographic imagers or SPIMs) and four ultraviolet and visible multispectral imagers. These nine imaging sensors provided hyperspectral and multispectral capabilities in wavelength range 110–900 nm. The UVISI records data in a total of 1360 spectral bands simultaneously, with a GSD of 770 m at nadir and a swath about 15 km wide. It is not well known because of its coarse spatial resolution and the nature as a military satellite.

### 2.2.2.2 HyperSpectral Imager (HSI) for the LEWIS Mission

The second earliest spaceborne hyperspectral imager was HSI for the LEWIS mission launched in 1997 [23] as a technology demonstration under the NASA's SSTI program. Unfortunately, HSI was unsuccessful and lost as described in Section 2.1.2 [5].

### 2.2.2.3 MODIS Onboard Terra and Aqua Satellites

The Moderate Resolution Imaging Spectroradiometers (MODIS) were deployed onboard NASA's Terra (EOS AM) satellite and Aqua (EOS PM) satellite, which were launched in 1999 and 2002, respectively. MODIS uses band-pass filters to disperse the spectrum of the object scene. This is the technology used for multispectral imagers (e.g. Landsat and SPOT). From this perspective, MODIS should be classified as a multispectral imager. However, many of its bands are narrow enough (10 nm), which falls in the definition of a hyperspectral imager "the acquisition of many images of contiguous, narrow, registered spectral bands such that for each pixel a radiance spectrum can be derived" described in Section 2.1. That is why it is counted as a hyperspectral imager here. Detailed description of MODIS instruments and their applications can be found in the literature [5, 26].

#### 2.2.2.4 Hyperion Onboard NASA's EO-1 Satellite

Hyperion is well known in the remote sensing community and is often regarded as the first spaceborne hyperspectral imager in the remote sensing community for the reasons described in the paragraphs above. It was onboard NASA's EO-1 satellite launched on 21 November 2000 as part of the New Millennium Program to develop and validate new technologies for future Earth imaging observatories.

Hyperion is a pushbroom hyperspectral imager with a swath of 7.65 km. The ground footprint size is 30 × 30 m. The 30 m size in the along-track direction was obtained by basing the frame rate on the velocity of the spacecraft for a 705 km orbit. The entire 7.65 km swath is obtained in a single frame. Each image cube contains data for a 7.65 km wide in cross-track direction by 185 km long in along-track direction [27]. Hyperion instrument consisted of a VNIR spectrometer and a SWIR spectrometer covering a wavelength range from 0.4 to 2.5 μm with a spectral sampling interval of 10 nm. This resulted in 242 raw spectral bands and 198 bands after being processed. Hyperion had been continuously acquiring hyperspectral data for scientific research and user community until its retirement in 2017 although it was designed for a one-year life. This was a great success story of spaceborne hyperspectral imagers.

#### 2.2.2.5 CHRIS Onboard ESA's PROBA Satellite

The Compact High-Resolution Imaging Spectrometer (CHRIS) onboard ESA's Project for On-Board Autonomy (PROBA) satellite was launched on 22 October 2001. The primary objective of PROBA was to test a number of innovations in platform design, such as attitude control and recovery from errors, autonomous operation with minimal intervention from the ground [28].

The CHRIS is hyperspectral imager working in the VNIR region at combined GSD of 17–20 m or 34–40 m with a multi-angle viewing capability and programmable spectral bands. It acquires up to 62 spectral bands at 5–15 nm spectral sampling interval in a wavelength range of 415–1050 nm. CHRIS has five operating modes, each of which has varied nominal number of bands, wavelength range, spectral bandwidth, and the nominal GSD, with GSD decreasing as spectral bandwidth increases. At perigee, CHRIS provides a GSD of 17 m, over typical image areas 13 km square. The instrument provides sets of images of selected target areas, at different viewing angles, forming up to five images of each target in a single overpass. The longevity of PROBA satellite is quite impressive. It is still running after over 20 years orbiting (as of July 2022).

#### 2.2.2.6 MERIS Onboard ESA's ENVISAT Satellite

The MERIS onboard ESA's ENVISAT satellite was launched on 1 March 2002 and demissioned in 2012. MERIS was the first large-swath spaceborne hyperspectral imager with high spectral and radiometric accuracy developed for observing the color of oceans, both in the open ocean and in coastal zones to study the oceanic biology and marine water quality of the global carbon cycle and the productivity of these regions and the atmosphere and land surface-related processes. It was a hyperspectral imager working in the VNIR region and operated in a pushbroom mode. MERIS had a wide FOV with a swath width of 1150 km measuring the solar radiation reflected by the Earth. By design,

MERIS could record 520 spectral bands within the wavelength range from 390 to 1040 nm at the instrument native spectral sampling interval of 1.25 nm. However, the MERIS was restricted by its downlink capability and transmitted to ground only 15 channels, where each channel is an average taken over 8–10 elements (native spectral samplings) of the detector array. The GSD varies in the cross-track direction, between 260 m at nadir and 390 m at swath extremities [5].

#### 2.2.2.7 VIRTIS for ESA's Rosetta, Venus-Express, and NASA-Dawn Planetary Missions

The VIRTIS was originally built for ESA's Rosetta cometary mission [29]. It was the third cornerstone mission of the ESA's Horizon 2000 program. Rosetta was a space probe, including an orbiter and a lander Philae, launched on 2 March 2004 for studying comet 67P/Churyumov–Gerasimenko (67P). The spacecraft reached the comet on 6 August 2014. During its journey to the comet, the spacecraft flew by Mars and the asteroids 21 Lutetia and 2867 Šteins. VIRTIS was one of the scientific payloads of the Rosetta orbiter to detect and characterize the evolution of specific signatures – such as the typical spectral bands of minerals and molecules – arising from surface components and from materials dispersed in the coma.

VIRTIS was also selected as a key instrument for ESA's Venus Express and NASA-Dawn missions. This makes the VIRTIS a great success story of one payload design for three planetary missions. Venus Express mission was to study the Venusian atmosphere and clouds in detail, and to study the plasma environment and the surface characteristics of Venus from orbit. It was launched on 9 November 2005 and entered the target Venus orbit at apoapsis on 7 May 2006 [30]. Dawn is a space probe launched on 27 September 2007 for studying two of the three known proto-planets of the asteroid belt, Vesta and Ceres. The space probe entered orbit around Vesta on 16 July 2011, and completed a 14-month survey mission before leaving for Ceres in late 2012. It then entered orbit around Ceres on 6 March 2015 [31].

VIRTIS is a hyperspectral imager with three focal planes in two channels. The mapping channel, referred to as VIRTIS-M, has two 2D focal planes covering the visible region from 0.28 to 1.1 μm and the infrared region from 1.05 to 5.13 μm. The spectroscopic channel, referred to as VIRTIS-H, has a single aperture covering a wavelength range from 1.84 to 4.99 μm with high spectral sampling interval. VIRTIS-M generates a total of 432 spectral band images of size 256 × 256 pixels in the wavelength range 0.28–4.99 μm with spectral sampling interval of 1.89 and 9.49 nm in the visible region and infrared region, respectively.

#### 2.2.2.8 CRISM Aboard Mars Reconnaissance Orbiter

The Compact Reconnaissance Imaging Spectrometer for Mars (CRISM) is a VNIR and infrared (IR) hyperspectral imager onboard the NASA spacecraft of Mars Reconnaissance Orbiter (MRO), which was launched on 12 August 2005 and attained Martian orbit on 10 March 2006. CRISM is one of the six major science instruments on MRO. It is used to produce detailed maps of the surface mineralogy of Mars and to identify minerals and chemicals indicative of the past or present existence of water on the surface of Mars [32].

The CRISM acquires electromagnetic radiation from 362 to 3920 nm with a spectral sampling interval of 6.55 nm. It operates in three modes: multispectral mode, targeted mode, and atmospheric mode. In the multispectral mode, it points at planet nadir to reconnoiter Mars with 72 of its 544 spectral bands at a footprint size of 100–200 m per pixel. Nearly the entire planet can be mapped in this fashion. The objective of this mode is to identify new scientifically interesting locations that could be further investigated. In targeted mode, the CRISM measures radiation reflected from Mars surface in all 544 spectral bands. When the MRO spacecraft is at an altitude of 300 km, the CRISM detects a scene of interest at full spatial (15–19 m/pixel) and spectral resolution (362–3920 nm at 6.55 nm/ band). In atmospheric mode, it acquires only the emission phase function. Global grids of the resulting lower data volume observations are taken repeatedly throughout the Martian year to measure seasonal variations in atmospheric properties.

#### 2.2.2.9 Moon Mineralogy Mapper for Mapping Lunar Surface

The Moon Mineralogy Mapper (M3) is a NASA-contributed hyperspectral imager to India's first mission to the Moon, Chandrayaan-1, launched on 22 October 2008. The M3 is the first high-resolution imaging spectrometer operating in pushbroom mode to map the entire lunar surface spatially and spectrally. It generates images of moon surfaces in long narrow strips in a wavelength range from 400 to 3000 nm with a spectral sampling interval of 10 nm. This forms 260 spectral images for a scene of the lunar surface. The swath width (i.e. width of a line scene) is 40 km on the moon's surface at a moment. This line scene is imaged onto 600 detector pixels, with each pixel representing a footprint size of 70 × 70 m on the surface. The second spatial dimension of the scene is obtained by the flight of the spacecraft along the flight direction. The circumference of the moon is 10,930 km. With overlap, it takes more than 274 image swaths to completely map the moon [33].

The Chandrayaan-1 mission was cut short at 10 months in August 2009, when contact was lost with the spacecraft. Despite the abbreviated mission, M3 was able to meet its mission requirements: collecting more than 95% of the Moon in Global Mode along with a small number of Target Mode images [34].

#### 2.2.2.10 Fourier Transform Hyperspectral Imager Onboard Chinese Environment Satellite

Fourier transform hyperspectral imager (FTHSI) onboard Chinese Huan Jing Satellite-1A (HJ-1A) is a spatially modulated imaging interferometer. Unlike dispersive element-based hyperspectral imagers, a FT-based hyperspectral imager produces interferometric data in FT domain, which need to be processed before obtaining radiometric data. The FTHSI instrument has a total of 115 bands covering a spectral range from 0.45 to 0.95 μm after processing of the raw FT data and returning to spectral domain. The width of the swath is 50 km with the ground footprint size 100 m [5].

The mini-satellite constellation Huan Jing (means *environment* in Chinese) is a China's national program for disaster monitoring and mitigation using remote sensing technology and improvement of the efficiency of disaster mitigation and relief. The HJ-1 constellation includes three mini-satellites referred to as HJ-1A, HJ-1B, and HJ-1C. HJ-1A and B satellites were launched on 6 September 2008.

#### 2.2.2.11 HySI Onboard Indian Mini Satellite-1

HySI onboard IMS-1 launched on 28 April 2008 was a 64-band VNIR imaging spectrometer with spatial sampling distance of about 500 m and swath width of about 130 km. It was aimed at validating the design of hyperspectral imager and providing hands-on experience for users and scientists of the hyperspectral applications [35].

HySI was an optical filter-based hyperspectral imager. Unlike the traditional imaging spectrometers that use dispersive elements (e.g. grating or prism) to disperse the spectrum, HySI used a wedge filter, also known as a linear variable filter (LVF), to disperse the spectrum. It is the first spaceborne hyperspectral imager that uses a LVF as the dispersive element. HySI consisted of only a telescope, a 2D detector array with the wedge filter mounted very close to it, and the associated front-end electronics. This is because an LVF-based hyperspectral imager omits a bulky spectrometer for dispersion of spectrum; thus, it is much compact in volume and light in weight. The operational principle of an LVF-based hyperspectral imager is different from that of a dispersive element-based hyperspectral imager. Its FOV is not limited in the along-track direction by a slit. The 2D detector array with an LVF mounted very close to it in front sees the full FOV (i.e. scene) via the telescope. Each row of the detector array senses a corresponding cross-track line at a different wavelength. As the satellite flies over the scene, each detector row passes over the scene. Once the complete scene has been passed over, the spectrum of each pixel can be reconstructed by reorganizing the acquired data. Detailed description of LVF-based hyperspectral imager can be found in Qian [5].

#### 2.2.2.12 ARTEMIS Onboard TacSat-3

The Advanced Responsive Tactically Effective Military Imaging Spectrometer (ARTEMIS) was on board TacSat-3 satellite launched on 19 May 2009, which was the third in a series of U.S. DoD military reconnaissance satellites. The TacSat satellites are all designed to demonstrate the ability to provide real-time data collected from space to combatant commanders in the field [36].

The ARTEMIS hyperspectral imager used a single 2D detector array covering both VNIR and SWIR spectral regions from 0.4 μm to 2.5 μm at a uniform spectral sampling interval of 5 nm with 4 m GSD and a swath width of 4 km. The detector array used was a substrate-removed mercury-cadmium-telluride (MCT) 2D detector array that extends its sensitivity into the blue wavelengths to cover the full spectral range. This allowed a simple instrument design that uses only a single spectrometer to cover both VNIR and SWIR spectral regions and significant reduction of instrument complexity and cost [37].

#### 2.2.2.13 HICO Onboard the International Space Station

The Hyperspectral Imager for the Coastal Ocean (HICO) was deployed on the ISS on 23 September 2009. It was the first spaceborne imaging spectrometer designed for coastal ocean research [38]. HICO was built by using non-space-hardened, commercial-off-the-shelf (COTS) parts to demonstrate improved coastal remote sensing products including bathymetry, bottom types, water optical properties, and on-shore vegetation maps. During its five years life, it collected approximately 10,000 hyperspectral scenes of the earth. The data have enabled ocean color scientists and managers to assess data quality and apply the imagery to a variety of scientific and societal problems.

HICO was a pushbroom imaging spectrometer working in the VNIR region. It acquired imagery in wavelength range of 380–960 nm with a spectral sampling interval of 5.7 nm. It had a swath width of 51 km when the ISS altitude was 420 km altitude. Its ground footprint size was 100 m, which was much smaller than that of other spaceborne ocean color imagers (300 m of MERIS and 1000 m of MODIS ocean color bands). Even with such small footprint compared to other ocean color imagers, it still achieved reasonably high SNR: maximum SNR 470 : 1 at 480 nm, SNR > 200 : 1 in spectral range 400–600 nm [39].

#### 2.2.2.14 Visible and Near-infrared Imaging Spectrometer Aboard Chang'E 3 Spacecraft

The Visible and Near-infrared Imaging Spectrometer (VNIS) is one of the main scientific payloads of China's lunar rover Yutu ("Jade Rabbit" in Chinese) of the Chang'E 3 mission, which reached lunar orbit on 6 December 2013 and landed on the moon on 14 December 2013. After soft landing on the

Moon, Chang'E 3 carried out a lunar survey and scientific exploration activities. The Chang'E 3 spacecraft included a lander and a lunar rover Yutu, each of them carried scientific payloads. The objective of the VNIS was to make in-situ measurements of the composition and resources of the lunar surface via imaging and spectrometry in the VNIR and SWIR regions. Mounted on the Yutu in the front, the VNIS measured the radiance diffusely reflected from the Moon's surface of the solar illumination with a 45° view angle and acquired spectral and geometric data for determining the lunar surface mineral composition and performing comprehensive analysis of the chemical composition [40]. VNIS is the first in-situ spectral imaging detection on the lunar surface. The high resolution and effective spectral imaging data obtained by the VNIS has provided valuable hyperspectral data for lunar scientific applications.

The VNIS instrument includes a VNIR imaging spectrometer covering a wavelength range of 0.45–0.95 μm and a SWIR spectrometer covering a wavelength range of 0.9–2.4 μm. Acousto-optic tunable filters (AOTFs) were used as the dispersive elements of the two spectrometers. When a radio-frequency (RF) signal is applied to the transducer of the AOTF crystal, the electrical signal is converted into an ultrasonic vibration. Then a coupling, quasi-monochromatic wavelength is diffracted at a given separation angle by momentum matching between the collimated light and ultrasound vibration at a given frequency. The spectrally tuned light reached the surface of the detector for spectral imaging. The VNIR imaging spectrometer had a FOV of 8.5° × 8.5° for 256 × 256 pixels in the scene with 100 spectral bands of 2–7 nm spectral resolution, whereas the SWIR spectrometer had a FOV of 3.6° × 3.6° for a single pixel in the scene with 300 spectral bands of 3–12 nm spectral resolution [41].

#### 2.2.2.15 Ocean and Land Color Imager (OLCI) on Sentinel-3A

The Ocean and Land Color Imager (OLCI) is a VNIR pushbroom imaging spectrometer. It is one of the seven instruments onboard ESA's Sentinel-3A satellite launched on 16 February 2016. OLCI is the successor of MERIS onboard ESA's ENVISAT, which was out of service since April 2012. OLCI was designed to provide global and regional measurements of ocean and land surface with high radiative accuracy based on a heritage design from MERIS.

OLCI has a ground footprint size of 300 m over both land and water surfaces and slightly wide swath of 1270 km. Same as MERIS, the OLCI instrument has five identical fan-arranged Dyson spectrometers (also called cameras) with five focal plane arrays covering the wide swath. Its revisit time with global coverage has been reduced to three days, instead of around 15 days of MERIS. OLCI transmits to ground 21 spectral bands, compared to 15 on MERIS. A lesson learned from MERIS is the negative impact of the direct solar reflection at sea surface to the sensor, which is referred to as sun-glint. To minimize the impact of sun-glint, OLCI adopted an asymmetric swath with respect to the satellite ground track to avoid sun-glint. A cross-track tilt of 12.6° of the overall FOV is used that results in a maximum observation zenith angle (OZA) slightly above 55° [42].

#### 2.2.2.16 Miniature High-Resolution Imaging Spectrometer on GHGSat-D

The miniature high-resolution imaging spectrometer (MHRIS) is on board the Greenhouse Gas Demonstration Satellite (GHGSat-D) launched on 22 June 2016 as a secondary payload to ISRO's CartoSat-2C spacecraft. GHGSat is a commercial venture for measurement of greenhouse gas and air-quality gas emissions from industrial sites using satellite technology [43].

The GHGSat is a nanosatellite based on a low-cost and high-performance nanosat bus NEMO-AM made by University of Toronto, Institute for Aerospace Studies/Space Flight Laboratory

(UTIAS/SFL) [44]. It has a launch mass of 15 kg and a volume of approximately 25U (20 cm × 30 cm × 42 cm) plus a mezzanine of size 7 cm × 18 cm × 42 cm on one side (−X).

MHRIS is an electronically tunable filter (ETF)-based SWIR hyperspectral imager covering a wavelength range 1600–1700 nm with over 300 spectral bands of 0.1 nm bandwidth. It has a spatial resolution of under 50 m, covering a 15 km ground swath. The MHRIS instrument uses InGaAs SWIR detector array with enhanced dynamic range. The detector array format is 640 × 512 pixels, of which GHGSat-D masks the area outside the central 512 × 512 array. The selected InGaAs array has heritage on a NASA mission.

#### 2.2.2.17 Aalto-1 Spectral Imager (AaSI) on a 3U Nanosatellite

Aalto-1 Spectral Imager (AaSI) is based on a spectrally tunable Fabry–Pérot interferometer developed by VTT Technical Research Centre of Finland. It is the main payload of the Aalto-1 nanosatellite that was launched on 23 June 2017 by PSLV-C38 rocket from India. The Aalto-1 nanosatellite was built mainly by students at Aalto University in Finland. It has a volume of 34 cm × 10 cm × 10 cm and a mass of ~ 4 kg. The design life is 2 years. It has an average power production of 4.8 W. AaSI is able to record 2D spatial images of selected spectral bands in a spectral range from 500 to 900 nm with a spectral resolution 10–30 nm by electronically tuning the Fabry–Pérot interferometer. Filter apertures of 7 or even 19 mm can be reached with the piezo-actuated Fabry–Pérot interferometer [45].

#### 2.2.2.18 DLR Earth Sensing Imaging Spectrometer on the International Space Station

The DLR Earth Sensing Imaging Spectrometer (DESIS) is a pushbroom hyperspectral imager in the VNIR region from 400 to 1000 nm with a minimum spectral sampling interval of 2.55 nm. It has a ground swath width of 30 km with a ground footprint size of 30 m. It is hosted on the Multi-User System for Earth Sensing (MUSES) mounted on the ISS. The DESIS instrument was launched on 29 June 2018 as part of the SpaceX CRS-15 logistics flight to the ISS and was installed to the exterior of the ISS on 27 August 2018 [46].

The main difference between DESIS design and most of the hyperspectral designs is that DESIS is equipped with a pointing mirror in front of the entrance slit. It can point in forward direction and back direction up to ±15°. It can operate in either the static mode with 3° angle steps or the dynamic mode with up to 1.5° change in viewing direction per seconds. When operating in the static mode, it allows acquiring hyperspectral data to produce BRDF products or stereo images. When operating in the dynamic mode, it allows continuous observations of the same targets with ground motion compensation to further improve the SNR of the acquired hyperspectral data.

#### 2.2.2.19 HyperScout Hyperspectral Camera on ESA's Nanosatellite GomX-4B

HyperScout is a miniaturized hyperspectral camera of size 1U (10 cm × 10 cm × 10 cm) developed by Cosine Research in the Netherlands. It is onboard ESA's nanosatellite GomX-4B, which is one of a pair of two 6U nanosatellites (GomX-4 A and GomX-4B). They were launched at the same time as secondary payloads on 2 February 2018 on a Long March 2D vehicle from Jiuquan Satellite Launch Center (JSLC), China.

HyperScout hyperspectral camera is one of the five demonstration payloads onboard GomX-4B. It is a spectral filter-based hyperspectral imager covering a spectral range from 400 to 1000 nm with a spectral resolution of 15 nm. An LVF is used to separate the different wavelengths before the radiance reaching on a 2D CMOS detector array. Thanks to the omission of the spectrometer, HyperScout instrument consists of only a telescope, a focal plane array, an instrument control unit, and onboard data handling unit, which fit in a volume of 1U. HyperScout has a swath width on ground of 200 km with a footprint size 70 m [5, 47].

#### 2.2.2.20 Advanced Hyperspectral Imager (AHSI) on Chinese Gaofen-5 Satellite

The Advanced Hyperspectral Imager (AHSI) is the main payload of the Chinese Gaofen-5 (GF-5) satellite, which was launched on 8 May 2018 for scientific research on the Earth's atmosphere and terrestrial observation. The satellite carries six payloads, a hyperspectral payload, and a multispectral payload for terrestrial Earth observation, along with four atmospheric observation payloads. The objectives of AHSI are to address many key science questions and operational needs using remote sensing technology.

AHSI has 330 spectral bands covering a wavelength range from 0.4 to 2.5 μm. The spectral sampling interval is 5 nm in the VNIR (0.4~1.0 μm) region and 10 nm in the SWIR (1.0~2.5 μm) region. The GSD of AHSI is 30 m, which is the same as that of Hyperion, whereas the swath width of AHSI is 60 km, which is about eight times wider than that of Hyperion [48].

#### 2.2.2.21 Italian Hyperspectral Satellite PRISMA

PRISMA (PRecursore IperSpettrale della Missione Applicativa) is an Italian hyperspectral satellite aiming to qualify the technology, contribute to development of applications, and provide products to institutional and scientific users for environmental observation and risk management. It was launched on 22 March 2019 on a Vega launch vehicle from the European base of Kourou in French Guyana, into a sun synchronous orbit [49].

PRISMA instrument is the core of the PRISMA mission and consists of a hyperspectral imager and panchromatic (PAN) camera. The hyperspectral imager operates in pushbroom mode. It is made up of a VNIR spectrometer and a SWIR spectrometer to cover spectral bands ranging from 400 to 1010 nm and from 920 to 2505 nm. It provides hyperspectral images of the Earth with 30 m GSD, 30 km swath width, and spectral bands at an spectral sampling interval of 12 nm. PAN camera provides images at spatial resolution of 5 m within a wavelength range of 400–700 nm, spatially co-registered to the hyperspectral images, so as to allow images fusion to sharpen the spatial resolution of the hyperspectral images [50].

#### 2.2.2.22 Hyperspectral Imager Suite Onboard the International Space Station

Hyperspectral Imager Suite (HISUI) includes a hyperspectral imager (HSI) and a multispectral imager (MSI) deployed onboard the Japan Experiment Module (JEM) in the ISS. The HISUI-Exposed Payload (HISUI-ExP) and HISUI Mission Data Recorder – Pressurized Module (MDR-PM) were launched to the ISS from US Cape Canaveral Air Force Station by SpaceX's Falcon-9 Dragon cargo rocket on 6 December 2019 [51].

The HSI and MSI were fabricated in two separate boxes and operate independently or simultaneously. The HSI covers a wavelength range from 0.40 to 2.5 μm with a total of 185 spectral bands (57 VNIR + 128 SWIR). The spectral sampling interval is 10.0 nm in the VNIR region and 12.5 nm in the SWIR region. The swath of HSI is 20 km, which is one-third of that of the MSI (60 km) due to the constraint of optical design and the availability of large form 2D detector arrays for the HSI. To fill the gap between the swaths of two imagers, the HSI is equipped with a pointing mechanism, which can tilt the HSI for ±5° in cross-track direction to match the swaths of the imagers [52].

Due to the constraint of data downlink capacity of the ISS, the hyperspectral and multispectral data generated by HISUI are partially transmitted to ground stations ($\approx$ 10 GB/day $\approx$ 30,000 km$^2$). The rest ($\approx$ max. 300 GB/day $\approx$ 900,000 km$^2$) is recorded in removable mass memory storages that are shipped back to Earth by cargo ships three or four times a year.

#### 2.2.2.23 German Spaceborne Hyperspectral Imager EnMAP

The EnMAP is a German hyperspectral satellite mission launched on April 1, 2022. It aims at monitoring and characterizing the Earth's environment on a global scale by providing high-quality hyperspectral data that are not achievable by the currently available spaceborne hyperspectral sensors.

EnMAP is a prism-based hyperspectral imager operating in pushbroom mode. It has 242 spectral bands covering a wavelength range from 420 to 2450 nm with a spectral sampling interval of 6.5 nm for VNIR bands and 10 nm for SWIR bands. Its ground swath width is 30 km with a GSD of 30 × 30 m [53].

#### 2.2.2.24 ESA's Moons and Jupiter Imaging Spectrometer (MAJIS)

The MAJIS was selected as one of the scientific payloads by ESA in February 2013 for its Jupiter Icy Moons Explorer (JUICE) mission intended to explore Jupiter and three of its icy moons: Europa, Callisto, and Ganymede. It is scheduled to be launched in 2023. The spacecraft of the JUICE mission is targeted to fly by Callisto, Ganymede, and Europa, then a one-year orbital phase around Ganymede [54].

The MAJIS hyperspectral imager is a dual grating-spectrometer design with the VNIR spectrometer covering a spectral range from 0.5 to 2.35 μm and the IR spectrometer covering a spectral range from 2.25 to 5.54 μm. The two spectrometers share a common telescope. A dichroic beam-splitter separates the light between the two spectrometers. The spectral and spatial resolution of MAJIS takes advantage of up-to-date developments of detector technology with 2 × 508 spectral bands to achieve a wide wavelength range of 0.5 to 5.5 μm at spectral sampling interval of 3.6 nm for VNIR and 6.4 nm for IR over 400 spatial pixels in a cross-track line. The IFOV of 150 μrad of the instrument corresponds to a footprint size of 75 m on Ganymede from a 500 km circular orbit over Ganymede and to a footprint size of 150 km for observations of the atmosphere of Jupiter when flies by it [55].

## 2.3 Conclusion

Hyperspectral imaging is a new generation of remote sensing technology, thanks to its expanded capability of acquiring hundreds of contiguous and narrow spectral bands for each pixel in the scene. Hyperspectral imaging has been widely used in various disciplines including agriculture, forestry, environment, geology, ocean, atmosphere, climate change, defense and security, and law enforcement. Fascinating detailed spectral information provided by hyperspectral imagery often provides results not possible with multispectral or other types of imaging technologies.

As an introduction of hyperspectral imaging remote sensing from satellites, this chapter described how hyperspectral imaging technology evolves from aircraft to spacecraft. It reviewed the history of the development of hyperspectral imaging technology starting from the world's first airborne hyperspectral imager in the late 1970s, the development of hyperspectral technologies between 1980s and 1990s, early development of hyperspectral imagers in Canada, and the NASA's planned orbiting hyperspectral imagers in the 1990s. A few well-known airborne hyperspectral sensors, such as the AIS, AVIRIS, FLI, CASI, and SFSI, are reviewed as representative examples of the early development history of the airborne hyperspectral imagers. The development of spaceborne hyperspectral imagers began in the early 1980s. The SISEX for space shuttle flight, the free-flyer HIRIS, and the HSI for LEWIS mission are the examples of the earliest planned spaceborne hyperspectral imaging sensors.

This chapter briefly reviewed spaceborne hyperspectral sensors based on author's thorough survey. A total of 27 spaceborne hyperspectral sensors are reviewed. It clarified the UVISI onboard the MSX mission launched in 1996 as the first spaceborne hyperspectral sensor, although Hyperion onboard NASA's EO-1 satellite launched in 2000 is often regarded as the first spaceborne hyperspectral sensor due to its popularity and longevity. The review of these spaceborne hyperspectral sensors is from the perspective of payload and its performance and capability. All the 27 spaceborne hyperspectral sensors use 2D detector arrays and operate in pushbroom mode, except MODIS that uses 1D linear detector arrays and operates in whiskbroom mode. With respect to spectral dispersion means, 20 of them use dispersive elements, either gratings or prisms, 2 of them use LVFs, 1 of them uses band-pass filters, 3 of them utilize ETFs, and 1 uses FT interferometer to disperse the spectrum. In terms of platforms and orbits of these hyperspectral sensors, majority of them are aboard satellites on LEOs. Three of them are deployed on the ISS. Six of them are out of Earth orbits: one (CRISM) on a Mars orbit, one (M3) on a lunar orbit, one (VNIS) on a lunar rover for in-situ observation, the VIRTIS and its two slight variants being deployed onboard the space probes of three planetary missions on orbits of a comet, Venus, and two proto-planets.

## References


1. Chang, C.-I (2013). *Hyperspectral Data Processing*. Hoboken: Wiley.
2. Grahn, H. and Geladi, P. (2007). *Techniques and Applications of Hyperspectral Image Analysis*. Hoboken: Wiley.
3. Qian, S.-E. (2021). Hyperspectral satellites, evolution, and development history. *IEEE Journal of Selected Topics in Earth Observations and Remote Sensing*. 14: 7032–7056.
4. Goetz, A.F.H., Vane, U., Solomon, J., and Rock, B.N. (1985). Imaging spectrometry for earth remote sensing. *Science* 228: 1147–1153.
5. Qian, S.-E. (2020). *Hyperspectral Satellites and System Design*. Boca Raton: CRC Press/Taylor & Francis.
6. Vane, G. and Goetz, A.F.H. (1988). Terrestrial imaging spectroscopy. *Remote Sensing of Environment* 24: 1–29.
7. Goetz, A.F.H. (1995). Imaging spectrometry for remote sensing: vision to reality in 15 years. *Proceedings of SPIE* 2480: 1–13.
8. Green, R.O., Eastwood, M.L., Sarture, C.M. et al. (1998). Imaging spectroscopy and the Airborne Visible/Infrared Imaging Spectrometer (AVIRIS). *Remote Sensing of Environment* 65 (3): 227–248.
9. Vane, G., Green, R.O., Chrien, T.G. et al. (1993). The airborne visible/infrared imaging spectrometer (AVIRIS). *Remote Sensing of Environment* 44: 127–143.
10. Staenz, K. (1992). A decade of imaging spectrometry in Canada. *Canadian Journal of Remote Sensing* 18 (4): 187–197.
11. Gower, J.F.R., Borstad, G.A., Gray, L.H., and Edel, H. R. (1987). The fluorescence line imager: imaging spectroscopy over water and land. *Proceedings of the Eleventh Canadian Symposium on Remote Sensing,* Waterloo, Ontario, June 22–25, 689–697.
12. Gower, J.F.R., Borstad, G.A., Anger, C.D., and Edel, H.R. (1992). CCD-based imaging spectroscopy for remote sensing: he FLI and CASI programs. *Canadian Journal of Remote Sensing* 18 (4): 199–208.
13. Babey, S.K. and Anger, C.D (1989). A Compact Airborne Spectrographic Imager (CASI). *Proceedings of IEEE IGARSS'89/Twelfth Canadian Symposium on Remote Sensing* 2:1028–1031.
14. Weale, G.P., Achal, S.B., and Anger, C.D. (1990). Application of CASI, a commercial remote-sensing instrument, to astronomy. *Proceedings of SPIE* 1235: 799–800.

15. Neville, R.A. and Powell, I. (1992). Design of SFSI: an imaging spectrometer in the SWIR. *Canadian Journal of Remote Sensing* 18 (4): 210–223.
16. Barducci, A. and Pippi, I. (1995). MIVIS evaluation for hyperspectral sensing of the environment. *Proceedings of SPIE* 2585: 241–246.
17. Rickard, L.J., Basedow, R., Zalewske, E. et al. (1993). HYDICE: an airborne system for hyperspectral imaging. *Proceedings of SPIE* 1937: 173–179.
18. Richter, R. (1996). Atmospheric correction of DAIS hyperspectral image data. *Computers and Geosciences* 22: 785–793.
19. Cocks, T., Jenssen, R., Stewart, A. et al. (1998). The HyMap airborne hyperspectral imager: the system, calibration and performance. *Proceedings 1st EARSeL Workshop on Imaging Spectroscopy (M. Schaepman, D. Schläpfer, and K.I. Itten, Eds.)*, 6-8 October 1998, Zurich, EARSeL, Paris, p. 37–43.
20. Goetz, A.F.H., Rowan, L.C., and Kingston, M.J. (1982). Mineral identification from orbit: initial results from the shuttle multispectral infrared radiometer. *Science* 218: 1020–1024.
21. Dozier, J. (1988). HIRIS - The high resolution imaging spectrometer. *Proceedings SPIE 0924 Recent Advances in Imagers, Radiometry, and Data Processing for Remote Sensing*: 23–30.
22. Goetz, A.F.H. and Herring, M. (1989). The high resolution imaging spectrometer (HIRIS) for EOS. *IEEE Transactions on Geoscience and Remote Sensing* 27: 136–144.
23. De Long, R. K., J. Marmo, Bowler, D.P., and Romesser, T.E. (1995). Airborne and satellite imaging spectrometer development at TRW. *Proceedings of SPIE international Symposium on Aerospace/Defense Sensing & Control and Dual-Use Photonics*. 2480: 287–294.
24. Lewis (2014). http://space.skyrocket.de/doc_sdat/lewis.htm (accessed 5 March 2020).
25. Ungar, S.G., Pearlman, J.S., Mendenhall, J.A., and Reuter, D. (2003). Overview of the Earth Observing One (EO-1) mission. *IEEE Transactions on Geoscience and Remote Sensing* 41: 1149–1153.
26. Xiong, X.-X., King, M.D., Salomonson, V.V. et al. (2016). Moderate resolution imaging spectroradiometer on Terra and Aqua missions. In: *Optical Payloads for Space Missions* (ed. S.-E. Qian), 53–90. West Sussex, UK: Wiley.
27. Pearlman, J., Barry, P., Segal, C. et al. (2003). Hyperion, a Space-Based Imaging Spectrometer. *IEEE Transactions on Geoscience and Remote Sensing* 41 (6): 1160–1173.
28. Barnsley, M.J., Settle, J.J., Cutter, M.A. et al. (2004). The PROBA/CHRIS mission: a low-cost smallsat for hyperspectral, multi-angle, observations of the earth surface and atmosphere. *IEEE Transactions on Geosciences and Remote Sensing* 42: 1512–1520.
29. Coradini, A., Capaccioni, F., Drossart, P. et al. (2007). VIRTIS: an imaging spectrometer for the Rosetta Mission. *Space Science Reviews* 128 (1–4): 529–559.
30. Piccioni, G., Drossart, P., Suetta, E. et al. (2007). VIRTIS: The visible and infrared thermal imaging spectrometer. *ESA-SP-1295, ESA Publications Division*, vol. SP-1295, Noordwijk, The Netherlands.
31. Russell, C.T., Capaccioni, F., Coradini, A. et al. (2007). Dawn mission to Vesta and Ceres. *Earth Moon Planet* 101: 65–91.
32. Murchie, S., Arvidson, R., Bedini, P. et al. (2007). Compact Reconnaissance Imaging Spectrometer for Mars (CRISM) on Mars Reconnaissance Orbiter (MRO). *Journal of Geophysical Research* 112, E05S03: 1–57.
33. Mouroulis, P., Green, R.O., and Chrien, T.G. (2000). Design of pushbroom imaging spectrometers for optimum recovery of spectroscopic and spatial information. *Applied Optics* 39 (13): 2210–2220.
34. Boardman, J.W. C. M. Pieters, R. Green, S. R. Lundeen, P. Varanasi, J. Nettles, N. Petro, P. Isaacson, S. Besse, and L. A. Taylor (2011). Full-mission selenolocation progress for the moon mineralogy mapper on Chandrayaan-1. *Proceedings of 42nd Lunar and Planetary Science Conference*.
35. Kumar, A.S.K. and Samudraiah, D.R.M. (2016). Hyperspectral imager onboard Indian Mini Satellite-1. In: *Optical Payloads for Space Missions* (ed. S.-E. Qian), 141–160. West Sussex, UK: Wiley.
36. Lockwood, R. B. T. Cooley, R. Nadile, J. Gardner, P. Armstrong, A. Payton, T. Davis, S. Straight, T. Chrien, E. Gussin, and D. Makowski (2006). Advanced responsive tactically-effective military imaging spectrometer (ARTEMIS) design. *Proceedings of IEEE Geoscience and Remote Sensing Symposium IGARSS'2006*: 1628–1630.

37. Straight, S.D. Christina Doolittle, Thomas Cooley, James Gardner, Peter Armstrong, Richard Nadile, Thomas Davis (2010). Tactical Satellite-3 mission overview and initial lessons learned. *Proceedings of the 24th Annual AIAA/USU Conference on Small Satellites.*
38. Lucke, R.L., Corson, M., McGlothlin, N.R. et al. (2011). Hyperspectral imager for the coastal ocean: instrument description and first images. *Applied Optics* 50: 1501–1516.
39. Kappus, M.E., Ackleson, S.G., Bowles, J. et al. (2016). Hyperspectral imager for the coastal ocean on the international space station. In: *Optical Payloads for Space Missions* (ed. S.-E. Qian), 29–49. West Sussex, UK: Wiley.
40. He, Z.P., Shu, R., and Wang, J.Y. (2011). Imaging spectrometer based on AOTF and its prospects in deep-space exploration application. *Proceedings of SPIE International Symposium on Photoelectronic Detection and Imaging* 8196: 819625-1–819625-7.
41. Wang, J.-Y., He, Z., Shu, R. et al. (2016). Visible and near-infrared imaging spectrometer aboard Chinese Chang'E 3 spacecraft. In: *Optical Payloads for Space Missions* (ed. S.-E. Qian), 121–139. West Sussex, UK: Wiley.
42. Nieke, J., Mavrocordatos, C., Donlon, C. et al. (2016). Ocean and land color imager on Sentinel-3. In: *Optical Payloads for Space Missions* (ed. S.-E. Qian), 223–245. West Sussex, UK: Wiley.
43. Germain, S. (2016). GHGSat — Global emissions monitoring GHGSat Inc., January 2016. https://www.ghgsat.com/who-we-are/our-satellites/claire.
44. Zee, R.E. (2016). Space Flight Laboratory: Disruptively Capable Smaller Satellites. *Proceedings of Canadian SmallSat Symposium*, Toronto, Canada, February 2–3, 2016.
45. Praks, J., Kestilä, A., Tikka, T., Leppinen, H., Khurshid, O., and Hallikainen, M. (2015). Aalto-1 earth observation CubeSat mission educational outcomes. *Proceedings of IEEE International Geoscience and Remote Sensing Symposium IGARSS'15* (https://ieeexplore.ieee.org/document/7326023).
46. Perkins, R., Krutz, D., Mueller, R. et al. (2016). Teledyne's Multi-User System for Earth Sensing (MUSES). Joint Agency Commercial Imagery Evaluation (JACIE) Workshop, Fort Worth, Texas, USA, 11–15 April (https://calval.cr.usgs.gov/wordpress/wp-content/uploads/2016_JACIE-TDY_MUSES-3.pdf).
47. Contocello, S.S., Foglia Manzillo, P., van Dijk, C.N. et al. (2016). Hyperspectral imaging for real time land and vegetation inspection. Proceedings of Small Satellites, System & Services Symposium (4S Symposium 2016).
48. Liu, Y.-N., Sun, D.-X., Hu, X.-N. et al. (2019). The advanced hyperspectral imager. *IEEE Geoscience and Remote Sensing Magazine* 7 (4): 23–32.
49. Candela, L., Formaro, R., Guarini, R., Loizzo, R., Longo, F., and Varacalli, G. (2016). The PRISMA mission. *Proceedings of IEEE Geoscience & Remote Sensing Symposium 2016 (IGARSS2016)*, 253–256.
50. Meini, M., Bini, A., Giunti, L. et al. (2016). Hyperspectral payload for Italian PRISMA programme. In: *Optical Payloads for Space Missions* (ed. S.-E. Qian), 183–213. West Sussex, UK: Wiley.
51. Matsunaga, T., Iwasaki, A.; Tsuchida, S. et al. (2019). HISUI status toward 2020 launch. *2019 IEEE International Geoscience and Remote Sensing Symposium (IGARSS 2019)*, Yokohama, Japan, pp. 4495–4498.
52. Matsunaga, T., Iwasaki, A., Tsuchida, S. et al. (2016). Hyperspectral imager suite (HISUI): Japanese spaceborne hyperspectral imager for resource and environmental mapping. In: *Optical Payloads for Space Missions* (ed. S.-E. Qian), 215–222. West Sussex, UK: Wiley.
53. Kaufmann, H., Sang, B., Storch, T. et al. (2016). Environmental mapping and analysis program – A German hyperspectral mission. In: *Optical Payloads for Space Missions* (ed. S.-E. Qian), 161–181. West Sussex, UK: Wiley.
54. Langevin, Y., G. Piccioni, P. Eng, G. Filacchione, F. Pouletand the MAJIS team 2014. The MAJIS VIS-NIR imaging spectrometer for the JUICE mission. *Proceedings of 45th Lunar and Planetary Science Conference*, The Woodlands, TX, 1777:2493.
55. Guerri, I., Fabbri, A., Tommasi, L. et al. (2018). The optical design of the MAJIS instrument on board of the JUICE mission. *Proceedings of SPIE* 10690: 1–21 (1L).

# 3

# Efficient Hardware Implementation for Hyperspectral Anomaly and Target Detection

Jie Lei, Weiying Xie, Jiaojiao Li, Keyan Wang, Kai Liu, and Yunsong Li
*State Key Laboratory of Integrated Services Networks, Xidian University, Xi'an, China*

## 3.1 Introduction

Hyperspectral images (HSIs) provide a wealth of spectral information [1] that is provided by the visible, near-infrared, and shortwave infrared bands in remotely sensed scenes [2]. These spectra are divided into hundreds or even thousands of narrow contiguous spectral bands to capture more accurate measurements of the components of each pixel [3, 4]. Hyperspectral imaging has been widely applied to fields such as target detection [5, 6], anomaly detection [7], change detection [8], image classification [9, 10], band selection [11, 12], and hyperspectral unmixing [13, 14] due to its high spectral resolution. Among these applications, hyperspectral anomaly detection is of particular interest due to its capability of uncovering many unknown material substances [15, 16].

Real-time target detection aiming to find timely interesting targets has drawn much attention because of its significance in military and civilian applications [17, 18]. Although the presence of targets in HSI provides critical information in data analysis, it is rather rare and may be difficult to find because many of the HSI-detected targets are relatively small and weak. Consequently, real-time target detection in HSI presents a great challenge and has become increasingly important. Anomaly detection is a special category of target detection. It can be considered to be a binary classification problem that is designed to classify each pixel in an HSI as either an anomaly or background [19]. Different from the supervised target detection problems that were considered in Zhang et al. [20] and Zhang et al. [21], anomaly detection is an unsupervised target detection technique that detects anomalies without any prior information of the scene [22]. For HSIs, the goal of anomaly detection is to identify unusual objects that significantly deviate from the typical background with respect to either space or spectrum. For example, the airplanes in an airport background and tanks in a forest background can be judged as anomalies. There are several challenges in hyperspectral anomaly detection. First, there is no prior knowledge of anomalies or backgrounds that are available for data analysis. Second, complex backgrounds hinder detection [23]. Third, the HSI data are usually

*Advances in Hyperspectral Image Processing Techniques*, First Edition. Edited by Chein-I Chang.

corrupted by unknown interference due to significantly improved spectral resolutions [24]. Owing to the complexity and dimensionality of HSI, the target detection and anomaly detection algorithms are still computationally intensive [25]. This often leads to the requirement of hardware accelerators to speed up computations [26]. The algorithm could be implemented on one of the widely used platforms like CPUs, GPUs [27], and field-programmable gate arrays (FPGAs) [28]. Among them, FPGAs have significant advantages in supporting parallel computation and pipeline operation with the cost of low power consumption.

The constrained energy minimization (CEM) imposes a constraint on the targets or backgrounds and builds a finite-impulse response filter, which minimizes the output energy of the filter. The CEM algorithm has attracted the attention of scholars at home and abroad due to its outstanding detection performance, and has resulted in a series of improved algorithms.Wang et al. [29] proposed the two-time detection (TD) method, which uses PCCEM and MTCEM algorithms through two superposition to improve CEM detection efficiency. Zou et al. [30] proposed the hierarchical suppression method in 2016, which layers the CEM algorithm and uses a nonlinear suppression function to suppress the background to improve performance. In order to reduce the amount of computation of the algorithm, Bin Yang et al. [31] use the idea of continuous updating to solicit inverse matrices as well as autocorrelation matrices to optimize the CEM algorithm by streaming background statistical structures.

The Reed-Xiaoli (RX) detection algorithm [32] is the most classical anomaly detection algorithm. It calculates the Mahalanobis distance between the measured pixels and the background mean by assuming that the probability density function of the background can be modeled as a multivariate Gaussian distribution [33]. By comparing the Mahalanobis distance to a preset threshold, the pixel is determined as a background or an anomaly. The RX detection algorithm has high computational efficiency and excellent detection performance when the background distribution of the data set satisfies Gaussian distribution [34]. Depending upon how the background is described, the RX detection algorithm can be implemented in two versions [35]. One is the local RX (LRX) [36], which regards the region surrounding the measured pixel as the background and then uses a dual-window method to detect anomaly targets. The other is the global RX (GRX), which considers the entire image scene as the background. Since all the pixels in an entire image scene or a local region are used to compute the background statistics, the mean and the covariance matrix of the background are easily contaminated by anomalies [37, 38], which may lead to the performance of the RX algorithm being unstable. Because of that, many variants of the RX anomaly detection algorithm have been proposed. For example, a kernel RX detection (KRX) algorithm [39] expands a linear non-Gaussian model in the low-dimensional original data space to a nonlinear Gaussian domain in higher dimensional feature space [40]. The subspace RX (SSRX) detection algorithm [41] employs principal component analysis (PCA) and then applies the RX detection algorithm on limited numbers of PCA bands.

Currently, many algorithms have been developed for multiple target detection. For example, Boardman proposed the pixel purity index (PPI) algorithm [42]. The approach searches for a set of vertices of a convex hull in a given data set, which is supposed to be pure signatures present in the data. It has been widely used because it is available in the Environment for Visualizing Images (ENVI) software system and has found multiple applications in different areas [43]. The other one is the famous N-FINDR algorithm, developed by Winter et al., which assumes that, in $L$ spectral dimensions, the $L$-dimensional volume formed by a simplex with vertices specified by target pixels are always larger than those formed by any other combination of pixels [44]. Most importantly, both PPI and N-FINDR have the same drawback, which is the use of random initial conditions to initialize their algorithms [45]. In order to cope with this issue, the automatic target-generation process (ATGP) algorithm [46] was invented, which finds a set of spectrally distinct target pixel vectors using the concept of orthogonal subspace projection (OSP) [47] in the spectral domain. Quantitative and comparative assessments of target detection algorithms reveal good performance of ATGP with regard to other approaches in terms of detection accuracy and computational performance [48].

Currently, many algorithms have been developed for this purpose and designed based on various criteria. For example, OSP algorithm is proposed by Harsanyi and Chang. The approach can be used to detect the abundance score of a specified desired signature for a set of vertices of a convex hull in a given data set, which is supposed to be pure signatures present in the data. However, obtaining the prior information of the target to be tested required by the OSP is very difficult and the matrix inversion operation is complicated. In order to cope with this issue, a lot of scholars have carried out further algorithm research on the basis of OSP algorithm. For example, Wang et al. developed an unsupervised OSP algorithm (UOSP) [49], which combines the target generation algorithm TGP with the OSP target detection algorithm, and the TGP algorithm is responsible for finding undesired interference targets according to the expected target. Although the UOSP algorithm has got rid of the dependence on the prior information to some extent, this algorithm does not optimize the complex matrix inversion of the OSP algorithm. In 2015, Meiping Song and Chein-I Chang proposed ROSP [47] which used recursive formulas instead of matrix inversion to detect known target signatures. Furthermore, they put forward UROSP which makes a great contribution to the development of OSP.

## 3.2 Hyperspectral Anomaly and Target Detection

### 3.2.1 DPBS-CEM

Let $\mathbf{X} \in \mathbf{R}^{W \times H \times L}$ denote a HSI with $W \times H$ pixels (row of $\mathbf{X}$) and $L$ spectral bands (column of $\mathbf{X}$). We may interpret $\mathbf{X}$ either as a collection of $L$ 2D images (or bands) of size $N$ ($N = W \times H$), or as a collection of $W \times H$ spectral vectors of size $L$. The entire data matrix $\mathbf{X} = [\mathbf{x}_1, \mathbf{x}_2, \mathbf{x}_3, \ldots, \mathbf{x}_N]$, where $\mathbf{x}_i$ is the $i$th sample pixel vector $\mathbf{x}_i = (x_{i1}, x_{i2}, \ldots, x_{iL})^T$ for $1 \leq i \leq N$ and the signature $\mathbf{d} = (d_1, d_2, \ldots, d_L)^T$ of target is known. The basic purpose of CEM is to design a linear finite impulse response (FIR) filter with $L$ filter coefficients denoted by an $L$-dimensional vector $\mathbf{w} = (w_1, w_2, \ldots, w_L)^T$ that minimizes the energy of the output $y_i (1 \leq i \leq N)$ with the following constraint:

$$\min_{\mathbf{w}} \left(\mathbf{w}^T \mathbf{R} \mathbf{w}\right) \tag{3.1}$$

$$\text{s.t.} \mathbf{d}^T \mathbf{w} = 1 \tag{3.2}$$

where $\mathbf{R} = \frac{1}{N}\left[\sum_{i=1}^{N} \mathbf{x}_i \mathbf{x}_i^T\right]$ is the global correlation matrix of $\mathbf{X}$. The weighting vector $\mathbf{w}$ solved for Eqs. (3.1) and (3.2) is given by

$$\mathbf{w} = \frac{\mathbf{R}^{-1}\mathbf{d}}{\mathbf{d}^T \mathbf{R}^{-1} \mathbf{d}} \tag{3.3}$$

which yields the CEM described by

$$y_i = \delta_{\text{CEM}}(\mathbf{x}_i) = (\mathbf{w}_{\text{CEM}})^T \mathbf{x}_i \tag{3.4}$$

DPBS-CEM takes the inverse of the correlation matrix of the $K$-group pixel vectors to replace the entire pixel vectors. More specifically, DPBS-CEM can be described as follows:

$$\mathbf{R}_n = (1/K)\left[\sum_{i=n-K}^{n-1} \mathbf{x}_i \mathbf{x}_i^T\right] \tag{3.5}$$

$$\mathbf{S}_n = \left[ \sum_{i=n-K}^{n-1} \mathbf{x}_i \mathbf{x}_i^T \right] \tag{3.6}$$

Now, $\mathbf{S}_n^{-1}$ is the inverse of the correlation matrix of the $K$-group pixel vectors. The Sherman–Morrison formula is used to derive the following Eq. (3.7)

$$\mathbf{S}_{n+1}^{-1} = \left(\mathbf{S}_n + \mathbf{x}_n\mathbf{x}_n^T\right)^{-1} = \mathbf{S}_n^{-1} - \frac{\mathbf{S}_n^{-1}\mathbf{x}_n\mathbf{x}_n^T\mathbf{S}_n^{-1}}{\mathbf{x}_n^T\mathbf{S}_n^{-1}\mathbf{x}_n + 1} \tag{3.7}$$

Since data dependency exists between $\mathbf{S}_{n+1}^{-1}$ and $\mathbf{S}_n^{-1}$, there is a great increase in processing time. We solve the data dependency by means of data segmentation. As a result, the current input pixel can be processed directly with no need of waiting for the previous pixel to be completed. By making the inverse calculations between neighboring pixels independent, we are able to carry out a deep pipelined architecture.

The detailed procedure of DPBS-CEM algorithm is shown as Algorithm 3.1.

**Algorithm 3.1** The Deep Pipelined Background Statistics (DPBS) Target Detection CEM Algorithm

**Input:** Initialize the following parameters.

(1) HSI data size: $W \times H \times L = N \times L$;
(2) the value of $\beta$;
(3) the desired signature $\mathbf{d}$;
(4) the number of inverse matrices: $M = 4$;
(5) bn indicates the index of number;
(6) $K$ indicates the number of pixel vectors collected before starting target detection;

**Output:** the final target detection results.
define an initial inverse matrix $S_0^{-1}$: $S_0^{-1} = \beta \cdot \mathbf{I}$
data segmentation:
for $i = 1$ ; $i \leq N + K$ ; $i++$ do
  bn = $i$ % $M$
  calculate the inverse matrix:
  if $i \leq N$ then
    $(\mathbf{S}^{-1})^{\text{bn}} = (\mathbf{S}^{\text{bn}} + \mathbf{x}_i\mathbf{x}_i^T)^{-1} = (\mathbf{S}^{-1})^{\text{bn}} - \frac{(\mathbf{S}^{-1})^{\text{bn}}\mathbf{x}_i\mathbf{x}_i^T(\mathbf{S}^{-1})^{\text{bn}}}{\mathbf{x}_i^T(\mathbf{S}^{-1})^{\text{bn}}\mathbf{x}_i + 1}$
  endif
  calculate the target detection results:
  if $i \geq K$ then
    $\text{DPBS} - \text{CEM}\ (\mathbf{x}_{i-K}) = \frac{\mathbf{x}_{i-K}^T(\mathbf{S}^{-1})^{\text{bn}}}{\mathbf{d}^T(\mathbf{S}^{-1})^{\text{bn}}\mathbf{d}}$
  **endif**
**endfor**

### 3.2.2 *DBN-RXD*

Deep brief network (DBN) can effectively produce different layers of nonlinear latent features in an unsupervised and layer-by-layer manner where restricted Boltzmann machines (RBM) is commonly used as a layer-wise training model. This network can be employed as a tool for feature extraction and dimensionality reduction. We first review the main component in the construction of a DBN, i.e. the RBM.

RBM is a layer-wise network, which consists of visible and hidden units. A symmetrical weight matrix **W** can connect the visible units **v** and hidden units **h**. It is noted that there is no connection between two visible units or between two hidden units. The joint distribution of the visible and hidden units can be given by

$$p(\mathbf{v},\mathbf{h};\theta) = \frac{e^{-E(\mathbf{v},\mathbf{h};\theta)}}{Z(\theta)} \tag{3.8}$$

$$Z(\theta) = \sum_{\mathbf{v}}\sum_{\mathbf{h}} E(\mathbf{v},\mathbf{h};\theta) \tag{3.9}$$

where $Z(\theta)$ is a normalization constant defined as the sum of $E(\mathbf{v}, \mathbf{h}; \theta)$ over all possible values of the visible units **v** and hidden units **h**. $E(\mathbf{v}, \mathbf{h}; \theta)$ is the energy function of a joint distribution given by

$$\begin{aligned} E(\mathbf{v},\mathbf{h};\theta) &= -\sum_{i=1}^{K} a_i v_i - \sum_{j=1}^{J} b_j h_j - \sum_{i=1}^{K}\sum_{j=1}^{J} W_{ij} v_i h_j \\ &= -\mathbf{a}^T\mathbf{v} - \mathbf{b}^T\mathbf{h} - \mathbf{v}^T\mathbf{W}\mathbf{h} \end{aligned} \tag{3.10}$$

where $\theta = \{a_i, b_j, W_{ij}\}$; $W_{ij}$ denotes the weight between the visible unit $i$ and hidden unit $j$; $a_i$ and $b_j$ represent the bias terms of visible and hidden layers, respectively; $K$ and $J$ are the total numbers of units in visible and hidden layers, respectively.

It is clear from 1.10 that the hidden units are independent with the given visible units, and so are the visible units. The conditional distributions of the input data and hidden unit are specified by the following logistic functions:

$$p(v_i = 1|\mathbf{h},\theta) = g\left(a_i + \sum_{i=1}^{K} W_{ij} v_i\right) \tag{3.11}$$

$$p(h_j = 1|\mathbf{v},\theta) = g\left(b_j + \sum_{j=1}^{J} W_{ij} h_j\right) \tag{3.12}$$

where $g(\cdot)$ represents the logistic function given by

$$g(x) = \frac{1}{1 + e^{-x}} \tag{3.13}$$

The weight matrix **W** can be optimized by the contrastive divergence (CD) [50]. It is clear that RBM can reconstruct the input data by using the information in the hidden units. The information in the hidden units is the learned features from the input data. The reconstruction performance depends on whether the hidden units retain enough information of the input data.

Learning one layer at a time is relatively limited. The hidden units of RBM can be fed into another RBM to model the significant dependencies between the hidden units of the earlier RBM. More specifically, stacking RBMs one atop another can progressively capture the hierarchical representation of the input data, which is called DBN. In this case, DBN can be employed as a means for dimensionality reduction on high-dimensional data. Figure 3.1 illustrates an implementation of a two-layer DBN.

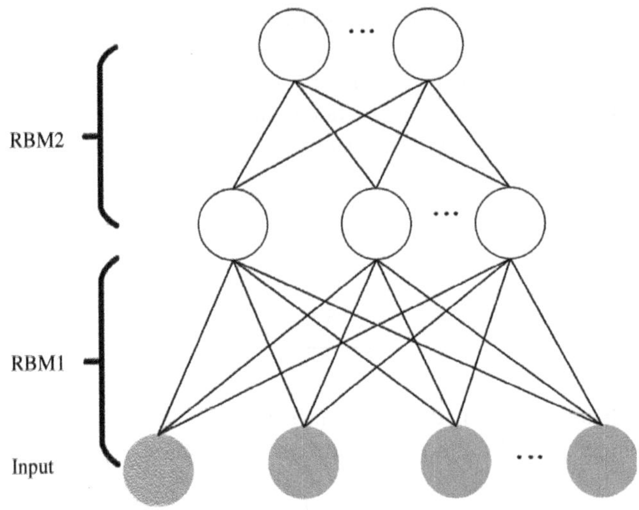


**Figure 3.1** An illustration of a two-layer DBN.

Let $\mathbf{X} \in \mathbf{R}^{L \times M}$ denote an HSI with $M(M = W \times H)$ pixels and $L$ spectral bands. The entire data matrix $\mathbf{X} = [\mathbf{x}_1, \mathbf{x}_2, \mathbf{x}_3, \ldots, \mathbf{x}_M]$, where $\mathbf{x}_i$ is the $i$th sample vector, $\mathbf{x}_i = [x_{i1}, x_{i2}, x_{i3}, \ldots, x_{iL}]^T$. The two competing hypotheses that the RX-algorithm must distinguish are given by

$$\begin{cases} \mathbf{H}_0 : x = \mathbf{n}, & \text{Target absent} \\ \mathbf{H}_1 : x = a\mathbf{s} + \mathbf{n}, & \text{Target present} \end{cases} \tag{3.14}$$

where $\mathbf{n}$ is a vector, which represents the background clutter noisy process, and $\mathbf{s}$ is the spectral signature of the target. The target signature $\mathbf{s}$ and background covariance $\mathbf{C}_b$ are assumed to be unknown. Under $\mathbf{H}_0$, the data (background clutter) is modeled as $N(0, \mathbf{C}_b)$ and under $\mathbf{H}_1$, it is modeled as $N(s, \mathbf{C}_b)$. The background covariance $\mathbf{C}_b$ is given by

$$\mathbf{C}_b = \frac{1}{M}\sum_{i=1}^{M}(\mathbf{x}_i - \mu_b)(\mathbf{x}_i - \mu_b)^T \tag{3.15}$$

where $\mu_b$ is the estimated background clutter sample mean given by

$$\mu_b = \frac{1}{M}\sum_{i=1}^{M}\mathbf{x}_i \tag{3.16}$$

Assuming a single pixel target is the observation test vector, the expression for the RX algorithm is given by

$$\mathrm{RX}(\mathbf{x}) = (\mathbf{x} - \mu_b)^T \mathbf{C}_b^{-1}(\mathbf{x} - \mu_b) \tag{3.17}$$

An important variant of the RX algorithm was introduced by Chang which replaces the sample covariance matrix with sample correlation matrix and removes the mean vector from each sample vector. As a result, the RX algorithm can be described as

$$\mathrm{RX}(\mathbf{x}) = \mathbf{x}^T\mathbf{R}^{-1}\mathbf{x} \tag{3.18}$$

which implements $\mathbf{x}$ and $\mathbf{R}^{-1}$ in place of $(\mathbf{x}-\mu_b)$ and $\mathbf{C}_b^{-1}$, respectively. This version of RX algorithm has been widely used in HIS processing.

For the RX anomaly detection algorithm, in order to obtain the anomaly detection results of each sample, we need to calculate the inverse of the correlation matrix. And for the $n$th testing pixel $\mathbf{x}_n$, let the correlation matrix $\mathbf{R}_n$ is computed by formerly collected $(n-1)$ pixels, then $\mathbf{R}_n$ can be expressed as

$$\mathbf{R}_n = \sum_{i=1}^{n-1} \mathbf{x}_i\mathbf{x}_i^T \tag{3.19}$$

Then, for the next testing pixel $\mathbf{x}_{n+1}$, its correlation matrix is given by

$$\mathbf{R}_{n+1} = \mathbf{R}_n + \mathbf{x}_n\mathbf{x}_n^T \tag{3.20}$$

Importantly, we can perform an inverse matrix real-time processing by using a Sherman–Morrison formula. We will introduce the Sherman–Morrison formula, first. Assuming $\mathbf{A}$ is an invertible square matrix, and $u$ and $v$ are two different column vectors. Then, the Sherman–Morrison formula is given by

$$\left(\mathbf{A} + \mathbf{u}\mathbf{v}^T\right)^{-1} = \mathbf{A}^{-1} - \frac{\mathbf{A}^{-1}\mathbf{u}\mathbf{v}^T\mathbf{A}^{-1}}{1 + \mathbf{v}^T\mathbf{A}^{-1}\mathbf{u}} \tag{3.21}$$

Here, $\mathbf{u}\mathbf{v}^T$ is the outer product of two vectors $\mathbf{u}$ and $\mathbf{v}$. The general form shown here is the one presented by Bartlett. If the inverse of $\mathbf{A}$ is already known, the formula provides a numerically cheap way to compute the inverse of $\mathbf{A}$ corrected by the matrix $\mathbf{u}\mathbf{v}^T$. The computation is relatively cheap because the inverse of $\mathbf{A} + \mathbf{u}\mathbf{v}^T$ does not have to be computed from scratch which in general is expensive but can be computed by correcting $\mathbf{A}^{-1}$ instead. Obviously, by using the Sherman–Morrison formula, the inverse of $\mathbf{R}_{n+1}$ can be computed as

$$\mathbf{R}_{n+1}^{-1} = \left(\mathbf{R}_n + \mathbf{x}_n\mathbf{x}_n^T\right)^{-1} = \mathbf{R}_n^{-1} - \frac{\mathbf{R}_n^{-1}\mathbf{x}_n\mathbf{x}_n^T\mathbf{R}_n^{-1}}{1 + \mathbf{x}_n^T\mathbf{R}_n^{-1}\mathbf{x}_n} \tag{3.22}$$

At this time, we still lack the inverse correlation matrix of the first sample to complete the inverse of the global correlation matrix. In order to avoid this problem, we set an identity matrix $E$ as the initial inverse matrix. More specifically, the calculation of the correlation matrix can be expressed as follows:

$$\mathbf{R}_{n+1} = \mathbf{E} + \mathbf{x}_1\mathbf{x}_1^T + \mathbf{x}_2\mathbf{x}_2^T + \cdots + \mathbf{x}_n\mathbf{x}_n^T \tag{3.23}$$

### *3.2.3 Fast-ATGP*

The original ATGP is based on OSP concepts and will be referred to hereinafter as ATGP-OSP. The basic concept of OSP is to project each pixel vector onto a subspace which is orthogonal to the obtained signatures. It is an iterative process in which orthogonal projections are applied to find a set of spectrally distinct pixels. In ATGP-OSP, $\mathbf{U}$ is a matrix of spectral signatures, $\mathbf{U}^T$ is the transpose of it, and $\mathbf{I}$ is the identity matrix. It should be emphasized that the ATGP-OSP primarily comes from a need of finding targets of interest in HSI only with the prior knowledge of $t$, where $t$ denotes the number of targets to be detected. In general, the value of $t$ can be obtained by the virtual dimensionality (VD) method.

In ATGP-OSP, $\mathbf{F} \in \mathbf{R}^n$ denote an HSI with $r(r = W \times H)$ pixels and $L$ spectral bands. The ATGP-OSP algorithm begins by an orthogonal projection operator specified by the following expression:

$$\mathbf{P}_{\mathbf{U}}^{\perp} = \mathbf{I} - \mathbf{U}\mathbf{U}^{\#} \tag{3.24}$$

Orthogonal projection operator $\mathbf{P}_{\mathbf{U}}^{\perp}$ has the same structure with the orthogonal complement projector, where $\mathbf{U}^{\#} = (\mathbf{U}^T\mathbf{U})^{-1}\mathbf{U}^T$ is the pseudoinverse of $\mathbf{U}$. The operator $\mathbf{P}_{\mathbf{U}}^{\perp}$ is first applied to the original image, with $\mathbf{U} = [\mathbf{m}_0]$, where $\mathbf{m}_0$ is an initial target signature (i.e. the pixel vector with maximum length). It then finds a target signature $\mathbf{m}_1$, with the maximum projection in $<\mathbf{m}_0>^{\perp}$, where $<\mathbf{m}_0>^{\perp}$ is the orthogonal projection operator $\mathbf{P}_{\mathbf{U}}^{\perp}$ composed with $\mathbf{U} = [\mathbf{m}_0]$. $<\mathbf{m}_0>^{\perp}$ is the orthogonal complement space linearly spanned by $\mathbf{m}_0$. The following target signature $\mathbf{m}_2$ is obtained by another $\mathbf{P}_{\mathbf{U}}^{\perp}$ with $\mathbf{U} = [\mathbf{m}_0, \mathbf{m}_1]$ and selected by the maximum projection in $<\mathbf{m}_0, \mathbf{m}_1>^{\perp}$. Then, the above procedure is repeated until the target pixels $\{\mathbf{m}_0, \mathbf{m}_1, \ldots, \mathbf{m}_{t-1}\}$ are detected.

---

**Algorithm 3.2** Pseudocode of Fast-ATGP

---

1:**Input:** $\mathbf{F} \in \mathbf{R}^n$, and $t$;

% $\mathbf{F}$ denotes an $n$-dimensional hyperspectral image with $r$ pixels and $t$ denotes the number of targets to be detected

2:**Initialized:** $\mathbf{P}_{\mathbf{U}_0}^{\perp} = \mathbf{I}_{L \times L}$, $\mathbf{P}_{\mathbf{V}_0}^{\perp} = [1, 1, \ldots, 1]$, $\omega = [1, 1, \ldots, 1]$;

% $\mathbf{I}$ is the identity matrix and $L$ denotes spectral bands of HSI, $\mathbf{P}_{\mathbf{U}_0}^{\perp}$ and $\mathbf{P}_{\mathbf{V}_0}^{\perp}$ are the initialized orthogonal projection operators

3:**for** $i = 0$ to $t - 1$**do**

4: $\mathbf{v} = \mathbf{P}_{\mathbf{V}_0}^{\perp} \cdot \mathbf{F}$;

% HSI data $\mathbf{F}$ is projected onto the direction indicates by $\mathbf{P}_{\mathbf{V}_i}^{\perp}$

5: $index = argmax_{\{1, \ldots, r\}}\mathbf{v}$;

% The maximum projection value is found, where $r$ denotes the total number of pixels

6: $\mathbf{m}_i \equiv \mathbf{U}[:, i + 1] = \mathbf{F}[:, index]$;

% The target $\mathbf{m}_i$ is detected and the target matrix $\mathbf{U}$ is updated, where the operator ":" denotes "all elements"

7: $\tilde{\mathbf{m}} = \mathbf{P}_{\mathbf{U}_i}^{\perp} \cdot \mathbf{m}_i$

% $\mathbf{m}_i$, used for updating the operator $\mathbf{P}_{\mathbf{U}_i+1}^{\perp}$ and $\mathbf{P}_{\mathbf{V}_i+1}^{\perp}$, is a vector obtained by the target signature $\mathbf{m}_i$ projected onto the direction indicated by $\mathbf{P}_{\mathbf{U}_i}^{\perp}$

8: $\mathbf{P}_{\mathbf{U}_i+1}^{\perp} = \mathbf{P}_{\mathbf{V}_i+1}^{\perp} - \frac{\tilde{\mathbf{m}}_i \tilde{\mathbf{m}}_i^T}{\tilde{\mathbf{m}}_i^T \tilde{\mathbf{m}}_i}$

% The projection operator of matrix $\mathbf{P}_{\mathbf{U}}^{\perp}$is updated

9: $\mathbf{P}_{\mathbf{V}_i+1}^{\perp} = \mathbf{P}_{\mathbf{V}_i}^{\perp} - \frac{\omega \cdot \tilde{\mathbf{m}}_i}{\tilde{\mathbf{m}}_i^T \tilde{\mathbf{m}}_i} \tilde{\mathbf{m}}_i^T$;

% The projection operator of vector $\mathbf{P}_{\mathbf{V}}^{\perp}$ is updated

10: end for $i$

11: **Output:** $\mathbf{U} = \{\mathbf{m}_0, \mathbf{m}_1, \ldots, \mathbf{m}_{t-1}\}$;

---

Through analyzing calculations of ATGP-OSP, an increasing operation problem occurs during growing matrix inversion process and a huge matrix multiplication problem also arises in the OSP process. For the first problem, a fixed operation scale is introduced to replace the continuously increasing computations when updating the OSP operator matrix, which ensures that the consumption of hardware resources does not change as the number of detected targets increases. Moreover, a vectorization approach is also developed for the operator matrix to avoid the latter problem, which updates the operator in a vector form only by one step and decreases computation to a great extent.

The Fast-ATGP calculates the orthogonal projection without using the pseudoinverse operation and updates the operator in one step, which is described in Algorithm 3.2.

### *3.2.4 Fast-MGD*

The proposed Fast-morphological reconstruction and the simplified guided filtering detector (MGD) approach consists of four steps. First, average fusion is adopted to reduce the dimension of the HSIs. Second, morphological opening and closing reconstruction are utilized for feature location. Third, a differential operation is conducted to extract anomalies in the HSIs. Finally, the final detection map is therefore obtained by means of the simplified self-guided filtering for feature clustering.

The high dimensionality of the original HSIs will undoubtedly and inevitably incur extremely high computational complexity, throughout which massive storage capability is therefore required. What's more, the adjacent bands are usually highly correlated with each other and contain redundant information. The original HSI is therefore first partitioned into $Q$ subsets of adjacent bands as follows:

$$\mathbf{C}^q = \begin{cases} \left(\mathbf{g}_{\lceil R/Q\rceil\,(q-1)+1}, \ldots, \mathbf{g}_{\lceil R/Q\rceil q}\right) & 1 \le q \le Q-1 \\ \left(\mathbf{g}_{\lceil R/Q\rceil(q-1)+1}, \ldots, \mathbf{g}_R\right) & q = Q \end{cases} \tag{3.25}$$

In the above expression, $\mathbf{C}^q$ ($q = 1, 2, \ldots, Q$) indicates the $q$th hyperspectral subset. $\mathbf{g} = (\mathbf{g}_1, \mathbf{g}_2, \mathbf{g}_3, \ldots, \mathbf{g}_R)$ denotes an HSI with $R$ spectral bands and $M \times N$ pixels in the spatial domain. $\lceil R/Q \rceil$ represents the smallest integer greater than or equal to $R/Q$.

Then, a simple and effective average fusion method [51] is adopted to reduce spectral dimension by combining the complementary information of adjacent bands in each subset. Specifically, the fused band $\mathbf{S}^q$ is obtained by the following equation:

$$\mathbf{S}^q = \frac{\sum_{i=1}^{N_q} \mathbf{C}_i^q}{N_q} \tag{3.26}$$

where $\mathbf{C}_i^q$ represents the $i$th band in the $q$th hyperspectral subset, and $N_q$ is the total number of spectral bands in the $q$th subset.

In this stage, the opening and closing reconstruction are applied for feature localization, in which a basic operand, a rectangular SE $\mathbf{b}$ of size $r_a \times r_a$ is therefore defined. The erosion and dilation of

**S** by **b** are first calculated as the marker images for dilation and erosion reconstruction, respectively. Let $\gamma$ and $\phi$ denote the opening and closing reconstruction, respectively. The operations performed at this stage can be expressed by the following equations:

$$\begin{aligned} \gamma^q &= R_{\mathbf{S}^q}^D(\mathbf{e}^q) = D_{\mathbf{S}^q}^{(k)}(\mathbf{S}^q \ominus \mathbf{b}) \\ &= D_{\mathbf{S}^q}^{(1)}\left(D_{\mathbf{S}^q}^{(k-1)}(\mathbf{S}^q \ominus \mathbf{b})\right) \\ \phi^q &= R_{\mathbf{S}^q}^E(\mathbf{d}^q) = E_{\mathbf{S}^q}^{(k)}(\mathbf{S}^q \oplus \mathbf{b}) \\ &= E_{\mathbf{S}^q}^{(1)}\left(E_{\mathbf{S}^q}^{(k-1)}(\mathbf{S}^q \oplus \mathbf{b})\right) \end{aligned} \quad (3.27)$$

where $q = 1, \ldots, Q$ and SE **b** is also used to perform erosion and dilation reconstruction. However, the morphological reconstruction following its definition is an iterative procedure that is performed until reaching idempotence, which results in unpredictable time performance on hardware implementation.

Considering that anomaly objects in HSIs usually appear as small-area objects compared with the background, the anomaly objects are extracted through the differential operation, which merges the dark and light features from the above stage. For $q = 1, \ldots, Q$, the process can be expressed as the following equation:

$$\mathbf{I}^q = |\phi^q - \mathbf{S}^q| + |\mathbf{S}^q - \gamma^q| = \phi^{\mathrm{q}} - \gamma^{\mathrm{q}} \quad (3.28)$$

Although morphological reconstruction can effectively extract small-sized bright and dark objects in the image, it is difficult for the reconstruction to filter out some background interference and random noise, to which may increase the false alarm rate of anomaly detection. However, as a local linear model that sufficiently utilizes the local spatial information of the image, guided filtering is a better alternative in solving the aforementioned problems.

The guided filter computes the filtering output by considering the content of a guidance image, which can be the input image itself or another different image. In this paper, we consider the special case in which the input image is used as the guide. The results of self-guided filtering can be defined as follows:

$$\mathbf{V}_i^q = a_j^q \mathbf{I}_i^q + b_j^q, \forall i \in \omega_j^q \quad (3.29)$$

The coefficients can be calculated as

$$\begin{aligned} a_j^q &= \frac{\frac{1}{|\omega^q|}\sum_{i\in\omega_j^q}\left(\mathbf{I}_j^q\right)^2 - \left(\mu_j^q\right)^2}{\left(\sigma_j^q\right)^2 + \varepsilon} \\ b_j^q &= \mu_j^q - a_j^q \mu_j^q \end{aligned} \quad (3.30)$$

where $\mu_j^q$ and $\sigma_j^{q2}$ are the mean and variance of $\mathbf{I}^q$ in the local window $\omega_j^q$ of size $(2r_b + 1) \times (2r_b + 1)$, $\epsilon$ is a parameter that controls the smoothness of the filtering, $q = 1, \ldots, Q$. The self-guided filter applied in this section is described in Algorithm 3.3.

**Algorithm 3.3** Pseudocode of Self-Guided Filter

**Input: I**
Parameters: $r_b$, $\epsilon$
$meanI = fmean(\mathrm{I})(1)$
$CorrI = fmean(\mathrm{I}.{*}\mathrm{I})$ (2)
$VarI = CorrI - meanI. * meanI$ (3)
$a = VarI/(VarI + \epsilon)$ (4)
$b = meanI - a. * meanI$ (5)
$\mathbf{V} = a. * \mathbf{I} + b$ (6)
**Output:V**

Finally, we perform the following operation to obtain the detection map:

$$\mathbf{O} = \sum_{q=1}^{Q} \omega_q \mathbf{V}^q \tag{3.31}$$

where $\omega_q = 1/Q$ ($q = 1, \ldots Q$) is the set, meaning that the processed result of each fused band contributes equally to finding anomaly objects.

## 3.3 Model-Based Design

### 3.3.1 What is Model-Based Design?

Model-Based Design is a model-centric approach to the development of control, signal processing, communications, and other dynamic systems. Different from the traditional development method of using text specifications and frequent prototype verification, the model-based design uses models throughout the entire development process. The model includes every component relevant to system behavior algorithms, control logic, physical components, and intellectual property (IP). Once the model is developed (elaborated), it becomes the source of many outputs, including reports, C code, and HDL code. Model-Based Design enables system-level and component-level design and simulation, automatic code generation, and continuous test and verification. Significant advantages of Model-Based Design include the fact that it facilitates rapid design iterations and it moves the verification process to the beginning of the design cycle. This helps detect system specification-related errors, design errors, and implementation errors early.

The model-based design process, also known as "*V*" mode, which takes the model as the core, each step in the process depends on the model, where "*V*" stands for "Verification" and "Validation". Such a complete system development process generally includes the following steps:

1. Definition of functional requirements. Traditional requirements definitions use a lot of text to describe specific functions. In order to avoid ambiguity and misinterpretation of text descriptions, in modern requirements definition methods, detailed descriptions will adopt model methods. An executable specification is a model that encapsulates all design information, including requirements, system components, IP, and test scenarios. It can be a model of the environment with use cases that the embedded software needs to manage or a high-level algorithm model that specifies the implementations' exact behavior.
2. Rapid prototyping verification. In the modern system design method, after the scheme design is completed, there is no need to wait for the software engineer to program and integrate the code into

the hardware. Instead, the computer-aided design tool is used to convert the scheme model into code and automatically download it to the hardware development platform, thereby quickly Prototype verification of the system. In this way, you can use the computer-aided test management tool software for various debugging and testing to verify the control effect of the program on the actual object and optimize the control parameters online. At this time, even if the model needs to be modified on a large scale, it only takes a relatively short time to reform the test prototype. In this way, before the actual product processing starts, the final plan and effect can be confirmed, which greatly reduces the waste of resources and time.

3. Generate code. Converting the model to code is the key to the model-based development process. Traditional manual programming is not only slow, but it is also easy to introduce defects. The model-based development method is relatively reliable. Most of the product code is automatically generated by the computer, but it may lose part of the real-time running efficiency of the code. However, for most engineers, if the development speed can be accelerated, these losses are acceptable.
4. Hardware in the loop simulation. After the product code is completed, a comprehensive test must also be conducted to verify the product and actual index requirements, especially under extreme conditions and failure conditions. In the model-based development method, the hardware-in-the-loop test method is used to perform tests under various conditions, especially fault conditions and extreme conditions. This is not possible with traditional development methods.
5. System integration test. After the product is completed, it needs to be connected with related subsystems to form a completely closed-loop system. The next system-level test can be carried out to confirm whether the product meets the design specifications and function definitions. At this stage, a difficult problem often arises. During the parallel development process, other related subsystems are not ready for complete system integration. Using HLS, you can replace parts of the closed-loop system that are not in place in time, use a mathematical model to simulate their characteristics, and provide corresponding feedback signals to the controller through the I/O port. Nesting other model systems integrate a complete set of closed loops.

### *3.3.2 FPGA Development Based on MBD*

The traditional FPGA development process usually uses a waterfall development model. First, the algorithm is described using Matlab Simulink or C language, and then a system-level model is established. After verification and testing, the algorithm is converted into a registered transmission level description using VHDL or Verilog HDL hardware description language. After that, it is necessary to carry out the functional simulation, onboard verification, and other processes on the register transmission-level circuit. However, this design method has many disadvantages. For example, for large-scale and complex systems, everyone has a different understanding of the text description of the system; the underlying hardware description process is error-prone and requires a lot of time and effort; it needs to be verified in steps Results and final results. Such a processing model is increasingly difficult to deal with increasingly complex designs.

In order to solve the above shortcomings, the model-based development process has attracted extensive attention. The model-based development tools on FPGA mainly include Xilinx's System Generator for DSP, Altera's DSP Builder, and MathWorks Simulink HDL Coder. Simulink HDL Coder can generate Simulink model, Stateflow block diagram, and MATLAB code to generate bit-true, cycle-accurate, synthesizable Verilog and VHDL codes, but the generated HDL code is independent of the target device. And the development efficiency is relatively low, there are still many problems to be solved from the model to the physical implementation. System Generator for DSP is Xilinx's system-level modeling development tool, which can be embedded into the Matlab Simulink modeling development platform of MathWorks, enabling developers to use model-based design methods in a graphical interactive environment. In this environment, developers can also make intuitive module calls and visually process system models and design modules of various subsystems. These models have a hierarchical structure that can divide the system into independent function modules and text-based design methods. For FPGA system-level hardware design, System Generator for DSP provides an environment for the establishment, simulation, and implementation of algorithm models for digital signal processing, which accelerates and simplifies product development and greatly improves the efficiency of designers implementing DSPs on FPGAs.

In recent years, some new development tools and hardware platforms support model-based FPGA design. PYNQ (Python Productivity for Zynq), based on the original Zynq architecture, added support for python. PYNQ fully encapsulates the bottom interaction logic of the ARM processor and the FPGA device. The top encapsulation uses python. The user can focus on the design of the algorithm model. Only the corresponding module name needs to be imported to import the corresponding hardware module. Interaction with upper-layer data or provide hardware acceleration for the system. Xilinx's new generation unified software platform Vitis has Vitis AI library, Vitis BLAS (Basic Linear Algebra Subprogram) library, Vitis data compression library, Vitis vision library, Vitis solver library, Vitis metrology financial library, Vitis database set, and Vitis security library eight Vitis libraries. These eight libraries provide a total of more than 400 open-source applications and are optimized for Xilinx hardware platforms, which can take full advantage of the model-based development process.

High-level Synthesis Tool (HLS) launched by Xilinx enables developers to use high-level languages for FPGA-based algorithm modeling and implementation. The research of HLS aims to allow designers to use commonly used high-level languages such as C or C++ to describe hardware circuits, and at the same time can automatically convert them into RTL-level circuits to achieve the purpose of hardware implementation. More than this, the advanced synthesis tool will also provide some constraints for users to choose, which limits the automatic synthesis of the tool to a certain extent, so that the finally realized system can meet various requirements of users. The use of advanced synthesis tools can greatly shorten the development cycle of hardware design, which greatly improves the efficiency of FPGA development, and can also reduce the probability of errors when using HDL for the code design. In addition to generating code, HLS can automatically generate algorithm IP cores. Xilinx's Vivado development tool supports the transfer of generated IP cores from HLS to connect with other modules and IP for more complex system construction.

In summary, the model-based development process further overcomes the shortcomings of the traditional progressive development process, making the system development difficult and reducing the development cycle. And, HLS is a more flexible and practical model-based development tool and means. Therefore, in this paper, we use HLS to implement a model-based design method to build a system algorithm processing hardware architecture based on the FPGA platform.

## *3.3.3 Examples of IP Design Based on HLS*

### 3.3.3.1 Efficient Off-Chip Storage Access IP

Due to the limitation of memory resources on FPGA, the large capacity off-chip memory (DDR3 SDRAM) is usually utilized to cache high-dimensional hyperspectral data. Multiple pixels need to be read from the DDR3 SDRAM concurrently to support multiple processing elements (PEs), thereby improving data processing speed. A general and standardized off-chip storage access IP is designed, which can access off-chip memory with high-bit width and split high-bit width data into low-bit width data for PEs. As shown in Figure 3.2, the data width of DDR3 SDRAM is set to 512 bits, which is the maximum bit width of the device available to use. The batch Data Loader and Distributor reads HSIs

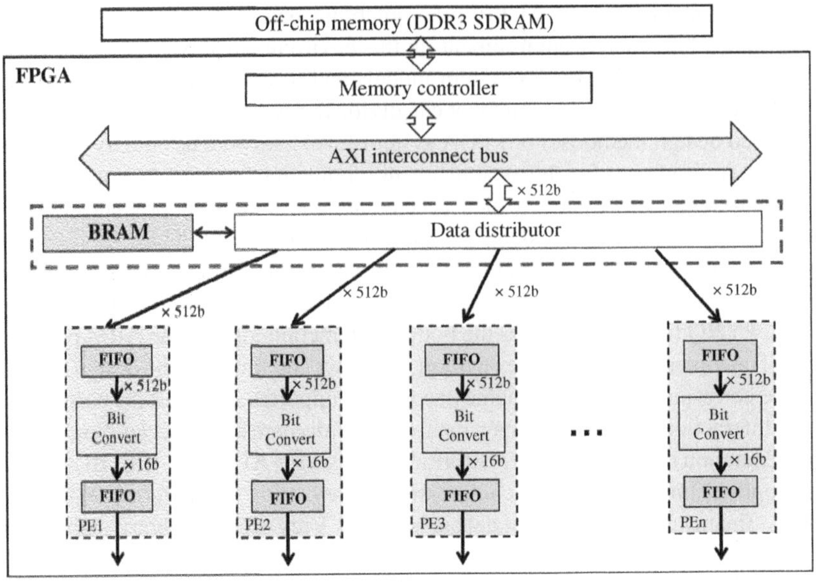


**Figure 3.2** Hardware structure of storage access IP.

from the DDR3 SDRAM through the AXI interconnect bus and loads pixel vectors for PEs in parallel. Specifically, given an HSI with a pixel width of 16 bits, 32 successive bands of a pixel vector at each address can be saved. In other words, when the number of PEs is set to 32, the processing unit can make full use of the bandwidth of DDR3 SDRAM.

#### 3.3.3.2 Parallel Matrix Multiplication IP

Figure 3.3 shows the specific hardware circuit of matrix multiplication IP. This module, which has two inputs including the matrix $L(N \times L)$ and $p(L \times M)$, is used to calculate the product $h(N \times M)$. In order to improve the capability of parallel computation with fewer logic resources, $L$ multipliers are allocated to realize the fully parallel or semi-parallel matrix calculation.

#### 3.3.3.3 Matrix Dot-Product-Plus IP

The main function of this IP is to get the result matrix through multiplying matrix $f$ by the fixed-point coefficient $Q$ and adding it to the matrix $S$. As shown in Figure 3.4, in each cycle, one element of each row in matrix $f$ and $S$ stored in two BRAMs is read in parallel for multiplication and sum, and then the resulting element is written into BRAM. There exists only $L$ multiply accumulate for the sake of cutting down resource consumption.

#### 3.3.3.4 Erosion/Dilation IP

Both erosion and dilation are window-based operations, aiming to find the maximum or minimum among the pixels belonging to the SE. As shown in Figure 3.5, the deep pipeline structure based on row and column decomposition is designed, which not only reduces the number of comparisons, but

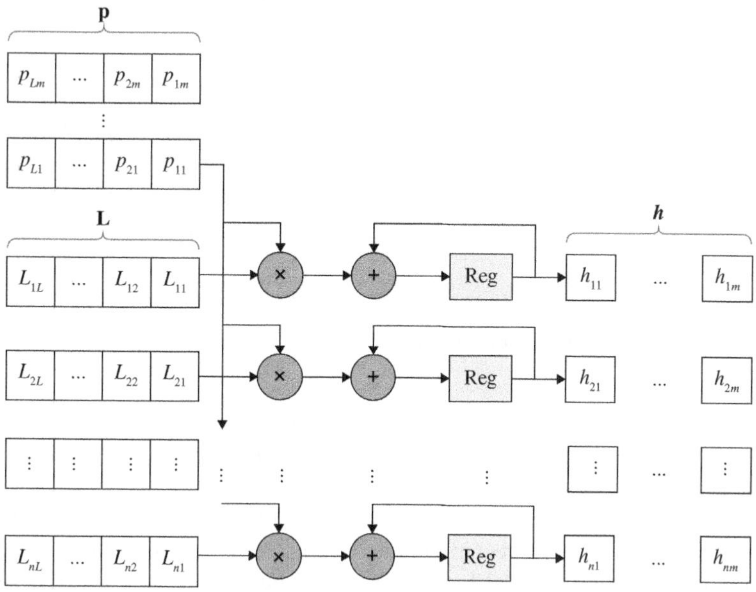


**Figure 3.3** Hardware structure of parallel matrix multiplication IP.

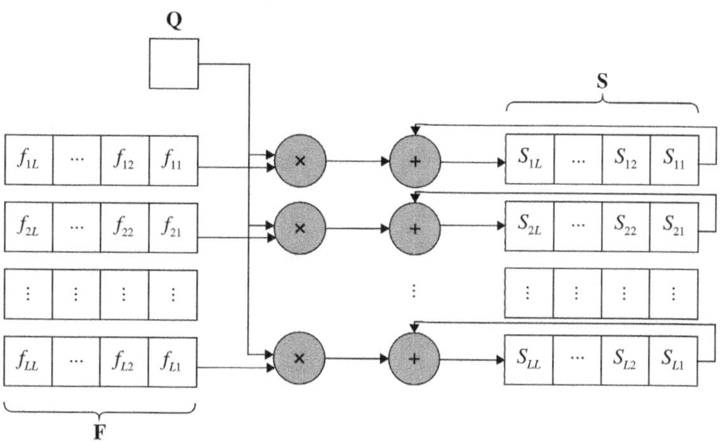


**Figure 3.4** Hardware structure of Matrix dot-product-plus IP.

also exhibits excellent performance in terms of latency and memory through keeping and reusing partial result generated during the computation process. Moreover, we solve the problem of image boundary processing by extending the row and column data.

## 3.4 System Integration Framework Design

The model-based multi-FPGA hardware processing hardware framework supports a single MPSoC or multiple FPGA accelerator cards to build a larger scale target detection algorithm prototype implementation system in parallel. The framework consists of an MPSoC system equipped with a single or

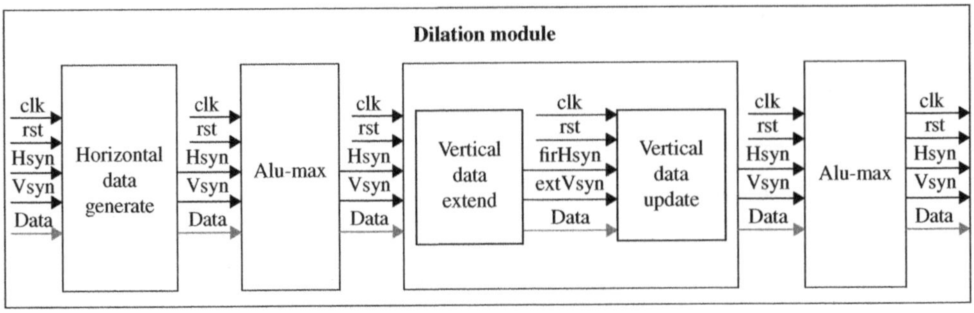


**Figure 3.5** Hardware structure of erosion.

multiple FPGAs. The entire system consists of the PS end processor APU in the MPSoC system as the main control host; the AXI DMA logic module is built inside the PL end of the MPSoC system, which leads the image to the memory address space AXI-FULL bus and AXI-LITE bus; AXI-FULL bus is equipped with AXI DDR4 MIG controller for controlling external DDR4 memory; AXI-LITE bus is equipped with user-defined dynamic reconfigurable Algorithm IP core. In addition, inside the FPGA connected to the PS end, the AXI-FULL and AXI-LITE buses derived from AXI DMA are also extended to other FPGA cards through the AXI CHIP2CHIP method; therefore, the user-defined dynamic reloadable mounted inside other FPGAs. The logic IP core is constructed, and the effect is equivalent to mounting inside the first FPGA.

The overall hardware architecture is shown in Figure 3.6. The FPGA is divided into several dynamic reconstruction blocks: reconfigurable partitions (RP). Each RP can be reconstructed as needed to accelerate the IP of the custom-designed algorithm. The processor APU is called by software API. Implement adaptive hardware IP acceleration of algorithm software. The API library plays the role of connecting software and hardware, so that we can access the processing unit on the FPGA

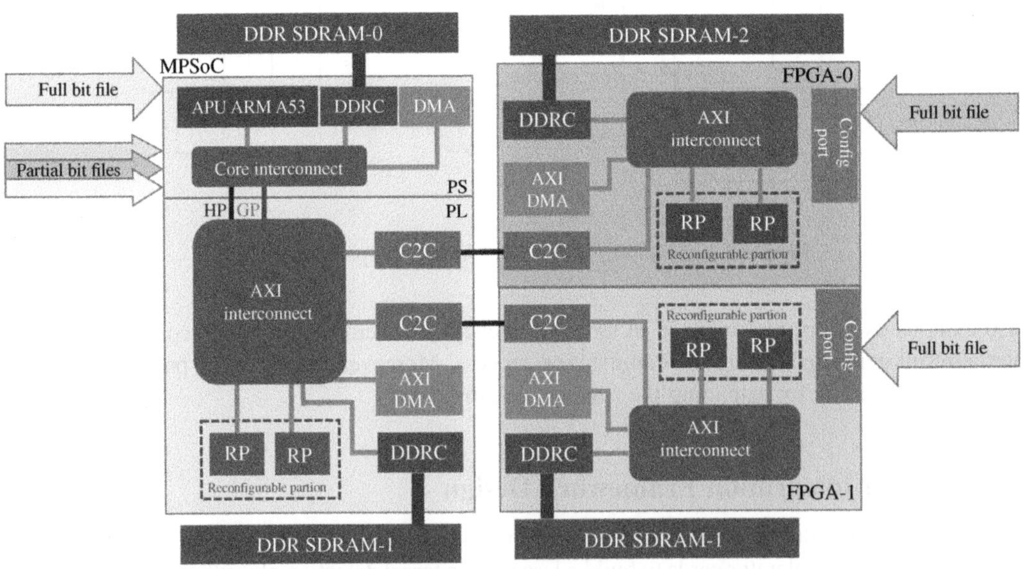


**Figure 3.6** System integration framework design.

side through the upper-layer Application. The acceleration library API includes official library functions and custom IP driver functions, which can be implemented through high-level languages such as C/C++ or Python. All kinds of target detection algorithm applications build the basic algorithm IP designed by Vivado HLS or RTL by calling the acceleration library API, implement different algorithm models as a whole, and generate Full Bitstream File for hardware deployment after verification; it can also be on the top of the software. Use C language or Python and other high-level languages to build the overall algorithm model, and then dynamically call different hardware acceleration IP, generate Partial Bitstream Files to achieve local hardware acceleration. The design of the entire FPGA part can generate different Bitstream files for multitasking and multi-applications, support calls through software APIs, and dynamically reconfigure the logical functions on the FPGA.

Such a model-based development platform can greatly improve development efficiency. The common structure of each algorithm is encapsulated into various IPs, generating part of the bit stream, which can be used by different algorithms to implement different algorithm functions. Next, this book will introduce the specific hardware implementation architecture of each target detection algorithm.

## *3.4.1 Efficient FPGA Implementation*

### 3.4.1.1 FPGA Implementation of DPBS-CEM

As shown in Figure 3.7, the framework of DPBS CEM mainly consists of three components including an off-chip memory and a scheduler. The off-chip memory (DDR3 SDRAM) is utilized to cache the hyperspectral image pixels. The processor core is responsible for the data processing of DPBS-CEM, which involves three modules: the first module is an inverse matrix updater, dedicated to update the

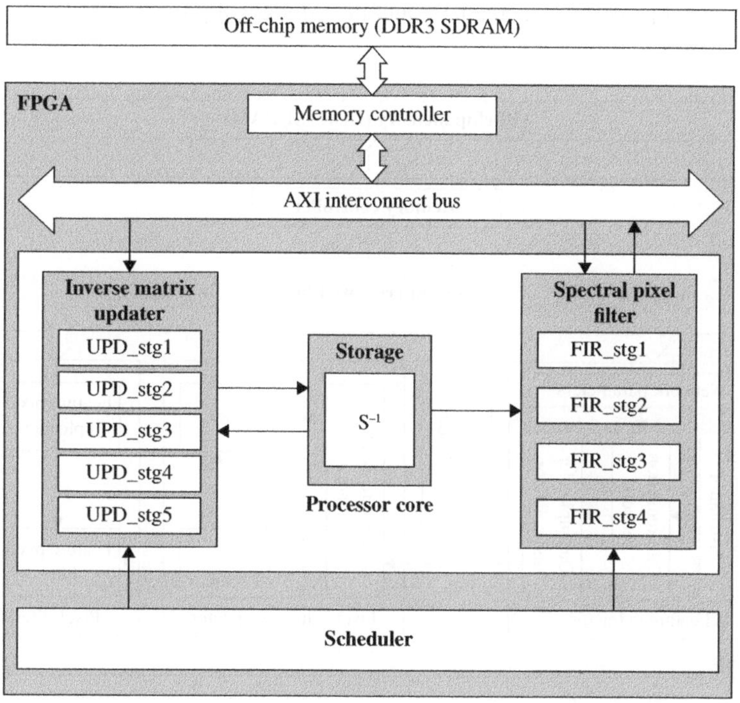


**Figure 3.7** FPGA implementation framework of DPBS-CEM.

inverse matrix in five stages; the second module is a spectral pixel filter, applied to filter pixels in four stages; and the last module is a storage component, utilized to cache the inverse matrix. Finally, the third component is scheduler which is designed to schedule the two modules of inverse matrix updater and spectral pixel filter.

#### 3.4.1.2 FPGA Implementation of DBN-RXD

As shown in Figure 3.8, the framework of DBN-RXD mainly consists of two components including an off-chip memory (DDR3 SDRAM) and a processor core. The off-chip memory is utilized to cache the HSI data. The processor core of DBN-RXD is mainly responsible for the data processing, which involves three modules. The first module is a feature extractor, dedicated to processing an HSI and extract spectral feature matrix for anomaly detection. The second module is an inverse matrix updater, applied to update the inverse matrix by extracted spectral feature matrix. And the last module is a pixel filter, utilized to filter pixels by means of filtering operator $\mathbf{S}_n^{-1}$.

#### 3.4.1.3 FPGA Implementation of Fast-ATGP

As shown in Figure 3.9, the framework of Fast-ATGP mainly consists of two components including an off-chip memory (DDR3 SDRAM) and a processor core. The off-chip memory is utilized to cache the HSI data. The processor core of Fast-ATGP is mainly responsible for the data processing, which involves three modules. The first module is a target searcher, dedicated to processing an HSI and find the target pixel's location. The second module is a subspace projector, applied to calculate the orthogonal projection vector $\tilde{\mathbf{m}}_i$ of the target pixel vector $\mathbf{m}_i$. And the last module is an updater, utilized to update the orthogonal projection operator $\mathbf{P}_{\mathbf{U}}^{\perp}$ and the orthogonal projector $\mathbf{P}_{\mathbf{V}}^{\perp}$.

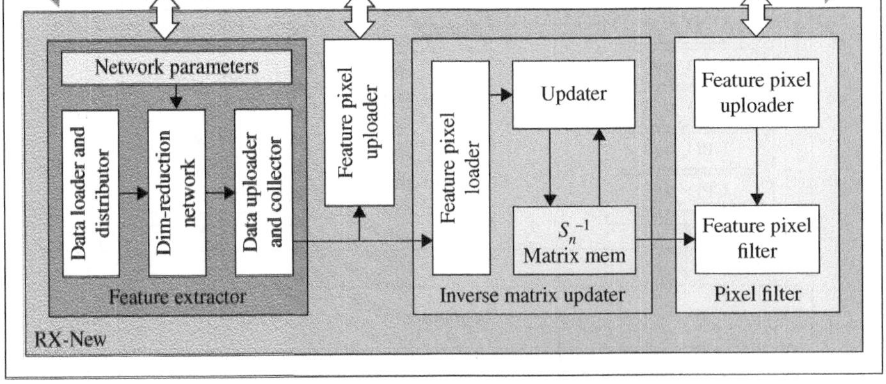


**Figure 3.8** FPGA implementation framework of DBN-RXD.

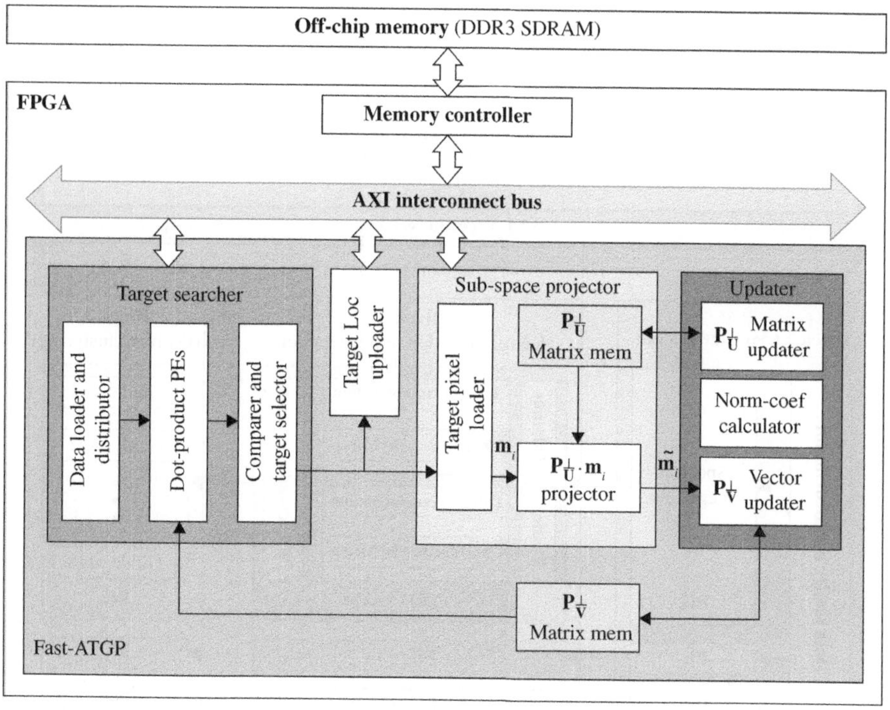


**Figure 3.9** FPGA implementation framework of Fast-ATGP.

#### 3.4.1.4 FPGA Implementation of Fast-MGD

As shown in Figure 3.10, the framework of Fast-MGD is mainly composed of two components including an off-chip memory (DDR3 SDRAM) for caching the HSI data and a processor core, which can be directly used for onboard real-time processing. The processor core that encompasses three modules is used for the major purpose of data processing. The first module is the average fusion unit (AFU) which is in charge of reducing the spectral dimensions of a given HSI. The second module is the feature location and extraction unit (FLEU), which is designed to extract anomalies based on morphological reconstruction. The last module is the feature clustering unit (FCU) being applied to further rectify the detection map through the self-guided filtering. Each module contains $Q$ identical PEs that work in parallel. It is emphasized that only deep pipelined design highlighted by gray in the figure is concerned while the rest of the figure only helps us to verify the function and performance of the proposed Fast-MGD hardware implementation. The highlights of the proposed hardware structure are summarized as follows:

First, it is worth noting that different optimization strategies concerning arrays, loops, latency, and throughput are utilized to achieve a trade-off between area and processing speed through adding directives provided in HLS. For instance, the entire hardware implementation employs global deep pipelining, which takes advantage of the *dataflow* directive to raise the concurrency of the RTL implementation and overall throughput. The three units are connected by FIFO memories with a depth of 1. The C++ template class *hls* :: *stream* <> is adopted for mapping these data to FIFOs.

Second, we design a sharing architecture for the core units of FCU and FLEU (erosion, dilation, and mean filter), which only requires reconfiguration of the RDU and CDU. The general pipelined

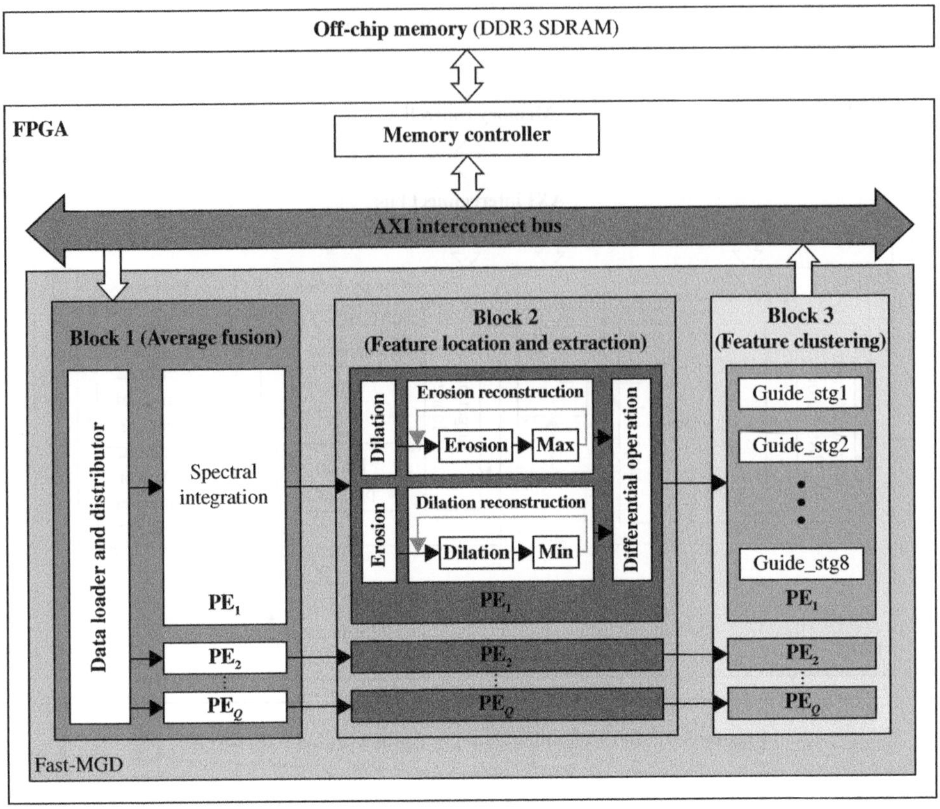


**Figure 3.10** FPGA implementation framework of Fast-MGD.

architecture based on row and column decomposition further simplifies the overall hardware implementation, thereby achieving high-throughput performance.

Third, the type of the input data is 16-bit unsigned fixed-point (15 bits fractional part), and the type of intermediate data needs to be taken into account seriously due to its significant impact on detection accuracy and resource consumption. In particular, a large number of arithmetic units are designed in the FCU. The appropriate data types provided by HLS are applied to different intermediate data by balancing detection accuracy and resource consumption.

## 3.5 Experiments and Discussions

### 3.5.1 Hyperspectral Image Data Set

#### 3.5.1.1 TE1 Data Set

As shown in Figure 3.11(a), 25 panels created with five United States Geological Survey (USGS, Reston, VA, USA) reflectance hyperspectral signatures: alunite (A), buddingtonite (B), calcite (C), kaolinite (K), and muscovite (M). Each row of the five panels in Figure 3.11(b) is simulated by the same mineral signature and each column of five panels has the same size [52, 53]. Among 25 panels are: five $4 \times 4$-pure pixel panels, $px^i_{4 \times 4}$ for $i = 1, \ldots, 5$ in the first column; five $2 \times 2$-pure pixel panels, $px^i_{2 \times 2}$ for $i = 1, \ldots, 5$ in the second column; five $2 \times 2$-mixed pixel panels, $\left\{px^i_{3,jk}\right\}^{2,2}_{j=1,k=1}$

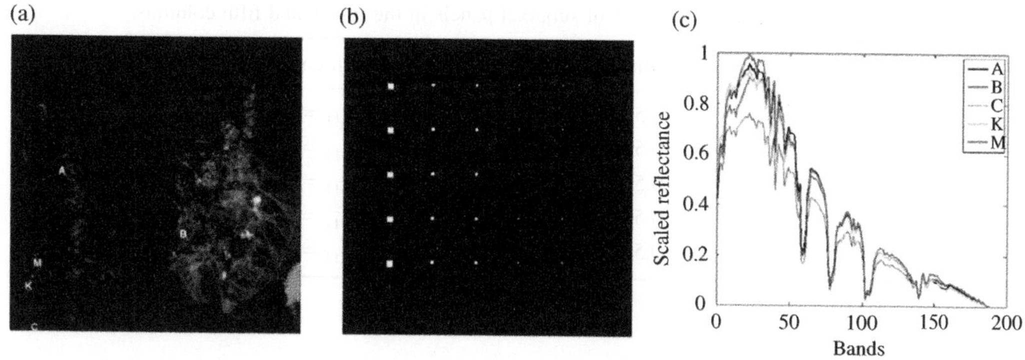


**Figure 3.11** (a) Cuprite Airborne Visible/Infrared Imaging Spectrometer (AVIRIS) image scene with spatial positions of five pure pixels corresponding to minerals: alunite (A), buddingtonite (B), calcite (C), kaolinite (K), and muscovite (M), (b) synthetic image simulated by Scenario TE1, (c) five reflectance USGS ground-truth mineral spectra.

**Table 3.1** Simulated 20 mixed panel pixels in the third column

| | | |
|---|---|---|
| Row1 | $px_{3,11}^{1} = 0.5A + 0.5B$ | $px_{3,12}^{1} = 0.5A + 0.5C$ |
| | $px_{3,21}^{1} = 0.5A + 0.5K$ | $px_{3,22}^{1} = 0.5A + 0.5M$ |
| Row2 | $px_{3,11}^{2} = 0.5B + 0.5A$ | $px_{3,12}^{2} = 0.5B + 0.5C$ |
| | $px_{3,21}^{2} = 0.5B + 0.5K$ | $px_{3,22}^{2} = 0.5B + 0.5M$ |
| Row3 | $px_{3,11}^{3} = 0.5C + 0.5A$ | $px_{3,12}^{3} = 0.5B + 0.5C$ |
| | $px_{3,21}^{3} = 0.5C + 0.5K$ | $px_{3,22}^{3} = 0.5C + 0.5M$ |
| Row4 | $px_{3,11}^{4} = 0.5K + 0.5A$ | $px_{3,12}^{4} = 0.5K + 0.5B$ |
| | $px_{3,21}^{4} = 0.5K + 0.5C$ | $px_{3,22}^{4} = 0.5K + 0.5M$ |
| Row5 | $px_{3,11}^{5} = 0.5M + 0.5A$ | $px_{3,12}^{5} = 0.5M + 0.5B$ |
| | $px_{3,21}^{5} = 0.5M + 0.5C$ | $px_{3,22}^{5} = 0.5M + 0.5K$ |

for $i = 1, \ldots, 5$ in the third column; five subpixel panels, $px_{4,11}^{i}$ for $i = 1, \ldots, 5$ in the fourth column; and five subpixel panels, $px_{5,11}^{i}$ for $i = 1, \ldots, 5$ in the fifth column. Table 3.1 tabulates the mixing details of mineral composition in the 20 panels in the third column, while subpixel panels in the fourth and fifth columns are simulated with their abundance fractions tabulated in Table 3.2, where the background (BKG) is simulated by the sample mean of the real cuprite image scene in USGS [52]. The Synthetic image TE1 is $200 \times 200$ pixels, 189 bands from 0.4 to 2.5 μm.

#### 3.5.1.2 HyMap Data Set

The hyperspectral data set is provided by the Digital Imaging and Remote Sensing Group, Center for Imaging Science, Rochester Institute of Technology [54]. Figure 3.12 shows the HyMap reflection map of Cook City, Montana, USA, with a resolution of $280 \times 800$ and a total of 126 bands distributed between 0.4 and 2.4 μm. There is a grass area and four real panels of fabric in the data set as shown in Table 3.3, where the area of interest is highlighted with a a hollow circle.

**Table 3.2** Abundance fractions of subpixel panels in the fourth and fifth columns

| Row | Fourth column | Fifth column |
|---|---|---|
| 1 | $px^{1}_{4,11} = 0.5\mathrm{A} + 0.5\mathrm{BKG}$ | $px^{1}_{5,11} = 0.25\mathrm{A} + 0.75\mathrm{BKG}$ |
| 2 | $px^{2}_{4,11} = 0.5\mathrm{B} + 0.5\mathrm{BKG}$ | $px^{2}_{5,11} = 0.25\mathrm{B} + 0.75\mathrm{BKG}$ |
| 3 | $px^{3}_{4,11} = 0.5\mathrm{C} + 0.5\mathrm{BKG}$ | $px^{3}_{5,11} = 0.25\mathrm{B} + 0.75\mathrm{BKG}$ |
| 4 | $px^{4}_{4,11} = 0.5\mathrm{K} + 0.5\mathrm{BKG}$ | $px^{4}_{5,11} = 0.25\mathrm{K} + 0.75\mathrm{BKG}$ |
| 5 | $px^{5}_{4,11} = 0.5\mathrm{M} + 0.5\mathrm{BKG}$ | $px^{5}_{5,11} = 0.25\mathrm{M} + 0.75\mathrm{BKG}$ |

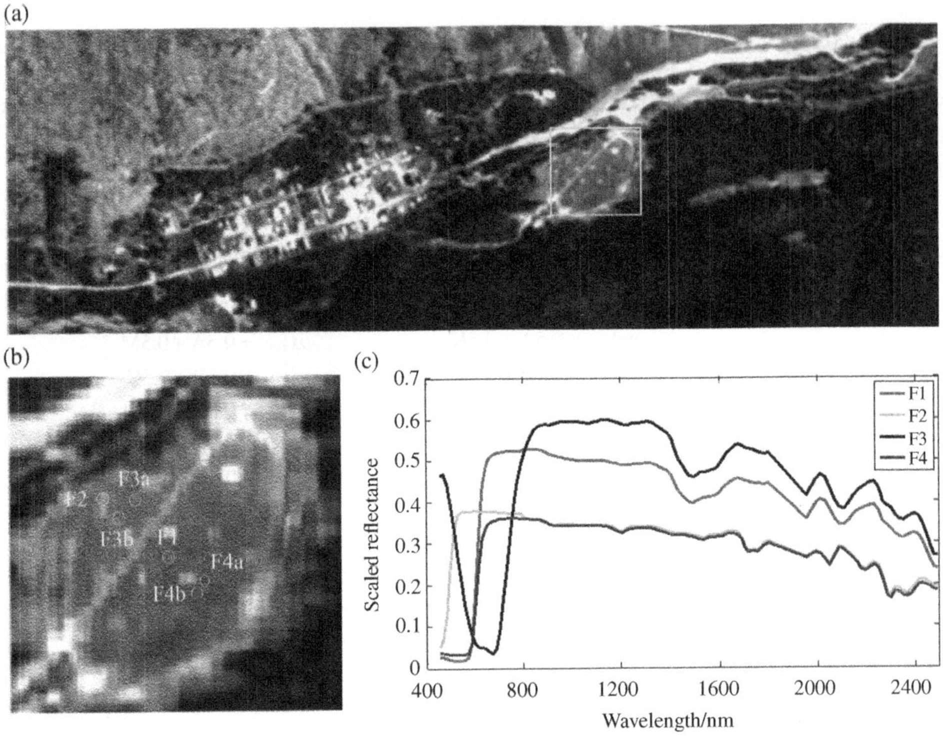


**Figure 3.12** (a) HyMap reflectance image of Cook City in Montana, USA, and locations of the real targets, (b) enlarged figure of red box area, (c) spectral signatures of four targets.

**Table 3.3** The characteristics of targets in the real scene of HyMap

| Name | F1 | F2 | F3a |
|---|---|---|---|
| Size (m) | 3 × 3 | 3 × 3 | 2 × 2 |
| Fabric type | Red cotton | Yellow nylon | Blue cotton |
| **Name** | **F3b** | **F4a** | **F4b** |
| Size (m) | 1 × 1 | 2 × 2 | 1 × 1 |
| Fabric type | Blue cotton | Red nylon | Red nylon |

#### 3.5.1.3 Airport-Beach-Urban (ABU) Data Set

Airport-Beach-Urban (ABU) Data Set contains a lot of data, which contain four airport and beach images and five urban images, and here we use one airport image, two beach images, and three urban images. These sample images contain 100 × 100 or 150 × 150 pixels with the corresponding references, which are manually extracted from large images downloaded from the Airborne Visible/Infrared Imaging Spectrometer (AVIRIS) website. The airplane and urban scenes were captured by the AVIRIS sensor, and most of the beach scenes are also captured by the AVIRIS sensor. One of the beach scenes was captured by the Reflective Optics System Imaging Spectrometer (ROSIS-03) sensor. Furthermore, noisy bands in the original images have been removed using a recently published noisy level estimation method. The reference maps of the sample images are manually labeled with the help of the ENVI software. Some features of these images are shown in Table 3.4. Moreover, due to the different heights of the flights, the spatial resolutions of the images are also different.

#### 3.5.1.4 Cuprite Data Set

Cuprite, as shown in Figure 3.13(a), is the most benchmarked data set for the hyperspectral detection research that covers the Cuprite in Las Vegas, NV, USA. The well-known AVIRIS Cuprite data set is considered as a reference within the hyperspectral remote sensing research field, which is available on the website http://aviris.jpl.nasa.gov/. This scene is composed by 250 × 191 pixels and 224 spectral bands distributed between 0.4 and 2.5 μm, with a spectral resolution of 10 nm. As a result, a total of 188 bands were used for experiments after removing the noisy channels (1–2 and 221–224) and water absorption channels (104–113 and 148–167). Reference ground-signatures of the above minerals (see Figure 3.13(b)), available in the form of a U.S. Geological Survey library (USGS), will be used to assess Fast-ATGP algorithm in this paper.

#### 3.5.1.5 San Diego Data Set

This real hyperspectral data set was also captured by the AVIRIS sensor, which has a spatial resolution of 3.5 m and the spectral resolution of 10 nm. The data set covers the San Diego airport area, CA, USA, and 134 pixels in three airports embedded in the background are regarded as anomalies to be detected. The original spatial size of the data set is 100 × 100, and the spectral bands are 224. After eliminating

**Table 3.4** Some features of the ABU data set

| Images | Captured place | Resolution | Sensor | Fight time |
|---|---|---|---|---|
| airport-1 | Los Angeles | 7.1 m | AVIRIS | 11/9/2011 |
| airport-2 | Los Angeles | 7.1 m | AVIRIS | 11/9/2011 |
| airport-3 | Los Angeles | 7.1 m | AVIRIS | 11/9/2011 |
| airport-4 | Gulfport | 3.4 m | AVIRIS | 7/7/2010 |
| beach-1 | Cat Island | 17.2 m | AVIRIS | 11/16/2011 |
| beach-2 | San Diego | 7.5 m | AVIRIS | Unknown |
| beach-3 | Bay Champagne | 4.4 m | AVIRIS | 7/4/2010 |
| beach-4 | Pavia | 1.3 m | ROSIS-03 | Unknown |
| urban-1 | Texas Coast | 17.2 m | AVIRIS | 8/29/2010 |
| urban-2 | Texas Coast | 17.2 m | AVIRIS | 8/29/2010 |
| urban-3 | Gainesville | 3.5 m | AVIRIS | 9/4/2010 |
| urban-4 | Los Angeles | 7.1 m | AVIRIS | 11/9/2011 |
| urban-5 | Los Angeles | 7.1 m | AVIRIS | 11/9/2011 |

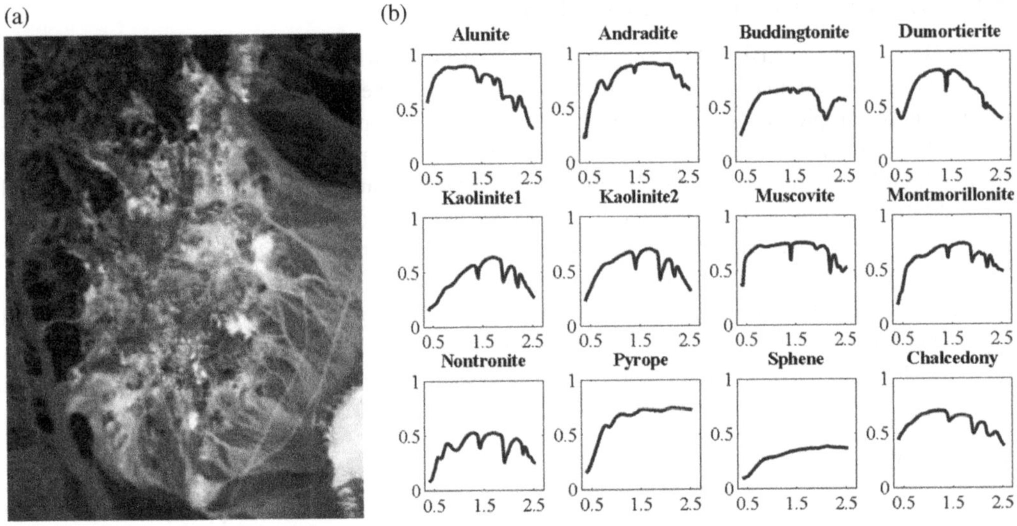


**Figure 3.13** (a) False color composition of the AVIRIS hyperspectral over the Cuprite mining district in Nevada, (b) U.S. Geological Survey mineral spectral signatures used for validation purposes. *Source:* Shao and Lan [55]/MDPI/CC BY 4.0.

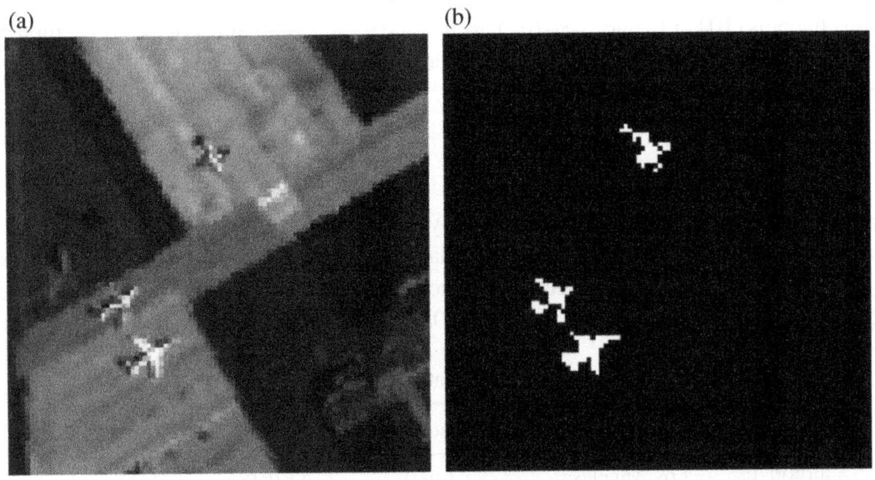


**Figure 3.14** The visualization (a) and the corresponding ground-truth, (b) maps of San Diego.

several noisy bands (1–6, 33–35, 97, 107–113, 153–166, and 221–224) by considering the water vapor and atmospheric effects, the other 189 bands are preserved. The visualization picture and its corresponding ground-truth map of this AVIRIS data set are shown in Figure 3.14(a) and (b), respectively.

#### 3.5.1.6 HYDICE Data Set

This data set describes an urban scene in the United States, which was captured by the Hyperspectral Digital Imagery Collection Experiment (HYDICE) airborne sensor. The original image is of size $307 \times 307$, with 162 spectral channels ranging from 400 to 2500 nm. A sub-image of the spatial size

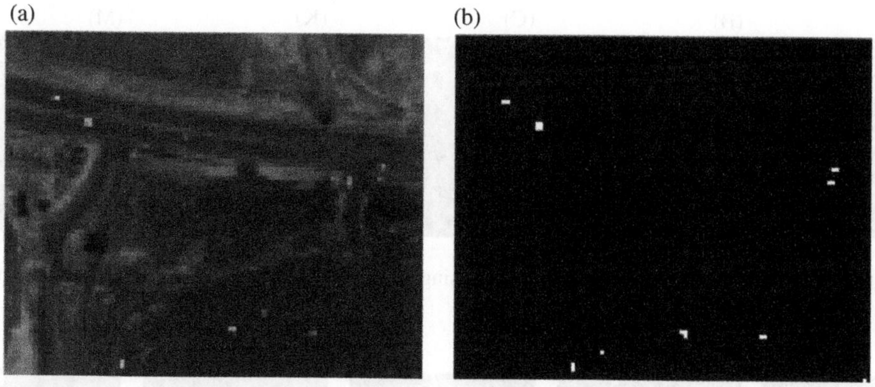

Figure 3.15 The visualization (a) and the corresponding ground-truth, (b) maps of HYDICE.

$80 \times 100$ is cropped from the whole original image for hyperspectral anomaly detection. As for the spectral bands, 162 bands are preserved for retaining useful information while other bands are removed for eliminating redundant information. The total anomalous pixels are 19 and account for 0.2375%, which are cars and roofs. The color composite of this data set and its corresponding ground truth are shown in Figure 3.15(a) and (b), respectively.

### 3.5.2 Experiments of DPBS-CEM

A Virtex7 FPGA board (Alpha-Data ADM-PCIE-7V3) is chosen as our development platform, which has more logic resources than the Kintex-7 board. Besides the FPGA implementation of DPBS-CEM, the simulation versions were also conducted using the MATLAB and C++ languages. The code of MATLAB and C++ are executed on Win7 operation system equipped with the Intel Core (TM) quad CPU @3.2 GHz and 4 GB main memory. We compare the performance of DPBS-CEM with SBS-CEM under the same condition of FPGA implementation.

#### 3.5.2.1 Detection Accuracy

In this part, we evaluate the detection accuracy of the FPGA implementations of DPBS-CEM by using the simulation/real HSI data sets described earlier. CEM and SBS-CEM are evaluated as well for comparison. The detection accuracy can be evaluated via Receiver Operating Characteristics (ROC). However, the ROC curves of different algorithms may be too close to determine which algorithm has better performance. Therefore, in this paper, we choose another way commonly used in medical diagnosis to calculate the area under a ROC curve, referred to as the area under the curve (AUC). The AUC values corresponding to the detection results can further quantify the differences in the accuracy of the algorithms. The higher the AUC, the better the detection results.

Figure 3.16 shows five detection maps produced by DPBS-CEM using the five-panel signatures A, B, C, K, and M as the desired target signatures. The two-dimensional (2-D) results of real-time detection of target A are illustrated in Figure 3.17. And the experimental results show that all the AUC values of five desired targets detected by DPBS-CEM are one, indicating that our detection results are extremely satisfactory.

In order to further measure the performance of DPBS-CEM, we also focus on the detection results of HyMap data set. Figure 3.18 shows the results of the target F4 obtained by Global-CEM,

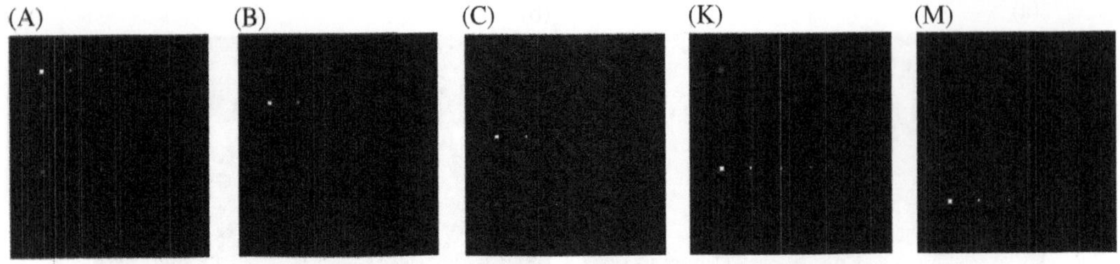


**Figure 3.16** Detection maps of DPBS-CEM using A, B, C, K, and M as desired target signature.

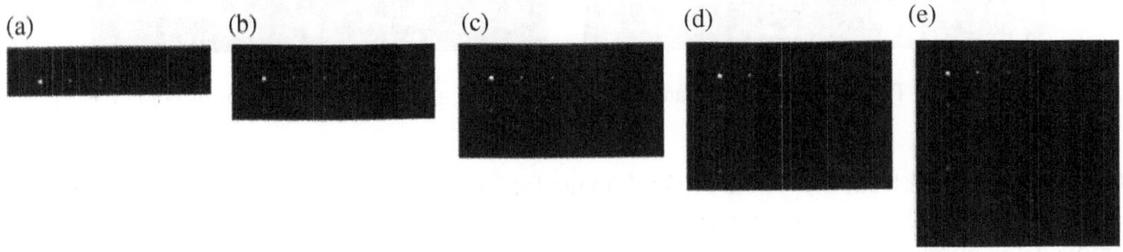


**Figure 3.17** (a–e) Real-time detection results with A used as the desired target signature.

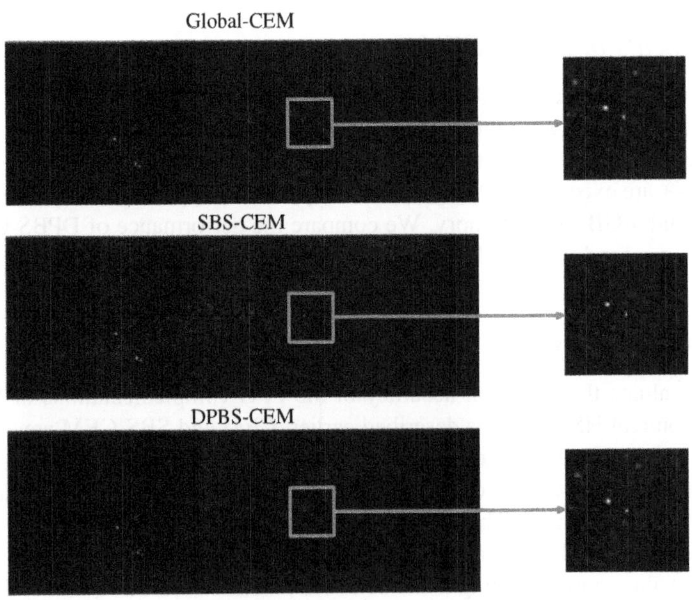


**Figure 3.18** Detection result for target F4 obtained by different algorithms.

SBS-CEM, and DPBS-CEM, respectively. For a more accurate representation of the detection results, we have an enlarged target region of interest, as shown in red boxes of target F4 and Figure 3.19 of target F1, F2, and F3. As we expected, in comparison to the target detection accuracy of SBS-CEM, the DPBS-CEM have the same or even better performance. This conclusion is further verified by the AUC values in Table 3.5.

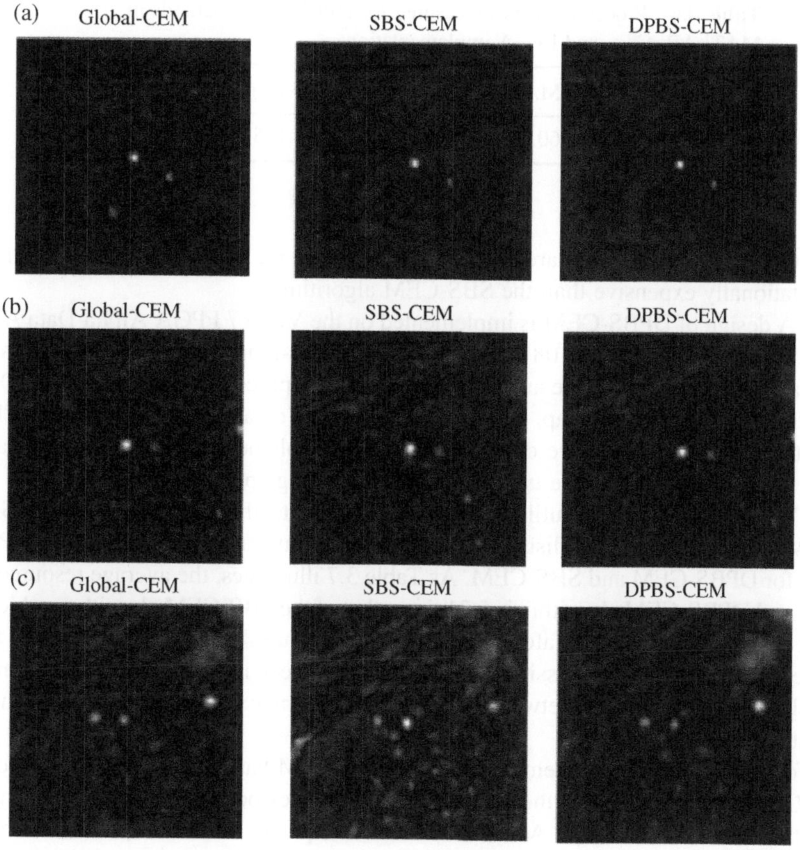


**Figure 3.19** Detection results obtained by different algorithms for targets. (a) F1, (b) F2, (c) F3.

**Table 3.5** AUC obtained by different algorithms for the targets

| | F1 | F2 | F3 | F4 |
|---|---|---|---|---|
| Global-CEM | 0.9107 | 1 | 0.9067 | 0.9987 |
| SBS-CEM | 0.9783 | 1 | 0.9862 | 0.9972 |
| DPBS-CEM | 0.9997 | 0.9999 | 0.9992 | 0.9994 |

#### 3.5.2.2 Acceleration Performance

From the previous section, we can see that the proposed DPBS-CEM is very close to SBS-CEM in detection accuracy; some detection results of DPBS-CEM are even superior to the latter one. Table 3.6 shows the processing time comparison of the proposed DPBS-CEM on different platforms (such as MATLAB, C++, and FPGA). The version of MATLAB used here is R2014a. The C++ environment directly uses the software simulation environment of Vivado HLS 2017.3. The working clock frequency of FPGA is set at 200 MHz. As shown in Table 3.6, the processing time of DPBS-CEM implemented on FPGA has achieved significant improvements compared to MATLAB and C++ implementations. On the other hand, the processing time of our software versions are also

**Table 3.6** Processing times measured for DPBS-CEM methods in MATLAB, C++, and FPGA implementations

| Platform | MATLAB(s) | C++(s) | FPGA(s) |
|---|---|---|---|
| HyMap | 60.7378 | 58.135 | 0.1568 |

superior to that of SBS-CEM software implementations since the proposed DPBS-CEM algorithm is less computationally expensive than the SBS-CEM algorithm.

The FPGA design of DPBS-CEM is implemented on the Virtex7 FPGA Alpha-Data ADM-PCIE-7V3 board. This FPGA contains 108 300 slices, 433 200 six-input LUTs, 1470 BRAMs, and 3600 DSPs. To facilitate the comparative analysis of resource and processing speed between DPBS-CEM and SBS-CEM, we selected HyMap, the same hyperspectral data source used by SBS-CEM, as our input HSI to be detected. Next, we compare the FPGA implementation of SBS-CEM and DPBS-CEM from two aspects of resource utilization and processing speed.

Table 3.7 shows the resource utilization corresponding to the two algorithms SBS-CEM and DPBS-CEM. The right-hand side lists the unit's ratios and average ratios of the two FPGA implementations for DPBS-CEM and SBS-CEM. As Table 3.7 illustrates, the average resource utilization of the proposed DPBS-CEM algorithm is 5.21 times that of the SBS-CEM algorithm, which is caused by the deep pipelined structure. As aforementioned, resource utilization is closely related to data processing precision, and data processing precision has a direct impact on the detection accuracy. Table 3.8 shows the relationship between the detection accuracy represented by AUC and the intermediate data precision.

The performance of FPGA implementation of DPBS-CEM has been greatly improved compared with SBS-CEM. Table 3.9 shows the number of clock cycles occupied by SBS-CEM and DPBS-CEM and the ratio between them. At the same clock frequency of 200MHz, the number of clock

**Table 3.7** Comparison of resource utilization for the FPGA-based implementation of SBS-CEM and DPBS-CEM

| | SBS-CEM | DPBS-CEM | Ratio |
|---|---|---|---|
| | **Units(G)** | **Units(Z)** | $(\frac{Z}{G})$ |
| Number of DSP48Es | 265 | 1396 | 5.27 |
| Number of block RAM | 120 | 379 | 3.16 |
| Number of slices | 12 088 | 58 167 | 4.81 |
| Number of flip flops | 28 245 | 217 958 | 7.72 |
| Number of LUTs | 21 730 | 111 073 | 5.11 |
| Average ratio | – | – | 5.21 |

**Table 3.8** Corresponding AUC values with different intermediate data accuracy of algorithm (we set F1 in HyMap image as desired target)

| Precision (bit) | 32 | 34 | 36 | 38 | 40 | 48 |
|---|---|---|---|---|---|---|
| AUC | 0.2530 | 0.4779 | 0.5463 | 0.9997 | 0.9997 | 0.9997 |

**Table 3.9** Comparison of processing speed for FPGA implementations of SBS-CEM and DPBS-CEM

| | SBS-CEM | DPBS-CEM | Speedup |
|---|---|---|---|
| Frequency (MHz) | 200 | 200 | 7.3× |
| Number of clock periods | 229 607 996 | 31 360 557 | |

cycles consumed by SBS-CEM is nearly 7.3 times that of DPBS-CEM. This leads to the following conclusion. When processing the same image, the processing speed of DPBS-CEM proposed by us is 7.3 times faster than that of SBS-CEM. The cost is only 5.21 times the resource than that of SBS-CEM. It is worthwhile to mention that our work is conducted by mainly using HLS.

### 3.5.3 *Experiments of DBN-RXD*

We implemented the proposed structure of DBN-RXD on a Virtex7 FPGA board (Alpha-Data ADM-PCIE-7V3), which is equipped with two independent channels of DDR3 memory capable of 1333MT/s (fitted with two 8GB SODIMMs). The hardware architecture described in Section 3.4.1.2 has been implemented using OpenCL framework. The implemented system contains two parts: the host API and the OpenCL device kernel. The communication between CPU and FPGA is completed by SDAccel which supports standard OpenCL APIs. The kernel of DBN-RXD is achieved through HLS tools in Vivado HLS. Besides the FPGA implementation of DBN-RXD, the simulation versions were also implemented using the MATLAB and C++ platforms. The code of MATLAB and C++ are executed on Windows 7 operating system equipped with the Intel Core (TM) quad CPU @3.2 GHz and 4 GB main memory.

#### 3.5.3.1 Detection Accuracy

In this section, we evaluate the detection accuracy of the proposed implementation of DBN-RXD by using the real HSI data sets described earlier. RXD and SSFE-Spe were evaluated together with DBN-RXD for comparison. The detection accuracy can be evaluated via ROC and the AUC. The AUC values corresponding to the detection results can further quantify the differences in the accuracy of the algorithms. The higher the AUC, the better the detection accuracy.

Figure. 3.20 shows the anomaly detection results of RXD algorithm, SSFE-Spe algorithm, and DBN-RXD algorithm and the ground-truth map in the ABU data set, indicating that the detection results were both similar to the ground-truth map. The AUC value detected by three different anomaly detectors is illustrated in Table 3.10. For all ABU scenes, the AUC values detected by DBN-RXD algorithm are higher than the RXD and SSFE-Spe. The experimental results show that all AUC values detected by DBN-RXD are better than the RXD, and a few are infinite, indicating that the detection results are extremely satisfactory.

In order to further measure the performance of DBN-RXD, we also focus on the detection results of San Diego data set. Figure 3.21 shows the anomaly detection results obtained by DBN-RXD, SSFE-Spe, and RXD, respectively. The AUC value detected by three different anomaly detectors is shown in Table 3.11. As we expected, in comparison to the anomaly detection accuracy of RXD and SSFE-Spe, DBN-RXD has even better performance. The conclusion in the last experimental result is further verified.

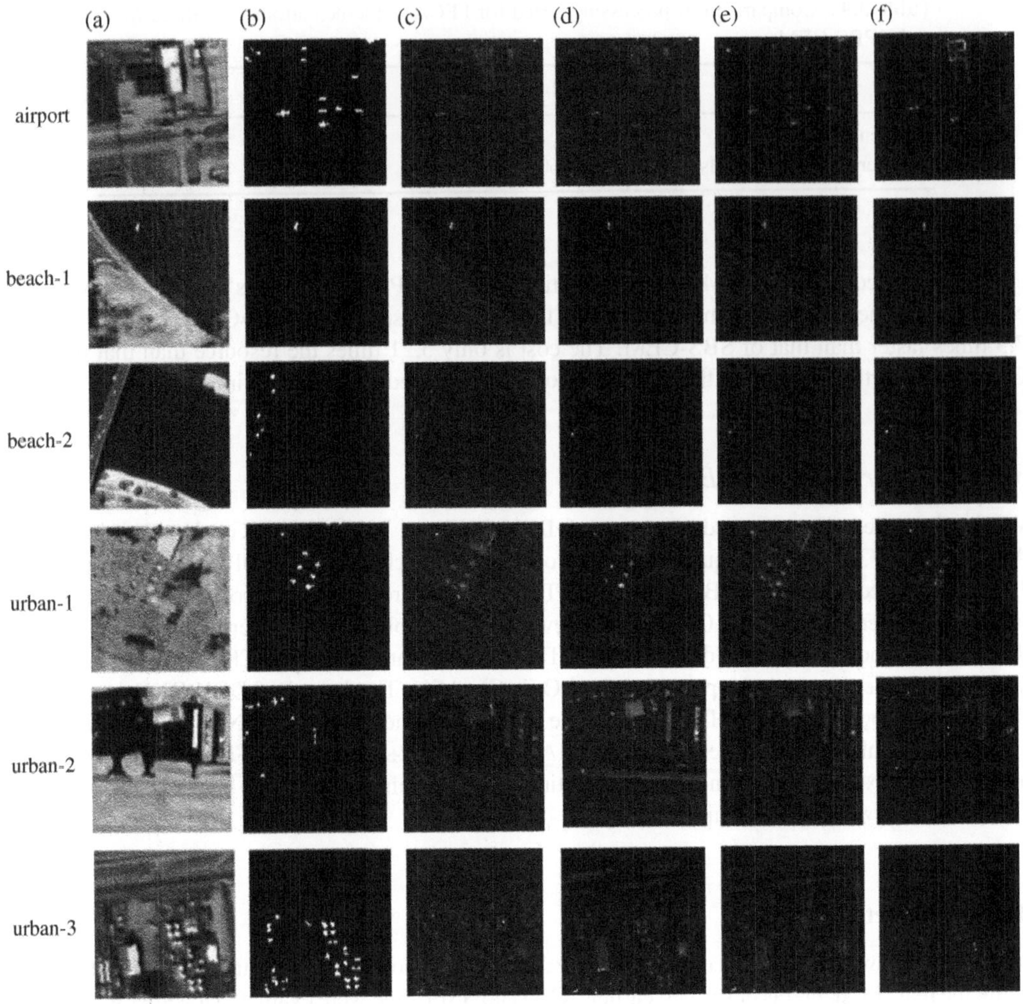


**Figure 3.20** Detection maps of the compared methods. (a) color composites of airport, beach and urban scene, (b) reference map, (c) RXD, (d) DBN-RXD by Matlab, (e) SSFE-Spe, (f) DBN-RXD by HLS.

**Table 3.10** The AUC value detected by three different anomaly detectors of urban scene in TE1

| AUC | airport | beach-1 | beach-2 | urban-1 | urban-2 | urban-3 |
|---|---|---|---|---|---|---|
| **RXD** | 0.8374 | 0.9998 | 0.9534 | 0.9906 | 0.9508 | 0.9693 |
| **SSFE-Spe** | 0.9427 | 0.9999 | 0.9940 | 0.9934 | 0.9732 | 0.9977 |
| **DBN-RXD** | 0.9503 | 0.9999 | 0.9950 | 0.9942 | 0.9804 | 0.9699 |

#### 3.5.3.2 Acceleration Performance

From Section 3.4.1.2, we can see that the proposed DBN-RXD algorithm is very close to RXD in detection accuracy; some detection results of DBN-RXD are even superior to the latter one. Table 3.12 shows the processing time comparison of the proposed DBN-RXD on C++, MATLAB,

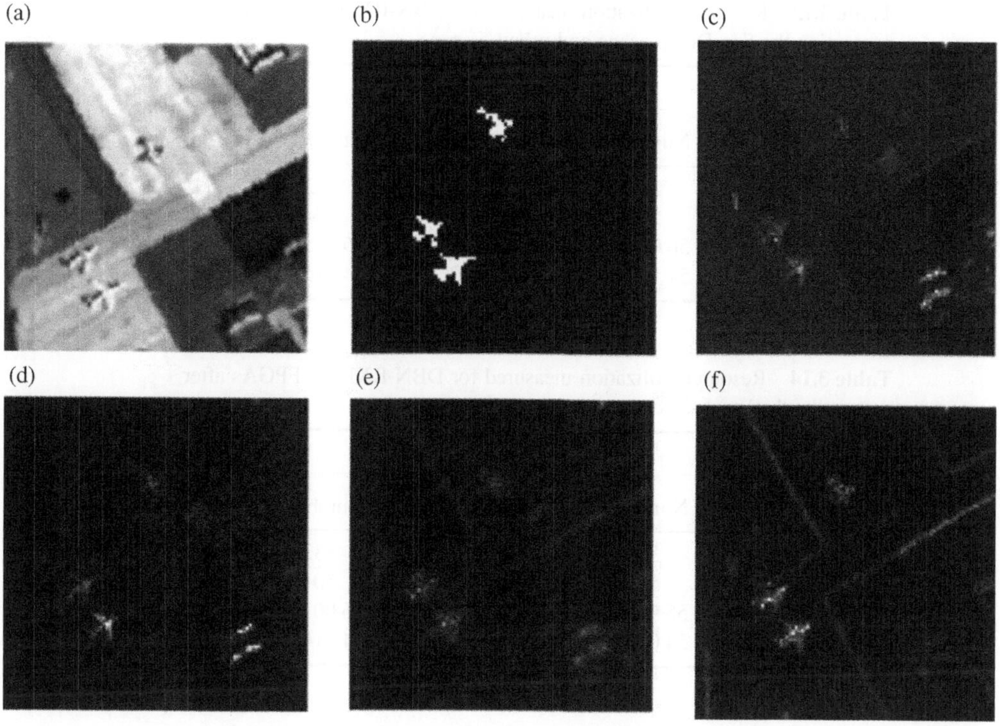


**Figure 3.21** Detection maps of the compared methods. (a) color composites of San Diego image, (b) reference map, (c) RXD, (d) DBN-RXD by Matlab, (e) SSFE-Spe, (f) DBN-RXD by HLS.

**Table 3.11** The AUC value detected by three different anomaly detectors in TE2

| | DBN-RXD | SSFE-Spe | RXD |
|---|---|---|---|
| AUC | 0.77852 | 0.9515 | 0.97532 |

**Table 3.12** Processing time measured for DBN-RXD methods in MATLAB, C++, and FPGA implementations

| | MATLAB (s) | C++ (s) | FPGA (s) |
|---|---|---|---|
| airport | 0.0432 | 0.0238 | 0.0086 |
| beach-1 | 0.0402 | 0.0238 | 0.0086 |
| beach-2 | 0.0852 | 0.0238 | 0.0193 |
| urban-1 | 0.0396 | 0.0238 | 0.0086 |
| urban-2 | 0.0411 | 0.0238 | 0.0086 |
| urban-3 | 0.0390 | 0.0238 | 0.0861 |
| San Diego | 0.0413 | 0.0131 | 0.0086 |

**Table 3.13** Resource utilization measured for DBN-RXD on FPGAs after processing the San Diego scene and urban-5 scene

| | San Diego | | urban-5 | |
|---|---|---|---|---|
| | Number | Proportion | Number | Proportion |
| BRAMs | 332 | 11% | 332 | 11% |
| DSP48E1s | 504 | 14% | 504 | 14% |
| Slice LUTs | 56 007 | 12% | 56 011 | 12% |
| Slice registers | 54 114 | 6% | 54 114 | 6% |

**Table 3.14** Resource utilization measured for DBN-RXD on FPGAs after processing the beach-2 scene and urban-2 scene

| | beach-2 | | urban-2 | |
|---|---|---|---|---|
| | Number | Proportion | Number | Proportion |
| BRAMs | 332 | 11% | 332 | 11% |
| DSP48E1s | 504 | 14% | 504 | 14% |
| Slice LUTs | 55 956 | 12% | 56 007 | 12% |
| Slice registers | 54 084 | 6% | 54 114 | 6% |

and FPGA platforms separately. The version of MATLAB used here is R2018a. The C++ environment directly uses the software simulation environment of Vivado HLS 2017.3. The clock frequency of FPGA is set at 200 MHz. Because the beach-2 contains 150 × 150 pixels and the rest of images contain 100 × 100 pixels, beach-2 will take more time and resources. As shown in Table 3.12, the processing time of DBN-RXD implemented on FPGA has achieved a speedup of 5282.5× on average, thus making significant improvements compared to C++ implementations.

Because the band number and pixel number of these images are not the same, so the consumption of hardware resources are also very different. Images with the same number of pixels and bands consume the same amount of resources. Tables 3.13 and 3.14 show the hardware resource utilization corresponding to DBN-RXD algorithm. Our hardware design is implemented on an FPGA, which consists of 2940 block RAMs, 3600 DSP48E1s, 433200 slice look-up tables (LUTs), and 866400 slice registers. As Table 3.13 illustrates, because the two images have different number of bands, they consume different Slice LUTs. As Table 3.14 shows, they have different number of pixels and bands, so the slice LUTs and slice registers are also different. But anyway because of the optimal architecture we have adopted, the resources such as block RAMs, DSP48E1s, and slice LUTs are obviously reduced. Although the usage of slice registers is increased, the proposed hardware structure occupies fewer resources than RXD in general.

## *3.5.4 Experiments of Fast-ATGP*

### 3.5.4.1 Detection Accuracy

In this section, we evaluate the detection accuracy of the proposed implementations of Fast-ATGP algorithm by using the real hyperspectral data sets for which ground-truth information is available. ATGP-OSP algorithms are evaluated together with our approach. It is worthwhile to emphasize that our hardware version of the Fast-ATGP algorithm provides exactly the same results as a software version of the same algorithm, implemented using the Intel C/C++ compilers.

The detection accuracy can be evaluated via spectral angle mapper (SAM) values (in degrees) between the detected target and the reference spectral signature, which reflects the similarity of pixels in an HSI. The SAM between two pixel vectors $\mathbf{x}_i$ and $\mathbf{x}_j$ is defined by the following expressions:

$$\mathrm{SAM}(\mathbf{x}_i, \mathbf{x}_j) = \cos^{-1}\left(\frac{\mathbf{x}_i \cdot \mathbf{x}_j}{\|\mathbf{x}_i\|_2 \cdot \|\mathbf{x}_j\|_2}\right) r \tag{3.32}$$

It should be noted that SAM is invariant in the presence of illumination interferes, which can provide advantages in terms of target detection in complex backgrounds.

### 3.5.4.2 Results for the AVIRIS Cuprite Scene

Table 3.15 shows the SAM values (in degrees) between the most similar target pixels detected by two different versions of ATGP (ATGP-OSP and Fast-ATGP) at the known target positions in the AVIRIS Cuprite image. In addition, the number of target pixels to be detected was set to $t = 19$ after calculating the VD.

Table 3.16 shows the processing time comparison of the proposed Fast-ATGP on C++ and FPGA platforms separately, where exists 12 targets in Cuprite and 4 targets in Urban data. As shown in Table 3.16, the processing time of Fast-ATGP implemented on FPGA has achieved a speedup of

**Table 3.15** Spectral angle distance values (in degrees) between the target pixels extracted by ATGP-OSP and Fast-ATGP and the known ground targets in the AVIRIS Cuprite scene

| | ATGP-OSP | Fast-ATGP |
|---|---|---|
| Alunite | 4.86° | 5.48° |
| Andradite | 7.56° | 4.80° |
| Buddingtonite | 6.00° | 4.67° |
| Dumortierite | 4.56° | 4.56° |
| $\text{Kaolinite}_1$ | 4.79° | 4.31° |
| $\text{Kaolinite}_2$ | 5.40° | 4.42° |
| Muscovite | 9.35° | 5.11° |
| Montmorillonite | 5.98° | 4.47° |
| Nontronite | 4.19° | 6.01° |
| Pyrope | 7.50° | 8.78° |
| Sphene | 3.74° | 9.37° |
| Chalcedony | 4.73° | 6.80° |
| **Average** | 5.72° | 5.73° |

**Table 3.16** Processing time measured for Fast-ATGP methods in C++ and FPGA implementations

| HSIs | C++ (s) | FPGA (s) |
|---|---|---|
| Cuprite | 132.668 | 0.0238 |

**Table 3.17** Processing time of the AVIRIS Cuprite scene in nanosecond measured for ATGP-OSP and Fast-ATGP on FPGAs

| | ATGP-OSP | Fast-ATGP | Speedup |
|---|---|---|---|
| Maximum frequency (MHz) | 72 | 200 | 34.2× |
| Number of clock periods | 102.24 | 8.31 | |

**Table 3.18** Resource utilization measured for ATGP-OSP and Fast-ATGP on FPGAs after processing the AVIRIS Cuprite scene

| | ATGP-OSP | | Fast-ATGP | |
|---|---|---|---|---|
| | Number | Proportion | Number | Proportion |
| BRAMs | 994 | 67.62% | 632.5 | 43.03% |
| DSP48E1s | 2847 | 79.08% | 1040 | 28.89% |
| Slice LUTs | 13 2802 | 30.66% | 80 080 | 18.49% |
| Slice registers | 22 962 | 2.65% | 140 143 | 16.18% |

5282.5× on average, thus making significant improvements compared to C++ implementations. Furthermore, the processing time of our software versions is much superior to that of ATGP-OSP software implementations since the proposed Fast-ATGP algorithm is less computationally expensive than the ATGP-OSP algorithm. Consequently, it is clear that the proposed FPGA implementation is even faster than the ATGP-OSP algorithm.

We evaluate the performance of ATGP-OSP and Fast-ATGP in the considered HLS tools. Table 3.17 reports that the maximum frequency of Fast-ATGP with 32 PEs is higher than that of ATGP-OSP, and the clock periods occupied by Fast-ATGP is fewer than ATGP-OSP. Thus, the processing time of clock cycles consumed by ATGP-OSP is about 34.2 times as much as that of Fast-ATGP.

Table 3.18 shows the hardware resource utilization corresponding to ATGP-OSP and Fast-ATGP algorithm with 32 PEs. Our hardware design is implemented on an FPGA, which consists of 1470 block RAMs, 3600 DSP48E1s, 433 200 slice LUTs, and 866 400 slice registers. As Table 3.18 illustrates, because of the optimal architecture we have adopted, the resources such as block RAMs, DSP48E1s, and slice LUTs are obviously reduced compared to the ATGP-OSP algorithm. Although the usage of slice registers is increased, the proposed hardware structure occupies fewer resources than ATGP-OSP in general.

### 3.5.5 *Experiments of Fast-MGD*

#### 3.5.5.1 Detection Accuracy

For San Diego data set, the AUC score of (TPR, FPR) is listed in Table 3.19. The AUC score is 0.98432, which is much higher than RX (0.94094), CRD (0.96263), and LRX (0.91430). Compared with the AED method in which the AUC score of (TPR, FPR) is ranked the first, the detection map of the Fast-MGD method contains less background information in Figure 3.22(a).

For Urban-Beach data sets, the proposed method in Table 3.19 shows the best AUC scores of (TPR, FPR) in all scenes. Compared with the RX, LRX, CRD, and AED, the detection results

**Table 3.19** Evaluation AUC scores of (TPR, FPR) obtained by the Fast-MGD (Matlab), Fast-MGD (FPGA), RX, AED, CRD, and LRX

| | Fast-MGD (Matlab) | Fast-MGD (FPGA) | RX | AED | CRD | LRX |
|---|---|---|---|---|---|---|
| (a) | 0.98432 | 0.98432 | 0.94041 | 0.98499 | 0.96263 | 0.91430 |
| (b) | 0.99942 | 0.99942 | 0.99463 | 0.95585 | 0.93042 | 0.91571 |
| (c) | 0.99745 | 0.99745 | 0.95119 | 0.99554 | 0.93970 | 0.93420 |
| (d) | 0.99470 | 0.99470 | 0.98870 | 0.99131 | 0.95156 | 0.97054 |
| (e) | 0.99776 | 0.99776 | 0.98281 | 0.99740 | 0.96720 | 0.97495 |

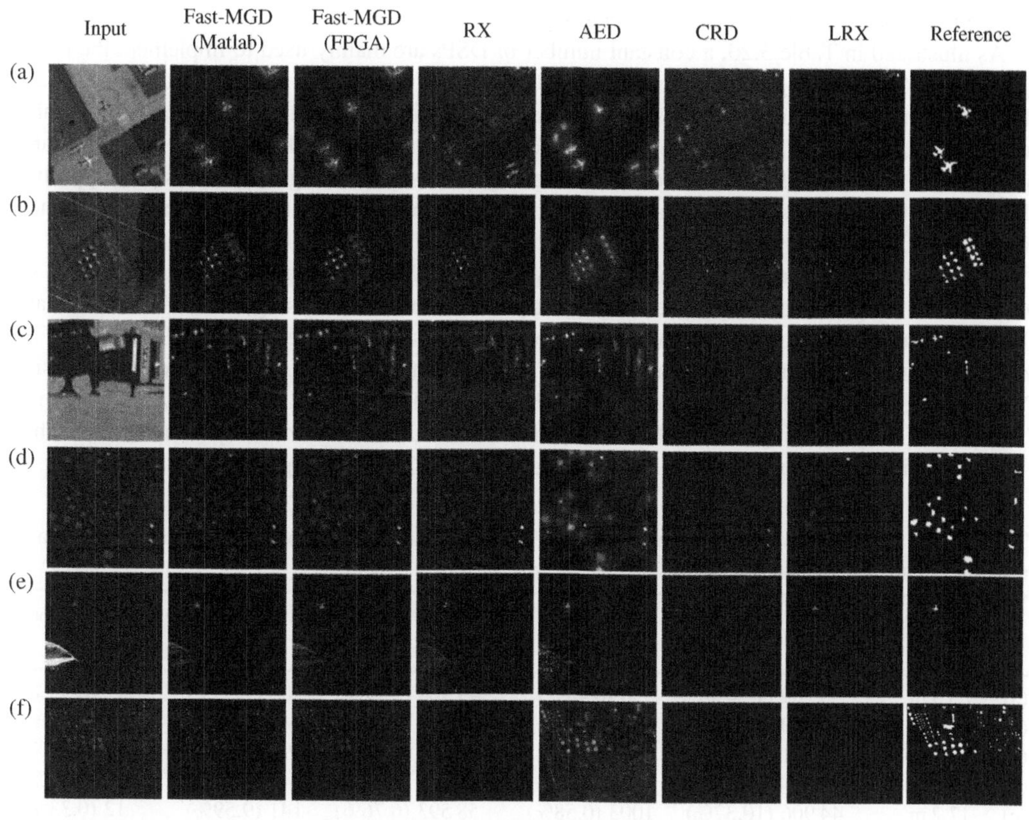


**Figure 3.22** Detection maps of the compared methods on different scenes. (a) San Diego, (b) urban-2, (c) urban-3, (d) urban-4, (e) beach-1, (f) EI Segundo.

obtained by Fast-MGD achieve better visual inspection as shown in Figure 3.22(b)–(e). Most of the anomalies in urban scenes (Figure 3.22(b)–(d)) are unable to be detected by the LRX and CRD methods, and the lousy AUC scores of (TPR, FPR) also support this observation. Nevertheless, these two methods display inadequate performance throughout with respect to computing time, not to mention the inconvenience induced in optimal settings of the inner and outer window. The RX method is merely suitable for sample images with simple background distribution and for large area anomalies, which is unsuitable for complex HSIs as shown in Figure 3.22(c).

It should be pointed out that the parameter settings in the FPGA implementation of Fast-MGD are identical with that in the software. As shown in Figure 3.22, the proposed FPGA implementation shares almost the same visual effects with the software version. Hence, the proposed deep pipelined hardware architecture is proven to be far more than satisfactory.

#### 3.5.5.2 Performance Evaluation

A Virtex7 XC7VX690T FPGA used to evaluate the proposed Fast-MGD hardware architecture contains 433 200 LUTs, 174 200 LUTRAMs, 866 400 FFs, 1470 Block RAMs, and 3600 DSPs. Table 3.20 shows the resource utilization corresponding to Fast-MGD for diverse images with different sizes and resolutions using Vivado 2018.3. Optimal configurable parameters are set for different HSIs to ensure excellent detection accuracy.

As illustrated in Table 3.20, a constant number of DSPs are mainly used to implement the multiplication of Stage 2, Stage 3, and Stage 6 in FCU. Most FIFOs of depth 1 are inferred as LUTs and FFs. The consumption of BRAMs is mainly due to COPU in the sharing architecture. The number of LUTs, LUTRAMs, FFs, and BRAMs is related to the image size and the setting of configurable parameters. Our cost-effective hardware designing philosophy using HLS aims at supporting HSIs with different sizes and spectral bands through adjustment of several configurable parameters including $Q$, $k$, $r_a$, and $r_b$.

Three different platforms including C++ using OpenMP, MATLAB, and FPGA for the above six data sets are adopted in our experiments. The code of software version using MATLAB R2017b and Visual studio 2017 is executed on Window 10 operation system equipped with the Intel Core (TM) quad CPU @2.50 GHz and 8 GB main memory. The hardware architecture is implemented by using Vivado HLS 2018.3. The clock frequency of the Virtex7 FPGA is set at 200 MHz. As shown in Table 3.21, the processing time of FPGA-based implementation achieved a speedup of more than 161 times higher than that of the MATLAB and more than 72 times higher than that of C++.

**Table 3.20** Summary of resource utilization for the FPGA-based implementation in HSIs with different spatial resolutions and spectral bands

| | Resolution | LUTs (433 200) | LUTRAMs (174 200) | FFs (866 400) | BRAMs (1470) | DSPs (3600) |
|---|---|---|---|---|---|---|
| (a) | 3.5 m | 33 969 (7.84%) | 836 (0.48%) | 49 964 (5.77%) | 108 (7.35%) | 12 (0.33%) |
| (b) | 17.2 m | 43 890 (10.13%) | 1001 (0.57%) | 57 962 (6.69%) | 141 (9.59%) | 12 (0.33%) |
| (c) | 3.5 m | 31 681 (7.31%) | 287 (0.16%) | 46 370 (5.35%) | 99 (6.73%) | 12 (0.33%) |
| (d) | 7.1 m | 4387 4(10.13%) | 1001 (0.57%) | 57 929 (6.69%) | 141 (9.59%) | 12 (0.33%) |
| (e) | 17.2 m | 44 906 (10.37%) | 1003 (0.58%) | 58 593 (6.76%) | 141 (9.59%) | 12 (0.33%) |

**Table 3.21** Processing time measured for Fast-MGD methods in MATLAB, C++, and FPGA implementations

| | MATLAB (ms) | C++(ms) | FPGA (ms) |
|---|---|---|---|
| (a) | 224 | 45 | 0.528 |
| (b) | 232 | 55 | 0.553 |
| (c) | 212 | 45 | 0.513 |
| (d) | 232 | 54 | 0.553 |
| (e) | 408 | 122 | 1.176 |

It must be emphasized that the processing time achieved by the FPGA implementation of Fast-MGD is strictly in real-time for all the above scenarios. In specific, the scanning time of cross-track line in AVIRIS being referred to as a push-broom instrument is quite fast (8.3 ms to collect 512 full-pixel vectors) [56], indicating that the processing time of the scene must be limited within 0.16s to achieve real-time performance thoroughly. From Table 3.21, it can be observed that the processing times are lower than 8 ms, including the load time and the data transfer time from the CPU to FPGA device.

## 3.6 Conclusion

In this chapter, we first introduce four effective target and anomaly detection algorithms, including DPBS-CEM, DBN-RXD, Fast-ATGP, and Fast-MGD. In order to perform an efficient FPGA implementation of them, we consider introducing the popular model-based design methods and tools to overcome the drawbacks of high difficulty and long period in the development of traditional hardware systems. Then, we use the flexible and practical model-based development tools and means, HLS, to implement and build four relevant algorithm IPs and system integration framework based on FPGA platform. The system integration framework supports a single MPSoC or multiple FPGA accelerator cards to build larger scale prototype implementations of target detection algorithms in parallel. Moreover, various detection algorithm applications can be dynamically reconfigured on FPGAs to implement different algorithm models via software APIs to call the basic algorithm IP designed by Vivado HLS or RTL. After that, we completed FPGA implementations of four algorithms based on this framework. Finally, on six different data sets, we effectively demonstrate that the hardware implementation of our algorithms using the model-based design approach and tools achieves good performance in terms of detection accuracy, computational speed, and resource utilization on six different data sets.

## References


1. Zhu, L. and Wen, G. (2018). Hyperspectral anomaly detection via background estimation and adaptive weighted sparse representation. *Remote Sens.* 10 (2): 272.
2. Yuan, Y., Zheng, X., and Xiaoqiang, L. (2015b). Spectral–spatial kernel regularized for hyperspectral image denoising. *IEEE Trans. Geosci. Remote Sens.* 53 (7): 3815–3832.
3. Wang, Q., Lin, J., and Yuan, Y. (2016). Salient band selection for hyperspectral image classification via manifold ranking. *IEEE Trans. Neural Netw. Learn. Syst.* 27 (6): 1279–1289.
4. Zhao, L., Lin, W., Wang, Y., and Li, X. (2018). Recursive local summation of RX detection for hyperspectral image using sliding windows. *Remote Sens.* 10 (1): 103.
5. Alarcon-Ramirez, A., Rwebangira, M.R., Chouikha, M.F., and Manian, V. (2016). A new methodology based on level sets for target detection in hyperspectral images. *IEEE Trans. Geosci. Remote Sens.* 54 (9): 5385–5396.
6. Zhang, Y., Wu, K., Du, B. et al. (2017). Hyperspectral target detection via adaptive joint sparse representation and multi-task learning with locality information. *Remote Sens.* 9 (5): 482.
7. Stein, D.W.J., Beaven, S.G., Hoff, L.E. et al. (2002). Anomaly detection from hyperspectral imagery. *IEEE Signal Process. Mag.* 19 (1): 58–69.
8. Wang, Q., Yuan, Z., Qian, D., and Li, X. (2019). Getnet: a general End-To-End 2-D CNN framework for hyperspectral image change detection. *IEEE Trans. Geosci. Remote Sens.* 57 (1): 3–13.
9. Fang, L., Liu, G., Li, S. et al. (2019). Hyperspectral image classification with squeeze multibias network. *IEEE Trans. Geosci. Remote Sens.* 57 (3): 1291–1301.
10. Li, S., Song, W., Fang, L. et al. (2019). Deep learning for hyperspectral image classification: an overview. *IEEE Trans. Geosci. Remote Sens.* 57 (9): 6690–6709.

11. Liu, K.-H., Chen, S.-Y., Chien, H.-C., and Meng-Han, L. (2018). Progressive sample processing of band selection for hyperspectral image transmission. *Remote Sens.* 10 (3): 367.
12. Wang, Q., Li, Q., and Li, X. (2019). Hyperspectral band selection via adaptive subspace partition strategy. *IEEE J. Sel. Top. Appl. Earth Obs. Remote Sens.* 12 (12): 4940–4950.
13. Rizkinia, M. and Okuda, M. (2017). Joint local abundance sparse unmixing for hyperspectral images. *Remote Sens.* 9 (12): 1224.
14. Zhang, X., Li, C., Zhang, J. et al. (2018). Hyperspectral unmixing via low-rank representation with space consistency constraint and spectral library pruning. *Remote Sens.* 10 (2): 339.
15. Jiang, T., Li, Y., Xie, W., and Du, Q. (2020). Discriminative reconstruction constrained generative adversarial network for hyperspectral anomaly detection. *IEEE Trans. Geosci. Remote Sens.*, to be published, doi: https://doi.org/10.1109/TGRS.2020.2965961.
16. Manolakis, D. and Shaw, G. (2002). Detection algorithms for hyperspectral imaging applications. *IEEE Signal Process. Mag.* 19 (1): 29–43.
17. Nasrabadi, N. (2014). Hyperspectral target detection: an overview of current and future challenges. *IEEE Signal Process. Mag.* 31 (1): 34–44.
18. Sun, W., Tian, L., Xu, Y. et al. (2018). A randomized subspace learning based anomaly detector for hyperspectral imagery. *Remote Sens.* 10 (3): 417.
19. Shaw, G. and Manolakis, D. (2002). Signal processing for hyperspectral image exploitation. *IEEE Signal Process. Mag.* 19 (1): 12–16.
20. Zhang, L., Zhang, L., Tao, D. et al. (2013b). Hyperspectral remote sensing image subpixel target detection based on supervised metric learning. *IEEE Trans. Geosci. Remote Sens.* 52 (8): 4955–4965.
21. Zhang, L., Zhang, L., Tao, D., and Huang, X. (2013a). Sparse transfer manifold embedding for hyperspectral target detection. *IEEE Trans. Geosci. Remote Sens.* 52 (2): 1030–1043.
22. Nasrabadi, N.M. (2013). Hyperspectral target detection: an overview of current and future challenges. *IEEE Signal Process. Mag.* 31 (1): 34–44.
23. Yuan, Y., Ma, D., and Wang, Q. (2015a). Hyperspectral anomaly detection by graph pixel selection. *IEEE Trans. Cybern.* 46 (12): 3123–3134.
24. Ying, Q., Wang, W., Guo, R. et al. (2018). Hyperspectral anomaly detection through spectral unmixing and dictionary-based low-rank decomposition. *IEEE Trans. Geosci. Remote Sens.* 56 (8): 4391–4405.
25. Chang, C.-I, Gao, C., and Chen, S.-Y. (2015). Recursive automatic target generation process in subpixel detection. *IEEE Geosci. Remote Sens. Lett.* 12 (9): 1848–1852.
26. Plaza, A.J. (2009). Special issue on architectures and techniques for real-time processing of remotely sensed images. *J. Real Time Image Process.* 4 (3): 191–193.
27. Vetter, J.S. and Vetter, J.S. (2015). *A Survey of CPU-GPU Heterogeneous Computing Techniques*. ACM.
28. Plaza, A. and Chang, C.-I (2008). Clusters versus FPGA for parallel processing of hyperspectral imagery. *Int. J. High Perform. Comput. Appl.* 22 (4): 366–385.
29. Wang, Y., Huang, S., Liu, D., and Wang, H. (2017). A target detection method for hyperspectral imagery based on two-time detection. *J. Ind. Soc. Remote Sens.* 45 (2): 239–246.
30. Zou, Z. and Shi, Z. (2016). Hierarchical suppression method for hyperspectral target detection. *IEEE Trans. Geosci. Remote Sens.* 54 (1): 330–342.
31. Yang, B., Yang, M., Plaza, A. et al. (2015). Dual-mode FPGA implementation of target and anomaly detection algorithms for real-time hyperspectral imaging. *IEEE J. Sel. Top. Appl. Earth Obs. Remote Sens.* 8 (6): 2950–2961.
32. Reed, I.S. and Yu, X. (1990). Adaptive multiple-band CFAR detection of an optical pattern with unknown spectral distribution. *IEEE Trans. Acoust.* 38 (10): 1760–1770.
33. Imani, M. (2017). Rx anomaly detector with rectified background. *IEEE Geosci. Remote Sens. Lett.* 14 (8): 1313–1317.
34. Huber-Lerner, M., Hadar, O., Rotman, S.R., and Huber-Shalem, R. (2016). Hyperspectral band selection for anomaly detection: the role of data gaussianity. *IEEE J. Sel. Topics Appl. Earth Obs. Remote Sens.* 9 (2): 732–743.

35. Li, W., Wu, G., and Du, Q. (2017). Transferred deep learning for anomaly detection in hyperspectral imagery. *IEEE Geosci. Remote Sens. Lett.* 14 (5): 597–601.
36. Molero, J.M., Garzón, E.M., García, I., and Plaza, A. (2013). Analysis and optimizations of global and local versions of the RX algorithm for anomaly detection in hyperspectral data. *IEEE J. Sel. Top. Appl. Earth Obs. Remote Sens.* 6 (2): 801–814.
37. Chang, C.-I (2002). Target signature-constrained mixed pixel classification for hyperspectral imagery. *IEEE Trans. Geosci. Remote Sens.* 40 (5): 1065–1081.
38. Du, B. and Zhang, L. (2010). Random-selection-based anomaly detector for hyperspectral imagery. *IEEE Trans. Geosci. Remote Sens.* 49 (5): 1578–1589.
39. Kwon, H. and Nasrabadi, N.M. (2005). Kernel rx-algorithm: a nonlinear anomaly detector for hyperspectral imagery. *IEEE Trans. Geosci. Remote Sens.* 43 (2): 388–397.
40. Khazai, S. and Mojaradi, B. (2015). A modified kernel-rx algorithm for anomaly detection in hyperspectral images. *Arab. J. Geosci.* 8 (3): 1487–1495.
41. A Schaum. Joint subspace detection of hyperspectral targets. *2004 IEEE Aerospace Conference Proceedings (IEEE Cat. No. 04TH8720)*, volume 3, page 1824. IEEE, 2004.
42. Zhou, Q., Jing, Z., Qi, T. et al. (2016). Automatic endmember extraction using pixel purity index for hyperspectral imagery. *LNCS* 9517: 207–217.
43. Xing, Y. and Gomez, R.B. (2001). Hyperspectral image analysis using envi (environment for visualizing images). *Proc.SPIE – Int. Soc. Opt. Eng.* 15 (38): 79–86.
44. Chang, C.-I, Wu, C.C., and Tsai, C.T. (2011). Random n-finder (n-findr) endmember extraction algorithms for hyperspectral imagery. *IEEE Trans. Image Process. Publ. IEEE Signal Process. Soc.* 20 (3): 641–656.
45. Chang, C.-I, Chen, S.Y., Li, H.C. et al. (2016). Comparative study and analysis among atgp, vca, and sga for finding endmembers in hyperspectral imagery. *IEEE J. Sel. Top. Appl. Earth Obs. Remote Sens.* 9 (9): 4280–4306.
46. Ren, H. and Chang, C. (2003). Automatic spectral target recognition in hyperspectral imagery. *IEEE Trans. Aerosp. Electron. Syst.* 39 (4): 1232–1249.
47. Song, M. and Chang, C. (2015). A theory of recursive orthogonal subspace projection for hyperspectral imaging. *IEEE Trans. Geosci. Remote Sens.* 53 (6): 3055–3072.
48. Yu, C., Lee, L.-C., Chang, C.-I et al. (2018). Band-specified virtual dimensionality for band selection: an orthogonal subspace projection approach. *IEEE Trans. Geosci. Remote Sens.* 56 (5): 1–11.
49. Wang, C., Chen, C.C., Yang, S.C. et al. (2002). Unsupervised orthogonal subspace projection approach to magnetic resonance image classification. *Opt. Eng.* 41 (7): 1546–1557.
50. Hinton, G.E. (2002). Training products of experts by minimizing contrastive divergence. *Neural Comput.* 14 (8): 1771–1800.
51. Zhang, Y., Kang, X., Li, S. et al. (2019). Feature extraction from hyperspectral images using learned edge structures. *Remote Sens. Lett.* 10 (3): 244–253.
52. Chang, C.-I (2013). *Hyperspectral Data Processing: Algorithm Design and Analysis*. Wiley.
53. Wang, J. and Chang, C.-I (2006). Applications of independent component analysis in endmember extraction and abundance quantification for hyperspectral imagery. *IEEE Trans. Geosci. Remote Sens.* 44 (9): 2601–2616.
54. David K. Snyder, John P. Kerekes, Ian Fairweather, Robert Crabtree, and Stacey Hager. Development of a web-based application to evaluate target finding algorithms. *IEEE International Geoscience & Remote Sensing Symposium, IGARSS 2008, July 8–11, 2008, Boston, Massachusetts, USA, Proceedings*, 2008.
55. Shao, Y. and Lan, J. (2019). A spectral unmixing method by maximum margin criterion and derivative weights to address spectral variability in hyperspectral imagery. *Remote Sens.* 11: 1045.
56. Lei, J., Lingyun, W., Li, Y. et al. (2019). A novel FPGA-based architecture for fast automatic target detection in hyperspectral images. *Remote Sens.* 11 (2): 146.

# Part II

## Band Selection for Hyperspectral Imaging

# 4

# Constrained Band Selection for Hyperspectral Imaging

Chein-I Chang
*Center for Hyperspectral Imaging in Remote Sensing (CHIRS), Information and Technology College, Dalian Maritime University, Dalian, China*
*Remote Sensing Signal and Image Processing Laboratory, Department of Computer Science and Electrical Engineering, University of Maryland Baltimore County, Baltimore, MD, USA*

## 4.1 Introduction

With significantly refined spectral resolution, a hyperspectral image (HSI) generally contains abundant spectral information that can be used to uncover many subtle material substances present in a single pixel at subscale. As a consequence, HSI has emerged as an advanced remote sensing technique that can resolve many issues which cannot be solved by multispectral imaging. In many practical applications, HSI has found its great potential in data exploitation, such as missile/vehicle detection in combat, rare materials in geology, pesticide residual detection in food safety and inspection, soil contamination in agriculture, agriculture, drug traffic in law enforcement, rescue and search in disaster area, and water/toxic pollution for environmental monitoring. While HSI has proved to be a very valuable remote sensing technology, it also suffers from so-called curse of dimensionality. In order to mitigate this dilemma, data dimensionality reduction (DR) is generally implemented to reduce data volume prior to data processing [1]. In general, two approaches to DR are of particular interest. One is data compression which utilizes transformations to reduce a high-dimensional hyperspectral data cube to a data set in a low-dimensional data space. The main issue resulting from this approach is that all spectral information is combined and compacted by transformations and the spectral data integrity in the original spectral information is lost. As an alternative, a second approach is band selection (BS), which only selects a partial subset of full bands for data processing. Unfortunately, while the selected bands preserve their original band information, it is also traded for the price that the spectral information of unselected bands is completely discarded. Nevertheless, BS is generally preferred to data compression in HSI applications because BS can retain the spectral band information of the selected bands to represent data faithfully. This is crucial to HSI. For an HSI, each spectral band has its own spectral properties and characteristics which cannot be compromised

*Advances in Hyperspectral Image Processing Techniques*, First Edition. Edited by Chein-I Chang.

by other bands. On the other hand, due to very high inter-band correlation, if one band is selected, those bands which may share similar spectral information such as its adjacent bands can be considered as redundant bands and can be removed without sacrificing much spectral information. Accordingly, BS is particularly suitable and useful for HSI.

Traditional BS is usually performed by several approaches. One is band prioritization (BP) which ranks all spectral bands by a BP criterion such as variance, entropy, signal-to-noise ratio (SNR), and then selects bands in accordance with their ranked priority scores [2]. Another is band clustering, which groups highly correlated bands as a number of separate clusters and then for each cluster, a most informative band is selected to represent the cluster [3]. A third one is BS which implements a BS search strategy to select an optimal set of bands by one band at a time according to a number of bands to be selected, $n_{\mathrm{BS}}$ [4]. A fourth one is band subset selection (BSS) [5], also known as band combinations [6], which must first determine $n_{\mathrm{BS}}$ and then selects an optimal band subset of size $n_{\mathrm{BS}}$ via a designed BSS search strategy. Since many BS techniques have been reported in the literature, it is impossible to discuss them all in a limited space. Accordingly, the main theme of this chapter is focused on constrained BS (CBS) [7], which has been shown to be effective.

There are two approaches to CBS depending upon how to impose the required constraints. The first approach, to be called constrained energy minimization-constrained band selection (CEM-CBS), was proposed by Chang and Wang [7] who took advantage of CEM [8] by constraining a band to be selected for BS. More specifically, instead of constraining a desired target signature to be detected as the way that CEM does, BCBS constrains a band and further calculates a priority score used as its ranking for BS. It considers each band image as a band vector as if it is a target signature so that CEM is readily applied. The resulting BCBS is called band vector-based BCBS. Unfortunately, a HSI generally has a very large size. The band vector to represent an HSI generally has a very large vector dimensionality which results in excessive computing time. To resolve this issue, as an alternative, a linearly constrained minimum variance-based CBS (LCMV-CBS) is developed to replace the single band vector-based BCBS by taking advantage of LCMV to constrain a band image to be selected as a matrix as to reduce significant computational complexity. The resulting LCMV-CBS is referred to as band image-based BCBS. Since CEM-CBS and LCMV-CBS do not specify a particular application, they are considered as application-independent BS.

In contrast to BCBS, another type of CBS is an application-specified BS, referred to as target-constrained BS (TCBS) which imposes constraints on targets to be detected for target detection [9]. Interestingly, CEM-CBS and LCMV-CBS developed for BCBS can be also developed in parallel as counterparts for TCBS. One is a single target-based TCBS which performs CEM by constraining the desired target to be detected to select bands. In order to make the single target-based TCBS work, a theorem is further proved to show that the minimum variance (MV) resulting from CEM using band subsets is monotonically decreasing as more bands are included in CEM. Once the full theory of the single target-based TCBS is established, it can be further extended to a multiple target-based TCBS to impose multiple targets as a target matrix as LCME-CBS does. Following a similar treatment developed for the single target-based TCBS, a theorem to that derived for the single target-based TCBS can be also proved for the multiple target-based TCBS.

## 4.2 Constrained BS

This section presents CEM-CBS [7], which is derived from the concept of CEM. Its idea is to consider each band image as a signal source and treats a selected band as a desired target signature. It then designs a linear finite impulse response filter which passes the selected band while minimizing the variance of the filter output caused by other bands.

### 4.2.1 Band Vector-Constrained BS

CEM was originally designed by constraining a signal source of interest arriving from a desired direction. Now, if we interpret all bands as signal sources and a selected band as a desired signal source, CEM can be used to lock the selected band in a desired direction and in the meantime, to minimize the least squares error (LSE) as variance incurred by all other bands. This variance can be used to rank the selected band according to the magnitude of its generated MV for BP. The resulting CEM is called BCBS.

#### 4.2.1.1 Band Correlation Minimization (BCM)

Assume that $\{\mathbf{B}_l\}_{l=1}^{L}$ is the set of all band images in a HSI cube where $L$ is the total number of bands. Assume that the size of all the band images is $M \times N$. Each band image $\mathbf{B}_l$ can be represented as a column vector of dimensions, $MN$, denoted by $\mathbf{b}_l$. Let $\mathbf{w}_l$ be an $MN$-dimensional column vector that is used to specify a finite impulse response (FIR) filter designed for the $l$th band image vector, $\mathbf{b}_l$, and $y_l$ be the filter output specified by

$$y_l = \mathbf{w}_l^T \mathbf{b}_l \tag{4.1}$$

The averaged LSE or MV resulting from the filter output (4.1) is given by

$$\frac{1}{L}\sum_{l=1}^{L} y_l^2 = \frac{1}{L}\sum_{l=1}^{L}\left(\mathbf{w}_l^T\mathbf{b}_l\right)\left(\mathbf{w}_l^T\mathbf{b}_l\right)^T = \mathbf{w}_l^T\left(\frac{1}{L}\sum_{l=1}^{L}\mathbf{b}_l\mathbf{b}_l^T\right)\mathbf{w}_l \tag{4.2}$$

Let $\mathbf{Q} = (1/L)\sum_{l=1}^{L}\mathbf{b}_l\mathbf{b}_l^T$ denote the band image correlation matrix. A similar optimization problem to CEM can be formulated by a CBS problem as

$$\min_{\mathbf{w}_l}\left\{\mathbf{w}_l^T\mathbf{Q}\mathbf{w}_l\right\} \text{ subject to } \mathbf{b}_l^T\mathbf{w}_l = 1 \tag{4.3}$$

The solution to (4.3), $\mathbf{w}_l^{\text{CEM-CBS}}$ is given by

$$\mathbf{w}_l^{\text{CEM-CBS}} = \left(\mathbf{b}_l^T\mathbf{Q}^{-1}\mathbf{b}_l\right)^{-1}\mathbf{Q}^{-1}\mathbf{b}_l \tag{4.4}$$

and the LSE resulting from $\mathbf{w}_l^{\text{CEM-CBS}}$ is

$$\rho_l = \left(\mathbf{w}_l^{\text{CEM-CBS}}\right)^T\mathbf{Q}\mathbf{w}_l^{\text{CEM-CBS}} \tag{4.5}$$

which represents the minimal correlation of the $l$th band image vector $\mathbf{b}_l$ with all other bands in the least squares sense by constraining $\mathbf{b}_l$. This implies that the larger the $\rho_l$, the higher correlated the $l$th band image vector $\mathbf{b}_l$ with the HSI, and thus the more important the $l$th band image vector $\mathbf{b}_l$. For example, a noisy band image generally has little correlation with the entire HSI, and thus has smaller CEM-LSE. The use of $\{\rho_l\}_{l=1}^{L}$ to select the band images $\{\mathbf{B}_l\}_{l=1}^{L}$ is called band correlation minimization (BCM).

#### 4.2.1.2 Band Dependence Minimization (BDM)

Alternatively, we can also exclude the $l$th band image $\mathbf{b}_l$ from the band image correlation matrix $\mathbf{Q}$ and further define $\tilde{\mathbf{Q}} = (1(L-1))\sum_{j=1,j\neq l}^{L}\mathbf{b}_j\mathbf{b}_j^T$ as the band image dependence matrix. Replacing $\mathbf{Q}$ in (4.3) with $\tilde{\mathbf{Q}}$ results in a similar CBS problem

$$\min_{\mathbf{w}_l}\left\{\mathbf{w}_l^T\tilde{\mathbf{Q}}\mathbf{w}_l\right\} \text{ subject to } \mathbf{b}_l^T\mathbf{w}_l = 1 \tag{4.6}$$

The solution to (4.6), $\tilde{\mathbf{w}}_l^{\text{CEM-CBS}}$ is the same as the one in (4.3) with the $\mathbf{Q}$ replaced by $\tilde{\mathbf{Q}}$ which is given by

$$\tilde{\mathbf{w}}_l^{\text{CEM-CBS}} = \left(\mathbf{b}_l^T\tilde{\mathbf{Q}}^{-1}\mathbf{b}_l\right)^{-1}\tilde{\mathbf{Q}}^{-1}\mathbf{b}_l \tag{4.7}$$

Analogous to (4.5) as BCM, the LSE resulting from $\tilde{\mathbf{w}}_l^{\text{CEM-CBS}}$ is

$$\tilde{\rho}_l = \left(\tilde{\mathbf{w}}_l^{\text{CEM-CBS}}\right)^T\tilde{\mathbf{Q}}\tilde{\mathbf{w}}_l^{\text{CEM-CBS}} \tag{4.8}$$

which indicates the dependence of the band image $\mathbf{b}_l$ on all other band images. The larger the $\tilde{\rho}_l$, the more dependent the band image $\mathbf{b}_l$ on other band images, and thus the more significant the band. Therefore, using $\{\tilde{\rho}_l\}_{l=1}^{L}$ for BS is called band dependence minimization (BDM).

#### 4.2.1.3 Band Correlation Constraint (BCC)

The BCM described above is a criterion to measure the LSE incurred by CEM. However, we can also measure the degree of a band image vector $\mathbf{b}_k$ deviated from the constrained $l$th band image vector $\mathbf{b}_l$ imposed by CEM, $\mathbf{b}_l^T\mathbf{w}_l = 1$, which is $\mathbf{b}_k^T\mathbf{w}_l$. With this interpretation, a new criterion can be defined by calculating the sum of the deviations of all bands $\{\mathbf{b}_k\}_{k=1,k\neq l}^{L}$, $\left(\mathbf{w}_l^{\text{CEM-CBS}}\right)^T\mathbf{b}_k$ other than $\mathbf{b}_l$ from CEM-imposed constraint $\mathbf{b}_l^T\mathbf{w}_l = 1$ via $\mathbf{w}_l^{\text{CEM-CBS}}$ as follows:

$$\begin{aligned}\eta_l &= \sum_{k=1,k\neq l}^{L}\left(\mathbf{w}_l^{\text{CEM-CBS}}\right)^T\mathbf{b}_k \\ &= \left[\sum_{k=1}^{L}\left(\mathbf{w}_l^{\text{CEM-CBS}}\right)^T\mathbf{b}_k\right] - 1 = \left(\mathbf{w}_l^{\text{CEM-CBS}}\right)^T\left[\sum_{k=1}^{L}\mathbf{b}_k\right] - 1\end{aligned} \tag{4.9}$$

It should be noted that (4.9) is the result of $\mathbf{b}_l^T\mathbf{w}_l = 1$ from (4.3). Additionally, due to the constraint $\mathbf{b}_l^T\mathbf{w}_l = 1$ in (4.3), a band image vector $\mathbf{b}_k$ has less correlation with band image vector $\mathbf{b}_l$ if its band constraint $\mathbf{w}_l^T\mathbf{b}_k$ is far away from 1. In other words, the closer the $\mathbf{w}_l^T\mathbf{b}_k$ to 1, the higher the correlation of $\mathbf{b}_k$ to $\mathbf{b}_l$. In light of (4.9), $\eta_l$ represents the degree of the band constraint of a particular band $\mathbf{b}_l$ on band correlation in regard to the entire HSI. The use of $\{\eta_l\}_{l=1}^{L}$ for BS is called band correlation constraint (BCC).

#### 4.2.1.4 Band Dependence Constraint (BDC)

Similar to BCC, we can also define a criterion based on BDM via $\tilde{\mathbf{w}}_l^{\text{CEM-CBS}}$ by

$$\tilde{\eta}_l = \sum_{k=1,k\neq l}^{L} \left(\tilde{\mathbf{w}}_l^{\text{CEM-CBS}}\right)^T \mathbf{b}_k$$
$$= \sum_{k=1}^{L} \left(\tilde{\mathbf{w}}_l^{\text{CEM-CBS}}\right)^T \mathbf{b}_k - 1 = \left(\tilde{\mathbf{w}}_l^{\text{CEM-CBS}}\right)^T \left[\sum_{k=1}^{L} \mathbf{b}_k\right] - 1 \quad (4.10)$$

The use of $\{\tilde{\eta}_l\}_{l=1}^{L}$ to measure the degree of the band constraint of a particular band $\mathbf{b}_l$ on band dependence upon all other band images $\{\mathbf{b}_k\}_{k=1,k\neq l}^{L}$ is called band dependence constraint (BDC).

### 4.2.2 *Band Image-Constrained BS*

One disadvantage of CEM-CBS is the enormous size of vectors converted from a band image that causes tremendous computing time. For example, it requires a vector with $4 \times 10^4$ dimensions to represent a band image with size of $200 \times 200$. In order to mitigate this dilemma, a LCMV-CBS is developed in this section. Instead of constraining a band image as a band vector, LCMV-CBS constrains a band image as a band image matrix without vector conversion. More specifically, assume that $\mathbf{r}_{l,C_1}, \mathbf{r}_{l,C_2}, \ldots, \mathbf{r}_{l,C_N}$ are $N$ columns of the $l$th band image $\mathbf{B}_l$ which has $M$ rows and $N$ columns and each column $\mathbf{r}_l$, $C_n$ is represented by $M$-dimensional column vector, $\mathbf{r}_{l,C_n} = (r_{l,1n}, r_{l,2n}, \ldots, r_{l,Mn})$. In this case, the $l$th band image $\mathbf{B}_l$ can be further expressed by a matrix given by

$$\mathbf{B}_l = \begin{bmatrix} r_{l,11} & r_{l,12} & \cdots & r_{l,1(N-1)} & r_{l,1N} \\ r_{l,21} & r_{l,22} & \ddots & r_{l,2(N-1)} & r_{l,2N} \\ \vdots & \ddots & \ddots & \vdots & \vdots \\ r_{l,(M-1)1} & \ddots & \ddots & r_{l,(M-1)(N-1)} & r_{l,(M-1)N} \\ r_{l,M1} & r_{l,M2} & \cdots & r_{l,M(N-1)} & r_{l,MN} \end{bmatrix} = \left[\mathbf{r}_{l,C_1}\mathbf{r}_{l,C_2}\cdots\mathbf{r}_{l,C_N}\right] \quad (4.11)$$

Like CEM, the goal is to design a constrained FIR linear filter with an $M$-dimensional weight column vector $\mathbf{v}_l = (v_{l,1}, v_{l,2}, \ldots, v_{l,M})^T$ specified by a set of $M$ filter coefficients $\{v_{l,1}, v_{l,2}, \ldots, v_{l,M}\}$ that minimizes the filter output energy subject to the following simultaneous $N$ multiple constraints, $\mathbf{r}_{l,C_n}^T \mathbf{v}_l = \sum_{m=1}^{M} r_{l,nm} v_{l,m} = 1,\ 1 \leq n \leq N$ which is equivalent to

$$\mathbf{B}_l^T \mathbf{v}_l = \mathbf{1}_N \quad (4.12)$$

where $\mathbf{1}_N$ is an $N$-dimensional column vector with all 1's in its $N$ components. It should be noted that since the weight vector $\mathbf{v}_l$ in (4.12) is used to constrain a column vector of a band image, its dimensionality $M$ is compared to the $MN$-dimensional weight vector $\mathbf{w}_l$ used in CEM-CBS that constrains a band image as a band vector with dimensionality $MN$. Using the $N$ multiple constraints in (4.12), CEM-CBS problem in (4.3) can be re-derived as the following optimization problem:

$$\min_{\mathbf{v}_l} \left\{\mathbf{v}_l^T \boldsymbol{\Sigma} \mathbf{v}_l\right\} \text{ subject to } \mathbf{B}_l^T \mathbf{v}_l = \mathbf{1}_N \quad (4.13)$$

where $\boldsymbol{\Sigma} = (1/L)\sum_{l=1}^{L} \mathbf{B}_l \mathbf{B}_l^T$ is the sample band correlation matrix. So, (4.13) is referred to as an LCMV-based CBS optimization problem. The solution to (4.13) can be solved by

$$\mathbf{v}_l^{\text{LCMV-CBS}} = \mathbf{\Sigma}^{-1}\mathbf{B}_l\left(\mathbf{B}_l^T\mathbf{\Sigma}^{-1}\mathbf{B}_l\right)^{-1}\mathbf{1}_N \tag{4.14}$$

and the $\tau_l$ given by

$$\tau_l = \left(\mathbf{v}_l^{\text{LCMV-CBS}}\right)^T\mathbf{\Sigma}\mathbf{v}_l^{\text{LCMV-CBS}} \text{ for LCMV-BCM} \tag{4.15}$$

plays the same role that $\rho_l$ does for CEM-BCM. Derivations similar to CEM-BDM can be also obtained for $\tilde{\mathbf{v}}_l^{\text{LCMV-CBS}} = \tilde{\mathbf{\Sigma}}^{-1}\mathbf{B}_l\left(\mathbf{B}_l^T\tilde{\mathbf{\Sigma}}^{-1}\mathbf{B}_l\right)^{-1}\mathbf{1}_N$ and

$$\tilde{\tau}_l = \left(\tilde{\mathbf{v}}_l^{\text{LCMV-CBS}}\right)^T\tilde{\mathbf{\Sigma}}^{-1}\tilde{\mathbf{v}}_l^{\text{LCMV-CBS}} \text{ for LCMV-BDM} \tag{4.16}$$

In analogy with CEM-based band correlation/dependence constraint criteria (BCC/BDC) developed in Section 4.2.1,

$$\zeta_l = \sum_{k=1, k\neq l}^{L} \mathbf{1}_N^T\left(\mathbf{B}_k^T\left(\mathbf{v}_l^{\text{LCMV-CBS}}\right)\right) \text{ for LCMV-BCC} \tag{4.17}$$

and

$$\tilde{\zeta}_l = \sum_{k=1, k\neq l}^{L} \mathbf{1}_N^T\left(\mathbf{B}_k^T\left(\tilde{\mathbf{v}}_l^{\text{LCMV-CBS}}\right)\right) \text{ for LCMV-BDC} \tag{4.18}$$

can be also derived for LCMV-based band correlation/dependence constraint criteria by replacing band image vector $\mathbf{b}_l$ and CEM-based weight vectors with band image $\mathbf{B}_l$ and LCMV-based weight vectors, respectively.

It should be noted that for CEM-CBS, the dimensions of the band image correlation/dependence matrices $\mathbf{Q}$ and $\tilde{\mathbf{Q}}$ are both $MN \times MN$, while for LCMV-CBS, the size of the band image correlation/dependence matrices $\mathbf{\Sigma}$ and $\tilde{\mathbf{\Sigma}}$ are $M \times M$. This is exactly the reason why we develop the LCMC-CBS to dramatically reduce the computational complexity caused by CEM-CBS.

One comment is worthwhile. The difference between CEM-CBS and LCMV-CBS lies in the fact that the former method considers a band image as a column vector via a scalar constraint, while the latter constraints each column vector of a band image through a vector constraint. In other words, if we impose each column vector in $\mathbf{B}_l$, $\mathbf{r}_{l,C_n}$ for $1 \leq n \leq N$ with the same scalar constraint $1/N$, i.e. $\mathbf{r}_{l,C_n}\mathbf{v}_l = \sum_{m=1}^{M} r_{l,mn}v_{l,m} = 1/N$, we can further concatenate all the column vectors $\{\mathbf{r}_{l,C_n}\}_{n=1}^{N}$ of the $l$th band image $\mathbf{B}_l$ to a vector to represent $\mathbf{B}_l$ as an $MN$-dimensional column vector $\mathbf{b}_l = \left[\mathbf{r}_{l,C_1}^T, \mathbf{r}_{l,C_2}^T, \ldots, \mathbf{r}_{l,C_N}^T\right]^T$ and the weight vectors $\mathbf{v}_l$ $N$ times as an $MN$-dimensional weight vector $\left[\underbrace{\mathbf{v}_l^T, \mathbf{v}_l^T, \ldots, \mathbf{v}_l^T}_{N}\right]^T$. The resulting $\mathbf{b}_l^T\left[\underbrace{\mathbf{v}_l^T, \mathbf{v}_l^T, \ldots, \mathbf{v}_l^T}_{N}\right] = \sum_{n=1}^{N}\sum_{m=1}^{M} r_{l,mn}v_{l,m} = \sum_{n=1}^{N} 1/N = 1$ is exactly the same constraint $\mathbf{b}_l^T\mathbf{w}_l = 1$ imposed in (4.3).

The advantage of using LCMV-CBS over CEM-CBS is its significant reduction of computational complexity. The only difference between these two approaches is that the weight vector, $\mathbf{v}_l$, generated by LCMV-CBS has dimensionality of $M$ as opposed to the weight vector, $\mathbf{w}_l$, produced by CEM-CBS which has the dimensionality of $MN$. Despite the fact that LCMV-CBS and CEM-CBS share the same design concept, both perform quite differently. According to our conducted experimental results, CEM-CBS would only perform better than LCMV-CBS when the number of selected bands is small such as fewer than 10 bands, in which case CEM-CBS is practically implementable and preferred to LCMV-CBS. Otherwise, both performed very similarly in which case LCMV-CBS is always preferable.

## 4.3 BCBS Experiments

### *4.3.1 HYDICE Data*

The first image data to be studied is an image scene shown in Figure 4.1(a), which has a size of $64 \times 64$ pixel vectors with 15 panels in the scene and the ground-truth map in Figure 4.1(b), where their color maps can be found in Figure 1.15 of Ref. [1]. It was acquired by 210 spectral bands with a spectral coverage from 0.4 μm to 2.5 μm. Low signal/high noise bands: bands 1–3 and bands 202–210; and water vapor absorption bands: bands 101–112 and bands 137–153 were removed. So, a total of 169 bands were used in experiments. The spatial resolution is 1.56 m and spectral resolution is 10 nm.

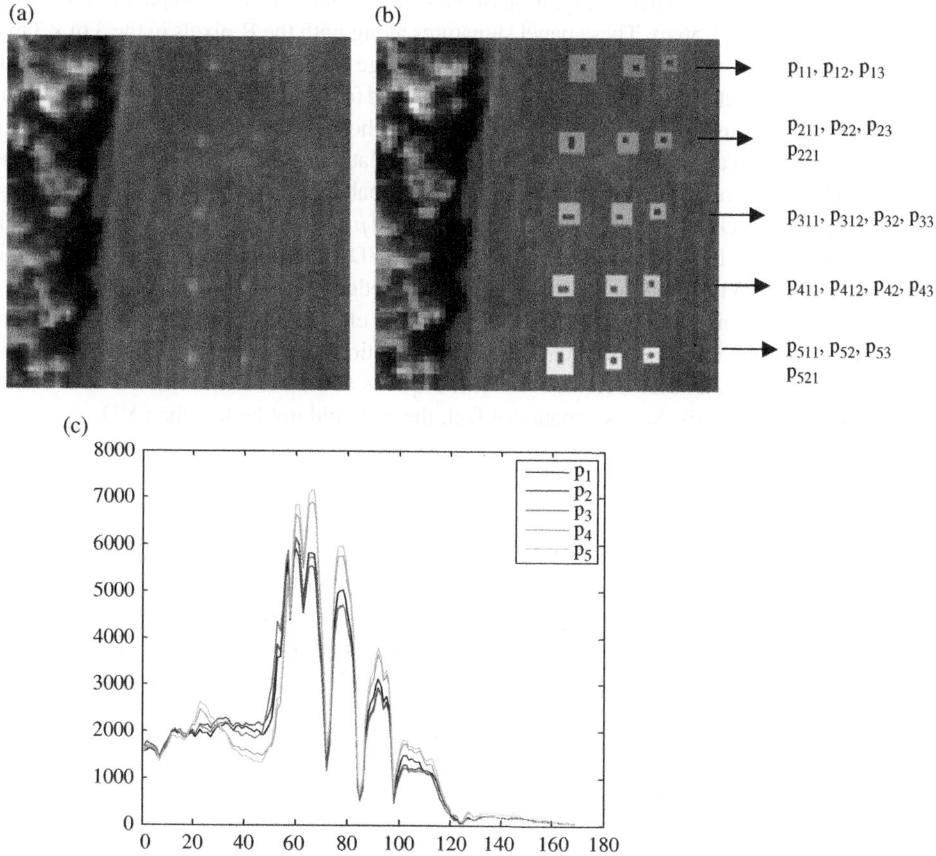


**Figure 4.1** (a) A HYDICE panel scene which contains 15 panels; (b) ground-truth map of spatial locations of the 15 panels; and (c) spectral signatures of $\mathbf{p}_1$, $\mathbf{p}_2$, $\mathbf{p}_3$, $\mathbf{p}_4$, and $\mathbf{p}_5$.

Within the scene in Figure 4.1(a), there is a large grass field background and a forest on the left edge. Each element in this matrix is a square panel and denoted by $p_{ij}$ with rows indexed by $i$ and columns indexed by $j = 1, 2, 3$. For each row $i = 1, 2, \ldots, 5$, there are three panels $p_{i1}, p_{i2}, p_{i3}$, painted by the same paint but with three different sizes. The sizes of the panels in the first, second, and third columns are 3 m × 3 m, 2 m × 2 m, and 1 m × 1 m, respectively. Since the size of the panels in the third column is 1 m × 1 m, they cannot be seen visually from Figure 4.1(a) due to the fact that its size is less than the 1.56 m pixel resolution. For each column $j = 1, 2, 3$, the five panels in the same column have the same size but with five different paints. However, it should be noted that the panels in rows 2 and 3 were made by the same material with two different paints and so are the panels in rows 4 and 5. Nevertheless, they were still considered as different panels. Our experiments will demonstrate that detecting panels in row 5 (row 4) may also have effect on detection of panels in row 2 (row 3). The 1.56 m-spatial resolution of the image scene suggests that most of the 15 panels are one pixel in size except the panels in the first column along with the second, third, fourth, and fifth rows which are two-pixel panels, denoted by $p_{211}, p_{221}, p_{311}, p_{312}, p_{411}, p_{412}, p_{511}, p_{521}$. Figure 4.1(b) shows the precise spatial locations of these 15 panels where dark gray pixels (DG pixels) are the panel center pixels and the pixels in light gray pixels (LG pixels) are panel pixels mixed with the background. Figure 4.1(c) plots the five panel spectral signatures $\mathbf{p}_i$ for $i = 1, 2, \ldots, 5$ obtained by averaging R pixels in the 3 m × 3 m and 2 m × 2 m panels in row $i$ in Figure 4.1(b). It should be noted the R pixels in the 1m × 1 m panels are not included in averaging spectral signatures in Figure 4.1(c) because they are not pure pixels in the sense that the spatial resolution of the R pixels in the 1 m × 1 m panels is 1m smaller than the pixel resolution is 1.56 m. These panel signatures along with the R pixels in the 3 m × 3 m and 2 m × 2 m panels were used as required prior target knowledge for the following comparative studies.

First of all, we need to determine how many bands are needed for BS. Let $p$ denote the number of bands required for BS. VD was used to estimate the $p$ where the method developed by Harsanyi, Farrand and Chang in [10], referred to as HFC method, was used to calculate VD for the HYDICE image scene in Figure 4.1(a). The values of VD with various false alarm probabilities of $P_F$ are tabulated in Table 4.1.

For our experiments, VD was chosen to be 9. The selection of $p = 9$ is empirical based on the false alarm fixed at probabilities $P_F = 10^{-3}, 10^{-4}$. As noted, the value of VD varies with the false alarm probability $P_F$. This makes sense. With no availability of prior knowledge, the false alarm probability via the Neyman–Pearson detection theory [11] is probably one of most effective criteria to determine the number of signals detected in the data. As interpreted in the introduction, one dimension can be only used to accommodate one single signal source. Therefore, it requires at least VD-determined dimensions to separate distinct signal sources. So, as a matter of fact, the $p$ should not be less than VD.

#### 4.3.1.1 Target Detection

In order to evaluate the impact of CBS on target detection, the commonly used CEM was implemented for performance analysis. The selection of CEM was purely based on our preference which only uses desired target signature provided a priori. It can be used to compare results for unsupervised mixed pixel classification in the following section where target knowledge is not provided a priori, rather obtained directly from the image scene. Table 4.2 tabulates the nine bands selected

**Table 4.1** VD estimates for the HYDICE scene in Figure 4.1(a) by the HFC method with various false alarm probabilities

| | $P_F = 10^{-1}$ | $P_F = 10^{-2}$ | $P_F = 10^{-3}$ | $P_F = 10^{-4}$ | $P_F = 10^{-5}$ |
|---|---|---|---|---|---|
| VD | 14 | 11 | 9 | 9 | 7 |

**Table 4.2** Comparison between selected bands using different techniques

| BP criteria | Nine selected bands |
|---|---|
| CEM-CBS BCM/BDM | 18/85/112/10/109/74/115/114/96 |
| CEM-CBS BCC/BDC | 115/96/84/86/105/13/103/87/24 |
| LCMV-CBS BCM/BDM | 125/169/168/164/146/124/165/128/167 |
| LCMV-CBS BCC/BDC | 125/169/124/164/167/161/163/168/146 |
| Uniform | 1/19/37/55/73/91/109/127/145 |
| MVPCA | 60/61/67/66/65/59/57/68/62 |
| ID | 154/157/156/153/150/158/145/164/163 |

by LCMV-CBS and CEM-CBS with nine highest band correlations along with the nine bands selected by uniform band and MVPCA in [2] and ID defined by (4.17) where a backslash "/" is used to separate two selected bands. Since the criteria BCM and BDM yielded IDentical bands for both LCMV-CBS and CEM-CBS, they are included in the same row in Table 4.2. Similarly, it was true for BCC and BDC which also produced IDentical bands for LCMV-CBS and CEM-CBS.

Figure 4.2(a) shows the 15-panel detection results by CEM using the complete set of full bands, 169 bands, whereas Figure 4.2(b–h) shows detection results of the 15 panels by CEM using nine bands selected by various criteria, BCM/BDM CEM-CBS, BCC/BDC CEM-CBS, BCM/BDM LCMV-CBS, BCC/BDC LCMV-CBS, uniform, MVPCA, and ID where the desired target signature used in CEM was selected from five panel signatures, $\mathbf{p}_1$, $\mathbf{p}_2$, $\mathbf{p}_3$, $\mathbf{p}_4$, and $\mathbf{p}_5$ in Figure 4.1(c).

As demonstrated in Figure 4.2(b–h), ID seemed to perform better than other BS algorithms. However, this was due to the fact that the desired target signature used in CEM was assumed to be known a priori. The results would be different and demonstrated in the following section where the unsupervised mixed pixel classification is considered as an application. Furthermore, if we compare the results in Figure 4.2(b–h) to that in Figure 4.2(a), CEM using full bands performed significantly better than BS-based techniques in Figure 4.2(b–h). This may be due to three reasons. One is targets to be detected. Since the 15 panels are small targets, their detection requires as many as bands to provide sufficient sample spectral correlation to capture their subtle spectral properties. A second reason is the algorithm used for detection. CEM only used one signature for detection while viewing other eight signatures as interfering signatures. As a result, it did not take advantage of the knowledge provided by these eight signatures. Instead, it used sample spectral correlation provided by the entire image to suppress these eight signatures. In this case, the sample correlation plays a key role in its detection. Like the first reason, such small panels can be only captured by subtle sample spectral correlation. BS may result in loss of some crucial information. A third reason is that VD only provides an estimate of the least number of bands to be selected. It does not imply that the $p$ must be VD. In this particular experiment, $p = 9$ was not sufficiently large for CEM to detect small panels. The results were significantly improved with all the 15 panels detected when the $p$ was increased to be twice VD, 18 or greater according to the results in [12, chapter 17]. Their results are not included here since this experiment was not designed to show target detection, but rather demonstrated the performance in target detection for all the BS methods that were compared fairly under the same conditions. Nevertheless, as will be also demonstrated in the next section, if the application is mixed pixel classification and the algorithm to be used is the fully constrained least squares method (FCLS) developed in [12, 13], the performance can be significantly improved using the same number of bands selected, $p = 9$.

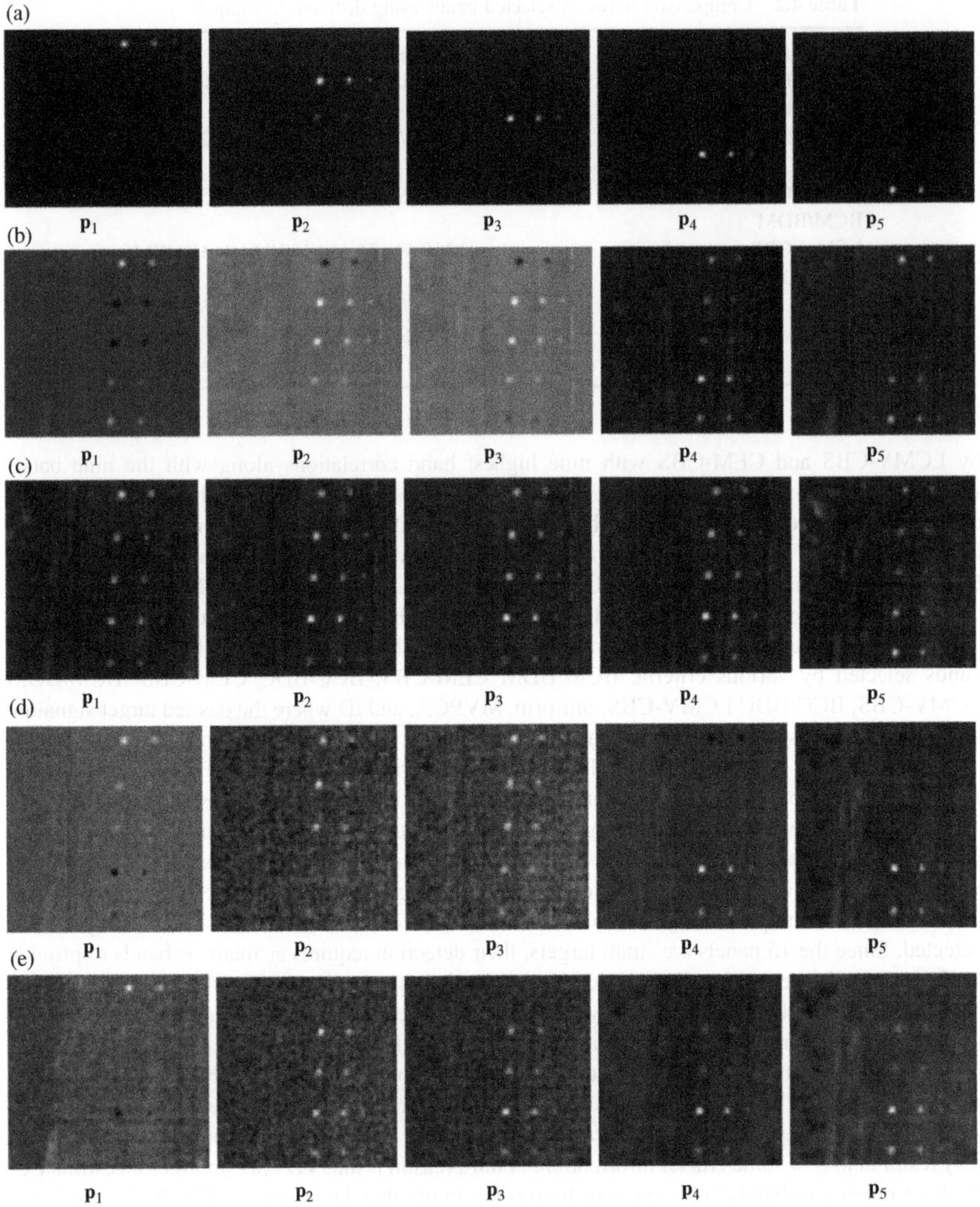


**Figure 4.2** Detection results of 15 panels by CEM with nine selected bands by various criteria. (a) full bands, (b) CEM-CBS BCM/BDM, (c) CEM-CBS BCC/BDC, (d) LCMV-CBS BCM/BDM, (e) LCMV-CBS BCC/BDC, (f) uniform, (g) MVPCA, and (h) ID.

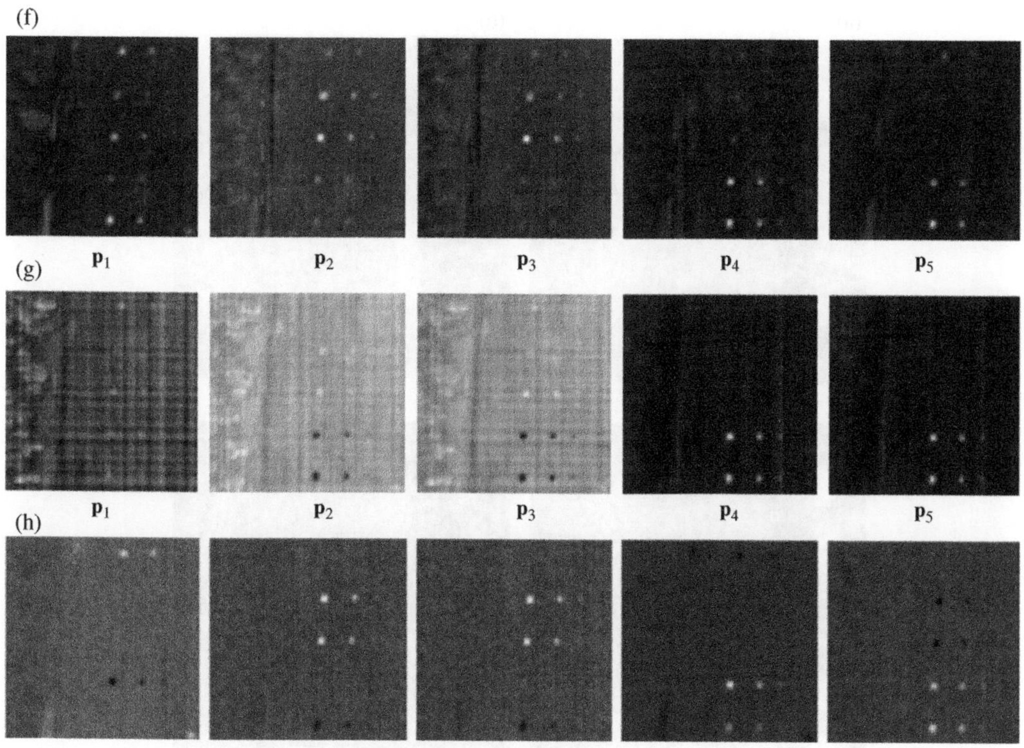


**Figure 4.2** (*Continued*).

#### 4.3.1.2 Unsupervised Mixed Pixel Classification

In the previous section, target detection was considered as an application for BS, where CEM was used to perform partially supervised target detection with only desired target signature required. In this section, unsupervised mixed pixel classification is considered as another application for BS. Since it must be performed unsupervisedly, an unsupervised algorithm which was proved to be effective, the automatic target generation process (ATGP) developed in [12, 14], was used to generate nine target pixels directly from the original image with full bands shown in Figure 4.3(a) and nine target pixels from nine-band selected images shown in Figure 4.3(b–h) to represent the required target knowledge for unsupervised classification with those target pixels corresponding to dark gray panel pixels marked by green triangles. It should be noted that only nine target pixels were generated by ATGP since a nine-band image has only nine dimensions that can be used for orthogonality projection.

An interesting finding from Figure 4.3 is that the nine target pixels generated by ATGP with full bands and nine selected bands by six different BS methods were quite different. Table 4.3 tabulates those ATGP-generated target pixels that were found to be actually R panel pixels in Figure 4.1(b). It is interesting to note that only LCMV-CBS produced four R panel pixels that were generated by ATGP based on the nine selected bands and LCMV-CBS with BCM/BDM was the only one finding an R panel pixel in the second column, $p_{22}$. The MVPCA was the worst which could only find one panel pixel, $p_{521}$.

For each BS method, the nine ATGP-generated target pixels were used to form the desired signature matrix for FCLS to classify the 15 panels. FCLS used in the experiments was preferred to the orthogonal subspace projection (OSP) in [12, 15] because the target signature knowledge provided by

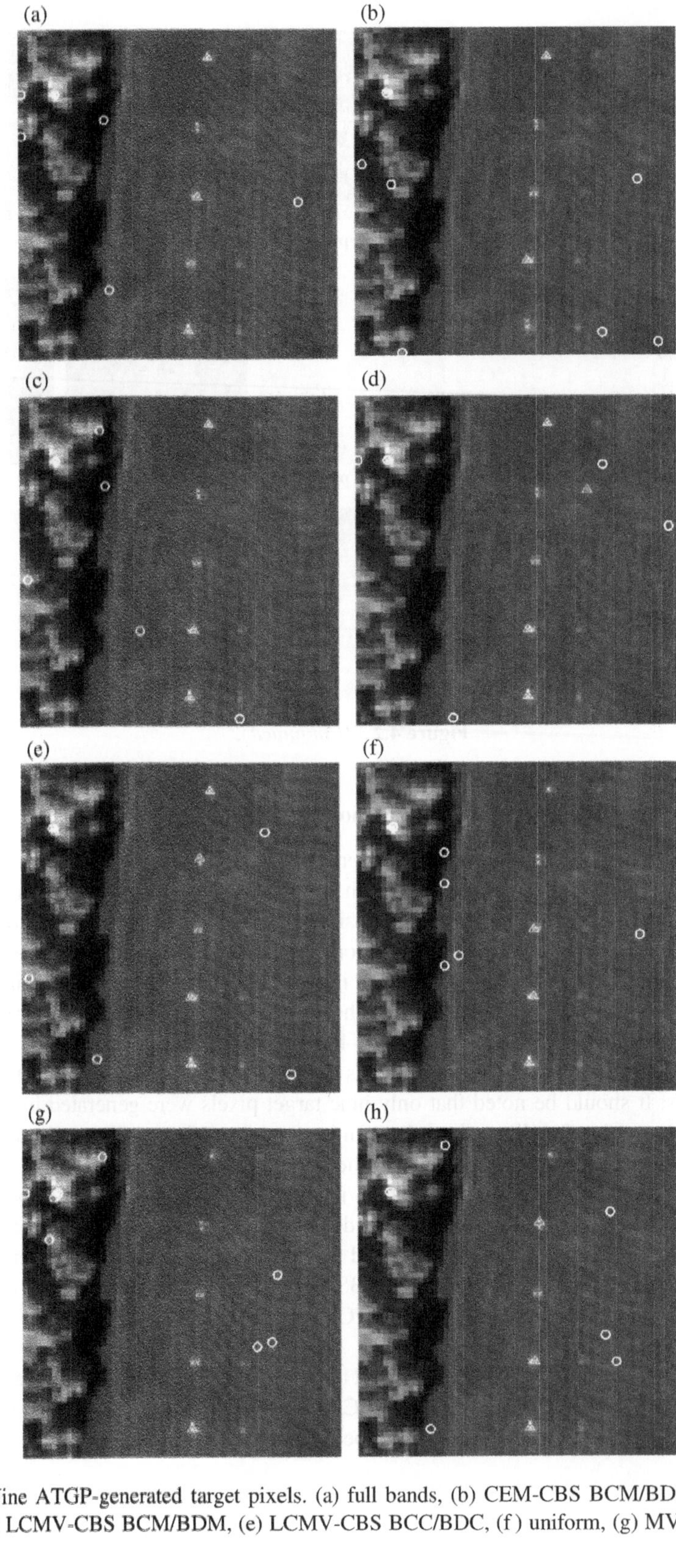


**Figure 4.3** Nine ATGP-generated target pixels. (a) full bands, (b) CEM-CBS BCM/BDM, (c) CEM-CBS BCC/BDC, (d) LCMV-CBS BCM/BDM, (e) LCMV-CBS BCC/BDC, (f) uniform, (g) MVPCA, and (h) ID.

**Table 4.3** ATGP-generated R panel pixels using full bands and six different band selection methods

| | ATGP-generated R panel pixels |
|---|---|
| Full bands | $p_{11}$, $p_{312}$, $p_{521}$ |
| CEM-CBS BCM/BDM | $p_{11}$, $p_{411}$ |
| CEM-CBS BCC/BDC | $p_{11}$, $p_{412}$, $p_{521}$ |
| LCMV-CBS BCM/BDM | $p_{11}$, $p_{22}$, $p_{411}$, $p_{521}$ |
| LCMV-CBS BCC/BDC | $p_{11}$, $p_{211}$, $p_{411}$, $p_{521}$. |
| Uniform | $p_{311}$, $p_{412}$, $p_{521}$ |
| MVPCA | $p_{521}$ |
| ID | $p_{211}$, $p_{412}$, $p_{521}$ |

single pixels was sensitive to spectral variability unless fully abundance constraints imposed on the OSP which is FCLS. Figure 4.4(a–h) shows FCLS-classification results of the 15 panels obtained by full bands and various BS methods, CEM-CBS, LCMV-CBS, uniform band selection (UBS), MVPCA, and ID where the same bands selected in Table 4.2 by each BS method were used for the FCLS method. It is very obvious that the classification performance was determined by the R panel pixels found by ATGP. According to Table 4.3, LCMV-CBS found four R panel pixels among the nine ATGP-generated target pixels compared to other methods which could only find three or fewer R panel pixels. Therefore, it is not surprising to see that LCMV-CBS yielded the best performance and it even performed better than those produced by using full bands. Additionally, FCLS-based mixed pixel classification performance in Figure 4.4 was improved significantly compared to Figure 4.3 produced by CEM which used the known panel signatures as the desired signatures. The experiments in Figure 4.4 showed otherwise where LCMV-CBS performed better than ID. So, this experiment demonstrated that the same nine selected bands used for target detection in Figure 4.3 produced significantly better performance for unsupervised mixed pixel classification shown in Figure Most interestingly, according to Figure 4.3, ID performed better than LCMV-CBS in CEM detection. The above two experiments showed that with the same number of selected bands, different applications may be very likely to result in different performance.

A remark on this example is noteworthy. The results in Figure 4.4 were only used to evaluate the performance of seven BS methods with fair comparison, in which case, the same ATGP and FCLS were implemented exactly the same way for full bands and the bands selected by each of BS methods.

### *4.3.2 AVIRIS Cuprite Data*

Another real image is a well-known Airborne Visible/InfraRed Imaging Spectrometer (AVIRIS) Cuprite image scene shown in Figure 4.5 which has been used to study endmember extraction extensively. It is available at website [16] and was collected by 224 spectral bands with 10 nm spectral resolution over the Cuprite mining site, Nevada, in 1997. The image in Figure 4.5 has size of $350 \times 350$ pixels and is well understood mineralogically where bands 1–3, 105–115, and 150–170 have been removed prior to the analysis due to water absorption and low SNR in those bands. As a result, a total of 189 bands were used for experiments. The ground truth also provides the spatial locations of the five minerals, alunite (A), buddingtonite (B), calcite (C), kaolinite (K), and muscovite (M) circled and labeled by A, B, C, K, and M, respectively, which can be used to verify endmembers

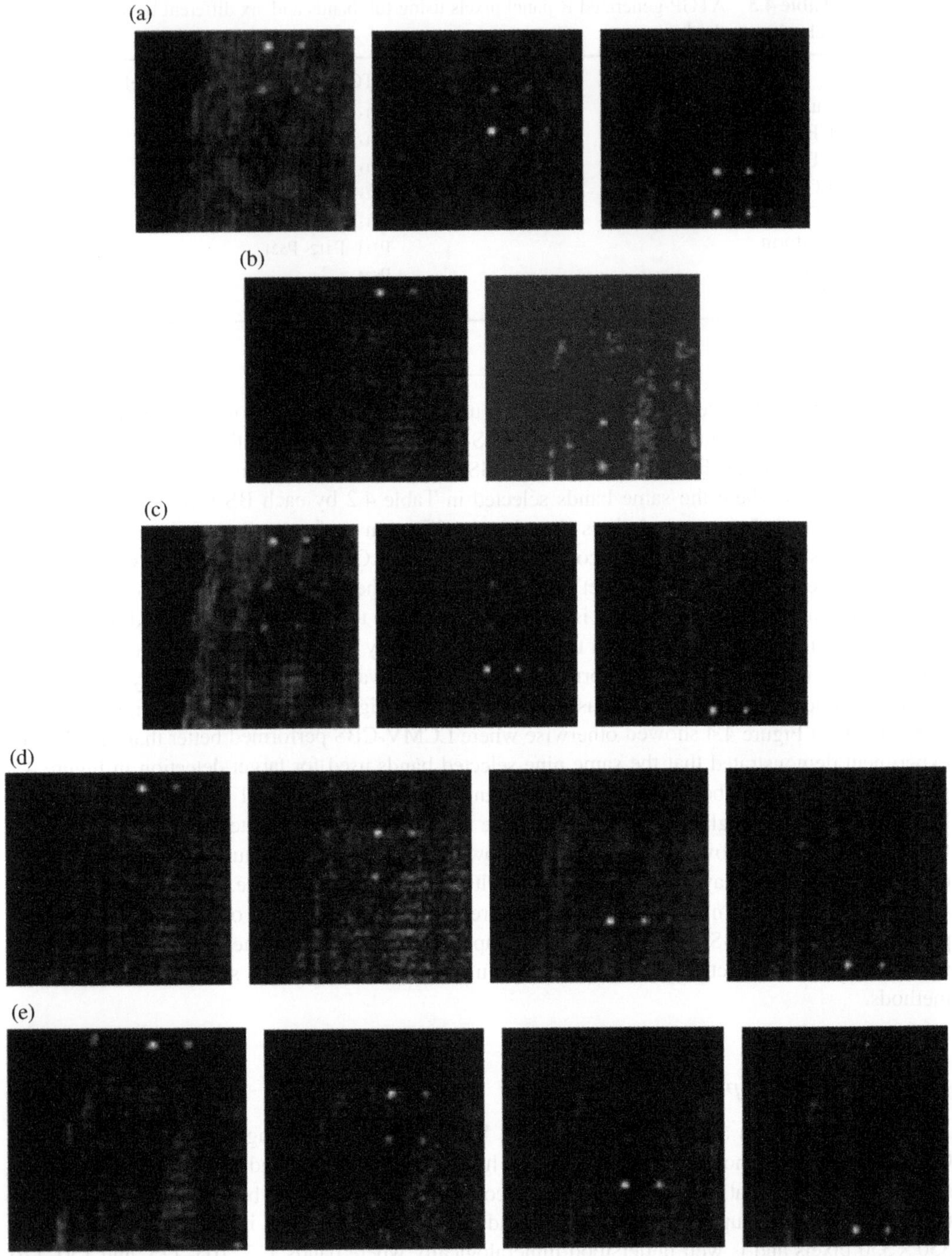


**Figure 4.4** Mixed pixel classification with endmembers resulted from ATGP obtained by various band selection methods using nine bands in Table 4.3: (a) full bands, (b) CEM-CBS BCM/BDM, (c) CEM-CBS BCC/BDC, (d) LCMV-CBS BCM/BDM, (e) LCMV-CBS BCC/BDC, (f) uniform, (g) MVPCA, and (h) ID.

(f)

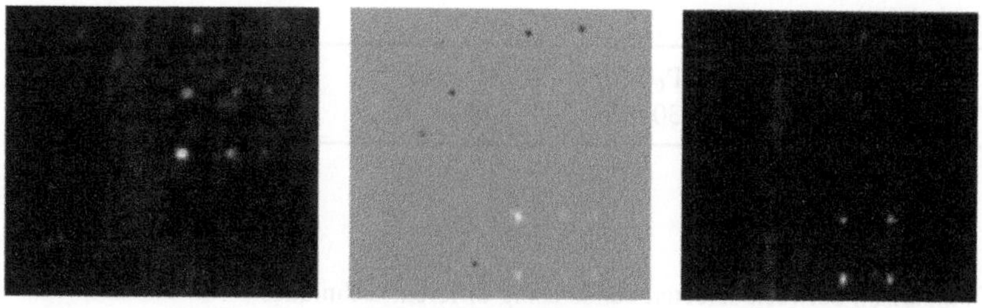

(g)

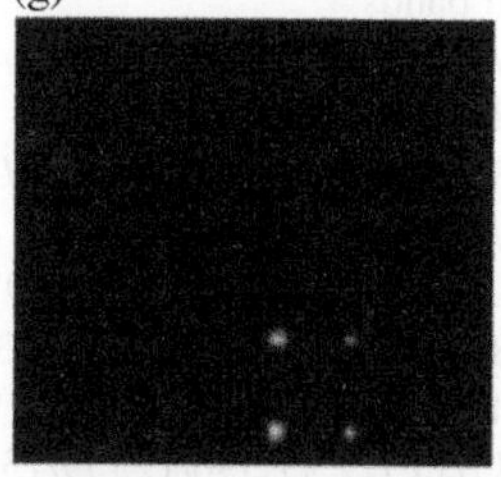

(h)

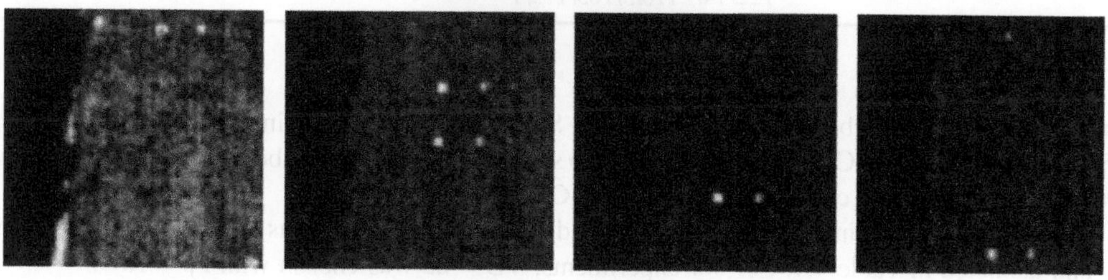

**Figure 4.4** (*Continued*).

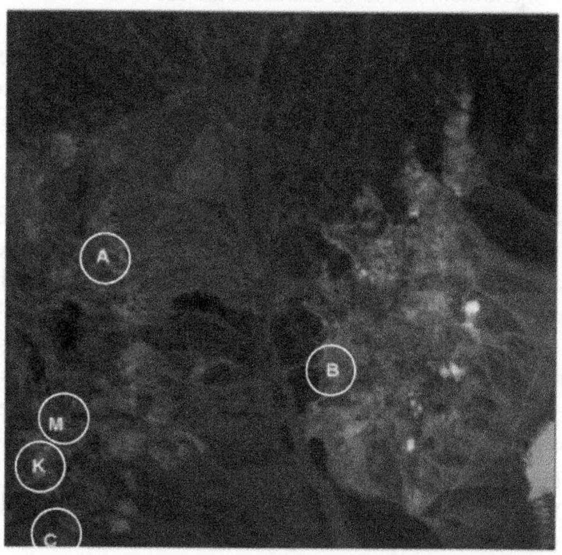


**Figure 4.5** Spatial positions of five pure pixels corresponding to minerals: alunite (A), buddingtonite (B), calcite (C), kaolinite (K), and muscovite (M).

**Table 4.4** VD estimates for the AVIRIS scene in Figure 4.5 with various false alarm probabilities

| | $P_F = 10^{-1}$ | $P_F = 10^{-2}$ | $P_F = 10^{-3}$ | $P_F = 10^{-4}$ | $P_F = 10^{-5}$ |
|---|---|---|---|---|---|
| VD | 34 | 30 | 24 | 22 | 20 |

**Table 4.5** Comparison between selected bands using different techniques

| Criteria | 22 selected bands |
|---|---|
| LCMV-CBS | 26/117/48/37/189/64/1/185/10/172 |
| BCM/BDM | 47/4/60/28/165/17/5/2/151/158/3/94 |
| LCMV-CBS | 185/37/2/3/5/64/8/9/6/7/10/165/4/11/12 |
| BCC/BDC | 14/151/13/28/15/16/153 |
| Uniform | 1/9/17/25/33/41/49/57/65/73/81/89/97/105/113/121/129/137/145/153/161/169 |
| MVPCA | 87/85/88/86/89/84/91/80/78/90/92/83/82/79/ 93/81/98/99/97/189/77/76 |
| ID | 6/2/124/117/11 /123/12/96/125 /9/13/15/10/126/ 122/14/7 /16/5/165/112/1 |

extracted by an endmember extraction algorithm. Since the size of a band image is huge with 350 × 350 pixels, only LCMV-CBS was implemented to select bands for endmember extraction to avoid the intensive computational complexity required by CEM-CBS.

VD estimated for this image scene was tabulated in Table 4.4 with various false alarm probabilities $P_F$ and tabulated in Table 4.4. For our experiments, VD = 22 was chosen with $P_F = 10^{-4}$.

It has been shown in [12, 17, 18] that the value of VD could be also used as the number of endmembers required to be generated. Therefore, in our experiments, both the number of bands to be selected and the number of endmembers to be extracted are set to 22. As noted previously, VD only provides the least number of bands for BS and is not necessary to be the exact number of bands to be selected. Table 4.5 tabulates 22 bands selected by LCMV-CBS in accordance with 22 highest band correlations along with the uniformly selected 22 bands and the MVPCA-selected 22 bands where a backslash "/" is used to separate two selected bands. It should be noted that the value of VD was only used as an estimate and not necessarily accurate which is almost impossible to know for real data.

For the purpose of endmember extraction, the fast algorithms developed in [19] to implement the widely used Winter's N-finder algorithm (N-FINDER) was used to extract 22 endmembers directly from the image scene. It should be noted that there is no particular reason to select N-FINDR to perform endmember extraction. Any other endmember extraction algorithm such as pixel purity index (PPI) [20] can be also used for experiments as well.

Figure 4.6(a) shows the 22 endmembers extracted by N-FINDR using full bands with Figure 4.6(b–f) showing the 22 endmembers extracted by the N-FINDR using 22 selected bands tabulated in Table 4.5 for comparison where the 22 N-FINDR-extracted endmember pixels are marked by open circles with pixels marked by the lower cases of "a,b,c,k,m" with triangles indicating that these pixels were found in correspondence to the five ground-truth mineral endmembers marked by the upper cases of "A,B,C,K,M" with crosses "x". It should be noted that N-FINDR-found endmembers "a,b,c,k,m" in Figure 4.6 were compared against the ground-truth endmember pixels, "A,B,C,K,M" in Figure 4.5 by the spectral angle mapper (SAM) [12] with their values tabulated in Table 4.6(a–f) where the coordinates included in the brackets for both "a,b,c,k,m" and "A,B,C,K,M" indicate the locations in the image scene.

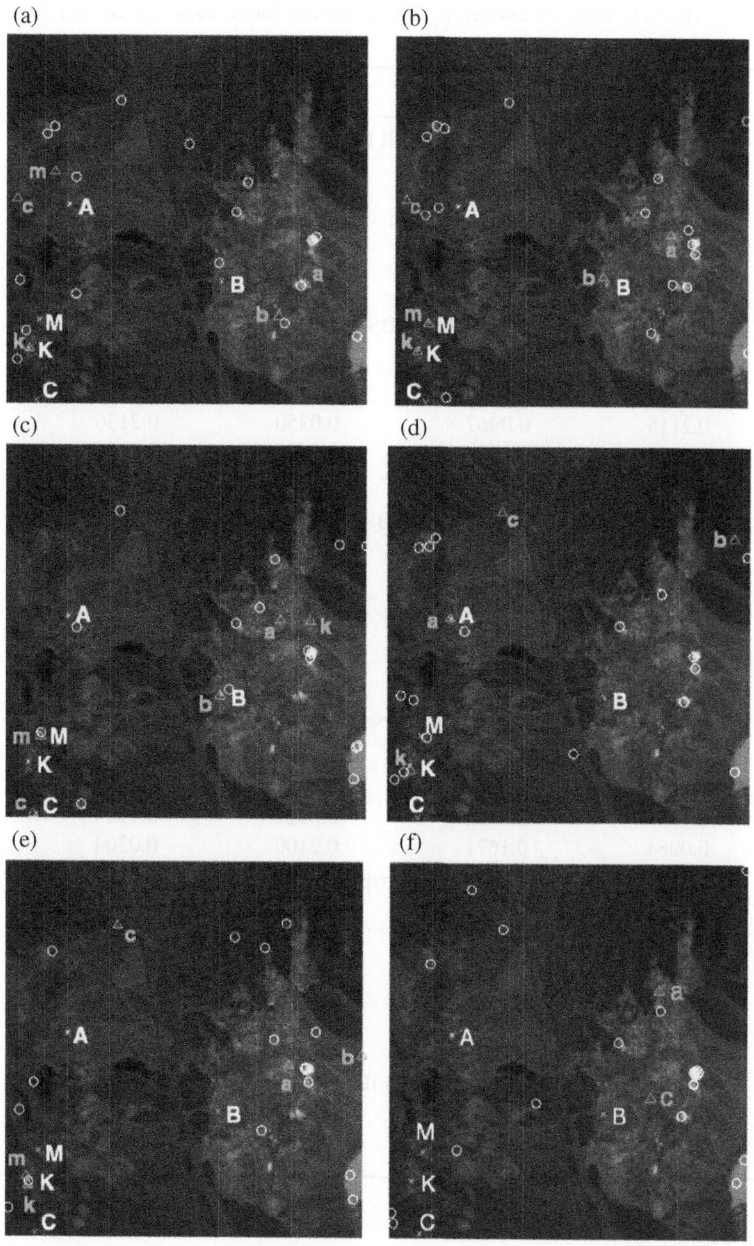


**Figure 4.6** Twenty-two endmembers extracted by N-FINDR using bands in Table 4.5. (a) full bands, (b) LCMV-CBS BCM/BDM, (c) LCMV-CBS BCC/BDC, (d) uniform, (e) MVPCA, and (f) ID.

As we can see from the above experiments, LCMV-CBS performed very well to identify all the five mineral endmembers. Interestingly, ID which produced good results for HYDICE data yielded the worst performance for Cuprite data in Figure 4.6(f) and Table 4.6(f) where N-FINDR could only find two mineral signatures, A and C. This result is the complete opposite of the results in the CEM-based target detection in Figure 4.3 where ID produced the best result.

According to Table 4.6, it is worth noting that none of ground-truth mineral pixels were found by N-FINDR even though full bands were used. Nevertheless, the spectral similarity values produced by

**Table 4.6** Spectral similarly values measured by SAM between found endmembers and the ground-truth endmembers

| | A (62,161) | B (209,234) | C (30,347) | K (22,298) | M (33,271) |
|---|---|---|---|---|---|
| | | | **(a)** Full bands | | |
| a (270,273) | 0.0575 | 0.1581 | 0.2045 | 0.0869 | 0.1381 |
| b (265,267) | 0.1608 | 0.0685 | 0.0803 | 0.1613 | 0.0884 |
| c (11,156) | 0.2115 | 0.0967 | 0.0350 | 0.2136 | 0.1170 |
| k (23,298) | 0.1098 | 0.1524 | 0.1918 | 0.0613 | 0.1079 |
| m (48,130) | 0.1786 | 0.0881 | 0.0681 | 0.1664 | 0.0643 |
| | | | **(b)** LCMV-CBS BCM/BDM | | |
| a (273,188) | 0.0567 | 0.1991 | 0.2427 | 0.0982 | 0.1742 |
| b (205,228) | 0.1247 | 0.0396 | 0.1123 | 0.1452 | 0.0866 |
| c (11,156) | 0.2115 | 0.0967 | 0.0350 | 0.2136 | 0.1167 |
| k (22,299) | 0.1076 | 0.1789 | 0.2177 | 0.0220 | 0.1305 |
| m (33,272) | 0.1519 | 0.0964 | 0.1040 | 0.1413 | 0.0264 |
| | | | **(c)** LCMV-CBS BCC/BDC | | |
| a (267,164) | 0.0493 | | 0.1813 | 0.0870 | 0.1230 |
| b (208,235) | 0.1459 | 0.0281 | 0.1110 | 0.1637 | 0.0968 |
| c (27,346) | 0.2208 | 0.1106 | 0.0466 | 0.2327 | 0.1330 |
| k (296,165) | 0.0727 | 0.1385 | 0.1844 | 0.0647 | 0.1124 |
| m (34,274) | 0.1681 | 0.0859 | 0.0898 | 0.1594 | 0.0441 |
| | | | **(d)** Uniform band selection | | |
| a (61,161) | 0.0172 | 0.1645 | 0.2115 | 0.0962 | 0.1476 |
| b (336,86) | 0.1525 | 0.0711 | 0.0792 | 0.1752 | 0.0978 |
| c (111,62) | 0.1783 | 0.1002 | 0.0725 | 0.1938 | 0.0974 |
| k (23,303) | 0.0984 | 0.1671 | 0.2100 | 0.0304 | 0.1276 |
| | | | **(e)** MVPCA | | |
| a (278,191) | 0.0582 | 0.1956 | 0.2385 | 0.1101 | 0.1765 |
| b (350,181) | 0.1344 | 0.0597 | 0.0817 | 0.1476 | 0.0692 |
| c (111,61) | 0.1873 | 0.0934 | 0.0550 | 0.196 | 0.0944 |
| k (23,303) | 0.0984 | 0.1671 | 0.2100 | 0.0304 | 0.1276 |
| m (21,295) | 0.1282 | 0.1036 | 0.1355 | 0.0956 | 0.0656 |
| | | | **(f)** ID | | |
| a (263,119) | 0.0348 | 0.1564 | 0.2007 | 0.0953 | 0.1392 |
| c (255,219) | 0.1657 | 0.0729 | 0.0567 | 0.1823 | 0.0933 |

SAM in Table 4.6 should provide information about closeness between N-FINDR-found mineral pixel and the ground-truth mineral pixels. Using the values in Table 4.6 as a measure of effectiveness, LCMV-CBS with BCM/BDM in Table 4.6(b) was the best among all BS methods and LCMV-CBS was also the best on the average and even better than the one produced by the full bands.

Three remarks are summarized as follows:

1. On some occasions, an appropriate BS may be more effective than one with full bands as demonstrated by the endmember extraction experiments. This could be due to the fact that spectral information from real data may be distorted by unknown signal sources and too much such contaminated information may eventually hinder data analysis. As a consequence, a judicious selection of appropriate bands can avoid such dilemma and further improve performance.

2. Moreover, the applications presented in this section only serve as an illustrative purpose to demonstrate the utility of CBS. The performance of CBS is also determined by the algorithms used for applications. Therefore, when a comparative analysis was conducted, the same algorithm was applied to various BS methods to make sure that all conditions were held to the same situation so that the only difference is selected bands used for analysis.
3. It should be noted that in order to make fair comparison among different BS criteria, the same algorithm must be implemented in conjunction with VD which determines the same number of bands to be selected for all the criteria. Only the same algorithm coupled with the same VD-determined number of bands can ensure that the difference in performance only comes from a specific BS criterion. In this section, we demonstrated the effectiveness of BS criteria with the number of bands determined by VD in three different applications where three algorithms, CEM for target detection, UFCLS for unsupervised mixed pixel classification, and N-FINDR for endmember extraction, were used for performance evaluation. According to our conducted experiments, CBS outperformed other BS criteria in all the case.
4. Our experimental results showed that CBS performed robustly and uniformly well compared to the other BS methods. Although only three applications, target detection, mixed pixel classification, and endmember extraction, were considered in this paper, some other applications such as mixed pixel quantification and data compression were also investigated for CBS. Similar conclusions can be also drawn. So, in order to avoid replication, their results are not included.

## 4.4 Target-Constrained BS

Unlike BCBS which constrains bands to be selected in Section 4.2, this section develops a rather new specific application-based supervised BS approach, to be called target-constrained band selection (TCBS) for multiple-target detection where only the knowledge of targets of interest is assumed to be known a priori and no other prior knowledge such as background (BKG) information is required. It was originally developed as constrained-target BS (CTBS) in [9] and referred to as TCBS in this chapter.

Recall that $\{\mathbf{b}_l\}_{l=1}^{L}$ is a set of band images representing a HSI cube where $\mathbf{b}_l$ is the $l$th spectral band represented by a column vector, $\mathbf{b}_l = (b_{l1}, b_{l2}, \ldots, b_{lN})^T$, and $\{b_{li}\}_{i=1}^{N}$ is the set of all $N$ pixels in the $l$th band image, $\mathbf{b}_l$.

### 4.4.1 *Target-Constrained Band Prioritization*

By taking advantage of CEM error, $\rho_l$ derived from (4.5), we can define a new measure as a criterion for TCBS given by

$$\mathrm{V}(\Omega_{\mathrm{BS}}) = \left(\mathbf{d}_{\Omega_{\mathrm{BS}}}^T \mathbf{R}_{\Omega_{\mathrm{BS}}}^{-1} \mathbf{d}_{\Omega_{\mathrm{BS}}}\right)^{-1} \tag{4.19}$$

which is the minimal variance specified by (4.2) using only those bands in $\mathbf{\Omega}_{\mathrm{BS}}$. Most importantly, we can prove in the following theorem that

$$\{\mathrm{V}(\Omega_j)\}_{j=1}^{L} \tag{4.20}$$

is in fact a monotonically decreasing sequence where $\Omega_j = \left\{\mathbf{b}_{l_1}, \mathbf{b}_{l_2}, \ldots, \mathbf{b}_{l_j}\right\}$ is any $j$-band subset containing $j$ bands, $\mathbf{b}_{l_1}, \mathbf{b}_{l_2}, \ldots, \mathbf{b}_{l_j}$.

Despite that $V(\boldsymbol{\Omega}_{BS})$ in (4.19) and $\left(\mathbf{b}_l^T\tilde{\mathbf{Q}}^{-1}\mathbf{b}_l\right)^{-1}$ in (4.7) have similar forms with $\mathbf{d}_{\Omega_{BS}}$ and $\mathbf{R}_{\Omega_{BS}}$ in (4.19) corresponding to $\mathbf{b}_l$ and $\tilde{\mathbf{Q}}$ in (4.6), respectively, they are quite different in design rationale. In particular, the former is specified by target signature $\mathbf{d}$ and sample correlation matrix as opposed to the latter specified by a desired band $\mathbf{b}_l$ and the band image correlation matrix $\tilde{\mathbf{Q}}$. The following theorem (Theorem 4.1) is crucial to the development of TCBS. There is no such corresponding theorem as a counterpart in BCBS.

**Theorem 4.1** Let $\Omega = \{\mathbf{b}_l\}_{l=1}^{L}$ be a full band set. For any band subset $\Omega_k = \left\{\mathbf{b}_{l_j}\right\}_{j=1}^{k} \subset \Omega$ and an arbitrary band image vector $\mathbf{b}_{k+1} \notin \Omega_k$.

$$\frac{1}{\mathbf{d}^T_{\Omega_k \cup \{\mathbf{b}_{k+1}\}}\mathbf{R}^{-1}_{\Omega_k \cup \{\mathbf{b}_{k+1}\}}\mathbf{d}} < \frac{1}{\mathbf{d}^{-1}_{\Omega_k}\mathbf{R}^{-1}_{\Omega_k}\mathbf{d}_{\Omega_k}} \tag{4.21}$$

**Proof.**
For simplicity of notations, we assume that $\Omega_k = \{\mathbf{b}_1, \mathbf{b}_2, \ldots, \mathbf{b}_k\}$ without loss of generality. Also, let the total number of data sample vectors be $\left\{\mathbf{r}_i^{(k)}\right\}_{i=1}^{N}$ using $k$ bands with $\mathbf{r}_i^{(k)} = \left(r_{i1}^{(k)}, r_{i2}^{(k)}, \ldots, r_{ik}^{(k)}\right)^T$. Let the data matrix be denoted by $\mathbf{X}_k = \left[\mathbf{r}_1^{(k)}\mathbf{r}_2^{(k)}\cdots\mathbf{r}_N^{(k)}\right] = \begin{bmatrix} r_{11} & r_{21} & \cdots & r_{N1} \\ r_{12} & r_{22} & \cdots & r_{N2} \\ \vdots & \vdots & \ddots & \vdots \\ r_{1k} & r_{2k} & \cdots & r_{Nk} \end{bmatrix}$ using $k$ bands. Then, the data matrix using ($k$+1) bands is given by

$$\mathbf{X}_{k+1} = \begin{bmatrix} r_{11} & r_{21} & \cdots & r_{(N-1)1} & r_{N1} \\ r_{12} & r_{22} & \cdots & r_{(N-1)2} & r_{N2} \\ \vdots & \vdots & \ddots & \vdots & \vdots \\ r_{1k} & r_{2k} & \cdots & r_{(N-1)k} & r_{Nk} \\ r_{1(k+1)} & r_{2(k+1)} & \cdots & r_{(N-1)(k+1)} & r_{N(k+1)} \end{bmatrix} = \begin{bmatrix} \mathbf{X}_k \\ \mathbf{r}^T(k+1) \end{bmatrix} \tag{4.22}$$

where $\mathbf{r}(k+1) = (r_{1(k+1)}, r_{2(k+1)}, \ldots, r_{N(k+1)})^T$. Let $\mathbf{R}_{\Omega_k} = \frac{1}{N}\sum_{i=1}^{N}\left(\mathbf{r}_i^{(k)}\left(\mathbf{r}_i^{(k)}\right)^T\right) = \frac{1}{N}\mathbf{X}_k\mathbf{X}_k^T$. We have

$$\mathbf{R}_{\Omega_k \cup \{\mathbf{b}_{k+1}\}} = \frac{1}{N}\begin{bmatrix} \mathbf{X}_k \\ \mathbf{r}^T(k+1) \end{bmatrix}\left[\mathbf{X}_k^T \ \mathbf{r}(k+1)\right] = \frac{1}{N}\begin{bmatrix} \mathbf{X}_k\mathbf{X}_k^T & \mathbf{X}_k\mathbf{r}(k+1) \\ \mathbf{r}^T(k+1)\mathbf{X}_k^T & \mathbf{r}^T(k+1)\mathbf{r}(k+1) \end{bmatrix}$$

$$= \begin{bmatrix} \mathbf{R}_{\Omega_k} & (1/N)\mathbf{X}_k\mathbf{r}(k+1) \\ (1/N)\mathbf{r}^T(k+1)\mathbf{X}_k^T & (1/N)\mathbf{r}^T(k+1)\mathbf{r}(k+1) \end{bmatrix}. \tag{4.23}$$

Using the following matrix identity in [1] and [17, appendix A]

$$\begin{bmatrix} \mathbf{U}^T\mathbf{U} & \mathbf{U}^T\mathbf{d} \\ \mathbf{d}^T\mathbf{U} & \mathbf{d}^T\mathbf{d} \end{bmatrix}^{-1} = \begin{bmatrix} \left(\mathbf{U}^T\mathbf{U}\right)^{-1} + \beta\mathbf{U}^{\#}\mathbf{d}\mathbf{d}^T\left(\mathbf{U}^{\#}\right)^T & -\beta\mathbf{U}^{\#}\mathbf{d} \\ -\beta\mathbf{d}^T\left(\mathbf{U}^{\#}\right)^T & \beta \end{bmatrix} \quad (4.24)$$

with $\mathbf{U} = \mathbf{X}_k^T$ and $\mathbf{d} = \mathbf{r}(k+1)$, we can derive

$$\begin{aligned} \mathbf{R}^{-1}_{\Omega_k\cup\{\mathbf{b}_{k+1}\}} &= \frac{1}{N}\begin{bmatrix} N\mathbf{R}_{\Omega_k} & \mathbf{X}_k\mathbf{r}(k+1) \\ \mathbf{r}^T(k+1)\mathbf{X}_k^T & \mathbf{r}^T(k+1)\mathbf{r}(k+1) \end{bmatrix}^{-1} \\ &= \frac{1}{N}\begin{bmatrix} N^{-1}\mathbf{R}^{-1}_{\Omega_k} + \beta\left(\mathbf{X}_k^T\right)^{\#}\mathbf{r}(k+1)\mathbf{r}^T(k+1)\mathbf{X}_k^{\#} & \left(\mathbf{X}_k^T\right)^{\#}\mathbf{r}(k+1) \\ \mathbf{r}^T(k+1)\mathbf{X}_k^{\#} & \beta \end{bmatrix} \end{aligned} \quad (4.25)$$

where $\left(\mathbf{X}_k^T\right)^{\#} = \left(\mathbf{X}_k\mathbf{X}_k^T\right)^{-1}\mathbf{X}_k^T$ and $\beta = \left(\mathbf{r}^T(k+1)\left(\mathbf{I}-\mathbf{X}_k^T\left(\mathbf{X}_k^T\right)^{\#}\right)\mathbf{r}(k+1)\right)^{-1} = \left(\mathbf{r}^T(k+1)P^{\perp}_{\mathbf{X}_k^T}\mathbf{r}(k+1)\right)^{-1}$ due to the fact that $P^{\perp}_{\mathbf{X}_k^T} = \mathbf{I}-\mathbf{X}_k^T\left(\mathbf{X}_k^T\right)^{\#}$ is idempotent, i.e. $\left(P^{\perp}_{\mathbf{X}_k^T}\right)^2 = P^{\perp}_{\mathbf{X}_k^T}$.

Using (4.25) and letting $\mathbf{d}^T_{\Omega_k\cup\{\mathbf{b}_{k+1}\}} = \begin{bmatrix} \mathbf{d}_k \\ d_{k+1} \end{bmatrix}$ and $\mathbf{d}_k = (d_1, d_2, \ldots, d_k)^T$, we can derive

$$\begin{aligned} & N^{-1}d^T_{\Omega_k\cup\{b_{k+1}\}}R^{-1}_{\Omega_k\cup\{b_{k+1}\}}d_{\Omega_k\cup\{b_{k+1}\}} \\ &= \begin{bmatrix} d_k^T & d_{k+1} \end{bmatrix}\begin{bmatrix} N^{-1}R^{-1}_{\Omega_k} + \beta\left(X_k^T\right)^{\#}r(k+1)r^T(k+1)X_k^{\#} & -\beta\left(X_k^T\right)^{\#}r(k+1) \\ -\beta r^T(k+1)X_k^{\#} & \beta \end{bmatrix}\begin{bmatrix} d_k \\ d_{k+1} \end{bmatrix} \\ &= N^{-1}d_k^TR^{-1}_{\Omega_k}d_k + \beta d_k^T\left(X_k^T\right)^{\#}r(k+1)r^T(k+1)X_k^{\#}d_k - 2\beta d_{k+1}r^T(k+1)X_k^{\#}d_k + d^2_{k+1}\beta \\ &= N^{-1}d_k^TR^{-1}_{\Omega_k}d_k + \beta\left[d_k^T\left(X_k^T\right)^{\#}r(k+1) - d_{k+1}\right]^2 > N^{-1}d_k^TR^{-1}_{\Omega_k}d_k \end{aligned} \quad (4.26)$$

Since $\left[\mathbf{d}_k^T\left(\mathbf{X}_k^T\right)^{\#}\mathbf{r}(k+1) - d_{k+1}\right]^2 > 0$ and $\beta > 0$, the second term, $\beta\left[\mathbf{d}_k^T\left(\mathbf{X}_k^T\right)^{\#}\mathbf{r}(k+1) - d_{k+1}\right]^2 > 0$. Therefore, (4.21) is proved.

■

From (4.21), we can also show that

$$\frac{1}{\mathbf{d}^T_{\Omega_1}\mathbf{R}^{-1}_{\Omega_1}\mathbf{d}_{\Omega_1}} < \frac{1}{\mathbf{d}^T_{\Omega_2}\mathbf{R}^{-1}_{\Omega_2}\mathbf{d}_{\Omega_2}} \text{ if } \Omega_2 \subset \Omega_1 \text{ and } \Omega_2 \neq \Omega_1 \quad (4.27)$$

Let $\Omega_1 = \Omega_2 \cup \{\mathbf{b}_j\}_{j=1}^{p}$. Then, (4.27) can be proved by implementing (4.21) $p$ times repeatedly as follows:

$$\frac{1}{\mathbf{d}_{\Omega_1}^T \mathbf{R}_{\Omega_1}^{-1} \mathbf{d}_{\Omega_1}} < \frac{1}{\mathbf{d}_{\Omega_2 \cup \{\mathbf{b}_1,\ldots,\mathbf{b}_{p-1}\}}^T \mathbf{R}_{\Omega_2 \cup \{\mathbf{b}_1,\ldots,\mathbf{b}_{p-1}\}}^{-1} \mathbf{d}_{\Omega_2 \cup \{\mathbf{b}_1,\ldots,\mathbf{b}_{p-1}\}}} < \cdots < \frac{1}{\mathbf{d}_{\Omega_2 \cup \{\mathbf{b}_1\}}^T \mathbf{R}_{\Omega_2 \cup \{\mathbf{b}_1\}}^{-1} \mathbf{d}_{\Omega_2 \cup \{\mathbf{b}_1\}}} < \frac{1}{\mathbf{d}_{\Omega_2}^T \mathbf{R}_{\Omega_2}^{-1} \mathbf{d}_{\Omega_2}}. \tag{4.28}$$

Furthermore, (4.21) also allows us to derive the following inequalities:

$$\frac{1}{\mathbf{d}_{\Omega_1 \cup \Omega_2}^T \mathbf{R}_{\Omega_1 \cup \Omega_2}^{-1} \mathbf{d}_{\Omega_1 \cup \Omega_2}} < \frac{1}{\mathbf{d}_{\Omega_1}^T \mathbf{R}_{\Omega_1}^{-1} \mathbf{d}_{\Omega_1}} \tag{4.29}$$

and

$$\frac{1}{\mathbf{d}_{\Omega_1 \cup \Omega_2}^T \mathbf{R}_{\Omega_1 \cup \Omega_2}^{-1} \mathbf{d}_{\Omega_1 \cup \Omega_2}} < \frac{1}{\mathbf{d}_{\Omega_2}^T \mathbf{R}_{\Omega_2}^{-1} \mathbf{d}_{\Omega_2}} \tag{4.30}$$

Of particular interest is the following inequality derived in [22]:

$$\frac{1}{\mathbf{d}_{\Omega}^T \mathbf{R}_{\Omega}^{-1} \mathbf{d}_{\Omega}} < \frac{1}{\mathbf{d}_{\tilde{\Omega}}^T \mathbf{R}_{\tilde{\Omega}}^{-1} \mathbf{d}_{\tilde{\Omega}}} \tag{4.31}$$

which can be also shown to be a special case of (4.27) with $\boldsymbol{\Omega}_1$ replaced by the full band set $\boldsymbol{\Omega}$ and $\boldsymbol{\Omega}_2$ replaced by $\tilde{\Omega}$ which is a proper subset of $\boldsymbol{\Omega}$.

Finally, by virtue of (4.21), it shows that $\{\mathrm{V}(\Omega_j)\}_{j=1}^{L}$ is indeed a monotonically decreasing sequence and we can further develop two types of BP criteria, single band minimum variance BP (Min*V*-BP) and leave-one-out maximum variance BP (Max*V*-BP).

#### 4.4.1.1 Single Band Minimum Variance Band Prioritization by TCBS

According to (4.5), $(\mathbf{d}^T\mathbf{R}^{-1}\mathbf{d})^{-1}$ is the variance resulting from CEM which can be considered as the interfering effect caused by signal sources including BKG signal sources which do not pass through CEM filter (4.3). This quality is obtained by full band set $\boldsymbol{\Omega}$, i.e. $\mathrm{V}(\Omega) = \left(\mathbf{d}_{\Omega}^T\mathbf{R}_{\Omega}^{-1}\mathbf{d}_{\Omega}\right)^{-1} = \left(\mathbf{d}^T\mathbf{R}^{-1}\mathbf{d}\right)^{-1}$. On the other hand, from (4.19), $\mathrm{V}(\Omega_{\mathrm{BS}}) = \left(\mathbf{d}_{\mathbf{b}_l}^T\mathbf{R}_{\mathbf{b}_l}^{-1}\mathbf{d}_{\mathbf{b}_l}\right)^{-1}$ is calculated by using a single band $\mathbf{b}_l$ with $\Omega_{\mathrm{BS}} = \{\mathbf{b}_l\}$. So, for each single band image $\mathbf{b}_l$, we calculate the variance

$$V(\mathbf{b}_l) = \left(\mathbf{d}_{\mathbf{b}_l}^T\mathbf{R}_{\mathbf{b}_l}^{-1}\mathbf{d}_{\mathbf{b}_l}\right)^{-1} \tag{4.32}$$

which can be used as a criterion to measure the variance caused by data sample vectors not specified by $\mathbf{d}$ using only one single band, $\mathbf{b}_l$. Consequently, the value of $V(\mathbf{b}_l)$ can be further used as a priority score of $\mathbf{b}_l$. With this interpretation, a single band-based band prioritization (SB-BP) criterion can be derived to rank bands according to increasing values of $V(\mathbf{b}_l)$ in (4.32). That is, the smaller the $V(\mathbf{b}_l)$ is, the higher priority the $l$th band, $\mathbf{b}_l$ is. Since it ranks bands by starting the minimal variance in (4.32) as its first selected band, it is called MinV-BP. A detailed description of implementing MinV-BP is provided in the following algorithm:

***MinV-BP***

1. Initial condition:

   Let $n_{\text{BS}}$ be the number of bands needed to be selected, which can be determined by a criterion such as virtual dimensionality (VD) in [1, chapter 17].
   Find the first band by

$$\mathbf{b}_{l_1} = \arg\{\min_{\mathbf{b}_l \in \Omega} \mathrm{V}(\mathbf{b}_l)\} \tag{4.33}$$

2. Rank all bands for $1 \leq l \leq L$ according to $\{\mathrm{V}(\mathbf{b}_l)\}_{l=1}^{L}$ by (4.19)

$$\mathbf{b}_{l_j} \succ \mathbf{b}_{l_k} \Leftrightarrow \mathrm{V}(\mathbf{b}_{l_j}) < \mathrm{V}(\mathbf{b}_{l_k}) \tag{4.34}$$

   i.e. $\mathbf{b}_{l_1} \succ \mathbf{b}_{l_2} \succ \cdots \succ \mathbf{b}_{l_L}$ where the notation "$\succ$" is used to indicate "superior to".

3. The final set of selected bands is given by the first $n_{\text{BS}}$ bands by $\{\mathbf{b}_{l_1}, \mathbf{b}_{l_2}, \cdots, \mathbf{b}_{l_{n_{BS}}}\}$.

It is worth noting that $\mathbf{R}_{\mathbf{b}_l}$ is a scalar and so is $\mathbf{R}_{\mathbf{b}_l}^{-1}$. Interestingly, according to [23], $\mathbf{R}_{\mathbf{b}_l}$ should remove all data sample vectors specified by $\mathbf{d}_{\mathbf{b}_l}$ because the functionality of $\mathbf{R}_{\mathbf{b}_l}^{-1}$ is to suppress the interfering effect caused by all data sample vectors other than those specified by $\mathbf{d}_{\mathbf{b}_l}$. So, let $\boldsymbol{\Delta}$ be such data set excluding from data sample vectors specified by $\mathbf{d}_{\mathbf{b}_l}$. We can define $\tilde{\mathbf{R}}_{\mathbf{b}_l} = \sum_{r_{\mathbf{b}_l} \in \Delta} r_{\mathbf{b}_l}^2$ and $(s_l^{\text{BKG}})^2$ as the BKG energy given by $(s_l^{\text{BKG}})^2 = \tilde{\mathbf{R}}_{\mathbf{b}_l}$. In this case, $\mathbf{d}_{\mathbf{b}_l}^T \tilde{\mathbf{R}}_{\mathbf{b}_l}^{-1} \mathbf{d}_{\mathbf{b}_l}$ becomes the signal-to-background (SBR) ratio, i.e. $\text{SBR}_l = \mathbf{d}_{\mathbf{b}_l}^T \tilde{\mathbf{R}}_{\mathbf{b}_l}^{-1} \mathbf{d}_{\mathbf{b}_l} = \frac{d_{\mathbf{b}_l}^2}{(s_l^{\text{BKG}})^2}$. Consequently, the larger the $\text{SBR}_l$ is, the smaller the $\left(\mathbf{d}_{\mathbf{b}_l}^T \tilde{\mathbf{R}}_{\mathbf{b}_l}^{-1} \mathbf{d}_{\mathbf{b}_l}\right)^{-1}$ is. Since the size of the data sample vectors specified by $\mathbf{d}_{\mathbf{b}_l}$ is generally small, $\tilde{\mathbf{R}}_{\mathbf{b}_l} \approx \mathbf{R}_{\mathbf{b}_l}$. In this case, $V(\mathbf{b}_l) = \left(\mathbf{d}_{\mathbf{b}_l}^T \mathbf{R}_{\mathbf{b}_l}^{-1} \mathbf{d}_{\mathbf{b}_l}\right)^{-1}$ in (4.32) can be also represented by $\text{Min}V(\mathbf{b}_l) = \left(\mathbf{d}_{\mathbf{b}_l}^T \tilde{\mathbf{R}}_{\mathbf{b}_l}^{-1} \mathbf{d}_{\mathbf{b}_l}\right)^{-1} = (\text{SBR}_l)^{-1}$ which is the reciprocal of $\text{SBR}_l$.

#### 4.4.1.2 Leave-One-Out Maximum Variance Band Prioritization by TCBS

As an alternative to Min$V$-BP, we can also replace the full band set $\boldsymbol{\Omega}$ in (4.19) with $\Omega_{\text{BS}} = \Omega - \mathbf{b}_l$ to derive

$$V(\Omega - \mathbf{b}_l) = \left(\mathbf{d}_{\Omega-\{\mathbf{b}_l\}}^T \mathbf{R}_{\Omega-\{\mathbf{b}_l\}}^{-1} \mathbf{d}_{\Omega-\{\mathbf{b}_l\}}\right)^{-1} \tag{4.35}$$

Like (4.19) which is $(\text{SBR}_l)^{-1}$ resulting from a single band $\mathbf{b}_l$, (4.35) can be also considered as $\left(\text{SBR}_{\Omega-\{\mathbf{b}_l\}}\right)^{-1}$ resulting from ($L$–1) bands, $\Omega - \{\mathbf{b}_l\}$. Its idea is to first remove any single band,

$\mathbf{b}_l$, from the full band set $\mathbf{\Omega}$ and then to calculate variance by (4.35). The band which yields the maximal variance should be most significant one since its removal from $\mathbf{\Omega}$ causes the maximal variance. With this in mind, $V(\Omega - \mathbf{b}_l)$ in (4.35) can be used as a priority score to rank $\mathbf{b}_l$ for BP. That is, the higher the $V(\Omega - \mathbf{b}_l)$ is, the more significant the band $\mathbf{b}_l$ is. As a result, all bands can be ranked by decreasing values of $V(\Omega - \mathbf{b}_l)$ as opposed to MinV-BP which ranks all bands by increasing values of $V(\mathbf{b}_l)$ in (4.32). Because of that, it is called maximum variance band prioritization (Max$V$-BP) criterion. A detailed step-by-step implementation of Max$V$-BP is given as follows:

***MaxV-BP***

1. Initial condition:

   Let $n_{\mathrm{BS}}$ be the number of bands needed to be selected, which can be determined by a criterion such as virtual VD.

$$\mathbf{b}_{l_1} = \arg\{\max_{\mathbf{b}_l \in \Omega} V(\Omega - \mathbf{b}_l)\} \tag{4.36}$$

2. Rank all bands $1 \le l \le L$ according to $\{V(\Omega - \mathbf{b}_l)\}_{l=1}^{L}$ by (4.35)

$$\mathbf{b}_{l_j} \succ \mathbf{b}_{l_k} \Leftrightarrow V(\Omega - \mathbf{b}_{l_k}) < V(\Omega - \mathbf{b}_{l_j}) \tag{4.37}$$

   i.e. $\mathbf{b}_{l_1} \succ \mathbf{b}_{l_2} \succ \cdots \succ \mathbf{b}_{l_L}$ where the notation "$\succ$" is used to indicate "superior to".

3. The final set of selected bands is given by the first $n_{\mathrm{BS}}$ bands by $\left\{\mathbf{b}_{l_1}, \mathbf{b}_{l_2}, \ldots, \mathbf{b}_{l_{n_{\mathrm{BS}}}}\right\}$.

### 4.4.2 *Constrained-Target Band Selection*

In this section, the two TCBS-based BP criteria designed in Section 4.41 are further used as search strategies to derive their corresponding BS algorithms, respectively, described in the following two subsections.

#### 4.4.2.1 Sequential Feed-Forward TCBS

For each single band $\mathbf{b}_l$, we can replace the full band set $\mathbf{\Omega}$ used in CEM-produced variance $(\mathbf{d}^T\mathbf{R}^{-1}\mathbf{d})^{-1} = (\mathbf{d}_\Omega^T\mathbf{R}_\Omega^{-1}\mathbf{d}_\Omega)^{-1}$ with the single band $\mathbf{b}_l$ to become $\left(\mathbf{d}_{\mathbf{b}_l}^T\mathbf{R}_{\mathbf{b}_l}^{-1}\mathbf{d}_{\mathbf{b}_l}\right)^{-1}$ which can be used as the priority score assigned to the $l$th band, $\mathbf{b}_l$, to rank the priority of $\mathbf{b}_l$. As a result, the band, which yields the minimal variance

$$\mathbf{b}_{l_1}^* = \arg\{\min_{\mathbf{b}_l \in \Omega} V(\mathbf{b}_l)\} = \arg\left\{\min_{\mathbf{b}_l \in \Omega}\left(\mathbf{d}_{\mathbf{b}_l}^T\mathbf{R}_{\mathbf{b}_l}^{-1}\mathbf{d}_{\mathbf{b}_l}\right)^{-1}\right\} \tag{4.38}$$

will be selected as the first band, denoted by $\mathbf{b}_{l_1}^*$. Then, the second band which yields the minimal variance again

$$\mathbf{b}_{l_2}^* = \arg\left\{\min_{\mathbf{b}_l \in \Omega - \{\mathbf{b}_{l_1}\}} V(\mathbf{b}_l)\right\} = \arg\left\{\min_{\mathbf{b}_l \in \Omega - \{\mathbf{b}_{l_1}\}}\left(\mathbf{d}_{\mathbf{b}_l}^T\mathbf{R}_{\mathbf{b}_l}^{-1}\mathbf{d}_{\mathbf{b}_l}\right)^{-1}\right\} \tag{4.39}$$

is then selected as the second band, $\mathbf{b}_{l_2}^*$. The same process is repeated over and over again by continuously adding one new band at a time via repeatedly implementing (4.39). The resulting algorithm is called sequential forward TCBS (SF-TCBS).

Now, if we consider a band as a feature vector, sequential feed-forward floating search (SFFS) developed in [4] can be used to develop an SFFS-based BS to augment bands to be selected one at a time sequentially by (4.19) instead of ranking all bands by their priority scores $\{V(\mathbf{b}_l)\}_{l=1}^{L}$ as MinV-BP does. The resulting BS is referred to as sequential feed forward TCBS (SF-TCBS).

***SF-TCBS***

1. Initial condition:

   Let $n_{\mathrm{BS}}$ be the number of bands needed to be selected, which can be determined by a criterion such as virtual VD.

   Find

$$\mathbf{b}_{l_1} = \arg\left\{ \min_{\mathbf{b}_l \in \Omega} \left( \mathbf{d}_{\mathbf{b}_l}^T \mathbf{R}_{\mathbf{b}_l}^{-1} \mathbf{d}_{\mathbf{b}_l} \right)^{-1} \right\} \tag{4.40}$$

   $\Omega_1 = \{\mathbf{b}_{l_1}\}$.

2. Band augmentation

$$\mathbf{b}_{l_j} = \arg\left\{ \min_{\mathbf{b}_l \in \Omega_{j-1}^c} \left( \mathbf{d}_{\Omega_{j-1} \cup \{\mathbf{b}_l\}}^T \mathbf{R}_{\Omega_{j-1} \cup \{\mathbf{b}_l\}}^{-1} \mathbf{d}_{\Omega_{j-1} \cup \{\mathbf{b}_l\}} \right)^{-1} \right\} \tag{4.41}$$

   where $\Omega_{j-1} = \left\{\mathbf{b}_{l_1}, \mathbf{b}_{l_2}, \ldots, \mathbf{b}_{l_{j-1}}\right\}$ and $\Omega_{j-1}^c = \Omega - \Omega_{j-1}$.

3. If $j < n_{\mathrm{BS}}$,

$$\Omega_j = \left\{\mathbf{b}_{l_1}, \mathbf{b}_{l_2}, \cdots, \mathbf{b}_{l_j}\right\} = \Omega_{j-1} \cup \left\{\mathbf{b}_{l_j}\right\} \tag{4.42}$$

   and go step 2. Otherwise, BS is terminated. The final set of selected bands is given by the first selected $n_{\mathrm{BS}}$ bands, i.e. $\left\{\mathbf{b}_{l_1}, \mathbf{b}_{l_2}, \ldots, \mathbf{b}_{l_{n_{\mathrm{BS}}}}\right\}$.

It should be pointed out that SF-TCBS does not have to run through all bands as MinV-BP does. It can terminate the augmentation process as long as the number of bands to be selected, $n_{\mathrm{BS}}$, is reached.

### 4.4.2.2 Sequential Backward TCBS

In contrast to SF-TCBS, we can also derive a sequential backward search (SBS), which uses leave-one-out as a technique to select optimal feature vectors. The resulting BS is called sequential backward TCBS (SB-TCBS).

More specifically, for each single band, say $l$th band, $\mathbf{b}_l$, we consider the band subset $\Omega - \mathbf{b}_l$ by removing $\mathbf{b}_l$ from the full band set $\mathbf{\Omega}$ and then replace the full band set $\mathbf{\Omega}$ used in CEM-produced variance $\left(\mathbf{d}^T \mathbf{R}^{-1} \mathbf{d}\right)^{-1} = \left(\mathbf{d}_\Omega^T \mathbf{R}_\Omega^{-1} \mathbf{d}_\Omega\right)^{-1}$ with the single band $\Omega - \mathbf{b}_l$ to become

$V(\Omega-\mathbf{b}_l)=\left(\mathbf{d}^T_{\Omega-\{\mathbf{b}_l\}}\mathbf{R}^{-1}_{\Omega-\{\mathbf{b}_l\}}\mathbf{d}_{\Omega-\{\mathbf{b}_l\}}\right)^{-1}$ which can be used as the priority score assigned to the $l$th band, $\mathbf{b}_l$, to rank the priority of $\mathbf{b}_l$. In particular, the band which yields the maximal variance

$$\mathbf{b}^*_{l_1}=\arg\{\max{}_{\mathbf{b}_l\in\Omega}V(\Omega-\{\mathbf{b}_l\})\}=\arg\left\{\max{}_{\mathbf{b}_l\in\Omega}\left(\mathbf{d}^T_{\Omega-\{\mathbf{b}_l\}}\mathbf{R}^{-1}_{\Omega-\{\mathbf{b}_l\}}\mathbf{d}_{\Omega-\{\mathbf{b}_l\}}\right)^{-1}\right\} \tag{4.43}$$

will be selected as the first band, denoted by $\mathbf{b}^*_{l_1}$ to be the most significant band since the variance in (4.43) produced the maximal variance if $\mathbf{b}^*_{l_1}$ is removed from $\mathbf{\Omega}$. Now, let $\Omega_1=\Omega-\{\mathbf{b}_{l_1}\}$. Then the second band which yields the maximal variance again

$$\mathbf{b}^*_{l_2}=\arg\{\min{}_{\Omega_1-\{\mathbf{b}_l\}}V(\Omega_1-\mathbf{b}_l)\}=\arg\left\{\min{}_{\Omega_1-\{\mathbf{b}_l\}}\left(\mathbf{d}^T_{\Omega_1-\{\mathbf{b}_l\}}\mathbf{R}^{-1}_{\Omega_1-\{\mathbf{b}_l\}}\mathbf{d}_{\Omega_1-\{\mathbf{b}_l\}}\right)^{-1}\right\} \tag{4.44}$$

is then selected as the second band, $\mathbf{b}^*_{l_2}$. The same process is repeated over and over again by continuously removing selected bands one at a time from the full band set $\mathbf{\Omega}$. The resulting algorithm is called SB-TCBS.

***SB-TCBS***

1. Initial condition:

   Let $n_{BS}$ be the number of bands needed to be selected, which can be determined by a criterion such as virtual VD.

   Find

$$\mathbf{b}_{l_1}=\arg\left\{\max{}_{\mathbf{b}_l\in\Omega}\left(\mathbf{d}^T_{\Omega-\{\mathbf{b}_l\}}\mathbf{R}^{-1}_{\Omega-\{\mathbf{b}_l\}}\mathbf{d}_{\Omega-\{\mathbf{b}_l\}}\right)^{-1}\right\} \tag{4.45}$$

   $\Omega_1=\{l_1\}$.

2. Band reduction

$$\mathbf{b}_{l_j}=\arg\left\{\max{}_{\mathbf{b}_l\in\Omega^c_{j-1}}\left(\mathbf{d}^T_{\Omega-(\Omega_{j-1}\cup\{\mathbf{b}_l\})}\mathbf{R}^{-1}_{\Omega-(\Omega_{j-1}\cup\{\mathbf{b}_l\})}\mathbf{d}_{\Omega-(\Omega_{j-1}\cup\{\mathbf{b}_l\})}\right)^{-1}\right\} \tag{4.46}$$

   where $\Omega_{j-1}=\{\mathbf{b}_{l_1},\mathbf{b}_{l_2},\ldots,\mathbf{b}_{l_{j-1}}\}$ and $\Omega^c_{j-1}=\Omega-\Omega_{j-1}$.

3. If $j<n_{BS}$,

$$\Omega_j=\{\mathbf{b}_{l_1},\mathbf{b}_{l_2},\ldots,\mathbf{b}_{l_j}\}=\Omega_{j-1}\cup\{\mathbf{b}_{l_j}\} \tag{4.47}$$

   and go step 2. Otherwise, BS is terminated. The final set of selected bands is given by the first selected $n_{BS}$ bands, i.e. $\{\mathbf{b}_{l_1},\mathbf{b}_{l_2},\ldots,\mathbf{b}_{l_{n_{BS}}}\}$.

It should be noted that SB-TCBS is different from SF-TCBS in two different aspects. First of all, SB-TCBS ranks all bands individually according to (4.46) compared to SF-TCBS which augments selected bands one at a time using (4.41). Second, SB-TCBS uses the sample correlation matrix formed by data sample vectors using all bands except the bands already selected, $\mathbf{R}_{\Omega-\Omega_j}$, while

SF-TCBS only uses the selected bands in $\mathbf{\Omega}_j$ to form the sample correlation matrix $\mathbf{R}_{\Omega_j}$. Nevertheless, both SF-TCBS and SB-TCBS do not have to run through all bands as Max*V*-BP does. Their augmentation processes can be terminated as long as the number of bands to be selected, $n_{\text{BS}}$, is reached.

## 4.5 TCBS Experiments

VD of this scene was estimated by the HFC method in [10] as nine. However, it was demonstrated in [24, 25] that 18 is a more appropriate value than 9. So, in the following experiments, 18 was used as the number of bands to be selected. Table 4.7 lists 18 bands selected by UBS, CBS [7], Min*V*-BP, Max*V*-BP, SF-TCBS, SB-TCBS, and their corresponding BFS methods where UBS was selected because it is the simplest BS method and also generally performs reasonably well. There are also two reasons to choose CBS for comparison. The most important one is that its designed rationale is also based on CEM. A second one is that CBS was compared and shown to perform better than many other BS methods in [2] in which case there is no need to include other methods here for comparison as long as we can show that TCBS indeed performs better than CBS.

Figures 4.7(a), 4.8(a), 4.9(a), and 4.10(a) show detection results of CEM using 18 bands selected in Table 4.7 by four TCBS-based methods, MinV-BP, MaxV-BP, SF-TCBS, and SB-TCBS, while each

**Table 4.7** Eighteen bands prioritized by BF-Min*V*-BP, BF-Max*V*-BP, BF-Min*V*-BP, and BF-Max*V*-BP, and 18 bands selected by UBS, SF-TCBS, SB-TCBS, BF-SF-TCBS, and BF-SB-TCBS

| | |
|---|---|
| Min*V*-BP | $\mathbf{p}_1$: 169/122/123/168/167/166/124/162/164/165/161/160/158/132/163/157/159/133 |
| | $\mathbf{p}_2$: 122/169/123/132/133/131/134/168/135/126/124/130 /125/127/136/98/167/137 |
| | $\mathbf{p}_3$: 122/123/132/133/124/131/134/135/52/169/167/130/168/53//51/136/127/125 |
| | $\mathbf{p}_4$: 123/122/124/125/127/128/126/129/130/131/132/133//98/134/99/135/100/168 |
| | $\mathbf{p}_5$: 122/123/124/12/126/168/127/169/128/129/130/167/166/131/165/132/164/133 |
| BF-Min*V*-BP | 122/123/124/132/133/168/169/125/127/167/131/130/126/134/135/128/166//129 |
| Max*V*-BP | $\mathbf{p}_1$: 109/6/110/128/17/142/122/116/57/24/28/130/106/132/140/51/111/112 |
| | $\mathbf{p}_2$: 133/44/50/129/143/118/45/46/154/47/71/144/29/101/38/153/115/66 |
| | $\mathbf{p}_3$: 123/44/45/51/50/52/19/99/131/98/38/46/5/169/113/60/88/94 |
| | $\mathbf{p}_4$: 158/152/154/61/155/63/98/138/52/105/118/25/27/107/153/82/18/ |
| | $\mathbf{p}_5$: 61/57/136/52/67/124/27/28/73/114/113/128/38/25/ 63/ 86/65/101 |
| BF-Max*V*-BP | 52/38/61/44/57/45/50/154/51/128/63/118/27/98/2/46/113/25 |
| SF-TCBS | $\mathbf{p}_1$: 169/127/122/105/112/149/130/116/128/123/148/17/ 22/2/106/109/142/132 |
| | $\mathbf{p}_2$: 122/128/133/154/143/137/142/148/153/11/27/19/6/138/46/49/42/129 |
| | $\mathbf{p}_3$: 122/126/53/41/118/111/15/52/45/51/38/107/44/5/49/26/37/98 |
| | $\mathbf{p}_4$: 123/166/50/128/158/101/152 141/149/46/153/25/134/ 9/38/112 /104/140 |
| | $\mathbf{p}_5$: 122/49/25/17/39/154/136/28/153/18/27/38/52/132/155/26/151/14 |
| BF-SF-TCBS | 12/49/128/153/38/123/25/17/154/112/149/142/52/148/46/27/132/26 |
| SB-TCBS | $\mathbf{p}_1$: 109/110/111/112/122/113/132/137/138/140/123/139 /108/114/136/142/141/120 |
| | $\mathbf{p}_2$: 133/129/143/144/118/44/46/47/45/43/115/78/154/153/142/33/35/32 |
| | $\mathbf{p}_3$: 123/45/44/46/47/43/51/52/50/ 49/53/54/55/56/57/48/70/39 |
| | $\mathbf{p}_4$: 158/152/154/155/153/63/156/151/150/149/159/148/147/57/161/163/166/165 |
| | $\mathbf{p}_5$: 61/62/73/67/124/58/94/114/27/28/25/26/24/113/136/23/21/29 |
| BF-SB-TCBS | 123/45/44/154/46/47/153/43/113/114/136/142/61/109/133/158/62/110 |
| CBS | 62/77/63/61/13/91/30/69/76/56/38/45/16/20/39/34/24/47 |
| UBS | 1/10/19/28/37/46/55/64/73/82/91/100/109/118/127/136/145/154 |

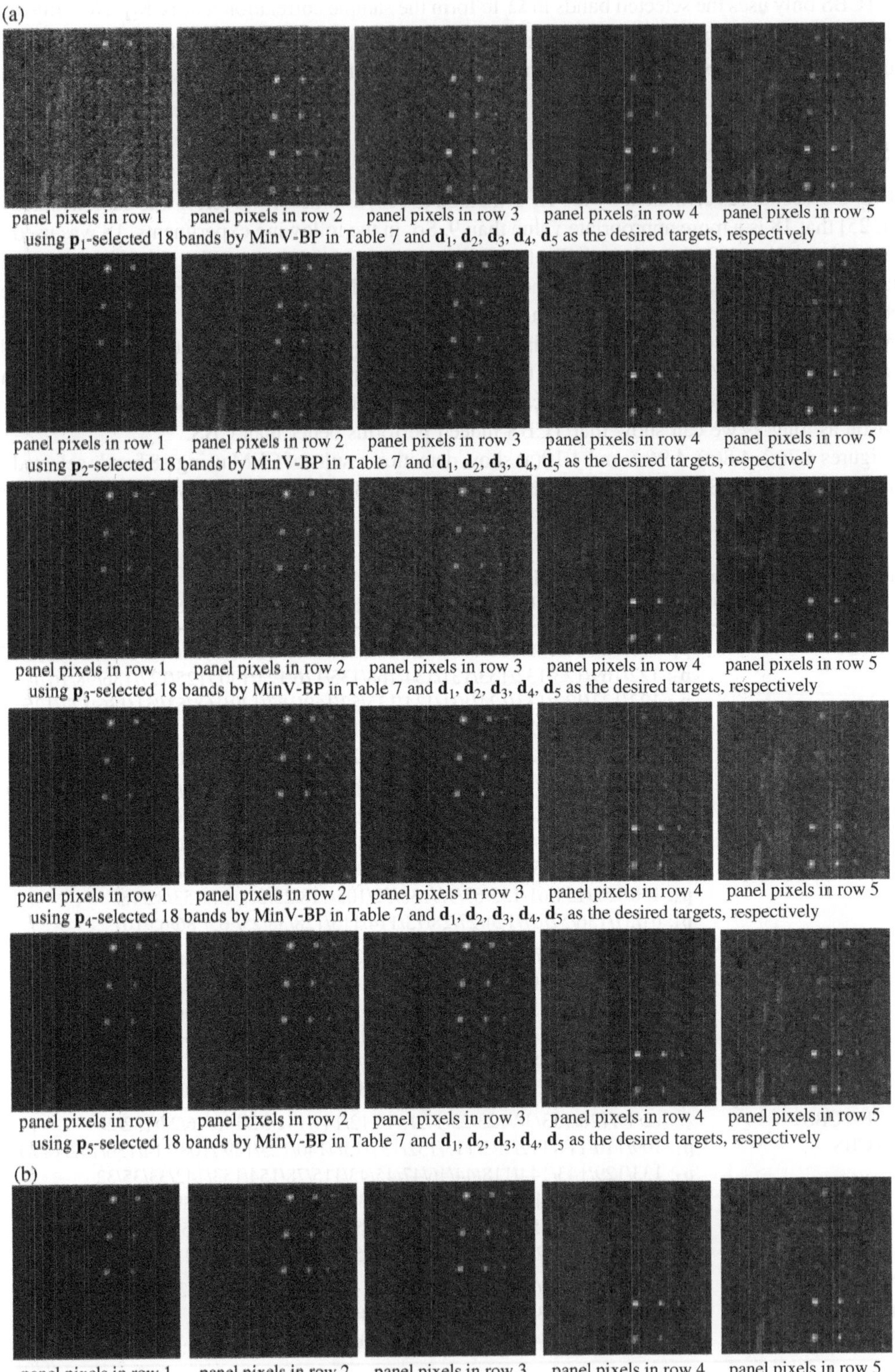


**Figure 4.7** CEM detection maps of 19 R panel pixels using 18 bands selected by Min*V*-BP and BF-Min*V*-BP in Table 4.7. (a) Min*V*-BP and (b) BF-Min*V*-BP.

(a)

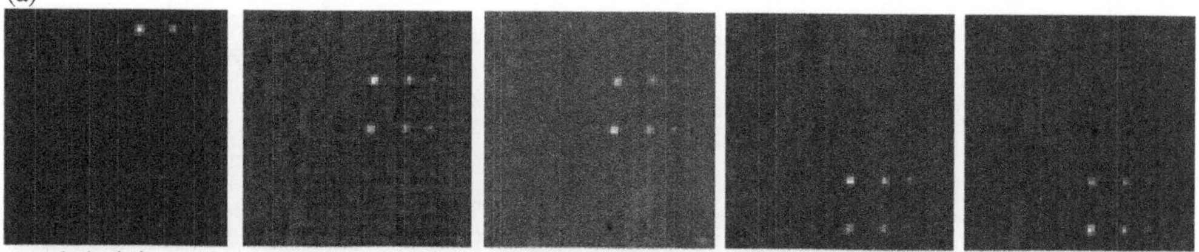

panel pixels in row 1 panel pixels in row 2 panel pixels in row 3 panel pixels in row 4 panel pixels in row 5

using $\mathbf{p}_1$-selected 18 bands by MaxV-BP in Table 1 and $\mathbf{d}_1$, $\mathbf{d}_2$, $\mathbf{d}_3$, $\mathbf{d}_4$, $\mathbf{d}_5$ as the desired targets, respectively

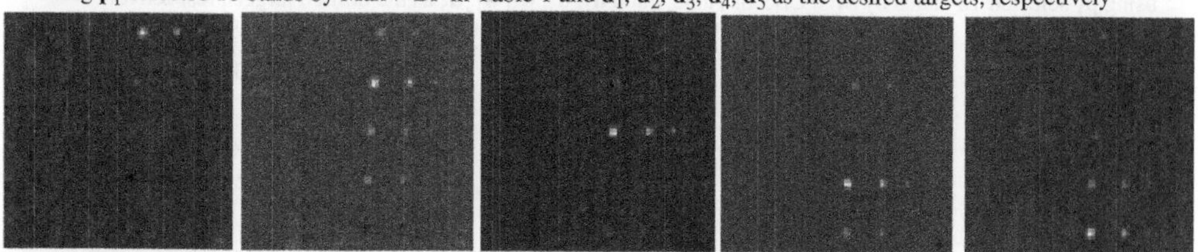

panel pixels in row 1 panel pixels in row 2 panel pixels in row 3 panel pixels in row 4 panel pixels in row 5

using $\mathbf{p}_2$-selected 18 bands by MaxV-BP in Table 1 and $\mathbf{d}_1$, $\mathbf{d}_2$, $\mathbf{d}_3$, $\mathbf{d}_4$, $\mathbf{d}_5$ as the desired targets, respectively

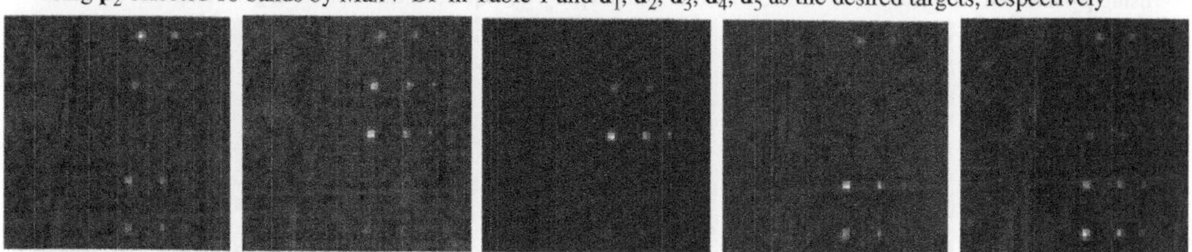

panel pixels in row 1 panel pixels in row 2 panel pixels in row 3 panel pixels in row 4 panel pixels in row 5

using $\mathbf{p}_3$-selected 18 bands by MaxV-BPin Table 1 and $\mathbf{d}_1$, $\mathbf{d}_2$, $\mathbf{d}_3$, $\mathbf{d}_4$, $\mathbf{d}_5$ as the desired targets, respectively

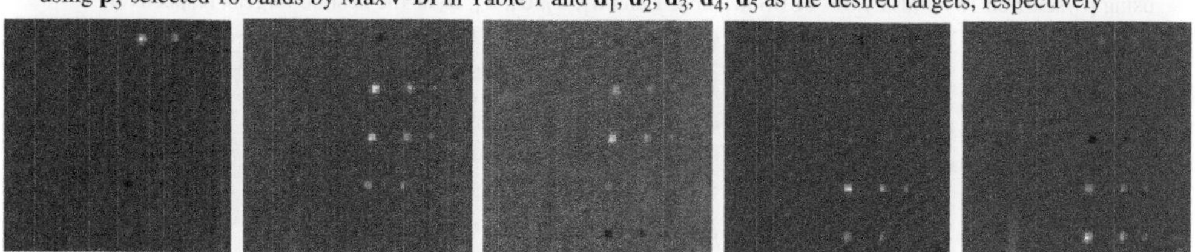

panel pixels in row 1 panel pixels in row 2 panel pixels in row 3 panel pixels in row 4 panel pixels in row 5

using $\mathbf{p}_4$-selected 18 bands by MaxV-BP in Table 1 and $\mathbf{d}_1$, $\mathbf{d}_2$, $\mathbf{d}_3$, $\mathbf{d}_4$, $\mathbf{d}_5$ as the desired targets, respectively

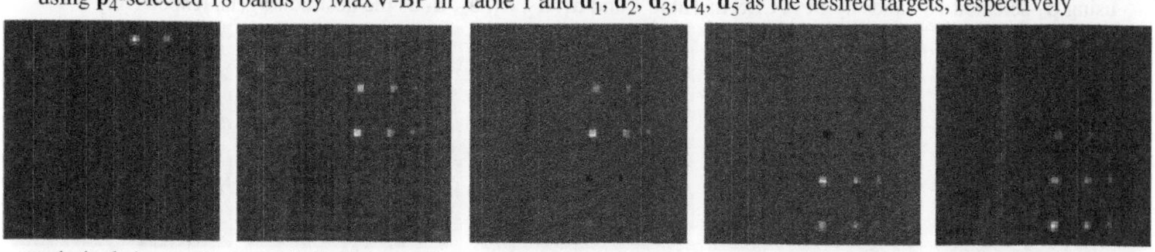

panel pixels in row 1 panel pixels in row 2 panel pixels in row 3 panel pixels in row 4 panel pixels in row 5

using $\mathbf{p}_5$-selected 18 bands by MaxV-BP in Table 1 and $\mathbf{d}_1$, $\mathbf{d}_2$, $\mathbf{d}_3$, $\mathbf{d}_4$, $\mathbf{d}_5$ as the desired targets, respectively

(b)

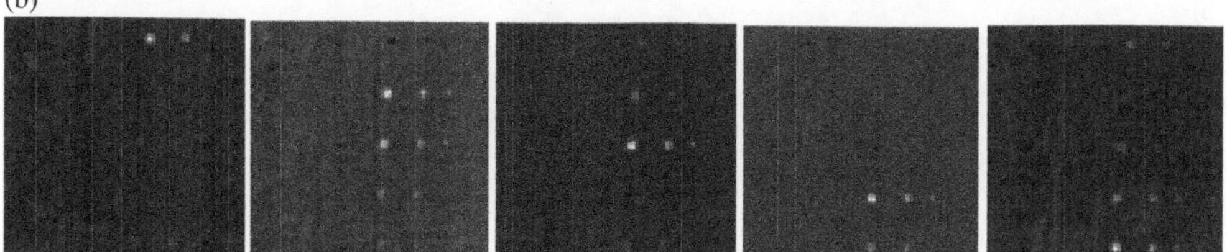

panel pixels in row 1 panel pixels in row 2 panel pixels in row 3 panel pixels in row 4 panel pixels in row 5

**Figure 4.8** CEM detection maps of 19 R panel pixels using 18 bands selected by Max*V*-BP and BF-Max*V*-BP in Table 4.7. (a) Max*V*-BP and (b) BF-Max*V*-BP.

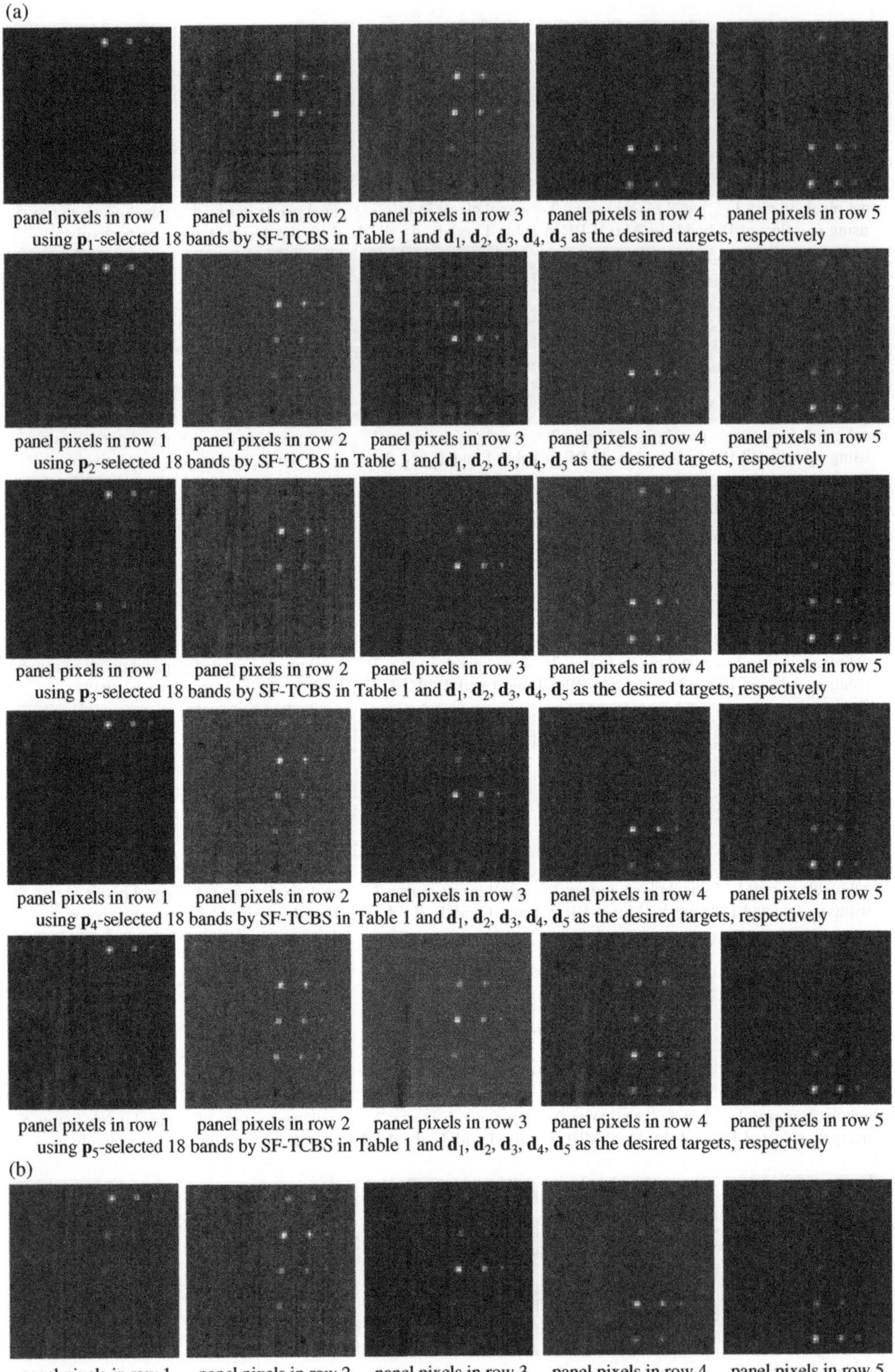


**Figure 4.9** CEM detection maps of 19 R panel pixels using 18 bands selected by SF-TCBS and BF-SF-TCBS in Table 4.7. (a) SF-TCBS and (b) BF-SF-TCBS.

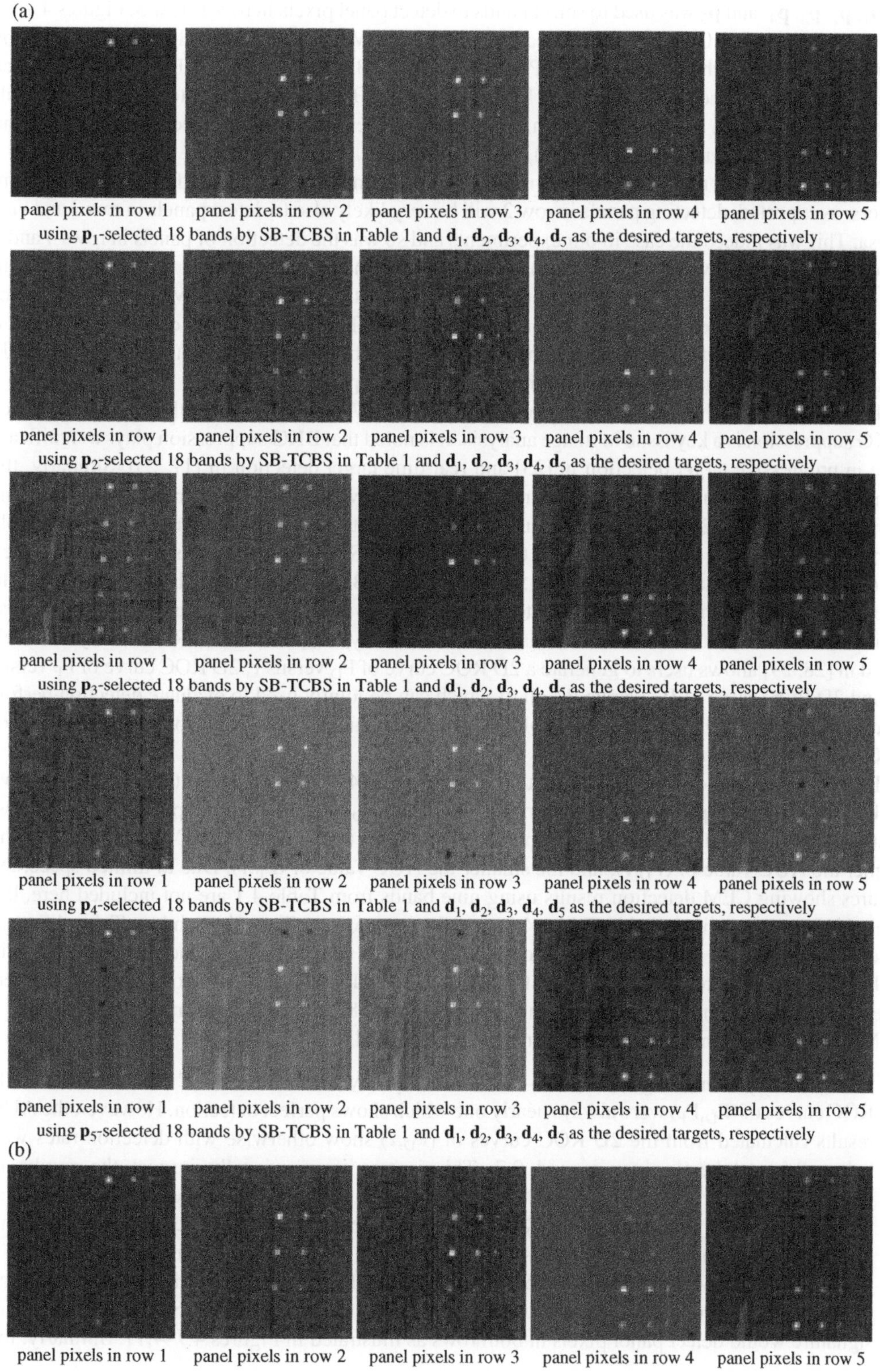


**Figure 4.10** CEM detection maps of 19 R panel pixels using 18 bands selected by SB-TCBS and BF-SB-TCBS in Table 4.7. (a) SB-TCBS and (b) BF-SB-TCBS.

of $\mathbf{p}_1$, $\mathbf{p}_2$, $\mathbf{p}_3$, $\mathbf{p}_4$, and $\mathbf{p}_5$ was used to select bands to detect panel pixels in row 1–row 5. Figures 4.7(b), 4.8(b), 4.9(b), and 4.10(b) also show detection results of CEM using 18 bands selected in Table 4.7 by four BFS-TCBS methods, BF-Min*V*-BP, BF-Max*V*-BP, BF-SF-TCBS, and BF-SB-TCBS which used fused bands selected by Min*V*-BP, Max*V*-BP, SF-TCBS, and SB-TCBS to detect panel pixels in row 1–row 5. For a further comparison, CBS and UBS were also implemented using 18 bands and their results are shown in Figure 4.11(a), (b).

Since the panels in rows 2 and 3 were made by the same material and painted by slightly different colors, as expected, detecting panels in row 2 will be very likely also to detect panels in row 3 and vice versa. This was indeed the case in our experiments and so was the detection of panels in rows 4 and 5 since the panels in rows 4 and 5 were made by the same material and painted by slightly different colors. According to visual inspection of Figures 4.7–4.11, it is difficult to see which one would produce best results since there was not much appreciable difference to make a call. Nevertheless, it seemed that CEM detection results using bands selected by particular signatures were generally not as good as the results produced by their fusion counterparts.

It is very obvious that the visual inspection of Figures 4.7–4.11 did not tell the whole story since BKG suppression is a key issue in image analysis. It is noted that BKG suppression plays a significant role in progressive target detection in [23] and real-time target detection in [17, 23]. Similarly, the same issue of BKG suppression is encountered in progressive anomaly detection [26] and real-time anomaly detection [22, 27]. This is because anomaly detection also uses the inverse of the sample correlation matrix, i.e. $\mathbf{R}^{-1}$, as CEM does to suppress BKG. Such issue of BKG suppression has been discussed in great detail in [23, chapter 5]. Accordingly, BKG suppression should be included to account for the performance of target detection. Apparently, plots of $P_D$ versus $P_F$ used by the traditional ROC analysis cannot address this issue. Fortunately, a 3D ROC (3D ROC) analysis developed in [28, 29] allows users to generate a 2D ROC curve of $P_D$ versus $\tau$, 2D ROC curve of $P_F$ versus $\tau$, and 2D ROC curve of $P_D$ versus $P_F$, all of which can be jointly used to evaluate detection performance. Specifically, the 2D ROC curve of $P_F$ versus $\tau$ is particularly designed to assess the impact of BKG suppression on target detection.

In order to further conduct a quantitative study, Tables 4.8–4.10 calculate the area under curve (AUC) of the three types of 2D ROC curves of $(P_D,P_F)$, $(P_D,\tau)$, and $(P_F,\tau)$, respectively, where best results are boldfaced. In other words, the better the detection, the higher the AUC value of $(P_D,P_F)$, the higher the AUC value of $(P_D,\tau)$, and the smaller the AUC value of $(P_F,\tau)$. Due to limited space, the figures showing CEM detection results using nine bands from Table 4.1 are not included here, but their AUC results are included in Tables 4.8–4.10 for comparison. Since the bands in Table 4.7 were selected sequentially, the nine bands used by TCBS were simply to select the first nine bands in Table 4.7 to produce results. In addition, the last column in Tables 4.8–4.10 also calculated the average AUC of panels in the five rows using selected 9 and 18 bands for comparison.

Many intriguing findings can be observed from Tables 4.8–4.10.

1. First of all, by examining Tables 4.8–4.10, it is very obvious that the AUC values based on 2D ROC curves of $(P_D,P_F)$ were always nearly one which shows perfect detection. However, the AUC results calculated from the 2D ROC curves of $(P_D,\tau)$ show otherwise with detection rates well below 0.5 and best values around 0.7. This is true if we visually inspect the results in Figures 4.7–4.11. The results shown in Tables 4.8–4.10 provide evidence that the traditional measure using 2D ROC curve of $(P_D,P_F)$ to evaluate detection performance is not reliable and could be misleading since it only tells half of a story in detection performance analysis.
2. As shown in Table 4.9, the detection results were not good. This is primarily due to the fact that the spectral signatures of panel pixels in rows 2 and 3, $\mathbf{p}_2$ and $\mathbf{p}_3$, are very close. So, using either panel signature would detect panel pixels in both rows as illustrated in Figures 4.7–4.11. Similarly, the

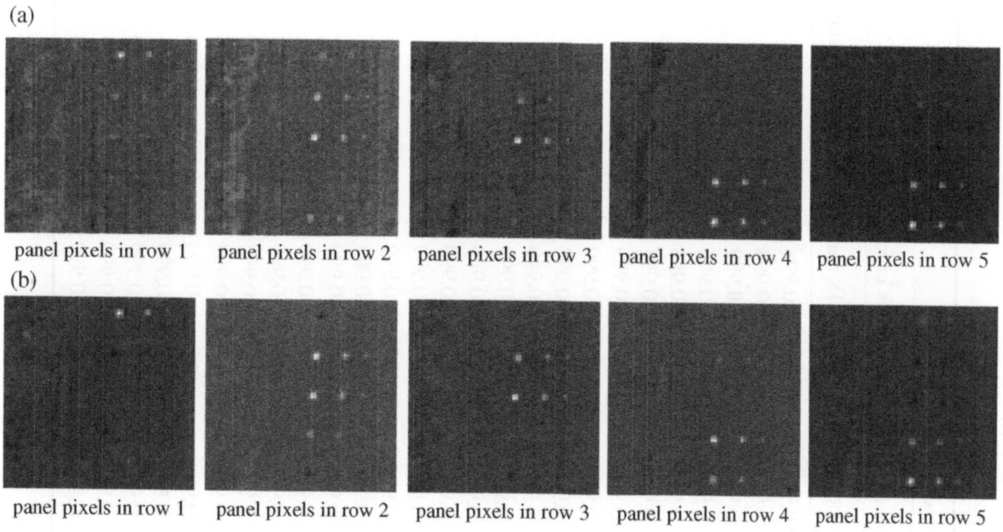


**Figure 4.11** CEM detection maps of 19 R panel pixels using 18 bands selected by UBS and CBS in Table 4.7. (a) CBS and (b) UBS.

spectral signatures of panel pixels in rows 4 and 5, $\mathbf{p}_4$ and $\mathbf{p}_5$, are also very close. So, detecting panel pixels in one row will also detect panel pixel in the other row.

3. To evaluate BKG suppression, a best way is suggested to visually inspect the results in Figures 4.7–4.11 in conjunction with comparison of the AUC values in Tables 4.8–4.10 since the AUC value calculated from 2D ROC curve of ($P_F$,$\tau$) reflects the degree of BKG suppression. That is, the smaller the AUC value of 2D ROC curve of ($P_F$,$\tau$), the better the BKG suppression. For example, the best performance of CEM using $\mathbf{p}_1$ as the desired target signature to detect the panel pixels in row 1 was SF-TCBS which yielded the smallest value of AUC value of 2D ROC curve of ($P_F$,$\tau$), 0.041, among all other AUC values. If we examine the top row of Figure 4.9(a) produced by SF-TCBS, its $\mathbf{p}_1$-detection map was very clean compared to $\mathbf{p}_1$-detection map produced by other BS methods. Due to the lack of prior knowledge on BKG, it is crucial to include the detection maps in Figures 4.7–4.11 for visual inspection. Unfortunately, the BKG suppression issue has been overlooked in the traditional 2D ROC analysis.
4. As expected, BF-BS methods may not be the best in detecting individual panel pixels but generally work reasonably well over all the panel pixels.
5. The band fusion results using nine bands were generally better than their corresponding counterparts using 18 bands. However, the conclusions are reversed for using individual bands without fusion. That is, for individual target signature, results using 18 bands were generally better than those using nine bands for all test BS methods. This makes sense. Since the band fusion methods integrate bands from different BS methods, the number of bands is expected to be smaller.
6. Although the TI and TE scenarios designed in [1, chapter 4] can be also used for experiments with complete ground truth of simulated panel pixels, the results are nearly perfect and not very interesting for analysis. Accordingly, their results are not included here. On the other hand, HYDICE data set provides precise knowledge of panel pixels in Figure 4.11(b). Such ground truth can be used for verification and validation. To the authors' best knowledge, this scene may be one of best data sets that can be used to evaluate the effectiveness of target detection.

**Table 4.8** Area under curve (AUC) of ($P_D$,$P_F$) for CEM detection results using 9 and 18 bands prioritized by Min*V*-BP, BF-Min*V*-BP, Max*V*-BP, and BF-Max*V*-BP, and selected by SF-TCBS, BF-SF-TCBS, SB-TCBS, and BF-SB-TCBS along with 9 and 18 bands by CBS and UBS

| BS algorithms | Desired targets for BS | Detection of panel pixels in rows | | | | | | | | | | | |
|---|---|---|---|---|---|---|---|---|---|---|---|---|---|
| | | Panel pixels in row 1 | | Panel pixels in row 2 | | Panel pixels in row 3 | | Panel pixels in row 4 | | Panel pixels in row 5 | | Average AUC | |
| | | 9 bands | 18 bands | 9 bands | 18 bands | 9 bands | 18 bands | 9 bands | 18 bands | 9 bands | 18 bands | 9 bands | 18 bands |
| Min*V*-BP | $\mathbf{p}_1$ | 0.997 | 0.961 | 0.996 | 0.987 | 0.996 | 0.981 | 0.997 | 0.997 | 0.998 | 0.998 | 0.997 | 0.985 |
| | $\mathbf{p}_2$ | 0.922 | 0.990 | 0.903 | 0.992 | 0.865 | 0.983 | 0.971 | 0.997 | 0.995 | 0.998 | 0.931 | 0.992 |
| | $\mathbf{p}_3$ | 0.994 | 0.992 | 0.988 | 0.992 | 0.911 | 0.772 | 0.992 | 0.998 | 0.996 | **0.999** | 0.976 | 0.951 |
| | $\mathbf{p}_4$ | 0.987 | 0.989 | 0.985 | 0.966 | 0.982 | 0.750 | 0.995 | 0.957 | 0.996 | 0.996 | 0.989 | 0.932 |
| | $\mathbf{p}_5$ | 0.994 | 0.979 | 0.994 | 0.970 | 0.993 | 0.944 | 0.991 | 0.995 | 0.997 | 0.998 | 0.994 | 0.977 |
| BF-Min*V*-BP | Fusion | 0.982 | 0.976 | 0.985 | 0.986 | 0.955 | 0.981 | 0.990 | 0.996 | 0.997 | 0.997 | 0.982 | 0.987 |
| Max*V*-BP | $\mathbf{p}_1$ | 0.997 | 0.905 | 0.786 | 0.591 | 0.571 | 0.983 | 0.995 | 0.998 | 0.997 | 0.999 | 0.869 | 0.895 |
| | $\mathbf{p}_2$ | 0.938 | 0.929 | 0.637 | 0.778 | 0.991 | 0.962 | 0.996 | 0.997 | 0.998 | 0.999 | 0.912 | 0.933 |
| | $\mathbf{p}_3$ | 0.997 | **0.995** | 0.995 | 0.987 | 0.994 | 0.960 | 0.997 | 0.996 | 0.997 | 0.998 | 0.996 | 0.987 |
| | $\mathbf{p}_4$ | 0.765 | 0.818 | 0.760 | 0.946 | 0.998 | 0.997 | 0.997 | 0.997 | **0.999** | 0.999 | 0.904 | 0.951 |
| | $\mathbf{p}_5$ | 0.864 | 0.845 | 0.848 | 0.528 | 0.959 | 0.991 | 0.997 | 0.998 | 0.998 | 0.999 | 0.933 | 0.872 |
| BF-Max*V*-BP | Fusion | 0.992 | 0.971 | 0.992 | 0.920 | 0.986 | 0.984 | **0.999** | 0.997 | 0.999 | 0.999 | 0.994 | 0.974 |
| SF-TCBS | $\mathbf{p}_1$ | 0.995 | 0.985 | 0.778 | 0.573 | 0.826 | 0.941 | 0.997 | 0.997 | 0.998 | 0.999 | 0.919 | 0.899 |
| | $\mathbf{p}_2$ | 0.991 | 0.989 | 0.875 | 0.857 | 0.850 | 0.986 | 0.997 | 0.997 | 0.998 | 0.999 | 0.942 | 0.966 |
| | $\mathbf{p}_3$ | 0.997 | 0.925 | 0.936 | 0.778 | 0.991 | 0.981 | 0.995 | 0.997 | 0.998 | 0.999 | 0.983 | 0.936 |
| | $\mathbf{p}_4$ | 0.757 | 0.786 | 0.711 | 0.879 | 0.631 | 0.975 | 0.998 | 0.997 | 0.999 | 0.999 | 0.819 | 0.927 |
| | $\mathbf{p}_5$ | 0.969 | 0.983 | 0.916 | 0.895 | 0.818 | 0.977 | 0.996 | 0.990 | 0.999 | 0.999 | 0.940 | 0.969 |
| BF-SF-TCBS | Fusion | 0.906 | 0.943 | 0.986 | 0.914 | 0.980 | 0.981 | 0.997 | 0.996 | 0.999 | 0.999 | 0.974 | 0.967 |
| SB-TCBS | $\mathbf{p}_1$ | **0.999** | 0.913 | 0.998 | 0.984 | 0.998 | **0.999** | 0.998 | 0.998 | 0.999 | 0.999 | **0.998** | 0.979 |
| | $\mathbf{p}_2$ | 0.999 | 0.962 | 0.544 | 0.569 | 0.617 | 0.623 | 0.996 | 0.997 | 0.997 | 0.998 | 0.831 | 0.830 |
| | $\mathbf{p}_3$ | 0.999 | 0.991 | 0.995 | 0.989 | 0.994 | 0.993 | 0.995 | 0.998 | 0.997 | 0.998 | 0.996 | **0.994** |
| | $\mathbf{p}_4$ | 0.999 | 0.797 | **0.999** | **0.996** | **0.999** | 0.996 | 0.932 | 0.994 | 0.998 | 0.999 | 0.985 | 0.956 |
| | $\mathbf{p}_5$ | 0.999 | 0.925 | 0.547 | 0.896 | 0.684 | 0.659 | 0.998 | 0.998 | 0.999 | 0.999 | 0.845 | 0.895 |
| BF-SB-TCBS | Fusion | 0.999 | 0.962 | 0.992 | 0.872 | 0.978 | 0.920 | 0.997 | 0.997 | 0.997 | 0.999 | 0.993 | 0.950 |
| CBS | CBS | 0.936 | 0.819 | 0.642 | 0.983 | 0.843 | 0.817 | 0.998 | **0.999** | 0.998 | 0.999 | 0.883 | 0.923 |
| UBS | UBS | 0.989 | 0.984 | 0.963 | 0.984 | 0.870 | 0.984 | 0.998 | 0.997 | 0.997 | 0.999 | 0.963 | 0.990 |

**Table 4.9** Area under curve (AUC) of ($P_D$,$\tau$) for CEM detection results using 9 and 18 bands prioritized by Min*V*-BP, BF-Min*V*-BP, Max*V*-BP, and BF-Max *V*-BP, and selected by SF-TCBS, BF-SF-TCBS, SB-TCBS, and BF-SB-TCBS along with 9 and 18 bands by CBS and UBS

| BS algorithms | Desired targets for BS | Detection of panel pixels in rows | | | | | | | | | | | |
|---|---|---|---|---|---|---|---|---|---|---|---|---|---|
| | | Panel pixels in row 1 | | Panel pixels in row 2 | | Panel pixels in row 3 | | Panel pixels in row 4 | | Panel pixels in row 5 | | Average AUC | |
| | | 9 bands | 18 bands | 9 bands | 18 bands | 9 bands | 18 bands | 9 bands | 18 bands | 9 bands | 18 bands | 9 bands | 18 bands |
| Min*V*-BP | $p_1$ | **0.747** | 0.475 | **0.778** | 0.461 | 0.625 | **0.450** | 0.470 | 0.420 | 0.650 | 0.642 | **0.654** | **0.490** |
| | $p_2$ | 0.442 | 0.185 | 0.491 | 0.357 | 0.522 | 0.258 | 0.493 | 0.511 | 0.699 | 0.713 | 0.529 | 0.405 |
| | $p_3$ | 0.535 | 0.239 | 0.572 | 0.281 | 0.375 | 0.157 | **0.704** | 0.562 | 0.695 | 0.700 | 0.576 | 0.388 |
| | $p_4$ | 0.281 | 0.135 | 0.296 | 0.173 | 0.292 | 0.104 | 0.678 | 0.554 | 0.608 | **0.757** | 0.431 | 0.345 |
| | $p_5$ | 0.461 | 0.139 | 0.597 | 0.184 | 0.641 | 0.183 | 0.489 | 0.413 | 0.732 | 0.658 | 0.584 | 0.315 |
| BF-Min*V*-BP | Fusion | 0.252 | 0.147 | 0.290 | 0.166 | 0.330 | 0.149 | 0.498 | 0.443 | 0.729 | 0.688 | 0.420 | 0.318 |
| Max*V*-BP | $p_1$ | 0.367 | 0.110 | 0.306 | 0.101 | 0.383 | 0.115 | 0.563 | 0.458 | 0.673 | 0.677 | 0.458 | 0.292 |
| | $p_2$ | 0.271 | 0.178 | 0.104 | 0.108 | 0.179 | 0.140 | 0.403 | 0.344 | 0.688 | 0.666 | 0.329 | 0.287 |
| | $p_3$ | 0.718 | 0.355 | 0.345 | 0.289 | 0.211 | 0.118 | 0.615 | 0.467 | 0.670 | 0.666 | 0.512 | 0.379 |
| | $p_4$ | 0.340 | 0.192 | 0.241 | 0.123 | 0.259 | 0.120 | 0.499 | 0.415 | 0.726 | 0.709 | 0.413 | 0.312 |
| | $p_5$ | 0.506 | 0.128 | 0.388 | 0.128 | 0.205 | 0.251 | 0.679 | 0.522 | 0.670 | 0.672 | 0.489 | 0.341 |
| BF-Max*V*-BP | Fusion | 0.511 | 0.159 | 0.195 | 0.239 | 0.174 | 0.147 | 0.681 | 0.360 | 0.664 | 0.648 | 0.445 | 0.311 |
| SF-TCBS | $p_1$ | 0.153 | 0.092 | 0.100 | 0.106 | 0.114 | 0.068 | 0.433 | 0.419 | 0.636 | 0.677 | 0.287 | 0.272 |
| | $p_2$ | 0.265 | 0.153 | 0.092 | 0.065 | 0.077 | 0.181 | 0.431 | 0.340 | 0.699 | 0.655 | 0.313 | 0.279 |
| | $p_3$ | 0.406 | 0.088 | 0.258 | 0.161 | 0.195 | 0.141 | 0.656 | 0.620 | 0.667 | 0.643 | 0.436 | 0.331 |
| | $p_4$ | 0.190 | 0.076 | 0.194 | 0.081 | 0.128 | 0.134 | 0.473 | 0.265 | 0.654 | 0.662 | 0.328 | 0.244 |
| | $p_5$ | 0.385 | 0.147 | 0.163 | 0.102 | 0.427 | 0.294 | 0.460 | 0.333 | 0.660 | 0.657 | 0.419 | 0.307 |
| BF-SF-TCBS | Fusion | 0.187 | 0.160 | 0.427 | 0.064 | **0.717** | 0.111 | 0.429 | 0.313 | 0.654 | 0.656 | 0.483 | 0.261 |
| SB-TCBS | $p_1$ | 0.299 | 0.168 | 0.150 | 0.068 | 0.124 | 0.070 | 0.554 | 0.501 | 0.684 | 0.671 | 0.362 | 0.296 |
| | $p_2$ | 0.234 | 0.221 | 0.095 | 0.109 | 0.176 | 0.149 | 0.657 | 0.431 | 0.688 | 0.651 | 0.370 | 0.312 |
| | $p_3$ | 0.722 | **0.501** | 0.410 | 0.318 | 0.221 | 0.168 | 0.674 | 0.664 | 0.693 | 0.673 | 0.544 | 0.465 |
| | $p_4$ | 0.612 | 0.382 | 0.152 | 0.154 | 0.143 | 0.138 | 0.701 | 0.448 | **0.741** | 0.706 | 0.470 | 0.366 |
| | $p_5$ | 0.613 | 0.312 | 0.249 | 0.318 | 0.352 | 0.319 | 0.695 | 0.690 | 0.715 | 0.719 | 0.525 | 0.472 |
| BF-SB-TCBS | Fusion | 0.689 | 0.224 | 0.453 | 0.055 | 0.532 | 0.060 | 0.478 | 0.439 | 0.689 | 0.666 | 0.568 | 0.289 |
| CBS | CBS | 0.635 | 0.146 | 0.510 | **0.511** | 0.637 | 0.219 | 0.579 | **0.695** | 0.590 | 0.642 | 0.590 | 0.443 |
| UBS | UBS | 0.664 | 0.202 | 0.450 | 0.136 | 0.341 | 0.063 | 0.697 | 0.480 | 0.664 | 0.647 | 0.563 | 0.306 |

**Table 4.10** Area under curve (AUC) of ($P_F$,$\tau$) for CEM detection results using 9 and 18 bands prioritized by Min*V*-BP, BF-Min*V*-BP, Max*V*-BP, and BF-Max*V*-BP, and selected by SF-TCBS, BF-SF-TCBS, SB-TCBS, and BF-SB-TCBS along with 9 and 18 bands by CBS and UBS

| BS algorithms | Desired targets for BS | Detection of panel pixels in rows | | | | | | | | | | | |
|---|---|---|---|---|---|---|---|---|---|---|---|---|---|
| | | Panel pixels in row 1 | | Panel pixels in row 2 | | Panel pixels in row 3 | | Panel pixels in row 4 | | Panel pixels in row 5 | | Average AUC | |
| | | 9 bands | 18 bands | 9 bands | 18 bands | 9 bands | 18 bands | 9 bands | 18 bands | 9 bands | 18 bands | 9 bands | 18 bands |
| Min*V*-BP | $\mathbf{p}_1$ | 0.283 | 0.224 | 0.240 | 0.181 | 0.192 | 0.175 | 0.175 | 0.132 | 0.206 | 0.166 | 0.219 | 0.176 |
| | $\mathbf{p}_2$ | 0.159 | 0.065 | 0.161 | 0.116 | 0.209 | 0.088 | 0.175 | 0.113 | 0.214 | 0.118 | 0.183 | 0.100 |
| | $\mathbf{p}_3$ | 0.167 | 0.073 | 0.166 | 0.126 | 0.204 | 0.128 | 0.121 | 0.110 | 0.096 | 0.102 | 0.151 | 0.108 |
| | $\mathbf{p}_4$ | 0.077 | 0.052 | 0.121 | 0.108 | 0.111 | 0.087 | 0.226 | 0.173 | 0.158 | 0.229 | 0.139 | 0.130 |
| | $\mathbf{p}_5$ | 0.144 | 0.053 | 0.196 | 0.098 | 0.193 | 0.090 | 0.172 | 0.122 | 0.203 | 0.184 | 0.182 | 0.109 |
| BF-Min*V*-BP | Fusion | 0.087 | 0.055 | 0.103 | **0.089** | 0.119 | 0.079 | 0.191 | 0.125 | 0.217 | 0.158 | 0.143 | 0.101 |
| Max*V*-BP | $\mathbf{p}_1$ | 0.141 | 0.058 | 0.272 | 0.112 | 0.392 | 0.246 | 0.109 | 0.115 | 0.090 | 0.114 | 0.201 | 0.129 |
| | $\mathbf{p}_2$ | 0.168 | 0.105 | 0.155 | 0.187 | 0.090 | 0.075 | 0.142 | 0.143 | 0.102 | 0.126 | 0.132 | 0.127 |
| | $\mathbf{p}_3$ | 0.213 | 0.122 | **0.084** | 0.134 | 0.094 | **0.063** | 0.159 | 0.113 | 0.085 | 0.075 | 0.127 | 0.101 |
| | $\mathbf{p}_4$ | 0.257 | 0.123 | 0.294 | 0.198 | 0.422 | 0.295 | 0.298 | 0.156 | 0.213 | 0.220 | 0.297 | 0.198 |
| | $\mathbf{p}_5$ | 0.186 | 0.078 | 0.266 | 0.130 | 0.401 | 0.147 | 0.149 | 0.143 | 0.111 | 0.076 | 0.222 | 0.115 |
| BF-Max*V*-BP | Fusion | 0.132 | 0.083 | 0.088 | 0.175 | **0.089** | 0.075 | 0.118 | 0.138 | 0.092 | 0.081 | 0.104 | 0.111 |
| SF-TCBS | $\mathbf{p}_1$ | **0.048** | **0.041** | 0.158 | 0.110 | 0.196 | 0.144 | 0.082 | **0.081** | 0.131 | 0.100 | 0.123 | 0.095 |
| | $\mathbf{p}_2$ | 0.138 | 0.087 | 0.186 | 0.158 | 0.152 | 0.092 | 0.177 | 0.137 | 0.152 | 0.114 | 0.161 | 0.118 |
| | $\mathbf{p}_3$ | 0.100 | 0.060 | 0.182 | 0.136 | 0.129 | 0.085 | **0.073** | 0.144 | 0.069 | 0.063 | 0.111 | 0.097 |
| | $\mathbf{p}_4$ | 0.115 | 0.044 | 0.156 | 0.171 | 0.168 | 0.076 | 0.153 | 0.090 | 0.078 | 0.087 | 0.134 | 0.094 |
| | $\mathbf{p}_5$ | 0.282 | 0.082 | 0.254 | 0.192 | 0.395 | 0.208 | 0.127 | 0.115 | 0.085 | 0.071 | 0.229 | 0.134 |
| BF-SF-TCBS | Fusion | 0.153 | 0.117 | 0.235 | 0.145 | 0.219 | 0.066 | 0.185 | 0.124 | 0.065 | 0.079 | 0.171 | 0.106 |
| SB-TCBS | $\mathbf{p}_1$ | 0.095 | 0.065 | 0.388 | 0.166 | 0.363 | 0.195 | 0.096 | 0.130 | 0.127 | 0.098 | 0.214 | 0.131 |
| | $\mathbf{p}_2$ | 0.080 | 0.136 | 0.092 | 0.149 | 0.163 | 0.136 | 0.107 | 0.199 | 0.092 | 0.081 | 0.107 | 0.140 |
| | $\mathbf{p}_3$ | 0.157 | 0.162 | 0.131 | 0.198 | 0.093 | 0.083 | 0.100 | 0.099 | 0.108 | 0.080 | 0.118 | 0.124 |
| | $\mathbf{p}_4$ | 0.265 | 0.157 | 0.382 | 0.364 | 0.383 | 0.348 | 0.220 | 0.206 | 0.323 | 0.252 | 0.315 | **0.265** |
| | $\mathbf{p}_5$ | 0.193 | 0.193 | 0.273 | 0.363 | 0.253 | 0.337 | 0.116 | 0.103 | 0.142 | 0.130 | 0.196 | 0.225 |
| BF-SB-TCBS | Fusion | 0.146 | 0.077 | 0.196 | 0.127 | 0.203 | 0.113 | 0.167 | 0.205 | 0.104 | 0.098 | 0.163 | 0.124 |
| CBS | CBS | 0.466 | 0.220 | 0.495 | 0.258 | 0.487 | 0.160 | 0.168 | 0.109 | 0.171 | **0.054** | **0.357** | 0.160 |
| UBS | UBS | 0.140 | 0.116 | 0.306 | 0.244 | 0.241 | 0.140 | 0.130 | 0.160 | 0.116 | 0.097 | 0.219 | 0.176 |

## 4.6 Conclusion

This chapter develops two types of CBS methods, BCBS and TCBS, both of which are derived from the well-known subpixel target detection algorithm, CEM. There are three key fundamental differences between BCBS and TCBS in terms of how to constrain the signal source used by CEM. One is the signal sources to be constrained by CEM. In BCBS, the signal sources to be constrained are the bands to be selected, which can be implemented by two versions, single-band constrained BS and multiple band-constrained BS. On the other hand, in TCBS, the signal sources to be constrained are the targets to be detected using a band subset selected by BS instead of full bands used by CEM. Like BCBS, there are also two versions that can be implemented for TCBS, single-target constrained BS and multiple-target constrained BS. A second difference is their used criteria. While BCBS uses the band-constrained minimum variance to prioritize all spectral bands with no reference to specific targets to be detected, TCBS uses the target-constrained minimum variance using selected band subsets. A third difference is that BCBS is application-independent, but TCBS is a target-specified BS method. Nevertheless, both BCBS and TCBS use the same BS criteria, minimum/maximum variance band prioritization and same BS search strategies, sequential forward/backward search algorithms. Accordingly, BCBS and TCBS can be considered as parallel twin CEM-CBS methods.

## References


1. Chang, C.-I (2013). *Hyperspectral Data Processing: Signal Processing Algorithm Design and Analysis*. NJ: Wiley.
2. Chang, C.-I, Du, Q., Sun, T.S., and Althouse, M.L.G. (1999). A joint band prioritization and band decorrelation approach to band selection for hyperspectral image classification. *IEEE Transactions on Geoscience and Remote Sensing* 37 (6): 2631–2641.
3. Wang, Q., Zhang, F., and Li, X. (2018). Optimal clustering framework for hyperspectral band selection. *IEEE Transactions on Geoscience and Remote Sensing* 56 (10): 5910–5922.
4. Pudil, P., Novovicova, J., and Kittler, J. (1994). Floating search methods in feature selection. *Pattern Recognition Letters* 15: 1119–1125.
5. Wang, L., Chang, C.-I, Lee, L.C. et al. (2017). Band subset selection for anomaly detection in hyperspectral imagery. *IEEE Transactions on Geoscience and Remote Sensing* 55 (9): 1–12.
6. Wang, Q., Zhang, F., and Li, X. (2020). Hyperspectral band selection via optimal neighborhood reconstruction. *IEEE Transactions on Geoscience and Remote Sensing* 58 (4): 8465–8476.
7. Chang, C.-I and Wang, S. (2006). Constrained band selection for hyperspectral imagery. *IEEE Transactions on Geoscience and Remote Sensing* 44 (6): 1575–1585.
8. Harsanyi, J.C. (1993). Detection and classification of subpixel spectral signatures in Hyperspectral image sequences. PhD. Dissertation. Department of Electrical Engineering, University of Maryland, Baltimore County, MD.
9. Wang, Y., Wang, L., Yu, C. et al. (2019). Constrained-target band selection for multiple-target detection. *IEEE Transactions on Geoscience and Remote Sensing* 57 (8): 6079–6103.
10. Harsanyi, J.C., Farrand, W. and Chang, C.-I (1994). Detection of subpixel spectral signatures in hyperspectral image sequences. *Proceedings of American Congress on Surveying & Mapping (ACSM)/American Society of Photogrammetry & Remote Sensing (ASPRS) Annual Convention and Exposition*, Baltimore, 1, 236–247.
11. Poor, H.V. (1994). *An Introduction to Detection and Estimation Theory*, 2e. New York: Springer-Verlag.
12. Chang, C.-I (2003). *Hyperspectral Imaging: Techniques for Spectral detection and Classification*. Kluwer Academic/Plenum Publishers.
13. Heinz, D. and Chang, C.-I (2001). Fully constrained least squares linear mixture analysis for material quantification in hyperspectral imagery. *IEEE Transactions on Geoscience and Remote Sensing* 39 (3): 529–545.

14. Ren, H. and Chang, C.-I (2003). Automatic spectral target recognition in hyperspectral imagery. *IEEE Transactions on Aerospace and Electronic Systems* 39 (4): 1232–1249.
15. Harsanyi, J.C. and Chang, C.-I (1994). Hyperspectral image classification and dimensionality reduction: an orthogonal subspace projection approach. *IEEE Transactions on Geoscience and Remote Sensing* 32 (4): 779–785.
16. http://aviris.jpl.nasa.gov.
17. Chang, C.-I (2017). *Real-Time Recursive Hyperspectral Sample and Band Processing: Algorithm Architecture and Implementation*. Berlin: Springer.
18. Chang, C.-I (2018). A review of virtual dimensionality for hyperspectral imagery. *IEEE Journal of Selected Topics in Applied Earth Observation and Remote Sensing* 11 (4): 1285–1305.
19. Xiong, W., Wu, C.-C., Chang, C.-I et al. (2011). Fast algorithms to implement N-FINDR for hyperspectral endmember extraction. *IEEE Journal of Selected Topics in Applied Earth Observation and Remote Sensing* 4 (3): 545–564.
20. Boardman, J.W. (1994). Geometric mixture analysis of imaging spectrometry data. *International Geoscience and Remote Sensing Symposium* 4: 2369–2371.
21. Yu, C., Song, M., and Chang, C.-I (2019). Class signature-constrained background suppressed approach to band selection for hyperspectral image classification. *IEEE Transactions on Geoscience and Remote Sensing* 57 (1): 14–31.
22. Li, H.C. and Chang, C.-I (2016). Real time hyperspectral anomaly detection via band-interleaved by line, *Remotely Sensed Data Compression, Communications, and Processing XII, part of SPIE Commercial + Scientific Sensing and Imaging*, Baltimore, MD, 17–21.
23. Chang, C.-I (2016). *Real-Time Progressive Hyperspectral Image Processing: Endmember Finding and Anomaly Detection*. New York: Springer.
24. Xiong, W., Liu, W., Wu, C.C., and Chen, C.C.C. (2010). Linear spectral mixture analysis-based approaches to estimation of virtual dimensionality in hyperspectral imagery. *IEEE Transactions on Geoscience and Remote Sensing* 48 (11): 3960–3979.
25. Chang, C.-I, Jiao, X., Du, Y., and Chen, H.M. (2011). Component-based unsupervised linear spectral mixture analysis for hyperspectral imagery. *IEEE Transactions on Geoscience and Remote Sensing* 49 (11): 4123–4137.
26. Chang, C.-I, Li, Y., Hobbs, M. et al. (2015). Progressive band processing of anomaly detection. *IEEE Journal of Selected Topics in Signal Processing* 8 (7): 3558–3571.
27. Chen, S.Y., Wang, Y., Wu, C.C. et al. (2014). Real time causal processing of anomaly detection in hyperspectral imagery. *IEEE Transactions on Aerospace and Electronics Systems* 50 (2): 1511–1534.
28. Chang, C.-I (2010). Multiple-parameter receiver operating characteristic analysis for signal detection and classification. *IEEE Sensors Journal* 10 (3): 423–442.
29. Chang, C.-I (2021). An effective evaluation tool for hyperspecral target detection: 3D receiver operating characteristic analysis. *IEEE Transactions on Geoscience and Remote Sensing* 59 (6): 5131–5153.

# 5

# Band Subset Selection for Hyperspectral Imaging

Chein-I Chang
*Center for Hyperspectral Imaging in Remote Sensing (CHIRS), Information and Technology College, Dalian Maritime University, Dalian, China*
*Remote Sensing Signal and Image Processing Laboratory, Department of Computer Science and Electrical Engineering, University of Maryland Baltimore County, Baltimore, MD, USA*

## 5.1 Introduction

A brief review on band selection (BS) is already given in Chapter 4. This chapter begins with how multiple bands are selected via search algorithms which are absent in BS. Since simultaneous multiple band selection (SMBS) selects multiple bands "*simultaneously*," it requires the prior knowledge of bands to be selected, $n_{\mathrm{BS}}$, in advance. Suppose that the full band set is given by $\Omega = \{\mathbf{B}_1, \mathbf{B}_2, \ldots, \mathbf{B}_L\}$ where $L$ is the total number of bands, $\mathbf{B}_{l_j}$ is the selected $j$th band. Let $n_{\mathrm{BS}} = p$ and $\Omega_p = \{\mathbf{B}_{l_1}, \mathbf{B}_{l_2}, \ldots, \mathbf{B}_{l_p}\}$ be a band subset selected from $\mathbf{\Omega}$. In order to find a set of optimal multiple bands, $\Omega_p^*$, it must exhaust all possible $p$-band combinations among $L$ bands, which are $\binom{L}{p} = \frac{L!}{p!(L-p)!}$ possibilities. Practically speaking, this is impossible in reality. To mitigate this dilemma, some SMMBS approaches select multiple bands as a band group/combination and then employ a search strategy to find an optimal band group/combination such as particle swarm optimization (PSO) [1], evolutional-like algorithm [2], multi-graph determinantal point process (DPP) [3], and optimal neighborhood reconstruction (ONR) in [4]. A major issue with these approaches is that they do not exhaust all the $p$-band combinations to find $\Omega_p^*$ but rather randomly select a very small portion of $\mathbf{\Omega}$ to make finding an optimal set of band group/combinations possible by a specific search algorithm.

This chapter develops band subset selection (BSS), which is also an SMMBS method but is quite different from the SMMBS methods described above. It considers multiple bands to represent a band subset as an entity, in which case a $p$-band combination is considered as a band subset. In particular, BSS distinguishes itself from the aforementioned SMMBS methods in terms of its search strategy by taking a direct approach to finding $\Omega_p^*$ from $\mathbf{\Omega}$ where the number of selected bands, $p$, is estimated by

*Advances in Hyperspectral Image Processing Techniques*, First Edition. Edited by Chein-I Chang.

virtual dimensionality (VD) developed in [5–8]. Its idea is originated from the concept of endmember finding. That is, if we interpret finding $p$ desired bands from a full band set $\mathbf{\Omega}$ as finding $p$ endmembers, $\mathbf{E}_p$, from an entire data space $\mathbf{X}$, then finding $\Omega_p^*$ is similar to finding an optimal set of $p$ endmembers, $\mathbf{E}_p^*$, among data sample vectors in $\mathbf{X}$. With this interpretation, an immediate discovery is that the search strategy for finding $\Omega_p^*$ is essentially equivalent to finding $\mathbf{E}_p^*$. It is well known that the N-finder algorithm (N-FINDR) developed by Winter [9] has been widely used for finding endmembers. But it also suffers from a similar computational issue by exhausting all possible $p$-endmember combinations. This issue was investigated by [10–13] where two sequential algorithms, called sequential N-FINDR (SQ N-FINDR) and successive N-FINDR (SC N-FINDR), were developed to execute N-FINDR sequentially and successively without exhausting all $p$-endmember combinations. By taking advantage of SQ N-FINDR and SC N-FINDR, two similar sequential algorithms can be therefore designed for BSS, referred to as sequential BSS (SQBSS) and successive BSS (SCBSS), to find $\Omega_p^*$. A crucial difference between SQ/SC N-FINDR and SQBSS/SCBSS is their used criteria. For N-FINDR, maximal simplex volume (MSV) is a criterion to find an optimal set of endmembers. Apparently, MSV is not applicable to BSS. To make BSS work, the concept of channel capacity (CC) [14] is particularly selected as a general criterion for BSS corresponding to MSV used for endmember finding. Nevertheless, if a particular application is specified for BSS, CC can be replaced by a custom-designed criterion. For example, if an application is specified by endmember finding, MSV can be used instead of CC for BSS for this purpose.

It is natural that BSS can be further considered as a generalization of traditional single band selection (SBS). Specifically, when $p = 1$, all possible 1–band combinations are then reduced to $L$. In this case, BSS is reduced to SBS where each single band can be selected according to its specifically designed criterion for SBS.

## 5.2 Simultaneous Multiple Band Selection

Since different spectral bands provide various levels of information of interest, the primary goal of multiple band selection (MBS) is to select multiple bands from the original band set to represent the original data in some sense of optimality. In this case, the information preserved by MBS has significant impact on data analysis. This is because the information of un-selected bands will be completely discarded after MBS. Therefore, a key success to MBS is how to design different effective criteria for MBS to meet various applications.

Assume that $J(\Omega_{\text{MBS}})$ is a generic objective function of $\Omega_{\text{MBS}}$ for MBS to be optimized. For a given number of selected bands, $n_{\text{BS}}$, a BS technique is to find an optimal band subset, $\Omega_{\text{MBS}}^*$ with $|\Omega_{\text{MBS}}^*| = n_{\text{MBS}}$ which satisfies the following optimization problem:

$$\Omega_{\text{MBS}}^* = \arg\left\{ \max / \min_{\Omega_{\text{MBS}} \subset \Omega, |\Omega_{\text{MBS}}| = n_{\text{MBS}}} J(\Omega_{\text{MBS}}) \right\}. \tag{5.1}$$

Depending upon how the objective function $J(\Omega_{\text{MBS}})$ is designed, the optimization in (5.1) can be performed by either maximization or minimization over all possible band subsets $\Omega_{\text{MBS}}$ in $\Omega$ with $|\Omega_{\text{MBS}}| = n_{\text{MBS}}$.

Solving a general BS problem (5.1) requires two processes. One is to develop an optimization algorithm to solve $J(\mathbf{\Omega}_{\text{MBS}})$ in (5.1) for a particular band set $\mathbf{\Omega}_{\text{MBS}}$, and then followed by an exhaustive search for all possible $\binom{L}{|\Omega_{\text{MBS}}|} = \frac{L!}{(L - |\Omega_{\text{MBS}}|)!\,|\Omega_{\text{MBS}}|!}$ combinations with $L$ and $|\Omega_{\text{MBS}}|$ being the total number of spectral bands in $\Omega$ and the number of bands to be selected in $\Omega_{\mathrm{MBS}}$, respectively. However, such exhaustive search is generally infeasible if $L$ is relatively large even though $|\Omega_{\mathrm{MBS}}|$ is small. Accordingly, to avoid directly solving $\mathbf{\Omega}_{\mathrm{MBS}}$ for $J(\Omega_{\mathrm{MBS}})$ in (5.1), many MBS techniques in the literature are developed according to (i) how to design $J(\mathbf{\Omega}_{\mathrm{MBS}})$ in (5.1) and (ii) how to develop an effective search algorithm for finding appropriate bands based on a specific designed $J(\mathbf{\Omega}_{\mathrm{MBS}})$ derived from (i).

To address the issue of (i) a general approach is to design various BP criteria to rank each of bands in accordance with priority scores calculated by a particular BP criterion for all bands using $J(\mathbf{\Omega}_{\mathrm{MBS}})$ such as SNR, variance, entropy, and information divergence (ID) [15]. Such a BP criterion is generally designed based on data statistics and can be performed in an unsupervised manner. But a disadvantage is that it has nothing to do with applications. In other words, once bands are selected by a BP criterion, these bands will be fixed and used for one-fit-all applications such as anomaly detection (AD), target detection, endmember finding, linear spectral unmixing (LSU), and classification. To address this issue, another approach is driven by applications as well as the concept of sparsity. However, in this case, BS requires prior knowledge about the applications. For example, the number of classes of interest to be classified include training samples, targets of interest with $J(\mathbf{\Omega}_{\mathrm{MBS}})$ in (5.1) measured by classification rate such as Jeffreys–Matusita (JM) distance or Bhattacharyya distance, minimum estimated abundance covariance (MEAC) in [16], and detection rate such as receiver operating characteristics (ROC). Accordingly, such BS methods are supervised.

As for the issue of (ii), developing an efficient search algorithm to find a desired set of bands is crucial. There are two ways to design a search strategy. One is SQMBS. For BP-based MBS methods, all the bands are already ranked by their priority scores. So, if there are $p$ bands needed to be selected, the bands can be selected with the $p$ highest priority scores and MBS is carried out by SQMBS in which case no search strategy is needed. However, this is traded for another issue needed to be taken care of. Since a band is selected with a higher priority score, its adjacent bands may be also selected due to the fact that these bands also have higher priority scores. Moreover, because these bands are selected individually without accounting for inter-band correlation, de-correlation (BD) must be also implemented in conjunction with BP to remove bands which are highly correlated to bands being selected. However, this also comes with an issue of how to choose an appropriate threshold for BD. Obviously, such a threshold cannot be fixed at a constant and must vary with different applications. To resolve these dilemmas, recent works on BS have been directed to SMMBS which finds a group of bands as band groups or combinations. In this case, their main focus is placed on the design of various search strategies, for example, sequential feed-forward floating search (SFFS) [17], PSO algorithm [1], firefly algorithm (FA) [18], evolutional algorithm in [2], MDPP [3], and dominant set extraction BS (DSEBS) in [19]. Unfortunately, all these search algorithms were carried out by randomly selecting a set of band groups/combinations, such as total number of particles in a swarm, $M = 25$ in [1], total number of fireflies, $m = 10$ in [17], antibody populations was fixed at a predetermined $N = 10$ in [2], and number of spectral subspaces, $\{m\}_{m=1}^{30}$ in [3], from which an optimal band group/combination can be found. Despite that all these methods can be considered as SMMBS methods, they only run for very small specific values of $M$, $m$, and $N$ with $1 \leq M, m, N = \binom{L}{p} = \frac{L!}{p!(L-p)!}$ and do not exhaust all $p$-band combinations for optimizing $J(\Omega_{\mathrm{MBS}})$ in (5.1). So, technically speaking, they are not optimal SMMBS methods. Figure 5.1 depicts a diagram of MBS methods.

In what follows, we will first focus on the second issue of developing algorithms for search strategy for BSS in Section 5.3 and investigate the first issue to Section 5.4.

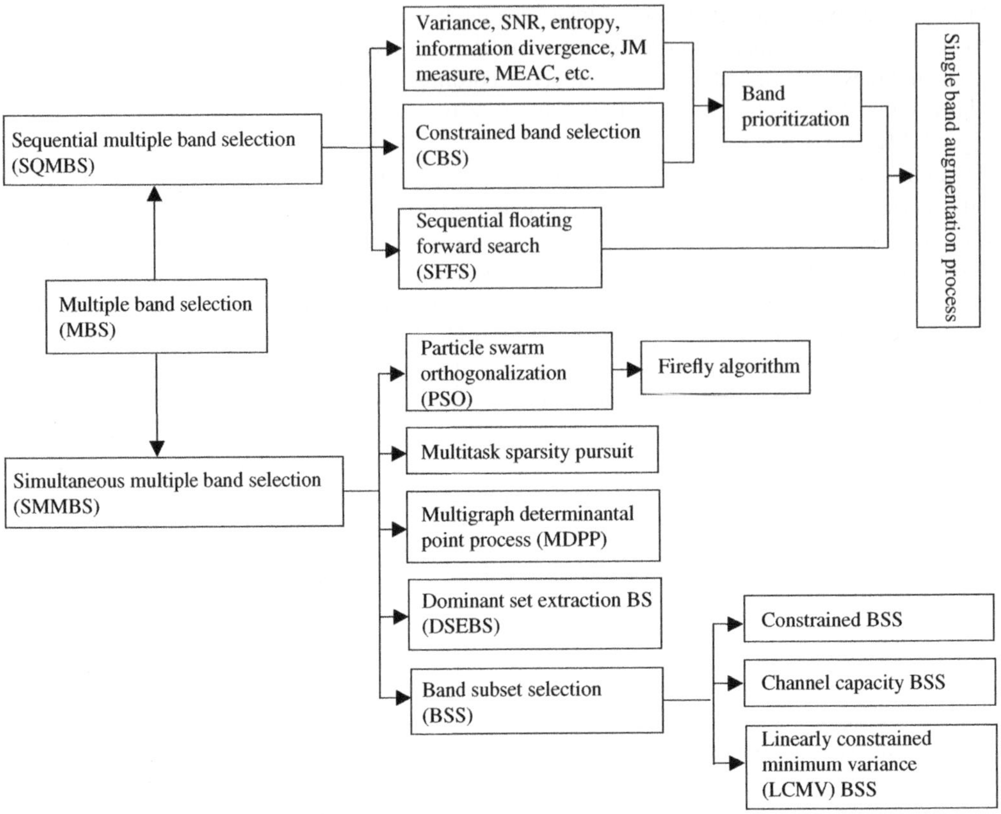


**Figure 5.1** Diagram of MBS methods.

## 5.3 Search Strategies for BSS

As noted above, there is a close relationship between searching for finding endmembers and finding optimal bands where two sequential versions of N-FINDR, SQ N-FINDR and SC N-FINDR developed in [10–13], can be further modified and re-derived for finding an optimal band subset.

### *5.3.1 Sequential Band Subset Selection*

The first algorithm presented in this section is derived from SQ N-FINDR and is called SQBSS which can be described as follows:

***SQBSS***

1. Initialization:
   a. Let $p$ be the number of selected bands determined by VD.
   b. Let $\Omega_p^{(0)} = \left\{\mathbf{B}_1^{(0)}, \mathbf{B}_2^{(0)}, \ldots, \mathbf{B}_p^{(0)}\right\}$ be a set of initial bands uniformly selected from the entire band set $\mathbf{\Omega}$. Set $l = 1$.
2. Outer Loop: (using index $l$ as a counter to keep track band $\mathbf{B}_l$)

   Check $l = L$. If it is, the algorithm terminated. Otherwise, let $l \leftarrow l + 1$ and continue.

3. Input the $l$th band, $\mathbf{B}_l$.

4. Inner Loop: (using $j$ as a counter to keep track the $j$th band $\mathbf{B}_j$)

   For$1 \leq j \leq p$, let $\Omega_p^{(l)} = \left\{\mathbf{B}_1^{(l)}, \ldots, \mathbf{B}_{j-1}^{(l)}, \mathbf{B}_j^{(l)}, \mathbf{B}_{j+1}^{(l)}, \ldots, \mathbf{B}_p^{(l)}\right\}$ and $\Omega_{j|p}^{(l)} = \left\{\mathbf{B}_1^{(l)}, \ldots, \mathbf{B}_{j-1}^{(l)}, \mathbf{B}_l, \mathbf{B}_{j+1}^{(l)}, \ldots, \mathbf{B}_p^{(l)}\right\}$ for the band $\mathbf{B}_l$ with the $j$th component and calculate $J_{\text{BSS}}\left(\Omega_{1|p}^{(l)}\right), J_{\text{BSS}}\left(\Omega_{2|p}^{(l)}\right), \ldots, J_{\text{BSS}}\left(\Omega_{p|p}^{(l)}\right)$.

5. If any $J_{\text{BSS}}\left(\Omega_{j|p}^{(l)}, \Omega\right) \geq J_{\text{BSS}}\left(\Omega_p^{(l)}, \Omega\right)$ for all $1 \leq j \leq p$, go to step 2. Otherwise, continue.

6. Replacement rule:

   Find an index $j^*$ by

$$j^* = \arg\left\{\min_{1 \leq j \leq p} J_{\text{BSS}}\left(\Omega_{j|p}^{(l)}\right)\right\} \tag{5.2}$$

   where the band specified by the $j^*$ band, i.e. $\mathbf{B}_{j^*}^{(l)}$ will be replaced by the $l$th band $\mathbf{B}_l$. As a result, a new set of bands for next iteration is then produced by letting $\mathbf{B}_{j^*}^{(l+1)} = \mathbf{B}_l$ and $\mathbf{B}_j^{(l+1)} = \mathbf{B}_j^{(l)}$ for $j \neq j^*$ and go to step 3.

### 5.3.2 *Successive Band Subset Selection*

A second algorithm is developed from SC N-FINDR and is called SCBSS with detailed implementation given as follows:

***SCBSS***

1. Initialization:

   Let $p$ be the number of endmembers required to generate and $\left\{\mathbf{B}_1^{(0)}, \mathbf{B}_2^{(0)}, \ldots, \mathbf{B}_p^{(0)}\right\}$ be a set of initial bands uniformly selected from $\Omega = \{\mathbf{B}_l\}_{l=1}^{L}$ which is the set of all spectral bands.

2. Outer Loop

   For $1 \leq j \leq p$ find $\mathbf{B}_j^{(*)}$ which will replace $\mathbf{B}_j^{(0)}$.

3. Inner Loop for a given $j$ in the outer loop:

   For $1 \leq l \leq L$ let $\Omega_{j-1|p}^{(l)} = \left\{\mathbf{B}_1^{(*)}, \ldots, \mathbf{B}_{j-1}^{(*)}, \mathbf{B}_j^{(0)}, \mathbf{B}_{j+1}^{(0)}, \ldots, \mathbf{B}_p^{(0)}\right\}$ and $\Omega_{j=\mathbf{B}_l|p}^{(l)} = \left\{\mathbf{B}_1^{(*)}, \ldots, \mathbf{B}_{j-1}^{(*)}, \mathbf{B}_l, \mathbf{B}_{j+1}^{(0)}, \ldots, \mathbf{B}_p^{(0)}\right\}$ for the band $\mathbf{B}_l$ with the $j$th component and calculate $J_{\text{BSS}}\left(\Omega_{j=\mathbf{B}_l|p}^{(l)}\right)$ for all $\{\mathbf{B}_l\}_{l=1}^{L}$, while fixing other bands $\mathbf{B}_i^{(*)}$ with $i<j$ and $\mathbf{B}_i^{(0)}$ with $i>j$.

4. If $J_{\text{BSS}}\left(\Omega_{j=\mathbf{B}_l|p}^{(l)}\right) \geq J_{\text{BSS}}\left(\Omega_{j-1|p}^{(l)}\right)$ for all $\mathbf{B}_l$ with $1 \leq l \leq L$, then $j \leftarrow j+1$ and go step 2. Otherwise, continue.

5. Find $\mathbf{B}_j^{(*)} = \arg\left\{\min_{\mathbf{B}_l} J_{\text{BSS}}\left(\Omega_{j=\mathbf{B}_l|p}^{(l)}\right)\right\}$ and set a new band subset $\Omega_{j|p}^{(l)} = \left\{\mathbf{B}_1^{(*)}, \ldots, \mathbf{B}_{j-1}^{(*)}, \mathbf{B}_j^{(*)}, \mathbf{B}_{j+1}^{(0)}, \ldots, \mathbf{B}_p^{(0)}\right\}$.

6. Stopping rule:

   If $j \leq p$, then Otherwise, the final set of $\Omega_p^* = \Omega_{p|p}^{(L)} = \left\{\mathbf{B}_1^{(*)}, \mathbf{B}_2^{(*)}, \ldots, \mathbf{B}_p^{(*)}\right\}$ is the desired $p$ bands.

It is worth noting that unlike SQ N-FINDR and SC N-FINDR which have the same issue of random initial conditions, SQBSS/SCBSS do not have this problem because they can use bands selected by uniform band selection (UBS) as their initial bands. Therefore, SQBSS/SCBSS always performs better than UBS. However, according to our experimental results, using UBS has also a potential drawback to have solutions trapped in local optima. In this case, randomly selecting band subsets can avoid such dilemma, but this is also traded for inconsistent final results.

## 5.4 Channel Capacity BSS

This section intends to deal with the first issue regarding $J(\Omega_{\text{MBS}})$ in (5.1). An approach which used a general criterion, CC [14] for BSS without specifying a particular application was proposed in [20].

To be more specific, let the original band set be denoted by $\Omega_L = \{\mathbf{B}_l\}_{l=1}^{L}$ and the selected multiple-band set be denoted by $\Omega_p = \{\tilde{\mathbf{B}}_j\}_{j=1}^{p}$ where the selected $\tilde{\mathbf{B}}_j$ is not necessarily the same as the original $\mathbf{B}_j$. Furthermore, we re-express each spectral band image $\mathbf{B}_l$ as a band vector by concatenating line by line, denoted by $\mathbf{b}_l$ for $1 \leq l \leq L$. Also, let $\mathbf{b}_l = (b_{l1}, b_{l2}, \ldots, b_{lN})^T$ represent the band image $\mathbf{B}_l$ as its band vector and $\tilde{\mathbf{b}}_j = (\tilde{b}_{j1}, \tilde{b}_{j2}, \ldots, \tilde{b}_{jN})^T$ represent the band image $\tilde{\mathbf{B}}_j$ as its band vector where $N$ is the total number of pixels in a band image. Then, we define their respective probability vectors, $\mathbf{p}_l = (p_{l1}, p_{l2}, \ldots, p_{lN})^T$ and $\tilde{\mathbf{p}}_j = (\tilde{p}_{j1}, \tilde{p}_{j2}, \ldots, \tilde{p}_{jN})^T$ as follows:

$$p_{li} = \frac{b_{li}}{\sum_{n=1}^{N} b_{ln}} \text{ and } \tilde{p}_{ji} = \frac{\tilde{b}_{ji}}{\sum_{n=1}^{N} \tilde{b}_{jn}}. \tag{5.3}$$

By virtue of (5.3), the band discrimination between a band vector, $\mathbf{b}_j$, and a selected band, $\tilde{\mathbf{b}}_j$, can be measured by spectral information divergence (SID) [21, 22] defined as

$$\text{D}_{\text{SID}}\left(\mathbf{B}_l, \tilde{\mathbf{B}}_j\right) = \text{SID}(\mathbf{p}_l, \tilde{\mathbf{p}}_j) = \text{D}\left(\mathbf{p}_j \| \tilde{\mathbf{p}}_j\right) + \text{D}\left(\tilde{\mathbf{p}}_j \| \mathbf{p}_j\right) \tag{5.4}$$

where

$$\text{D}\left(\mathbf{p}_j \| \tilde{\mathbf{p}}_j\right) = \sum_{n=1}^{N} p_{ln} \log \frac{p_{ln}}{\tilde{p}_{jn}} \tag{5.5}$$

is the directed divergence, Kullback–Leibler information distance or relative entropy [23]. Of course, another spectral measure, called spectral angle mapper (SAM) in [21] can be also used for this purpose. Using (5.4), we define a channel transition probability matrix from the original band set $\Omega_L = \{\mathbf{B}_l\}_{l=1}^{L}$ to the selected band set $\Omega_p = \{\tilde{\mathbf{B}}_j\}_{j=1}^{p}$ by

$$Q_{j|l} = \frac{\text{SID}(\mathbf{p}_l, \tilde{\mathbf{p}}_j)}{\sum_{i=1}^{N} \text{SID}(\mathbf{p}_l, \tilde{\mathbf{p}}_j)} \tag{5.6}$$

which measures correlation between a band and a selected band. If band $\mathbf{b}_l$ and $\tilde{\mathbf{b}}_j$ have less correlation, then $\tilde{\mathbf{b}}_j$ will be more likely to be selected. This implies that their transition probability

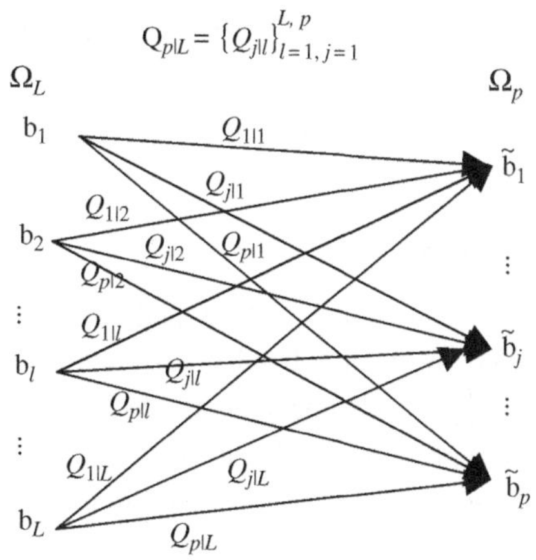


**Figure 5.2** Channel specified by $\boldsymbol{\Omega}_L$, $\Omega_p$, and $\mathbf{Q}_{p|L} = \{Q_{j|l}\}_{l=1,j=1}^{L,p}$.

will be higher. Now, we can construct a channel $\left(\Omega_L, \Omega_p, \mathbf{Q}_{p|L} = \{Q_{j|l}\}_{l=1,j=1}^{L,p}\right)$ as shown in Figure 5.2 specified by the input space given by $\boldsymbol{\Omega}$, the output space by $\Omega_p$, and the channel transition probability matrix given by $\mathbf{Q}_{p|L}(\Omega_L, \Omega_p) = \{Q_{j|l}\}_{l=1,j=1}^{L,p}$.

For a given band channel specified by $\left(\Omega_L, \Omega_p, \mathbf{Q}_{p|L} = \{Q_{j|l}\}_{l=1,j=1}^{L,p}\right)$, the mutual information is defined by

$$I(\Omega_L; \Omega_p) = \sum_{l=1}^{L} p_l \sum_{j=1}^{p} Q_{j|l} \log \frac{Q_{j|l}}{q_j} \tag{5.7}$$

where $q_j = \sum_{l=1}^{J} p_l Q_{j|l}$. According to information theory in [14], we can further define CC via (5.7) as a general criterion for BSS as follows:

$$J_{\mathrm{BSS}}\left(\Omega_p^{\mathrm{CC}}\right) = \max_{\{p_l\}_{l=1}^{L}} I(\Omega_L; \Omega_p) \tag{5.8}$$

which can be solved by the well-known Blahut algorithm [24]. Now we can use (5.8) to find an optimal band subset $\Omega_p^*$ which satisfies

$$\Omega_p^* = \arg\left\{\max_{\Omega_p^{\mathrm{CC}}}\left\{J_{\mathrm{BSS}}\left(\Omega_p^{\mathrm{CC}}\right)\right\}\right\} \tag{5.9}$$

that is,

$$\max_{\Omega_p^{\mathrm{CC}}}\left\{J_{\mathrm{BSS}}\left(\Omega_p^{\mathrm{CC}}\right)\right\} = J_{\mathrm{BSS}}\left(\Omega_p^*\right). \tag{5.10}$$

## 5.5 Multiple Band-Constrained Band Subset Selection

In this section, we extend the single band-based constrained band selection (CBS) in Chapter 4 to constrained BSS (CBSS) using the CC-based BSS criterion specified by (5.10).

### 5.5.1 Constrained BSS (CBSS)

For each band $\mathbf{B}_l$ in $\Omega_L = \{\mathbf{B}_l\}_{l=1}^L$ its size of $M \times N$ and can be represented by either a column vector of dimension *MN*, denoted by $\mathbf{b}_l$ or a band matrix denoted by $\mathbf{B}_{\Omega_L} = [\mathbf{b}_1\mathbf{b}_2\cdots\mathbf{b}_L]$ formed by all band image vectors, $\{\mathbf{b}_l\}_{l=1}^L$. Assume that a selected band subset is denoted by $\Omega_p = \{\mathbf{b}_{l_1}, \mathbf{b}_{l_2}, \ldots, \mathbf{b}_{l_p}\}$. For clarity, we represent $\mathbf{b}_{l_j} = \tilde{\mathbf{b}}_j$ and $\mathbf{B}_{\Omega_p} = [\mathbf{b}_1\mathbf{b}_2\cdots\mathbf{b}_p]$ as a band matrix formed by selected bands in $\boldsymbol{\Omega}_p$ where $p = n_{\text{BSS}}$. By taking advantage of the CEM-based CBS in Chapter 4, we can derive

$$\min_{\mathbf{w}} \mathbf{w}^T \tilde{\mathbf{Q}}_{\Omega-\Omega_{\text{BS}}} \mathbf{w} \text{ subject to}$$
$$[\mathbf{b}_1\mathbf{b}_2\cdots\mathbf{b}_p]^T \mathbf{w} = \mathbf{w}^T \mathbf{B}_{\Omega_{\text{BS}}} = \mathbf{1}_{p\times 1} = \Big(\underbrace{1, 1, \ldots, 1}_{p}\Big)^T \tag{5.11}$$

and

$$\tilde{\mathbf{Q}}_{\Omega-\Omega_{\text{BS}}} = (1/\,|\,\Omega-\Omega_{\text{BS}}\,|) \sum_{\mathbf{b}_j\in\Omega-\Omega_{\text{BS}}} \mathbf{b}_j\mathbf{b}_j^T. \tag{5.12}$$

Since $\tilde{\mathbf{Q}}_{\Omega-\Omega_{\text{BS}}}$ is not of full rank due to size of $\Omega-\Omega_{\text{BS}}$, $|\Omega-\Omega_{\text{BS}}|$ less than *MN*, it is ill-ranked. The solution to (5.11) is given by

$$\tilde{\mathbf{w}}_{\Omega_{\text{BS}}}^{\text{CBSS}} = \mathbf{Q}^{\#}_{\Omega-\Omega_{\text{BS}}} \mathbf{B}_{\Omega_{\text{BS}}} \left(\mathbf{B}^T_{\Omega_{\text{BS}}} \mathbf{Q}^{\#}_{\Omega-\Omega_{\text{BS}}} \mathbf{B}_{\Omega_{\text{BS}}}\right)^{-1} \mathbf{1}_{p\times 1} \tag{5.13}$$

with $\mathbf{Q}^{\#}_{\Omega-\Omega_{\text{BS}}}$ being the pseudo-inverse of $\tilde{\mathbf{Q}}_{\Omega-\Omega_{\text{BS}}}$. Using (5.13), we can define the following measure for BS:

$$\tilde{\rho}_{\Omega_{\text{BS}}} = \left(\tilde{\mathbf{w}}_{\Omega_{\text{BS}}}^{\text{CBSS}}\right)^T \tilde{\mathbf{Q}}_{\Omega-\Omega_{\text{BS}}} \tilde{\mathbf{w}}_{\Omega_{\text{BS}}}^{\text{CBSS}}. \tag{5.14}$$

The larger the $\tilde{\rho}_{\Omega_{\text{BS}}}$ is, the less dependent the band image $\boldsymbol{\Omega}_{\text{BS}}$ on other band images is, thus the more significant the $\boldsymbol{\Omega}_{\text{BS}}$ is. Therefore, $\tilde{\rho}_{\Omega_{\text{BS}}}$ can be used as a priority score measure to rank all band subsets. Its idea is the same as the priority score used in CBS except that the latter is used to rank single individual bands not band subsets. These priority scores allow users to select better band subsets rather than a set of individual bands for applications such as target detection and classification. If $p = 1$ in (5.17), then $[\mathbf{b}_1\mathbf{b}_2\cdots\mathbf{b}_p]$ is replaced by $\mathbf{b}_l$ and CBSS is reduced to CBS. Accordingly, CBS can be considered as a special case of CBSS.

### 5.5.2 *Search Algorithms for CBSS*

According to CBSS, one major issue in implementing CBSS, which is not encountered in CBS, is to search for an optimal band subset, $\Omega^*_{\mathrm{BS}}$, in (5.1). Interestingly, this turns out to be exactly the same issue arising in N-FINDR for finding $p$ endmembers simultaneously [9] which searches for $p$ endmembers among all the $MN$ data sample vectors with all possible $p$-endmember combinations, $\binom{MN}{p} = \frac{MN!}{p!(MN-p)!}$. Our recognition of the similarity between CBSS and N-FINDR allows us to modify and re-derive two sequential versions of CBSS, called sequential BSS (SQ CBSS) and successive CBSS (SC CBSS), as follows.

#### 5.5.2.1 Sequential CBSS (SQ CBSS)

The first algorithm is sequential CBSS (SQ CBSS) derived from SQ N-FINDR [10–13] and can be described as follows:

***SQ CBSS***

1. Initialization:

   a. Let $p$ be the number of selected bands determined by virtual dimensionality (VD).

   b. Let $\Omega^{(0)}_p = \left\{\mathbf{B}^{(0)}_1, \mathbf{B}^{(0)}_2, \ldots, \mathbf{B}^{(0)}_p\right\}$ be a set of initial bands selected from the entire band set $\boldsymbol{\Omega}$ by uniform band selection (UBS). Set $l = 1$.

2. Outer Loop: (using index $l$ as a counter to keep track of band $\mathbf{B}_l$) and $\Omega^{(l)}_p = \left\{\mathbf{B}^{(l)}_1, \mathbf{B}^{(l)}_2, \ldots, \mathbf{B}^{(l)}_p\right\}$

   Check $l = L$. If it is, the algorithm terminates. Otherwise, let $l \leftarrow l + 1$ and continue.

3. Input the $l$th band, $\mathbf{B}_l$.

4. Inner Loop: (using $j$ as a counter to keep track of the $j$th band $\mathbf{B}_j$)

   For $1 \leq j \leq p$, we recalculate $\tilde{\rho}_{\Omega_{\mathrm{BS}}}\left(\mathbf{B}^{(l)}_1, \ldots, \mathbf{B}^{(l)}_{j-1}, \mathbf{B}_j, \mathbf{B}^{(l)}_{j+1}, \ldots, \mathbf{B}^{(l)}_p\right)$ for the band $\mathbf{B}_j$. If all these $p$ recalculated the performance measure, $\tilde{\rho}_{\Omega_{\mathrm{BS}}}\left(\mathbf{B}_k, \mathbf{B}^{(l)}_2, \ldots, \mathbf{B}^{(l)}_p\right)$, $\tilde{\rho}_{\Omega_{\mathrm{BS}}}\left(\mathbf{B}^{(l)}_1, \mathbf{B}_k, \mathbf{B}^{(l)}_3, \ldots, \mathbf{B}^{(l)}_p\right), \ldots, \tilde{\rho}_{\Omega_{\mathrm{BS}}}\left(\mathbf{B}^{(l)}_1, \ldots, \mathbf{B}^{(l)}_{p-1}, \mathbf{r}_k\right)$, is greater than $\tilde{\rho}_{\Omega_{\mathrm{BS}}}\left(\mathbf{B}^{(l)}_1, \mathbf{B}^{(l)}_2, \ldots, \mathbf{B}^{(l)}_p\right)$, go to step 2. Otherwise, continue.

5. Find an index $j^*$ by

$$j^* = \arg\left\{\min_{1 \leq j \leq p} \tilde{\rho}_{\Omega_{\mathrm{BS}}}\left(\mathbf{B}^{(l)}_1, \ldots, \mathbf{B}^{(l)}_{j-1}, \underbrace{\mathbf{B}_l}_{j}, \mathbf{B}^{(l)}_{j+1}, \ldots, \mathbf{B}^{(l)}_p\right)\right\} \tag{5.15}$$

   which specifies the band to be replaced by the $l$th band $\mathbf{B}_l$. Such a band is now denoted by $\mathbf{B}^{(l+1)}_j$. A new set of bands is then produced by letting $\mathbf{B}^{(l+1)}_{j^*} = \mathbf{B}_l$ and $\mathbf{B}^{(l+1)}_j = \mathbf{B}^{(l)}_j$ for $j \neq j^*$ to form $\Omega^{(l+1)}_p = \left\{\mathbf{B}^{(l+1)}_1, \mathbf{B}^{(l+1)}_2, \ldots, \mathbf{B}^{(l+1)}_p\right\}$ and go to step 3.

### 5.5.2.2 Successive CBSS (SC CBSS)

The second algorithm is successive CBSS (SC CBSS) derived from SC N-FINDR [10–13] with its detailed implementation given as follows:

***SC CBSS***

1. Initialization:

   Let $p$ be the number of endmembers we need to generate and $\Omega_p^{(0)} = \left\{\mathbf{B}_1^{(0)}, \mathbf{B}_2^{(0)}, \ldots, \mathbf{B}_p^{(0)}\right\}$ be a set of initial bands uniformly selected from $\boldsymbol{\Omega}$ by UBS where $\Omega = \{\mathbf{B}_l\}_{l=1}^{L}$ is the set of all spectral bands.

2. For $1 \le j \le p$ find $\mathbf{B}_j^{(*)}$ which will replace $\mathbf{B}_j^{(0)}$.

3. Inner Loop for a given $j$ in the outer loop:

   For $1 \le l \le L$, calculate $\tilde{p}_{\Omega_{\mathrm{BS}}}\left(\mathbf{B}_1^{(*)}, \ldots, \mathbf{B}_{j-1}^{(*)}, \mathbf{B}_l, \mathbf{B}_{j+1}^{(0)}, \ldots, \mathbf{B}_p^{(0)}\right)$ for all $\{\mathbf{B}_l\}_{l=1}^{N}$, while fixing bands $\mathbf{B}_i^{(*)}$ with $i<j$ and $\mathbf{B}_i^{(0)}$ with $i>j$.

4. Find $\mathbf{B}_j^{(*)} = \arg\left\{\max_{\mathbf{B}_l} \tilde{p}_{\Omega_{\mathrm{BS}}}\left(\mathbf{B}_1^{(*)}, \ldots, \mathbf{B}_{j-1}^{(*)}, \mathbf{B}_l, \mathbf{B}_{j+1}^{(0)}, \ldots, \mathbf{B}_p^{(0)}\right)\right\}$ and form

   $\Omega_p^{(j)} = \left\{\mathbf{B}_1^{(*)}, \ldots, \mathbf{B}_j^{(*)}, \mathbf{B}_{j+1}^{(0)} \ldots, \mathbf{B}_p^{(0)}\right\}$.

5. Stopping rule:

   If $j \le p$, then $j \leftarrow j+1$ and go step 2. Otherwise, $\Omega_p^{(p)} = \left\{\mathbf{B}_1^{(*)}, \mathbf{B}_2^{(*)}, \ldots, \mathbf{B}_p^{(*)}\right\}$ is the desired $p$ bands.

## 5.6 Application-Specified BSS (AS-BSS)

As noted, the concept of BSS is derived from endmember finding. It formulates the problem of finding an optimal $p$-band set as the problem of finding an optimal $p$-endmember set where the purity of endmembers is used as a criterion in $J(\Omega_{\mathrm{BSS}})$ where $\Omega_{\mathrm{BSS}}$ is to specifically emphasize the band subset selected by BSS. In Section 5.4, BSS is developed using CC as a general criterion to select an optimal band subset. In Section 5.5, BSS is further extended to CBSS by imposing constraints on band subsets to be selected. While BSS can be made application-independent in Sections 5.4 and 5.5, on many occasions we can also make BSS application-dependent by specifying an application-dependent criterion for $J(\Omega_{\mathrm{MBS}})$ and the resulting BSS is referred to as application-specified BSS (AS-BSS). For example, if an application is LSU, $J_{\mathrm{A}}(\Omega_{\mathrm{BSS}})$ will be the minimal unmixed error (MUE), $J_{\mathrm{UME}}(\Omega_{\mathrm{BSS}})$. On the other hand, if endmember finding is designated as an application, $J_{\mathrm{A}}(\Omega_{\mathrm{BSS}})$ will be $J_{\mathrm{MSV}}(\Omega_{\mathrm{BSS}})$ where the MSV is used as a criterion for BSS. Furthermore, if AD is used for an application [25], $J_{\mathrm{A}}(\Omega_{\mathrm{BSS}})$ can be $J_{\mathrm{ROC}}(\Omega_{\mathrm{BSS}})$ that measures errors between the maximal areas under ROC produced by the full band set and a selected band subset if the ground truth is provided. As for classification, MEAC [16] can be used to replace $J_{\mathrm{A}}(\Omega_{\mathrm{BSS}})$ with $J_{\mathrm{MEAC}}(\Omega_{\mathrm{BSS}})$.

Figure 5.3 describes various BSS approaches and extensions with two extensions yet to be explored, compressive sensing BSS [26] and band sampling BSS [27].

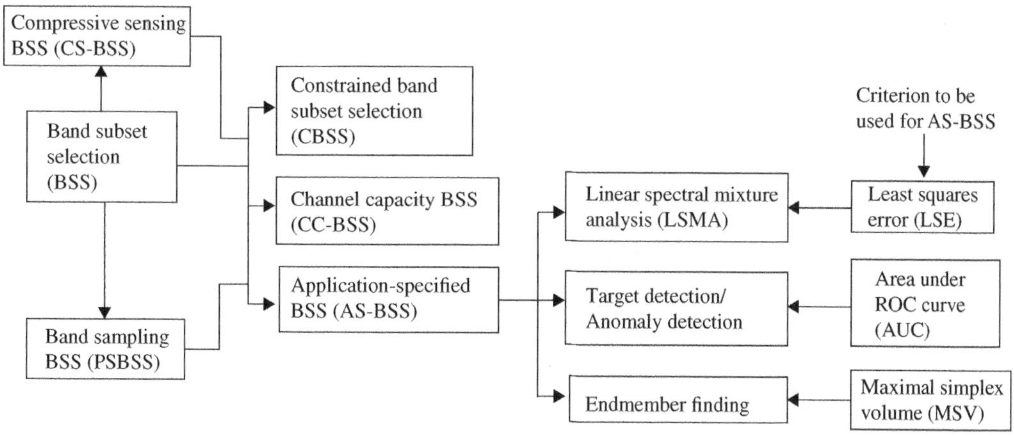


**Figure 5.3** Diagram of various BSS approaches and extensions.

### *5.6.1 Application to Hyperspectral Classification*

This section applies BSS to hyperspectral image classification [28, 29]. First and foremost is the criterion used for BSS, which is the minimum variance resulting from a linearly constrained minimum variance (LCMV) finite impulse response (FIR) filter arising in adaptive beamforming in array signal processing [30]. This LCMV-based BSS interprets signal sources as class signature vectors by linearly constraining the class signature vectors and finds an optimal band subset for classification. It is very different from constrained energy minimization (CEM)-based BS [31] which constrained a single selected band and also different from constrained multiple band selection (CMBS) [32], which extended CEM-BS by constraining multiple bands as a band subset not as class signature vectors as LCMV-CBSS does. Second, two new SQ BSS and SC BSS algorithms are developed for LCMV-CBSS, specifically for classification, referred to as SQ LCMV-CBSS and SC LCMV-CBSS. Third, the classifier used to evaluate BS performance is also an LCMV classifier which is particularly designed to best utilize bands selected by LCMV-CBSS. Fourth, despite the fact that LCMV-CBSS may not exhaust all possible band combinations, to the authors' best knowledge, LCMV-CBSS is probably the only BSS algorithm to numerically search band subsets among all possible band combinations compared to other SMMBS algorithms such as PSO, FA, MTSP, MDPP, and DSEBS which are indeed designed to run through only a very small set of selected band subsets. Last but most importantly, the proposed LVMV-BSS is very easy to be implemented because there are no parameters needed to be tuned as many BS methods do. This is a tremendous advantage since such parameters must be adaptive to various applications.

### *5.6.2 LCMV Criterion for BSS*

Suppose that there are $M$ classes of interest and each class is specified by a class signature vector, denoted by $\mathbf{d}_1, \mathbf{d}_2, \ldots, \mathbf{d}_M$. We can now form a class signature matrix, denoted by $\mathbf{D} = [\mathbf{d}_1\mathbf{d}_2\cdots\mathbf{d}_M]$. The goal is to design an FIR linear filter with $L$ filter coefficients $\{w_1, w_2, \ldots, w_L\}$, denoted by an $L$-dimensional vector $\mathbf{w} = (w_1, w_2, \ldots, w_L)^T$ that minimizes the filter output energy subject to the following constraint:

$$\mathbf{D}^T\mathbf{w} = \mathbf{c} \text{ where } \mathbf{d}_j^T\mathbf{w} = \sum_{l=1}^{L} w_l d_{jl} \text{ for } 1 \leq j \leq M \quad (5.16)$$

where $\mathbf{c} = (c_1, c_2, \ldots, c_k)^T$ is a constraint vector. Using (5.16), we derive the following linearly constrained optimization problem:

$$\min_{\mathbf{w}}\left\{\mathbf{w}^T\mathbf{R}\mathbf{w}\right\} \text{ subject to } \mathbf{D}^T\mathbf{w} = \mathbf{c} \tag{5.17}$$

where $\mathbf{R} = (1/N)\sum_{i=1}^{N}\mathbf{r}_i\mathbf{r}_i^T$ is the autocorrelation sample matrix of the image. The solution to (5.17) is called LCMV-based classifier and can be obtained in [33] by

$$\delta^{\text{LCMV}}(\mathbf{r}) = \left(\mathbf{w}^{\text{LCMV}}\right)^T\mathbf{r} \tag{5.18}$$

with

$$\mathbf{w}^{\text{LCMV}} = \mathbf{R}^{-1}\mathbf{D}\left(\mathbf{D}^T\mathbf{R}^{-1}\mathbf{D}\right)^{-1}\mathbf{c}. \tag{5.19}$$

Substituting (5.19) into (5.17) yields

$$\begin{aligned}
&\left(\mathbf{w}^{\text{LCMV}}\right)^T\mathbf{R}^{-1}\mathbf{w}^{\text{LCMV}} \\
&= \left[\mathbf{R}^{-1}\mathbf{D}\left(\mathbf{D}^T\mathbf{R}^{-1}\mathbf{D}\right)^{-1}\mathbf{c}\right]^T\mathbf{R}^{-1}\left[\mathbf{R}^{-1}\mathbf{D}\left(\mathbf{D}^T\mathbf{R}^{-1}\mathbf{D}\right)^{-1}\mathbf{c}\right] \\
&= \mathbf{c}^T\left(\mathbf{D}^T\mathbf{R}^{-1}\mathbf{D}\right)^{-1}\mathbf{D}^T\mathbf{R}^{-1}\mathbf{D}\left(\mathbf{D}^T\mathbf{R}^{-1}\mathbf{D}\right)^{-1}\mathbf{c} = \mathbf{c}^T\left(\mathbf{D}^T\mathbf{R}^{-1}\mathbf{D}\right)^{-1}\mathbf{c}
\end{aligned} \tag{5.20}$$

which is the minimum variance weighted by $(\mathbf{D}^T\mathbf{R}^{-1}\mathbf{D})^{-1}$. As a matter of fact, (5.20) can be also viewed as the minimal $(\mathbf{D}^T\mathbf{R}^{-1}\mathbf{D})^{-1}$-weighted LSE caused by misclassification errors from operating $\delta^{\text{LCMV}}$ on the entire image cube. For those who would like to learn more about LCMV, its details can be found in [21, 33].

### 5.6.3 LCMV-BSS Algorithms

Now, if we replace the full band set $\mathbf{\Omega}$ and class signature matrix, $\mathbf{D}$ in $(\mathbf{D}^T\mathbf{R}^{-1}\mathbf{D})^{-1}$ of (5.20) with a selected band subset, $\mathbf{\Omega}_{\text{BS}}$ and $\mathbf{D}_{\text{BS}}$, then (5.20) becomes

$$\mathbf{MV}(\mathbf{\Omega}_{\mathbf{BS}}) = \mathrm{c}^T\left(\mathrm{D}_{\Omega_{\mathbf{BS}}}^T\mathrm{R}_{\Omega_{\mathbf{BS}}}^{-1}\mathrm{D}_{\Omega_{\mathbf{BS}}}\right)^{-1}\mathrm{c} \tag{5.21}$$

which is the minimum variance weighted by $\left(\mathrm{D}_{\Omega_{\mathbf{BS}}}^T\mathrm{R}_{\Omega_{\mathbf{BS}}}^{-1}\mathrm{D}_{\Omega_{\mathbf{BS}}}\right)^{-1}$ resulting from the LCMV filter using a partial band subset specified by $\mathbf{\Omega}_{\text{BS}}$. There is another interpretation of (5.11) which can be also considered as the least $\left(\mathrm{D}_{\Omega_{\mathbf{BS}}}^T\mathrm{R}_{\Omega_{\mathbf{BS}}}^{-1}\mathrm{D}_{\Omega_{\mathbf{BS}}}\right)^{-1}$-weighted squares error. It should be noted that the constraint vector, $\mathbf{c}$, is specifically designed to take care of $M$ class signatures, $\mathbf{d}_1, \mathbf{d}_2, \ldots, \mathbf{d}_M$, not bands. Accordingly, $\mathbf{c}$ has nothing to do with the selected band subset, $\mathbf{\Omega}_{\text{BS}}$ and thus, it remains a constant in (5.21) for any selected band subset, $\mathbf{\Omega}_{\text{BS}}$.

Using the $\text{MV}(\mathbf{\Omega}_{\text{BS}})$ in (5.21), a criterion can be designed to find an optimal band subset, $\Omega_{\mathbf{BS}}^*$, which solves

$$\Omega_{\mathbf{BS}}^* = \arg\left\{\min_{\Omega_{\mathbf{BS}}\subset\Omega}\mathbf{MV}(\mathbf{\Omega}_{\mathbf{BS}})\right\}. \tag{5.22}$$

By virtue of (5.22), two types of algorithms from SQ N-FINDR and SC N-FINDR, called sequential LCMV-CBSS (SQ LCMV-CBSS) algorithm and successive LCMV-CBSS (SC LCMV-CBSS) algorithm, can be further developed as follows.

#### 5.6.3.1 SQ LCMV-CBSS

The idea of SQ LCMV-CBSS is to use two loops to iterate band subsets $\mathbf{\Omega}_{\text{BS}}$ in an outer loop and compute MV($\mathbf{\Omega}_{\text{BS}}$) in (5.21) in an inner loop. Depending upon how to computer MV($\mathbf{\Omega}_{\text{BS}}$) in the inner loop, two versions can be also developed. The first one is called SQ LCMV-CBSS-1 which finds the minimum variance $\text{MV}\left(\Omega_{\text{BS}}^{(j)}\right)$ currently being iterated for $1 \leq j \leq n_{\mathbf{BS}}$ in the inner loop compared to the minimum variance $\text{MV}\left(\Omega_{\text{BS}}^{(l)}\right)$ obtained at the $l$th iteration in the outer loop. Detailed of step-by-step implementation is described below.

***SQ LCMV-BSS-1***

Step 1: Initial conditions

(i) $n_{\text{BS}} = p$ which is the number of selected multiple bands determined by VD.

(ii) Let $\Omega_p^{(0)} = \left\{\mathbf{B}_1^{(0)}, \mathbf{B}_2^{(0)}, \ldots, \mathbf{B}_p^{(0)}\right\}$ with $\mathbf{B}_1^{(0)} = \mathbf{B}_1, \mathbf{B}_2^{(0)} = \mathbf{B}_2, \ldots, \mathbf{B}_p^{(0)} = \mathbf{B}_p$ uniformly selected from the band set $\mathbf{\Omega}$.

(iii) Calculate

$$\text{MV}\left(\Omega_p^{(0)}\right) = \mathbf{c}^T\left(\mathbf{D}_{\Omega_p^{(0)}}^T \mathbf{R}_{\Omega_p^{(0)}}^{-1} \mathbf{D}_{\Omega_p^{(0)}}\right)^{-1} \mathbf{c} \tag{5.23}$$

Step 2: Outer loop

For $l = 1, \ldots, L$ do

Step 3: Inner loop

compute$\text{MV}\left(\Omega_p^{(l)}\right)$

For $j = 1, \ldots, p$ do

find an index $j^*$ by

$$j^* = \arg\left\{\left(\min_{1 \leq j \leq p} \text{MV}\left(\Omega_p^{(j)}\right)\right) < \text{MV}\left(\Omega_p^{(l)}\right)\right\} \tag{5.24}$$

with

$$\text{MV}\left(\Omega_p^{(j)}\right) = \mathbf{c}^T\left(\mathbf{D}_{\Omega_p^{(j)}}^T \mathbf{R}_{\Omega_p^{(j)}}^{-1} \mathbf{D}_{\Omega_p^{(j)}}\right)^{-1} \mathbf{c} \tag{5.25}$$

which specifies the band to be replaced by the $l$th band $\mathbf{B}_l$. Such an band is now denoted by $\mathbf{B}_j^{(l+1)}$. A new set of bands is then produced by letting $\mathbf{B}_{j^*}^{(l+1)} = \mathbf{B}_l$ and $\mathbf{B}_j^{(l+1)} = \mathbf{B}_j^{(l)}$ for $j \neq j^*$

A second version of SQ LCMV-CBSS, referred to as SQ LCMV-CBSS-2, which always finds the minimum variance $\text{MV}\left(\Omega_{\text{BS}}^{(j)}\right)$ currently being iterated for $1 \leq j \leq n_{\mathbf{BS}}$ at each iteration in the inner loop with its detailed step-by-step implementation is summarized as follows:

***SQ LCMV-BSS-2***

Step 1: Initial conditions

(i) $n_{BS} = p$ which is the number of selected multiple bands determined by VD.

(ii) Let $\Omega_p^{(0)} = \left\{\mathbf{B}_1^{(0)}, \mathbf{B}_2^{(0)}, \ldots, \mathbf{B}_p^{(0)}\right\}$ with $\mathbf{B}_1^{(0)} = \mathbf{B}_1, \mathbf{B}_2^{(0)} = \mathbf{B}_2, \ldots, \mathbf{B}_p^{(0)} = \mathbf{B}_p$ uniformly selected from the band set $\mathbf{\Omega}$.

(iii) Calculate

$$\mathrm{MV}\left(\Omega_p^{(0)}\right) = \mathbf{c}^T\left(\mathbf{D}_{\Omega_p^{(0)}}^T \mathbf{R}_{\Omega_p^{(0)}}^{-1} \mathbf{D}_{\Omega_p^{(0)}}\right)^{-1} \mathbf{c} \tag{5.26}$$

Step 2: Outer loop

For $l = 1, \ldots, L$ do

Step 3: Inner loop

For $j = 1, \ldots, p$ do

find an index $j^*$ by

$$j^* = \arg\left\{\min_{1 \leq j \leq p} \mathrm{MV}\left(\Omega_p^{(j)}\right)\right\} \tag{5.27}$$

with

$$\mathrm{MV}\left(\Omega_p^{(j)}\right) = \mathbf{c}^T\left(\mathbf{D}_{\Omega_p^{(j)}}^T \mathbf{R}_{\Omega_p^{(j)}}^{-1} \mathbf{D}_{\Omega_p^{(j)}}\right)^{-1} \mathbf{c} \tag{5.28}$$

which specifies the band to be replaced by the $l$th band $\mathbf{B}_l$. Such an band is now denoted by $\mathbf{B}_j^{(l+1)}$. A new set of bands is then produced by letting $\mathbf{B}_{j^*}^{(l+1)} = \mathbf{B}_l$ and $\mathbf{B}_j^{(l+1)} = \mathbf{B}_j^{(l)}$ for $j \neq j^*$

#### 5.6.3.2 SC LCMV-CBSS

A second type of LCMV-CBSS algorithm is SC LCMV-CBSS which reverses the two loops implemented in SQ LCMV-CBSS by iterating computation of $\mathrm{MV}(\mathbf{\Omega}_{BS})$ in (5.27) in an outer loop, while iterating band subsets $n_{BS}$ in an inner loop. Its detailed step-by-step implementation is provided as follows:

***SC LCMV-CBSS***

Step 1: Initial conditions

(i) $n_{BS} = p$ which is the number of selected multiple bands determined by VD.

(ii) Let $\Omega_p^{(0)} = \left\{\mathbf{B}_1^{(0)}, \mathbf{B}_2^{(0)}, \ldots, \mathbf{B}_p^{(0)}\right\}$ with $\mathbf{B}_1^{(0)} = \mathbf{B}_1, \mathbf{B}_2^{(0)} = \mathbf{B}_2, \ldots, \mathbf{B}_p^{(0)} = \mathbf{B}_p$ uniformly selected from the band set $\mathbf{\Omega}$

(iii) Calculate

$$\mathrm{MV}\left(\Omega_p^{(0)}\right) = \mathbf{c}^T\left(\mathbf{D}_{\Omega_p^{(0)}}^T \mathbf{R}_{\Omega_p^{(0)}}^{-1} \mathbf{D}_{\Omega_p^{(0)}}\right)^{-1} \mathbf{c} \tag{5.29}$$

Step 2: Outer loop

For $j = 1, \ldots, p$ do

Step 3: Inner loop

For $l = 1, \ldots, L$ do

find

$$\mathbf{B}_j^{(*)} = \arg\left\{ \min_{\mathbf{B}_l \in \overline{\Omega}} \mathrm{MV}\left(\tilde{\Omega}_p^l\right) \right\} \quad (5.30)$$

where $\overline{\Omega} = \Omega - \left\{ \mathbf{B}_1^{(*)}, \ldots, \mathbf{B}_{j-1}^{(*)}, \mathbf{B}_{j+1}^{(p)}, \ldots, \mathbf{B}_p^{(p)} \right\}$

$\tilde{\Omega}_p^l = \left\{ \mathbf{B}_1^{(*)}, \ldots, \mathbf{B}_{j-1}^{(*)}, \mathbf{B}_l, \mathbf{B}_{j+1}^{(p)}, \ldots, \mathbf{B}_p^{(p)} \right\}$

Step 4: Output the final band subset, $\left\{ \mathbf{B}_1^{(*)}, \mathbf{B}_2^{(*)}, \ldots, \mathbf{B}_p^{(*)} \right\}$.

## 5.7 Experiments

The hyperspectral data to be studied is an image scene shown in Figure 5.4(a) with a total of 169 bands, 1.56 m spatial resolution, and 10 nm of spectral resolution. There are 15 panels of three different dimensions in the image, 3 m × 3 m, 2 m × 2 m, and 1 m × 1 m in size. These are marked in light gray in Figure 5.4(b) with 19 center pixels highlighted by dark gray, $p_{11}$, $p_{12}$, $p_{13}$ in row 1, $p_{211}$, $p_{221}$, $p_{22}$, $p_{23}$ in row 2, $p_{311}$, $p_{312}$, $p_{32}$, $p_{33}$ in row 3, $p_{411}$, $p_{412}$, $p_{32}$, $p_{33}$ in row 4, and $p_{511}$, $p_{521}$, $p_{52}$, $p_{53}$ in row 5. More details can be found in Figure 4.1 of Chapter 4 or [6, 21].

### *5.7.1 MBC-BSS*

It has been shown in [34] that automatic target generation process (ATGP) [35], vector component analysis (VCA) [36], and simplex growing algorithm (SGA) [37] are essentially the same algorithm and can be carried out by ATGP. Consequently, all the following experiments are performed based on ATGP. According to [21], VD was estimated as 9.

The experiments were conducted for various MBC-BSS methods by comparing three categories of BS methods. The first category consists of BP/band de-correlation (BD) criterion-based BS methods [15], variance, SNR, entropy, ID. The second category comprises CBS-based methods in [31], CBS band correlation minimization/band dependence minimization (CEM-BCM/BDM), CBS band correlation constraint/band dependence constraint (BDC) (CBS BCC/BDC), LCMV-based band correlation minimization/band dependence minimization (LCMV-BCM/BDM), and the LCMV-based band correlation constraint/band dependence constraint (LCMV-BCC/BDC) algorithms. The third category is made up of four CBSS methods, SC CBSS, SQ CBSS, PPI-SC CBSS, and PPI-SQ CBSS. Table 5.1 displays nine bands selected by BS methods from these three categories. It should be noted that SBS-based BP criteria generally require BD to remove redundant bands. In this case, a threshold $\varepsilon$ was used for this purpose and their values are also provided in the table. For ID, there was no BD

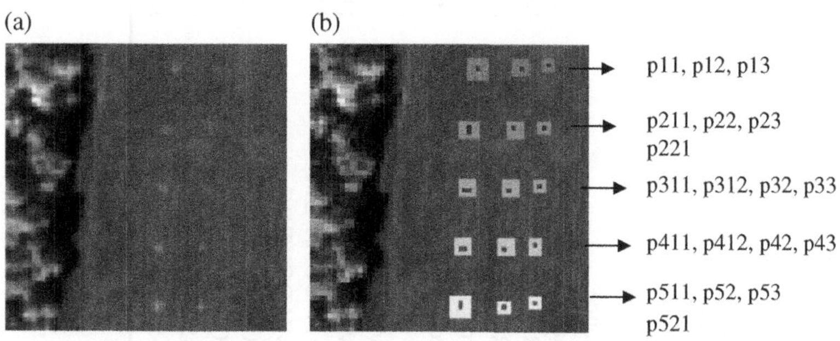


**Figure 5.4** (a) A HYDICE panel scene which contains 15 panels and (b) ground-truth map of spatial locations of the 15 panels, where its color maps can be found in [6], figure 1.15/with permission of John Wiley & Sons.

**Table 5.1** Nine bands selected by various BP criteria using SC CBSS and SQ CBSS along with their computing times

| BS methods | Selected bands | | | | | | | | | Computing time (s) |
|---|---|---|---|---|---|---|---|---|---|---|
| UBS | 1 | 20 | 39 | 58 | 77 | 96 | 115 | 134 | 153 | Not applicable |
| Variance | 52 | 40 | 36 | 46 | 47 | 48 | 44 | 45 | 41 | 0.0881 |
| Variance (BD with $\varepsilon = 0.4$) | 52 | 36 | 47 | 50 | 13 | 54 | 55 | 96 | 152 | 3.9663 |
| SNR | 102 | 104 | 119 | 139 | 117 | 90 | 140 | 134 | 101 | 0.5862 |
| SNR (BD with $\varepsilon = 0.45$) | 102 | 119 | 90 | 110 | 131 | 47 | 70 | 63 | 54 | 4.3532 |
| Entropy | 65 | 60 | 67 | 53 | 66 | 61 | 52 | 68 | 59 | 0.1247 |
| Entropy (BD with $\varepsilon = 0.4$) | 65 | 53 | 59 | 78 | 77 | 49 | 56 | 47 | 58 (BD) | 4.0278 |
| ID | 154 | 157 | 156 | 153 | 150 | 158 | 145 | 164 | 163 | 0.2258 |
| CEM-BCM/BDM | 18 | 85 | 112 | 10 | 109 | 74 | 115 | 114 | 96 | 18.5399/2.6442e + 03 |
| CEM-BCC/BDC | 115 | 96 | 84 | 86 | 105 | 13 | 103 | 87 | 24 | 18.2337/2.7017e + 03 |
| LCMV-BCM/BDM | 125 | 169 | 168 | 164 | 146 | 124 | 165 | 128 | 167 | 0.1548/0.2042 |
| LCMV-BCC/BDC | 125 | 169 | 124 | 164 | 167 | 161 | 163 | 168 | 146 | 0.6356/0.7193 |
| SC CBSS | 164 | 168 | 138 | 118 | 169 | 1 | 102 | 48 | 85 | 2.2205e+04 |
| SQ CBSS | 1 | 110 | 60 | 118 | 169 | 102 | 168 | 139 | 164 | 2.2330e+04 |
| PPI-SC CBSS | 140 | 168 | 145 | 166 | 169 | 125 | 130 | 134 | 42 | 16.1288 |
| PPI-SQ CBSS | 168 | 140 | 145 | 125 | 42 | 169 | 166 | 134 | 130 | 16.4393 |

implemented because our experiments showed that there was no appreciable difference between using and not using BD. This may be due to the fact that ID already takes care of BD by measuring discrepancy between two probability distributions. On the other hand, CBSS does not require BP and BD because it constrains selected bands simultaneously as a band subset. However, it does require SQ/SC CBSS running through over possible band subsets via a SQ/SC process. The computing time is documented in the last column of Table 5.1 where the computer environment was a double CPU, model: E5-2650 V2, main clock speed of 2.6 GHz and the capacity of memory is 48G, DDR3, frequency is 1600 MHz.

Despite the fact that BCM and BDM produced the same bands in [15], their computing times were actually different in Table 5.1. This holds for BCC and BDC. As shown in Table 5.1, the CBSS requires tremendous computing time to carry out an iterative processes search for best possible band subsets. However, this computational disadvantage is offset greatly by avoiding the exhaustive search which is practically infeasible in real applications. As will be also demonstrated by the following experiments, ATGP using CBSS selected bands performs better than other tested BS methods and CBS.

Table 5.2 also shows panel pixels found by ATGP with the numbers in parentheses indicating their found order and the number of panel pixels calculated in the last column.

As shown in Table 5.2, the best results as measured by SNR (using a best selected threshold $\varepsilon = 0.45$ for BD) were PPI-SC CBSS and PPI-SQ CBSS which were the only two BSS methods to find all the five panel pixels, each of which corresponded to one of the five rows. This is followed by the second best results of four ATGP-found panel pixels by SNR, entropy using a best selected threshold $\varepsilon = 0.4$ for BD, LCMV-CBS BCM/BDM, SC CBSS, and SQ CBSS. Interestingly, ATGP using full bands and UBS could only find three panel pixels. As a matter of fact, in order for ATGP using the full number of bands to find five panel pixels corresponding to each of five rows, it would require 18 target pixels [19, 32, 33] such as SNR with BD = 0.45 to accomplish the task where the five panel pixels were $p_{11}(6)$, $p_{212}(17)$, $p_{312}(5)$, $p_{412}(18)$, and $p_{521}(3)$. It should be noted that the 17th ATGP-found target pixel was a Y pixel $p_{212}$ instead of an R panel pixel $p_{211}$. This is because $p_{212}$ is more spectrally distinct than $p_{211}$ as noted in [6].

The results in Table 5.2 reveal an important fact that finding five panel pixels as endmember pixels to represent five different panel signatures requires only nine bands provided that a BP criterion was selected appropriately such as SNR in contrast to ATGP which requires 18 target pixels using full

**Table 5.2** Panel pixels found by ATGP using nine bands selected in Table 5.1

| BS method | Found panel pixels | Number of found panels |
|---|---|---|
| Full bands | $p_{11}(6)$, $p_{312}(5)$ $p_{521}(3)$ | 3 |
| UBS | $p_{311}(7)$ $p_{412}(9)$, $p_{521}(4)$ | 3 |
| Variance | $p_{312}(3)$ | 1 |
| Variance (BD) | $p_{11}(9)$, $p_{312}(3)$ $p_{521}(5)$ | 3 |
| SNR | $p_{11}(3)$, $p_{312}(6)$, $p_{411}(7)$, $p_{521}(1)$ | 4 |
| SNR (BD 0.45) | $p_{11}(6)$, $p_{211}(9)$, $p_{312}(4)$, $p_{411}(8)$, $p_{521}(3)$ | 5 |
| Entropy | $p_{211}(9)$, $p_{312}(4)$, $p_{521}(3)$ | 3 |
| Entropy | $p_{11}$ (2), $p_{211}(3)$, $p_{312}(5)$, $p_{521}(1)$ | 4 |
| ID | $p_{11}(2)$, $p_{211}(3)$, $p_{412}(5)$, $p_{521}(1)$ | 4 |
| CEM-BCM/BDM | $p_{11}(5)$, $p_{411}(7)$ | 2 |
| CEM-BCC/BDC | $p_{11}(3)$ $p_{412}(7)$, $p_{521}(1)$ | 3 |
| LCMV-BCM/BDM | $p_{11}(3)$, $p_{22}(9)$, $p_{411}(4)$ $p_{521}(1)$ | 4 |
| LCMV-BCC/BDC | $p_{10}(4)$, $p_{211}(5)$, $p_{411}(3)$, $p_{521}(1)$ | 3 |
| SC CBSS | $p_{11}(6)$, $p_{211}(7)$, $p_{412}(5)$, $p_{521}(3)$ | 4 |
| SQ CBSS | $p_{11}(6)$, $p_{211}(7)$, $p_{412}(5)$, $p_{521}(3)$ | 4 |

bands to do the same. This implied that using full bands did not necessarily perform better than using a partial band subset as long as the nine bands are selected wisely.

## *5.7.2 MTC-BSS*

Unlike MBC-BSS which imposes constraints on multiple bands to select an optimal band subset, MTC-BSS imposes constraints on multiple targets to select an optimal band subset. So, the key difference between MBC-BSS and MTC-BSS is that MTC-BSS requires the prior knowledge of multiple targets, while MBC-BSS does not. As a result, MBC-BSS can be considered as application-independent BSS as opposed to MTC-BSS which is application-dependent. In this case, to evaluate effectiveness of MTC-BSS, we need to specify targets of interest used by applications. Two applications are of particular interest, target detection [37, 38] and image classification [28, 29]. Due to limited space, only application to classification [28, 29] will be discussed in this chapter. Nevertheless, the same treatment can be applied to target detection [37, 38].

Three popular real hyperspectral images, Purdue University's Indiana Indian Pines, Salinas, and University of Pavia, were used for experiments.

### 5.7.2.1 Purdue Indiana Indian Pines Scene

The first image scene used for experiments is an airborne visible/infrared imaging spectrometer (AVIRIS) hyperspectral data set which is Purdue Indiana Indian Pines test site in Figure 5.5(a) with

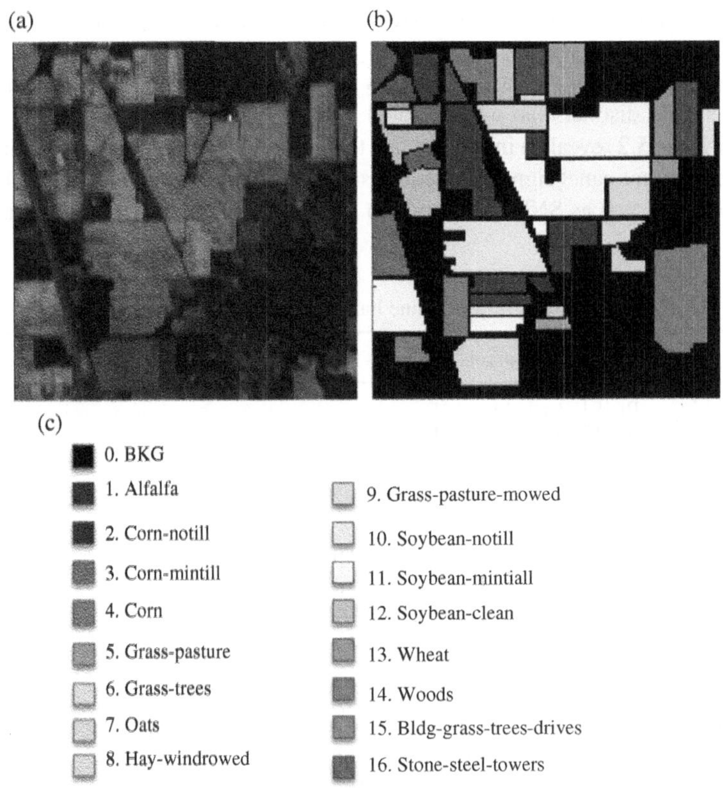


**Figure 5.5** Purdue's Indiana Indian Pines scene with 16 classes: (a) band 186, (b) gray ground-truth map, and (c) ground-truth class labels, where its color maps can be found in Ref. [28], figure 1/MDPI/CC BY 4.0.

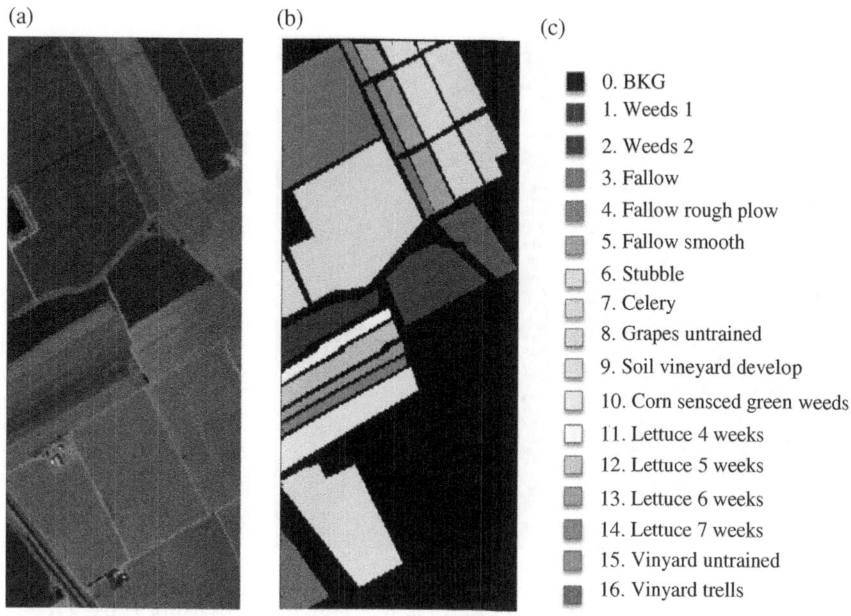


**Figure 5.6** Ground truth of Salinas scene with 16 classes: (a) band 126, (b) gray ground-truth map, and (c) class labels, where its color maps can be found in Ref. [28], figure 2/MDPI/CC BY 4.0.

its ground truth of 16 class maps along with background class as 17th class shown in Figure 5.5(b). It has size of 145 × 145 pixel vectors taken from an area of mixed agriculture and forestry in Northwestern Indiana, USA, with details of band and wavelength given in the caption. The data set is available at website https://purr.purdue.edu/publications/1947/serve/1?el=1. It was recorded in June 1992 with 220 bands which include water absorption bands (bands 104–108 and 150–163, 220).

#### 5.7.2.2 Salinas

A second AVIRIS data used for experiments was Salinas scene shown in Figure 5.6(a) which was captured by the AVIRIS sensor over Salinas Valley, California, and with a spatial resolution of 3.7-m per pixel with spectral resolution of 10 nm. It is of size 512 × 217 × 224. Figure 5.6(b) and (c) shows the color composite of the Salinas image along with the corresponding ground-truth class labels.

#### 5.7.2.3 ROSIS Data

The last hyperspectral image data used for experiments was the University of Pavia image shown in Figure 5.7 which is an urban area surrounding the University of Pavia, Italy. It was recorded by the ROSIS-03 satellite sensor. It is of size 610 × 340 × 115 with a spatial resolution of 1.3-m per pixel and a spectral coverage ranging from 0.43 to 0.86 μm with spectral resolution of 4 nm (12 most noisy channels were removed before experiments). Nine classes of interest plus BKG class, class 0 are considered for this image.

In the following experiments, four types of BS methods are tested for a comparative study and analysis.

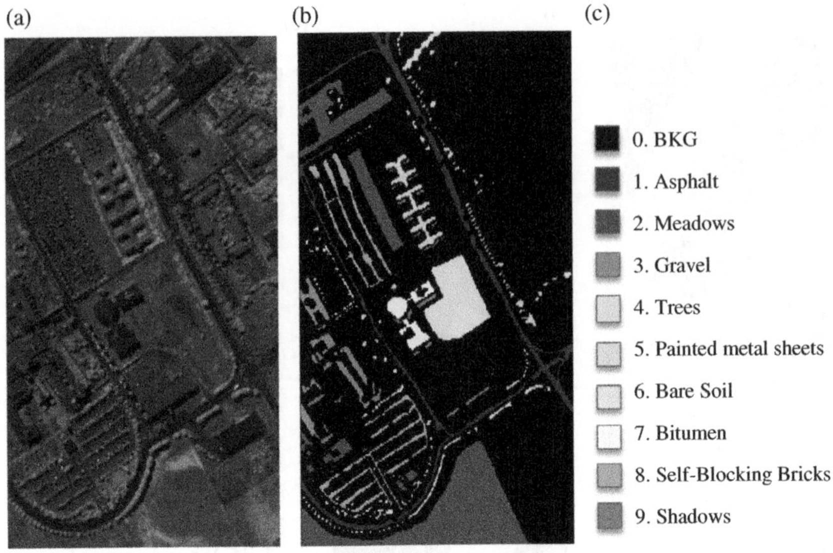


**Figure 5.7** Ground truth of University of Pavia scene with 9 classes: (a) band 95, (b) gray ground-truth image, and (c) class labels, where its color maps can be found in Ref. [28], figure 3/MDPI/CC BY 4.0.

1. UBS: According to our extensive experiments, UBS is a reasonably good BS method which is also reported in the literature. It does not require any prior knowledge or BS criterion. It is the simplest BS method.
2. MEAC: It uses the minimum covariance derived from the estimated abundance matrix, which is similar to the minimum variance in (5.5). In addition, it can also represent the category of SQMBS methods.
3. MDPP and DSEBS: Both represent the category of SMMBS methods. They make use of graph representations to specify band groups. Most importantly, these two methods also compared their results to CEM/LCMV-based methods in [24] which are also based on the LCMV formulation specified by (5.2).
4. LCMV-BSS developed in this chapter: It represents the category of BSS methods using the LCMV formulation in (5.2).

Table 5.3 tabulates the number of selected bands, $n_{BS}$, estimated for three scenes by HFC/NWHFC methods developed for VD in [6, 21] where $n_{BS}$ was determined by the false alarm probability $10^{-4}$ in which case, $n_{BS} = 18$ for Purdue's data, 21 for Salinas and, 14 for University of Pavia [29].

Table 5.4 lists the bands selected by 7 BS methods, UBS, MEAC, MDPP, DSEBS, SQ LCMV-BSS-1, SQ LCMV-BSS-2, and SC LCMV-BSS for the three scenes, Purdue's Indian Pines for $n_{BS} = 18$, Salinas for $n_{BS} = 21$, and University of Pavia for $n_{BS} = 14$.

**Table 5.3** $n_{BS}$ estimated by HySime and HFC//NWHFC

| | $P_F = 10^{-1}$ | $P_F = 10^{-2}$ | $P_F = 10^{-3}$ | $P_F = 10^{-4}$ | $P_F = 10^{-5}$ |
|---|---|---|---|---|---|
| Purdue | 73/21 | 49/19 | 35/18 | 27/18 | 25/17 |
| Salinas | 32/33 | 28/24 | 25/21 | 21/21 | 20/20 |
| University of Pavia | 25/34 | 21/27 | 16/17 | 14/14 | 13/12 |

**Table 5.4** Bands selected by UBS, SQ LCMV-BSS-1, SQ LCMV-BSS-2, and SC LCMV-BSS

| Data | Methods | Selected bands | | | | | | | | | | | | | | | | | | | | |
|---|---|---|---|---|---|---|---|---|---|---|---|---|---|---|---|---|---|---|---|---|---|---|
| Purdue Indian | UBS | 1 | 14 | 27 | 40 | 53 | 66 | 79 | 92 | 105 | 118 | 131 | 144 | 157 | 170 | 183 | 196 | 209 | 220 | | | |
| Pines (18 bands) | MEAC | 159 | 3 | 92 | 96 | 82 | 36 | 39 | 55 | 41 | 1 | 2 | 33 | 206 | 38 | 163 | 17 | 204 | 9 | | | |
| | MDPP | 10 | 39 | 59 | 75 | 79 | 85 | 92 | 130 | 140 | 146 | 147 | 149 | 150 | 152 | 160 | 164 | 175 | 193 | | | |
| | DSEBS | 42 | 129 | 97 | 131 | 174 | 16 | 176 | 177 | 172 | 43 | 192 | 193 | 98 | 171 | 99 | 132 | 40 | 33 | | | |
| | SQ LCMV-BSS-1 | 39 | 164 | 29 | 155 | 108 | 66 | 79 | 8 | 105 | 42 | 44 | 17 | 156 | 150 | 3 | 43 | 213 | 41 | | | |
| | SQ LCMV-BSS-2 | 38 | 109 | 29 | 52 | 163 | 66 | 158 | 8 | 164 | 219 | 43 | 78 | 157 | 220 | 3 | 49 | 218 | 2 | | | |
| | SC LCMV-BSS | 54 | 156 | 42 | 159 | 53 | 41 | 79 | 91 | 105 | 57 | 51 | 43 | 157 | 48 | 107 | 160 | 115 | 163 | | | |
| Salinas (21 bands) | UBS | 1 | 12 | 23 | 34 | 45 | 56 | 67 | 78 | 89 | 100 | 111 | 122 | 133 | 144 | 155 | 166 | 177 | 188 | 199 | 210 | 224 |
| | MEAC | 107 | 148 | 203 | 149 | 5 | 8 | 105 | 3 | 28 | 12 | 18 | 10 | 44 | 36 | 25 | 17 | 51 | 32 | 110 | 68 | 58 |
| | MDPP | 1 | 8 | 11 | 22 | 27 | 28 | 50 | 57 | 58 | 65 | 90 | 99 | 105 | 119 | 123 | 134 | 142 | 157 | 175 | 191 | 204 |
| | DSEBS | 99 | 101 | 16 | 119 | 177 | 112 | 44 | 46 | 120 | 47 | 131 | 175 | 196 | 121 | 17 | 102 | 174 | 180 | 187 | 135 | 42 |
| | SQ LCMV-BSS-1 | 7 | 50 | 23 | 48 | 45 | 73 | 65 | 15 | 40 | 19 | 80 | 122 | 38 | 41 | 42 | 46 | 78 | 47 | 200 | 37 | 2 |
| | SQ LCMV-BSS-2 | 7 | 42 | 56 | 28 | 45 | 58 | 67 | 15 | 41 | 19 | 50 | 122 | 38 | 34 | 36 | 47 | 224 | 46 | 183 | 37 | 172 |
| | SC LCMV-BSS | 18 | 39 | 41 | 31 | 45 | 44 | 67 | 78 | 90 | 101 | 40 | 91 | 42 | 141 | 46 | 48 | 102 | 185 | 47 | 86 | 50 |
| University of | UBS | 1 | 9 | 17 | 25 | 33 | 41 | 49 | 57 | 65 | 73 | 81 | 89 | 97 | 103 | | | | | | | |
| Pavia (14 bands) | MEAC | 1 | 23 | 24 | 40 | 42 | 58 | 56 | 59 | 48 | 31 | 47 | 83 | 25 | 54 | | | | | | | |
| | MDPP | 2 | 23 | 44 | 46 | 50 | 62 | 66 | 73 | 89 | 91 | 92 | 93 | 96 | 102 | | | | | | | |
| | DSEBS | 86 | 102 | 64 | 20 | 21 | 63 | 65 | 6 | 19 | 22 | 7 | 66 | 95 | 67 | | | | | | | |
| | SQ LCMV-BSS-1 | 1 | 4 | 55 | 16 | 95 | 83 | 84 | 93 | 39 | 77 | 91 | 102 | 92 | 103 | | | | | | | |
| | SQ LCMV-BSS-2 | 1 | 4 | 38 | 76 | 85 | 55 | 84 | 102 | 16 | 83 | 93 | 89 | 92 | 103 | | | | | | | |
| | SC LCMV-BSS | 1 | 4 | 84 | 16 | 38 | 102 | 85 | 92 | 83 | 72 | 95 | 91 | 96 | 103 | | | | | | | |

It has been shown in [39, 40] that using overall accuracy (OA), $P_{OA}$ may not be sufficient to evaluate the effectiveness of classification performance. To address this issue, an additional measure, called precision rate, $P_{precision}$, developed in [35, 36] was introduced for hyperspectral image classification where the precision rate has been widely used in pattern recognition such as medical imaging, handwritten character recognition, and biometric recognition. The precision definitions and details of $P_{OA}$ and $P_{precision}$ can be found in [41]. Specifically, [29] proposed a new classifier, called class-signature constrained background suppression classifier (CSCBSC), and further demonstrated that CSCBSC did perform well in both $P_{OA}$ and $P_{precision}$. Since CSCBSC was also derived from LCMV and implemented by LCMV in an iterative manner, the iterative linearly constrained minimum variance (ILCMV) is used in [40] to reflect its idea arising from LCMV and its iterative nature in algorithm implementation. Most importantly, ILCMV was adopted for two main reasons. One is because of the work in [40] which showed that ILCMV could perform at least comparably in $P_{OA}$ but significantly better than the work [41]. The other is that ILCMV is indeed derived from the LCMV criterion specified by (5.20). So, it is natural to use ILCMV to perform classification.

Tables 5.5–5.7 show OA and precision rates calculated by the ILCMV-classification results using the bands selected in Table 5.4 for Purdue's data, Salinas, and University of Pavia, respectively, where the best results with highest rates are boldfaced. Here, we would like to point out a crucial fact used in the experiments where the $P_{OA}$ and $P_{precision}$ rates were calculated by including background (BKG) for classification because LCMV is particularly designed to take care of BKG issue in classification as shown in [40]. This is quite different from many reports which calculate $P_{OA}$ excluding BKG from classification such as [42].

As we can see from these tables, SQ LCMV-BSS-2 and SC LCMV-BSS outperformed all other five BS methods in terms of $P_{OA}$ and $P_{precision}$ for Salinas and University of Pavia scenes but slightly

**Table 5.5** OA and precision rates calculated by the classification results using the bands selected in Table 5.4 for Purdue's data

| Class | Full bands | UBS | MEAC | MDPP | DSEBS | SQ LCMV-CBSS-1 | SQ LCMV-CBSS-2 | SC LCMV-CBSS |
|---|---|---|---|---|---|---|---|---|
| 1 | 95.65 | 95.65 | 93.48 | 95.65 | 95.65 | 95.65 | 97.83 | 100 |
| 2 | 96.01 | 97.13 | 93.07 | 96.08 | 96.99 | 95.59 | 93.78 | 94.89 |
| 3 | 96.99 | 96.51 | 96.27 | 97.35 | 97.23 | 96.39 | 95.67 | 94.10 |
| 4 | 98.73 | 98.73 | 98.31 | 99.58 | 98.31 | 97.89 | 98.31 | 98.31 |
| 5 | 89.44 | 90.68 | 91.51 | 92.34 | 93.58 | 91.93 | 92.34 | 92.96 |
| 6 | 97.12 | 97.67 | 97.40 | 96.71 | 97.12 | 96.44 | 97.95 | 95.75 |
| 7 | 100 | 100 | 100 | 100 | 100 | 100 | 100 | 100 |
| 8 | 98.78 | 98.54 | 99.16 | 97.49 | 97.91 | 97.91 | 99.16 | 98.95 |
| 9 | 100 | 100 | 90.00 | 100 | 100 | 100 | 100 | 100 |
| 10 | 93.93 | 91.98 | 93.31 | 94.65 | 93.00 | 94.24 | 91.98 | 91.98 |
| 11 | 94.70 | 96.13 | 94.55 | 95.48 | 95.85 | 95.48 | 96.17 | 95.93 |
| 12 | 95.45 | 94.94 | 96.29 | 96.80 | 97.30 | 95.95 | 95.11 | 96.63 |
| 13 | 98.54 | 98.54 | 99.02 | 97.56 | 96.59 | 97.56 | 98.54 | 98.54 |
| 14 | 93.52 | 94.15 | 94.78 | 94.70 | 94.55 | 95.89 | 96.05 | 96.13 |
| 15 | 90.67 | 95.60 | 92.49 | 96.89 | 93.52 | 94.82 | 94.56 | 96.11 |
| 16 | 98.92 | 98.92 | 98.92 | 98.92 | 98.92 | 98.92 | 97.85 | 95.70 |
| $P_{OA}$ | 95.09 | 95.69 | 94.91 | **95.89** | 95.88 | 95.67 | 95.48 | 95.46 |
| $P_{Precision}$ | 97.61 | 97.90 | 97.52 | **98.00** | 97.99 | 97.89 | 97.80 | 97.79 |

**Table 5.6** OA and precision rates calculated by the classification results using the bands selected in Table 5.4 for Salinas

| Class | Full bands | UBS | MEAC | MDPP | DSEBS | SQ LCMV-CBSS-1 | SQ LCMV-CBSS-2 | SC LCMV-CBSS |
|---|---|---|---|---|---|---|---|---|
| 1 | 95.52 | 97.16 | 97.71 | 97.76 | 97.16 | 96.37 | 97.01 | 96.91 |
| 2 | 98.42 | 98.85 | 98.44 | 97.99 | 99.17 | 98.79 | 98.36 | 98.71 |
| 3 | 93.78 | 95.50 | 94.03 | 93.98 | 95.65 | 90.44 | 95.14 | 95.95 |
| 4 | 95.62 | 94.69 | 94.33 | 97.49 | 94.74 | 96.56 | 95.91 | 92.04 |
| 5 | 96.90 | 96.45 | 95.19 | 95.22 | 96.90 | 95.87 | 95.94 | 90.78 |
| 6 | 98.79 | 98.59 | 98.56 | 98.79 | 98.56 | 97.95 | 98.91 | 97.75 |
| 7 | 98.63 | 98.21 | 98.18 | 97.99 | 97.65 | 98.35 | 98.44 | 98.32 |
| 8 | 96.69 | 95.81 | 97.40 | 95.23 | 96.11 | 95.84 | 97.47 | 96.61 |
| 9 | 95.87 | 95.60 | 94.74 | 95.29 | 95.73 | 94.79 | 94.89 | 95.44 |
| 10 | 96.67 | 96.37 | 96.34 | 96.46 | 97.25 | 95.73 | 96.58 | 96.77 |
| 11 | 97.75 | 97.85 | 91.10 | 97.75 | 98.31 | 95.79 | 97.38 | 97.66 |
| 12 | 97.15 | 96.16 | 95.54 | 97.46 | 97.66 | 96.32 | 95.39 | 95.43 |
| 13 | 96.51 | 96.94 | 93.35 | 96.40 | 95.63 | 87.77 | 97.38 | 94.00 |
| 14 | 95.89 | 98.14 | 97.66 | 97.01 | 98.04 | 97.76 | 97.20 | 96.93 |
| 15 | 94.00 | 95.27 | 96.52 | 95.42 | 95.25 | 95.42 | 96.27 | 95.84 |
| 16 | 93.30 | 96.07 | 93.86 | 95.07 | 95.02 | 95.68 | 95.13 | 95.41 |
| $P_{OA}$ | 96.37 | 96.49 | 96.45 | 96.25 | 96.63 | 95.93 | **96.81** | 96.21 |
| $P_{Precision}$ | 98.23 | 98.29 | 98.27 | 98.17 | 98.36 | 98.02 | **98.45** | 98.15 |

**Table 5.7** OA and precision rates calculated by the classification results using the bands selected in Table 5.4 for University of Pavia

| Class | Full bands | UBS | MEAC | MDPP | DSEBS | SQ LCMV-CBSS-1 | SQ LCMV-CBSS-2 | SC LCMV-CBSS |
|---|---|---|---|---|---|---|---|---|
| 1 | 77.24 | 75.86 | 76.32 | 74.01 | 76.17 | 80.21 | 78.22 | 77.08 |
| 2 | 86.42 | 87.67 | 86.44 | 87.97 | 87.71 | 84.44 | 88.05 | 88.67 |
| 3 | 73.34 | 84.38 | 83.33 | 82.14 | 84.63 | 84.14 | 85.21 | 86.76 |
| 4 | 79.85 | 78.90 | 76.66 | 76.17 | 79.22 | 76.49 | 74.71 | 78.56 |
| 5 | 98.81 | 97.84 | 98.88 | 96.95 | 97.99 | 98.14 | 97.77 | 97.70 |
| 6 | 91.49 | 89.93 | 91.33 | 93.32 | 87.11 | 93.50 | 90.57 | 90.77 |
| 7 | 89.10 | 91.35 | 82.78 | 87.53 | 87.44 | 86.19 | 90.00 | 91.13 |
| 8 | 81.10 | 83.32 | 76.26 | 75.64 | 76.34 | 82.84 | 82.92 | 82.46 |
| 9 | 78.46 | 79.09 | 79.51 | 79.83 | 79.30 | 77.09 | 77.09 | 79.30 |
| $P_{OA}$ | 84.32 | 85.19 | 83.85 | 84.33 | 84.25 | 84.45 | 85.41 | **85.92** |
| $P_{Precision}$ | 96.76 | 96.93 | 96.64 | 96.76 | 96.75 | 96.78 | **96.96** | 96.95 |

worse than MDPP in $P_{OA}$ and DSEBS in $P_{precision}$. Interestingly, both MDPP and DSEBS produced the best results in terms of $P_{OA}$ and $P_{precision}$, respectively, for the Purdue data. As also noted in Tables 5.5–5.7, the $P_{OA}$ and $P_{precision}$ using full bands were generally not as good as those produced by most of test BS methods, but also worse than that produced by UBS. These experiments showed

**Table 5.8** Computing time in seconds required by six test BS methods, MEAC, MDPP, DSEBS, SQ LCMV-CBSS-1, SQ LCMV-CBSS-2, and SC LCMV-CBSS

| | MEAC | MDPP | DSEBS | SQ LCMV-CBSS-1 | SQ LCMV-CBSS-2 | SC LCMV-CBSS |
|---|---|---|---|---|---|---|
| Purdue | 13.70 | 41.14 | 0.58 | 7.00 | 7.10 | 6.93 |
| Salinas | 83.64 | 44.66 | 5.27 | 43.43 | 46.63 | 43.55 |
| University of Pavia | 44.53 | 29.22 | 4.62 | 16.67 | 17.52 | 16.84 |

that hyperspectral image classification can be benefited greatly by judiciously selecting bands with appropriately determined $n_{BS}$. Table 5.8 tabulates the computing times in seconds for each of 6 BS methods where the computer environment is specified by processor: 1.6GHz Intel Core i5 with OS X EI Capitan and 4GB 1600 MHz DDR3 and the software used to run experiments was Matlab_R2014b. Apparently, the best time was DSEBS followed by SC LCMV-BSS and SQ LCMV-BSS. The worst time was MDPP for the Purdue data and MEAC for Salinas and University of Pavia.

As noted above, a classifier can also have significant impact on BS, specifically, when BKG is included for consideration. A recent work [39] developed four EPF-based techniques, EPF-B-c, EPF-G-c, EPF-B-g, and EPF-G-g, for hyperspectral image classification and also conducted a comprehensive comparative analysis to show that their methods indeed performed better than most recently developed spectral-spatial techniques. Therefore, in what follows, we conducted experiments to evaluate the performance of ILCMV in comparison with these four EPF-based techniques with BKG particularly included for classification. To see that, we also implemented these four EPF-based techniques with "B" and "G" used to specify bilateral filter and guided filter, respectively, and "g" and "c" indicate that the first principal component and color composite of three principal components are used as reference images [39].

Tables 5.9–5.17 tabulate their results in terms of $P_{OA}$ and $P_{precision}$ rates produced by the four EFP-based methods and ILCMV, all of which included BKG for classification and also used the bands selected in Table 5.4 to implement the three image scenes, Purdue's data in Tables 5.9–5.11 using bands selected SQ LCMV-BSS-1, SQ LCMV-BSS-2, and SC LCMV-BSS; Salinas in Tables 5.12–5.14 using bands selected SQ LCMV-BSS-1, SQ LCMV-BSS-2, and SC LCMV-BSS; and University of Pavia in Tables 5.15–5.17 using bands selected SQ LCMV-BSS-1, SQ LCMV-BSS-2, and SC LCMV-BSS. In addition, their computing times in seconds are also included in the tables for comparison.

There are several interesting findings derived from the results in Tables 5.9–5.17.

1. It is very obvious to note that BSS did improve ILCMV classification results. Such improvement cannot be applied to the four EPF-based methods where the classification results of the four EPF-based methods using band subsets could only get worse compared to the results using full bands. This may be due to the fact that the four EPF-based methods used principal components analysis (PCA) to compress the original data as a preprocessing which retains some crucial information provided by full bands.
2. The precision rates produced by the four EPF-based methods were very low as also noted in [38, 39]. But ILCMV using bands selected by LCMV-BSS consistently performed very well in both $P_{OA}$ and $P_{precision}$.
3. According to Tables 5.9–5.14, ILCMV performed slightly better than the 4 EPF-based methods in $P_{OA}$ but significantly better in $P_{precision}$ for Purdue's data and Salinas. The scene of University of

**Table 5.9** OA and precision rates calculated by the classification results using the bands selected by SQ LCMV-CBSS-1 for the Purdue data

| Class | EPF-B-g with full bands | EPF-B-c with full bands | EPF-G-g with full bands | EPF-G-c with full bands | ILCMV with full bands | EPF-B-g-BS | EPF-B-c-BS | EPF-G-g-BS | EPF-G-c-BS | ILCMV-BS |
|---|---|---|---|---|---|---|---|---|---|---|
| 1 | 100 | 100 | 97.83 | 100 | 95.65 | 100 | 100 | 100 | 100 | 95.65 |
| 2 | 76.47 | 87.82 | 80.60 | 82.98 | 96.01 | 74.79 | 90.76 | 71.99 | 79.90 | 95.59 |
| 3 | 93.49 | 83.98 | 79.64 | 65.42 | 96.99 | 63.37 | 75.66 | 84.10 | 72.77 | 96.39 |
| 4 | 99.16 | 100 | 100 | 96.20 | 98.73 | 100 | 100 | 100 | 97.89 | 97.89 |
| 5 | 93.79 | 94.00 | 97.10 | 94.82 | 89.44 | 89.86 | 97.52 | 96.07 | 94.41 | 91.93 |
| 6 | 100 | 99.59 | 99.59 | 99.45 | 97.12 | 98.22 | 99.86 | 99.73 | 99.32 | 96.44 |
| 7 | 92.86 | 92.86 | 96.43 | 96.43 | 100 | 92.86 | 64.29 | 92.86 | 89.29 | 100 |
| 8 | 100 | 100 | 100 | 100 | 98.78 | 100 | 100 | 100 | 100 | 97.91 |
| 9 | 80.00 | 65.00 | 100 | 100 | 100 | 65.00 | 95.00 | 65.00 | 10.00 | 100 |
| 10 | 90.53 | 91.46 | 87.14 | 93.00 | 93.93 | 75.31 | 73.97 | 84.88 | 76.54 | 94.24 |
| 11 | 90.67 | 92.67 | 86.27 | 88.88 | 94.70 | 78.98 | 70.79 | 64.73 | 80.94 | 95.48 |
| 12 | 98.31 | 96.46 | 93.93 | 92.07 | 95.45 | 47.55 | 42.50 | 79.76 | 57.67 | 95.95 |
| 13 | 99.02 | 99.51 | 99.51 | 99.51 | 98.54 | 100 | 100 | 100 | 100 | 97.56 |
| 14 | 97.71 | 97.00 | 97.87 | 98.26 | 93.52 | 94.86 | 94.86 | 87.67 | 92.57 | 95.89 |
| 15 | 100 | 100 | 99.74 | 82.90 | 90.67 | 95.85 | 96.37 | 95.85 | 97.41 | 94.82 |
| 16 | 97.85 | 100 | 100 | 100 | 98.92 | 98.92 | 100 | 98.92 | 97.85 | 98.92 |
| $P_{OA}$ | 92.27 | 93.45 | 90.32 | 89.79 | 95.09 | 81.62 | 82.94 | 81.17 | 84.15 | 95.67 |
| $P_{Precision}$ | 44.98 | 44.56 | 44.03 | 43.77 | 97.61 | 39.79 | 40.43 | 39.86 | 41.02 | 97.89 |
| Time (s) | 196.58 | 200.84 | 194.09 | 200.87 | 25.37 | 31.27 | 36.77 | 31.14 | 36.16 | 37.25 |

**Table 5.10** OA and precision rates calculated by the classification results using the bands selected by SQ LCMV-CBSS-2 for the Purdue data

| Class | EPF-B-g with full bands | EPF-B-c with full bands | EPF-G-g with full bands | EPF-G-c with full bands | ILCMV with full bands | EPF-B-g-BS | EPF-B-c-BS | EPF-G-g-BS | EPF-G-c-BS | ILCMV-BS |
|---|---|---|---|---|---|---|---|---|---|---|
| 1 | 100 | 100 | 100 | 100 | 95.65 | 100 | 100 | 100 | 100 | 97.83 |
| 2 | 86.83 | 82.49 | 86.62 | 84.52 | 96.01 | 85.78 | 81.86 | 64.22 | 66.60 | 93.78 |
| 3 | 90.84 | 85.90 | 82.53 | 80.72 | 96.99 | 81.93 | 74.58 | 63.98 | 64.82 | 95.67 |
| 4 | 99.16 | 99.16 | 100 | 99.58 | 98.73 | 99.16 | 100 | 99.58 | 100 | 98.31 |
| 5 | 95.03 | 92.96 | 97.72 | 92.34 | 89.44 | 93.17 | 95.03 | 91.72 | 92.13 | 92.34 |
| 6 | 100 | 99.86 | 99.59 | 99.73 | 97.12 | 99.73 | 97.81 | 100 | 99.59 | 97.95 |
| 7 | 89.29 | 89.29 | 96.43 | 96.43 | 100 | 92.86 | 78.57 | 96.43 | 92.86 | 100 |
| 8 | 100 | 100 | 100 | 100 | 98.78 | 100 | 100 | 100 | 100 | 99.16 |
| 9 | 60 | 75 | 70 | 50 | 100 | 100 | 95 | 65 | 35 | 100 |
| 10 | 92.59 | 90.33 | 90.74 | 92.49 | 93.93 | 70.27 | 66.26 | 64.71 | 54.42 | 91.98 |
| 11 | 89.12 | 89.33 | 88.51 | 86.44 | 94.70 | 64.89 | 70.26 | 82.12 | 82.24 | 96.17 |
| 12 | 96.46 | 98.31 | 98.15 | 98.99 | 95.45 | 67.45 | 52.78 | 51.10 | 34.74 | 95.11 |
| 13 | 99.02 | 99.02 | 99.51 | 99.51 | 98.54 | 99.51 | 99.51 | 99.51 | 99.51 | 98.54 |

*(continued)*

**Table 5.10** (*Continued*)

| Class | EPF-B-g with full bands | EPF-B-c with full bands | EPF-G-g with full bands | EPF-G-c with full bands | ILCMV with full bands | EPF-B-g-BS | EPF-B-c-BS | EPF-G-g-BS | EPF-G-c-BS | ILCMV-BS |
|---|---|---|---|---|---|---|---|---|---|---|
| 14 | 98.81 | 98.18 | 96.36 | 95.02 | 93.52 | 90.59 | 97.79 | 92.81 | 89.41 | 96.05 |
| 15 | 99.74 | 100 | 94.04 | 95.34 | 90.67 | 90.41 | 90.41 | 93.26 | 83.42 | 94.56 |
| 16 | 100 | 100 | 100 | 100 | 98.92 | 95.70 | 98.92 | 100 | 100 | 97.85 |
| $P_{OA}$ | 93.37 | 92.17 | 92.10 | 90.96 | 95.09 | 81.49 | 81.28 | 80.01 | 77.66 | 95.48 |
| $P_{Precision}$ | 45.52 | 44.93 | 44.89 | 44.34 | 97.61 | 39.72 | 39.61 | 39 | 37.85 | 97.80 |
| Time (s) | 194.14 | 199.37 | 194.13 | 200.36 | 25.37 | 31.16 | 37.93 | 32.56 | 36.51 | 41.58 |

**Table 5.11** OA and precision rates calculated by the classification results using the bands selected by SC LCMV-CBSS for the Purdue data

| Class | EPF-B-g with full bands | EPF-B-c with full bands | EPF-G-g with full bands | EPF-G-c with full bands | ILCMV with full bands | EPF-B-g-BS | EPF-B-c-BS | EPF-G-g-BS | EPF-G-c-BS | ILCMV-BS |
|---|---|---|---|---|---|---|---|---|---|---|
| 1 | 100 | 100 | 100 | 97.83 | 95.65 | 100 | 97.83 | 97.83 | 100 | 100 |
| 2 | 83.26 | 83.26 | 83.05 | 78.92 | 96.01 | 48.25 | 42.09 | 48.74 | 39.92 | 94.89 |
| 3 | 79.16 | 69.64 | 80.48 | 66.87 | 96.99 | 62.05 | 55.18 | 60.36 | 44.10 | 94.10 |
| 4 | 100 | 98.31 | 100 | 99.58 | 98.73 | 100 | 100 | 100 | 99.58 | 98.31 |
| 5 | 93.79 | 93.17 | 95.86 | 93.79 | 89.44 | 91.93 | 94 | 93.58 | 93.79 | 92.96 |
| 6 | 99.73 | 99.73 | 98.08 | 99.32 | 97.12 | 99.32 | 99.45 | 94.93 | 99.45 | 95.75 |
| 7 | 96.43 | 92.86 | 96.43 | 89.29 | 100 | 85.71 | 71.43 | 92.86 | 64.29 | 100 |
| 8 | 100 | 99.79 | 100 | 100 | 98.78 | 100 | 100 | 99.79 | 100 | 98.95 |
| 9 | 20 | 75 | 75 | 65 | 100 | 20 | 45 | 35 | 0 | 100 |
| 10 | 75.31 | 84.16 | 84.47 | 89.61 | 93.93 | 59.05 | 48.97 | 47.22 | 49.90 | 91.98 |
| 11 | 91.65 | 91 | 87.98 | 91.73 | 94.70 | 76.25 | 78.09 | 70.22 | 72.87 | 95.93 |
| 12 | 96.29 | 92.07 | 96.12 | 97.47 | 95.45 | 70.83 | 60.20 | 44.86 | 58.85 | 96.63 |
| 13 | 99.51 | 99.51 | 99.51 | 99.51 | 98.54 | 100 | 100 | 100 | 100 | 98.54 |
| 14 | 92.89 | 98.34 | 96.36 | 96.68 | 93.52 | 98.34 | 93.04 | 97.23 | 97.39 | 96.13 |
| 15 | 100 | 100 | 98.19 | 99.48 | 90.67 | 93.26 | 89.12 | 79.27 | 67.36 | 96.11 |
| 16 | 100 | 100 | 100 | 100 | 98.92 | 100 | 100 | 93.55 | 95.70 | 95.70 |
| $P_{OA}$ | 90.06 | 90.42 | 90.56 | 90.37 | 95.09 | 77.37 | 74.12 | 72.31 | 71.25 | 95.46 |
| $P_{Precision}$ | 43.90 | 44.08 | 44.15 | 44.05 | 97.61 | 37.72 | 36.13 | 35.25 | 34.73 | 97.79 |
| Time (s) | 187.76 | 203.60 | 195.20 | 201.21 | 25.37 | 32.01 | 38.78 | 31.92 | 38.07 | 42.73 |

Pavia is interesting as shown in Tables 5.15–5.17. The 4 EPF-based methods performed very well in $P_{OA}$ but did very poorly in $P_{precision}$ with about only 20%. Furthermore, $P_{OA}$ rates produced by ILCMV may not be as good as the 4 EPF-based methods with about 10% less but the $P_{precision}$ rates produced by ILCMV were around 96% which is nearly 4.8 times better than that with about 20% produced by the 4 EPF-based methods. These experiments demonstrated that the BKG issue is critical in data analysis of the University of Pavia and cannot be ignored or discarded in data processing. Unfortunately, this BKG issue has never been investigated in the past.

4. Unlike the 4 EPF-based methods which can only perform well in $P_{OA}$ but very poorly in $P_{precision}$, ILCMV consistently performs well in both $P_{OA}$ and $P_{precision}$, even better when it is implemented

**Table 5.12** OA and precision rates calculated by the classification results using the bands selected by SQ LCMV-CBSS-1 in Table 5.4 for Salinas

| Class | EPF-B-g with full bands | EPF-B-c with full bands | EPF-G-g with full bands | EPF-G-c with full bands | ILCMV with full bands | EPF-B-g-BS | EPF-B-c-BS | EPF-G-g-BS | EPF-G-c-BS | ILCMV-BS |
|---|---|---|---|---|---|---|---|---|---|---|
| 1 | 100 | 100 | 100 | 100 | 95.52 | 100 | 99.75 | 100 | 100 | 96.37 |
| 2 | 100 | 100 | 100 | 100 | 98.42 | 99.97 | 100 | 99.97 | 100 | 98.79 |
| 3 | 100 | 100 | 100 | 100 | 93.78 | 100 | 100 | 100 | 100 | 90.44 |
| 4 | 100 | 100 | 100 | 99.93 | 95.62 | 99.93 | 99.86 | 99.93 | 100 | 96.56 |
| 5 | 99.37 | 99.25 | 98.69 | 99.25 | 96.90 | 98.92 | 98.95 | 98.92 | 98.95 | 95.87 |
| 6 | 100 | 100 | 100 | 100 | 98.79 | 99.97 | 99.97 | 99.97 | 99.97 | 97.95 |
| 7 | 100 | 99.92 | 100 | 100 | 98.63 | 99.66 | 99.72 | 99.66 | 99.83 | 98.35 |
| 8 | 90.92 | 90.40 | 89.26 | 90.61 | 96.69 | 87.11 | 91.94 | 87.11 | 91.70 | 95.84 |
| 9 | 99.98 | 100 | 99.97 | 99.97 | 95.87 | 99.58 | 100 | 99.58 | 99.97 | 94.79 |
| 10 | 96.95 | 98.08 | 98.60 | 98.51 | 96.67 | 96.19 | 98.57 | 96.19 | 98.60 | 95.73 |
| 11 | 99.91 | 99.91 | 99.91 | 100 | 97.75 | 99.91 | 99.91 | 99.91 | 99.81 | 95.79 |
| 12 | 100 | 100 | 100 | 100 | 97.15 | 100 | 100 | 100 | 100 | 96.32 |
| 13 | 99.89 | 99.56 | 99.02 | 99.89 | 96.51 | 98.47 | 99.13 | 98.47 | 99.56 | 87.77 |
| 14 | 99.91 | 99.25 | 99.63 | 100 | 95.89 | 98.97 | 99.35 | 98.97 | 98.04 | 97.76 |
| 15 | 89.01 | 85.65 | 85.65 | 87.08 | 94 | 90.66 | 82.21 | 90.66 | 85.99 | 95.42 |
| 16 | 99.83 | 100 | 99.45 | 99.83 | 93.30 | 99.67 | 100 | 99.67 | 100 | 95.68 |
| $P_{OA}$ | 96.40 | 95.89 | 95.64 | 96.17 | 96.37 | 95.64 | 95.73 | 95.64 | 96.19 | 95.93 |
| $P_{Precision}$ | 46.97 | 46.26 | 46.95 | 46.85 | 98.23 | 46.60 | 46.64 | 46.60 | 46.86 | 98.02 |
| Time (s) | 1060.77 | 741.84 | 1082.51 | 1134.06 | 167.80 | 75.17 | 78.73 | 75.17 | 104.31 | 134.96 |

**Table 5.13** OA and precision rates calculated by the classification results using the bands selected by SQ LCMV-CBSS-2 in Table 5.4 for Salinas

| Class | EPF-B-g with full bands | EPF-B-c with full bands | EPF-G-g with full bands | EPF-G-c with full bands | ILCMV with full bands | EPF-B-g-BS | EPF-B-c-BS | EPF-G-g-BS | EPF-G-c-BS | ILCMV-BS |
|---|---|---|---|---|---|---|---|---|---|---|
| 1 | 100 | 100 | 100 | 100 | 95.52 | 100 | 100 | 100 | 100 | 97.01 |
| 2 | 100 | 100 | 100 | 100 | 98.42 | 100 | 100 | 100 | 99.97 | 98.36 |
| 3 | 100 | 100 | 100 | 100 | 93.78 | 100 | 100 | 99.95 | 100 | 95.14 |
| 4 | 100 | 100 | 100 | 99.78 | 95.62 | 100 | 99.93 | 100 | 100 | 95.91 |
| 5 | 99.48 | 98.95 | 98.58 | 99.14 | 96.90 | 98.66 | 98.73 | 99.14 | 98.84 | 95.94 |
| 6 | 100 | 100 | 100 | 100 | 98.79 | 99.95 | 100 | 100 | 100 | 98.91 |
| 7 | 100 | 99.89 | 100 | 99.89 | 98.63 | 99.92 | 99.97 | 99.89 | 99.80 | 98.44 |
| 8 | 88.63 | 91.98 | 87.90 | 90.86 | 96.69 | 87.84 | 89.50 | 90.61 | 91.86 | 97.47 |
| 9 | 99.90 | 99.98 | 99.95 | 100 | 95.87 | 99.97 | 99.94 | 99.60 | 100 | 94.89 |
| 10 | 97.56 | 98.90 | 97.04 | 97.50 | 96.67 | 99.33 | 99.51 | 99.21 | 98.29 | 96.58 |
| 11 | 100 | 99.91 | 99.91 | 99.72 | 97.75 | 100 | 99.91 | 99.72 | 100 | 97.38 |
| 12 | 100 | 100 | 100 | 100 | 97.15 | 100 | 100 | 100 | 100 | 95.39 |
| 13 | 99.24 | 99.89 | 98.80 | 99.13 | 96.51 | 99.02 | 100 | 98.36 | 99.13 | 97.38 |
| 14 | 99.91 | 99.91 | 99.91 | 99.25 | 95.89 | 100 | 100 | 99.91 | 99.35 | 97.20 |
| 15 | 88.66 | 85.43 | 94.25 | 91.43 | 94 | 79.35 | 93.04 | 83.94 | 86.13 | 96.27 |

(*continued*)

**Table 5.13** *(Continued)*

| Class | EPF-B-g with full bands | EPF-B-c with full bands | EPF-G-g with full bands | EPF-G-c with full bands | ILCMV with full bands | EPF-B-g-BS | EPF-B-c-BS | EPF-G-g-BS | EPF-G-c-BS | ILCMV-BS |
|---|---|---|---|---|---|---|---|---|---|---|
| 16 | 99.94 | 100 | 99.61 | 99.50 | 93.30 | 100 | 100 | 100 | 100 | 95.13 |
| $P_{OA}$ | 95.91 | 96.24 | 96.42 | 96.69 | 96.37 | 94.56 | 96.77 | 95.71 | 96.24 | 96.81 |
| $P_{Precision}$ | 46.73 | 46.89 | 46.97 | 47.11 | 98.23 | 46.07 | 47.15 | 46.63 | 46.89 | 98.45 |
| Time (s) | 1128.55 | 755.52 | 1050.34 | 722.76 | 167.80 | 73.49 | 101.80 | 71.57 | 98.16 | 159.26 |

**Table 5.14** OA and precision rates calculated by the classification results using the bands selected by SC LCMV-CBSS in Table 5.4 for Salinas

| Class | EPF-B-g with full bands | EPF-B-c with full bands | EPF-G-g with full bands | EPF-G-c with full bands | ILCMV with full bands | EPF-B-g-BS | EPF-B-c-BS | EPF-G-g-BS | EPF-G-c-BS | ILCMV-BS |
|---|---|---|---|---|---|---|---|---|---|---|
| 1 | 100 | 100 | 100 | 100 | 95.52 | 100 | 100 | 100 | 100 | 96.91 |
| 2 | 100 | 100 | 100 | 100 | 98.42 | 99.95 | 99.84 | 99.92 | 99.97 | 98.71 |
| 3 | 100 | 100 | 100 | 100 | 93.78 | 99.90 | 100 | 100 | 99.90 | 95.95 |
| 4 | 100 | 100 | 100 | 100 | 95.62 | 100 | 99.57 | 100 | 99.86 | 92.04 |
| 5 | 99.07 | 99.22 | 99.10 | 98.92 | 96.90 | 98.88 | 99.55 | 99.33 | 98.99 | 90.78 |
| 6 | 100 | 100 | 100 | 100 | 98.79 | 100 | 100 | 100 | 99.90 | 97.75 |
| 7 | 100 | 99.97 | 100 | 99.97 | 98.63 | 99.61 | 99.89 | 100 | 99.94 | 98.32 |
| 8 | 90.18 | 89.51 | 89.40 | 91.38 | 96.69 | 90.88 | 89.02 | 90.68 | 90.30 | 96.61 |
| 9 | 100 | 99.98 | 99.81 | 99.94 | 95.87 | 99.95 | 99.95 | 99.97 | 100 | 95.44 |
| 10 | 97.47 | 97.71 | 97.41 | 99.48 | 96.67 | 98.90 | 99.24 | 98.11 | 98.29 | 96.77 |
| 11 | 100 | 100 | 100 | 100 | 97.75 | 100 | 99.72 | 99.34 | 99.72 | 97.66 |
| 12 | 100 | 100 | 100 | 100 | 97.15 | 100 | 100 | 100 | 100 | 95.43 |
| 13 | 100 | 99.89 | 98.91 | 99.89 | 96.51 | 99.34 | 99.89 | 99.24 | 99.13 | 94 |
| 14 | 100 | 100 | 100 | 99.81 | 95.89 | 99.35 | 99.63 | 98.79 | 98.32 | 96.93 |
| 15 | 88.99 | 84.26 | 88.08 | 86.90 | 94 | 89.74 | 93.35 | 78.92 | 88.92 | 95.84 |
| 16 | 99.28 | 99.89 | 100 | 99.94 | 93.30 | 100 | 100 | 100 | 99.94 | 95.41 |
| $P_{OA}$ | 96.25 | 95.52 | 95.95 | 96.34 | 96.37 | 96.50 | 96.70 | 95.02 | 96.26 | 96.21 |
| $P_{Precision}$ | 46.89 | 46.53 | 46.75 | 46.94 | 98.23 | 47.02 | 47.11 | 46.29 | 46.90 | 98.15 |
| Time (s) | 1139.99 | 1106.95 | 1154.78 | 1089.61 | 167.80 | 73.06 | 94.17 | 69.92 | 91.50 | 147.30 |

in conjunction with BSS, a case that the EPF-based methods actually failed as shown in Tables 5.9–5.17.

5. Last but not least, BS is heavily determined by three factors, data to be processed, a BS method to be selected, and a classifier to be used. Unfortunately, most works on BS for hyperspectral image classification have been focused on design and development of BS methods but very little has been reported on performance evaluation of different classifiers which use the same set of bands selected by a BS method. For example, as shown in Tables 5.7–5.15, if the 4 EPF methods were implemented by BS, their classification results could not be improved, but ILCMV could.

**Table 5.15** OA and precision rates calculated by the classification results using the bands selected by SQ LCMV-CBSS-1 in Table 5.4 for University of Pavia

| Class | EPF-B-g with full bands | EPF-B-c with full bands | EPF-G-g with full bands | EPF-G-c with full bands | ILCMV with full bands | EPF-B-g-BS | EPF-B-c-BS | EPF-G-g-BS | EPF-G-c-BS | ILCMV-BS |
|---|---|---|---|---|---|---|---|---|---|---|
| 1 | 98.04 | 98.04 | 98.08 | 97.81 | 77.24 | 96.15 | 92.47 | 96.34 | 97.29 | 80.21 |
| 2 | 98.66 | 99.39 | 97.79 | 98.28 | 86.42 | 94.31 | 94.84 | 97.39 | 98.06 | 84.44 |
| 3 | 91.09 | 93.52 | 95 | 94.33 | 73.34 | 92.62 | 94.81 | 95.57 | 95.33 | 84.14 |
| 4 | 93.47 | 95.27 | 92.92 | 98.01 | 79.85 | 97.52 | 97.45 | 97 | 96.87 | 76.49 |
| 5 | 100 | 100 | 100 | 99.85 | 98.81 | 100 | 100 | 100 | 100 | 98.14 |
| 6 | 99.98 | 100 | 100 | 100 | 91.49 | 99.64 | 98.15 | 99.64 | 100 | 93.50 |
| 7 | 100 | 99.32 | 99.92 | 99.77 | 89.10 | 99.92 | 99.32 | 99.40 | 100 | 86.19 |
| 8 | 99.02 | 99 | 97.80 | 99.78 | 81.10 | 95.60 | 96.22 | 96.41 | 96.28 | 82.84 |
| 9 | 100 | 100 | 100 | 100 | 78.46 | 100 | 99.89 | 100 | 100 | 77.09 |
| $P_{OA}$ | 98.12 | 98.67 | 97.80 | 98.46 | 84.32 | 95.96 | 97.49 | 97.49 | 98.01 | 84.45 |
| $P_{Precision}$ | 20.24 | 20.35 | 20.17 | 20.31 | 96.76 | 19.79 | 19.71 | 20.11 | 20.21 | 96.82 |
| Time (s) | 225.93 | 265.79 | 232.05 | 252.50 | 401.08 | 51.16 | 83.46 | 52.90 | 78.32 | 1387.01 |

**Table 5.16** OA and precision rates calculated by the classification results using the bands selected by SQ LCMV-CBSS-2 in Table 5.4 for University of Pavia

| Class | EPF-B-g with full bands | EPF-B-c with full bands | EPF-G-g with full bands | EPF-G-c with full bands | ILCMV with full bands | EPF-B-g-BS | EPF-B-c-BS | EPF-G-g-BS | EPF-G-c-BS | ILCMV-BS |
|---|---|---|---|---|---|---|---|---|---|---|
| 1 | 94.89 | 97.80 | 97.84 | 98.21 | 77.24 | 95.43 | 95.96 | 88.76 | 92.01 | 78.22 |
| 2 | 98.58 | 97.63 | 99.25 | 98.01 | 86.42 | 96.43 | 94.88 | 92.59 | 93.13 | 88.05 |
| 3 | 94.90 | 93.57 | 92.57 | 95.19 | 73.34 | 92.19 | 92.47 | 95.52 | 93.62 | 85.21 |
| 4 | 95.63 | 94.97 | 93.02 | 98.83 | 79.85 | 98.69 | 96.02 | 98.37 | 98.47 | 74.71 |
| 5 | 100 | 100 | 100 | 99.85 | 98.81 | 100 | 100 | 100 | 100 | 97.77 |
| 6 | 100 | 100 | 100 | 100 | 91.49 | 100 | 99.52 | 99.72 | 99.94 | 90.57 |
| 7 | 100 | 100 | 99.40 | 99.77 | 89.10 | 100 | 99.92 | 100 | 100 | 90 |
| 8 | 96.93 | 98.89 | 98.94 | 98.48 | 81.10 | 93.97 | 94.70 | 95.22 | 93.56 | 82.92 |
| 9 | 100 | 100 | 100 | 100 | 78.46 | 100 | 100 | 100 | 100 | 77.09 |
| $P_{OA}$ | 97.76 | 97.85 | 98.36 | 98.39 | 84.32 | 96.74 | 95.97 | 94.25 | 94.78 | 85.41 |
| $P_{Precision}$ | 20.16 | 20.18 | 20.29 | 20.29 | 96.76 | 19.95 | 19.79 | 19.44 | 19.55 | 96.94 |
| Time (s) | 219.10 | 249.76 | 226.85 | 238.81 | 401.08 | 48.52 | 82.95 | 53.53 | 79.05 | 971.56 |

**Table 5.17** OA and precision rates calculated by the classification results using the bands selected by SC LCMV-CBSS in Table 5.4 for University of Pavia

| Class | EPF-B-g with full bands | EPF-B-c with full bands | EPF-G-g with full bands | EPF-G-c with full bands | ILCMV with full bands | EPF-B-g-BS | EPF-B-c-BS | EPF-G-g-BS | EPF-G-c-BS | ILCMV-BS |
|---|---|---|---|---|---|---|---|---|---|---|
| 1 | 99.43 | 97.16 | 97.95 | 97.98 | 77.24 | 95.93 | 92.73 | 98.27 | 97.99 | 77.08 |
| 2 | 98.19 | 99.15 | 98.80 | 98.79 | 86.42 | 95.44 | 94.31 | 97 | 96.23 | 88.67 |
| 3 | 99.24 | 95.09 | 94.85 | 93.57 | 73.34 | 93.52 | 91.04 | 92.38 | 94.14 | 86.76 |
| 4 | 93.93 | 94.65 | 94.61 | 98.27 | 79.85 | 97.52 | 96.74 | 98.27 | 96.96 | 78.56 |
| 5 | 100 | 99.85 | 100 | 99.93 | 98.81 | 100 | 100 | 100 | 100 | 97.70 |
| 6 | 99.96 | 99.34 | 99.68 | 100 | 91.49 | 99.70 | 98.11 | 99.05 | 100 | 90.77 |
| 7 | 100 | 100 | 99.70 | 99.77 | 89.10 | 99.62 | 99.47 | 99.92 | 99.55 | 91.13 |
| 8 | 96.85 | 98.13 | 97.77 | 99.51 | 81.10 | 92.88 | 92.61 | 92.99 | 92.97 | 82.46 |
| 9 | 100 | 100 | 100 | 100 | 78.46 | 100 | 100 | 100 | 100 | 79.30 |
| $P_{OA}$ | 98.37 | 98.32 | 98.28 | 98.67 | 84.32 | 96.22 | 94.85 | 97.21 | 96.92 | 85.92 |
| $P_{Precision}$ | 20.29 | 20.28 | 20.27 | 20.35 | 96.76 | 19.85 | 19.56 | 20.05 | 19.99 | 97.09 |
| Time (s) | 238.01 | 270.24 | 234.01 | 259.33 | 401.08 | 51.13 | 83.41 | 49.37 | 76.48 | 998.17 |

## 5.8 Conclusion

This chapter presents a new theory of BSS which can be considered as a generalization of SBS which considers a band subset as a singleton set. In particular, it derives constrained BSS (CBSS) as a counterpart of constrained BS (CBS) discussed in Chapter 4. There are several significant results derived from BSS. First of all, it introduces a new concept of BSS which selects multiple bands simultaneously as a band subset instead of selecting single bands at a time. This is not trivial since BSS extends a single band as a band vector to a band subset as a band matrix where matrix complexity is much more sophisticated than vector complexity. Second of all, no band prioritization and band decorrelation are required for BSS because BSS already takes care of both by selecting multiple bands as a band subset. Third of all, it interprets finding optimal BSS as finding optimal endmember sets where selecting an appropriate band subset is equivalent to finding an appropriate endmember set. With this interpretation, two sequential algorithms developed for N-FINDR, SQ N-FINDR and SC N-FINDR, can be redesigned for finding optimal BSS, called SQBSS and SCBSS, to avoid the exhaustive search for all possible band subsets. Fourth of all, BSS can be adapted according to various applications by specifying $J(\Omega_{MBS})$ in (5.1). Fifth of all, from the experiments conducted in this chapter, it clearly shows that on some cases, judiciously selecting band subsets can produce better results than using full bands. In other words, selecting right band subsets is more crucial than using full band set. BSS provides a better means of selecting bands for this purpose. Lastly and most importantly, BSS does not follow the traditional wisdom which either uses BP/BD to rank individual bands to select bands or application-based features such as detection, classification followed by a feature extraction algorithm such as SFFS to select bands. Instead, it directly solves the BS problem specified by $J(\Omega_{MBS})$ in (5.1) and further develops two sequential algorithms, SQBSS and SCBSS, to replace the exhaustive search for finding an optimal band subset.

## References


1. Su, H., Du, Q., Chen, G., and Du, P. (2014). Optimized hyperspectral band selection using particle swam optimization. *IEEE Journal Selected Topics in Applied Earth Observations and Remote Sensing* 7 (6): 2659–2670.
2. Yuan, Y., Zhu, G., and Wang, Q. (2015). Hyperspectral band selection by multitask sparisty pursuit. *IEEE Transactions on Geoscience and Remote Sensing* 53 (2): 631–644.
3. Yuan, Y., Zheng, X., and Lu, X. (2017). Discovering diverse subset for unsupervised hyperspectral band selection. *IEEE Transactions Image Processing* 26 (1): 51–64.
4. Wang, Q., Zhang, F., and Li, X. (2020). Hyperspectral band selection via optimal neighborhood reconstruction. *IEEE Transactions on Geoscience and Remote Sensing* 58 (4): 8465–8476.
5. Chang, C.-I and Du, Q. (2004). Estimation of number of spectrally distinct spectral signal sources in hyperspectral imagery. *IEEE Transactions on Geoscience and Remote Sensing* 42: 608–619.
6. Chang, C.-I (2013). *Hyperspectral Data Processing: Signal Processing Algorithm Design and Analysis.* Hoboken, NJ: Wiley.
7. Chang, C.-I (2018). A review of virtual dimensionality for hyperspectral imagery. *IEEE Journal of Selected Topics in Applied Earth Observation and Remote Sensing* 11 (4): 1285–1305.
8. Chang, C.-I (2017). *Real-Time Recursive Hyperspectral Sample and Band Processing: Algorithm Architecture and Implementation.* New York: Springer.
9. Winter, M.E. (1999). N-finder: an algorithm for fast autonomous spectral endmember determination in hyperspectral data. *Image Spectrometry V*, Proc. SPIE 3753: 266–277.
10. Wu, C.C., Chu, S., and Chang, C.-I (2008). Sequential N-FINDR algorithm. *SPIE Conference on Imaging Spectrometry XIII,* (10–14 August 2008). San Diego.
11. Xiong, W., Wu, C.-C., Chang, C.-I et al. (2011). Fast algorithms to implement N-FINDR for hyperspectral endmember extraction. *IEEE Journal of Selected Topics in Applied Earth Observation and Remote Sensing* 4 (3): 545–564.
12. Chang, C.-I (2013). Maximum simplex volume-based endmember extraction algorithms. US Patent 8,417,748 B2, 9 April 2013.
13. Chang, C.-I (2016). *Real-Time Progressive Hyperspectral Image Processing: Endmember Finding and Anomaly Detection.* New York: Springer.
14. McEliece, R.J. (2004). *Theory of Information and Coding.* New York, NY: Cambridge University Press.
15. Chang, C.-I, Du, Q., Sun, T.S., and Althouse, M.L.G. (1999). A joint band prioritization and band decorrelation approach to band selection for hyperspectral image classification. *IEEE Transactions on Geoscience and Remote Sensing* 37 (6): 2631–2641.
16. Yang, H., Du, Q., Su, H., and Sheng, Y. (2011). An efficient method for supervised hyperspectral band selection. *IEEE Geoscience and Remote Sensing Letters* 8: 138–142.
17. Pudil, P., Novovicova, J., and Kittler, J. (1994). Floating search methods in feature selection. *Pattern Recognition Letters* 15: 1119–1125.
18. Su, H., Yang, B., and Du, Q. (2016). Hyperspectral band selection using improved firefly algorithm. *IEEE Geoscience and Remote Sensing Letters* 13 (1): 68–72.
19. Zhu, G., Huang, Y., Lei, J. et al. (2016). Unsupervised hyperspectral band selection by dominant set extraction. *IEEE Transactions on Geoscience and Remote Sensing* 54 (1): 227–239.
20. Chang, C.-I, Lee, L.C., Xue, B. et al. (2017). Channel capacity approach to band subset selection for hyperspectral imagery. *IEEE Journal of Selected Topics in Applied Earth Observation and Remote Sensing* 10 (10): 4630–4644.
21. Chang, C.-I (2003). *Hyperspectral Imaging: Techniques for Spectral detection and Classification.* Kluwer Academic/Plenum Publishers.
22. Chang, C.-I (2000). An information theoretic-based approach to spectral variability, similarity and discriminability for hyperspectral image analysis. *IEEE Transactions on Information Theory* 46 (5): 1927–1932.
23. Fano, R.M. (1961). *Transmission of Information.* MIT Press.

24. Blahut, R.E. (1972). Computations of channel capacity and rate-distortion functions. *IEEE Transactions on Information Theory* 18 (4): 460–473.
25. Wang, L., Chang, C.-I, Lee, L.C. et al. (2017). Band subset selection for anomaly detection in hyperspectral imagery. *IEEE Transactions on Geoscience and Remote Sensing* 55 (9): 1–12.
26. Della Porta, C.J. and Chang, C.-I (2021). Esimating number of compressively sensed bands for hyperspectral image classification. *IEEE Journal of Selected Topics in Applied Earth Observation and Remote Sensing* 14: 11775–11788.
27. Chang, C.-I (2022). Band sampling for hyperspectral imaging. *IEEE Transactions on Geoscience and Remote Sensing* 60, 5514024: 1–24.
28. Yu, C., Song, M., and Chang, C.-I (2018). Band subset selection for hyperspectral image classification. *Remote Sensing* 10: 113. https://doi.org/10.3390/rs10010113.
29. Yu, C., Song, M., and Chang, C.-I (2019). Class signature-constrained background suppressed approach to band selection for hyperspectral image classification. *IEEE Transactions on Geoscience and Remote Sensing* 57 (1): 14–31.
30. Frost, O.L. III (1972). An algorithm for linearly constrained adaptive array processing. *Proceedings of the IEEE* 60: 926–935.
31. Chang, C.-I and Wang, S. (2006). Constrained band selection for hyperspectral imagery. *IEEE Transactions on Geoscience and Remote Sensing* 44 (6): 1575–1585.
32. Wang, L., Li, H.C., Xue, B., and Chang, C.-I (2017). Constrained band subset selection for hyperspectral imagery. *IEEE Geoscience and Remote Sensing Letters* 14 (11): 2032–2036.
33. Chang, C.-I (2002). Target signature-constrained mixed pixel classification for hyperspectral imagery. *IEEE Transactions on Geoscience and Remote Sensing* 40 (5): 1065–1081.
34. Chang, C.-I, Chen, S.Y., Li, H.C., and Wen, C.-H. (2016). A comparative analysis among ATGP, VCA and SGA for finding endmembers in hyperspectral imagery. *IEEE Journal of Selected Topics in Applied Earth Observation and Remote Sensing* 9 (9): 4280–4306.
35. Ren, H. and Chang, C.-I (2003). Automatic spectral target recognition in hyperspectral imagery. *IEEE Transactions on Aerospace and Electronic Systems* 39 (4): 1232–1249.
36. Nascimento, J.M.P. and Dias, J.M. (2005). Vertex component analysis: a fast algorithm to unmix hyperspectral data. *IEEE Transactions Geoscience and Remote Sensing* 43 (4): 898–910.
37. Wang, Y., Wang, L., Yu, C. et al. (2019). Constrained target band selection for target detection. *IEEE Transactions on Geoscience and Remote Sensing* 57 (8): 6079–6103.
38. Shang, X., Song, M., Wang, Y. et al. (2021). Target-constrained interference-minimized band selection for hyperspectral target detection. *IEEE Transactions on Geoscience and Remote Sensing* 59 (7): 6044–6064.
39. Xue, B., Yu, C., Wang, Y. et al. (2017). A subpixel target approach to hyperpsectral image classification. *IEEE Transactions on Geoscience and Remote Sensing* 55 (9): 5093–5114.
40. Yu, C., Xue, B., Song, M. et al. (2018). Iterative target–constrained interference-minimized classifier for hyperspectral image classification. *IEEE Journal of Selected Topics in Applied Earth Observation and Remote Sensing* 11 (4): 1095–1117.
41. Chang, C.-I (2019). A statistical detection theory approach to hyperspectral image classification: performance analysis. *IEEE Transactions on Geoscience and Remote Sensing* 57 (4): 2057–2074.
42. Kang, X., Li, S., and Benediktsson, J.A. (2014). Spectral-spatial hyperspectral image classification with edge-preserving filtering. *IEEE Transactions on Geoscience and Remote Sensing* 52 (5): 2666–2677.

# 6

# Progressive Band Selection Processing for Hyperspectral Image Classification

Chunyan Yu[1], Meiping Song[1], and Chein-I Chang[1,2]
[1]*Center for Hyperspectral Imaging in Remote Sensing (CHIRS), Information and Technology College, Dalian Maritime University, Dalian, China*
[2]*Remote Sensing Signal and Image Processing Laboratory, Department of Computer Science and Electrical Engineering, University of Maryland Baltimore County, Baltimore, MD, USA*

## 6.1 Introduction

Hyperspectral image classification (HSIC) has received considerable interest in recent years, just to name a few [1–15]. A general approach is to use a spectral classifier such as support vector machine (SVM) to perform spectral classification coupled with a spatial domain-based technique to capture spatial contextual information such as Markov random field (MRF) model [5], extended morphological profiles (EMP) [2], and edge preserving filter (EPF) [7] to take care of both spectral and spatial correlation. SVM is originally designed to perform binary classification. So, when it is used to perform multi-class classification, two common approaches are considered, one-against-one and one-against-rest [16]. No matter which one is adopted for multi-class classification, there are always inherent issues in misclassification that need to be addressed [16, pp. 182–184]. Most importantly, such multi-class classification is usually carried out in a one-shot operation in the sense that all multiple classes are classified by a classifier simultaneously at the same time using full band information. However, since various classes have their own spectral characteristics responding to different wavelengths for their classification, it is important to take advantage of those bands that are crucial to individual classes for their classification. As a result, when a misclassification error occurs, there is no clue about how it occurs and which classes cause such misclassification errors. This is particularly important for environmental

*Advances in Hyperspectral Image Processing Techniques*, First Edition. Edited by Chein-I Chang.

monitoring when contaminated areas may be slowly migrated. The HSIC currently being considered in the literature did not take this into account. This chapter develops a new approach, called progressive band selection processing of hyperspectral image classification (PBSP-HSIC), to address these issues.

To realize PBSP-HSIC, the main idea is to design a criterion that can measure the class classification priority (CCP) for each class of interest so that all the classes can be prioritized according to their scores calculated by CCP. Then, CCP can be used for two purposes. One is to find a desired band subset for each class for its classification. The other is to navigate classification being performed progressively based on CCP-ranked classes along with their corresponding band subsets selected by custom-designed algorithms. However, this is easily said than done because four challenging issues need to be addressed: (i) how to design CCP to measure the priority for each of the classes of interest, (ii) how to select band subset for each class, (iii) how to perform progressive classification according to CCP-ranked classes using their own selected band subsets, and (iv) how to classify all classes progressively stage by stage with each stage carried out by using a particular selected band subset. Most importantly, due to data spectral characteristics, data samples in each class responding to wavelengths quite differently, band selection (BS) becomes crucial when it comes to HSIC [17]. Therefore, including BS in PBSP-HSIC is critical to make HSIC more effective.

First of all, we define a correlation (CR)-based criterion to calculate CCP scores for all classes in terms of probabilities, referred to as CCP probabilities. That is, the higher is the CCP probability assigned to a class, the higher priority is the class to be classified. In order to classify various classes progressively, these CCP probabilities are further used to design a $p$-ary Huffman code that encodes all the classes of interest as codewords according to their CCP probabilities, where $p$ is the number of codeword alphabets used for encoding. So, the longer the codeword is assigned to a class, the lower priority the class is to be classified. These $p$-ary encoded Huffman codewords are then used to construct a CCP-based $p$-ary Huffman coding tree (HCT) that will be used as a guide to navigate progressive classification, where each layer of the designed $p$-ary HCT is considered as a stage that performs classification of particular classes specified by the codewords at that particular layer. As a result, classification can be carried out in a top-to-down fashion stage-by-stage using a $p$-ary HCT from the highest CCP level to the lowest CCP level. The depth of the tree is the number of stages required for progressive classification. Using such a $p$-ary HCT to represent classes with different levels of CCP is similar to using a graph network to select a band subset where each internal node specifies a particular band and each path represents a selection of desired bands such as MDPP in [18]. Since different classes are classified at different stages, a stage of classification requires a different set of selected bands to respond to the signatures of classes to be classified at the same stage. To address this issue, a recently developed BS for HSIC in [19], called class signature-constrained background suppressed (CSCBS) approach, is used to further develop class signature constrained band prioritization-based band selection (CSCBP-BS) to select appropriate bands for each stage. Finally, according to the results in [14], a new iterative classification technique, called iterative linearly constrained minimum variance (ILCMV), which is derived from the iterative MCBSC (IMCBSC) in [13] was shown to perform better than many other existing classifiers. Therefore, this ILCMV is used to derive progressive band selection processing of ILCMV (PBSP-ILCMV), which implements a CCP-constructed $p$-ary HCT using bands selected by CSCBP-BS to perform PBSP-HSIC. To evaluate the effectiveness of the proposed PBSP-HSIC, three real hyperspectral images including Purdue's Indiana Indian Pines, Salinas, and University of Pavia are used for performance analysis.

## 6.2 Measures of Class Classification Priority

Assume that there are $M$ classes of interest to be classified, $\{C_i\}_{i=1}^{M}$. Suppose that $\boldsymbol{\mu}_i$ is the sample mean of the $i$th class, $C_i$. We can define an index using the global sample correlation matrix $\mathbf{R}$ as

$$\rho_{\mathbf{R}}^{\mathrm{CCP}}(C_i) = \boldsymbol{\mu}_i^T \mathbf{R}^{-1} \boldsymbol{\mu}_i, \tag{6.1}$$

where $\mathbf{R} = \frac{1}{N}\sum_{i=1}^{N} \mathbf{r}\mathbf{r}_i^T$.

Using the index in (6.1), a criterion can measure CCP associated with different classes in terms of probabilities as follows:

$$p_i^{\mathrm{CCP}} = p^{\mathrm{CCP}}(C_i) = \frac{\rho_{\mathbf{R}}^{\mathrm{CCP}}(C_i)}{\sum_{i=1}^{M} \rho_{\mathbf{R}}^{\mathrm{CCP}}(C_i)}, \tag{6.2}$$

which can be considered as the priority probability assigned to class $C_i$. This implies that the higher the $p_i^{\mathrm{CCP}}$, the higher priority the class $C_i$ to be classified. From information theory [20], we can use (6.2) to further define the self-information of each class

$$I^{\mathrm{CCP}}(C_i) = -\log p_i^{\mathrm{CCP}}. \tag{6.3}$$

Based on the probabilities and self-information calculated by (6.2) and (6.3), we now introduce the entropy of CCP in terms of (6.3) defined by

$$H\left(\{C_i\}_{i=1}^{M}\right) = \sum_{i=1}^{M} p_i^{\mathrm{CCP}} I^{\mathrm{CCP}}(C_i) = -\sum_{i=1}^{M} p_i^{\mathrm{CCP}} \log p_i^{\mathrm{CCP}}. \tag{6.4}$$

Basically, the entropy of CCP by (6.4) can be used to measure the complexity of classifying hyperspectral data.

## 6.3 *p*-Ary Huffman Coding Tree Construction

The well-known optimal variable length coding method in information theory, Huffman coding [21], is used to construct a coding tree for progressive image classification. The idea can be described as follows.

Suppose that a source $S$ with the source alphabet space specified by the classes of interest, $\{C_i\}_{i=1}^{M}$, i.e. each class specified by a source alphabet, where $\{p_i^{\mathrm{CCP}} = p^{\mathrm{CCP}}(C_i)\}_{i=1}^{M}$ given in (6.2) is the source alphabet probabilities. Now, we assume that the codeword space used for encoding $S$ is given by $\Delta_p = \{0, 1, 2, \ldots, p-1\}$. Then, each of the codewords assigned to $\{C_i\}_{i=1}^{M}$ is actually a $p$-ary representation. As a special case with $p = 2$, the codewords are encoded by bits, i.e. binary digits. It should be noted that the "$p$" determines $p$-class classification at each level of a $p$-ary coding tree. In other words, for a layer of the constructed coding tree except for the last layer, there are $p$ internal nodes in that particular layer, each of which corresponds to one subset of classes ranked by CCP scores for $p$-class classification. The depth of the coding tree is given by $d = \lceil \log_p M \rceil$.

Constructing a $p$-ary Huffman code with $p > 2$ is not trivial. This is because the $p$-ary tree constructed by a $p$-ary Huffman code may not be complete in the sense that there may not be sufficient codewords to fill in all nodes and leaves of the tree. In this case, unencoded nodes must be leaves and should not be internal nodes.

In order to address this issue, we need to introduce dummy source alphabets so that these dummy alphabets will be treated as source alphabets with zero probability in such a way that they will have longest source codewords which do not exist. The key issue is how many dummy variables are required to do so when we construct a $p$-ary Huffman code for any given positive integer $p$.

Let $q$ be the number of the first set of source alphabets to be combined to reduce original probabilities $\{p_1, p_2, \ldots, p_n\}$ to $\left\{p_1^{(1)}, p_2^{(1)}, \ldots, p_{n^{(1)}}^{(1)}\right\}$. According to [21, eqs. (11.2) and (11.3), p. 338], $q$ must satisfy the following constraints:

$$q \in \{2, 3, \ldots, p\}, \tag{6.5}$$

$$n \equiv q \mod (p-1), \tag{6.6}$$

which means that $n = m(p-1) + q$ for some non-negative integer $m$. Then, the number of dummy variables is $\{x_1, x_2, \ldots, x_{p-q}\}$. In this case, $n_1^{(1)} = m(p-1) + 1$. As an example, for $n = 6$ and $p = 3$, $n$ can be expressed by (6.6) as $n = 2 \cdot 2 + 2$ with $m = 2$ and $q = 2$. This implies that the number of the dummy variable is $p - q = 3 - 2 = 1$.

## 6.4 Iterative LCMV

The classifier to be used for PBSP-HSIC is derived from adaptive beamforming in array signal processing [22, 23].

### 6.4.1 *Linearly Constrained Minimum Variance (LCMV)*

Suppose that $\boldsymbol{\mu}_1, \boldsymbol{\mu}_2, \ldots, \boldsymbol{\mu}_M$ are $M$ specific class signatures of interest. We can now form a class signature matrix, denoted by $\Lambda = [\boldsymbol{\mu}_1 \boldsymbol{\mu}_2 \cdots \boldsymbol{\mu}_M]$. The goal is to design an FIR linear filter with $L$ filter coefficients $\{w_1, w_2, \ldots, w_L\}$, denoted by an $L$-dimensional vector $\mathbf{w} = (w_1, w_2, \ldots, w_L)^T$ that minimizes the filter output energy subject to the following constraint:

$$\Lambda^T \mathbf{w} = \mathbf{c} \text{ where } \boldsymbol{\mu}_j^T \mathbf{w} = \sum\nolimits_{l=1}^{L} w_l \mu_{jl} = c_j \text{ for } 1 \le j \le M, \tag{6.7}$$

where $\mathbf{c} = (c_1, c_2, \ldots, c_p)^T$ is a constraint vector each of which constrains one particular class signature.

Now, let $y_i$ denote the output of the designed FIR filter resulting from the input $\mathbf{r}_i$. Then, $y_i$ can be expressed by

$$y_i = \sum_{l=1}^{L} w_l r_{il} = \mathbf{w}^T \mathbf{r}_i = \mathbf{r}_i^T \mathbf{w}. \tag{6.8}$$

Using (6.7) and (6.8), a linearly constrained minimum variance (LCMV)-based classifier can be designed by minimizing the average energy of the filter outputs

$$\left(\frac{1}{N}\right)\sum_{i=1}^{N} y_i^2 = \left(\frac{1}{N}\right)\sum_{i=1}^{N} \left(\mathbf{r}_i^T\mathbf{w}\right)^T\left(\mathbf{r}_i\mathbf{w}\right) = \mathbf{w}^T\left(\left(\frac{1}{N}\right)\sum_{i=1}^{N}\mathbf{r}_i\mathbf{r}_i^T\right)\mathbf{w} = \mathbf{w}^T\mathbf{R}_{L\times L}\mathbf{w}, \qquad (6.9)$$

where $\mathbf{R}_{L\times L} = (1/N)\sum_{i=1}^{N}\mathbf{r}_i\mathbf{r}_i^T$ is the autocorrelation sample matrix of the image data. The LCMV classifier can be obtained by solving the following linearly constrained optimization problem:

$$\min_{\mathbf{w}}\left\{\mathbf{w}^T\mathbf{R}_{L\times L}\mathbf{w}\right\} \text{ subject to } \Lambda^T\mathbf{w} = \mathbf{c}. \qquad (6.10)$$

The solution of $\mathbf{w}$ to (6.10) is given by

$$\mathbf{w}^{\text{LCMV}} = \mathbf{R}_{L\times L}^{-1}\Lambda\left(\Lambda^T\mathbf{R}_{L\times L}^{-1}\Lambda\right)^{-1}\mathbf{c} \qquad (6.11)$$

and the LCMV classifier $\delta^{\text{LCMV}}(\mathbf{r})$ for each hyperspectral image pixel vector $\mathbf{r}$ can be expressed as

$$\delta^{\text{LCMV}}(\mathbf{r}) = \left(\mathbf{w}^{\text{LCMV}}\right)^T\mathbf{r}. \qquad (6.12)$$

The minimum variance resulting from $\delta^{\text{LCMV}}(\mathbf{r})$ in (6.12) is therefore

$$\left(\mathbf{w}^{\text{LCMV}}\right)^T\mathbf{R}_{L\times L}\mathbf{w}^{\text{LCMV}} = \mathbf{c}^T\left((\Lambda)^T\mathbf{R}_{L\times L}^{-1}\Lambda\right)^{-1}\mathbf{c}. \qquad (6.13)$$

It should be noted that in adaptive beamforming of array signal processing, $\mathbf{c}$ is only used to lock signal arrivals in $M$ desired directions. For simplicity, we can choose $\mathbf{c}$ to be the $M$-dimensional unity vector with ones in all the $M$ components, $\mathbf{1} = \left(1, \ldots, \underbrace{1}_{j}, \ldots, \underbrace{1}_{M}\right)^T$. Substituting $\mathbf{1}$ for $\mathbf{c}$ in (6.13) yields

$$\left(\mathbf{w}^{\text{LCMV}}\right)^T\mathbf{R}_{L\times L}\mathbf{w}^{\text{LCMV}} = \mathbf{1}^T\left(\Lambda^T\mathbf{R}_{L\times L}^{-1}\Lambda\right)^{-1}\mathbf{1}, \qquad (6.14)$$

which is the minimum variance using the $M$-dimensional unity vector $\mathbf{1}$ to constrain $M$ class signatures, $\boldsymbol{\mu}_1, \boldsymbol{\mu}_2, \ldots, \boldsymbol{\mu}_M$.

### 6.4.2 *Iterative Linearly Constrained Minimum Variance (ILCMV)*

One major issue of using LCMV for classification is that LCMV is a mixed pixel-based technique and does not take into account the spatial information of the mixed pixel to be processed. So, blindly applying LCMV to HSIC will result in unsatisfactory results. In order to resolve this issue, we need to add spatial information to LCMV. The ILCMV presented in this section takes advantage of a Gaussian filter to capture spatial contextual information and then feeds back the Gaussian-filter data to be added to the data, which is currently being processed to create a new data cube to be used for reprocessing ILCMV. Its detailed step-by-step implementation is described in the following along with its graphic implementation illustrated in Figure 6.1.

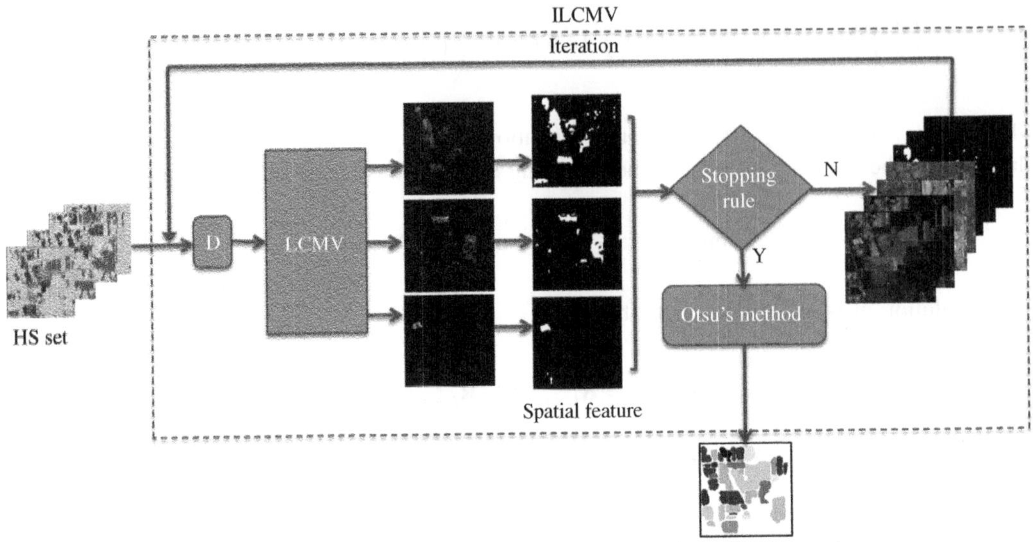


**Figure 6.1** Graphic implementation of ILCMV.

***ILCMV***

1. Initial condition: Let $\Omega^{(0)}$ be the original band set, $\mathbf{T}^{(0)}$ be the original desired class signatures. Set an initial value for the error threshold $\varepsilon$. Let $k = 1$.
2. Implement $\delta_k^{\text{LCMV}}$ on $\Omega^{(k)}$ using $\mathbf{T}^{(k)}$ and $\mathbf{R}^{(k)}$ to produce$\mathbf{B}_k^{\text{LCMV}}$ which is the resulting LCMV-classification map.
3. Use a Gaussian filter to blur $|\ \mathbf{B}_k^{\text{LCMV}}\ |$ where $|\ \mathbf{B}_k^{\text{LCMV}}\ |$ is the absolute value of $\mathbf{B}_k^{\text{LCMV}}$. The resulting image is denoted by a Gaussian-filtered LCMV-classification map, $|\ \mathbf{GB}_k^{\text{LCMV}}\ |$.
4. Form a new hyperspectral image cube $\Omega^{(k)}$ by augmenting the hyperspectral image cube $\Omega^{(k-1)}$ obtained at $k$-1 iteration with adding new $|\ \mathbf{GB}_k^{\text{LCMV}}\ |$ obtained at the $k$th iteration, i.e. $\Omega^{(k)} = \Omega^{(k-1)} \cup \left|\mathbf{GB}_k^{\text{LCMV}}\right|$, where the desired signature matrix $\mathbf{T}^{(k)}$ is obtained by finding new LCMV-classified class sample means from $\Omega^{(k)}$.
5. Calculate the following stopping rule,

$$\text{TI}^{(k)} = \frac{\text{size}\left(|\ \mathbf{B}_k^{\text{LCMV}}\ | \cap |\ \mathbf{B}_{k-1}^{\text{LCMV}}\ |\right)}{\text{size}\left(|\ \mathbf{B}_k^{\text{LCMV}}\ | \cup |\ \mathbf{B}_{k-1}^{\text{LCMV}}\ |\right)}, \tag{6.15}$$

where TI is Tanimoto index (TI) defined in [24] and size(S) is the size of a set $S$, $|\ \mathbf{B}_k^{\text{LCMV}}\ |$ and $|\ \mathbf{B}_{k-1}^{\text{LCMV}}\ |$ are the binary image of the $k$th LCMV detection map and the binary image of the $k$-1st LCMV classification map thresholded by Ostu's method in [25], respectively.

6. If $\text{TI}^{(k)}$ is within an error threshold $\varepsilon$, go to step 7. Otherwise, continue.
7. Form $\Omega^{(k+1)} = \Omega^{(k)} \cup \{\ |\ \mathbf{GB}_k^{\text{LCMV}}\ |\}$. Let $k \leftarrow k + 1$ and go to step 2.
8. LCMV is terminated. Otsu's method is applied to threshold $|\ \mathbf{B}_k^{\text{LCMV}}\ |$ into multi-class classification map.

## 6.5 Class Signature Constrained Band Prioritization-Based Band Selection

Due to the use of hundreds of contiguous spectra bands, a hyperspectral image generally has a very high inter-band correlation. Besides, spectral properties of data samples in different classes respond to wavelengths quite differently and not all bands are useful for classification. Accordingly, BS becomes an effective means that can improve classification rates [17]. To be consistent with the ILCMV classifier developed in Section 6.4.2, this section presents a BS method which is also based on LCMV, to be called class signature constrained band prioritization band selection (CSCBP-BS).

Specifically, for any given band image, $\mathbf{B}_l$, it can be represented by a band vector $\mathbf{b}_l$. In this case, we can modify (6.14) and define a criterion by minimizing (6.14) with respect to $\mathbf{b}_l$ as follows:

$$\text{MinV}(\mathbf{b}_l) = \mathbf{1}^T\left(\mathbf{D}_{\mathbf{b}_l}^T\mathbf{R}_{\mathbf{b}_l}^{-1}\mathbf{D}_{\mathbf{b}_l}\right)^{-1}\mathbf{1}, \tag{6.16}$$

where $\mathbf{D}_{\mathbf{b}_l} = \left[\boldsymbol{\mu}_{\mathbf{b}_l}^1\boldsymbol{\mu}_{\mathbf{b}_l}^2\cdots\boldsymbol{\mu}_{\mathbf{b}_l}^M\right]$ with $\boldsymbol{\mu}_{\mathbf{b}_l}^i$ being the $i$th class signature using only one single band, $\mathbf{b}_l$.

Let $S^{(i)} = \left\{\overline{S}_1^{(i)}, \ldots S_j^{(i)}, \ldots \overline{S}_{m_i}^{(i)}\right\}$ be the set of $m_i$ subsets of classes at the $i$th layer of a Huffman tree where $S_j^{(i)} = \left\{C_{jl}^{(i)}\right\}_{l=1}^{n_j}$ is the set of classes of interest needed to be classified at the $i$th layer and $\overline{S}_k^{(i)} = \left\{\overline{C}_{k_i}^{(i)}\right\}_{i=1}^{n_k}$ with $1 \le k \ne j \le m_i$ is the $k$th subset of classes that will not be classified at the $i$th layer.

$$\mathbf{D}_{\mathbf{b}_l}^{(i)} = \left[\mathbf{D}_{\mathbf{b}_l}^{\overline{S}_1^{(i)}}\cdots\mathbf{D}_{\mathbf{b}_l}^{\overline{S}_{j-1}^{(i)}}\mathbf{D}_{\mathbf{b}_l}^{S_j^{(i)}}\mathbf{D}_{\mathbf{b}_l}^{\overline{S}_{j+1}^{(i)}}\cdots\mathbf{D}_{\mathbf{b}_l}^{\overline{S}_{m_i}^{(i)}}\right], \tag{6.17}$$

where $\mathbf{D}_{\mathbf{b}_l}^{S_j^{(i)}} = \left[\mathbf{d}_l^{C_{j1}^{(i)}}\mathbf{d}_l^{C_{j2}^{(i)}}\cdots\mathbf{d}_l^{C_{jn_j}^{(i)}}\right]$,

$$\mathbf{D}_{\mathbf{b}_l}^{\overline{S}_k^{(i)}} = \frac{1}{\overline{n}_k}\sum_{i=1}^{\overline{n}_k}\mathbf{d}_l^{\overline{C}_{k_1}^{(i)}}, \tag{6.18}$$

$n_j$ is the number of leaves corresponding to desired classes to be classified at the $i$th layer and $\overline{n}_k$ is the number of internal nodes corresponding to subtrees at the $i$th layer. The implementation of CBSCBP-BS is provided in the following.

***CSCBP-BS***

1. **Initial condition:**

   For each band, $\mathbf{B}_l$, find minimal priority scores, $\text{MinV}(\mathbf{B}_l)$, according to (6.1).

2. Prioritize bands according to $\{\text{MinV}(\mathbf{B}_l)\}_{l=1}^L$

$$\mathbf{B}_{l_j} \succ \mathbf{B}_{l_k} \Leftrightarrow \text{MinV}\left(\mathbf{B}_{l_j}\right) < \text{MinV}(\mathbf{B}_{l_k}), \tag{6.19}$$

   i.e. $\mathbf{B}_{l_1} \succ \mathbf{B}_{l_2} \succ \cdots \succ \mathbf{B}_{l_L}$ where the notation "$\succ$" is used to indicate "superior to".

3. The final set of selected bands is given by $\left\{\mathbf{B}_{l_1}, \mathbf{B}_{l_2}, \ldots, \mathbf{B}_{l_{n_{BS}}}\right\}$ where $n_{\text{BS}}$ is the number of bands needed to be selected.

## 6.6 Progressive Band Selection

In Section 6.2, CCP is designed to construct a $p$-ary HCT that can navigate HSIC to be performed at each internal node in a progressive manner where CSCBP-BS is used to find an appropriate band subset for each of layers in a $p$-ary HCT for stage-by-stage classification. The resulting HSIC is called PBSP-HSIC, which implements four key steps: (i) calculation of CCP, (ii) construction of a $p$-ary HCT, (iii) CSCBP-BS to select appropriate band subsets, and (iv) ILCMV to perform multi-class classification at each stage. Detailed step-by-step implementations of each of the four steps are described as follows:

***PBSP-HSIC***

1. Use virtual dimensionality (VD) to determine $n_{\mathrm{BS}}$.
2. Construction of a $p$-ary HCT
   (a) Use (6.3) to calculate probabilities of classification priority associated with different classes, $p_i^{\mathrm{CCP}}$, for all classes to be classified, $\{C_i\}_{i=1}^{M}$.
   (b) $\{p_i^{\mathrm{CCP}}\}_{i=1}^{M}$ to design a $p$-ary Huffman code.
   (c) Use the $p$-ary Huffman code obtained in step 3 to construct a $p$-ary Huffman coding tree.
3. At each layer, there are terminal nodes which are leaves and internal nodes which are roots of subtrees. So, in this layer, only leaves which correspond to single classes will be classified while internal nodes will be passed on to the next layer.
4. Find the mean of each class at each leave and denote desired class signatures $\mathbf{d}_l^{C_{j_1}^{(i)}}, \mathbf{d}_l^{C_{j_2}^{(i)}}, \ldots, \mathbf{d}_l^{C_{j_{n_j}}^{(i)}}$. In the meantime, also find mean of all classes in subtrees at the intermediate nodes and denote undesired class signatures $\mathbf{D}_{\mathbf{b}_l}^{\overline{S}_1^{(i)}} \cdots \mathbf{D}_{\mathbf{b}_l}^{\overline{S}_{j-1}^{(i)}} \mathbf{D}_{\mathbf{b}_l}^{\overline{S}_{j+1}^{(i)}} \cdots \mathbf{D}_{\mathbf{b}_l}^{\overline{S}_{m_i}^{(i)}}$ which are obtained in (6.17) and (6.18) by averaging all class means in the subtrees, $\overline{S}_k^{(i)}$, for $1 \leq k \neq j \leq m_i$.
5. For each $k$, use CSCBP-BS to find an appropriate band subset for the $k$th layer.
6. Apply ILCMV to perform multi-class classification at this layer as a stage process. Let the resulting LCMV-classification map be denoted by $\mathrm{LCMV}^{(k)}$.
7. Find

$$\begin{aligned}\text{ILCMV-map}^{(k)} \\ = \text{LCMV-map}^{(k)} \cup \left(\left(\text{LCMV-map}^{(k)}\right)^c \cap \text{LCMV-map}^{(k-1)}\right),\end{aligned} \tag{6.20}$$

   where $(\text{LCMV - map}^{(k)})^c$ is the complement set of $\text{LCMV - map}^{(k)}$ and $\text{ILCMV - map}^{(1)} = \text{LCMV - map}^{(1)}$.
8. Then, the same process is continued on until it reaches the last layer of the $p$-ary HCT.

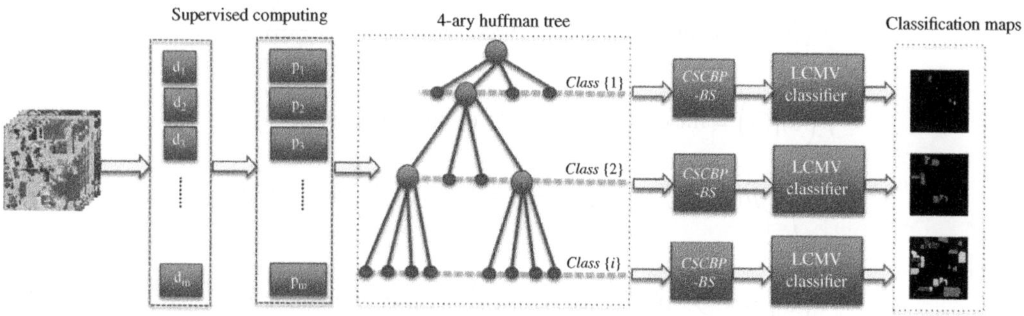


**Figure 6.2** A graphic diagram of implementing PBSP-HSIC.

It should be noted that the key idea is (6.20) which takes advantage of ILCMV-classification maps produced by progressive band selection processing to correct misclassification errors as will be illustrated by an example in Figure 6.7. Figure 6.2 depicts a graphic diagram of PBSP-HSIC by implementing ILCMV using a constructed $p$-ary HCT in conjunction with bands selected by CSCBP-BS. For example, at the first layer corresponding to the first stage, three leaves highlighted by dark gray circles are classified, while a large circle highlighted by light gray is a subtree representing undesired classes which will not be classified at this particular stage. At the last stage, all the leaves are single classes and will be classified all together to complete the classification.

## 6.7 Classification Measures

How to select an effective classification performance measure is also very crucial to validate a classifier. Over the past years, OA, average accuracy (AA), and kappa coefficient have been widely used for this purpose. Unfortunately, these three measures are very similar and do not take into account the misclassification error of one class resulting from other classes, specifically, background class. This section proposes several classification measures to be used for classification.

$M$ = the number of classes.
$n_i$ = the number of data samples in the $i$th class.
$n_{ij}$ = the number of signal samples in the $j$th class, $C_j$ to be classified into the $i$th class, $C_i$.
$\hat{n}_{ij}$ = the number of data samples classified in the $i$th class, which are supposed to be in the $j$th class, $C_j$.
$n_{ii}$ = the number of signal samples in the $i$th class correctly classified into the $i$th class.
$\tilde{n}_i = \sum_{j=1,j\neq i}^{M} n_{ji}$ = the number of signal samples in the $i$th class misclassified into the $i$th class with $j \neq i$
$n_i = \sum_{j=1}^{M} n_{ji}$.
$N$ = total number of data samples, $N = \sum_{i=1}^{M} n_i$.
$C_i$ = the number of data samples in the $i$th class.
$\tilde{C}_i$ = the number of data samples from classes $\{C_j\}_{j=1,j\neq i}^{M}$ classified into $C_j$.
$p^{\text{SR}}(C_i) = p_i^{\text{SR}} = \frac{n_i}{N}$, where SR is sample ratio defined as the ratio of $i$th class sample size $n_i$ to data sample size $N$.
$p^{\text{CCP}}(C_i) = p_i^{\text{CCP}}$.

$$p_{\mathrm{A}}(C_i) = \text{accuracy of the } i\text{th class} = p(C_{ii} \mid C_i) = \frac{n_{ii}}{\sum_{j=1}^{M} n_{ji}}. \tag{6.21}$$

$$p\left(\tilde{C}_i|C_i\right) = \text{misclassification rate of the } i\text{th class, } C_i$$
$$= \sum_{j=1}^{M} p\left(C_{ij}|C_i\right) = \frac{\sum_{j=1, j\neq i}^{M} n_{ij}}{\sum_{j=1}^{M} n_{ij}} = 1 - p(C_{ii}|C_i) = 1 - p_{\mathrm{A}}(C_i). \tag{6.22}$$

$$\mathrm{P}\left(\tilde{C}\right) == \sum_{i=1}^{M} p^{\mathrm{CCP}}(C_i) p\left(\tilde{C}_i \mid C_i\right) = \text{misclassification rate.} \tag{6.23}$$

$$\mathrm{P}_{\mathrm{PR}}(C_i) = p_{\text{precision}}(C_i \mid \text{classification}) = \frac{\hat{n}_{ii}}{\sum_{j=1}^{M} \hat{n}_{ij}}. \tag{6.24}$$

$$\mathrm{P}_{\mathrm{PR}}^{\mathrm{SR}} = \sum_{i=1}^{M} p^{\mathrm{SR}}(C_i)\mathrm{P}_{\mathrm{PR}}(C_i). \tag{6.25}$$

$$\mathrm{P}_{\mathrm{PR}}^{\mathrm{CCP}} = \sum_{i=1}^{M} p^{\mathrm{CCP}}(C_i)\mathrm{P}_{\mathrm{PR}}(C_i). \tag{6.26}$$

$$\mathrm{P}_{\mathrm{OA}}^{\mathrm{SR}} = p_{\mathrm{OA}}^{\mathrm{SR}}\left(\{C_i\}_{i=1}^{M}\right) = \sum_{i=1}^{M} p_i^{\mathrm{SR}} p(C_{ii} \mid C_i) = \sum_{i=1}^{M} \left(\frac{n_i}{N}\right)\left(\frac{n_{ii}}{n_i}\right) = \frac{1}{N}\sum_{i=1}^{M} n_{ii} = \mathrm{P}_{\mathrm{OA}}. \tag{6.27}$$

$$\mathrm{P}_{\mathrm{OA}}^{\mathrm{CCP}} = p_{\mathrm{OA}}^{\mathrm{CCP}}\left(\{C_i\}_{i=1}^{M}\right) = \sum_{i=1}^{M} p_i^{\mathrm{CCP}} p(C_{ii} \mid C_i). \tag{6.28}$$

$$\mathrm{P}_{\mathrm{AA}} = p_{\text{average-accuracy}}\left(\{C_i\}_{i=1}^{M}\right) = \frac{1}{M}\sum_{i=1}^{M} \frac{n_{ii}}{n_i}. \tag{6.29}$$

Three remarks on accuracy and precision are noteworthy.

1. In the HSIC literature, $\mathrm{P}_{\mathrm{OA}}$ has been widely used to measure the effectiveness of a classifier. It is actually $\mathrm{P}_{\mathrm{OA}}^{\mathrm{SR}}$ as shown in (6.28). Here, we particularly define three types of accuracy. The $\mathrm{P}_{\mathrm{OA}}^{\mathrm{SR}}$ defined in (6.27) and $\mathrm{P}_{\mathrm{AA}}$ in (6.29) are probably those that have been used in the literature. $\mathrm{P}_{\mathrm{OA}}^{\mathrm{CCP}}$ defined in (6.28) is a new measure which is only applied to progressive classification when CCP is introduced to account for priorities of classifying various classes. So, in the following experiments, these three measures will be used for classification evaluation.
2. If we let $p^{\mathrm{CCP}}(C_i) = 1/M$, then (6.28) is reduced to (6.29). If we let $p^{\mathrm{CCP}}(C_i) = n_{ii}/n_i$, then (6.28) becomes (6.27). Accordingly, $\mathrm{P}_{\mathrm{AA}}$ and $\mathrm{P}_{\mathrm{OA}}^{\mathrm{SR}}$ can be considered as special cases of (6.28).

The $\mathrm{P}_{\mathrm{PR}}$ defined in (6.24)–(6.26) is designed to measure the classification effectiveness produced by a classifier compared to the $\mathrm{P}_{\mathrm{OA}}$ defined in (6.27) which is used to measure classification accuracy given that the knowledge of $\{C_j\}_{j=1}^{M}$ is provided a priori. Accordingly, $\mathrm{P}_{\mathrm{PR}}$ is referred to a posteriori classification rate as opposed to $\mathrm{P}_{\mathrm{OA}}$ referred to as a priori classification rate [14].

## 6.8 Real Images to be Used for Experiments

Three real hyperspectral images were used for experiments, Purdue University's Indiana Indian Pines, University of Pavia, Italy, and Salinas, each of which has its own unique feature characteristics worth being explored.

### 6.8.1 *Purdue Indiana Indian Pines*

A real image to be used for experiments is a well-known airborne visible infrared imaging spectrometer (AVIRIS) image scene, Purdue Indiana Indian Pine test site with a band 186 shown in Figure 6.3(a), its ground truth in Figure 6.3(b), and an aerial view in Figure 6.3(c). Table 6.1 also tabulates all the specific types of 16 classes. It has the size of 145 × 145 pixel vectors taken from an

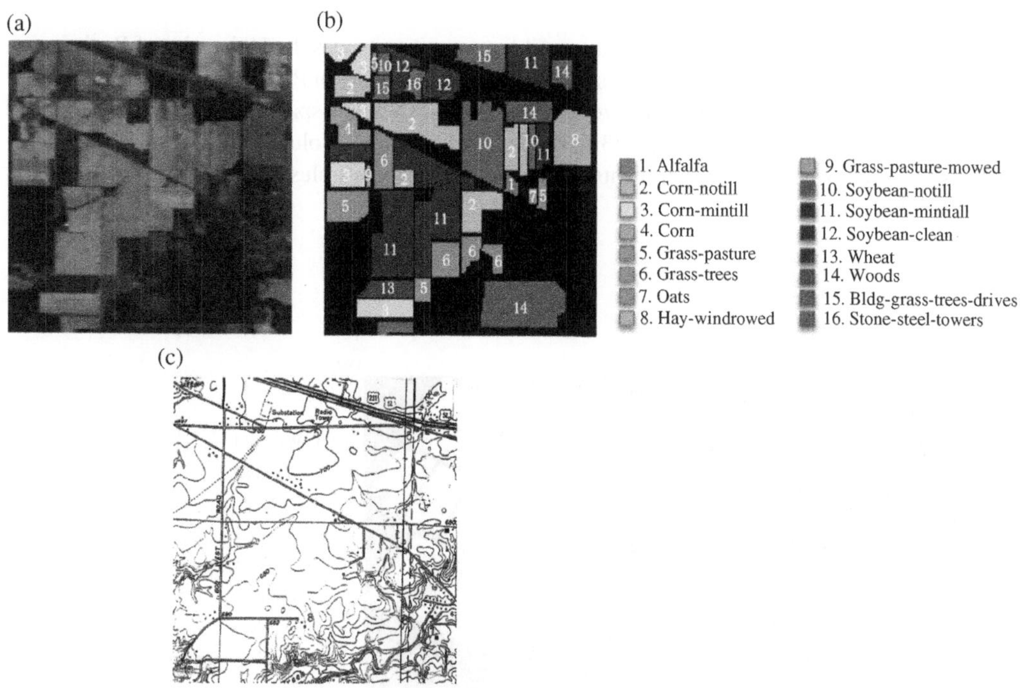


**Figure 6.3** Purdue's Indiana Indian Pines scene: (a) band 186 (2162.56 nm), (b) ground-truth map and classes by different degrees of gray intensities, and (c) A USGS Quadrangle map of the Indian Pines scene, where its color maps can be found in Ref. [14], figure 6/with permission of IEEE.

**Table 6.1** $n_{BS}$ estimated by HFC//NWHFC

| | $P_F = 10^{-1}$ | $P_F = 10^{-2}$ | $P_F = 10^{-3}$ | $P_F = 10^{-4}$ | $P_F = 10^{-5}$ |
|---|---|---|---|---|---|
| Purdue | 73/21 | 49/19 | 35/18 | 27/18 | 25/17 |
| Salinas | 32/33 | 28/24 | 25/21 | 21/21 | 20/20 |
| University of Pavia | 25/34 | 21/27 | 16/17 | 14/14 | 13/12 |

area of mixed agriculture and forestry in Northwestern Indiana, USA, with details of band and wavelength given in the caption. The data set is available at website https://purr.purdue.edu/publications/1947/serve/1?el=1. It was recorded in June 1992 with 220 bands including water absorption bands (bands 104–108, 150–163, 220).

### 6.8.2 *Salinas*

A second AVIRIS data used for experiments was Salinas scene shown in Figure 6.4(a) which was captured by the AVIRIS sensor over Salinas Valley, California, and with a spatial resolution of 3.7 m/pixel with spectral resolution of 10 nm. It has the size of 512 × 217 × 224 including 20 water absorption bands, 108–112, 154–167, 224. Figure 6.4(b) and (c) shows the color composite of the Salinas image along with the corresponding ground-truth class labels.

### 6.8.3 *ROSIS Data*

A third and the last hyperspectral image data used for experiments was the University of Pavia image shown in Figure 6.5 which is an urban area surrounding the University of Pavia, Italy. It was recorded by the ROSIS-03 satellite sensor. It has the size of 610 × 340 × 115 with a spatial resolution of 1.3 m/pixel and a spectral coverage ranging from 0.43 to 0.86 m with spectral resolution of 4 nm (12 most noisy channels were removed before experiments). Nine classes of interest plus BKG class, class 0 are considered for this image.

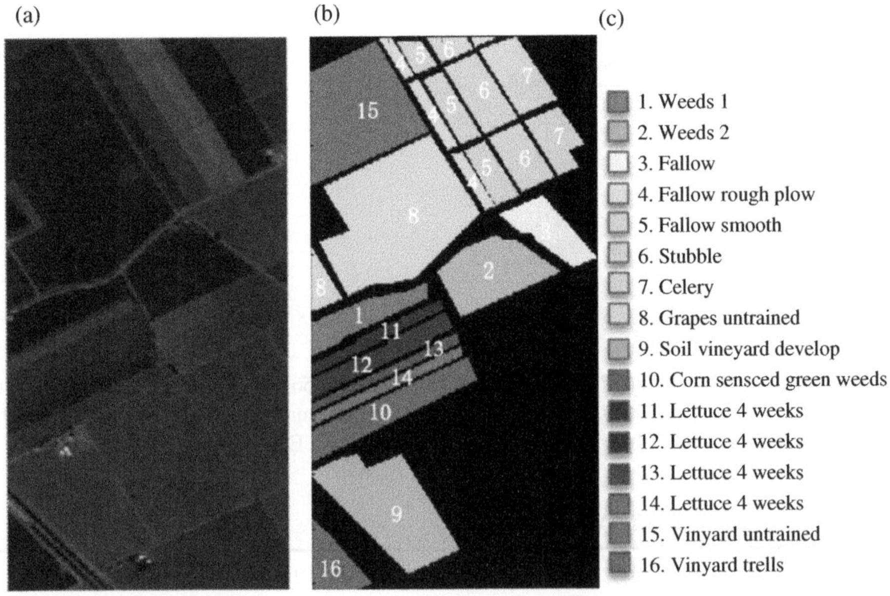


**Figure 6.4** Ground-truth of Salinas scene with 16 classes: (a) Salinas scene, (b) ground-truth map, and (c) ground-truth class labels, where its color maps can be found in Ref. [14], figure 10/with permission of IEEE.

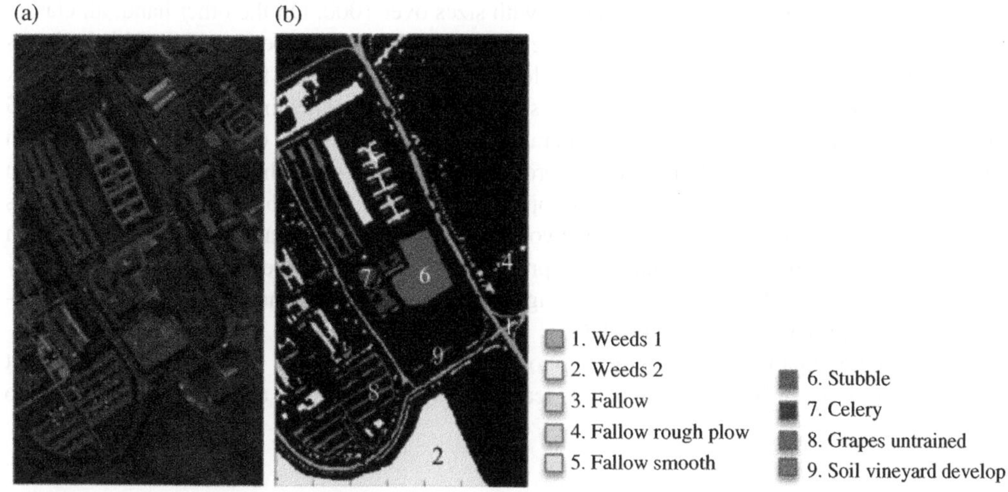


**Figure 6.5** Ground truth of University of Pavia scene with nine classes: (a) University of Pavia scene and (b) ground-truth map along with classes labeled by different degrees of gray intensities with colored ground truth map found in Ref. [14], figure 14/with permission of IEEE.

## 6.9 Experiments

According to [26–28], VD was used to estimate $n_{BS}$ for the three hyperspectral images using the Harsanyi-Farrand-Chang (HFC) method in [29] and noise-whitened HFC (NWHFC) method [26, 27]. Table 6.1 tabulates the values of $n_{BS}$, $n_{BS} = 18$ for the Purdue data, $n_{BS} = 21$ for Salinas, and $n_{BS} = 14$ for the University of Pavia with the false alarm probability $10^{-4}$ where a/b indicate the values of VD estimated by HFC/NWHFC methods.

### 6.9.1 *Purdue Indiana Indian Pines*

First of all, the sample means $\{\boldsymbol{\mu}_i\}_{i=1}^{16}$ of all the 16 classes are used to calculate CCP probabilities via (6.1). Table 6.2 tabulates the 16 class CCP probabilities $\{p_i^{CCP} = p^{CCP}(C_i)\}_{i=1}^{16}$ along with their assigned priorities. Based on the CCP probabilities in Table 6.2, the five most difficult classes to be classified ranked from priority 12 to priority 16 are class 11 (2468): soybeans-min followed by class 2 (1434): corn-notill, class 3 (384): corn-min, class 10 (968): soybean-notill, and class

**Table 6.2** Priority probabilities of CCP of 16 classes in Purdue's Indian Pines calculated by (1–2) with **R** including BKG

| Class | 1 | 2 | 3 | 4 | 5 | 6 | 7 | 8 |
|---|---|---|---|---|---|---|---|---|
| CCP | 0.1441 | 0.0121 | 0.0121 | 0.0308 | 0.0403 | 0.0313 | 0.1103 | 0.0562 |
| Priority | 2 | 15 | 14 | 10 | 7 | 9 | 3 | 6 |
| Class | 9 | 10 | 11 | 12 | 13 | 14 | 15 | 16 |
| CCP | 0.1045 | 0.0138 | 0.0076 | 0.0233 | 0.0956 | 0.0229 | 0.0359 | 0.2583 |
| Priority | 4 | 13 | 16 | 11 | 5 | 12 | 8 | 1 |

14 (1294): woods which include all the classes with sizes over 1000. On the other hand, all classes with five smallest sizes are all ranked by top five priorities for classes to be classified. This indicates that the CCP is closely related to the class sample size. It should be noted that the correlation matrix **R** used in (6.1) to calculate the CCP probabilities of 16 classes calculated in Table 6.2 includes BKG samples due to the fact that the BKG class generally has significant impact on classification as shown in [13, 14]. Interestingly, the CCP is inversely proportional to class size. That is, the four classes with the highest priorities are those classes with sample size less than 100 as opposed to the two classes with least priorities, which correspond to the two largest classes, class 11 and class 2 with more than 1000 data samples. Now, these calculated CCP probabilities in Table 6.2 were used to produce a 4-ary Huffman code shown at the left-hand side of Figure 6.6 and its corresponding 4-ary HCT was constructed at the right-hand side in Figure 6.6.

It is worth noting that the largest value of $p$ to implement a $p$-ary HCT for 16 classes is $p = 4$. At each layer of the 4-ary HCT in Figure 6.6, only leaves marked by dark gray are codewords assigned to

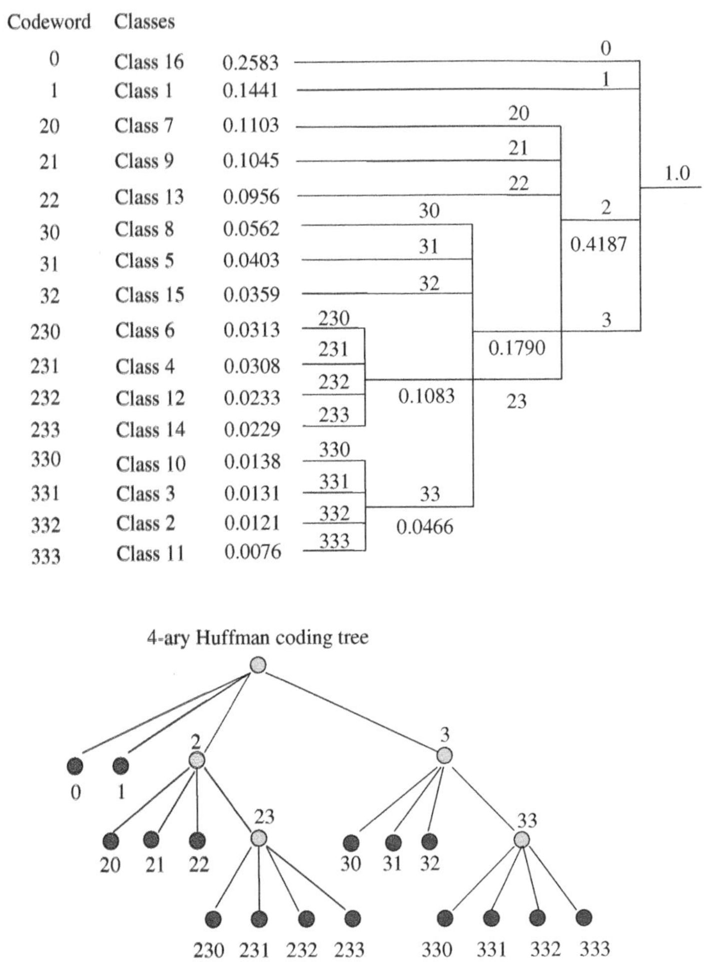


**Figure 6.6** 4-Ary Huffman code and coding tree of CCP with BKG for Purdue's Indian Pines with $p = 4$, $m = q = 4$, $p-q = 0$.

classes which will be classified at that particular layer, while the light gray internal nodes are codewords representing subtrees, which will be further processed at the next layer. In this case, each of the leaves corresponds to one particular class. On the other hand, the number of light gray internal nodes actually determines the number of times that the classification must be performed by ILCMV for $p$ classes. For example, in Figure 6.6, there are five light gray nodes which determine that ILCMV must be performed progressively five times and each time, classes at leaves highlighted by dark gray are classified. At the first layer, two leaves specified by class 1 and class 16 assigned by codewords 0 and 1 will be first classified and two light gray internal nodes assigned by codewords 2 and 3 represent two subtrees to specify two joint sets of classes, {7,9,13,8,5,15} and {6,4,12,14,10,3,2,11}, respectively, both of which will be further processed at the second layer. At the second layer, six leaves specified by six classes, 7,9,13,8,5,15 assigned by codewords 20, 21, 22, 30, 31, 32 will be classified, whereas the two light gray internal nodes assigned by codewords 23 and 33 represent two subtrees that specify two joint sets of classes, {6,4,12,14} and {10,3,2,11}, respectively, needed to be processed at the next layer. Finally, at the third layer, a total of eight leaves specified by eight classes, 12,6,4,14,10,3,2,11 assigned by codewords, 230, 231, 232, 233, 330, 331, 332, 333, with four leaves from each subtree yet to be classified to complete PBSP-HSIC. Then, CSCBP-BS was implemented to find desired bands listed in Table 6.3 with $n_{\text{BS}} = 18$.

Now, using the 4-ary HCT constructed in Figure 6.6 as well as bands selected by CSCBP-BS in Table 6.3, both ILCMV and PBSP-ILCMV were implemented with the parameters used by ILCMV listed in Table 6.4.

Figure 6.7(a) describes a three-stage PBSP-ILCMV using 4-ary HCT stage-by-stage and Figure 6.7(b) shows ILCMV without PBSP in one-shot operation. As shown in Figure 6.7(a), PBSP-ILCMV performed classification in three stages where classes 1 and 16 were classified in the first stage shown by $\text{ILCMV - map}^{(1)} = \text{LCMV - map}^{(1)}$, classes 7,9,13,8,5,15 at the second stage shown by LCMV-map$^{(2)}$ along with $\text{ILCMV - map}^{(2)} = \text{ILCMV - map}^{(1)} \cup ((\text{LCMV - map}^{(2)})^c \cap \text{LCMV - map}^{(1)})$ which combined classification results of eight classes, {1,16,7,9,13,8,5,15}, and classes 6,4,12,14,10,3,2,11 at the third and last stage, LCMV-map$^{(3)}$ along with the final classification ILCMV-map$^{(3)}$ given by

$$\text{ILCMV-map}^{(3)} = \text{ILCMV-map}^{(2)} \cup \left(\left(\text{LCMV-map}^{(3)}\right)^c \cap \text{LCMV-map}^{(2)}\right).$$

**Table 6.3** Bands selected by CSCBP-BS for each subset of class of the Purdue Indiana Indian Pines

| Number of subclasses | Selected bands for classification |
|---|---|
| {1,16} | 28,29,27,26,25,24,30,23,32,22,21,20,19,33,31,18,12,13 |
| {7,9,13,8,5,15} | 71,73,72,70,74,53,52,69,55,56,54,51,50,75,68,49,57,48 |
| {12,6,4,14,10,3,2,11} | 29,28,30,32,27,26,25,31,33,121,120,122,24,119,123,124,23,118 |

**Table 6.4** Specifications of parameters used by ILCMV

| D | 16 class sample means |
|---|---|
| Class number | 1–16 |
| $\sigma$ used in Gaussian filter | 0.5 with window size 5 × 5 |
| Thresholding method | Otsu |
| Stopping threshold (TI) | 0.99 |

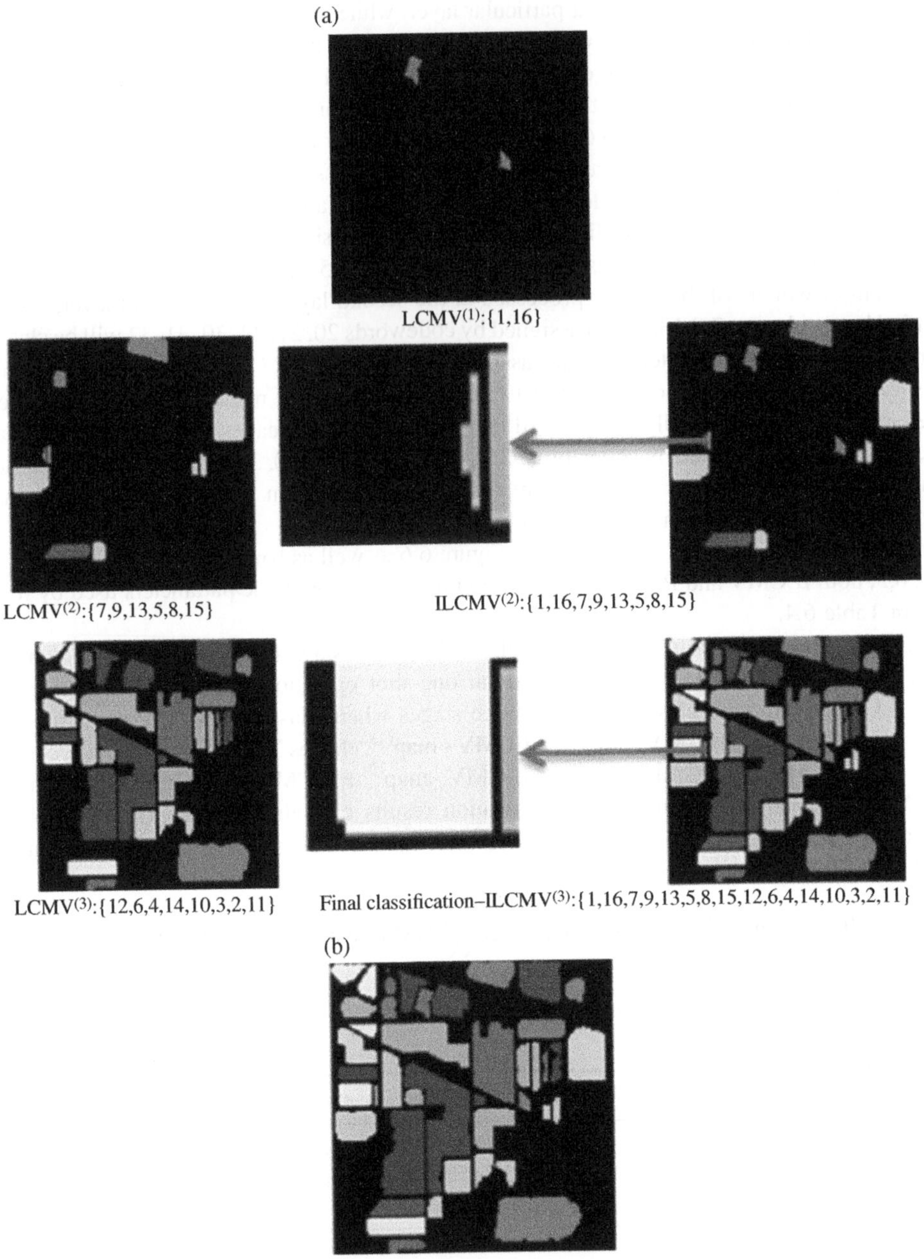


**Figure 6.7** (a) Three layers/stages processed by PBSP-ILCMV using 4-ary HCT and (b) classification of ILCMV without PBSP for Purdue Indian Pines.

**Table 6.5** $P_A$ calculated from the classification results of Purdue's data

| Class | 1 | 2 | 3 | 4 | 5 | 6 | 7 | 8 |
|---|---|---|---|---|---|---|---|---|
| ILCMV without PBSP | 95.65 | 96.01 | 96.99 | 98.73 | 89.44 | 97.12 | 100 | 98.78 |
| PBSP-ILCMV | 100 | 97.55 | 97.11 | 98.73 | 92.75 | 97.01 | 100 | 98.74 |
| Class | 9 | 10 | 11 | 12 | 13 | 14 | 15 | 16 |
| ILCMV without PBSP | 100 | 93.93 | 94.70 | 95.45 | 98.54 | 93.52 | 90.67 | 98.92 |
| PBSP-ILCMV | 100 | 95.47 | 96.78 | 96.63 | 97.07 | 94.23 | 92.49 | 98.92 |

**Table 6.6** $P_{OA}^{SR}$, $P_{AA}^{SR}$, $P_{PR}^{SR}$, $P_{OA}^{CCP}$, and $P_{PR}^{CCP}$ calculated by ILCMV without PBSP and progressive layer/stage-by-layer/stage PBSP-ILCMV for Purdue's data

| | $P_{OA}^{SR}$ | $P_{AA}^{SR}$ | $P_{PR}^{SR}$ | $P_{OA}^{CCP}$ | $P_{PR}^{CCP}$ |
|---|---|---|---|---|---|
| ILCMV without PBSP | 95.09 | 96.15 | 99.93 | — | — |
| | | | | Overall $P_{OA}^{CCP} = 98.19$ | Overall $P_{PR}^{CCP} = 99.91$ |
| PBSP-ILCMV | ---- | ---- | Layer 1/stage 1 | 99.31 | 100 |
| | | | Layer 2/stage 2 | 97.94 | 100 |
| | | | Layer 3/stage 3 | 96.58 | 99.99 |

Comparing Figure 6.7(a) and (b), there are at least two immediate benefits offered by PBSP-HSIC. One is to dictate changes in classified data samples. For example, in the second stage, there were 11 pixels in the third class $C_3$ misclassified in class 9 as shown in a zoom-in window of ILCMV$^{(2)}$, but these misclassified were later corrected as shown in the third stage classification by ILCMV$^{(3)}$. Another is to allow users to calculate classification accuracy and precision rates progressively. In order to see how it works, Table 6.5 further calculates $P_A$ in (6.21) of each of 16 classes produced by ILCMV without PBSP and PBSP-ILCMV using 4-ary HCT in Figure 6.6 where PBSP-ILCMV produced higher $P_A$ values for every class except class 13 than their corresponding $P_A$ values produced by ILCMV without PBSP. Table 6.6 also calculates $P_{OA}^{SR}$ in (6.27), $P_{OA}^{CCP}$ in (6.28), $P_{PR}^{SR}$ in (6.25), $P_{PR}^{CCP}$ in (6.26), and $P_{AA}$ in (6.29) where PBSP-ILCMV performed better than ILCMV without PBSP by producing higher values of $P_{OA}^{SR}$, $P_{AA}^{SR}$, and $P_{PR}^{SR}$.

Also, from Table 6.6, PBSP-ILCMV actually produced $P_{OA}^{CCP}$ better than $P_{OA}^{SR}$ by 2% with $P_{PR}^{CCP}$ nearly the same as $P_{PR}^{SR}$ by a difference of $8 \times 10^{-4}$. The most important results in Table 6.6 is the last two columns which calculate $P_{OA}^{CCP}$ and $P_{PR}^{CCP}$ progressively according to CCP for each of stages. For example, in the first stage $P_{OA}^{CCP}$ and $P_{PR}^{CCP}$ for classifying classes, 1 and 16 were 99.28 and 100%, respectively, and 95.25 and 100% for classes 7,9,13,5,8,15, respectively, for the second stage. At the final and third stage, $P_{OA}^{CCP}$ and $P_{PR}^{CCP}$ for classifying last 8 classes, 12,6,4,14,10,3,2,11 were 96.25 and 99.99%, respectively. For overall performance $P_{OA}^{CCP} = 98.19$ and $P_{PR}^{CCP} = 99.91$, which were higher than ILCMV by 3% in $P_{OA}$ and nearly the same in $P_{PR}$. These experiments demonstrated the benefits of using PBSP-ILCMV for HSIC.

### 6.9.2 *Salinas*

Similar experiments conducted for the Purdue data were also performed for Salinas. Table 6.7 calculates the CCP probabilities of 16 classes by (6.1) with **R** including BKG samples. Since a 4-ary

**Table 6.7** Priority probabilities of CCP of 16 classes in Salinas calculated by (1-2) with **R** including BKG

| Class | 1 | 2 | 3 | 4 | 5 | 6 | 7 | 8 |
|---|---|---|---|---|---|---|---|---|
| CCP | 0.1236 | 0.0672 | 0.0397 | 0.0493 | 0.0397 | 0.0773 | 0.0720 | 0.0205 |
| Priority | 2 | 7 | 11 | 9 | 10 | 5 | 6 | 15 |
| Class | 9 | 10 | 11 | 12 | 13 | 14 | 15 | 16 |
| CCP | 0.0249 | 0.0395 | 0.1295 | 0.0278 | 0.0642 | 0.0988 | 0.0189 | 0.1071 |
| Priority | 14 | 12 | 1 | 13 | 8 | 4 | 16 | 3 |

**Table 6.8** Bands selected for each subset of class of three hyperspectral images

| Data set | Number of subclass | Selected bands for classification |
|---|---|---|
| Salinas (21 bands) | {11} | 152,151,150,114,144,149,145,143,115,142,148,146,141,147, 140,139,138,137,136,134,135 |
| | {1,2,3,4,5,6,7,10,13,14,16} | 44,45,43,46,47,48,42,49,50,51,52,53,54,55,56,57,58,41,59,60,61 |
| | {8,9,12,15} | 223,221,222,220,224,218,219,217,215,214,211,208,216,210,212, 209,207,213,204,206,205 |

**Table 6.9** Specifications of parameters used by ILCMV

| D | 16 class sample means |
|---|---|
| Class number | 1–16 |
| $\sigma$ used in Gaussian filter | 1.0 with window size $9 \times 9$ |
| Thresholding method | Otsu |
| Stopping threshold (TI) | 0.99 |

HCT has shown to be better than binary and ternary HCTs for Purdue's data, only 4-ary HCT was constructed for Salinas. Then, CSCBP-BS was implemented to find desired bands listed in Table 6.8 with $n_{\text{BS}} = 21$.

Table 6.9 also lists the parameters used by ILCMV without PBSP when PBSP-ILCMV was implemented.

Figure 6.8 shows the progressive classification of PBSP-ILCMV using a 4-ary HCT in Figure 6.8 (a) along with the results produced by ILCMV without PBSP, which classified all classes in one-shot operation in Figure 6.8(b) where three layers of classes were processed progressively with classes in the parentheses indicating what classes were assigned to a particular layer. To see that is, Table 6.10 further calculates $\text{P}_{\text{A}}$ in (6.21) of each of 16 classes produced by ILCMV and PBSP-ILCMV using 4-ary HCT in Figure 6.8 where PBSP-ILCMV produced higher $\text{P}_{\text{A}}$ values for every class than their corresponding $\text{P}_{\text{A}}$ values produced by ILCMV without PBSP except class 13. Table 6.11 also calculates $\text{P}_{\text{OA}}^{\text{SR}}$ in (6.27), $\text{P}_{\text{OA}}^{\text{CCP}}$ in (6.28), $\text{P}_{\text{PR}}^{\text{SR}}$ in (6.25), $\text{P}_{\text{PR}}^{\text{CCP}}$ in (6.26), and $\text{P}_{\text{AA}}$ in (6.29) where PBSP-ILCMV performed better than ILCMV without PBSP by producing higher values of $\text{P}_{\text{OA}}^{\text{SR}}$, $\text{P}_{\text{AA}}^{\text{SR}}$, and $\text{P}_{\text{PR}}^{\text{SR}}$ than ILCMV.

These conclusions were consistent with conclusions drawn from the experiments conducted for Purdue's Indian Pines.

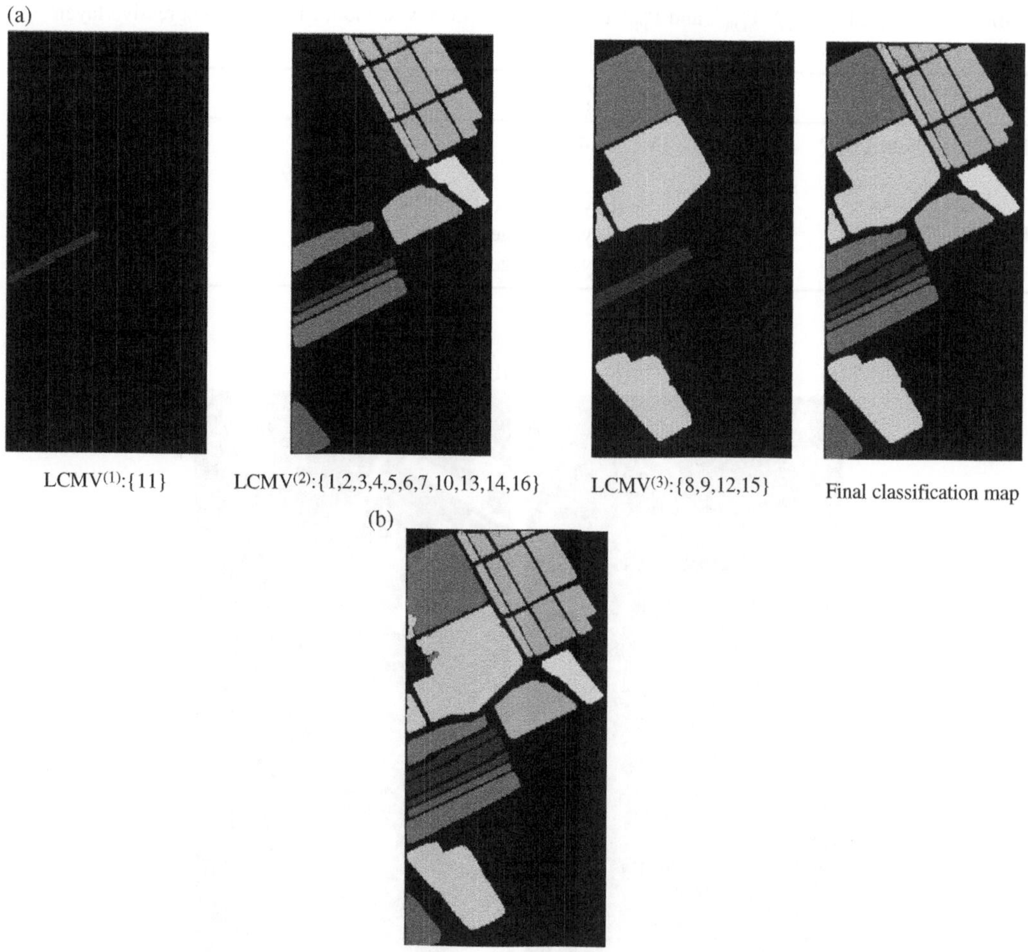


**Figure 6.8** (a) Three layers/stages processed by PBSP-ILCMV using 4-ary HCT and (b) classification of ILCMV without PBSP for Salinas.

**Table 6.10** $P_A$ calculated from the classification results of Salinas's data

| Class | 1 | 2 | 3 | 4 | 5 | 6 | 7 | 8 |
|---|---|---|---|---|---|---|---|---|
| ILCMV without PBSP | 95.52 | 98.42 | 93.78 | 95.62 | 96.90 | 98.79 | 98.63 | 96.69 |
| PBSP-ILCMV | 97.26 | 99.28 | 98.03 | 96.05 | 95.22 | 97.95 | 98.27 | 96.94 |
| Class | 9 | 10 | 11 | 12 | 13 | 14 | 15 | 16 |
| ILCMV without PBSP | 95.87 | 96.67 | 97.75 | 97.15 | 96.51 | 95.89 | 94.00 | 93.30 |
| PBSP-ILCMV | 97.07 | 96.67 | 97.75 | 97.87 | 95.52 | 97.85 | 96.67 | 98.92 |

**Table 6.11** $P_{OA}^{SR}$, $P_{AA}^{SR}$, $P_{PR}^{SR}$, $P_{OA}^{CCP}$, and $P_{PR}^{CCP}$ calculated by ILCMV without PBSP and progressive layer/stage-by-layer/stage by PBSP-ILCMV for Salinas

| | $P_{OA}^{SR}$ | $P_{AA}^{SR}$ | $P_{PR}^{SR}$ | $P_{OA}^{CCP}$ | $P_{PR}^{CCP}$ |
|---|---|---|---|---|---|
| ILCMV without PBSP | 96.37 | 96.34 | 99.45 | — | — |
| | | | | Overall $P_{OA}^{CCP}$ = 97.07 | Overall $P_{PR}^{CCP}$ = 99.91 |
| | | | Layer 1/stage 1 | 97.75 | 100 |
| PBSP-ILCMV | ---- | ---- | Layer 2/stage 2 | 96.93 | 99.88 |
| | | | Layer 3/stage 3 | 97.20 | 100 |

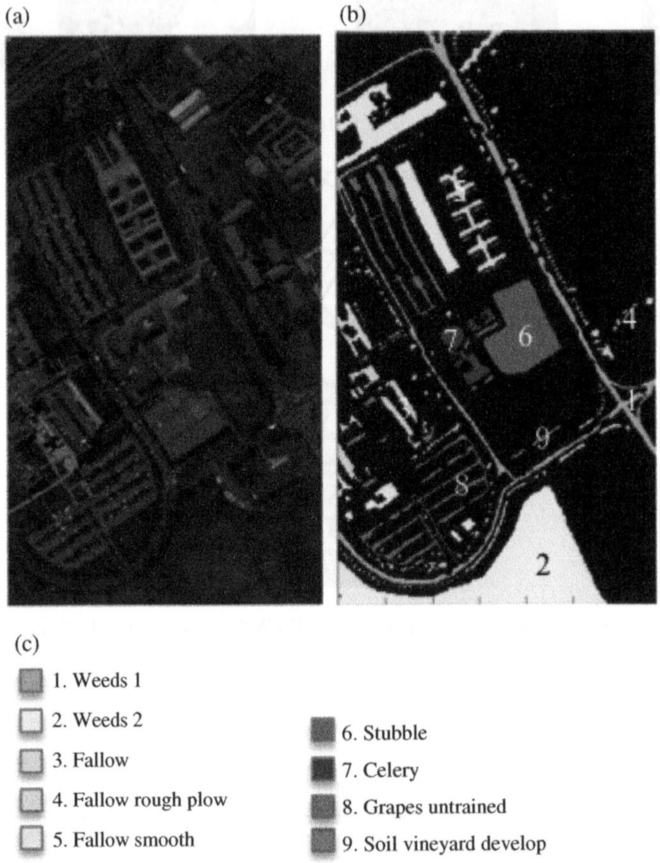


**Figure 6.9** Ground-truth of University of Pavia scene with 9 classes. (a) University of Pavia scene (b) Color ground-truth image, (c) classes by colors.

### *6.9.3 University of Pavia*

A third and the last hyperspectral image data used for experiments was the University of Pavia image shown in Figure 6.9 which is an urban area surrounding the University of Pavia, Italy. It was recorded by the ROSIS-03 satellite sensor. It has size of 610 × 340 × 115 with a spatial resolution of 1.3 m/pixel and a spectral coverage ranging from 0.43 to 0.86 μm with spectral resolution of 4 nm (12 most noisy

**Table 6.12** Specifications of parameters used by ILCMV

| D | Nine class sample means |
|---|---|
| Class number | 1–9 |
| $\sigma$ used in Gaussian filter | 0.5 with window size $5 \times 5$ |
| Thresholding method | Otsu |
| Stopping threshold (TI) | 0.98 |

**Table 6.13** Priority probabilities of CCP of nine classes in University of Pavia calculated by (1-2) with **R** including BKG

| Class | 1 | 2 | 3 | 4 | 5 | 6 | 7 | 8 | 9 |
|---|---|---|---|---|---|---|---|---|---|
| CCP | 0.0195 | 0.0237 | 0.0523 | 0.0903 | 0.6985 | 0.0183 | 0.0281 | 0.0288 | 0.0404 |
| Priority | 8 | 7 | 3 | 2 | 1 | 9 | 6 | 5 | 4 |

**Table 6.14** Bands selected for each subset of University of Pavia

| Data set | Number of subclass | Selected bands for classification |
|---|---|---|
| Pavia | {5,4} | 16,15,14,17,13,18,19,12,20,11,21,10,22,9 |
| (14 bands) | {3,9,8,7,2,1,6} | 58,59,48,60,47,57,49,44,61,51,52,43,50,45 |

channels were removed before experiments). Nine classes of interest plus BKG class, class 0 are considered for this image.

Table 6.12 lists the parameters used by ILCMV when PBSP-ILCMV was implemented.

Like Salinas, we also constructed a 4-ary HCT for University of Pavia. According to (6.8) and (6.9), $p = 4$, $m = 2$, $q = 3$, $p-q = 4-3 = 1$, there is a need of one dummy variable required for the 4-ary HCT. The CCP probabilities were calculated in Table 6.13 based on nine classes using **R** including BKG samples. Then, CSCBP-BS was implemented to find desired bands listed in Table 6.14 with $n_{BS} = 14$.

Figure 6.10 shows the progressive classification of PBSP-ILCMV using a 4-ary HCT in Figure 6.10(a) along with the results of all classes classified by ILCMV without PBSP in one-shot operation in Figure 6.10(b) where Figure 6.10(a) shows only two layers of classes being processed progressively with classes in the parentheses indicating what classes were assigned to a particular layer.

Table 6.15 further calculated $P_A$ in (6.21) of each of 9 classes produced by PBSP-ILCMV and ILCMV without PBSP in Figure 6.10 where PBSP-ILCMV produced higher $P_A$ values for every class except class 13 than their corresponding $P_A$ values produced by ILCMV. Table 6.16 also calculates $P_{OA}^{SR}$ in (6.27), $P_{OA}^{CCP}$ in (6.28), $P_{PR}^{SR}$ in (6.25), $P_{PR}^{CCP}$ in (6.26), and $P_{AA}$ in (6.29) where PBSP-ILCMV performed better than ILCMV by producing higher values of $P_{OA}^{SR}$, $P_{AA}^{SR}$, and $P_{PR}^{SR}$ than ILCMV.

It is worth noting from Table 6.16 that the OA rates produced by PBSP-ILCMV and ILCMV were nearly 13% better than that by PBSP-ILCMVB and ILCMV by taking BKG into consideration. This makes sense since the lower $P_{OA}$ classification rate was mainly caused by misclassifications from samples in the BK class. On the other hand, from Table 6.16, the precision rates produced by

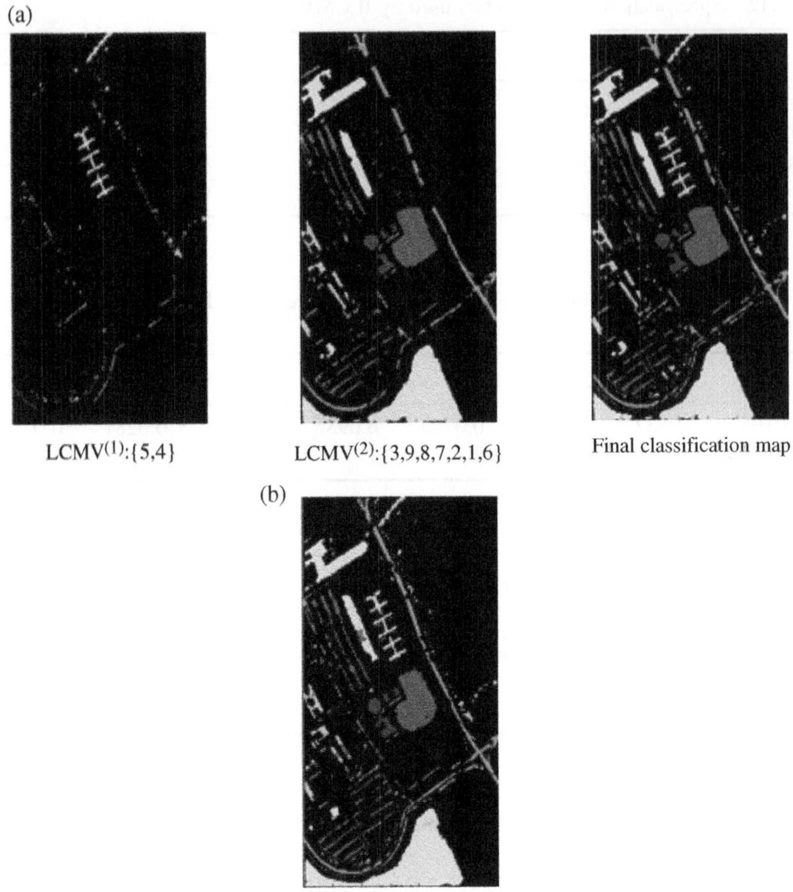


**Figure 6.10** (a) Two layers/stages processed by PBSP-ILCMV using 4-ary HCT and (b) classification of ILCMV without PBSP for University of Pavia.

**Table 6.15** $P_A$ calculated from the classification results of University of Pavia

| Class | 1 | 2 | 3 | 4 | 5 | 6 | 7 | 8 | 9 |
|---|---|---|---|---|---|---|---|---|---|
| ILCMV without PBSP | 77.24 | 86.42 | 73.34 | 79.85 | 98.81 | 91.49 | 89.10 | 81.10 | 78.46 |
| PBSP-ILCMV | 72.93 | 86.25 | 80.85 | 73.79 | 94.05 | 95.11 | 88.95 | 75.10 | 73.39 |

PBSP-ILCMV and ILCMV are nearly the same as those produced by PBSP-ILCMV and ILCMV. This is mainly due to the fact that the precision rate is calculated based on the classification results produced by PBSP-ILCMV and ILCMV, not based on ground truth as $P_{OA}$ does as illustrated in Figure 6.10. These experiments demonstrated that the BKG in the University of Pavia had a significant impact on classification. However, if we only looked at $\mathbf{P}_{OA}^{SR}$ and $\mathbf{P}_{OA}^{CCP}$ without factoring the BKG into classification, $P_{OA}$ was very high up to 97% that was also reported in many reports in the literature. Unfortunately, this only tells half of a story. The other half of a story is Table 6.16 which also demonstrated that BKG issue can be well taken care of by precision rates. This interesting finding

**Table 6.16** $P_{OA}^{SR}$, $P_{AA}^{SR}$, $P_{PR}^{SR}$, $P_{OA}^{CCP}$, and $P_{PR}^{CCP}$ calculated by ILCMV without PBSP and progressive layer/stage-by-layer/stage by PBSP-ILCMV for Salinas

| | $P_{OA}^{SR}$ | $P_{AA}^{SR}$ | $P_{PR}^{SR}$ | $P_{OA}^{CCP}$ | $P_{PR}^{CCP}$ |
|---|---|---|---|---|---|
| ILCMV without PBSP | 84.32 | 83.98 | 98.18 | — | — |
| PBSP-ILCMV | ---- | ---- | | Overall $P_{OA}^{CCP}$ = 89.42 | Overall $P_{PR}^{CCP}$ = 98.62 |
| | | | Layer 1/stage 1 | 91.73 | 98.00 |
| | | | Layer 2/stage 2 | 80.83 | 99.29 |

provides evidence that when it comes to supervised HSIC, the BKG issue must be carefully addressed. This dilemma can be resolved by including a precision rate that can effectively deal with the BKG issue.

## 6.10 Conclusion

This chapter develops a new approach to HSIC, called PBSP-HSIC, which makes use of a custom-designed $p$-ary HCT to perform classification stage-by-stage progressively. Several contributions are summarized as follows:

1. First and foremost is a new idea of developing PBSP-HSIC which allows users to process classes in multiple stages progressively where each stage performs classification of selective classes according to their priorities calculated by a CCP measure.
2. Second, to measure the priority of a class being classified, a new concept of CCP is further defined and a criterion is also designed to calculate a priority probability of each class in terms of CCP.
3. Third, these CCP probabilities are then used to construct a $p$-ary HCT with each internal node specified by a set of $p$ classes and each level consists of several internal nodes having similar levels of CCP. This CCP-constructed $p$-ary HCT can be used to navigate PBSP-HSIC to perform stage-by-stage classification progressively where each stage performs classification of $p$-class subsets located at its corresponding layer of the HCT.
4. Fourth, the CCP probabilities can be further used to define a new classification measure for OA. This new CCP-based OA and precision rate can be used as weighting factors to reflect the impact of CCP on classification. With these the CCP probabilities factored in the $P_{OA}$ as well as the precision rate, PBSP-HSIC provides progressive classification profiles of class subsets in multiple stages which cannot be offered by traditional HSIC.
5. Finally, with the CCP-constructed $p$-ary HCT, a progressive ILCMV (PBSP-ILCMV) is further developed to perform HSIC stage-by-stage progressively where each stage represents a layer of class–subset classification specified by their corresponding CCP in a HCT. All the abovementioned concepts are new and have never been investigated and explored in the literature.

## References


1. Melgani, F. and Bruzzone, L. (2004). Classification of hyperspectral remote sensing images with support vector machines. *IEEE Transactions on Geoscience and Remote Sensing* 42 (8): 1778–1790.
2. Benediktsson, J.A., Palmason, J.A., and Sveinsson, J.R. (2005). Classification of hyperspectral data from urban areas based on extended morphological profiles. *IEEE Transactions on Geoscience and Remote Sensing* 43 (3): 480–491.

3. Vamps-Valls, G. and Bruzzone, L. (2005). Kernel-based methods for hyperspectral image classiication. *IEEE Transactions on Geoscience and Remote Sensing* 43 (6): 1351–1362.
4. Tarabalka, Y., Benediktsson, J.A., and Chanussot, J. (2009). Spectral-spatial classification of hyperspectral imagery based on partitional clustering techniques. *IEEE Transactions on Geoscience and Remote Sensing* 47 (8): 2973–2987.
5. Tarabalka, Y., Fauvel, M., Chanussot, J., and Benediktsson, J.A. (2010). SVM- and MRF-based method for accurate classification of hyperspectral images. *IEEE Geoscience and Remote Sensing Letters* 9 (4): 736–740.
6. Fauvel, M., Tarabalka, Y., Benediktsson, J.A. et al. (2013). Advances in spectral-spatial classification of hyperspectral images. *Proceedings of IEEE* 101 (3): 652–675.
7. Kang, X., Li, S., and Benediktsson, J.A. (2014). Spectral-spatial hyperspectral image classification with edge-preserving filtering. *IEEE Transactions on Geoscience and Remote Sensing* 52 (5): 2666–2677.
8. Guo, X., Huang, X., Zhang, L. et al. (2016). Support tensor machines for classification of hyperspectral remote sensing imagery. *IEEE Transactions on Geoscience and Remote Sensing* 54 (6): 3248–3264.
9. Ma, X., Geng, J., and Wang, H. (2015). Hyperspectral image classification via contextual deep learning. *EURASIP Journal on Image and Video Processing* 2015: 20.
10. Ma, X., Wang, H., and Wang, J. (2016). Semisupervised classification for hyperspectral image based on multi-decision lableing and deep feature deep learning. *ISPRS Journal Photogrammetry and Renote Sensing* 120: 99–107.
11. Lu, T., Li, S., Fang, L. et al. (2017). From subpixel to superpixel: a novel fusion framework for hyperspectral image classification. *IEEE Transactions on Geoscience and Remote Sensing* 55 (9): 4398–4411.
12. Li, H., Song, Y., and Chen, C.L.P. Hyperspectral image classificatoion based on multiscale spatial iformation funsion. *IEEE Transactions on Geoscience and Remote Sensing*.
13. Yu, C., Xue, B., Wang, Y. et al. (2017). Multi-class constrained background suppression approach to hyperspectral image classification. *2017 IEEE/GRSS International Geoscience and Remote Sensing Symposium* (IGARSS) (23–28 July). TX, USA: Fort Worth.
14. Yu, C., Xue, B., Song, M. et al. (2018). Iterative target–constrained interference-minimized classifier for hyperspectral image classification. *IEEE Journal of Selected Topics in Applied Earth Observation and Remote Sensing* 11 (4): 1095–1117.
15. Xue, B., Yu, C., Wang, Y. et al. (2017). A subpixel target approach to hyperpsectral image classification. *IEEE Transactions on Geoscience and Remote Sensing* 55 (9): 5093–5114.
16. Bishop, C.M. (2006). *Machine Learning and Pattern Recognition*. Springer.
17. Yu, C., Song, M., and Chang, C.-I (2018). Band subset selection for hyperspectral image classification. *Remote Sensing* 10: 113. https://doi.org/10.3390/rs10010113.
18. Yuan, Y., Zheng, X., and Lu, X. (2017). Discovering diverse subset for unsupervised hyperspectral band selection. *IEEE Transactions Image Processing* 26 (1): 51–64.
19. Yu, C., Song, M., and Chang, C.-I (2019). Class signature-constrained background suppressed approach to band selection for hyperspectral image classification. *IEEE Transactions on Geoscience and Remote Sensing* 57 (1): 14–31.
20. Fano, R.M. (1961). *Transmission of Information*. MIT Press.
21. McEliece, R.J. (2004). *Theory of Information and Coding*. Cambridge University Press.
22. Frost, O.L. III (1972). An algorithm for linearly constrained adaptive array processing. *Proceedings of the IEEE* 60: 926–935.
23. Chang, C.-I (2002). Target signature-constrained mixed pixel classification for hyperspectral imagery. *IEEE Transactions on Geoscience and Remote Sensing* 40 (2): 1065–1081.
24. Theodoridis, S. and Koutroumbas, K. (1999). *Pattern Recognition*, 366. Academic Press.
25. Otsu, N. (1979). A threshold selection method from gray-level histgram. *IEEE Transactions on Systems, Man, and Cybernetics* SMC-9 (1): 62–66.
26. Chang, C.-I (2003). *Hyperspectral Imaging: Spectral Techniques for Detection and Classification*. NJ: Kluwer Academic Publishers.

27. Chang, C.-I and Du, Q. (2004). Estimation of number of spectrally distinct signal sources in hyperspectral imagery. *IEEE Transactions on Geoscience and Remote Sensing* 42 (3): 608–619.
28. Chang, C.-I (2013). *Hyperspectral Data Processing: Signal Processing Algorithm Design and Analysis.* NJ: Wiley.
29. Harsanyi, J.C., Farrand, W., and Chang, C.-I (1994). Detection of subpixel spectral signatures in hyperspectral image sequences. *Proceedings of American Congress on Surveying & Mapping (ACSM)/American Society of Photogrammetry & Remote Sensing (ASPRS) Annual Convention and Exposition, Baltimore*, vol. 1, pp. 236–247.

# Part III

## Compressive Sensing for Hyperspectral Imaging

# 7

# Restricted Entropy and Spectrum Properties for Hyperspectral Imaging

Chein-I Chang[1,2] and Bernard Lampe[1]
[1]*Remote Sensing Signal and Image Processing Laboratory, Department of Computer Science and Electrical Engineering, University of Maryland Baltimore County, Baltimore, MD, USA*
[2]*Center for Hyperspectral Imaging in Remote Sensing (CHIRS), Information and Technology College, Dalian Maritime University, Dalian, China*

## 7.1 Introduction

Restricted isometry property (RIP) plays a key role in developing the field of compressive sensing (CS) [1–3]. Concisely, when a matrix satisfies the RIP condition for a certain integer sparsity level $k$, the information of any signal vector with $k$ or fewer nonzero components transformed by the matrix will be preserved for nearly lossless recovery using nonlinear optimization [2]. Using RIP, the theory of CS has shown that using incoherent matrices allows for under-sampling sparse signals while preserving the sensed signal energy [4]. As a consequence, nonlinear reconstruction algorithms can be used to recover signals via CS [5–7]. Another property derived from RIP, known as restricted conformal property (RCP), states that the angle between two incoherently sensed sparse signals is preserved [8, 9]. Generally speaking, any signal vector can be completely specified by the length of the signal which corresponds to the total energy in terms of RIP and the angle between the signal vector and the representation basis in the sense of RCP. A hyperspectral signature vector (HSV) is acquired across a range of electromagnetic (EM) wavelengths using hundreds of contiguous spectral bands [10]. RIP and RCP preserve a signal vector length and the angle between two signal vectors in the original data space (ODS) and post-CS acquisition. In other words, all the information about an HSV can be preserved by similar properties in such a way that a compressively sensed signal vector can be reconstructed with high fidelity at a very low sample rate by RIP and RCP.

Interestingly, if we interpret the optical spectrum of an HSV in the EM domain as a histogram, we can normalize it to a probability vector [11]. An HSV can be indeed treated as a random variable with the EM wavelengths and the normalized spectrum as its sample space and probability distribution, respectively. This normalized spectrum-based probability distribution can be further used to define the entropy of an HSV [12]. Using entropy, the discrepancy between probabilistic distributions of two HSVs by spectral information divergence (SID) is developed in [11].

*Advances in Hyperspectral Image Processing Techniques*, First Edition. Edited by Chein-I Chang.

RIP is a length-preserving property of signal vectors when collected compressively. It can be applied to hyperspectral imaging (HSI) in many areas such as anomaly detection, target detection, and band selection and many HSI exploitation algorithms can be developed. Underpinning each of these algorithms are pixel and band discrimination as well as similarity measures which can be re-derived using the RIP. Using these modified measures, the algorithm performance can be preserved, while being processed with significantly reduced bands and data sets. Euclidean distance (ED), spectral angle mapper (SAM), and SID are ubiquitous measures to quantify the distance and angle between two HSVs where SAM is used to measure the spectral angle between two deterministic HSVs as opposed to SID which is used to measure the probabilistic profiles between two HSVs [10]. As a result, what the vector length is to a signal vector and the SAM-measured angle is to discrimination between two signal vectors can be directly translated to what entropy is to an HSV and SID-measured spectral similarity is to discrepancy between two HSVs in HSI. With this interpretation, it brings up an interesting conjecture, "Can RIP and RCP be also translated to their counterparts in hyperspectral imaging in CSBD?". This chapter derives two new properties associated to HSI, to be called restricted entropy property (REP) which preserves the entropy of an HSV in compressively sensed band domain (CSBD) and restricted spectrum property (RSP) which preserves spectral similarity between two HSVs in CSBD. The novel idea to derive REP is based on the entropy of an HSV that can be used to better describe spectral variability of an HSV than its vector length. It can be further shown that such defined entropy for an HSV is preserved in CSBD in a similar manner that the vector length of a signal vector is preserved by RIP. Moreover, REP can be also used to derive RSP, which states that spectral similarity between two HSVs measured by SID can be further preserved in CSBD, in a similar manner that the angle between two signal vectors measured by SAM is preserved by RCP after CS. More specifically, REP and RSP are derived as counterparts of RIP and RCP in HSI to preserve entropy of an HSV and spectral similarity between two HSVs in CSBD.

There are several great benefits of using REP and RSP over RIP and RCP. One is to extend deterministic properties (RIP and RCP) to probabilistic properties (REP and RSP) from an information theory point of view. Second, REP and RSP possess better properties to be used for HSI to characterize the spectral variability which is generally caused by many unknown factors such as interfering effects. Third, it has been shown in [11, 13, 14] that SID is the more effective measure than ED and SAM in hyperspectral data exploitation because SID incorporates both length and direction of a vector into a vector spectrum whose probabilistic profile can faithfully characterize spectral discrimination between two HSVs. Fourth and most importantly, the fact that SID-measured probabilistic profiles preserved in CSBD enables many HSI analysis algorithms to be employed directly in CSBD to produce nearly the same results as they are applied to the ODS.

## 7.2 Compressive Sensing Review

CS is a recent breakthrough in data acquisition and reconstruction of sparse signals which provides high reconstruction accuracy while data being sampled below the Nyquist rate under the assumption that the signal is sparse in a certain domain and also sensed incoherently [7].

Assume that there are a total number of $N$ hyperspectral image pixel vectors acquired by $L$ spectral bands as a spectral signature, $\{\mathbf{r}_i\}_{i=1}^N$ with $\mathbf{r}_i = (r_{i1}, r_{i2}, \ldots, r_{iL})^T \in \Re^L$ being an $L$-dimensional HSV where $r_{il}$ is the amount of energy collected in the $i$th pixel in the $l$th spectral band. Therefore, we can compress an HSV, $\mathbf{r}_i$, using the following linear representation model:

$$\mathbf{y}_i = \boldsymbol{\Phi}\mathbf{r}_i + \mathbf{n}_i \tag{7.1}$$

where $\mathbf{\Phi} \in \Re^{m \times L}$ is the under-determined, incoherent sensing matrix such that $m \leq L$.

Therefore, according to (7.1), specifying the sensing matrix $\mathbf{\Phi}$ and sensing dimensionality are required to apply CS to an HSV. In addition, an HSV must also have a sparse representation, denoted by $\mathbf{\Psi r}_i$ with $\mathbf{\Psi} \in \Re^{L \times L}$ being a sparse basis [15]. But an optimally sparse basis is hard to find [16]. However, due to universality [17], the sparse basis is only needed for reconstruction which is not necessary for REP [18]. This is because RIP is specified without $\mathbf{\Psi}$ by

$$(1-\delta_k)\|\mathbf{r}\|_2^2 \leq \|\mathbf{\Phi r}\|_2^2 \leq (1+\delta_k)\|\mathbf{r}\|_2^2 \tag{7.2}$$

where $\delta_k$ is defined as the restricted isometry constant (RIC) and is the smallest possible constant that satisfies (7.2) for all $k$-sparse vectors in $\mathbf{r}$. That is, RIP is a sufficient condition for design of an under-determined sensing matrix $\mathbf{\Phi}$ [19]. This property ensures that the length of each sampled vector is approximately preserved within the factor $\left\|\mathbf{\Phi}^T\mathbf{\Phi}-\mathbf{I}\right\|_F^2 < \delta_k$ with $\|.\|_F$ defined as Frobenius norm as $m \to L$.

Another sufficient property of the sensing matrix $\mathbf{\Phi}$ is the RCP property which is derived from the RIP as follows [20]:

$$\frac{(1-\delta_k)}{(1+\delta_k)}\cos\theta \leq \cos\alpha \leq \frac{(1+\delta_k)}{(1-\delta_k)}\cos\theta \tag{7.3}$$

where $\theta$ is the angle between any two HSVs, $\mathbf{r}_i$ and $\mathbf{r}_j$, and $\alpha$ is the angle between the corresponding pixels in the compressively sensed domain, $\mathbf{y}_i$ and $\mathbf{y}_j$.

Constructing the sensing matrix $\mathbf{\Phi}$ probabilistically is an efficient means to ensure that the RIP and RCP conditions are preserved [21]. In addition, using a probability distribution guarantees that the sensing/sampling basis will be incoherent with any structured representation basis. A widely used and well-studied way to generate the entries of $\mathbf{\Phi}$ is to use a Gaussian distribution and normalize the columns explicitly or scale the variance to the unit length in expectation [22]. This is a widely used sampling scheme. Under this design, the number of samples, $m$, is specified by

$$m \geq ck\log\left(\frac{L}{k}\right) \tag{7.4}$$

where $c$ is a constant generally considered to be 2 and $k$ is the maximum signal sparsity which can be recovered. This constraint requires estimating the maximum sparsity of the signal, $k$, which remains an open problem [20, 21].

A recent realization has been made that the Harsanyi–Farrand–Chang (HFC) method used to estimate virtual dimensionality (VD) of hyperspectral images is in fact a sparsity estimation algorithm [23–27]. It can be used to choose an optimal sensing/sampling rate for the following experiments which show results for a range of sensing/sampling rates to illustrate the performance of the HFC method.

The set of HSVs can be sensed using the model described by (7.1) and each sensed HSV $\{\mathbf{y}_i\}_{i=1}^N$ can be reconstructed individually using nonlinear reconstruction algorithms minimizing the $l_1$-norm such as orthogonal match pursuit (OMP) [5], basis pursuit (BP) [6], and iterative hard thresholding (IHT) [5] to solve the following constrained optimization problem:

$$(\mathbf{\Psi r}_i)^* = \arg\left\{\min_{\mathbf{\Psi r}_i}\|\mathbf{\Psi r}_i\|_1\right\}\text{subject to } \mathbf{y}_i = \mathbf{\Phi r}_i \tag{7.5}$$

It should be noted that the reconstruction algorithm solving (7.5) is not required to develop the properties, REP and RSP, as well as hyperspectral data processing algorithms in the CSBD. This is because the sparse representation matrix $\mathbf{\Psi}$ need not be specified due to universality [28].

An important consequence of the RIP condition in (7.2) is $\|\mathbf{r}_i\|_2^2 \approx \|\mathbf{\Phi r}_i\|_2^2$ as $m \to L$. A proof of the RIP condition can be found in [28]. However, a statistical treatment for such approximation can be made considerably simpler as follows.

Using (7.1), the sensing transformation from $\mathbf{r}_i$ to $\mathbf{y}_i$ can be expressed as

$$\mathbf{y}_i = \mathbf{\Phi r}_i = \frac{1}{\sqrt{m}} \begin{bmatrix} \phi_{11} & \cdots & \phi_{1L} \\ \vdots & \ddots & \vdots \\ \phi_{m1} & \cdots & \phi_{mL} \end{bmatrix} \begin{pmatrix} r_{i1} \\ \vdots \\ r_{iL} \end{pmatrix} \tag{7.6}$$

where $\phi_{ul} \sim N\left(\mu = 0, \sigma_u^2 = 1\right)$ for $1 \leq u \leq m$ is a standard Gaussian distribution and the $1/\sqrt{m}$ term is the column normalization term to define the columns of the sensing matrix as $\mathbf{\Phi} = (\mathbf{\phi}_1\mathbf{\phi}_2 \cdots \mathbf{\phi}_L)$ where $\mathbf{\phi}_l \in \Re^{m \times 1}$. The expectation of the $l_2$-norm of the $l$th column vector $\mathbf{\phi}_l$ under the standard Gaussian can be calculated by

$$E[\|\mathbf{\phi}_l\|_2] = \sqrt{\|\mathbf{\mu}\|_2 + \sum_{i=1}^{m} \sigma_i^2} = \sqrt{m} \tag{7.7}$$

Therefore, consider the HSV $\mathbf{r}_i$ to be constant and the sensing matrix $\mathbf{\Phi}$ to be random. Then, $\mathbf{y}_i = (y_{i1}, y_{i2}, \ldots, y_{im})^T$ and $y_{iu}$ is distributed Gaussian with $y_{iu} \sim N\left(\mu = 0, \sigma_u^2 = \frac{1}{m}\sum_{l=1}^{L} r_{il}\right)$. Knowing the entries of the HSV in the CSBD denoted by $\mathbf{y}_i$ are Gaussian-distributed, then the expectation of length of the sensed HSV $\|\mathbf{y}_i\|_2^2$ can be computed using the same form in (7.7) as

$$E\left[\|\mathbf{y}_i\|_2^2\right] = \sqrt{\|\mathbf{\mu}\|_2 + \sum_{i=1}^{m}\left(\frac{1}{m}\sum_{l=1}^{L} r_{il}\right)} = \sqrt{\sum_{l=1}^{L} r_{il}^2} = \|\mathbf{r}_i\|_2 \tag{7.8}$$

Therefore, the expected values of the HSV in the ODS and the HSV in the CSBD is $E\left[\|\mathbf{r}_i\|_2^2\right] = E\left[\|\mathbf{y}_i\|_2^2\right] = E\left[\|\mathbf{\Phi r}_i\|_2^2\right]$ and $\|\mathbf{r}_i\|_2^2 \approx \|\mathbf{y}_i\|_2^2 \approx \|\mathbf{\Phi r}_i\|_2^2$ as $m \to L$. This result is critical in developing REP and RSP.

Another important result in CS which is useful and will be used in the derivation of REP and RSP is $\mathbf{\Phi}^T\mathbf{\Phi} \to \mathbf{I}$ as $m \to L$. Using (7.6), we define the columns of the sensing matrix as $\mathbf{\Phi} = (\mathbf{\phi}_1\mathbf{\phi}_2 \cdots \mathbf{\phi}_L)$ where $\mathbf{\phi}_l \in \Re^{m \times 1}$. Then, the inner product of the sensing matrix results in an $L \times L$ matrix with entries defined by the inner product of the columns $\phi_{uv} = \mathbf{\phi}_u^T\mathbf{\phi}_v$ with the magnitude of the columns normalized to one. So, when $u = v$, the inner product becomes 1 and when $u \neq v$ and knowing the vectors are independent identically distributed Gaussian, the inner product is 0, i.e. $\mathbf{\phi}_u$ and $\mathbf{\phi}_v$ become orthogonal as $m \to L$. Therefore, as dimensionality, $m$ increases, the diagonal entries approach to 1 and the off-diagonal entries go to zero exponentially.

## 7.3 Restricted Entropy Property

For any HSV, $\mathbf{r}_i = (r_{i1}, r_{i2}, \ldots, r_{iL})^T \in \Re^L$, we follow the approach in [9, 10] and define the spectral information measure (SIM) as the probability measure computed by

$$p_{il} = \frac{|r_{il}|}{\sum_{l=1}^{L}|r_{il}|} \tag{7.9}$$

According to (7.9), $\mathbf{r}_i$ is considered as a random variable defined on a probability space $X = (\Omega, \Sigma, P)$ where the sample space $\mathbf{\Omega}$ consisting of $\{r_{il}\}_{l=1}^{L}$ corresponding to all individual wavelengths $\{\lambda_l\}_{l=1}^{L}$ of a hyperspectral image, the event space $\Sigma$ is a power set and the probability measure $P$ is defined by (7.9). Using this random HSV, we can simply measure the $p_{il} = p_i(\lambda_l)$ by (7.9) via SIM. Another way to interpret this probability is to say that $p_{il}$ is the probability that energy from the scene will be collected in wavelength $\lambda_l$.

Furthermore, suppose that $\mathbf{p}_i = (p_{i1}, p_{i2}, \ldots p_{iL})^T$ and $\mathbf{q}_j = (q_{j1}, q_{j2}, \ldots q_{jL})^T$ are probability vectors calculated by (7.9) from the $i$th and $j$th HSVs. The Kullback–Leibler (KL) divergence between two probability distributions $\mathbf{p}_i$ and $\mathbf{q}_j$ is defined as the difference between the cross entropy,

$$\mathrm{D_{KL}}\left(\mathbf{p}_i \| \mathbf{q}_j\right) = H(\mathbf{p}_i, \mathbf{q}_j) - H(\mathbf{p}_i) \tag{7.10}$$

where $H(\mathbf{p}_i, \mathbf{q}_j)$ is the joint entropy of $\mathbf{p}_i$ and $\mathbf{q}_j$. Therefore, if we can prove the entropy is preserved under a CS transform $\mathbf{\Phi} \in \Re^{m \times L}$, then the divergence $\mathrm{D_{KL}}(\mathbf{p}_i \| \mathbf{q}_j)$ can be also preserved. More specifically, we prove the following theorem.

**Theorem 7.1** Restricted Entropy Property (REP)

$$H(\mathbf{p}_i) \to H(\mathbf{\Phi}\mathbf{p}_i) \text{ as } m \to L \tag{7.11}$$

**Proof.**

The first step of the proof is to express the entropy in terms of the angles between the basis vectors and the pixel vector $\mathbf{r}_i$ which is denoted by $\theta_l = \angle(\mathbf{e}_l, \mathbf{r}_i)$ where the $\mathbf{e}_l = (0, \ldots, 0, 1, 0, \ldots, 0)^T$ is the basis vector with the $l$th position equal to 1 and all other positions equal to 0.

$$\begin{aligned} H(\mathbf{p}_i) &= -\sum_{l=1}^{L} p_{il}\log p_{il} = -\sum_{l=1}^{L}\frac{|r_{il}|}{\bar{r}_i}\log\frac{|r_{il}|}{\bar{r}_i} \\ &= -\sum_{l=1}^{L}\left(|\langle \mathbf{e}_l, \mathbf{r}_i\rangle| / \sum_{l=1}^{L}|\langle \mathbf{e}_l, \mathbf{r}_i\rangle|\right)\log\left(|\langle \mathbf{e}_l, \mathbf{r}_i\rangle| / \sum_{l=1}^{L}|\langle \mathbf{e}_l, \mathbf{r}_i\rangle|\right) \\ &= -\sum_{l=1}^{L}\left(|\mathbf{e}_l^T\mathbf{r}_i| / \sum_{l=1}^{L}|\mathbf{e}_l^T\mathbf{r}_i|\right)\log\left(|\mathbf{e}_l^T\mathbf{r}_i| / \sum_{l=1}^{L}|\mathbf{e}_l^T\mathbf{r}_i|\right) \end{aligned} \tag{7.12}$$

where $\bar{r}_i = \sum_{l=1}^{L}|r_{il}|$. Furthermore, (7.12) can be re-expressed as

$$\begin{aligned} H(\mathbf{p}_i) &= -\sum_{l=1}^{L}\left(\|\mathbf{e}_l\|\|\mathbf{r}_i\|\cos\theta_l / \sum_{l=1}^{L}\|\mathbf{e}_l\|\|\mathbf{r}_i\|\cos\theta_l|\right)\log\left(\|\mathbf{e}_l\|\|\mathbf{r}_i|\,\|\cos\theta_l| / \sum_{l=1}^{L}|\cos\theta_l|\right) \\ &= -\sum_{l=1}^{L}\left(|\cos\theta_l| / \sum_{l=1}^{L}|\cos\theta_l|\right)\log\left(|\cos\theta_l| / \sum_{l=1}^{L}|\cos\theta_l|\right) \end{aligned} \tag{7.13}$$

where the norm $\|\cdot\|$ is the $l_2$-norm, $\|\cdot\|_2$. It should be noted that the magnitudes of $\|\mathbf{r}_i\|$ and $\|\mathbf{e}_l\|$ are constant. So, they are cancelled in (7.12).

Now, we can express $\cos(\theta_l)$ as

$$\cos\theta_l = \frac{\|\mathbf{e}_l\|_2^2 - \|\mathbf{r}_i\|_2^2 - \|\mathbf{e}_l - \mathbf{r}_i\|_2^2}{2\|\mathbf{e}_l\|_2^2 \mid \|\mathbf{r}_i\|_2^2} \tag{7.14}$$

By taking advantage of the RIP condition in (7.2), then $\|\mathbf{r}_i\|_2^2 \approx \|\mathbf{\Phi r}_i\|_2^2$ and we can replace the magnitudes in the law of cosines with the transformed magnitudes.

$$\cos\theta_l \approx \cos\alpha_l = \frac{\|\mathbf{\Phi e}_l\|_2^2 - \|\mathbf{\Phi r}_i\|_2^2 - \|\mathbf{\Phi e}_l - \mathbf{\Phi r}_i\|_2^2}{2\|\mathbf{\Phi e}_l\|_2^2 \mid \|\mathbf{\Phi r}_i\|_2^2} \tag{7.15}$$

Now, we redefine the entropy of $\mathbf{\Phi r}_i$, $H(\mathbf{\Phi r}_i)$, in the compressed domain via $\mathbf{\Phi}$ with $\alpha_i = \angle(\mathbf{\Phi e}_l, \mathbf{\Phi r}_i)$. In addition, this redefinition allows the angle to be preserved by the RCP property in (7.3).

$$\begin{aligned} H(\mathbf{\Phi r}_i) &= -\sum_{l=1}^{m}\left(\left\|\mathbf{\Phi e}_l\right\|\left\|\mathbf{\Phi r}_i\right\|\left|\cos\alpha_l\right| / \sum_{l=1}^{m}\left\|\mathbf{\Phi e}_l\right\|\left\|\mathbf{\Phi r}_i\right\|\left|\cos\alpha_l\right|\right)\log\left(\left\|\mathbf{\Phi e}_l\right\|\left\|\mathbf{\Phi r}_i\right\|\left|\cos\alpha_l\right| / \sum_{l=1}^{L}\left|\cos\alpha_l\right|\right) \\ &= -\sum_{l=1}^{m}\left(\left|\langle\mathbf{\Phi e}_l, \mathbf{\Phi r}_i\rangle\right| / \sum_{l=1}^{m}\left|\langle\mathbf{\Phi e}_l, \mathbf{\Phi r}_i\rangle\right|\right)\log\left(\left|\langle\mathbf{\Phi e}_l, \mathbf{\Phi r}_i\rangle\right| / \sum_{l=1}^{m}\left|\langle\mathbf{\Phi e}_l, \mathbf{\Phi r}_i\rangle\right|\right) \\ &= -\sum_{l=1}^{m}\left(\left|\mathbf{e}_l^T\mathbf{\Phi}^T\mathbf{\Phi r}_i\right| / \sum_{l=1}^{m}\left|\mathbf{e}_l^T\mathbf{\Phi}^T\mathbf{\Phi r}_i\right|\right)\log\left(\left|\mathbf{e}_l^T\mathbf{\Phi}^T\mathbf{\Phi r}_i\right| / \sum_{l=1}^{m}\left|\mathbf{e}_l^T\mathbf{\Phi}^T\mathbf{\Phi r}_i\right|\right) \end{aligned} \tag{7.16}$$

where the norm $\|\cdot\|$ in (7.16) is the $l_2$-norm, $\|\cdot\|_2$.

Now, as $m \to L$, $\mathbf{\Phi}^T\mathbf{\Phi} \to \mathbf{I}$ and $\delta_k \to 0$ Then, we use (7.16) to derive

$$\begin{aligned} H(\mathbf{\Phi r}_i) &= -\sum_{l=1}^{L}\left(\left|\mathbf{e}_l^T\mathbf{r}_i\right| / \sum_{l=1}^{L}\left|\mathbf{e}_l^T\mathbf{r}_i\right|\right)\log\left(\left|\mathbf{e}_l^T\mathbf{r}_i\right| / \sum_{l=1}^{L}\left|\mathbf{e}_l^T\mathbf{r}_i\right|\right) \\ &= -\sum_{l=1}^{L}\frac{|r_{il}|}{\bar{r}_i}\log\frac{|r_{il}|}{\bar{r}_i} = -\sum_{l=1}^{L}p_{il}\log p_{il} = H(\mathbf{p}_i) \end{aligned} \tag{7.17}$$

which implies that $H(\mathbf{\Phi r}_i) = H(\mathbf{p}_i)$. Since $\mathbf{p}_i$ is the normalized probability vector of $\mathbf{r}_i$ resulting from (7.6), $H(\mathbf{\Phi r}_i) = H(\mathbf{\Phi p}_i)$. The theorem is proved. ■

***Corollary***

Let the cross entropy be defined by

$$H(\mathbf{p}_i, \mathbf{q}_j) = -\sum_{l=1}^{L}p_{il}\log q_{jl} \tag{7.18}$$

Then,

$$H(\mathbf{p}_i, \mathbf{q}_j) \to H(\mathbf{\Phi p}_i, \mathbf{\Phi q}_j) \text{ as } m \to L \tag{7.19}$$

and

$$D_{\mathrm{KL}}\left(\mathbf{p}_i \middle\| \mathbf{q}_j\right) + H(\mathbf{p}_i) = D_{\mathrm{KL}}\left(\mathbf{\Phi p}_i \middle\| \mathbf{\Phi q}_j\right) + H(\mathbf{\Phi p}_i) \text{ as } m \to L \tag{7.20}$$

**Proof of Corollary**

$$\begin{aligned} H\left(\mathbf{\Phi p}_i, \mathbf{\Phi q}_j\right) &= -\sum_{l=1}^{m}\left(|\langle\mathbf{\Phi e}_l, \mathbf{\Phi p}_i\rangle| / \sum_{l=1}^{m}|\langle\mathbf{\Phi e}_l, \mathbf{\Phi p}_i\rangle|\right)\log\left(|\langle\mathbf{\Phi e}_l, \mathbf{\Phi q}_j\rangle| / \sum_{l=1}^{m}|\langle\mathbf{\Phi e}_l, \mathbf{\Phi q}_j\rangle|\right) \\ &= -\sum_{l=1}^{m}\left(|\mathbf{e}_l^T\mathbf{\Phi}^T\mathbf{\Phi p}_i| / \sum_{l=1}^{m}|\mathbf{e}_l^T\mathbf{\Phi}^T\mathbf{\Phi p}_i|\right)\log\left(|\mathbf{e}_l^T\mathbf{\Phi}^T\mathbf{\Phi q}_j| / \sum_{l=1}^{m}|\mathbf{e}_l^T\mathbf{\Phi}^T\mathbf{\Phi q}_j|\right) \end{aligned} \tag{7.21}$$

Since $\mathbf{\Phi}^T\mathbf{\Phi} \to \mathbf{I}$ as $m \to L$, (7.21) becomes $H(\mathbf{p}_i, \mathbf{q}_j)$ which results in (7.19).

■

## 7.4 Restricted Spectrum Property

Using (7.14), we can further prove the following theorem which validates RSP.

**Theorem 7.2** Restricted Spectrum Property (RSP)

$$D_{\mathrm{SID}}\left(\mathbf{p}_i \middle\| \mathbf{q}_j\right) \to D_{\mathrm{SID}}\left(\mathbf{\Phi p}_i \middle\| \mathbf{\Phi q}_j\right) \text{ as } m \to L \tag{7.22}$$

where SID defined in [10, 11] is given by

$$D_{\mathrm{SID}}(\mathbf{p}_i, \mathbf{q}_j) = D_{\mathrm{SID}}(\mathbf{q}_j, \mathbf{p}_i) \equiv D\left(\mathbf{p}_i \middle\| \mathbf{q}_j\right) + D\left(\mathbf{q}_j \middle\| \mathbf{p}_i\right) \tag{7.23}$$

**Proof.**
By using $H(\mathbf{p}_i) \to H(\mathbf{\Phi p}_i)$ in (7.11) and $H(\mathbf{p}_i, \mathbf{q}_j) \to H(\mathbf{\Phi p}_i, \mathbf{\Phi q}_j)$ in (7.19) as $m \to L$, we can easily prove that as $m \to L$

$$D_{\mathrm{KL}}\left(\mathbf{p}_i \middle\| \mathbf{q}_j\right) \to D_{\mathrm{KL}}\left(\mathbf{\Phi p}_i \middle\| \mathbf{\Phi q}_j\right) \tag{7.24}$$

and

$$D_{\mathrm{KL}}\left(\mathbf{q}_i \middle\| \mathbf{p}_j\right) \to D_{\mathrm{KL}}\left(\mathbf{\Phi q}_i \middle\| \mathbf{\Phi p}_j\right) \tag{7.25}$$

Combining the definition of SID in (7.23) with (7.24) and (7.25) yields

$$D_{\mathrm{SID}}\left(\mathbf{p}_i \middle\| \mathbf{q}_j\right) \approx D_{\mathrm{SID}}\left(\mathbf{\Phi p}_i \middle\| \mathbf{\Phi q}_j\right) \tag{7.26}$$

■

Note that the above proof implies that (7.9) can be further modified in CSBD as

$$p\big((\mathbf{\Phi r}_i)_l\big) = p(y_{il}) = \frac{|\langle \mathbf{\Phi e}_l, \mathbf{\Phi r}_i \rangle|}{\sum_{l=1}^{m} |\langle \mathbf{\Phi e}_l, \mathbf{\Phi r}_i \rangle|} \tag{7.27}$$

The proof shows that SID is preserved and therefore SID is also preserved. However, it does not give explicit bounds on the compressed entropy. The assumption made during the proof was the $\cos\theta_l \approx \cos \alpha_l$ for all the angles of $\alpha_i = \angle(\mathbf{\Phi e}_l, \mathbf{\Phi r}_i)$. This simplification was useful for the proof but did not yield a specific bound. In reality, the angles are bounded as in (7.3) and each pair of vectors has a slightly different error bound as noted by $\delta_{kl}$. This error bound results from the norm of the difference term in (7.14). The sparsity of this vector can vary. We can rewrite the compressed entropy using the error to obtain a lower and upper bound as follows:

$$\frac{\left(\frac{1-\delta_{kl}}{1+\delta_{kl}}\right)|\cos\theta_l|}{\sum_{l=1}^{L}\left(\frac{1-\delta_{kl}}{1+\delta_{kl}}\right)|\cos\theta_l|} \leq \frac{|\cos\alpha_l|}{\sum_{l=1}^{m}|\cos\alpha_l|} \leq \frac{\left(\frac{1+\delta_{kl}}{1-\delta_{kl}}\right)|\cos\theta_l|}{\sum_{l=1}^{L}\left(\frac{1+\delta_{kl}}{1-\delta_{kl}}\right)|\cos\theta_l|} \tag{7.28}$$

## 7.5 REP and RSP Hyperspectral Measures

Hyperspectral measures have found many applications in hyperspectral data analysis [23, 26, 29] such as endmember finding and identification in [28], target detection, linear spectral mixture analysis, and band subset selection [30]. In particular, hyperspectral measures were used to construct a band channel whose capacity can be used to select bands subsets [31]. In this section, several commonly used hyperspectral measures, ED, SAM, and SID [9–11] will be shown to be preserved in a CS-compressed domain by RIP, RCP, and REP.

Assume that $\{\mathbf{r}_i\}_{i=1}^{N}$ and $\{\mathbf{y}_i = \mathbf{\Phi r}_i\}_{i=1}^{N}$ are a set of hyperspectral data sample vectors in the ODS and their corresponding data sample vectors in CS-compressed domain, respectively.

The simplest hyperspectral measure is ED which can be preserved by RIP, i.e.

$$D_{\text{ED}}(\mathbf{r}_i, \mathbf{r}_j) \approx D_{\text{ED}}(\mathbf{y}_i, \mathbf{y}_j) = \|\mathbf{y}_i - \mathbf{y}_j\|_2. \tag{7.29}$$

An alternative measure to ED is called SAM, which calculates the angle between two data sample vectors which can be preserved by RCP, i.e.

$$D_{\text{SAM}}(\mathbf{r}_i, \mathbf{r}_j) \approx D_{\text{SAM}}(\mathbf{y}_i, \mathbf{y}_j) = \cos^{-1}\left(\frac{|\langle \mathbf{y}_i, \mathbf{y}_j \rangle|}{\|\mathbf{y}_i\|_2 \|\mathbf{y}_j\|_2}\right) \tag{7.30}$$

and the absolute value is taken due to possible negative values caused by two vectors with opposite directions.

Using (7.23), SID can be preserved by REP as follows:

$$D_{\text{SID}}(\mathbf{r}_j, \mathbf{r}_k) \approx D_{\text{SID}}(\mathbf{y}_i, \mathbf{y}_j) = D(\mathbf{y}_i \| \mathbf{y}_j) + D(\mathbf{y}_j \| \mathbf{y}_i) \tag{7.31}$$

where

$$D(\mathbf{y}_i \| \mathbf{y}_j) = \sum_{l=1}^{m} p_{il} \log \frac{p_{il}}{p_{jl}} \tag{7.32}$$

and $p_{il} = |y_{il}| / \sum_{l=1}^{m} |y_{il}|$.

Finally, an interesting measure was developed in [32] which combines SID in (7.31) and SAM in (7.30), referred to as SIDAM which can be further preserved by both RCP and REP as follows:

$$D_{\text{SIDAM}}(\mathbf{y}_i, \mathbf{y}_j) = D_{\text{SID}}(\mathbf{y}_i, \mathbf{y}_j) \cdot \tan\left(D_{\text{SAM}}(\mathbf{y}_i, \mathbf{y}_j)\right) \tag{7.33}$$

## 7.6 Experiments

The first image data set is an airborne Hyperspectral Digital Imagery Collection Experiment (HYDICE) scene shown in Figure 7.1(a) which was collected in August 1995 from a flight altitude of 10 000 ft with the ground sampling spectral discrimination measure approximately 1.56 m. It has size of $64 \times 64$ pixel vectors shown in Figure 7.1(a) along with its ground truth provided in Figure 7.1(b), where the center and boundary pixels of objects are highlighted by dark gray and light gray, respectively.

This scene has been well studied and can be used for experiments where the center and boundary pixels of objects are highlighted by dark gray and light gray, respectively. Each element in Figure 7.1 (b) is a square panel and denoted by $p_{ij}$ with rows indexed by $i = 1, 2, \ldots, 5$ and columns indexed by $j = 1, 2, 3$. The 1.56 m spatial resolution of the image scene suggests that most of the 15 panels are one pixel in size except that the panels in the 1st column with the 2nd, 3rd, 4th, and 5th rows which are two-pixel panels, denoted by $\mathrm{p}_{211}$, $\mathrm{p}_{221}$, $\mathrm{p}_{311}$, $\mathrm{p}_{312}$, $\mathrm{p}_{411}$, $\mathrm{p}_{412}$, $\mathrm{p}_{511}$, and $\mathrm{p}_{521}$. As a result, there are a total of 19 panel pixels. Figure 7.1(b) shows the precise spatial locations of these 19 panel pixels where dark gray pixels are the panel center pixels and the pixels in light gray are panel pixels mixed with the background. Figure 7.1(c) also plots the signatures of five panels obtained by averaging all dark gray panel pixels in each row. That is, panel $j$ in Figure 7.1(c) is the average value of all dark gray panel pixels in row $j$. It should be noted that there is an interesting panel pixel $\mathrm{p}_{212}$ which is located next to the panel pixel $\mathrm{p}_{211}$ horizontally. A detailed description of this particular scene with color maps can be found in [10, 13, 29].

In order to perform practical experiments, we must find a suitable compressively sensed band ratio (CSBR), $m/L$. First of all, we used the HFC method developed in [33] to estimate the sparsity of order as VD, $k = 9$ with the false alarm probability $P_F$ set to $10^{-4}$. We then increased the number of compressively sensed bands (CSBs), $m$, up to 169. Figure 7.2 shows the convergence of the estimated values of VD at $k \approx 9$ versus $m$ where $m \approx 20$ seemed to be a reasonably good number of $m$ required for the sparsity of order, $k = 9$.

**Experiment 7.1** In order to validate REP for hyperspectral data analysis, four hyperspectral discrimination measures, $D_{\text{ED}}$, $D_{\text{SAM}}$, $D_{\text{SID}}$, and $D_{\text{SIDAM}}$ were investigated for HYDICE experiments.

Using one of five panel signatures in Figure 7.1(c) as a reference signature $\mathbf{d}$, we computed spectral discrimination values between $\mathbf{d}$ and each of all the data sample vectors, $\{\mathbf{r}_i\}_{i=1}^{4096}$ by $D_{\text{ED}}$, $D_{\text{SAM}}$, $D_{\text{SID}}$, and $D_{\text{SIDAM}}$ in the ODS and CSBD as shown in Figure 7.3 where the spectral discrimination measure is specified by $D_{\text{SID}}$.

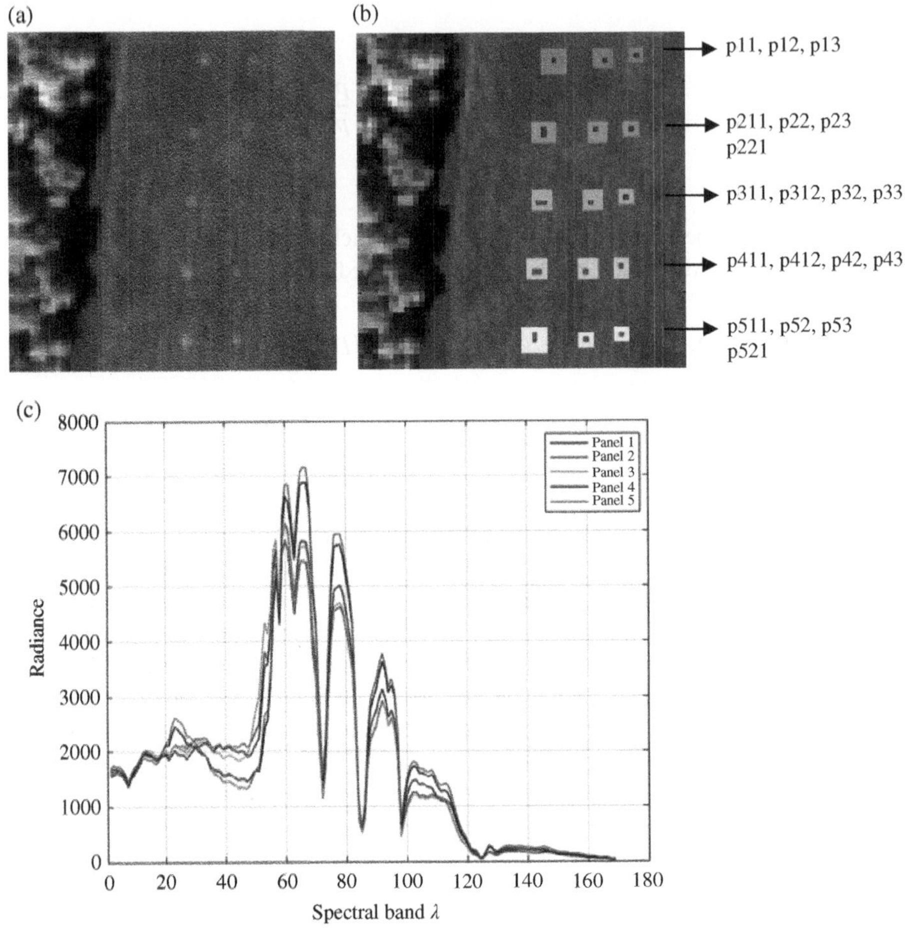


**Figure 7.1** (a) A HYDICE panel scene which contains 15 panels; (b) ground-truth map of spatial locations of the 15 panels; (c) spectra of five panels obtained by averaging all *R* panel pixels in each row.

Figures 7.4–7.8 plot the spectral discrimination values calculated by $D_{\mathrm{ED}}$, $D_{\mathrm{SAM}}$, $D_{\mathrm{SID}}$, and $D_{\mathrm{SIDAM}}$ between $\mathbf{d} = \mathbf{p}_1, \mathbf{p}_2, \mathbf{p}_3, \mathbf{p}_4, \mathbf{p}_5$, and $\{\mathbf{r}_i\}_{i=1}^{4096}$, respectively, with the number of CSBs, $m$, was set to 20 where the solid lines and dashed lines are used to specify the results calculated in the original full band domain and CSBD. In addition, Monte Carlo simulations were also used to reduce the variability introduced by the randomly generated sensing matrix. The experiments were run 100 times and the results were averaged to produce Figures 7.4–7.8 where solid and dashed lines were nearly identical one to another. This indicates that RIP, RCP, and RSP were preserved by ED, SAM, and SID in CSBD, respectively. In addition, both RCP and REP were also preserved by SIDAM in CSBD.

As noted in Figures 7.4–7.8, differences between the dashed and solid lines are not clearly visible. This is because the property of each of four measures, ED, SAM, SID, and SIDAM is preserved in ODS and CSBD resulting from RIP, RCP, REP, and RSP. In other words, the discrimination values of each measure in ODS and CSBD are expected to be very small. As a result, the discrepancy between dashed and solid lines produced by each measure cannot be recognized clearly. It must be this case. Otherwise, if their discrimination values are visibly recognizable, then it implies that the properties of

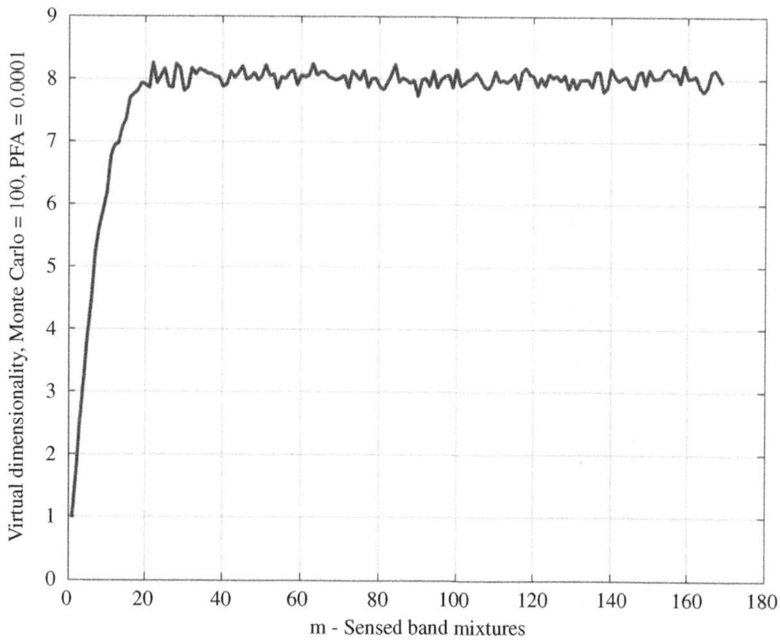


**Figure 7.2** Sparsity of order, $k \approx 9$ versus number of CSBs, $m$.

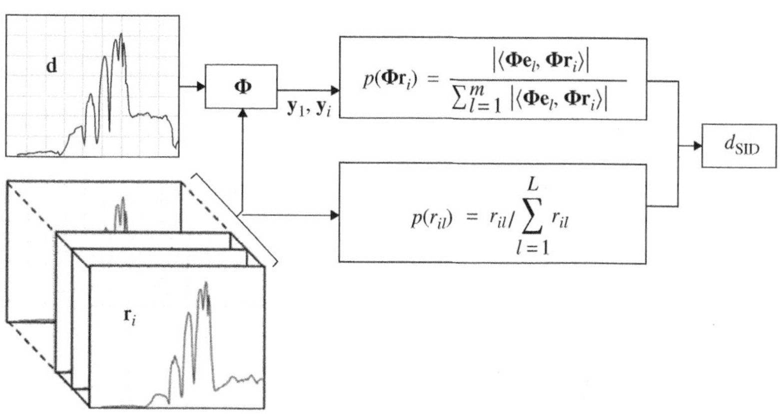


**Figure 7.3** Experiment 7.1 setup depicting the comparison of SID between compressively sensed and collected pixels.

each measure are not preserved. Nevertheless, Figures 7.4–7.8 demonstrate by the plots that each measure did have a different degree in error of property preservation. To further enlarge the differences between solid and dashed lines which are visually undetectable in Figures 7.4–7.8, Figures 7.9–7.13 provide zoom-in portions of $D_j(\mathbf{d}, \mathbf{r}_i)$ and $D_j(\mathbf{\Phi d}, \mathbf{\Phi r}_i)$ in Figures 7.4–7.8 of data samples, $\{\mathbf{r}_i\}_{i=2000}^{2050}$ for further inspection.

It is obvious from Figures 7.9–7.13 that both curves of $D_{\text{ED}}(\mathbf{d}, \mathbf{r}_i)$ in the original full band domain and $D_{\text{ED}}(\mathbf{\Phi d}, \mathbf{\Phi r}_i)$ in CSBD were nearly identical. This indicated that RIP was well preserved by ED

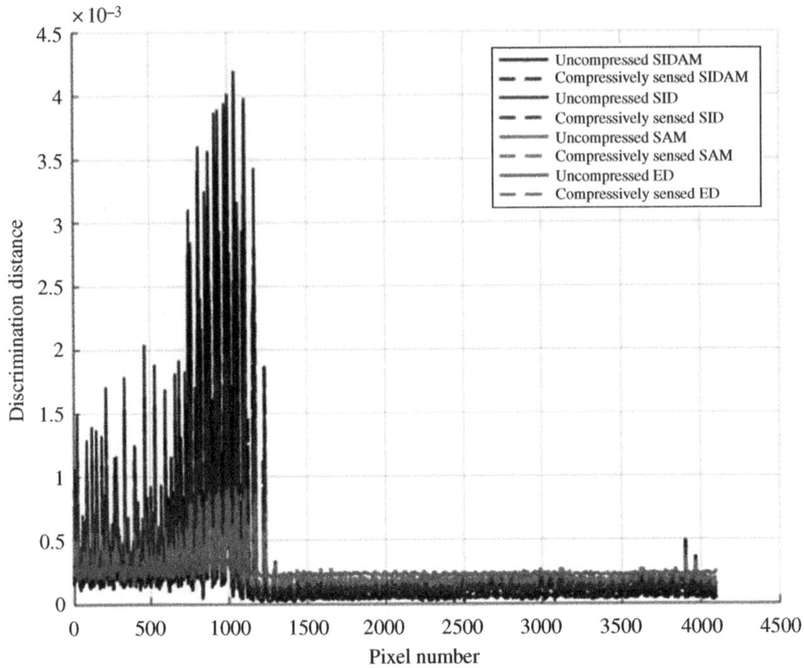


**Figure 7.4** Discrimination values calculated by $D_j(\mathbf{d}, \mathbf{r}_i)$ and $D_j(\mathbf{\Phi d}, \mathbf{\Phi r}_i)$ for $\mathbf{d} = p_1$ and $\{\mathbf{r}_i\}_{i=1}^{4096}$ and $j$ = ED, SAM, SID, SIDAM.

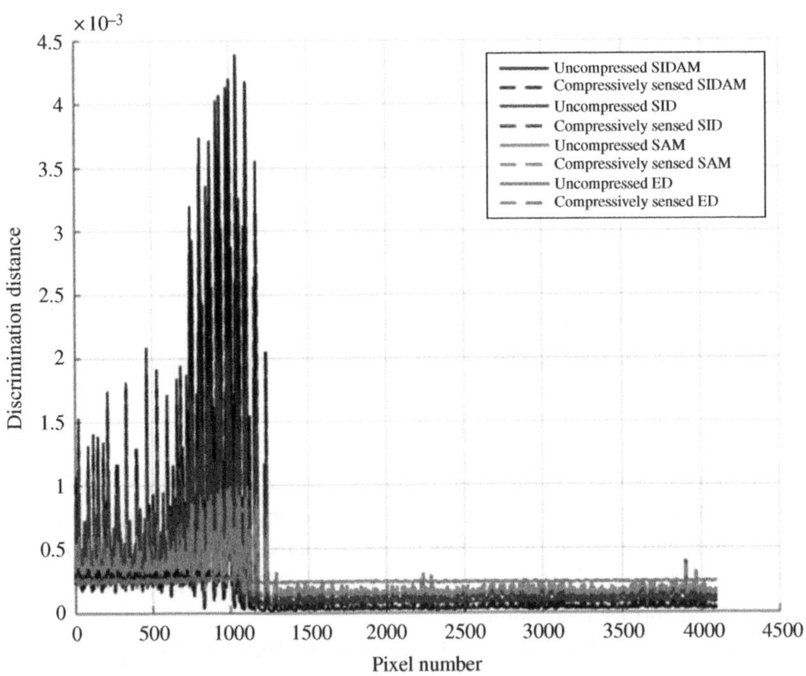


**Figure 7.5** Discrimination values calculated by $D_j(\mathbf{d}, \mathbf{r}_i)$ and $D_j(\mathbf{\Phi d}, \mathbf{\Phi r}_i)$ for $\mathbf{d} = p_2$ and $\{\mathbf{r}_i\}_{i=1}^{4096}$ and $j$ = ED, SAM, SID, SIDAM.

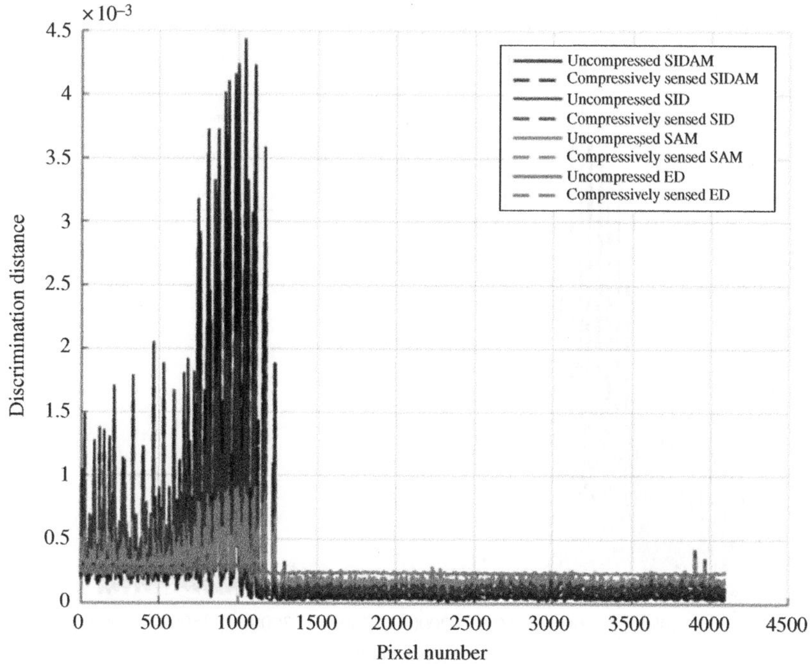


**Figure 7.6** Discrimination values calculated by $D_j(\mathbf{d}, \mathbf{r}_i)$ and $D_j(\mathbf{\Phi d}, \mathbf{\Phi r}_i)$ for $\mathbf{d} = p_3$ and $\{\mathbf{r}_i\}_{i=1}^{4096}$ and $j$ = ED, SAM, SID, SIDAM.

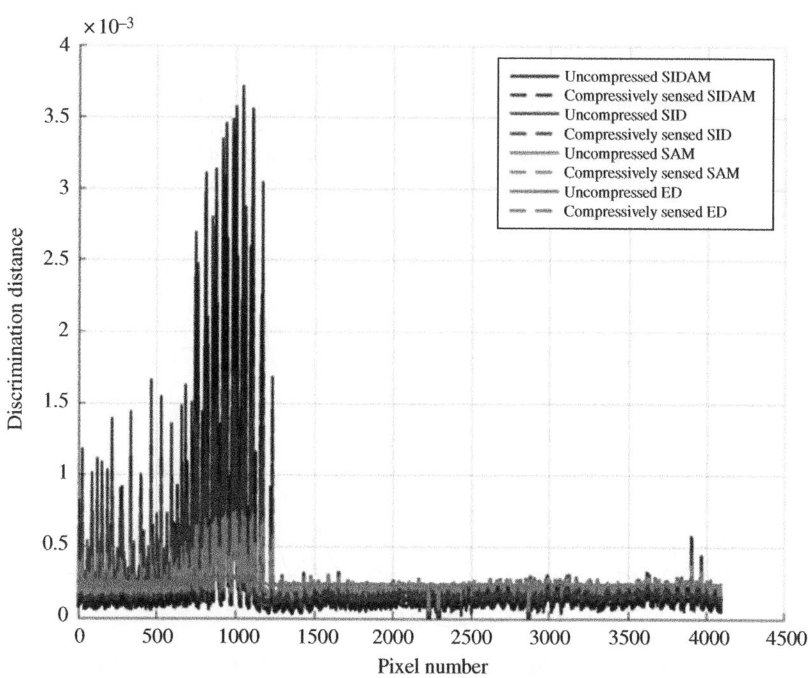


**Figure 7.7** Discrimination values calculated by $D_j(\mathbf{d}, \mathbf{r}_i)$ and $D_j(\mathbf{\Phi d}, \mathbf{\Phi r}_i)$ for $\mathbf{d} = p_4$ and $\{\mathbf{r}_i\}_{i=1}^{4096}$ and $j$ = ED, SAM, SID, SIDAM.

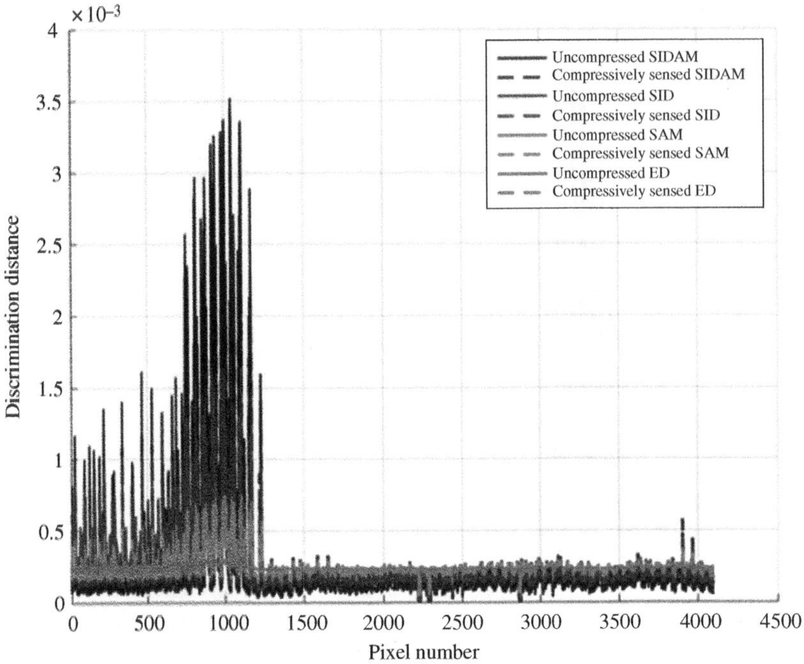


**Figure 7.8** Discrimination values calculated by $D_j(\mathbf{d}, \mathbf{r}_i)$ and $D_j(\mathbf{\Phi d}, \mathbf{\Phi r}_i)$ for $\mathbf{d} = p_5$ and $\{\mathbf{r}_i\}_{i=1}^{4096}$ and $j$ = ED, SAM, SID, SIDAM.

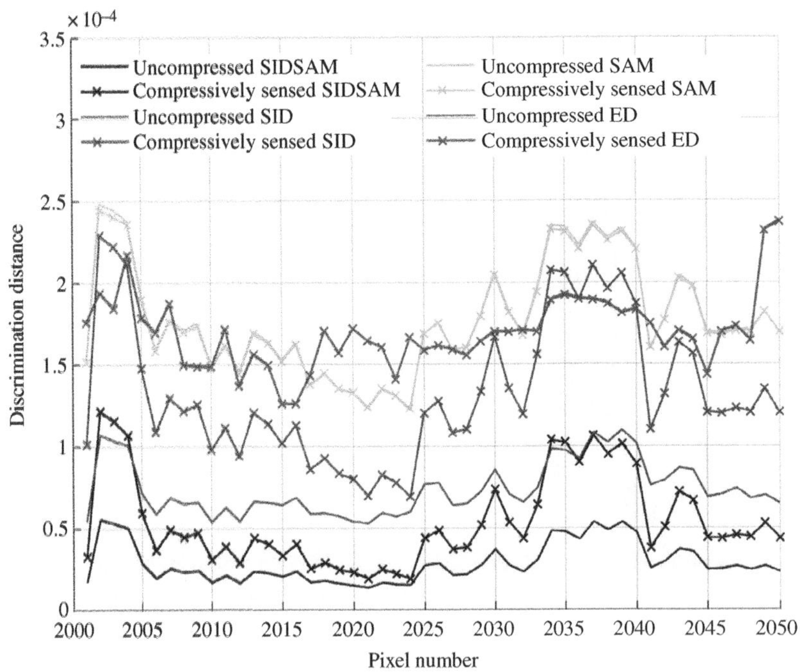


**Figure 7.9** Zoom-in portion of Figure 7.4 for $\mathbf{d} = p_1$ and $\{\mathbf{r}_i\}_{i=2000}^{2050}$.

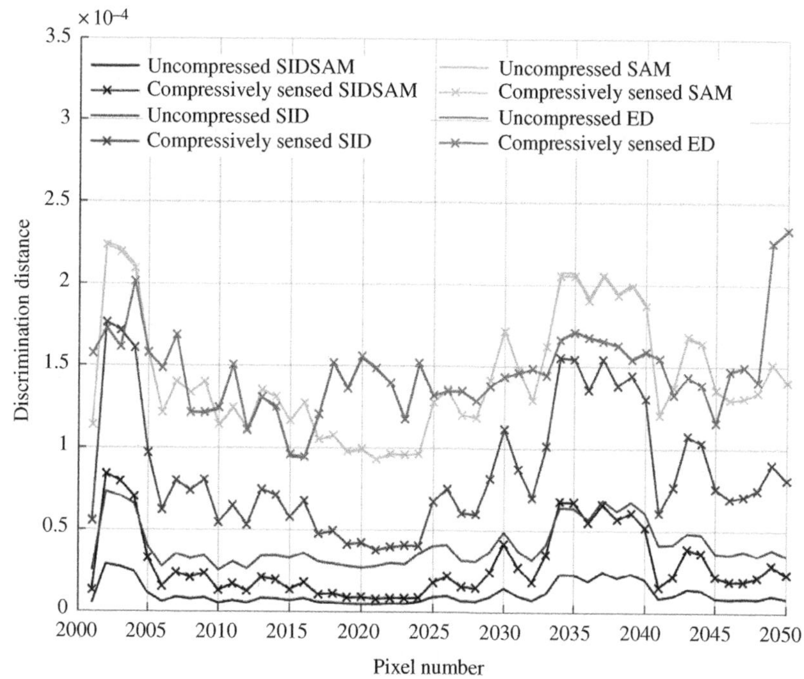


**Figure 7.10** Zoom-in portion of Figure 7.5 for $\mathbf{d} = \mathrm{p}_2$ and $\{\mathbf{r}_i\}_{i=2000}^{2050}$.

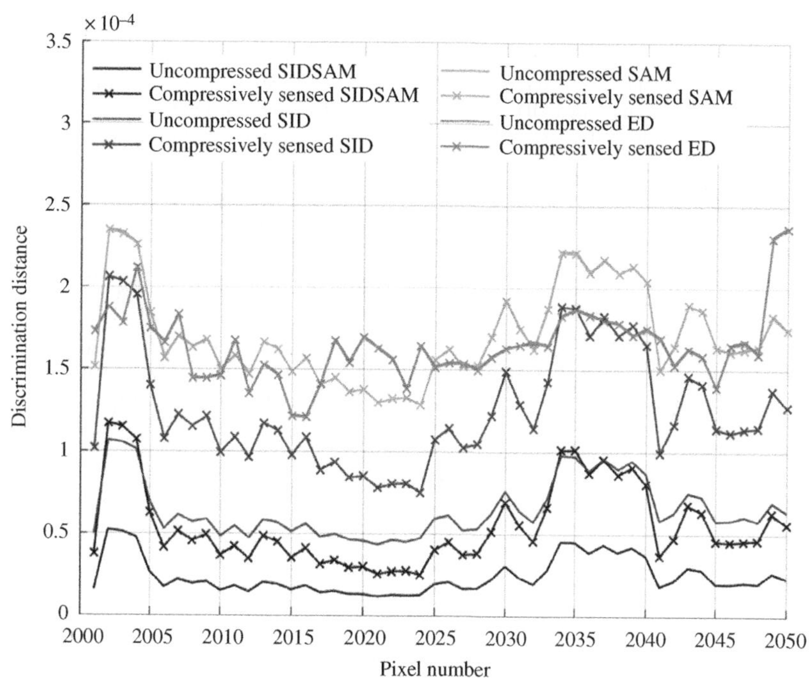


**Figure 7.11** Zoom-in portion of Figure 7.6 for $\mathbf{d} = \mathrm{p}_3$ and $\{\mathbf{r}_i\}_{i=2000}^{2050}$.

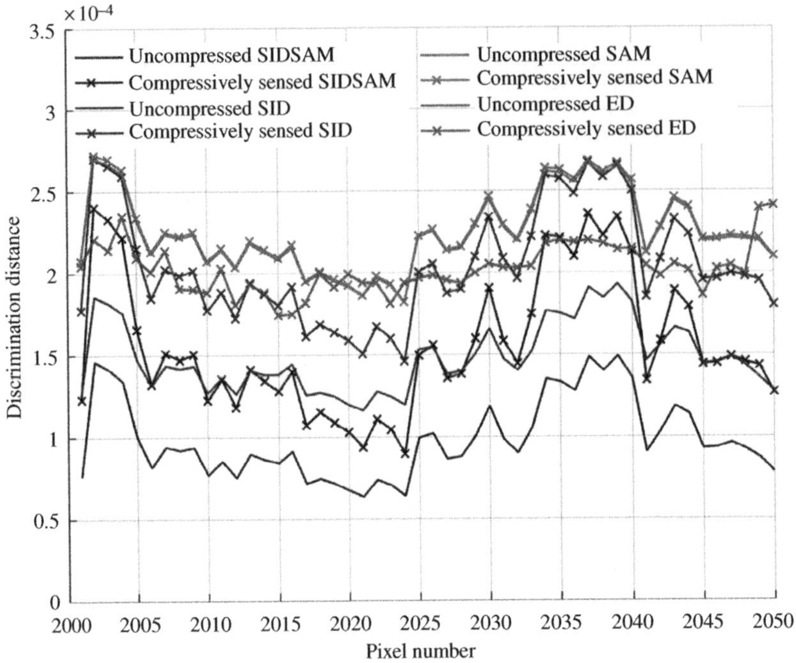


**Figure 7.12** Zoom-in portion of Figure 7.7 for $\mathbf{d} = p_4$ and $\{\mathbf{r}_i\}_{i=2000}^{2050}$.

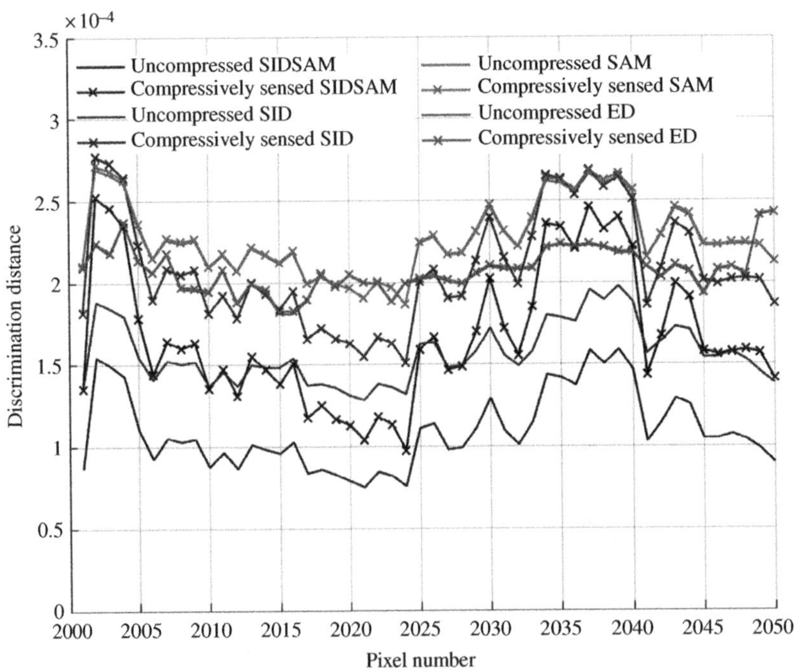


**Figure 7.13** Zoom-in portion of Figure 7.8 for $\mathbf{d} = p_5$ and $\{\mathbf{r}_i\}_{i=2000}^{2050}$.

with the very tight error bound in (7.2), $\delta_k \approx 0$, i.e. $1 - \delta_k \approx 1 + \delta_k$. Similarly, it is also very true for RCP which was persevered by SAM with $\delta_k$ approximating zero but not as good as the $\delta_k \approx 0$ by ED. However, although the $\delta_k$ in the error bound in (7.2) in Figures 7.9–7.13 produced by SID and SIDAM for REP was also very tight, i.e. $\delta_k$ nearly zero, its values were not as good as its counterparts produced by ED and SAM.

To further evaluate the comparisons in Figures 7.4–7.8 quantitatively, we introduce the Pearson correlation coefficient as a criterion to measure the correlation between $\{D_j(\mathbf{d}, \mathbf{r}_i)\}_{i=1}^N$ in the original full band domain and $\{D_j(\mathbf{\Phi d}, \mathbf{\Phi r}_i)\}_{i=1}^N$ in CSBD with $\{\mathbf{r}_i\}_{i=1}^N$ being the total number of data sample vectors. Specifically, let $\eta_i^j = D_j(\mathbf{d}, \mathbf{r}_i)$ and $\xi_i^j = D_j(\mathbf{\Phi d}, \mathbf{\Phi r}_i)$. Then, we can calculate $\mu_\eta^j = \frac{1}{N}\sum_{i=1}^N \eta_i^j$ and $\mu_\xi^j = \frac{1}{N}\sum_{i=1}^N \xi_i^j$ which are the means of $\{\eta_i^j\}_{i=1}^N$ and $\{\xi_i^j\}_{i=1}^N$, respectively, along with their corresponding variances, $\left(\sigma_\eta^j\right)^2 = \frac{1}{N}\sum_{i=1}^N \left(\eta_i^j - \mu_\eta^j\right)^2$ and $\left(\sigma_\xi^j\right)^2 = \frac{1}{N}\sum_{i=1}^N \left(\xi_i^j - \mu_\xi^j\right)^2$. Furthermore, let the covariance between $\{\eta_i^j\}_{i=1}^N$ and $\{\xi_i^j\}_{i=1}^N$ be calculated and denoted by $\mathrm{cov}\left(\eta^j, \xi^j\right) = \frac{1}{N}\sum_{i=1}^N \left(\eta_i^j - \mu_\eta^j\right)\left(\xi_i^j - \mu_\xi^j\right)$. We can now define the Pearson correlation coefficient between $\{\eta_i^j\}_{i=1}^N$ and $\{\xi_i^j\}_{i=1}^N$, denoted by $\rho$, as

$$\rho = \frac{\mathrm{cov}(\eta, \xi)}{\left(\sigma_\eta^j\right)\left(\sigma_\xi^j\right)} \in [-1, 1] \tag{7.34}$$

which indicates the discrepancy between $\{\eta_i^j\}_{i=1}^N$ and $\{\xi_i^j\}_{i=1}^N$. Table 7.1 calculates Pearson correlation coefficient, $\rho$, in (7.34) with $N = 4096$ for $j$ = ED, SAM, SID, and SIDAM and $\mathbf{d} = \mathrm{p}_1, \mathrm{p}_2, \mathrm{p}_3, \mathrm{p}_4, \mathrm{p}_5$, and $\{\mathbf{r}_i\}_{i=1}^{4096}$ where ED was the best followed by SAM, SIDAM, and concluded by SID with the worst. The results in Table 7.1 provide very important insights into the error bounds in (7.2), (7.3), and (7.8). Since RIP and RCP are well preserved by (7.2) and (7.3), it is expected that their error bounds are very tight and close to zero which indicates that the correlation between data samples in the original full band domain and CSBD is also close to 1.

Also, due to more complicated error bounds specified by (7.22), SIDAM and SID suffer from larger error bounds. This fact was also reflected in Figures 7.4–7.8 and Table 7.1 where the Pearson correlation coefficients between data samples in the original full band domain and CSBD might not be as close to one as ED and SAM were, their Pearson correlation correlations were still very high close to 0.95. Interestingly, according to Table 7.1, SIDAM resulted in better correlation than SID did because it included SAM as part of its measure which increased high correlation.

**Table 7.1** $\rho$ in (7.34) for $j$ = ED, SAM, SID, and SIDAM, and $\mathbf{d} = \mathrm{p}_1, \mathrm{p}_2, \mathrm{p}_3, \mathrm{p}_4, \mathrm{p}_5$, and $\{\mathbf{r}_i\}_{i=1}^{4096}$

| Measure **d** | ED | SAM | SID | SIDAM |
|---|---|---|---|---|
| $\mathrm{p}_1$ | 1.0000 | 0.9997 | 0.9779 | 0.9887 |
| $\mathrm{p}_2$ | 0.9998 | 1.0000 | 0.9832 | 0.9908 |
| $\mathrm{p}_3$ | 1.0000 | 0.9998 | 0.9826 | 0.9906 |
| $\mathrm{p}_4$ | 0.9999 | 0.9999 | 0.9503 | 0.9841 |
| $\mathrm{p}_5$ | 1.0000 | 0.9995 | 0.9515 | 0.9828 |

**Experiment 7.2** The second HSI used for the classification experiment is the well-known Pavia Center scene with band 80 shown in Figure 7.14(a) imaged by the Reflective Optics System Imaging Spectrometer (ROSIS) sensor of Pavia University of Northern Italy [34]. The sensor has 115 bands from 430 to 860 nm with spatial resolution of 1.3 and spectral resolution of 4 nm. The image consists of 715 × 1096 pixel vectors at 102 spectral bands after removal of the low SNR water absorption bands. The scene has 9 classes the ground truth shown in Figure 7.14(b) where the spectral classes consist of water, trees, asphalt, brick, bitumen, tiles, shadows, meadows, and soil highlighted by different degrees of gray level intensities. The classifier employed in this experiment utilizes the class means as a descriptor. Each class mean is computed from the image pixels corresponding to the classes the ground-truth image in Figure 7.14(b), where color image and the ground truth color map can be found in [34].

Finding the CSBR, $m$, for this data set is required to perform CS as in (7.1). Using the HFC method referenced earlier as a sparsity estimation method, the sparsity level is found to be $k = 21$ when the $P_{fa} = 0.0001$. Using (7.4), the CSBR is estimated at $m = 67$. Therefore, the classification experiment is run using data in the ODS and CSBD using a CSBR of $m/L = 67/102$.

A nearest centroid classifier is employed based on the class means. Let $M$ be the number of classes in the set $\{C_i\}_{i=1}^{M}$. Each class is a set of HSVs from the HSI such that $\mathbf{r}_i \in C_j$ denotes class membership. The class centroids are computed as the mean of all the pixels in the $j$th class creating a set of class means $\{\boldsymbol{\mu}_i\}_{i=1}^{M}$. Using this notation, class membership of each HSV in $\{\mathbf{r}_i\}_{i=1}^{N}$ is estimated as

$$\mathbf{r}_i \in C_{i^*} \text{ if } i^* = \arg\left\{ \min{}_k D_j(\mathbf{r}_i, \boldsymbol{\mu}_k)\right\} \tag{7.35}$$

While there are more accurate hyperspectral classifiers [13, 35], this simple classifier is useful to demonstrate the preservation of the HSV discrimination functions $D_j = \{\text{ED, SAM, SID, SIDAM}\}$. The evaluation criteria are the overall classification accuracy, the average accuracy per class, and the

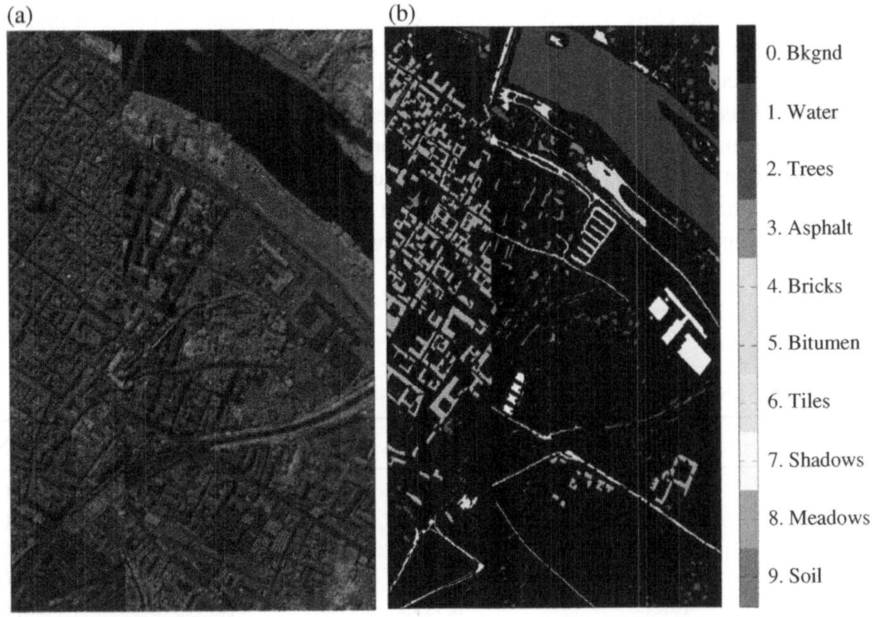


**Figure 7.14** (a) Rendering of band 80 of the Pavia Center scene. (b) Rendering of the ground-truth class labels.

**Table 7.2** Overall classification accuracy and average classification accuracy in ODS and CSBD using $D_j = \{\text{ED, SAM, SID, SIDAM}\}$

| | $D_{\text{ED}}$ | $D_{\text{SAM}}$ | $D_{\text{SID}}$ | $D_{\text{SIDAM}}$ |
|---|---|---|---|---|
| OA - ODS | 0.8217 | 0.9168 | 0.8482 | 0.9229 |
| OA - CSBD | 0.8081 | 0.9193 | 0.8675 | 0.9186 |
| AA - ODS | 0.7708 | 0.7862 | 0.7944 | 0.8393 |
| AA - CSBD | 0.7616 | 0.7904 | 0.7987 | 0.8420 |

**Table 7.3** Individual class accuracy in the ODS and CSBD using $D_j = \{\text{ED, SAM, SID, SIDAM}\}$

| | Water | Trees | Asphalt | Bricks | Bitumen | Tiles | Shadows | Meadows | Soil |
|---|---|---|---|---|---|---|---|---|---|
| $D_{\text{ED}}$ - ODS | 0.9901 | 0.7794 | 0.8754 | 0.7207 | 0.2792 | 0.9364 | 0.7240 | 0.6359 | 0.9965 |
| $D_{\text{ED}}$ - CSBD | 0.9897 | 0.7843 | 0.8761 | 0.7128 | 0.2672 | 0.9303 | 0.7028 | 0.5959 | 0.9958 |
| $D_{\text{SAM}}$ - ODS | 0.9763 | 0.7917 | 0.6838 | 0.8011 | 0.7278 | 0.8908 | 0.8275 | 0.9541 | 0.4230 |
| $D_{\text{SAM}}$ - CSBD | 0.9795 | 0.7934 | 0.7084 | 0.8037 | 0.7345 | 0.8902 | 0.8280 | 0.9549 | 0.4205 |
| $D_{\text{SID}}$ - ODS | 0.9890 | 0.8535 | 0.8961 | 0.7400 | 0.3068 | 0.9363 | 0.7302 | 0.7084 | 0.9892 |
| $D_{\text{SID}}$ - CSBD | 0.9925 | 0.7873 | 0.7314 | 0.7974 | 0.5627 | 0.8795 | 0.7386 | 0.7651 | 0.9340 |
| $D_{\text{SIDAM}}$ - ODS | 0.9888 | 0.8351 | 0.8311 | 0.7899 | 0.4304 | 0.9757 | 0.7765 | 0.9367 | 0.9892 |
| $D_{\text{SIDAM}}$ - CSBD | 0.9943 | 0.8003 | 0.7553 | 0.8466 | 0.6183 | 0.9441 | 0.7987 | 0.9003 | 0.9200 |

individual class accuracies (CA) when using data in ODS and the CSBD. Examining the differences in accuracies in ODS and CSBD demonstrates the preservation properties of RIP, RCP, REP, and RSP in addition to demonstrating utility on an important HSI exploitation algorithm class.

Using the Pavia Center data set, the class overall accuracy (OA) and class average accuracy (AA) in the ODS and CSBD is displayed in Table 7.2. This table depicts comparable classification rates using the ODS and data in the CSBD at the CSBR of 67/102.

In addition, the individual CA are examined to ensure the discrimination metrics are consistent. Table 7.3 displays the individual class accuracy for the Pavia Center data set of nine classes in the ODS and CSBD with CSBR of 67/102.

Tables 7.2 and 7.3 show comparable OA, AA, and CA using data in the ODS and CSBD. While the results may not be identical, they indeed support the derivations in Sections 7.3 and 7.4. In addition, this result also shows the utility of using RIP, RCP, and RSP for hyperspectral discrimination measures defined in Section 7.5.

## 7.7 Conclusion

In this chapter, two new properties of CS for HSI, REP and RSP, are coined and proved, both of which have never been explored in the past. It can be shown that the entropy of an HSV can be preserved in CSBD. With the help of REP, it can be further shown that SID between two HSVs can be also preserved in CSBD. As a result of preserving SID in CSBD, RSP is also validated. Since RSP not only preserves ED between two data samples but also their angle, RSP can be considered as a generalization of RIP and RCP. In other words, what ED and SAM are to RIP and RCP are also applied to what entropy and SID are to REP and RSP, respectively. Finally, experiments conducted based on real hyperspectral images show that all the four properties, RIP, RCP, REP, and RSP, are preserved in CSBD.

## References


1. Donoho, D.L. (2006). Compressed sensing. *IEEE Transactions on Information Theory* 52 (4): 1289–1306.
2. Candes, E.J. and Wakin, M.B. (2008). An introduction to compressive sampling. *IEEE Signal Processing Magazine* 21–30.
3. Baraniuk, R.G. (2007). Compressive sensing. *IEEE Signal Processing Magazine* 116–124.
4. Candes, E. and Romberg, J. (2007). Sparsity and incoherence in compressive sampling. *Inverse Problems* 23 (3): 969–985.
5. Tropp, J.A. and Gilbert, A.C. (2007). Signal recovery from random measurements via OMP. *IEEE Transactions on Information Theory* 53 (12): 4655–4666.
6. Blumensath, T. and Davies, M.E. (2009). Iterative hard thresholding for compressed sensing. *Applied and Computational Harmonic Analysis* 27 (3): 265–274.
7. Chen, S.S., Donoho, D.L., and Saunders, M.A. (2003). Atomic decomposition by basis pursuit. *SIAM Journal on Scientific Computing* 20 (1): 33–61.
8. Cheng, T. (2014). Restricted conformal property of compressive sensing, 2014 11th International Computer Conference on Wavelet Activ Media Technology and Information Processing(ICCWAMTIP), pp. 152–161.
9. Chang, L.H. and Wu, J.Y. (2013). Achievable angles between two compressed sparse vectors under norm/distance constraints imposed by the restricted isometry property: a plane geometry approach. *IEEE Transactions on Information Theory* 59 (4): 2059–2081.
10. Chang, C.-I (2013). Hyperspectral Data Processing: Algorithm Design and Analysis. Hoboken, NJ: Wiley-Interscience.
11. Chang, C.-I (2000). An information theoretic-based approach to spectral variability, similarity and discriminability for hyperspectral image analysis. *IEEE Transactions on Information Theory* 46 (5): 1927–1932.
12. Cover, T.M. and Thomas, J.A. (2006). Elements of Information Theory (Wiley Series in Telecommunications and Signal Processing). New York, NY, USA: Wiley-Interscience.
13. Chang, C.-I (2003). Hyperspectral Imaging: Techniques for Spectral Detection and Classification. US: Springer.
14. Chang, C.-I (2000). An information-theoretic approach to spectral variability, similarity, and discrimination for hyperspectral image analysis. *IEEE Transactions on Information Theory* 46 (5): 1927–1932.
15. Lopes, M.E. (2013). Estimating unknown sparsity in compressed sensing. *Proceedings of the 30th International Conference on Machine Learning, PMLR* 28 (3): 217–225.
16. Hurley, N. and Rickard, S. (2009). Comparing measures of sparsity. *IEEE Transactions on Information Theory* 55 (10): 4723–4741.
17. Haupt, J. and Nowak, R. (2007). A generalized restricted isometry property. *University of Wisconsin-Madison, Technology Representative* 1–16. https://citeseerx.ist.psu.edu/viewdoc/download?doi=10.1.1.536.3662&rep=rep1&type=pdf.
18. Baraniuk, R.G. (2007). Compressive sensing [lecture notes]. *IEEE Signal Processing Magazine* 24 (4): 118–121.
19. Foucart, S. and Rauhut, H. (2013). A Mathematical Introduction to Compressive Sensing. New York, NY: Springer.
20. Lopes, M.E. (2013). Estimating unknown sparsity in compressed sensing. *Proceedings of the 30th International Conference on International Conference on Machine Learning* 28 (ICML'13): 217–225.
21. Lopes, M.E. (2016). Unknown sparsity in compressed sensing: Denoising and inference. *IEEE Transactions on Information Theory* 62 (9): 5145–5166.
22. Harsanyi, J.C. (1993). Detection and classification of subpixel spectral signatures in hyperspectral image sequences. University of Maryland Baltimore County.
23. Chang, C.-I and Du, Q. (2004). Estimation of number of spectrally distinct signal sources in hyperspectral imagery. *IEEE Transactions on Geoscience and Remote Sensing* 42 (3): 608–619.
24. Chang, C.-I (2018). A review of virtual dimensionality for hyperspectral imagery. *IEEE Journal of Selected Topics in Applied Earth Observation and Remote Sensing* 11 (4): 1285–1305.

25. Lee, L.C., Ouyang, Y.C., Chen, S.Y., and Chang, C.-I (2016). An information theoretical approach to multiple-band selection for hyperspectral imagery. 2016 IEEE Geoscience and Remote Sensing Symposium, Beijing, China (10–15 July 2016).
26. Manolakis, D., Siracusa, C., and Shaw, G. (2001). Hyperspectral subpixel target detection using the linear mixing model. *IEEE Transactions on Geoscience and Remote Sensing* 39 (7): 1392–1409.
27. Chang, C.-I (2018). A review of virtual dimensionality for hyperspectral imagery. *IEEE Journal of Selected Topics in Applied Earth Observations and Remote Sensing* 11 (4): 1285–1305.
28. Baraniuk, R., Davenport, M., DeVore, R., and Wakin, M. (2008). A simple proof of the restricted isometry property for random matrices. *Constructive Approximation* 28 (3): 253–263.
29. Chang, C.-I (2016). Real-Time Progressive Hyperspectral Image Processing. New York, NY: Springer Berlin Heidelberg.
30. Chang, C.-I, Lee, L.C., Xue, B. et al. (2017). Channel capacity approach to hyperspectral band subset selection. *IEEE Journal of Selected Topics in Applied Earth Observations and Remote Sensing* 10 (10): 4630–4644.
31. Chang, C.-I (2018). Spectral inter-band discrimination capacity of hyperspectral imagery. *IEEE Transactions on Geoscience and Remote Sensing* 56 (3): 1749–1766.
32. Chang, C.-I (2004). New hyperspectral discrimination measure for spectral characterization. *Optical Engineering* 43 (8): 1777.
33. Harsanyi, J.C., Farrand, W., and Chang, C.-I (1994). Detection of subpixel spectral signatures in hyperspectral image sequences. *Proceedings of American Congress on Surveying & Mapping (ACSM)/American Society of Photogrammetry & Remote Sensing (ASPRS) Annual Convention and Exposition*, Baltimore 1: 236–247.
34. http://www.ehu.eus/ccwintco/index.php/Hyperspectral_Remote_Sensing_Scenes.
35. Kang, X., Li, S., and Benediktsson, J.A. (2014). Spectral – spatial hyperspectral image classification with edge-preserving filtering. *IEEE Transactions on Geoscience and Remote Sensing* 52 (5): 2666–2677.

# 8

# Endmember Finding in Compressively Sensed Band Domain

Chein-I Chang[1,2] and Adam Bekit[1]
[1]*Remote Sensing Signal and Image Processing Laboratory, Department of Computer Science and Electrical Engineering, University of Maryland Baltimore County, Baltimore MD, USA*
[2]*Center for Hyperspectral Imaging in Remote Sensing (CHIRS), Information and Technology College, Dalian Maritime University, Dalian, China*

## 8.1 Introduction

Endmembers are pure signatures generally used to identify spectral classes in hyperspectral imagery [1, 2]. As a result, it is a fundamental task prior to spectral unmixing. Unfortunately, finding endmembers is a very challenging since it must be carried out in an unsupervised manner due to the fact that endmembers are generally unknown and relatively small targets which cannot be identified by visual inspection or prior knowledge. Despite that many approaches have been developed and reported in the literature, there are several issues. One is how to determine the number of endmembers to be found, $p$, which is generally much smaller than the total number of spectral bands, $L$. Fortunately, a recently developed concept, virtual dimensionality (VD) [3], can be used for this purpose. Once the $p$ is determined, a following issue is finding a desired set of $p$ endmembers. One of most widely used approaches is simplex volume (SV)-based methods, which considers $p$ endmembers as vertices to form a $p$-vertex simplex and makes use of its SV as a criterion for finding a desired set of $p$ endmembers. There are two SV-based methods. One is maximal SV-based methods which inflate $p$-vertex simplexes from inside out the data space and find a $p$-vertex simplex with the maximal volume that is embedded in the data space. The most celebrated algorithm is the N-finder algorithm (N-FINDR) developed by Winter [4]. The other is minimum SV-based methods which deflate $p$-vertex simplexes from outside in the data space to find a $p$-vertex simplex with the minimal SV that embraces all data sample vectors. The key difference between these two is that the $p$ endmembers found by using the minimal SV are not necessarily in the data space as opposed to all $p$ endmembers found by the maximal SV which are indeed real data samples in the data space. For the minimal SV found endmembers, some of them may be very likely to be noisy samples or interferers rather than

*Advances in Hyperspectral Image Processing Techniques*, First Edition. Edited by Chein-I Chang.

true endmembers. Because of this reason, N-FINDR is selected in this chapter in preference to minimal SV-based endmember finding (EF) algorithms. For the purpose of simplicity and clarification, in discussions, the notations of $p$ and $N$ are used specifically to indicate the number of endmembers to be found and total number of data sample vectors, respectively, throughout this chapter. Also, N-FINDR is generally designed for EF with $N$ interpreted as $N$ endmembers. In this case, to avoid confusion, we use $p$-FINDR instead of N-FINDR by replacing "$N$" with "$p$" to emphasize the fact of finding $p$ endmembers rather than $N$ endmembers to distinguish $N$ being used for the total number of data sample vectors in this chapter.

Unfortunately, there are some practical issues in implementing N-FINDR. First and foremost is finding all $p$ endmembers altogether at the same time which requires $C_p^N = N!/(p!(N-p)!)$ $p$-combinations out of the total number of data samples in the data space, $N$. Realistically speaking, this exhaustive search is not feasible. To resolve this issue, two approaches have been developed. Instead of finding all $p$ endmembers simultaneously, an alternative way is to find $p$ endmembers one after another. This gives rise to a well-known EF algorithm, called simplex growing algorithm (SGA) [5]. Another way is to develop a numerical approach for searching potential $p$ endmembers among all possible $p$-combinations. This results in two algorithms developed in [6], called sequential N-FINDR (SQ N-FINDR) and successive N-FINDR (SC N-FINDR). Since a $p$-vertex simplex can be embedded in a ($p$–1)-dimensional data space, a second issue arising in EF is spectral dimensionality reduction (DR) which is generally required such as singular value decomposition (SVD) prior to its SV calculation. However, it has been shown in [5, 7] the techniques used for DR have significant impact on finding endmembers. As a consequence, various DR techniques [1] may lead to finding different sets of $p$ endmembers even if the same EF such as N-FINDR is used. Interestingly, very little work has been devoted to such an issue. This chapter deviates the conventional wisdom to develop a completely different approach using a recently developed concept, compressive sensing (CS) to finding $p$ endmembers by extending N-FINDR in compressively sensed band domain (CSBD) without DR.

A CS system is characterized by two matrices, sparse matrix $\mathbf{\Psi}$ and sensing matrix $\mathbf{\Phi}$. The novelty of the proposed CSBD-based $p$-FINDR is to explore applicability of the universality model [8] to N-FINDR in finding endmembers. Specifically, the universal model states that the sparse basis matrix, $\mathbf{\Psi}$, need not be known at the time of signal acquisition, but it is only required during sparse recovery. Since N-FINDR does not require signal reconstruction, the universality model allows N-FINDR to perform EF in CSBD via a sensing matrix $\mathbf{\Phi}$ without specifying the sparse matrix $\mathbf{\Psi}$. Most importantly, this approach avoids the need of DR to calculate the SV as required by the traditional $p$-FINDR. Furthermore, another novelty is to derive a new property, called restricted simplex volume property (RSVP), to ensure that SV calculated by $p$-FINDR in the original data space (ODS) is preserved in CSBD. Therefore, several great benefits can be gained from CS. First of all, there is no need of specifying techniques to perform DR, which is required for N-FINDR since DR can be simply accomplished by a sensing matrix $\mathbf{\Phi}$ randomly generated via CS. Second, N-FINDR can be performed in CSBD using the number of compressively sensed bands (CSBs), $p$. In other words, N-FINDR is carried out using $p$ CSBs that can be randomly sensed by $\mathbf{\Phi}$. Third, in order to further resolve the issue of using an exhaustive search for $p$ endmembers, SQ/SC N-FINDR can be used to replace N-FINDR in ODS to yield CSBD-based SQ/SC N-FINDR in CSBD. Finally and most importantly, SVD and matrix determinant generally used to calculate the $p$-vertex SV can be shown to be preserved in CSBD by the sensing matrix $\mathbf{\Phi}$. The new $\mathbf{\Phi}$-derived RSVP allows N-FINDR or SQ/SC N-FINDR to find nearly the same $p$ endmembers in both ODS and CSBD.

## 8.2 Compressive Hyperspectral Band Sensing

CS is a data acquisition and sampling technique which allows signals to be acquired by fewer samples than samples required by the conventional Nyquist–Shannon sampling rate. Since hyperspectral imaging sensors make use of hundreds of $L$ contiguous spectral bands to acquire data, a hyperspectral image generally contains a wealth of spectral information which results in very high correlation among bands. This is particularly true for SV-based EF where the dimensionality of a simplex with $p$ vertices formed by endmembers is $p$–1 which is much smaller than $L$.

Suppose that there are $p$ endmembers. A $p$-vertex simplex with all of its vertices specified by endmembers generally yields the maximal SV among all possible $p$-vertex simplexes. In addition, a $p$-vertex simplex with maximal volume is embedded in a ($p$–1)-dimensional data space. As a result, all the $p$ endmembers can be actually found by a band dimensionality-reduced ($p$–1)-dimensional band space. So, why do we need full bands, many of which will end up not being used eventually? One common approach to dealing with this issue is to develop data compression algorithms such as principal components analysis (PCA) in [9], independent component analysis (ICA) [10], which compress ODS into a $p$-dimensional data space before applying an EF algorithm. The problem with this approach is that the each spectral band information is distorted and compromised. The endmembers found from this data-compressed space may not be true endmembers. This chapter takes a rather different viewpoint to address this issue. That is, an EF algorithm is applied in CSBD using $p$ CSBs via CS, while a hyperspectral image is being sensed at the same time during acquisition. In other words, a CS-based EF algorithm is used to find all endmembers from a hyperspectral image cube using only a minimum number of sensed bands, i.e. $p$ CSBs without performing DR.

### *8.2.1 Compressive Sensing Framework*

In this section, we present a framework completely different from many existing sparse-based hyperspectral imaging methods which consider image pixels to be compressed sparsely. Instead, the proposed work considers the spectral bands of pixel vectors to be sparsely sampled from the $L$ original bands in ODS to $m$ linear mixtures of spectral bands as CSBs in CSBD.

More specifically, in the sparse acquisition process, an $L \times 1$ dimensional signal of interest $\mathbf{r}$ and an $m \times 1$ dimensional sparsely sampled signal $\mathbf{y}$ can be expressed mathematically as follows:

$$\mathbf{y} = \boldsymbol{\Phi}\boldsymbol{\Psi}\mathbf{r} + \mathbf{n} \tag{8.1}$$

where $\boldsymbol{\psi}$ is an $L \times L$ sparse transformation matrix, $\boldsymbol{\Phi}$ is an $m \times L$ sampling matrix, and $\mathbf{n}$ is an $L \times 1$ error vector. The sampling strategy in CS is determined by two key principles: sparsity and incoherence [11–13]. The signal of interest, $\mathbf{r}$, must be sparse that can be expressed as in an appropriate sparse matrix, $\boldsymbol{\psi}$, and must be also sampled in a sensing matrix, $\boldsymbol{\Phi}$, that is maximally incoherent with $\boldsymbol{\psi}$. Since the CS strategy is implemented without prior knowledge of signals to be processed, it ought to rely on probabilistic convergence guaranteed under specific conditions.

The restricted isometry property (RIP) [8] is a very important property that is typically imposed in CS to guarantee that certain distance measures are maintained and preserved in both ODS and CSBD. For a signal $\mathbf{r}$ with a sparsity order of $k$ determined by $\boldsymbol{\psi}$, the RIP condition is defined mathematically by

$$(1-\delta_k)\|\mathbf{r}\|_2^2 \leq \|\boldsymbol{\Phi}\mathbf{r}\|_2^2 \leq \|\mathbf{r}\|_2^2(1+\delta_k) \tag{8.2}$$

where $\delta_k$ is defined as the restricted isometry constant (RIC) and is the smallest possible constant that satisfies (8.2) for all $k$-sparse vectors in $\mathbf{r}$ [14]. As the dimensionality of the compressed vector, $m$, approaches the dimensionality of the original signal, $L$, the RIC approaches zero resulting in an almost exact preservation of lengths.

Under mild conditions, the original signal acquired in (8.1) can later be recovered by a sparse recovery process which solves the following optimization problem [14–17]:

$$\min_{\tilde{\mathbf{r}} \in \Re^L} \left\| \tilde{\mathbf{r}}_S \right\| \text{ subject to } \mathbf{\Phi} \tilde{\mathbf{r}}_S = \mathbf{y} \tag{8.3}$$

where $\tilde{\mathbf{r}}_s$ represents the sparse signal that is recovered from the compressed measurement in (8.1). It is very important to note that for a solution to be feasible, the original vector $\mathbf{r}$ must be sparse or compressible (approximately sparse). The complete CS model is fully described by the sparse acquisition model shown in (8.1) and the sparse recovery solved by (8.3).

### *8.2.2 Compressive Sensing of Hyperspectral Bands*

As described by (8.1), a CS system is completely characterized by two matrices, sparse matrix $\mathbf{\Psi}$ and sensing matrix $\mathbf{\Phi}$. One of greatest challenges in designing a CS system is to identify the domain in which the signal of interest is optimally sparse. Previous works on signal sparsity have suggested the use of the discrete cosine transform (DCT) [18] wavelet transforms [19] and dictionary learning approaches [20]. Fortunately, a random sampling strategy has been shown to be sufficient, if not optimal, for most signals. In other words, certain random distributions can be shown to satisfy both RIP and incoherent sampling with respect to the sparse representation, $\mathbf{\Psi}$. That is, for most practical systems, a random matrix is indeed a nearly optimal choice regardless of the sparse representation by $\mathbf{\Psi}$. This is critical for designing real-world systems as it makes the sampling system agnostic to the sensing modality and is exactly the main reason that random sampling strategies have become ubiquitous in the CS literature. A compressive sampling system proposed in this chapter takes advantage of this fact by subsampling true hyperspectral bands via the linear acquisition matrix, i.e. sensing matrix $\mathbf{\Phi}$ in (8.1). In this case, $m$ $\mathbf{\Phi}$-sensed bands can be obtained through the original $L$ spectral bands with $m \ll L$. Especially, the sparse acquisition model in (8.1) can be implemented by constructing the sampling matrix from any distribution that satisfies the concentration measure of inequalities [8]. For such a family of sampling matrices, it can be shown that RIP is valid if

$$m \geq C \cdot k \cdot \log(L/k) \tag{8.4}$$

is satisfied where $C$ is a constant that depends on each particular instance.

The relationship between the number of CSBs, $m$, and the true sparsity order of $k$ is clearly described by (8.4). It is important to note that for most applications in hyperspectral imaging, a user-selected sparsity order of $k$ is imposed directly. Specifically, let the lower bound of $m$ in (8.4) be denoted by

$$g(k) = C \cdot k \cdot \log(L/k) \tag{8.5}$$

Then, differentiating $g(k)$ with respect to $k$ yields

$$\frac{dg(k)}{dk} = C \cdot \log(L/k) - C = C \cdot (\log(L/k) - 1) > 0 \tag{8.6}$$

which shows that $g(k)$ is a monotonically increasing function of $k$ if

$$\log(L/k) > 1 \Leftrightarrow L/k > e \Leftrightarrow k < L/e \tag{8.7}$$

where $e \approx 2.71828$ is the base of the natural logarithm. Otherwise, $g(k)$ in (8.5) is monotonically decreasing if $2k > L$. So,

$$\left.\frac{dg(k)}{dk}\right|_{k^*} = 0 \Rightarrow k^* = L/e \tag{8.8}$$

which derives the maximal value of $k$, $k^* = L/e$. The implication of (8.6) and (8.7) indicates that the curve of (8.5) is actually convex cap. When $k < L/e$, the lower bound on $m$ is increasing as $k \to L/e$ and is decreasing as $k > L/e$ and $k \to Lk \to L$. In this case, we can choose $m$ to be the lower bound, i.e. $m = g(k)$ even if we do not know what the value of $k$ is.

This chapter takes a rather different perspective by looking into the sensing matrix to ensure that a sufficient number of CSBs, $m$, are collected. In doing so, the Gaussian distribution, $\mathbf{\Phi}$, is specifically chosen, where each element is distributed as $\phi_{ij} \sim N(0, 1/L)$ and the columns are normalized.

### 8.2.3 Universality Model

This subsection presents a very important property which is the key idea in our proposed CS approach to EF. It is universality [8], which states that the sparse matrix, $\mathbf{\Psi}$, need not be known at the time of signal acquisition, but it is only needed during sparse recovery. Specifically, by adopting a random sampling strategy, the sparse acquisition model in (8.1) can be simplified to

$$\mathbf{y} = \mathbf{\Phi}\mathbf{r} + \mathbf{n} \tag{8.9}$$

where $\mathbf{\Phi}$ represents a random sampling matrix. Furthermore, the constraint of the sparse recovery objective in (8.3) can be modified by including the sparse matrix $\mathbf{\Psi}$ to yield

$$\min_{\tilde{\mathbf{r}}_{\mathbf{S}} \in \Re^L} \|\tilde{\mathbf{r}}_{\mathbf{S}}\|_{l_1} \text{ subject to } \mathbf{\Phi}\mathbf{\Psi}\tilde{\mathbf{r}}_{\mathbf{S}} = \mathbf{y} \tag{8.10}$$

By combining (8.9) and (8.10), a new universal CS model is established to show that the knowledge of the sparse matrix is only required for the signal reconstruction process. This has an extremely important implication that for algorithms performed in the compressively sensed domain, an optimal performance can be actually achieved with no need of identifying the sparse matrix $\mathbf{\Psi}$.

## 8.3 Simplex Volume Calculation

In this section, we present two different but equivalent methods of calculating the volume of a $p$-vertex simplex. In ODS, if we let $p$ linearly independent endmember column vectors $\mathbf{m}_1$, $\mathbf{m}_2$, …, $\mathbf{m}_p$ make up an endmember matrix $\mathbf{M} \in \Re^{L \times p}$, then the $p$ columns of $\mathbf{M}$ will represent $p$ vertexes enclosing a $p$-vertex and vol($\mathbf{M}$) is its SV calculated by $\mathbf{M}$. With this in mind, the two SV methods described below can be used to find desirable $p$ endmembers in a hyperspectral image using SQ $p$-FINDR and SC $p$-FINDR described in Section 8.5, which can be used to replace the traditional N-FINDR.

**Table 8.1** Simplex volumes calculated by using SVD and determinant methods with random matrices

| | Log$_{10}$ [vol(**M**)] | | | | | | | | | |
|---|---|---|---|---|---|---|---|---|---|---|
| $L \times p$ | $152 \times 4$ | $199 \times 43$ | $180 \times 15$ | $92 \times 16$ | $148 \times 35$ | $64 \times 3$ | $153 \times 32$ | $117 \times 29$ | $62 \times 31$ | $134 \times 35$ |
| vol(**M**)$_{\text{SVD}}$ | 13.1519 | 144.1464 | 49.9970 | 46.1666 | 110.5461 | 8.0313 | 102.0655 | 87.3074 | 79.5317 | 108.2173 |
| vol(**M**)$_{\text{DET}}$ | 13.1519 | 144.1464 | 49.9970 | 46.1666 | 110.5461 | 8.0313 | 102.0655 | 87.3074 | 79.5317 | 108.2173 |

### *8.3.1 Simplex Volume via Singular Value Decomposition*

The first method used to calculate the SV is one using the well-known matrix decomposition technique, SVD. The endmember matrix is first decomposed as $\mathbf{M} = \mathbf{U\Sigma V}^T$ with $\mathbf{U} \in \Re^{L \times L}$ and $\mathbf{V} \in \Re^{p \times p}$ containing columns representing the left-singular vectors and the right-singular vectors of **M**, respectively. The $p$ nonzero diagonal entries of $\mathbf{\Sigma} \in \Re^{L \times p}$ represent the singular values **M**, where $\sigma_i$ is the $i$th singular value of **M**. The SV is obtained simply by

$$\text{vol}(\mathbf{M})_{\text{SVD}} = \frac{\prod_{i=1}^{p} \sigma_i}{(p-1)!} \tag{8.11}$$

### *8.3.2 Simplex Volume via Matrix Determinant*

A second method is to use the square root of the absolute value of the determinant (DET) of the so-called Gram matrix $\mathbf{M}^T\mathbf{M} \in \Re^{p \times p}$, which is a square, positive-definite, and symmetric matrix. Then, the SV specified by the endmember matrix **M** can be calculated by

$$\text{vol}(\mathbf{M})_{\text{DET}} = \frac{\sqrt{|\det(\mathbf{MM}^T)|}}{(p-1)!} \tag{8.12}$$

Considering $\prod_{i=1}^{p} \lambda_i = \det(\mathbf{MM}^T)$ where $\lambda_I$ represents the $i$th eigenvalue of the Gram matrix, $\mathbf{MM}^T$, then the term $\sqrt{|\det(\mathbf{MM}^T)|}$ in (8.12) is equivalent to $\prod_{i=1}^{p} \sigma_i$ in (8.11). As a result,

$$\text{vol}(\mathbf{M})_{\text{SVD}} = \text{vol}(\mathbf{M})_{\text{DET}} \tag{8.13}$$

Table 8.1 tabulates the SVs of 10 matrices with size $L \times p$ containing $p$ randomly generated and linearly independent column vectors calculated by (8.11)–(8.12). From this table, the two methods yield exactly the same SV.

Now, the main question remaining is that does CS preserve SVs by post-acquisition (sensing) in CSBD and allow EF algorithms find desired endmembers? This issue will be addressed in the following section.

## 8.4 Restricted Simplex Volume Property

In this section, a new property called RSVP is presented, which is directly derived from the RIP property [8]. It is this property that allows us to find the SV in CSBD upper and lower bounded by the SV in ODS with a negligible tolerance. More importantly, we will show that the lower and upper bounds imposed by RSVP are equally applicable to two SV calculation methods, SVD and determinant. In addition, RSVP also allows SQ $p$-FINDR and SC $p$-FINDR to find nearly the same endmembers in both ODS and CSBD.

Before we derive RSVP, we first need to understand how $\mathbf{\Phi}$ behaves as $m \rightarrow L$ and how it impacts SV calculation in CSBD. In particular, we are interested in knowing what happens to $\mathbf{\Phi}^T\mathbf{\Phi} \rightarrow \mathbf{I}$ as $m \rightarrow L$. As easily shown by the proof given in [8], $\mathbf{\Phi}^T\mathbf{\Phi} \rightarrow \mathbf{I}$, which is the identity matrix, as $m \rightarrow L$. Also, it is important to mention that the aforementioned methods of calculating SVs are directly applicable in CSBD via $\mathbf{\Phi}$. As a result, both SQ $p$-FINDR and SC $p$-FINDR can utilize both SV calculation methods by SVD and determinant in CSBD to find the desired $p$ endmembers in a hyperspectral image. To show this, we denote the p endmember vectors in CSBD by $\mathbf{M}_{\mathbf{\Phi}_1}, \mathbf{M}_{\mathbf{\Phi}_2}, \ldots, \mathbf{M}_{\mathbf{\Phi}_p}$ and its formed endmember matrix by $\mathbf{M}_{\mathbf{\Phi}} \in \Re^{m \times p}$. Finally, the corresponding SV enclosed by a $p$-vertex simplex in CSBD is then denoted by $\text{vol}(\mathbf{M}_{\mathbf{\Phi}})$. Therefore, the endmember matrix $\mathbf{M}_{\mathbf{\Phi}} \in \Re^{L \times p}$ in ODS is replaced by $\mathbf{M}_{\mathbf{\Phi}} \in \Re^{m \times p}$, which contains the $p$ corresponding compressively sensed endmembers. That is, $\mathbf{M}_{\mathbf{\Phi}} = \mathbf{\Phi}\mathbf{M}$. This simple relationship between $\mathbf{M}_{\mathbf{\Phi}}$ and $\mathbf{M}$ allows us to rewrite (8.11) in CSBD as

$$\text{vol}(\mathbf{M}_{\mathbf{\Phi}})_{\text{SVD}} = \frac{\prod_{i=1}^{p} \sigma_{\mathbf{\Phi}_i}}{(p-1)!} \tag{8.14}$$

where $\sigma_{\mathbf{\Phi}_i}$ is now the $i$th singular value of $\mathbf{M}_{\mathbf{\Phi}}$ and $\sigma_{\mathbf{\Phi}_i} = \sqrt{\lambda_{\mathbf{\Phi}_i}}$, which is the square root of the $i$th eigenvalue of the Gram matrix $\mathbf{M}_{\mathbf{\Phi}}^T\mathbf{M}_{\mathbf{\Phi}} = \mathbf{M}^T\mathbf{\Phi}^T\mathbf{\Phi}\mathbf{M}$. However, using the fact that $\mathbf{\Phi}^T\mathbf{\Phi} \rightarrow \mathbf{I}$ as $m \rightarrow L$, $\mathbf{M}_{\mathbf{\Phi}}^T\mathbf{M}_{\mathbf{\Phi}} \cong \mathbf{M}^T\mathbf{M}$. This suggests that $\text{vol}(\mathbf{M}_{\mathbf{\Phi}})_{\text{SVD}} \cong \text{vol}(\mathbf{M}_{\mathbf{\Phi}})_{\text{DET}}$ as $m \rightarrow L$. It worth mentioning that as $m \rightarrow L$, $\mathbf{\Phi}$ will have less impact on the SV in CSBD.

Again, using the direct relationship $\mathbf{M}_{\mathbf{\Phi}} = \mathbf{\Phi}\mathbf{M}$, we can also rewrite (8.12) as

$$\text{vol}(\mathbf{M}_{\mathbf{\Phi}})_{\text{DET}} = \frac{\sqrt{\left|\det\left(\mathbf{M}_{\mathbf{\Phi}}^T\mathbf{M}_{\mathbf{\Phi}}\right)\right|}}{(p-1)!} \tag{8.15}$$

Remembering the simple fact $\prod_{i=1}^{p} \lambda_{\mathbf{\Phi}_i} = \det\left(\mathbf{M}_{\mathbf{\Phi}}^T\mathbf{M}_{\mathbf{\Phi}}\right)$, which is also equivalent to $\det(\mathbf{M}^T\mathbf{\Phi}^T\mathbf{\Phi}\mathbf{M})$. If we let $m \rightarrow L$, however, $\det(\mathbf{M}^T\mathbf{\Phi}^T\mathbf{\Phi}\mathbf{M}) \cong \det(\mathbf{M}^T\mathbf{M})$. Consequently, $\text{vol}(\mathbf{M})_{\text{DET}} = \text{vol}(\mathbf{M})_{\text{SVD}}$ as $m \rightarrow L$.

So far, we have shown that $\text{vol}(\mathbf{M}_{\mathbf{\Phi}}) \cong \text{vol}(\mathbf{M})$ as $m \rightarrow L$. The remaining issue needs to be addressed is what are the lower and upper bounds imposed by RSVP on $\text{vol}(\mathbf{M}_{\mathbf{\Phi}})$?

The bounds imposed by RSVP on $\text{vol}(\mathbf{M}_{\mathbf{\Phi}})_{\text{SVD}}$ and $\text{vol}(\mathbf{M}_{\mathbf{\Phi}})_{\text{DET}}$ are shown by deriving the relationship between the eigenvalues of $\mathbf{M}^T\mathbf{M}$ and $\mathbf{M}_{\mathbf{\Phi}}^T\mathbf{M}_{\mathbf{\Phi}}$, respectively. We can easily write the eigenvalue equation for $\mathbf{M}^T\mathbf{M}$ as

$$\mathbf{M}^T\mathbf{M}\bar{\mathbf{v}} = \lambda\bar{\mathbf{v}} \tag{8.16}$$

where $\lambda$ is an eigenvalue with its associated normalized eigenvector defined by $\bar{\mathbf{v}} = \frac{\mathbf{v}}{\|\mathbf{v}\|_2}$. Taking the norm and squaring both sides of (8.16) yields in ODS

$$\|\mathbf{M}^T\mathbf{M}\bar{\mathbf{v}}\|_2^3 = \lambda^2\|\bar{\mathbf{v}}\|_2^2 \tag{8.17}$$

and the eigenvalue equation for $\mathbf{M}_{\mathbf{\Phi}}^T\mathbf{M}_{\mathbf{\Phi}}$ as

$$\|\mathbf{M}^T\mathbf{\Phi}^T\mathbf{\Phi}\mathbf{M}\bar{\mathbf{v}}_{\mathbf{\Phi}}\|_2^2 = \lambda_{\mathbf{\Phi}}^2\|\bar{\mathbf{v}}_{\mathbf{\Phi}}\|_2^2 \tag{8.18}$$

Using the RIP's upper bound condition, we can relate (8.17) and (8.18) by

$$\left\|\mathbf{M}^T\mathbf{\Phi}^T\mathbf{\Phi}\mathbf{M}\overline{\mathbf{v}}_{\mathbf{\Phi}}\right\|_2^2 \le \left\|\mathbf{M}^T\mathbf{M}\overline{\mathbf{v}}\right\|_2^2(1+\delta_k) \tag{8.19}$$

which directly implies that

$$\lambda_{\mathbf{\Phi}}^2\|\overline{\mathbf{v}}_{\mathbf{\Phi}}\|_2^2 \le \lambda^2\|\overline{\mathbf{v}}\|_2^2(1+\delta_k) \tag{8.20}$$

However, by the upper bound of RIP, $\|\overline{\mathbf{v}}_{\mathbf{\Phi}}\|_2^2 = \|\overline{\mathbf{v}}\|_2^2 = 1$, and

$$\lambda_{\mathbf{\Phi}} \le \lambda(1+\delta_k)^{1/2} \tag{8.21}$$

By using symmetry, we can apply the RIP's lower bound condition and write

$$\lambda(1-\delta_k)^{1/2} \le \lambda_{\mathbf{\Phi}} \tag{8.22}$$

Combining (8.21) and (8.22) together, we obtain the bounds on $\lambda_{\mathbf{\Phi}}$ as

$$\lambda(1-\delta_k)^{1/2} \le \lambda_{\mathbf{\Phi}} \le \lambda(1+\delta_k)^{1/2} \tag{8.23}$$

The result given in (8.23) is significant since it indicates that the eigenvalues of the Gram matrix are almost preserved in CSBD and are bounded in the same way that the length of a vector is bounded by RIP in (8.2).

Now, since $\sqrt{\lambda_{\mathbf{\Phi}_i}} = \sigma_{\mathbf{\Phi}_i}$, the bounds on $\lambda_{\mathbf{\Phi}}$ in (8.23) allow us to derive the bounds on $\sigma_{\mathbf{\Phi}}$ as

$$\sigma(1-\delta_k)^{1/4} \le \sigma_{\mathbf{\Phi}} \le \sigma(1+\delta_k)^{1/4} \tag{8.24}$$

By combining the bounds on $\lambda_{\mathbf{\Phi}}$ and $\sigma_{\mathbf{\Phi}}$, we can derive the RSVP for vol$(\mathbf{M}_{\mathbf{\Phi}})_{\text{SVD}}$ as

$$(1-\delta_k)^{p/4}\text{vol}(\mathbf{M}_{\mathbf{\Phi}})_{\text{SVD}} \le \text{vol}(\mathbf{M}_{\mathbf{\Phi}})_{\text{SVD}} \le (1+\delta_k)^{p/4}\text{vol}(\mathbf{M}_{\mathbf{\Phi}})_{\text{SVD}} \tag{8.25}$$

It should be noted that RSVP in (8.25) is applied to vol$(\mathbf{M}_{\mathbf{\Phi}})_{\text{DET}}$ since vol$(\mathbf{M}_{\mathbf{\Phi}})_{\text{SVD}}$ = vol$(\mathbf{M}_{\mathbf{\Phi}})_{\text{DET}}$. The significance of (8.25) resulting from RSVP demonstrates that SV in CSBD can be preserved within a tolerance $\delta_k$. The implication of (8.25) is that the SV of the $p$-vertex simplex constructed by using $p$ vectors with full $L$ bands can be approximated by the SV calculated by their counterparts using the corresponding $p$ vertices in compressively sensed vectors with $m$ CSBs. Most importantly, the SV preserved by RSVP allows SQ/SC $p$-FINDR to maintain performance results in CSBD that are comparable to that obtained in ODS.

## 8.5 Two Sequential Algorithms for *p*-FINDR

Many EF algorithms are designed to find endmembers one after another sequentially [21], such as vertex component analysis (VCA) [22], SGA [5, 7, 23], but not to find all endmembers simultaneously, such as simultaneous N-FINDR (SM N-FINDR) discussed in [1]. Although there are algorithms also proposed to implement $p$-FINDR in [24–26], they are either identical or very close to the SC $p$-FINDR. $p$-FINDR is a simplex-based method designed to look for a simplex of $p$ vertices specified by $p$-desired endmembers with maximal SV in ODS. Specifically, for a given value of $p$, the key idea of $p$-FINDR is to find a $p$-vertex simplex with maximal volume embedded in ODS in the sense that it embraces the maximal number of data sample vectors in the sense of SV. Theoretically, it can be done by exhausting all possible combinations of $p$-vertex simplexes. Practically, such an exhaustive search requires a formidable computational power and is nearly impossible in implementation. So, the algorithm developed by Winter in [27] seems to be the first numerical algorithm to address this issue by proposing a two-loop iterative algorithm to replace his original N-FINDR, where the outer loop was designated to run all data sample vectors in the data, while the inner loop was used to update each of the $p$ endmembers. Interestingly, the resulting $p$-FINDR turned out to be a variant of SQ $p$-FINDR. It should be noted that in this chapter, we modify SQ $p$-FINDR and SC N-FINDR as SQ $p$-FINDR and SC $p$-FINDR by replacing "$N$" with "$p$" to specifically emphasize the fact of finding $p$ endmembers instead of $N$ endmembers.

### *8.5.1 SeQuential* p-*FINDR (SQ* p-*FINDR)*

The SQ $p$-FINDR presented in this subsection implements two loops with outer loop designed to iterate data sample vectors while the inner loop is to update one of $p$ desired endmembers for each given data sample vector **r**.

---

**Algorithm 1: SQ *p*-FINDR**

---

1. Initial conditions:
   a. Let $p$ be the number of endmembers required to generate.
   b. Let $\left\{ \mathrm{m}_1^{(0)}, \mathrm{m}_2^{(0)}, \mathrm{m}_3^{(0)}, \ldots, \mathrm{m}_p^{(0)} \right\}$ be a set of initial vectors randomly selected from the data and let $\{\mathbf{r}_i\}_{i=1}^{N}$ be all data sample vectors. Set $i = 1$ and go to step 4.
2. Outer Loop: (using index i as a counter to keep track data sample vector $\mathbf{r}_{i}$). Check $i = N$. If it is, the algorithm terminated. Otherwise, let $i \leftarrow i+1$ and continue.
3. Input the $i$th data sample vector, $\mathbf{r}_i$.
4. Inner Loop: (using $j$ as a counter to keep track the $j$th endmember $\mathbf{m}_j$). For $1 \leq j \leq p$, we recalculate simplex volume $V\left(\mathbf{m}_1^{(i)}, \ldots, \mathbf{m}_{j-1}^{(i)}, \mathbf{r}_i, \mathbf{m}_{j+1}^{(i)}, \ldots, \mathbf{m}_p^{(i)}\right)$ for the data sample vector $r_i$. If any of these $p$ recalculated volumes,
   $V\left(\mathbf{r}_i\mathbf{m}_2^{(i)}, \ldots, \mathbf{m}_i^{(i)}\right)$, $V\left(\mathbf{m}_1^{(i)}, \mathbf{r}_i, \mathbf{m}_3^{(i)}, \ldots, \mathbf{m}_p^{(i)}\right), \ldots,$
   $V\left(\mathbf{m}_1^{(i)}, \ldots, \mathbf{m}_{p-}^{(i)}, \mathbf{r}_i\right)$ is greater than $V\left(\mathbf{m}_1^{(i)}, \mathbf{m}_2^{(i)}, \ldots, \mathbf{m}_p^{(i)}\right)$, go to step 6.
   Otherwise, go to step 3.
5. Replacement rule:
   The endmember which is absent in the largest simplex among the $p$ simplexes, $S\left(\mathbf{r}_i, \mathbf{m}_2^{(i)}, \ldots, \mathbf{m}_p^{(i)}\right)$, $S\left(\mathbf{m}_1^{(i)}, \mathbf{r}_i, \mathbf{m}_3^{(i)}, \ldots, \mathbf{m}_p^{(i)}\right), \ldots,$ $S\left(\mathbf{m}_1^{(i)}, \ldots, \mathbf{m}_{p-1}^{(i)}, \mathbf{r}_i\right)$ will be replaced by the $i$th sample vector $\mathbf{r}_i$. Assume that such an endmember is denoted by $\mathbf{m}_j^{i+1}$. A new set of endmembers is then produced by letting $\mathbf{m}_i^{i+1} = \mathbf{r}_i$ and $\mathbf{m}_i^{i+1} = \mathbf{m}_i^{i}$ for $i \neq j$ and go to step 3.

---

### 8.5.2 *SuCcessive* p-*FINDR (SC* p-*FINDR)*

The SC $p$-FINDR developed in this subsection reverses the two loops implemented in SQ $p$-FINDR by iterating endmembers in its outer loop while iterating data sample vectors in its inner loop.

---

**Algorithm 2: SC *p*-FINDR**

---

1. Initial conditions:
   Let the number of endmembers be $p$.
   Let $\left\{\mathbf{m}_1^{(0)}, \mathbf{m}_2^{(0)}, \mathbf{m}_3^{(0)}, \ldots, \mathbf{m}_p^{(0)}\right\}$ be a set of initial vectors randomly generated from the data. Let $\{\mathbf{r}_i\}_{i=1}^{N}$ be all data sample vectors.
2. Outer Loop:
   For $1 \le j \le p$ find $\mathbf{m}_j^{(*)}$ which will replace $\mathbf{m}_j^{(0)}$.
3. Inner Loop:
   For $1 \le j \le p$ calculate the simplex volume of $V\left(\mathbf{m}_1^{(*)}, \ldots, \mathbf{m}_{j-1}^{(*)}, \mathbf{r}_i \mathbf{m}_{j+1}^{(0)}, \ldots, \mathbf{m}_p^{(0)}\right)$ while fixing other endmembers $\mathbf{m}_i^{(*)}$ with $i<j$ and $\mathbf{m}_i^{(0)}$ with $i>j$. Let $\mathbf{m}_j^{(*)} = \max_{1 \le k \le N} V\left(\mathbf{m}_1^{(*)}, \ldots, \mathbf{m}_{j-1}^{(*)}, \mathbf{r}_k, \mathbf{m}_{j+1}^{(0)}, \ldots, \mathbf{m}_p^{(0)}\right)$
4. Stopping rule:
   If $j \le p$, then $j \leftarrow j+1$ and go step 3. Otherwise, the final set of $\left\{\mathbf{m}_1^{(*)}, \mathbf{m}_2^{(*)}, \mathbf{m}_3^{(*)}, \ldots, \mathbf{m}_p^{(*)}\right\}$ is the desired $p$ endmembers.

---

### 8.5.3 *SQ* p-*FINDR and SC* p-*FINDR in CSBD*

Our novel approach of using the CS model to sense bands allows $p$-FINDR to find a desired set of $p$ endmembers in CSBD without undermining $p$-FINDR in ODS. More specifically, SQ $p$-FINDR and SC $p$-FINDR can be directly applied to CSBD with SVs preserved. However, it should be noted that the input pixel vectors to SQ $p$-FINDR and SC $p$-FINDR should be in CSBD via the sensing matrix $\mathbf{\Phi}$ from full $L$ bands to $m$ CSBs. Namely, the CSBD-based random sample approach given in (8.9) is applied to the hyperspectral data vectors $\{\mathbf{r}_i\}_{i=1}^{N}$ prior to implementing SQ $p$-FINDR and SC $p$-FINDR. Furthermore, unlike the traditional $p$-FINDR, no additional spectral DR is required for the hyperspectral data vectors $\{\mathbf{r}_{\Phi i}\}_{i=1}^{N}$ pre- or post-compressive sensing. Most importantly, the found endmembers in CSBD are nearly same endmembers in ODS without performing any sparse recovery (reconstruction) process as in (8.3).

## 8.6 Experiments

The well-known Airborne Infrared Imaging Spectrometer (AVIRIS) Cuprite hyperspectral image shown in Figure 8.1 was used for experiments, where a detailed description of this scene along with its color maps is available in Figures 1.11 and 1.12 in [1]. This image scene has 350 × 350 pixels (i.e. $N = 122\,500$) with a spatial resolution of 20 m collected using 224 bands (only $L = 189$ used for experiment) and 10 nm spectral resolution in the range of 0.4–2.5 μm. According to the ground truth, the image contains five pure pixels that can be identified and located to represent five endmembers denoted by alunite (A), buddingtone (B), calcite (C), kaolinite (K), and muscovite (M). The VD was used to estimate the number of endmembers to be $p = 22$.

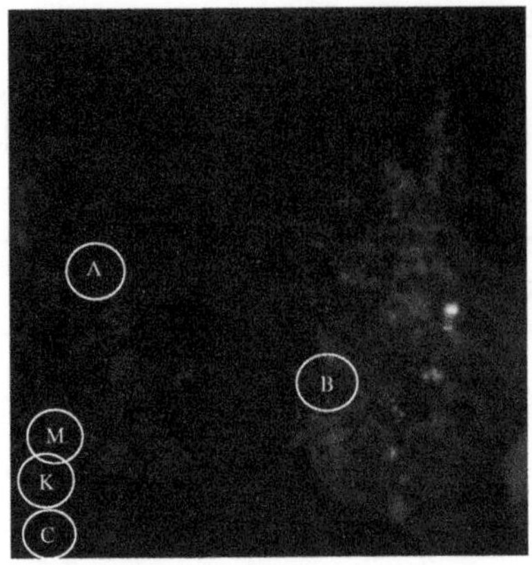


**Figure 8.1** Cuprite real image scene.

### 8.6.1 Experimental Setup

Our experimental setup consists of two parts. The first part is to implement SQ *p*-FINDR and SC *p*-FINDR in ODS without applying any CS technique (i.e. using $\{\mathbf{r}_i\}_{i=1}^{N}$ with full bands). The resulting endmember matrix in ODS is denoted by **M**. The second part is to implement CS acquisition process via a sensing matrix $\mathbf{\Phi}$ without performing any "sparsifying" by taking advantage of the universality concept discussed earlier. Therefore, the pixel vectors $\{\mathbf{r}_i\}_{i=1}^{N}$ are sensed by using a Gaussian sensing matrix $\mathbf{\Phi}$ to produce compressively sensed data sample vectors $\{\mathbf{r}_{\Phi i}\}_{i=1}^{N}$ which are then used as an input to SQ *p*-FINDR and SC *p*-FINDR. The resulting endmember matrix in CSBD is denoted by $\mathbf{M}_{\Phi}$. Figure 8.2 depicts such an experimental setup.

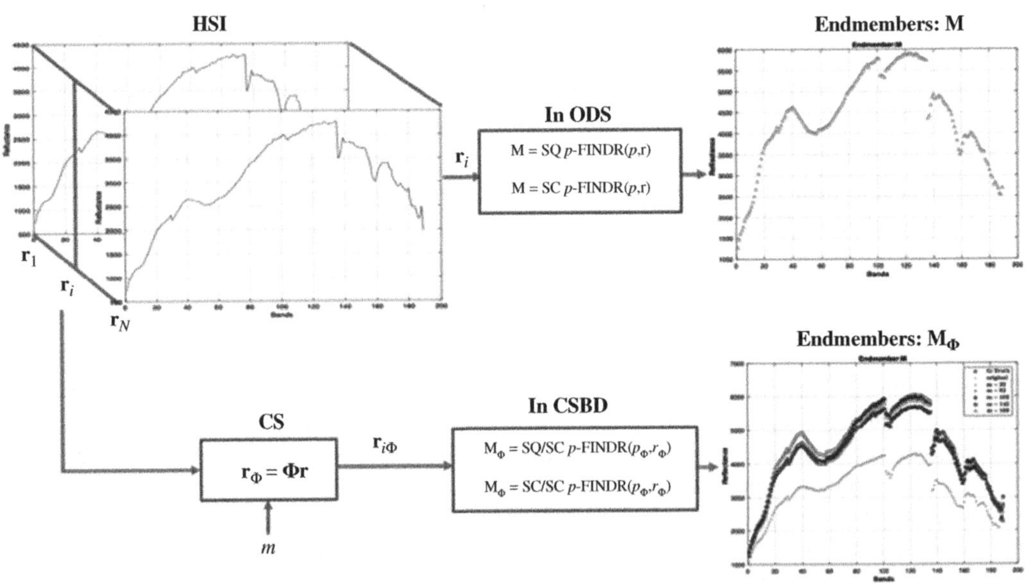


**Figure 8.2** Experimental setup.

Overall, our main experimental objectives are to compare the results produced by SQ $p$-FINDR and SC $p$-FINDR in CSBD with their counterparts produced in OSD as well as to demonstrate the RSVP concept experimentally. To accomplish these objectives, we implemented multiple Monte Carlo (MC) runs and collected experimental data in both ODS and CSBD for analysis.

In CSBD, for a single MC experimental run, the number of CSBs, $m$, varied from $p$ (number of desired endmembers) to $L$ (total number of bands in the hyperspectral image) with a step size of 40. In addition, each experimental run was repeated 10 times for MC size of 10. Therefore, a full MC run consisted of a set of 10 experimental runs; each of the runs was executed two times, one for each SV calculation method. We also conducted the same experiments in ODS without CS. All experiments were conducted using $p = 22$ for the Cuprite image, as it was estimated by the Harsanyi–Farrand–Chang (HFC) method [3, 28–32].

### 8.6.2 *Algorithm Analysis on Experimental Data*

For each MC set, we collected experimental data. The experimental data consisted of $p$ found endmembers and their spectral abundances as well as the SV calculated by the $p$ found endmembers. Furthermore, to obtain insights into the performances of SQ $p$-FINDR and SC $p$-FINDR, we also computed the spectral angle mapper (SAM) [29] between the found endmembers in ODS and the ground-truth endmembers to quantify similarity between the two vectors. For comparison reasons, we also computed the SAM values between the endmembers found in CSBD and the ground-truth endmembers. Finally, we measured the computational times for SQ $p$-FINDR and SC $p$-FINDR both in CSBD and ODS. For the CSBD case, the computational times were computed as a function of $m$.

## 8.7 Experimental Results and Discussions

We conducted performance analysis on our experimental results by focusing on the five materials shown in the Cuprite image, Figure 8.1, where the five ground-truth endmembers as (A), (B), (C), (K), and (M), are located, respectively. For our experiments, these five ground-truth endmembers were used to evaluate performance analysis of SQ $p$-FINDR and SC $p$-FINDR with $p = 22$ by measuring the smallest SAM values between their found endmembers and these five ground-truth endmembers. In other words, we were interested in analyzing how well the two algorithms performed in finding endmembers against ground-truth endmembers, (A), (B), (C), (K), and (M) both in CSBD and ODS. In addition, we determined the minimum number of CSBs, $m$, for the two algorithms required to achieve same or nearly the same performance as that obtained in ODS. Also, we investigated the impact of increasing the number of CSBs, $m$, on SV calculated by the $p$ found endmembers, and the role that RSVP played in imposing bounds on SV calculated using SVD in (8.11) and determinant method in (8.12). Finally, the computational times of these two algorithms were also analyzed both in ODS and CSBD.

### 8.7.1 *SQ* p-*FINDR Experimental Result Analysis*

To calculate SV, SQ $p$-FINDR used (8.11) and (8.12) to find $p$ endmembers in ODS and CSBD. Figures 8.3–8.4 show the results obtained by SQ $p$-FINDR using the $\mathrm{vol}(\mathbf{M}_{\Phi})_{\mathrm{SVD}}$-based SV method in (8.11). In particular, Figure 8.3(a) shows the spatial locations of its found $p$ endmembers in ODS without applying any CS model. As seen in the figure, SQ 22-FINDR was able to find five

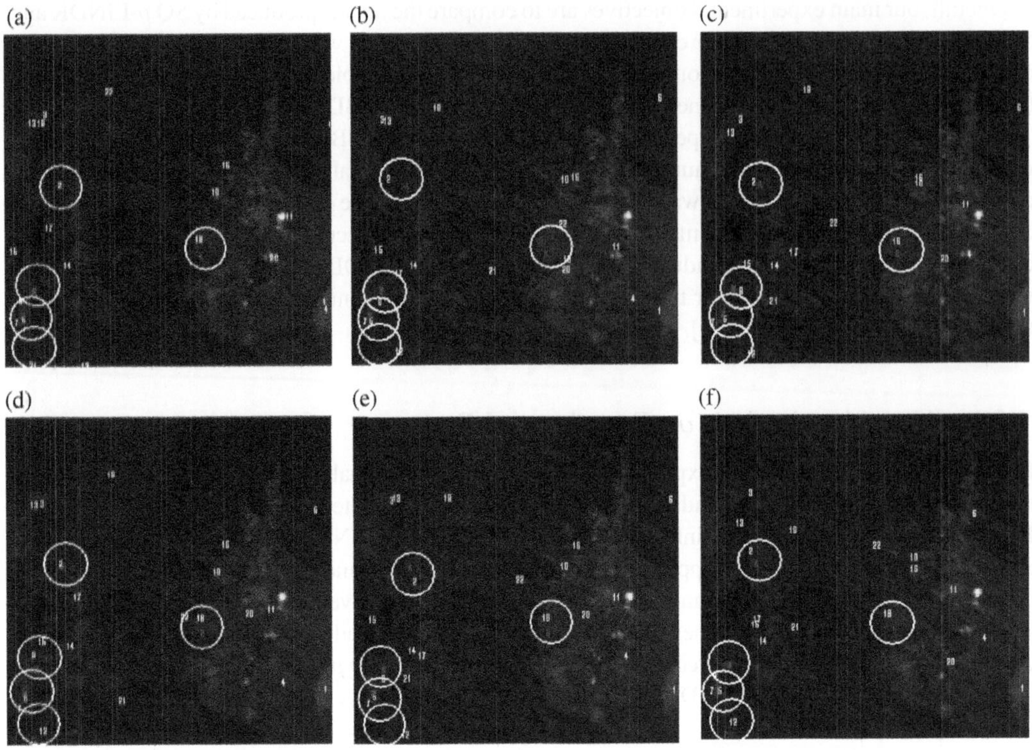

**Figure 8.3** SQ 22-FINDR: 22 found endmembers via $vol(\mathbf{M})_{SVD}$ (a) in ODS. (b)–(f) in CSBD for different values of $m$; (a) ODS (b) CSBD: $m$ = 22, (c) CSBD: $m$ = 62, (d) CSBD: $m$ = 102, (e) CSBD: $m$ = 142, and (f) CSBD: $m$ = 189.

endmembers that were very close to (A), (B), (C), (K), and (M) using all $L$ available bands shown in Figure 8.3(a). On the other hand, when the same SQ 22-FINDR was implemented in CSBD, it could find four endmembers that were very close to endmembers (A), (C), (K), and (M) with using only $m = p = 22$ CSBs. If we further define the compressively sensed band ratio (CSBR) as CSBR = $m/L$, then the number of CSBs, $m$, divided by the total number of bands, $L$. Then, with only CSBR = 22/189 = 0.12, SQ 22-FINDR found four endmembers that were very close to endmembers (A), (C), (K), and (M) as shown in Figure 8.3(b). Increasing the number of CSBs, $m$, beyond 22 helped SQ 22-FINDR perform even better in CSBD. For example, with the number of CSBs, $m = 62$, SQ 22-FINDR yielded nearly the same performance as observed in ODS where an additional endmember was found to be very close to the ground-truth endmember, (B) as shown in Figure 8.3(c). However, increasing the number of CSBs, $m$, beyond 62 did not provide much performance as observed in Figure 8.3(d) through Figure 8.3(f). Therefore, SQ 22-FINDR performed similarly both in ODS and CSBD by using only $m = 62$ with CSBR = 62/189 = 0.33 in CSBD. This represented a significant advantage in using the CS model over the traditional $p$-FINDR that used all $L$ available bands in ODS to provide similar performance results.

As shown in Figure 8.4(a), the SAM values for the five found endmembers that are closest to the ground-truth endmembers (A), (B), (C), (K), and (M) have similar values both in ODS and in

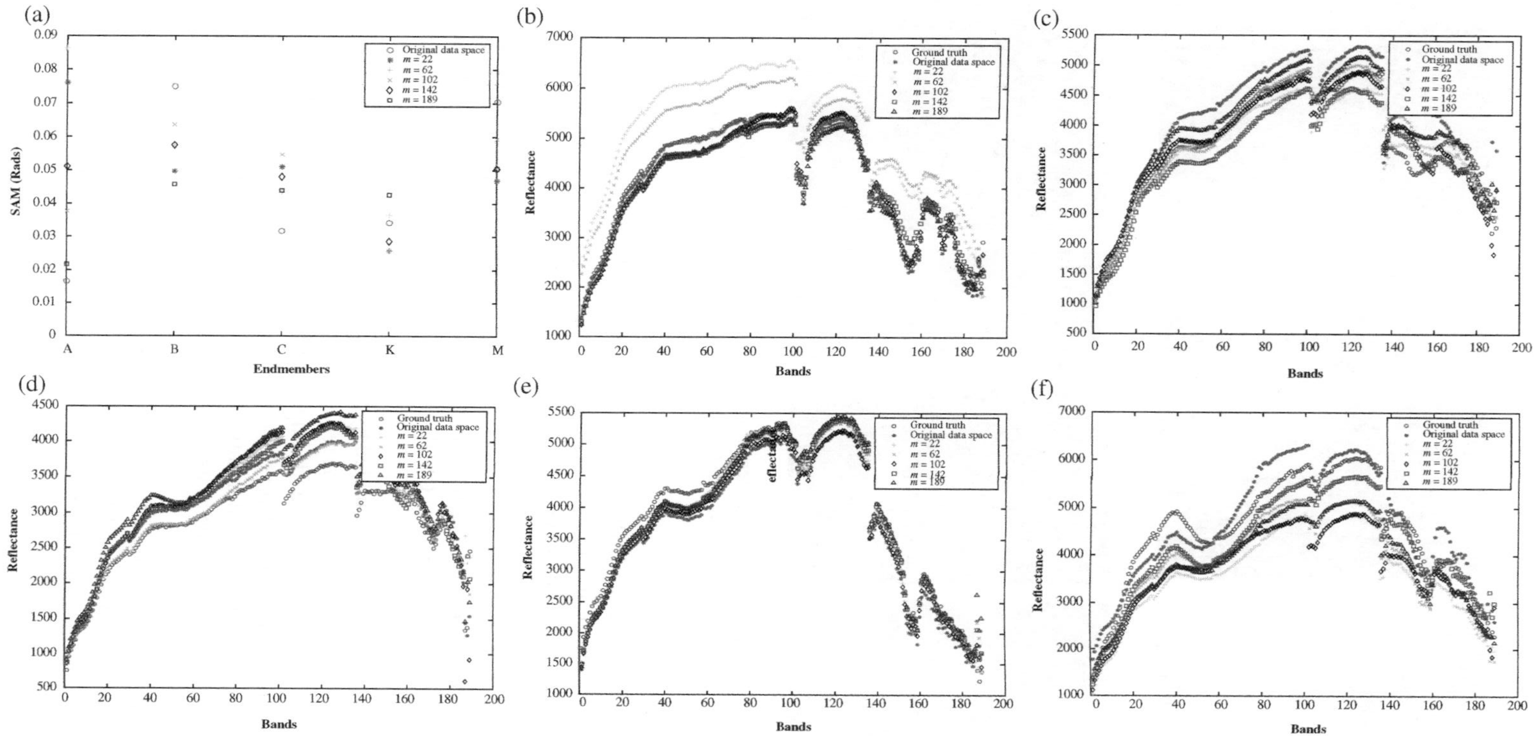


**Figure 8.4** SQ 22-FINDR: five endmembers found by vol($\mathbf{M}$)$_{SVD}$ that are closest to ground truth (A), (B), (C), (K), and (M); (a) SAMs. (b)–(f) Radiance; (a) SAM, (b) endmember (A), (c) endmember (B), (d) endmember (C), (e) endmember (K), and (f) endmember (M).

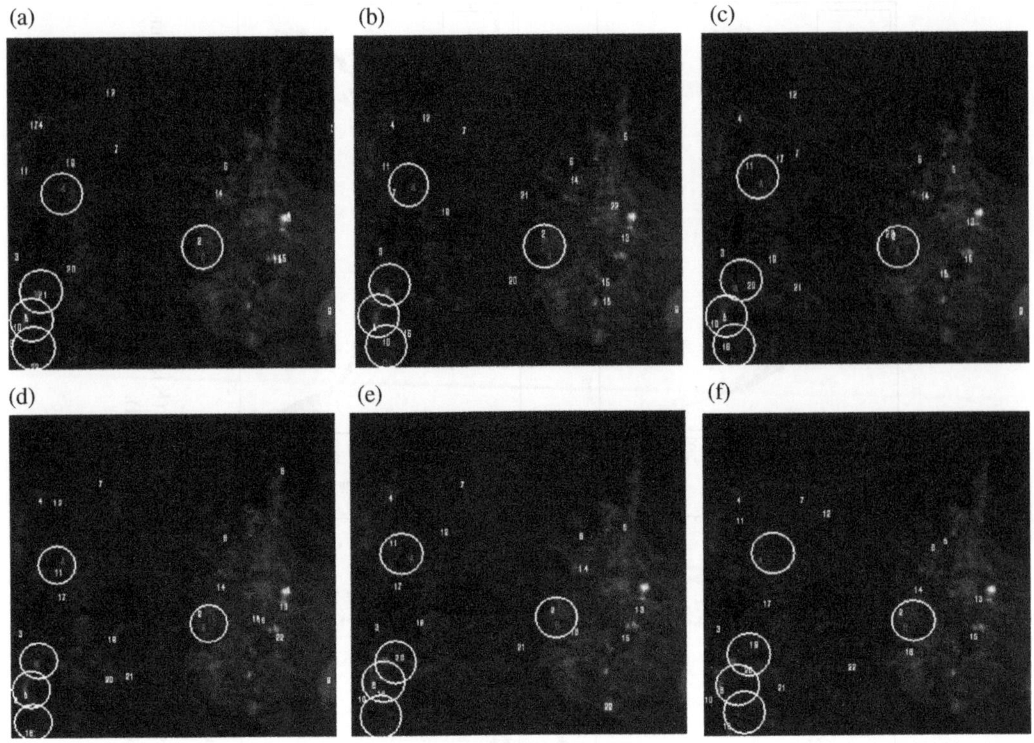

**Figure 8.5** SQ 22-FINDR: 22 found endmembers via vol($\mathbf{M}$)$_{DET}$ (a) in ODS. (b)–(e) in CSBD for different $m$; (a) ODS, (b) CSBD: $m = 22$, (c) CSBD: $m = 62$, (d) CSBD: $m = 102$, (e) CSBD: $m = 142$, and (f) CSBD: $m = 189$.

CSBD, even when the number of CSBs, $m = 22$. The same behavior was also observed for their spectral abundances as shown in Figure 8.4(b) through Figure 8.4(f). Despite using $m = 62$, the spectral values of the five found endmembers in CSBD had similar values as the five endmembers found in ODS.

We also observed similar performance results when SQ 22-FINDR used the vol($\mathbf{M}_{\Phi}$)$_{DET}$-based SV method to find endmembers in the Cuprite image. Once again, SQ 22-FINDR only needed $m = 62$ (i.e. CSBR = 0.33) to find endmembers in CSBD that were similar to endmembers obtained in ODS, as shown in Figures 8.5–8.6.

The SVs of 22 found endmembers calculated by SVD and determinant methods in ODS and CSBD are plotted in Figure 8.7(a), (b). Looking at the results, the estimated SVs in CSBD are bounded by the estimated SVs in ODS, confirming the bounds imposed by RSVP. Furthermore, the estimated SVs in CSBD increased as a function of $m$. However, it is interesting to note that the SVs did not increase much for $m$ beyond 62. This provided an additional evidence that the performance of SQ $p$-FINDR for $m \geq 62$ in CSBD was nearly the same as its performance in ODS (i.e. CSBR = 0.33 is sufficient for SQ $p$-FINDR).

The computing times of SQ 22-FINDR are plotted in Figure 8.7(c) through Figure 8.7(d) using the two different SV calculation methods, SVD and determinant, where the computing time of SQ 22-FINDR increased as a function of m in CSBD but they were bounded by its computing time in ODS.

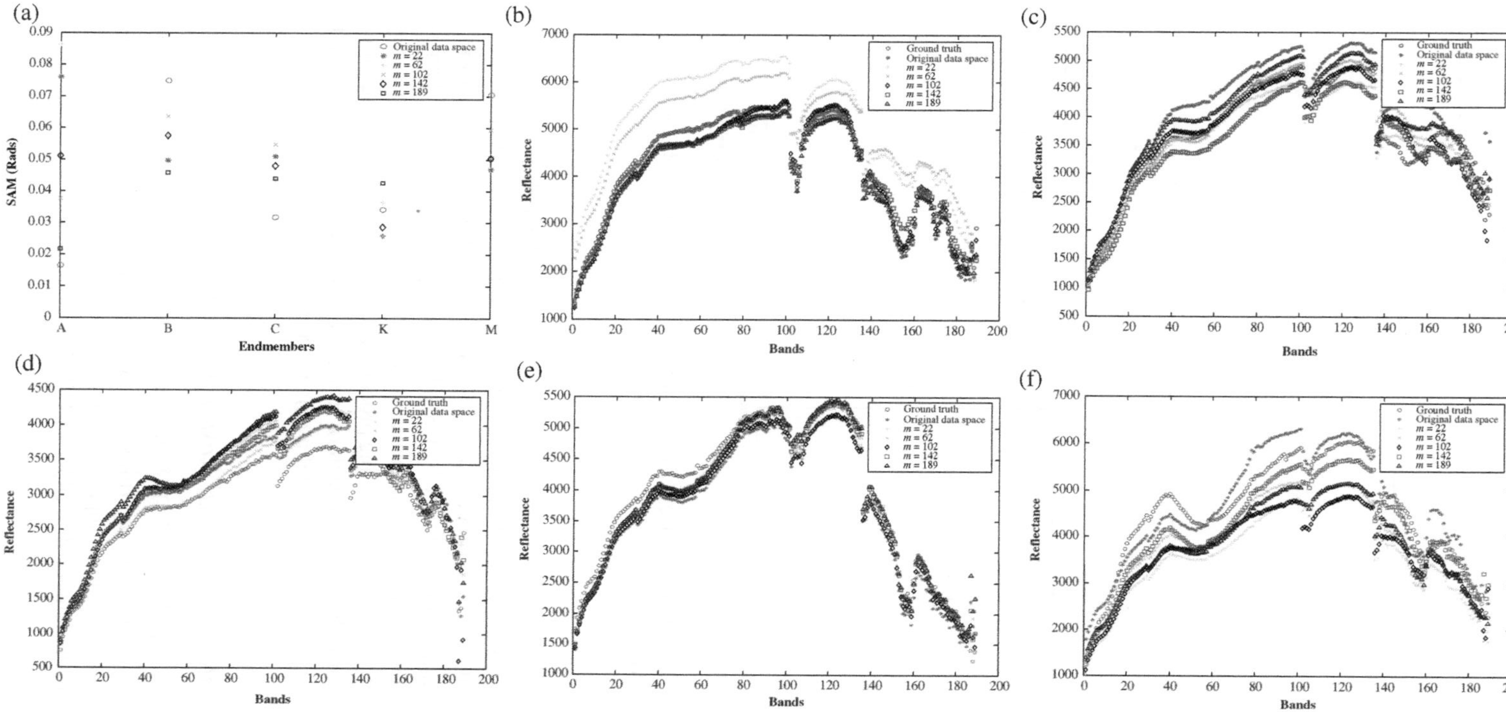


**Figure 8.6** SQ 22-FINDR: five endmembers found via vol$(\mathbf{M})_{\mathrm{DET}}$ that are close to ground truth (A), (B), (C), (K), and (M). (a) SAMs. (b)–(f) radiance; (a) SAM, (b) endmember (A), (c) endmember (B), (d) endmember (C), (e) endmember (K), and (f) endmember (M).

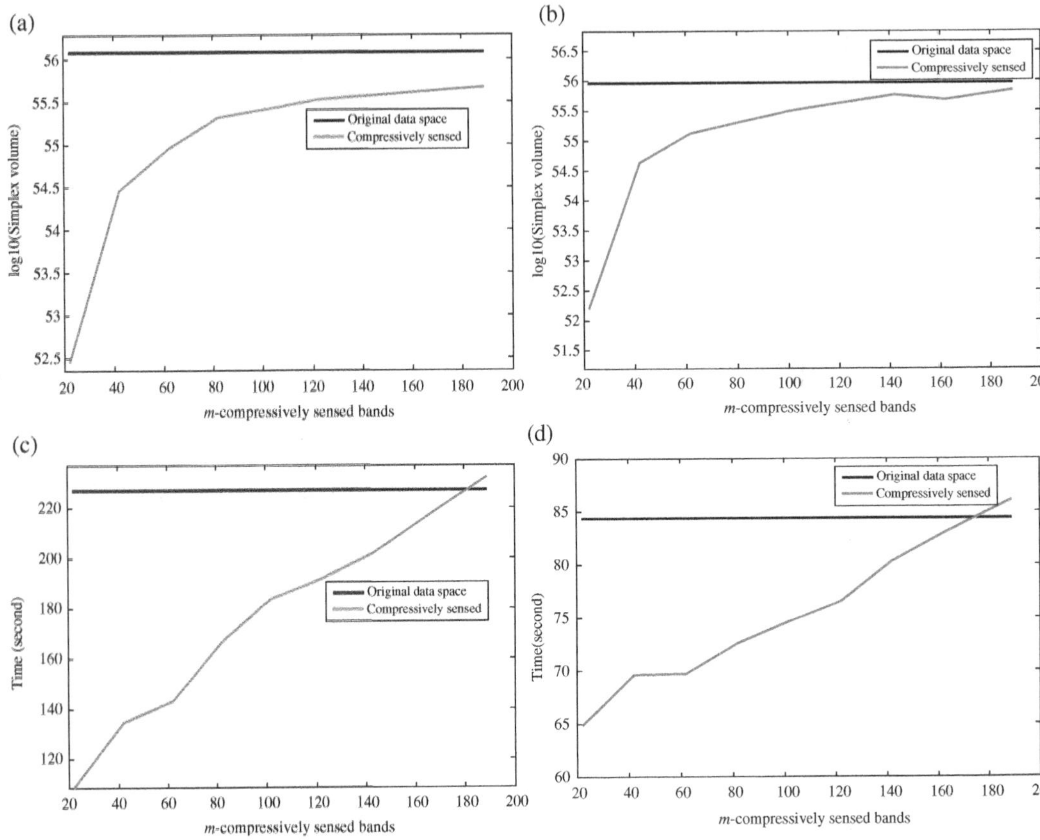


**Figure 8.7** SQ 22-FINDR: (a), (b) volumes calculated by $vol(\mathbf{M})_{SVD}$ and $vol(\mathbf{M})_{DET}$. (c), (d) Computational time $vol(\mathbf{M})_{SVD}$ and $vol(\mathbf{M})_{DET}$.

This was expected considering less amount of data was processed in CSBD for $m < L$ compared with the data processed in ODS utilizing $L$ full bands. A very important observation was that as mentioned earlier, when $m = 62$, the performance of SQ 22-FINDR in CSBD was similar to that observed in ODS. However, its computational times in CSBD were significantly lower than its counterpart observed in the ODS. However, the performance achieved in CSBD were comparable to the performance achieved in ODS despite less data being processed and with less computing time. These results presented further evidence that our approach of using the CS model was superior to the traditional method of data acquisition process using all available bands.

## *8.7.2 SC* p-*FINDR Experimental Result Analysis*

The performance results for SC ($p = 22$)-FINDR in ODS and in CSBD were expected to be similar to that of SQ 22-FINDR since the two algorithms are very similar in nature with the only difference in that their inner and outer loops are swapped. In other words, the main difference is that SC 22-FINDR reversed the two loops implemented in SQ 22-FINDR.

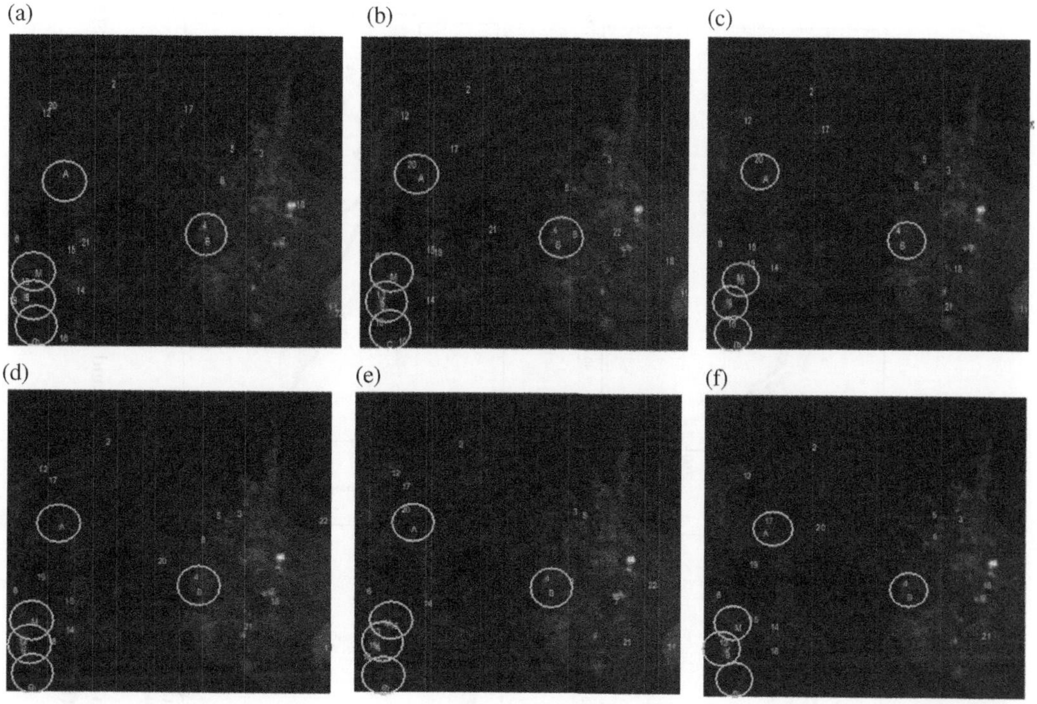

**Figure 8.8** SC 22-FINDR: 22 endmembers found via $\text{vol}(\mathbf{M})_{\text{SVD}}$. (a) In ODS. (b)–(e) In CSBD for different values of $m$; (a) ODS, (b) CSBD: $m$ = 22, (c) CSBD: $m$ = 62, (d) CSBD: $m$ = 102, (e) CSBD: $m$ = 142, and (f) CSBD: $m$ = 189.

As it was the case for SQ 22-FINDR, SC $p$-FINDR was also implemented using the two different SV calculation methods, SVD and determinant. Despite their computational time differences, their found endmembers were similar to that obtained by SQ 22-FINDR as shown in Figures 8.8–8.12 in comparison with Figures 8.3–8.7. So, similar conclusions drawn for SQ 22-FINDR were also applied to SC 22-FINDR where its performances in ODS and CSBD were nearly the same for $m = 62$ (CSBR = 0.33). Once again, these experimental results demonstrated the advantage of SC $p$-FINDR using the CS model over the traditional non-CS approaches.

In summary, both SQ 22-FINDR and SC 22-FINDR achieved nearly the same level of performance in both ODS and CSBD even though fewer number of CSBs for $m = 62$ or CSBR = 0.33 were used with significant less computational time. This was because the SVs were preserved by RSVP for the two types of SV calculation methods, SVD and determinant.

## 8.8 Conclusion

This chapter presents a CSBD-based $p$-FINDR to find $p$ endmembers in hyperspectral images. Since EF does not require signal reconstruction, the proposed CSBD-based $p$-FINDR takes the advantage of the universality derived in [8] without specifying a sparse matrix $\boldsymbol{\Psi}$ but rather uses a sensing matrix $\boldsymbol{\Phi}$ to sense bands in spectral domain. Most importantly, there is no need for CSBD-based $p$-FINDR to

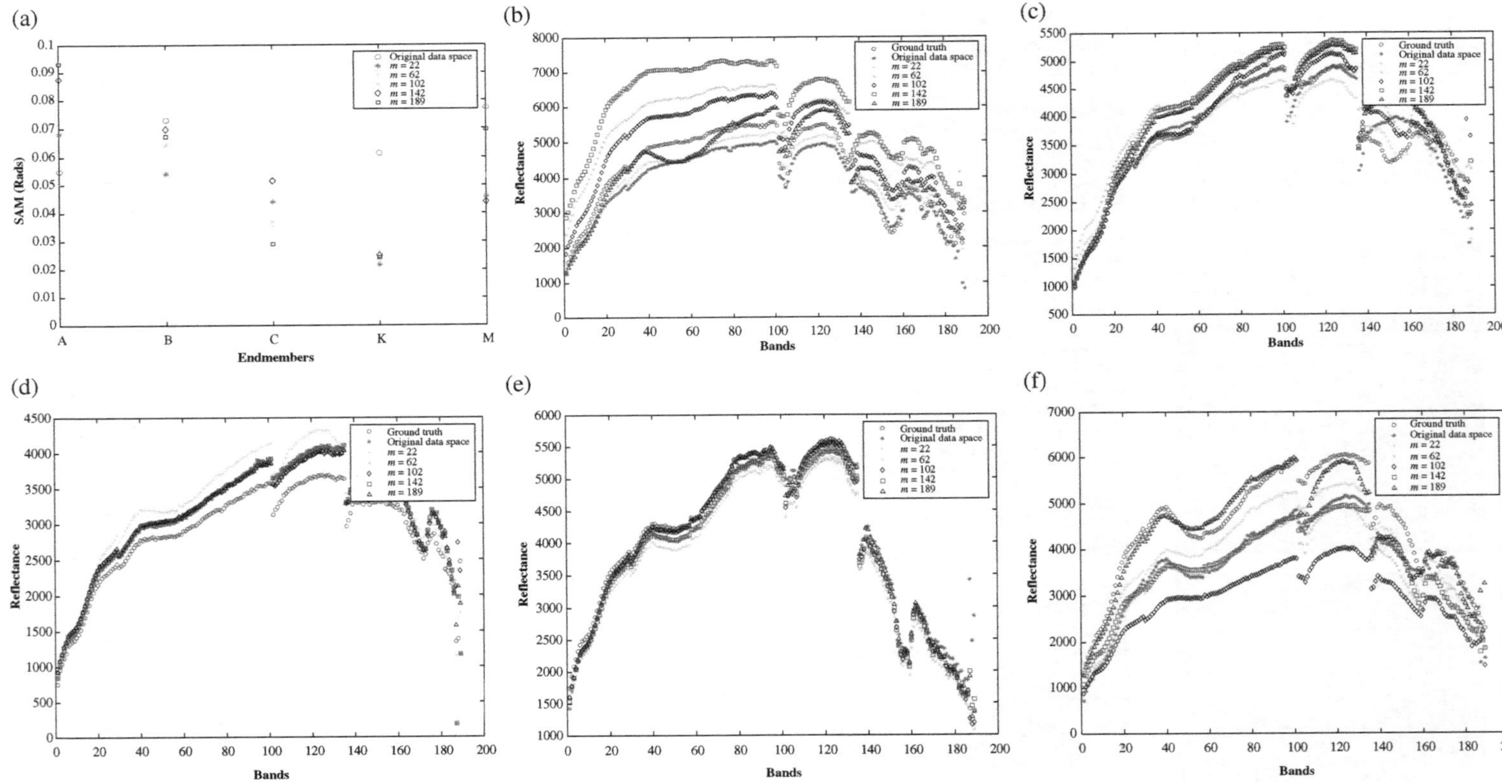


**Figure 8.9** SC 22-FINDR: five endmembers found via vol$(\mathbf{M})_{SVD}$ that are closest to ground truth (A), (B), (C), (K), and (M). (a) SAMs. (b)–(f) Radiance; (a) SAM, (b) endmember (A), (c) endmember (B), (d) endmember (C), (e) endmember (K), and (f) endmember (M).

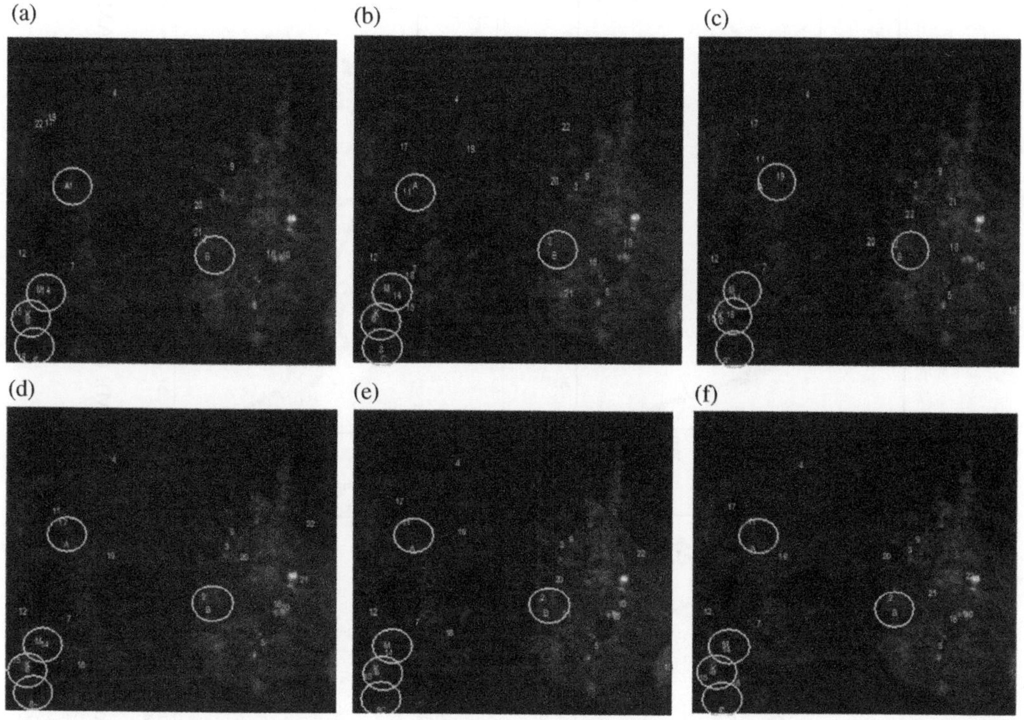

**Figure 8.10** SC 22-FINDR: 22 endmembers found via vol(**M**)$_{DET}$ (a) In ODS. (b)-(f) In CSBD for different values of Type equation here.; (a) ODS (b) CSBD: $m = 22$, (c) CSBD: $m = 62$, (d) CSBD: $m = 102$, (e) CSBD: $m = 142$, and (f) CSBD: $m = 189$.

perform spectral DR to calculate the SV as required by the traditional *p*-FINDR. In order to ensure that SV calculated by *p*-FINDR in CSBD is preserved in ODS, a new property, called RSVP, is further proved to show that SV calculated by *p*-FINDR in ODS can be also preserved in CSBD by *p*-FINDR. This nice property allows any SV-based EF algorithm to be performed in CSBD that can also yield nearly the same performance as it is applied to ODS. By virtue of RSVP, the two well-known EF algorithms, SQ *p*-FINDR and SC *p*-FINDR, can be further implemented in CSBD to avoid a computationally expensive process of finding endmembers via an exhaustive search as it is done by the traditional *p*-FINDR. In addition, unlike the traditional *p*-FINDR which requires DR, CSBD-based *p*-FINDR avoids such need. Experimental results demonstrate that both SQ *p*-FINDR and SC *p*-FINDR find nearly the same endmembers in the sense of producing the same SV in both OSD and CSBD with only a few numbers of CSBs, *m*.

As a final remark, despite that this chapter considers *p*-FINDR in CSBD, the treatment presented in this chapter can be also applicable to any EF algorithm implemented in CSBD, for example, minimum volume-based algorithms such as [33]. As a matter of fact, a CSBD-based SGA is also developed in [34] as a counterpart of SGA [5]. Interestingly, the work derived from CSBD may also lead to many other applications such as linear spectral mixture analysis in [34], hyperspectral image classification [35], and automatic target detection and band selection in [36].

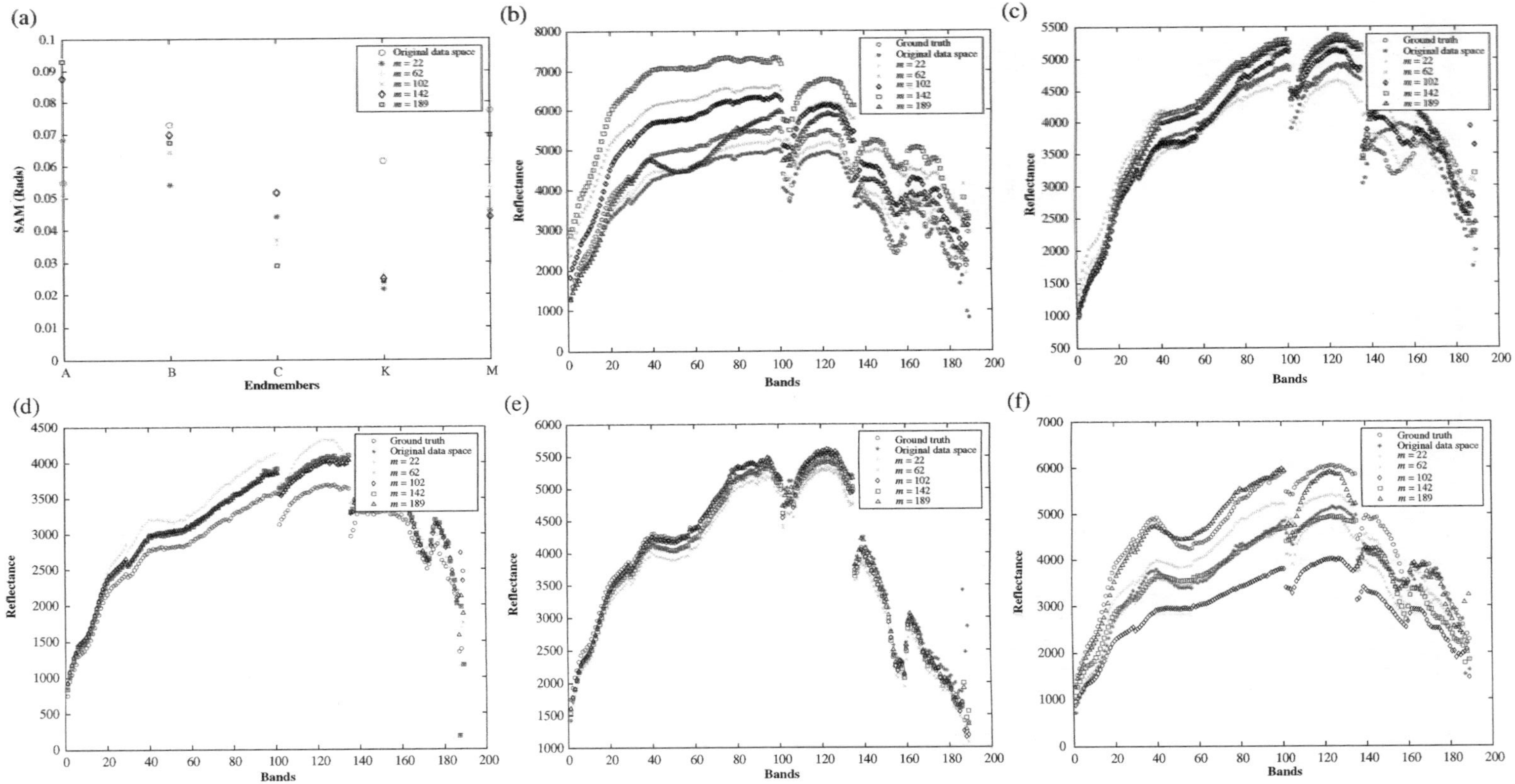


**Figure 8.11** SC 22-FINDR: five endmembers found via vol$(\mathbf{M})_{DET}$ that are close to (A), (B), (C), (K), and (M). (a) SAMs, (b)–(f) radiance; (a) SAM, (b) endmember (A), (c) endmember (B), (d) endmember (C), (e) endmember (K), and (f) endmember (M).

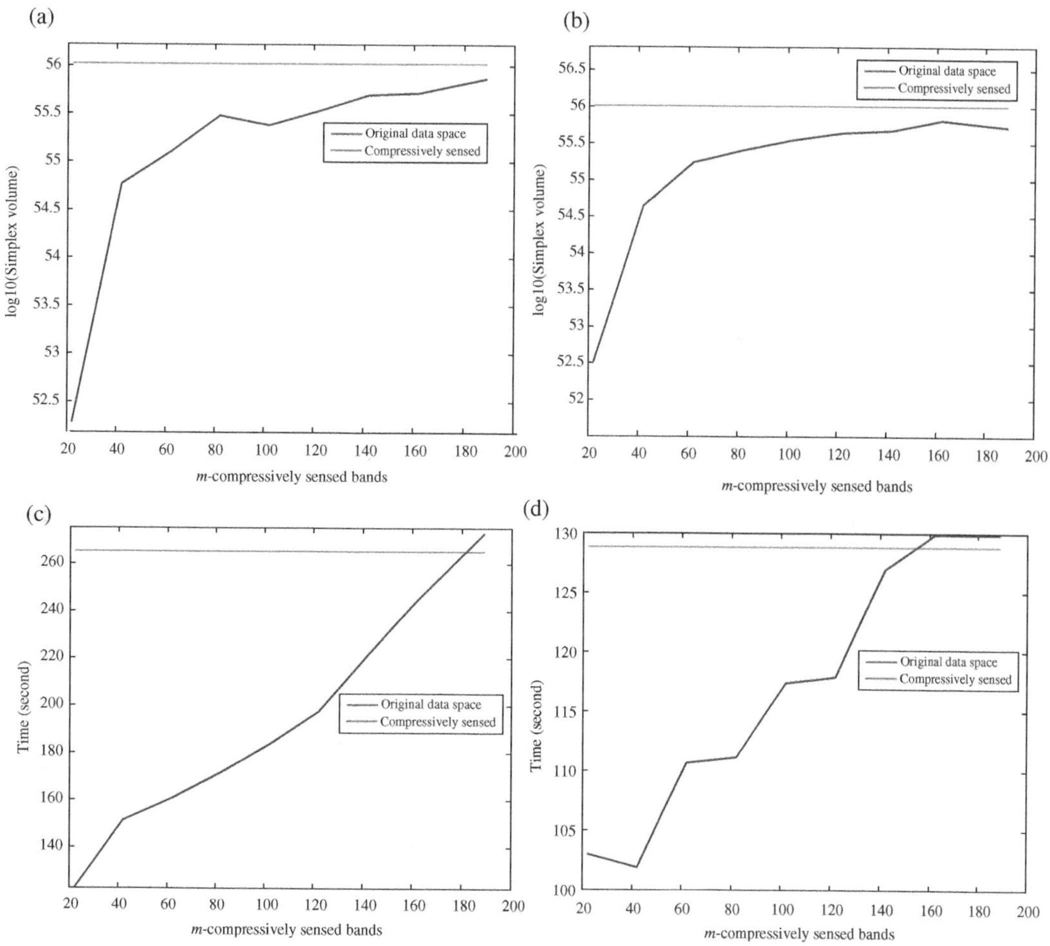


**Figure 8.12** SC 22-FINDR: (a), (b) SV calculated by $vol(\mathbf{M})_{SVD}$ and $vol(\mathbf{M})_{DET}$. (c), (d) Computational time $vol(\mathbf{M})_{SVD}$ and $vol(\mathbf{M})_{DET}$.

## References


1. Chang, C.-I (2013). *Hyperspectral Data Processing: Algorithm Design and Analysis*. Hoboken, NJ: Wiley.
2. Chang, C.-I (2016). *Real-Time Progressive Hyperspectral Image Processing*. New York: Springer New York.
3. Chang, C.-I and Du, Q. (2004). Estimation of number of spectrally distinct signal sources in hyperspectral imagery. *IEEE Trans. Geosci. Remote Sens.* 42 (3): 608–619.
4. Winter, M.E. (1999). N-FINDR: an algorithm for fast autonomous spectral end-member determination in hyperspectral data. Presented at the SPIE's International Symposium on Optical Science, Engineering, and Instrumentation, Denver, CO (27 October 1999), pp. 266–275.
5. Chang, C.-I, Wu, C.-C., Liu, W., and Ouyang, Y.-C. (2006). A new growing method for simplex-based endmember extraction algorithm. *IEEE Trans. Geosci. Remote Sens.* 44 (10): 2804–2819.
6. Wu, C.-C., Chu, S., and Chang, C.-I (2008). Sequential N-FINDR algorithms. SPIE Conference on Imaging Spectrometry XIII, San Diego, CA (10–14 August 2008).
7. Chang, C.-I, Wu, C.C., Lo, C.-S., and Chang, M.-L. (2010). Real-time simplex growing algorithms for hyperspecral endmember extraction. *IEEE Trans. Geosci. Remote Sens.* 40 (4): 1834–1850.

8. Baraniuk, R., Davenport, M., De Vore, R., and Wakin, M.B. (2008). A simple proof of the restricted isometry property for random matrices. *Constr. Approx.* 3: 253–263.
9. Wold, S., Esbensen, K., and Geladi, P. (1987). Principal component analysis. *Chemom. Intell. Lab. Syst.* 2 (1–3): 37–52.
10. Wang, J. and Chang, C.-I (2006). Independent component analysis-based dimensionality reduction with applications in hyperspectral image analysis. *IEEE Trans. Geosci. Remote Sens.* 44 (6): 1586–1600.
11. Davenport, M.A., Duarte, M.F., Eldar, Y.C., and Kutyniok, G. (2011). Introduction to compressed sensing. In: *Compressed Sensing: Theory and Applications* (ed. Y.C. Eldar and G. Kutyniok). Cambridge University Press.
12. Foucart, S. and Rauhut, H. (2013). *A Mathematical Introduction to Compressive Sensing*. New York: Springer New York.
13. Candes, E.J. and Wakin, M.B. (2008). An introduction to compressive sampling. *IEEE Signal Process. Mag.* 25 (2): 21–30.
14. Gally, T. and Pfetsch, M.E. (2016). Computing restricted isometry constants via mixed-integer semidefinite programming. Prepr. Submitt. http://www.optimization-online.org/DB_FILE/2016/04/5395.pdf.
15. Candes, E. and Romberg, J. (2005). l1-MAGIC: recovery of sparse signals via convex programming. https://candes.su.domains/software/l1magic/downloads/l1magic.pdf.
16. Willett, R.M., Duarte, M.F., Davenport, M.A., and Baraniuk, R.G. (2014). Sparsity and structure in hyperspectral imaging: sensing, reconstruction, and target detection. *IEEE Signal Process. Mag.* 31 (1): 116–126.
17. Tropp, J. and Gilbert, A.C. (2007). Signal recovery from partial information via orthogonal matching pursuit. *IEEE Trans. Inf. Theory* 53 (12): 4655–4666.
18. Amhed, N., Natarajan, T., and Rao, K.R. (1974). Discrete cosine transform. *IEEE Trans. Comput.* 100 (1): 90–93.
19. Antonini, M., Barlaud, M., Mathieu, P., and Daubechies, I. (1992). Image coding using wavelet trasform. *IEEE Trans. Image Process.* 1 (2): 205–220.
20. Li, C., Ma, L., Wang, Q. et al. (2013). Construction of sparse basis by dictionary training for compressive sensing hyperspectral imaging. Geoscience and Remote Sensing Symposium IGARSS 2013 IEEE International, Melbourne, VIC (21–26 July 2013), pp. 1442–1445.
21. Chang, C.-I (2017). *Real-Time Recursive Hyperspectral Sample and Band Processing*. Cham: Springer International Publishing.
22. Nascimento, J.M.P. and Dias, J.M.B. (2005). Vertex component analysis: a fast algorithm to unmix hyperspectral data. *IEEE Trans. Geosci. Remote Sens.* 43 (4): 898–910.
23. Chen, S.Y., Li, H.C., and Wen, C.-H. (2016). A comparative analysis among ATGP, VCA and SGA for finding endmembers in hyperspectral imagery. *IEEE J. Sel. Top. Appl. Earth Obs. Remote Sens.* 9 (9): 4280–4306.
24. Du, Q., Raksuntorn, N., Younan, N.H. and King, R.L. (2008). Variants of N-FINDR algorithm for endmember extraction. SPIE Remote Sensing, Cardiff, Wales, United Kingdom (10 October 2008), p. 71090G.
25. Wang, Y., Guo, L., and Liang, N. (2009). Using a new search strategy to improve the performance of N-FINDR algorithm for end-member determination. 2009 2nd International Congress on Image and Signal Processing, Tianjin, China, pp. 1–4.
26. Du, Q., Raksuntorn, N., Younan, N.H., and King, R. (2008). Endmember extraction algorithms for hyperspectral image analysis. *Appl. Opt.* 47 (28): F77–F84.
27. Winter, M.E. (2004). A proof of the N-FINDR algorithm for the automated detection of endmembers in a hyperspectral image. Proceedings Volume 5425, Algorithms and Technologies for Multispectral, Hyperspectral, and Ultraspectral Imagery X, Defense and Security, Orlando, FL (12 August 2004), p. 31.
28. Harsanyi, J.C., Farrand, W. and Chang, C.-I (1994). Detection of subpixel spectral signatures in hyperspectral image sequences. Proceedings of American Congress on Surveying & Mapping (ACSM)/American Society of Photogrammetry & Remote Sensing (ASPRS) Annual Convention and Exposition, Baltimore, MD, vol. 1, pp. 236–247.
29. Chang, C.-I and Du, Q. (2004). Estimation of number of spectrally distinct signal sources in hyperspectral imagery. *IEEE Trans. Geosci. Remote Sens.* 42 (3): 608–619.

30. Chang, C.-I (2003). *Hyperspectral Imaging: Techniques for Spectral Detection and Classification*. New York, N.Y.: Kluwer Academic/Plenum Publishers.
31. Chang, C.-I (2017). *Real-Time Recursive Hyperspectral Sample and Band Processing*. New York: Springer.
32. Chang, C.-I (2018). A review of virtual dimensionality for hyperspectral imagery. *IEEE J. Sel. Top. Appl. Earth Obs. Remote Sens.* 11 (4): 1285–1305.
33. Li, J., Agathos, A., Zaharie, D. et al. (2015). Minimum volume simplex analysis: a fast algorithm for linear hyperspvtral unmixing. *IEEE Trans. Geosci. Remote Sens.* 53 (9): 5067–5082.
34. Bekit, A. (2019). Compressive sensing for endmember finding and linear spectral unmixing for hyperspectral imagery. Ph.D. dissertation. Department of Computer Science and Electrical Engineering, University of Maryland, Baltimore County, MD.
35. Della Porta, C.J., Bekit, A., Lampe, B., and Chang, C.-I (2019). Hyperspectral image classification via compressive sensing. *IEEE Trans. Geosci. Remote Sens.* 57 (10): 8290–8303.
36. Lampe, B. (2019). Compressive sensing with applications to hyperspectral image processing. Ph.D. dissertation. Department of Computer Science and Electrical Engineering, University of Maryland, Baltimore County, MD.

# 9

# Hyperspectral Image Classification in Compressively Sensed Band Domain

Charles J. Della-Porta[1] and Chein-I Chang[1,2]
[1]*Remote Sensing Signal and Image Processing Laboratory, Department of Computer Science and Electrical Engineering, University of Maryland Baltimore County, Baltimore, MD, USA*
[2]*Center for Hyperspectral Imaging in Remote Sensing (CHIRS), Information and Technology College, Dalian Maritime University, Dalian, China*

## 9.1 Introduction

Hyperspectral sensing technology has found success in a variety of applications ranging from agricultural land cover/use mapping, food inspection, environmental monitoring to medical imaging, law enforcement, and military reconnaissance/surveillance. Although it has continued to improve over the years, its applications are still limited due to size, weight, and power (SWaP) constraints. One of the challenging requirements is a need of sampling a large number of very fine spectral bands which require very fast and expensive analog-to-digital converters, high-capacity onboard storage, and optimized computational hardware and software to allow for real-time processing. Such requirements limit the utility of many applications and preclude the use of hyperspectral technology in many hyperspectral data exploitation including future space hyperspectral data processing.

Compressive sensing (CS) [1] has recently developed as a promising approach in hyperspectral data acquisition and analysis. CS approaches allow for data to be acquired directly in a compressed format, providing savings in both onboard storage and communication bandwidth. These techniques involve two fundamental concepts, sparsity specified by a sparse matrix, $\boldsymbol{\Psi} \in \Re^{L \times L}$, and incoherence specified by a sensing matrix, $\boldsymbol{\Phi} \in \Re^{m \times L}$, where $L$ is the total number of bands and $m$ is the number of sensed samples. In this chapter, the sensed samples commonly considered in CS are actually the sensed bands. Mutual incoherence plays a significant role in CS. It is used to measure the correlation between $\boldsymbol{\Phi}$ and $\boldsymbol{\Psi}$. According to the CS theory, the lesser the coherence between $\boldsymbol{\Phi}$ and $\boldsymbol{\Psi}$ is, the better signal sensing and reconstruction is [1]. Mathematically, mutual coherence is defined as

$$1 \leq \mu(\boldsymbol{\Phi}, \boldsymbol{\Psi}) = \sqrt{L} \max_{1 \leq i,j \leq L} \left| \langle \phi_i, \psi_j \rangle \right| \leq \sqrt{L} \tag{9.1}$$

*Advances in Hyperspectral Image Processing Techniques*, First Edition. Edited by Chein-I Chang.

where $\phi_j$ and $\psi_j$ represent the columns of $\mathbf{\Phi}$ and $\mathbf{\Psi}$, respectively. Specifically, the parameter $m$ satisfies the following inequality:

$$m \geq C \cdot k \cdot \log(L/k) \tag{9.2}$$

where $C$ is a constant that depends on each particular instance and $k$ is the order of sparsity. It has been shown in [2] that the low coherence between $\mathbf{\Phi}$ and $\mathbf{\Psi}$ is guaranteed by choosing a sensing matrix with random entries and the incoherence can be directly achieved using a Gaussian sensing matrix. Therefore, it is not necessary to check for incoherence explicitly.

In hyperspectral imaging, sparsity has been used as an additional constraint to expand existing hyperspectral imaging algorithms in various applications, such as linear spectral unmixing [3, 4], band selection [5], feature extraction [6, 7], and classification [8, 9]. Rather than incorporating sparsity as a constraint, this chapter takes a different approach by looking into the data acquisition process which is fundamentally changed to account for sparsity by sampling data at sub-Nyquist rates without the loss of information. More specifically, by taking advantage of CS, the burdens imposed by high sampling rates (band rates in terms of band acquisition) can be significantly reduced, i.e. by leveraging the fact that the spectral bands are redundant. Such a compressive sampling approach can be applied to each hyperspectral pixel vector by reducing a very large number of spectral bands to a small number of compressively sensed bands (CSBs) that need to be collected. However, to fully exploit this benefit, hyperspectral processing should be able to be applied directly in the compressively sensed band domain (CSBD) without an appreciable loss of performance [10].

One of the earliest works using CS was the smashed filter [11] which was an extension of the matched filter to matched embeddings in CSBD. The use of more complicated classifiers such as the support vector machine (SVM) in CSBD were also investigated [12]; however, the work was limited to the linear kernel and focused on fairly trivial synthetic classification problems. The work in [13] introduced the concept of compressed classification to hyperspectral data; however, the sampling strategy was based on an adaptively optimized sensing paradigm which is not easily realizable in sensing hardware. Additionally, the effects of scene complexity on compressed performance were not addressed.

In this chapter, hyperspectral image classification (HSIC) in CSBD is investigated. First of all, a universal CS model is developed which can be implemented without creation of new, complex sensing hardware. While the concept of universal sampling is not new to CS, the implications of using such a model in the context of hyperspectral classification are in fact new and critical for achieving easily realizable hardware designs. Second, error expressions between the full band domain and CSBD are derived for the linear kernel and radial basis function (RBF)-based kernel-based SVM. These error expressions show analytically that performance using full bands is indeed achievable in CSBD for sufficient band sensing conditions. Third, a series of experiments are performed to demonstrate HSIC in CSBD. The experiments include multiple images of various scenes from different sensors and illustrate the effects of scene complexity.

## 9.2 Compressive Sensing Review

CS is a sensing paradigm that is able to recover signals of interest at sub-Nyquist sample rates using nonadaptive strategies under the right circumstances. CS is the process of acquiring information in an already compressed state and should be thought of as a modification to the typical analog-to-digital conversion process. Rather than sampling a full signal, a compressed form of the signal is directly acquired, allowing for reductions in both sampling speed and data storage. The full signal can later be uncompressed, or recovered, using one of many sparse recovery algorithms [1]. A complete CS model consists of a *sparse acquisition* process (signal encoding) and a *sparse recovery* process (signal decoding). It should be noted, that while a typical CS system includes both encoding and decoding, this work is based on a completely compressed system where only the encoded signal is used.

The following sections introduce the fundamental concepts of CS and their applicability to hyperspectral pixel vectors.

### 9.2.1 Compressive Sensing Framework

In this section, we present a framework completely different from many existing sparse-based hyperspectral imaging methods which consider image pixels to be compressed sparsely. It considers the spectral bands of pixel vectors to be sparsely sampled from $L$ original bands to $m$ linear mixtures of spectral bands as CSBs with $m << L$.

More specifically, in a sparse acquisition process, an $L \times 1$ dimensional signal of interest $\mathbf{r}$ and a sparsely sampled signal $\bar{\mathbf{r}} \in \Re^{m \times 1}$ can be related mathematically as follows:

$$\bar{\mathbf{r}} = \mathbf{\Phi\Psi r} + \mathbf{n} \tag{9.3}$$

where $\bar{\mathbf{r}}$ is the compressed measurement vector, $\mathbf{\Phi} \in \Re^{m \times L}$ is the compressed sampling matrix, $\mathbf{\Psi} \in \Re^{L \times L}$ is the sparse representation basis, and $\mathbf{n} \in \Re^{m \times 1}$ is a noise term. The sampling strategy in CS is determined by two key principles: sparsity and incoherence [1]. The signal of interest, $\mathbf{r}$, must be sparse that can be expressed as in an appropriate sparse matrix, $\mathbf{\Psi}$, and must be also sampled in a sensing matrix, $\mathbf{\Phi}$, that is maximally incoherent with $\mathbf{\Psi}$.

Contrary to dimensionality reduction methods such as principal components analysis (PCA) [14], the CS strategy is implemented without prior knowledge of signals to be processed. Thus, it ought to rely on probabilistic convergence guaranteed under specific conditions. The *restricted isometry property* (RIP) [2] is a very important property that is typically imposed in CS to guarantee that relative measures of distance are maintained and preserved between the original and compressed signal domains. For a signal $\mathbf{r}$ with true sparsity level of $k$ determined by $\mathbf{\Psi}$, RIP satisfies the following inequalities:

$$(1 - \delta_k)\|\mathbf{r}\|_2^2 \le \|\mathbf{\Phi r}\|_2^2 \le \|\mathbf{r}\|_2^2(1 + \delta_k) \tag{9.4}$$

where $\delta_k$ is defined as the restricted isometry constant (RIC) and is the smallest possible constant that satisfies (9.4) for all $k$-sparse vectors in $\mathbf{r}$. As the value of the compressed parameter, $m$, approaches the dimensionality of the original signal, $L$, RIC approaches zero which results in an exact preservation of lengths.

Under mild conditions, the original signal acquired in (9.1) can later be recovered by a sparse recovery process which solves the following optimization problem [14–16]:

$$\min_{\tilde{\mathbf{r}} \in \Re^L} \left\|\tilde{\mathbf{r}}_S\right\| \text{ subject to } \mathbf{\Phi}\tilde{\mathbf{r}}_S = \tilde{\mathbf{r}} \tag{9.5}$$

where $\tilde{\mathbf{r}}_s$ represents the sparse signal that is recovered from the compressed measurement in (9.1). It is very important to note that for a solution to be feasible, the original vector $\mathbf{r}$ must be sparse or compressible (approximately sparse). The complete CS model is fully described by the sparse acquisition model shown in (9.3) and the sparse recovery produced by (9.5).

### 9.2.2 Compressive Sensing of Hyperspectral Bands

As described by (9.3), a CS system is completely characterized by two matrices, sparse matrix $\boldsymbol{\Psi}$ and sensing matrix $\boldsymbol{\Phi}$. One of the greatest challenges in designing a CS system is to identify the domain in which the signal of interest is optimally sparse. Previous works on hyperspectral sparsity have suggested the use of the discrete cosine transform (DCT) [16], wavelet transforms [17], and dictionary learning approaches [18]. Each of these approaches has shown varying levels of success based on the specific objectives and particular image scenes examined; however, it is difficult to prove a single basis that is universally optimal for all cases. Fortunately, as it will be shown later, with the universal sampling model, optimal compressed classification performance can be achieved without identifying a specific sparse basis.

For a given sparse matrix $\boldsymbol{\Psi}$, a sampling matrix $\boldsymbol{\Phi}$ must be chosen such that the coherence is minimized and RIP is also satisfied. As it turns out, for most signals a random sampling strategy has been shown to be sufficient, if not optimal [2]. In other words, certain random distributions can be shown to satisfy both RIP and incoherent sampling with respect to any sparse representation, $\boldsymbol{\Psi}$. This is critical for designing real-world systems as it makes the sampling system agnostic to the sensing modality and is exactly the main reason that random sampling strategies have become ubiquitous in the CS literature. Especially, the sparse acquisition model in (9.3) can be implemented by constructing the sampling matrix from any distribution that satisfies the *concentration of measure inequalities* shown in Eq. (4.3) of [2]. For such a family of sampling matrices, it can be shown that RIP is valid provided that (9.2) is satisfied, where the number of sensed bands, $m$, is directly related to the true order of sparsity, $k$, and also satisfies (9.2).

Several distributions have been shown to satisfy the concentration of measure inequalities, making them good choices for designing a random sampling sensing matrix. The Gaussian and Bernoulli distributions are cited in [2] as being the most prominent examples of valid distributions that can be used to generate elements of the sampling matrix $\boldsymbol{\Phi}$. For this work, the Gaussian distribution, $\boldsymbol{\Phi}_G$, is specifically chosen, where each entry element is distributed as $\phi_{ij} \sim N(0, 1/L)$ and the columns are normalized to unit length. The Gaussian sampling matrix ensures low coherence with any arbitrary basis, removing the need to measure and verify coherence. Furthermore, the Gaussian sampling matrix is guaranteed to satisfy RIP. This is a common approach for modeling CS systems and has been used in other HSI works [10].

It is important to note that for most applications in hyperspectral imaging, a user-selected sparsity order of $k$ is imposed directly [3–9] and is tied to a very specific sparse basis. This chapter makes a rather different attempt by emphasizing the choice of the sensing matrix to ensure that a sufficient number of CSBs, $m$, are collected to preserve relevant properties such as RIP. Specifically, the random sensing matrix $\boldsymbol{\Phi}$ drawn from distributions that must satisfy RIP will be nearly optimally incoherent with any basis function. The Gaussian distribution turns out to be one of such distributions that satisfies RIP [2].

### 9.2.3 Universality (Universal Sensed Model)

This subsection presents a very important property which is the key idea in our proposed CS approach. It is a universal sensed model which states that the sparse basis matrix, $\boldsymbol{\Psi}$, need not be known at the time of signal acquisition, but only required during sparse recovery. More specifically, by adopting a random sampling strategy, the sparse acquisition model in (9.3) can be simplified to

$$\bar{\mathbf{r}} = \mathbf{\Phi}_G \mathbf{\Psi} \mathbf{r} + \mathbf{n} \tag{9.6}$$

where $\mathbf{\Phi}_G$ represents the Gaussian sampling matrix. Furthermore, the constraint of the sparse recovery in (9.5) can be modified by including the sparse matrix $\mathbf{\Psi}$ to yield

$$\min_{\tilde{\mathbf{r}}_S \in \Re^L} \|\tilde{\mathbf{r}}_S\|_{l_1} \text{ subject to } \mathbf{\Phi}_G \mathbf{\Psi} \tilde{\mathbf{r}}_S = \mathbf{r} \tag{9.7}$$

By combining (9.6) and (9.7), a new universal sensed model, referred to as universality, is established to show that knowledge of the sparse matrix is only required for the signal reconstruction process not for the compressed domain. This has extremely important implications that, for algorithms performed in CSBD, optimal performance can be achieved without the need of identifying the sparse matrix $\mathbf{\Psi}$. That is, the $m$-CSBs are selected by the sensing matrix, $\mathbf{\Phi}$, without specifically identifying the sparse matrix $\mathbf{\Psi}$. This fact was shown in (9.6) and proved by universality in [2] which ensures that all of the information is captured in CSBD, even though the exact sparse basis is not known.

The universality also has very important implications in design of physically realizable compressively sensed hyperspectral systems. Given the fact that the sparsity transform, $\mathbf{\Psi}$, is no longer a part of the acquisition process, there is no need of requiring a specific transform to be included in the hardware design. This means that new compressively sampled hyperspectral systems can be designed using a generic linear mixing hardware. As a result, it reduces the overall cost of new systems. In addition, it also facilitates the process of retrofitting existing hyperspectral sensors to be able to compressively sample bands.

Technically speaking, CSBD can be considered as a different type of dimensionality reduction (DR) from commonly used DR techniques in the sense that the former does not require the knowledge of its used specific transform compared to the latter which must specify the transforms they use such as PCA or independent component analysis (ICA) for DR. The reason why CSBD can do so is due to the universality resulting from CS in [2]. More specifically, the universality in [2] states that data can be compressively sensed without first transforming the data to a particular domain. But it should be noted that the data in CSBD is still sparse because the sparseness is included in $m$-CSBs. This is indeed a significant advantage in hardware design since the knowledge of universality derived from the universal model in [2] allows for any transform without specifically being designed in the hardware development, thus facilitating the development of a real system. The concept of universality, as applied to compressed hyperspectral processing, is described in more detail and illustrated with an experiment in [19].

## 9.3 Hyperspectral Image Classification

Classification is one of many tasks in hyperspectral data exploitation. The spectral diversity intrinsically available in hyperspectral data can be leveraged directly as classification features. HSIC continues to be a very active area of research in utilizing state-of-the-art classifiers, such as deep neural network architectures [20–22], kernel-based discriminators [23, 24], statistical learning theory [25], and complex feature strategies based on both spatial and spectral features [26–28].

This section presents a joint spectral-spatial classifier in CSBD based on an SVM and EPF, following the methods developed in [27].

### 9.3.1 Linear Support Vector Machines

The SVM [29] is one of most widely used and popular classifiers which has been successfully applied in hyperspectral applications and is well understood in the machine learning and pattern recognition communities. The SVM is a binary discriminant function that finds a hyperplane separating two classes. The linear SVM can be defined as the discriminant function by

$$f(\mathbf{r}) = \mathbf{w}^T\mathbf{r} + b \tag{9.8}$$

where the weight vector $\mathbf{w}$ and bias $b$ are selected by satisfying a *maximum margin* objective function, subject to $\|\mathbf{w}\| = 1$. A hard classification of test data samples is performed by observing the sign of the output, where positive value +1 belong to one class and negative value −1 belong to the other. Although, the SVM is natively a binary classifier, it is often extended to multi-class classification problems through the use of a binary extension strategy, such as one-vs-rest or one-vs-one [30]. The one-vs-one approach is specifically adopted in this work.

To make SVMs more mathematically convenient, the Representer Theorem [31] is often leveraged. The theorem states that, with appropriate constraints, $\boldsymbol{\omega}$ can always be written as a linear combination of the training data, $\mathbf{w} = \sum_{j=1}^{N} \alpha_j d_j \mathbf{r}_j^T$ where $d_j \in \{-1. + 1\}$is the training label, $\alpha_j \in [0, 1]$ is a linear coefficient, and $N$ is the total number of training samples. Using this relationship, the linear SVM discriminant function can also be written as

$$f(\mathbf{r}) = \sum_{j=1}^{N} \alpha_j d_j \mathbf{r}_j^T \mathbf{r}_j \tag{9.9}$$

where the bias term is implicitly included in the weight vector. In general, most values of $\alpha_j$ will be zero and the nonzero values are referred to as the support vectors. In this form, the SVM is completely described by the support vectors, which are typically solved for iteratively.

Classification in CSBD can be represented using a modification of the discriminant function that includes the sparse acquisition matrix. The compressed discriminant function then takes the following form:

$$f_s(\boldsymbol{\Phi}(\mathbf{r})) = \sum_{j=1}^{N} \alpha_{s_j} d_j (\boldsymbol{\Phi}(\mathbf{r}_j))^T \boldsymbol{\Phi}(\mathbf{r}_j) \tag{9.10}$$

where the subscript $s$ of $\alpha$ denotes the fact that the SVM in CSBD is not guaranteed to use the same support vectors or bias.

To examine the performance of the support vectors in the classified domain, an error function can be constructed between them in the original and compressed domains. Let $\varepsilon = |f(\mathbf{r}) - f_s(\boldsymbol{\Phi}(\mathbf{r}))|$ denote the absolute error between the fully sampled and the sparsely sampled classifiers.

$$\varepsilon = \left| \sum_{j=1}^{N} \alpha_j d_j \mathbf{r}_j^T \mathbf{r}_j - \sum_{j=1}^{N} \alpha_{s_j} d_j (\boldsymbol{\Phi}(\mathbf{r}_j))^T \boldsymbol{\Phi}(\mathbf{r}_j) \right| \tag{9.11}$$

In general, it is impossible to derive a closed-form solution for (9.9) since the support vectors are solved iteratively in practice. Additionally, there is no guarantee of a unique solution when solving for these model parameters. To simplify the analysis, the support vectors are chosen, without guarantee of optimality, to be the same in the sparsely sampled domain. This implies that $\alpha_{s_j} = \alpha_j$, $\forall j$ and the error function in (9.11) can be reduced to

$$\varepsilon = \left| \sum_{j=1}^{N} \alpha_j d_j \mathbf{r}_j^T \mathbf{r}_j - \left[ (\mathbf{\Phi}(\mathbf{r}_j))^T \mathbf{\Phi}(\mathbf{r}_j) \right] \right| \tag{9.12}$$

Furthermore, recognizing that $\mathbf{r}_j^T \mathbf{r}_j = \|\mathbf{r}_j\|^2$, the classification error can be defined as

$$\varepsilon = \left| \sum_{j=1}^{N} \alpha_j d_j \left[ \|\mathbf{r}_j\|_2^2 - \|\mathbf{\Phi}(\mathbf{r}_j)\|_2^2 \right] \right| \tag{9.13}$$

Given the fact that the sampling matrix, $\mathbf{\Phi}$, satisfies the RIP in (9.4), the error can be rewritten as an inequality in terms of the RIC, $\delta_k$, as

$$\varepsilon = \left| \sum_{j=1}^{N} \alpha_j d_j \left[ \|\mathbf{r}_j\|_2^2 - (1 - \delta_k) \|\mathbf{r}_j\|_2^2 \right] \right| \tag{9.14}$$

Expanding the RIC term and simplifying (9.14) yields the final error

$$\varepsilon = \left| \sum_{j=1}^{N} \alpha_j d_j \left[ \delta_k \|\mathbf{r}_j\|_2^2 \right] \right| \tag{9.15}$$

which indicates that the classification error is directly bounded by the RIC and will go to zero for all $m$ that satisfy the condition (9.2) as $\delta_k$ goes to zero. This derivation demonstrates that it is indeed possible to approach full classification performance in CSBD provided that sufficient sampling conditions are satisfied, i.e. $m$ satisfies (9.2) and $\mathbf{\Phi}$ satisfies RIP in (9.4).

### *9.3.2 Kernel Support Vector Machines*

In many practical classification problems, the data are not linearly separable in the original feature space. A solution to this problem is to transform the feature vectors into a high dimensionality feature space where the data is linearly separable. For an arbitrary transformation, $\phi(\mathbf{r})$, the linear discriminant function can be rewritten, more generally, as

$$f_\phi(\mathbf{r}) = \sum_{j=1}^{N} \alpha_j d_j (\phi(\mathbf{r}_j))^T \phi(\mathbf{r}_j) \tag{9.16}$$

In practice, transforming all of the data samples to the new feature space is computationally prohibitive but can be avoided using what is referred to as the kernel trick. A property of Mercer's theorem [32] is used to directly calculate the required dot products without ever mapping the data into the nonlinear feature space. The dot products in (9.16) are replaced with an arbitrary kernel operator, $K(\mathbf{r}_j, \mathbf{r}) = (\phi(\mathbf{r}_j))^T \phi(\mathbf{r})$ where the kernel is carried out on the original data.

$$f_\phi(\mathbf{r}) = \sum_{j=1}^{N} \alpha_j d_j K(\mathbf{r}_j, \mathbf{r}) \tag{9.17}$$

Many different types of kernels have been proposed for use with support vector machines; however, one of the most common kernels is the Gaussian RBF kernel. The RBF is defined as

$K(\mathbf{r}_j, \mathbf{r}) = \exp\left(-\frac{\|\mathbf{r}_j-\mathbf{r}\|^2}{2\sigma^2}\right)$ where $\sigma$ is a tunable hyperparameter. Substituting the kernel definition into (9.17), the RBF SVM discriminant function is written as

$$f_\phi(\mathbf{r}) = \sum_{j=1}^{N} \alpha_j d_j \exp\left(-\frac{\|\mathbf{r}_j-\mathbf{r}\|^2}{2\sigma^2}\right) \quad (9.18)$$

Following a similar approach to the linear SVM, the performance of the compressed RBF SVM can be analyzed by observing the error between the original and compressed domains. Choosing, again, to maintain the same support vectors in CSBD, the error function for the RBF is described as

$$\varepsilon = \sum_{j=1}^{N} \alpha_j d_j \left| \exp\left(-\frac{\|\mathbf{r}_j-\mathbf{r}\|^2}{2\sigma^2}\right) - \exp\left(\frac{\|\phi(\mathbf{r}_j)-\phi(\mathbf{r})\|^2}{2\sigma^2}\right) \right| \quad (9.19)$$

For simplicity, let $\mathbf{z}_j = \mathbf{r}_j - \mathbf{r}$ and substitute it into (9.19) to yield

$$\varepsilon = \sum_{j=1}^{N} \alpha_j d_j \left| \exp\left(-\frac{\|\mathbf{z}_j\|^2}{2\sigma^2}\right) - \exp\left(\frac{\|\phi(\mathbf{z}_j)\|^2}{2\sigma^2}\right) \right| \quad (9.20)$$

Knowing that the sampling matrix, $\mathbf{\Phi}$, satisfies the RIP in (9.4), the error can be rewritten as an inequality in terms of the RIC, $\delta_k$.

$$\varepsilon \leq \sum_{j=1}^{N} \alpha_j d_j \left| \exp\left(-\frac{\|\mathbf{z}_j\|^2}{2\sigma^2}\right) - \exp\left(\frac{\|(1-\delta_k)\mathbf{z}_j\|^2}{2\sigma^2}\right) \right| \quad (9.21)$$

Recognizing that $(1-\delta_k)$ is a scalar quantity, the expression can be further reduced by expanding the RIC term and factoring out the exponential.

$$\varepsilon \leq \sum_{j=1}^{N} \alpha_j d_j \left| \exp\left(-\frac{\|\mathbf{z}_j\|^2}{2\sigma^2}\right) - \exp\left(\frac{\|\mathbf{z}_j\|^2}{2[\sigma/(1-\delta_k)]^2}\right) \right| \quad (9.22)$$

Interestingly, the error is no longer bounded linearly by the RIC, but rather by a difference in the kernel space. In this particular case, both terms are identical with the exception of the RBF hyperparameter, which is scaled by $(1-\delta_k)^{-2}$. Just as in the linear case, the RBF classification error will go to zero for all $m$ that satisfy the condition shown in (9.2), as $\delta_k$ goes to zero.

### 9.3.3 Edge-Preserving Filters

Spatial correlation between neighboring pixels must be taken into account to improve hyperspectral classification performance. To capture the spatial contextual information, edge-preserving filters (EPFs) are included in the spectral-spatial classification process [28]. In this approach, the initial pixel class probabilities are determined by an SVM classifier and then the SVM-classification map is further refined by a follow-up EPF. It should be noted that there were four EPF-based methods developed in [28]. As an illustrative purpose, the guided EPF filter based on a 3-channel principal component image, referred to as EPF-G-c, is specifically selected and used for all the following experiments. Nevertheless, the proposed CSBD-based hyperspectral classification can be also applicable to other classification methods including the other three EPF-based methods as well. A description of the full classification algorithm is shown as follows.

---

***EPF-G-c Classification Algorithm***

---

**Input**: A hyperspectral image $R \in \Re^{N_x \times N_y \times L}$ and $M$ classes

1. Generate a classification map by an SVM classifier, $\boldsymbol{C}_{SVM} \in \Re^{N_x \times N_y}$.
2. Generate $M$ initial binary probability maps, $\boldsymbol{P} \in \Re^{N_x \times N_y}$, by assigning the pixelwise class to each respective classification map as

$$P_n = \begin{cases} 1; \text{if } C_i = n \\ 0; \text{otherwise} \end{cases}; \text{for } n = 1, \ldots, M \tag{9.23}$$

3. Optimize the probability map by applying an edge-preserving filter,

$$\overline{P}_n = \sum_j w_{ij}(I) P_n$$

   where the weights, $w_{ij}$, depend on the choice of EPF, pixel location $(i,j)$, and the guidance image $I$.
4. Generate the final classification map by choosing the class with the highest probability $C_{\text{EPF}} = \arg\{\max_{1 \leq n \leq M} \overline{P}_n\}$

**Output**: A final classification map, $C_{EPF} \in \Re^{N_x \times N_y}$.

---

## 9.4 Classification Measures

It has been a common practice in hyperspectral classification that overall accuracy (OA), average accuracy (AA), and kappa coefficient are major measures to be used to evaluate classification performance. Unfortunately, it has been shown in [33, 34] that OA and AA were not sufficient enough to evaluate the effectiveness of a classifier. There is a more important measure, called precision rate (PR), arising in pattern classification that can be used for this purpose.

Consider an $M$-class confusion matrix constructed in Figure 9.1.

Based on Figure 9.1, several definitions are described as follows:

- $M$ = the number of classes
- $n_{ij}$ = the number of samples in the $i$th class that have been classified in the $j$th class
- $N$ = total number of data samples, $N = \sum_{i=1}^{M} n_{ii}$ In the following, we define several classification measures that are used to assess performance. Specifically, the class accuracy, AA, OA, and average precision are all defined.
- Accuracy of the $i$th class

$$P_{\mathrm{A}}(C_i) = \frac{n_{ii}}{\sum_{i=1}^{M} n_{ii}} \tag{9.24}$$

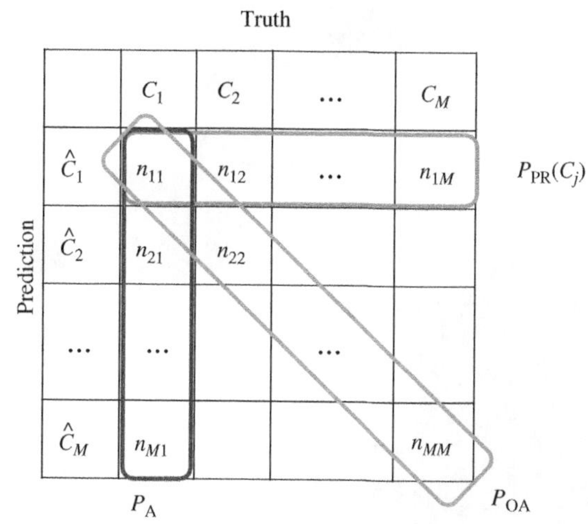


**Figure 9.1** *M*-class confusion matrix.

- Overall accuracy (OA)

$$P_{OA}(C_i) = \frac{\sum_{i=1}^{M} n_{ii}}{N} \tag{9.25}$$

- Average accuracy (AA)

$$P_{AA}(C_i) = \frac{1}{M}\sum_{i=1}^{M} P_A(C_i) \tag{9.26}$$

- Precision (PR)

$$P_{PR}(C_i) = \frac{n_{ii}}{\sum_{j=1}^{M} n_{ij}} \tag{9.27}$$

- Average precision rate (APR)

$$P_{APR}(C_i) = \frac{1}{M}\sum_{i=1}^{M} P_{PR}(C_i) \tag{9.28}$$

It should be noted that $P_{PR}$ in (9.27) is generally calculated for the entire image scene including background (BKG). The ability to include BKG pixels is the key advantage to the precision metric, since it is the only performance measure that can include all of the pixel data.

## 9.5 Experiments

Four images were used for experiments. Two scenes were acquired by Airborne Visible/Infrared Imaging Spectrometer (AVIRIS) [35] and two other scenes were acquired by the reflective optics system imaging spectrometer (ROSIS) [36]. The AVIRIS is an aircraft-based hyperspectral sensor with 224

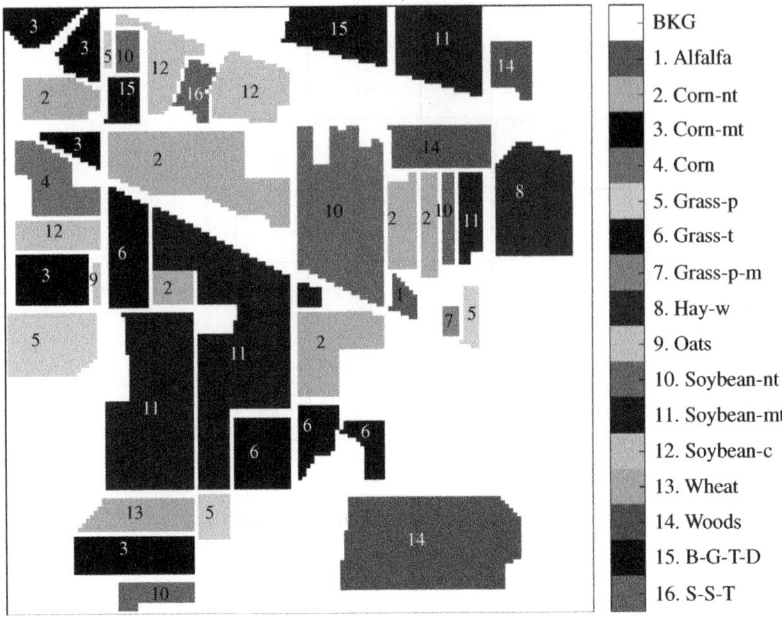


**Figure 9.2** Indian Pines ground truth.

contiguous bands over the 400–2500 nm spectrum with 10 nm spectral resolution. The first AVIRIS image shown in Figure 9.2 was collected over Purdue's Indiana Indian Pines test site and is composed of a mixture of agriculture and forestry with 20 m spatial resolution [37]. It is a well-known benchmark data set for hyperspectral classification and consists of 200 spectral channels after removal of the water absorption bands. The provided ground truth includes 16 classes and BKG samples. A second AVIRIS data set is the Salinas image [37] which has 3.7 m spatial resolution and consists of 200 spectral channels after removal of the water absorption bands. The image was collected over Salinas Valley, California, shown in Figure 9.3 and is composed of a mixture of agriculture signatures. The ground truth consists of 17 classes, including the BKG.

ROSIS is an aircraft-based hyperspectral sensor with 115 contiguous bands over the 430–860 nm spectrum with a spatial resolution of 1.3-meter per pixel and spectral resolution of 4 nm (12 most noisy channels were removed before experiments) [38]. The sensor was originally designed for resolving fine spectral structures and has produced several images that have been widely used within the literature.

Two ROSIS images were used for the experiments, the Pavia University and Pavia Centre image scenes [37], which are well-known benchmark data sets for hyperspectral classification. The images have 1.3 m spatial resolution and consists of 102 spectral channels after removal of the water absorption bands. The images were collected over Pavia, northern Italy, and are composed of urban scenes. The provided ground truths, for each image, consists of nine different classes and are shown in Figures 9.4 and 9.5.

### *9.5.1 Experimental Setup*

The sparse hyperspectral pixel vectors were modeled using random Gaussian sensing matrices with $m$-CSBs taken directly in the original full band domain, as shown in Eq. (9.6) (i.e. no sparsifying

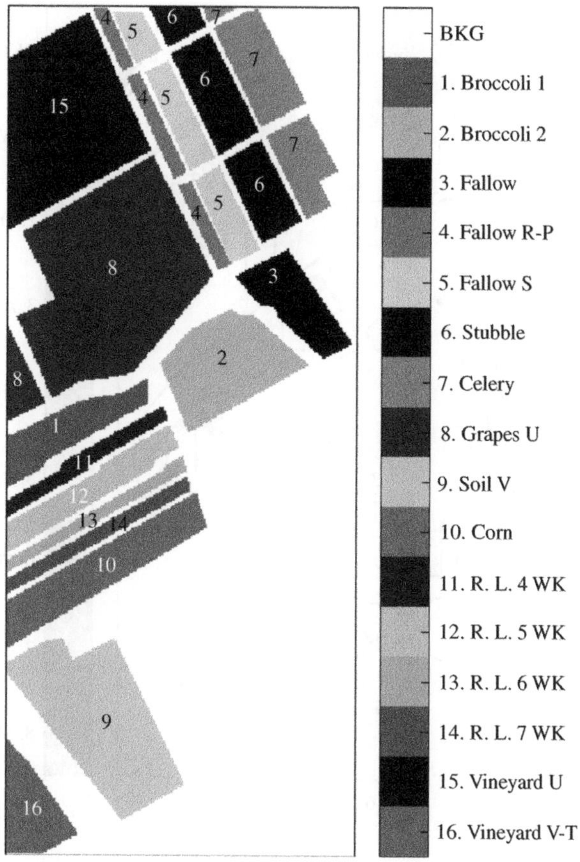


**Figure 9.3** Salinas ground truth.

transformation was applied before sampling). Additionally, the sampling noise, **n**, has been omitted for simplicity. The number of CSBs, *m*, was varied from five up to the full band number of bands for each image. The RBF-based SVM was used as the spectral classifier, with the tunable kernel parameter set to 0.72, 0.64, 0.75, and 0.59 for Indian Pines, Salinas, Pavia University, and Pavia Centre, respectively. The edge-preserving filtering process was based on a guided filter with a color guidance image, based on the first three principal components, i.e. EPF-G-c. Algorithm performance is measured by three common metrics: OA, AA, and average precision (APR) as described in Section 9.4.

For each experiment, the pixel vectors were randomly partitioned into training and test sets following the same setup described in [28] for Indian Pines, Salinas, and Pavia University. The Pavia Centre scene was partitioned in a similar fashion to Pavia University with 285 training samples from each class. The exact number of training and test samples is tabulated in Tables 9.1–9.4.

The same randomly selected set of training samples in Tables 9.1–9.4 was used for experiments in both the original data space (ODS) and CSBD for EPF-G-c. Each experiment was repeated 20 times and the average results were calculated. For all of the plots in this section, the dashed and solid lines show the results of each classification measure produced by EPF-G-c for the ODS, i.e. data without CS and CSBD, respectively.

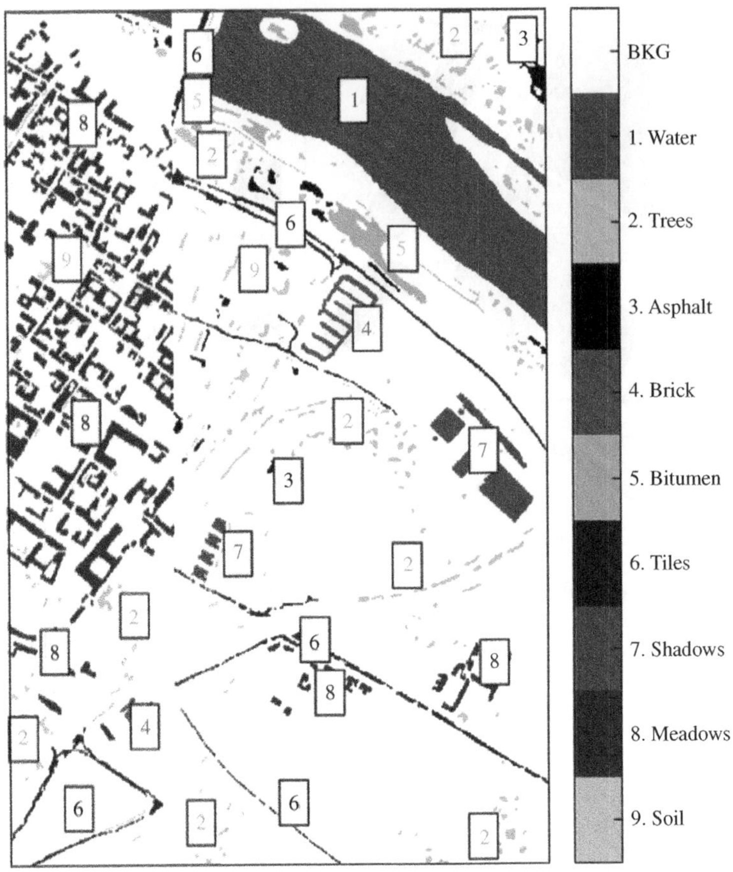


**Figure 9.4** Pavia Centre ground truth.

## *9.5.2 Classification Accuracy Analysis*

### 9.5.2.1 Purdue Indian Pines Scene

$P_{OA}$ and $P_{AA}$ are calculated for Purdue's Indian Pines and plotted in Figure 9.6 versus *m*-CSBs with *m* ranging from 1 to 220 where the plots had a large jump around 30 CSBs, and then began flat after 100 CSBs with near-full band performance. As can be seen in the figure, the maximum performance was achieved using approximately 50% of the total number of CSBs. The range of accuracy for this image is significant with just over a 20% difference in accuracy between the minimum number of CSBs and using full CSBs. This particular image has most imbalanced classes and was also the most difficult one among the four images that were tested. The presented results are in line with many of the performance measures reported in the literature.

### 9.5.2.2 Salinas Scene

$P_{OA}$ and $P_{AA}$ for Salinas are calculated and plotted in Figure 9.7. Similar to the experiments conducted for the Purdue Indian Pines, the largest improvement in classification accuracy occurs within the first 30 CSBs and continues to improve slowly until the plots become flat after 100 CSBs.

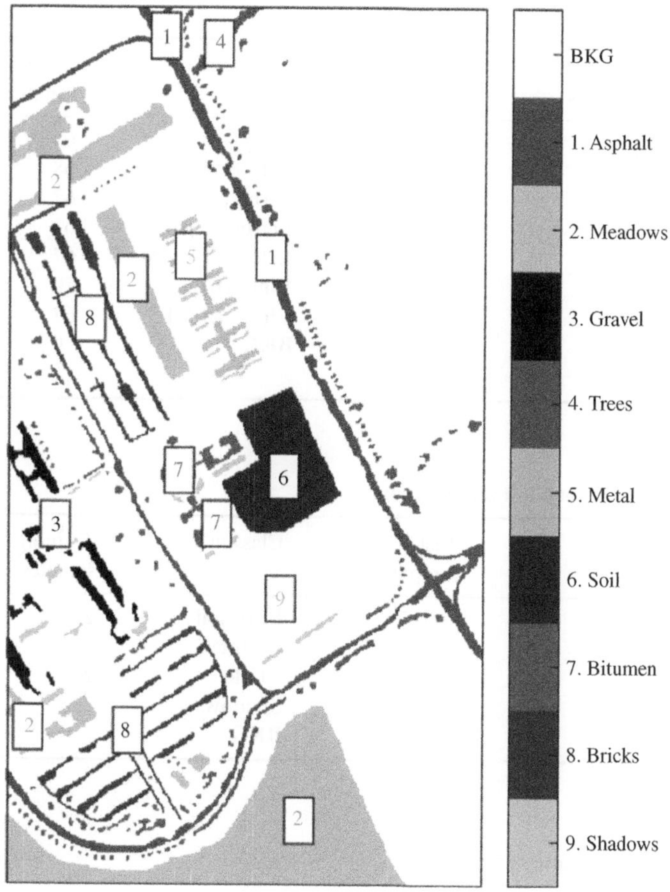


**Figure 9.5** Pavia University ground truth.

**Table 9.1** Indian Pines training and test sample counts

| Class name | Training samples | Test samples | Class name | Training samples | Test samples |
|---|---|---|---|---|---|
| **Alfalfa** | 25 | 21 | **Soybean-notil** | 81 | 891 |
| **Corn-notil** | 83 | 1345 | **Soybean-mintil** | 99 | 2356 |
| **Corn-mintil** | 78 | 752 | **Soybean-clean** | 73 | 520 |
| **Corn** | 68 | 169 | **Wheat** | 70 | 135 |
| **Grass-pasture** | 79 | 404 | **Woods** | 90 | 1175 |
| **Grass-trees** | 78 | 652 | **Building** | 65 | 321 |
| **Grass-mowed** | 14 | 14 | **Towers** | 46 | 47 |
| **Hay-windrow** | 66 | 412 | **BKG** | 0 | 10,076 |
| **Oates** | 10 | 10 | | | |

**Table 9.2** Salinas training and test sample counts

| Class name | Training samples | Test samples | Class name | Training samples | Test samples |
|---|---|---|---|---|---|
| **Broc. 1** | 67 | 1942 | **Corn** | 68 | 3210 |
| **Broc. 2** | 67 | 3659 | **Lettuce 4 week** | 68 | 1000 |
| **Fallow** | 67 | 1909 | **Lettuce 5 week** | 67 | 1860 |
| **Fallow rough** | 69 | 1325 | **Lettuce 6 week** | 67 | 849 |
| **Fallow smooth** | 67 | 2611 | **Lettuce 7 week** | 67 | 1003 |
| **Stubble** | 67 | 3892 | **Vineyard U** | 70 | 7198 |
| **Celery** | 68 | 3511 | **Vineyard VT** | 67 | 1740 |
| **Grapes** | 69 | 11202 | **BKG** | 0 | 56 975 |
| **Soil vineyard** | 68 | 6135 | | | |

**Table 9.3** Pavia University and test sample counts

| Class name | Training samples | Test samples | Class name | Training samples | Test samples |
|---|---|---|---|---|---|
| **Asphalt** | 286 | 6345 | **Bare soil** | 285 | 4744 |
| **Meadow** | 286 | 18363 | **Bitumen** | 285 | 1045 |
| **Gravel** | 285 | 1814 | **Bricks** | 285 | 3397 |
| **Trees** | 285 | 2779 | **Shadows** | 285 | 662 |
| **Metal** | 285 | 1060 | **BKG** | 0 | 164 624 |

**Table 9.4** Pavia Centre and test sample counts

| Class name | Training samples | Test samples | Class name | Training samples | Test samples |
|---|---|---|---|---|---|
| **Water** | 285 | 65686 | **Tile** | 285 | 8963 |
| **Trees** | 285 | 7313 | **Shadow** | 285 | 7002 |
| **Asphalt** | 285 | 2805 | **Meadow** | 285 | 42541 |
| **Brick** | 285 | 2400 | **Soil** | 285 | 2578 |
| **Bitumen** | 285 | 6299 | **BKG** | 0 | 635 488 |

Contrary to the Indian Pines image, the range of accuracy values is quite small with just about a 6% difference in performance. For the minimum case, it required only 5 CSBs for EPF-G-c to achieve an OA and an AA of 92 and 95%, respectively. Remarkably, even with just 2.5% of the total number of CSBs, the achieved accuracy level is acceptable for many practical applications.

#### 9.5.2.3 Pavia University

$P_{OA}$ and $P_{AA}$ are calculated for Pavia University and also plotted in Figure 9.8. Both accuracies converged quickly between 20 and 40 CSBs. This ROSIS image showed a similar trend to the AVIRIS Indian Pines image. There is an exponential increase in performance for the lower number of CSBs.

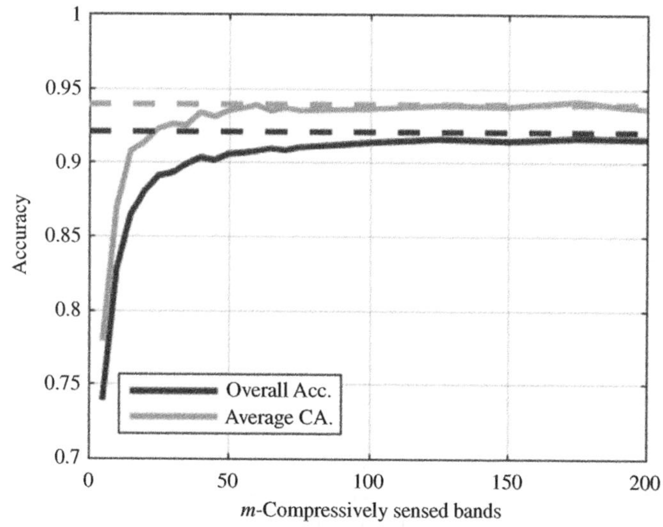


**Figure 9.6** Indian Pines accuracy.

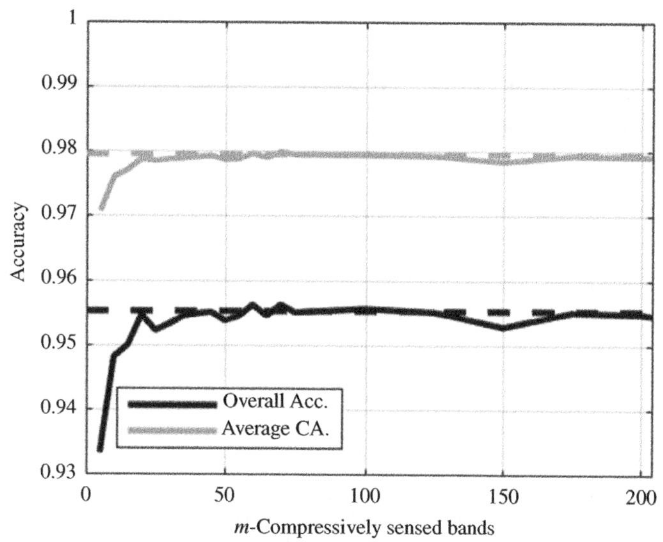


**Figure 9.7** Salinas accuracy.

Similarly, the range of accuracy values between the fewest CSBs and the full band performance is quite significant with a total difference of about 15% for the OA and 20% for the AA.

#### 9.5.2.4 Pavia Centre

The $P_{OA}$ and $P_{AA}$ for Pavia Centre are plotted in Figure 9.9. This image is somewhat unique because both accuracies appeared to converge at a slightly different number of CSBs where $P_{OA}$ converged with about 20 CSBs and the $P_{AA}$ converged with around 30–50 CSBs. The reason for this would

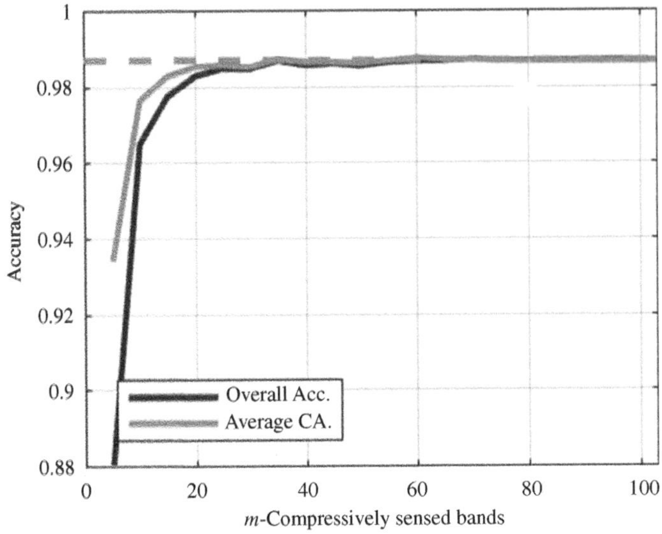


**Figure 9.8** Pavia University accuracy.

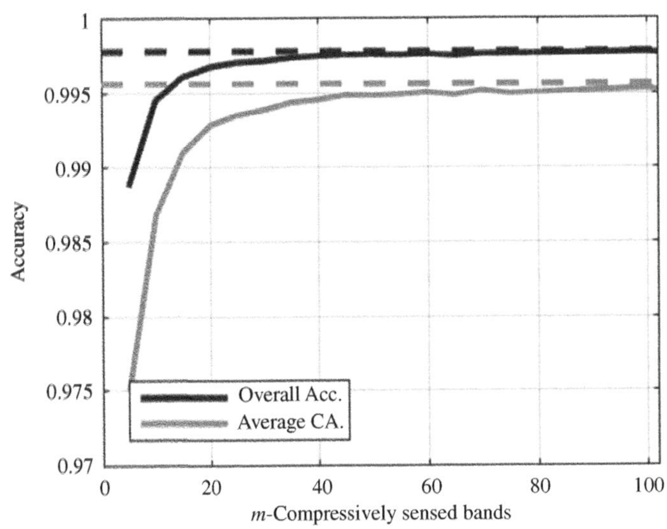


**Figure 9.9** Pavia Centre accuracy.

become more apparent when the individual class accuracies are discussed in later experiments. The range of accuracy for this image was exceptionally small. The difference in accuracy between using 5 CSBs and the full band image was just 2.5% for $P_{AA}$ and less than 1% for $P_{OA}$. This image is another example which shows accuracy required for practical applications can be achieved with small fractions of the total number of bands.

Interestingly, the classification accuracy achievable in the CSBD did not appear to be correlated with either the scene types (agricultural and urban) or the sensors (AVIRIS and ROSIS) used for experiments. However, for all images, acceptable accuracy can be achieved directly in CSBD with a significant reduction in the number of CSBs.

### *9.5.3 Classification Precision Analysis*

$P_{APR}$ of EPF-G-c is considered in this section and has received little interest in HSIC. However, its practical impact is quite important since it is the only measure that accounts for BKG data samples for classification. To address BKG issue in classification, $P_{APR}$ is calculated with and without inclusion of the BKG samples in the image scenes.

#### 9.5.3.1 Purdue Indian Pines Scene

The average precision for the Purdue Indian Pines is shown in Figure 9.10. Near-full precision performance is achieved with only 50 CSBs with and without the inclusion of the BKG samples. The range of precision values between the fewest CSBs and the full band algorithm is much smaller than the spread in accuracy, with only about a 10% difference. Following a similar trend as with accuracy, Indian Pines is again the most difficult of the four images tested.

#### 9.5.3.2 Salinas and Pavia Scenes

The results of $P_{APR}$ for Salinas, Pavia University, and Pavia Centre are shown in Figure 9.11(a–c), respectively, where the results of using full bands in the ODS could be achieved for the each of images with using only 5–10 CSBs! This is quite interesting given the fact that both Indian Pines and Salinas were collected from the same sensor and consist of similar agricultural classes. The main reason is that the Purdue data contains many imbalanced classes with 4 classes less than 100 data samples.

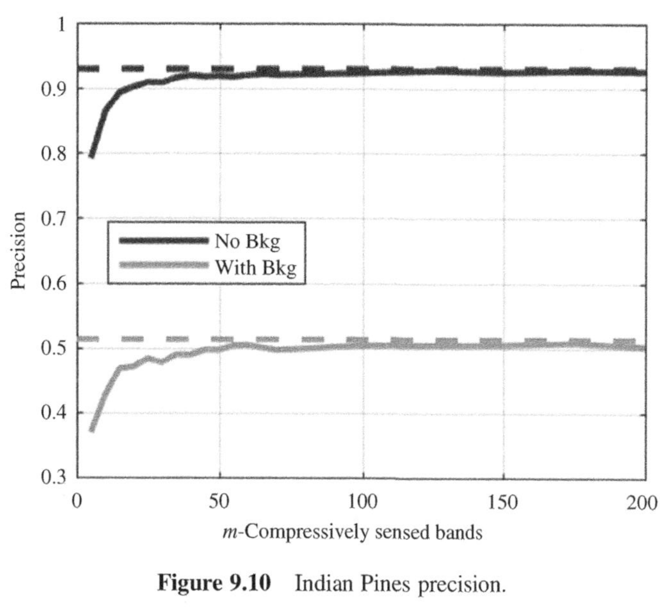


**Figure 9.10** Indian Pines precision.

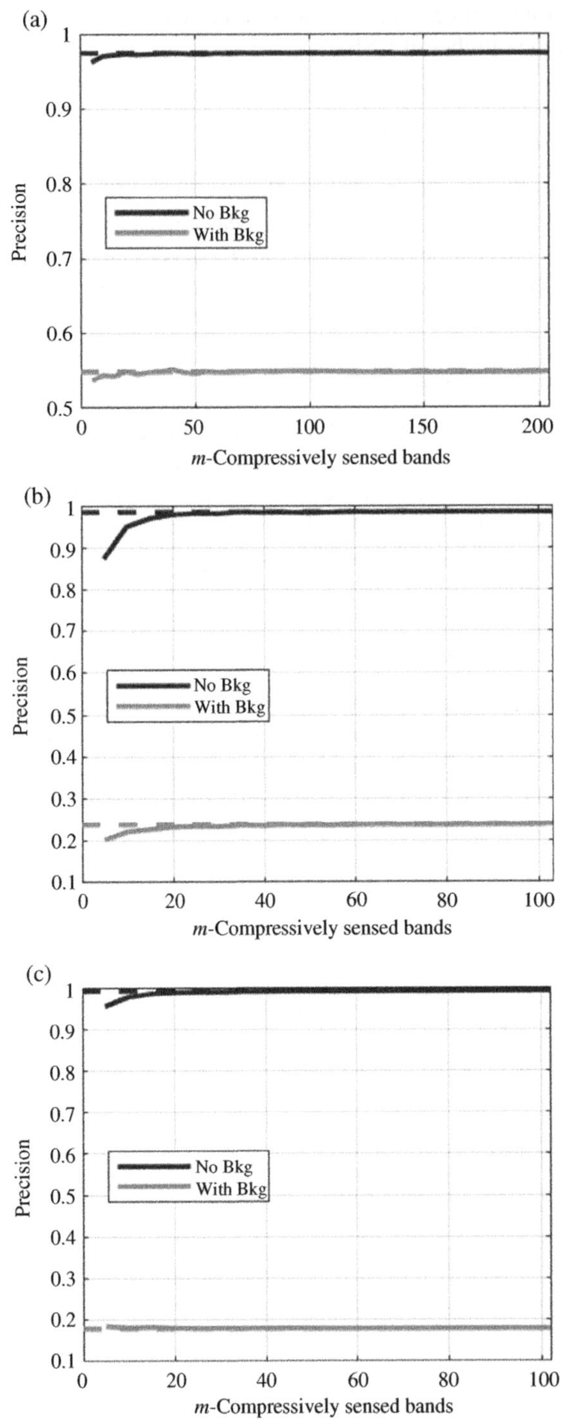


**Figure 9.11** Overall precision (OPR) of Salinas, Pavia University of Pavia Centre. (a) Salinas, (b) Pavia University, and (c) Pavia Centre.

In general, the precision is less affected by the compressive sampling for all images. In all cases, using full band performance could be achieved with 25% or less of CSBs. For Salinas and Pavia Centre, full band precision performance could be achieved while using only 5% of CSBs. The precision performance was also fairly consistent with and without the inclusion of BKG samples. In agreement with the accuracy performance results, the classification in the CSBD was capable of achieving acceptable precision performance with just fractions of the total number of CSBs.

### 9.5.4 *Discussions on Individual Class Accuracies*

In this section, the individual class accuracy is calculated. The individual class accuracies for Indian Pines are plotted in Figure 9.12 Interestingly, many of the classes converged to full band performance with as few as 10 CSBs, i.e. only 5% of the spectral bands! This is in stark contrast to the 100 CSBs needed before the OA converged. Accuracy performance in both the full band domain and CSBD is clearly limited by specific classes. The individual class accuracies for Salinas are shown in Figure 9.12. Remarkably, in comparison with the Purdue data, all of the classes with the exception of "vineyard untrained" converged to full band performance with only 10 CSBs or less! This explains why $P_{AA}$ is higher than $P_{OA}$. This suggested that the data set is almost completely separable. It is also very significant because it shows a clear correlation between class separability and the viability of compressed classification.

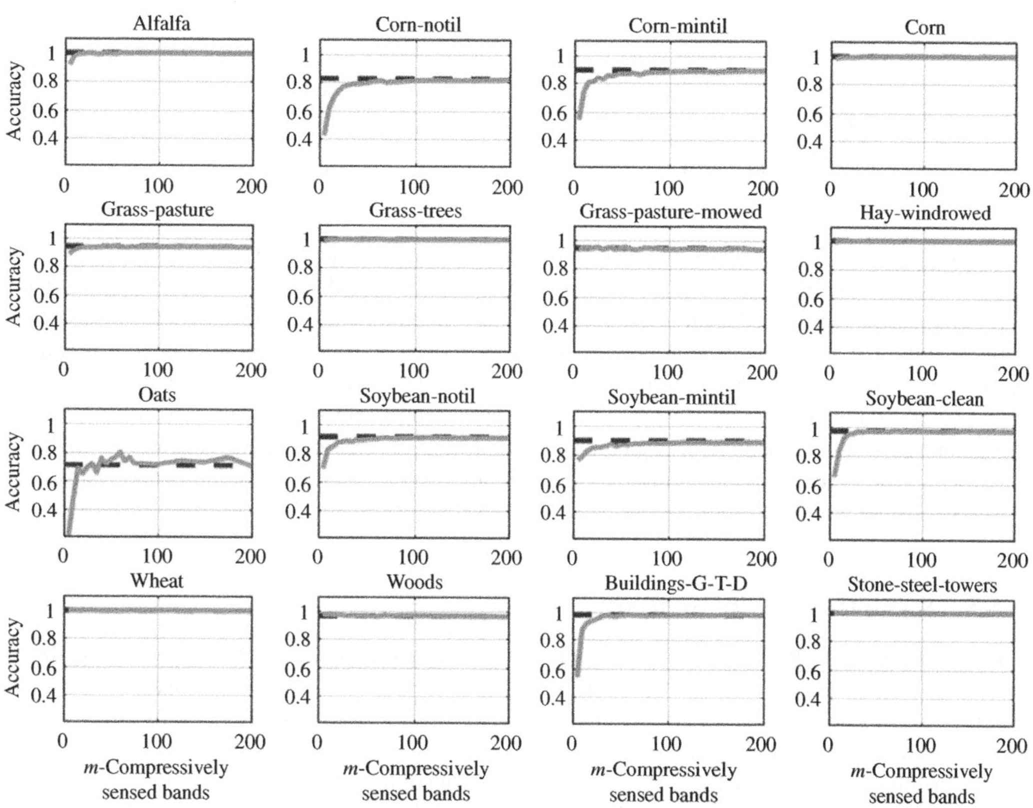


**Figure 9.12** Individual class accuracies of Indian Pines.

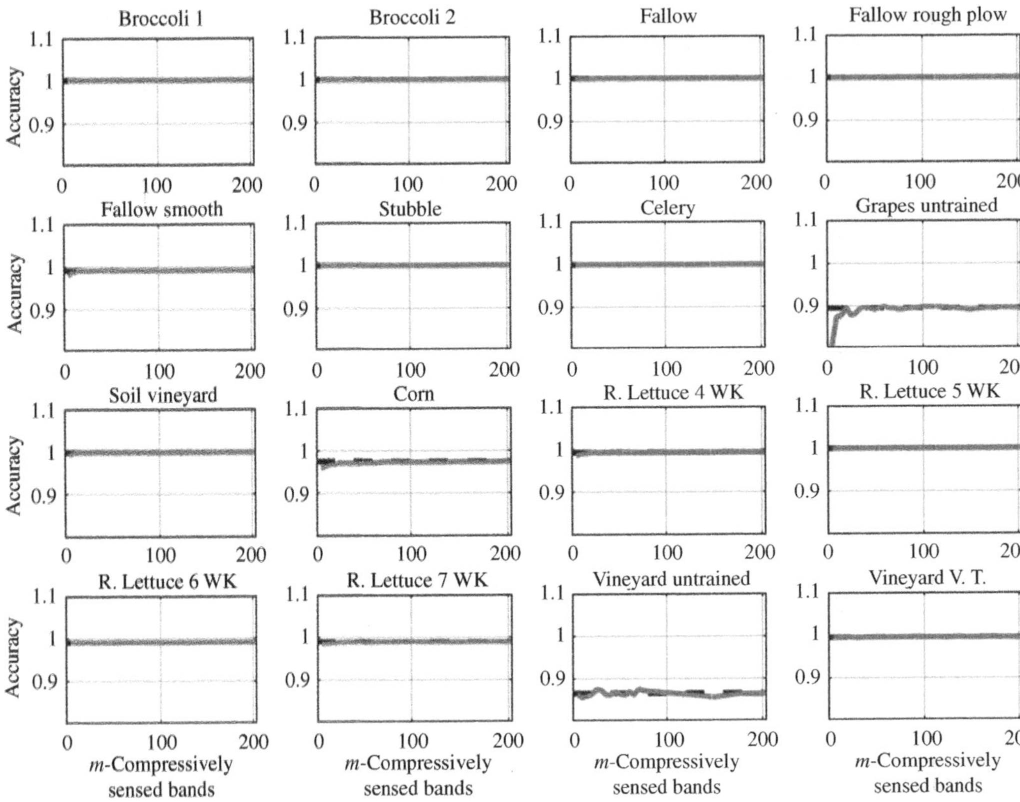


**Figure 9.13** Individual class accuracies of Salinas.

Similar performance was also observed for the ROSIS images. The individual class accuracies for the Pavia University and Pavia Centre are shown in Figures 9.13 and 9.14. For the Pavia University, half of the classes converged with just 5 CSBs and the remaining classes converged at various numbers of CSBs. All of the classes with the exception of "Gravel" converged to full band classification accuracy. For the Pavia Centre, a similar trend was observed with half of the classes converging with only 5 CSBs and the remaining classes converging at various numbers of CSBs. Two of the classes that did not converge immediately, "asphalt" and "shadows," are classes that never achieved full band classification accuracy. This also reinforces the correlation between class separability and viability of CS classification.

### *9.5.5 Scene Complexity Analysis*

One of interesting results from the above experiments is variability in the number of required CSBs to achieve maximum performance for each of the different scenes. This observation was true for both AVIRIS and ROSIS sensors, suggesting that it was indeed scene complexity that limits effectiveness of CSBs. To investigate this concept further, an efficacy criterion is defined for the OA, which calculates the ratio of $P_{OA}$ in CSBD to $P_{OA}$ in the original data domain, i.e. full band domain, that is, the compressively sensed band ratio (CSBR), CSBR = $m/L$, defined as the number of CSBs divided by the total number of bands, $L$. Using these criteria makes it possible to directly compare the

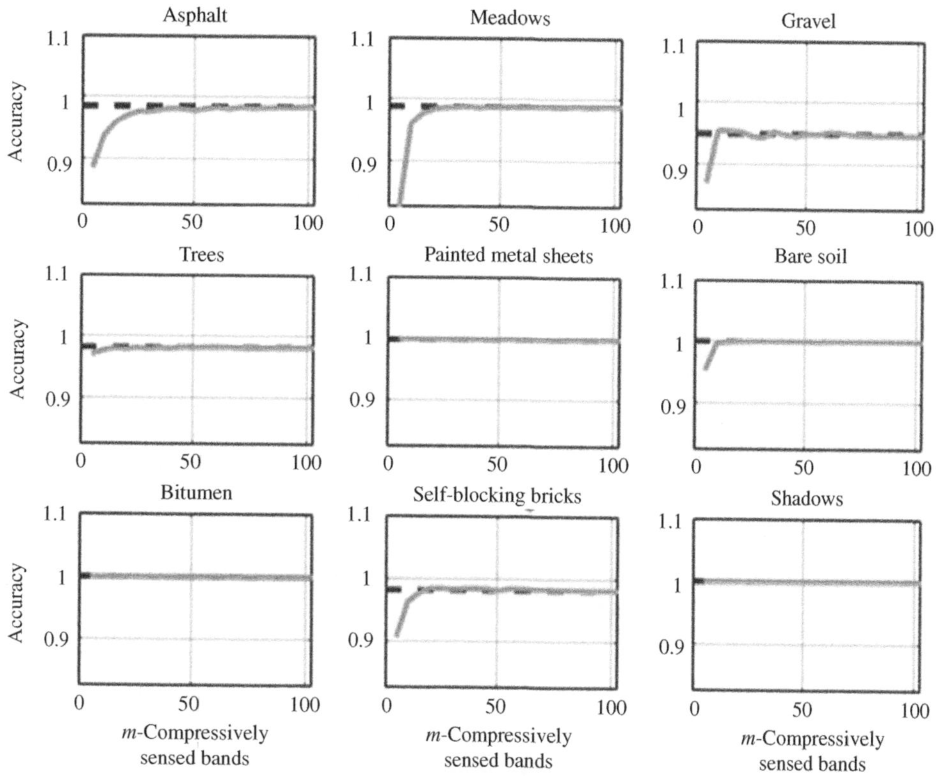


**Figure 9.14** Individual class accuracies of Pavia University.

performance for all four images. The OA efficiency is plotted as a function of CSBR in Figure 9.15 (a). For both Salinas and Pavia Centre, nearly full band performance can be achieved with CSBR of only 0.1. For the more difficult images, Pavia University required a CSBR of approximately 20% and Indian Pines required approximately 50%.

Similarly, the efficacy of precision calculates a ratio of precision in the CSBD to precision in the original full band domain and its results are plotted in Figure 9.15(b). As shown in the previous section, precision is generally better than accuracy. The efficacy of precision confirmed this fact with the worst-case performance never being lower than 90% of the full band precision performance after CSBR greater than 0.2. The efficacy of precision also showed the same trend as the efficacy of $P_{OA}$ with the Indian Pines and Pavia University images being more difficult than the other images.

### 9.5.6 *Selecting an Appropriate Number of CSBs*

For each real hyperspectral image scene, its complexity determines an appropriate number of CSBs based on the expected classes. Fortunately, a rather simple and practical approach for selecting the required number of CSBs can be adopted based on the training data. In other words, the number of CSBs can be selected based solely on the training data. Figure 9.16 shows an example of the training $P_{AA}$ and test $P_{AA}$ for the Indian Pines image scene.

Note that while there is approximately 2% difference between the training and test $P_{AA}$, the relative effect from varying the number of CSBs is consistent.

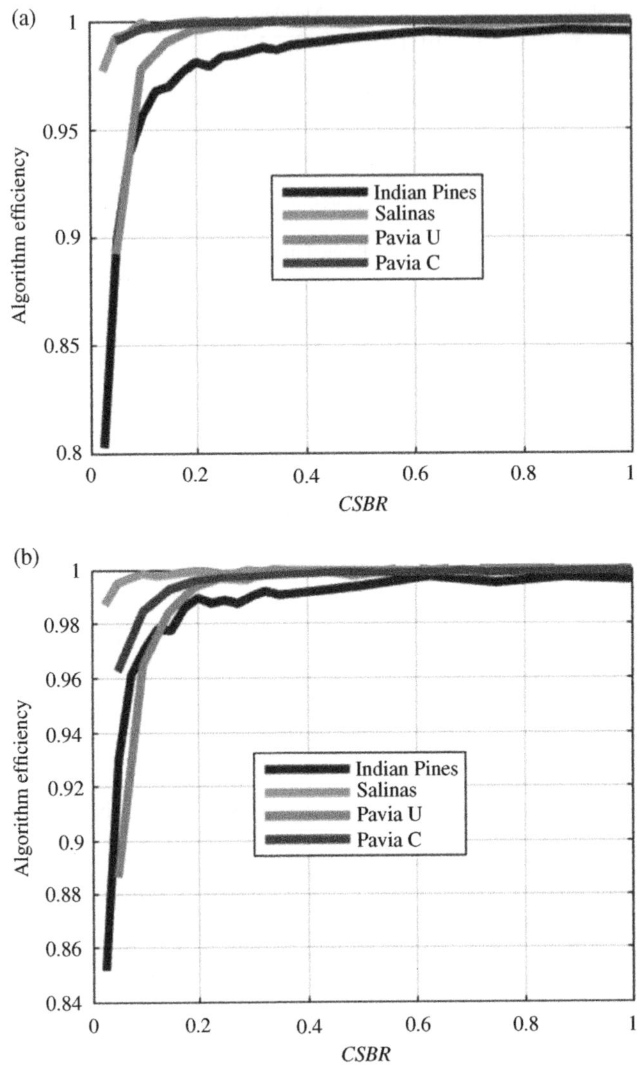


**Figure 9.15** Overall and accuracy efficacy and overall precision efficacy for all four images: (a) overall accuracy efficacy and (b) overall precision efficacy.

Furthermore, the correlation between the training and test performance can be easily quantified by using a measure such as the Pearson correlation coefficient (PCC), which is defined as

$$\frac{\text{cov}(a, b)}{\text{SD}(a)\text{SD}(b)} \tag{9.29}$$

where $a$ and $b$ are arbitrary vectors of equal length, cov($a$,$b$) represents the covariance between $a$ and $b$, and SD($a$) and SD($b$) represents the standard deviations of $a$ and $b$. The PCC between the training and test performance is equal to 0.9980. Similarly, the correlation coefficients for Salinas, Pavia University, and Pavia Centre are equal to 0.9811, 0.9996, and 0.9996, respectively. It is quite clear for these images that the relationship between the training and test data is highly correlated as a function

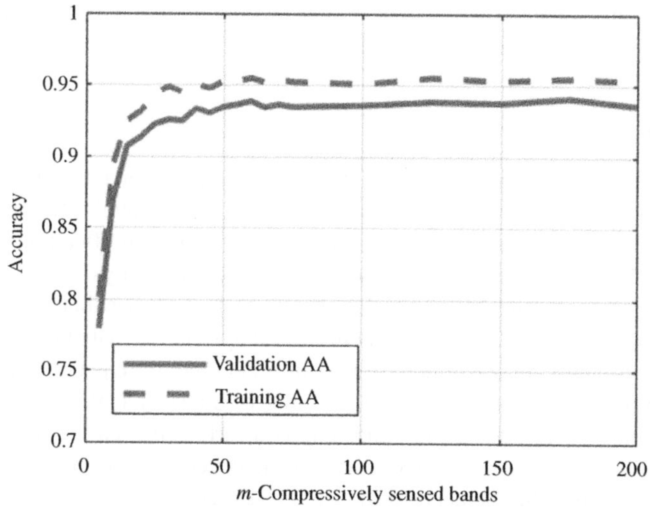


**Figure 9.16** Comparison of training and test $P_{AA}$ for the Indian Pines scene.

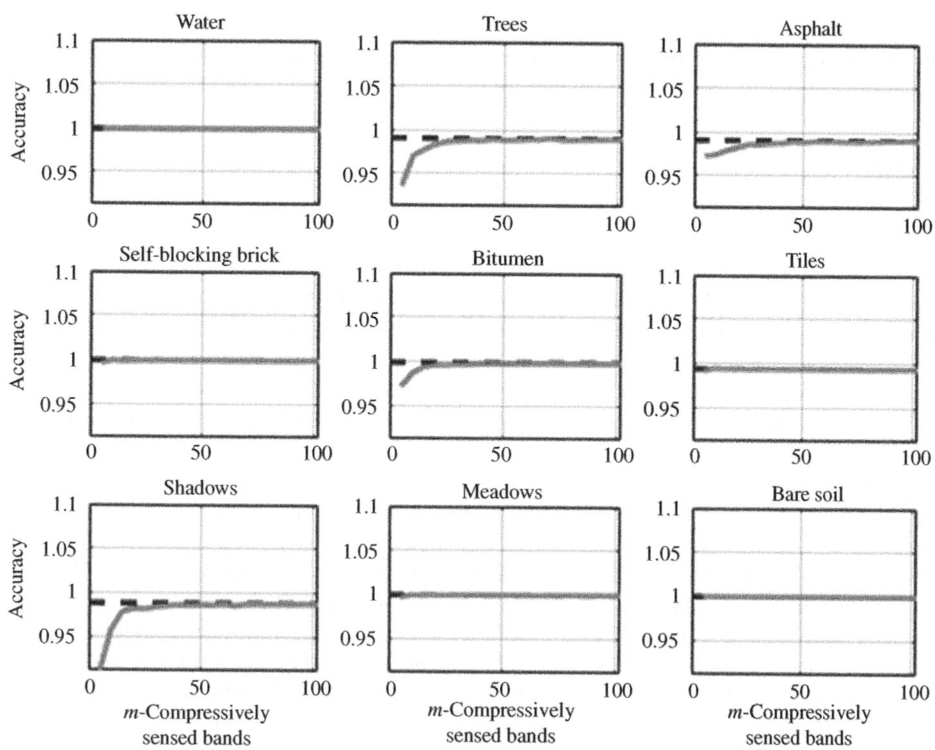


**Figure 9.17** Individual class accuracies of Pavia Centre.

of the number of CSBs and that the training data can be used reliably. Thus, a very simple, but effective, method for selecting the appropriate number of CSBs is to simply vary $m$ during the training process and to observe where performance begins to asymptote as illustrated in Figure 9.17. While this approach is not guaranteed to be optimal, it has been shown empirically to be effective. Future research will further investigate how to estimate a close bound on $m$ as well as the applicability of adaptive methods that can infer the appropriate $m$ *in-situ*.

## 9.6 Conclusion

HSIC in CSBD is explored in this chapter with the motivation of enabling low-cost, low-SWAP hyperspectral designs. A random sampling strategy is identified through the utility of universality which can be used to realize sensing hardware with no fundamental modification. Two new concepts of efficacy and CSBR are also introduced to conduct performance analysis on classification accuracies between using full bands and CSBs for classification which shows that full band classification performance can be achieved in CSBD for a compressed system that satisfies RIP.

Classification experiments are performed for two different types of hyperspectral sensors, AVIRIS and ROSIS. The experimental results demonstrated that with a sufficient subsampling, the classification performance in the ODS using full bands can be nearly achieved in a CSBD. The precision performance also shows the best performance can be achieved by only requiring a small number of CSBs before converging to full band performance. The experiments also highlight that the intrinsic sparsity of different images varies with requiring different levels of subsampling to achieve full band classification performance. These results also point to the important issue of choosing the subsampling level if the scene complexity is not known a priori. Figure 9.16 demonstrated a strong correlation between the training and test accuracy error, as a function of CSBs. By simply varying the number of CSBs during the training process, the appropriate choice for $m$ can be determined by observing where the performance begins to asymptote.

## References

1. Candes, E.J. and Wakin, M.B. (2008). An introduction to compressive sampling. *IEEE Signal Processing Magazine* 21–30.
2. Baraniuk, R., Davenport, M., De Vore, R., and Wakin, M.B. (2008). A simple proof of the restricted isometry property for random matrices. *Constructive Approximation* 28 (3): 253–263.
3. Iordache, M., Bioucas-Dias, J.M., and Plaza, A. (2011). Sparse unmixing of hyperspectral data. *IEEE Transactions on Geoscience and Remote Sensing* 49 (6): 2014–2039.
4. Bioucas-Dias, J.M. and Figueiredo, M. A. (2010). Alternating direction algorithms for constrained sparse regression: application to hyperspectral unmixing. Hyperspectral Image and Signal Processing: Evolution in Remote Sensing (WHISPERS), pp. 1–4.
5. Du, Q., Bioucas-Dias, J.M., and Plaza, A. (2012). Hyperspectral band selection using a collaborative sparse model. International Geoscience and Remote Sensing Symposium (IGARSS), Germany Munich (22–27 July 2012), pp. 3054–3057.
6. Zhong, P. and Wang, R. (2008). Learning sparse CRFs for feature selection and classification of hyperspectral imagery. *IEEE Transactions on Geoscience and Remote Sensing* 46 (12): 4186–4197.
7. Tuia, D., Flamary, R., and Courty, N. (2015). Multiclass feature learning for hyperspectral image classification: sparse and hierarchical solutions. *ISPRS Journal of Photogrammetry and Remote Sensing* 105: 272–285.
8. Chen, Y., Nasrabadi, N.M., and Tran, T.D. (2011). Hyperspectral image classification using dictionary-based sparse representation. *IEEE Transactions on Geoscience and Remote Sensing* 49 (10): 3973–3985.
9. Chen, Y., Nasrabadi, N.M., and Tran, T.D. (2013). Hyperspectral image classification via kernel sparse representation. *IEEE Transactions on Geoscience and Remote Sensing* 51: 217–231.
10. Martin, G., Bioucas-Dias, J.M., and Plaxa, A. (2015). HYCA: a new technique for hyperspectral compressive sensing. *IEEE Transactions on Geoscience and Remote Sensing* 53 (5): 2819–2831.
11. Davenport, M.A., Duarte, M.F., Wakin, M.B. et al. (2007). Smashed filter for compressive classification and target recognition. *Computational Imaging V, SPIE* 6498: 64980H.

12. Calderbank, R., Jafarpour, S., and Schapire, R. (2009). Compressed Learning: Universal Sparse Dimensionality Reduction and Learning in the Measurement Domain. Technical Report. https://citeseerx.ist.psu.edu/viewdoc/download?doi=10.1.1.481.8129&rep=rep1&type=pdf.
13. Hahn, J., Rosenkranz, S., and Zoubir, A.M. (2014). Adaptive compressed classification for hyperspectral imagery. 2014 IEEE International Conference on Acoustics, Speech and Signal Processing (ICASSP), Italy Florence (4–9 May 2014), pp. 1020–1024.
14. Wold, S., Esbensen, K., and Geladi, P. (1987). Principal component analysis. *Chemometrics and Intelligent Laboratory Systems* 2 (1–3): 37–52.
15. Candes, E. and Tao, T. (2005). Decoding by linear programming. *IEEE Transactions on Information Theory* 51 (12): 4203–4215.
16. Amhed, N., Natarajan, T., and Rao, K.R. (1974). Discrete cosine transform. *IEEE Transactions on Computers* 100 (1): 90–93.
17. Antonini, M., Barlaud, M., Mathieu, P., and Daubechies, I. (1992). Image coding using wavelet transform. *IEEE Transactions on Image Processing* 1 (2): 205–220.
18. Li, C., Ma, L., Wang, Q. et al. (2013). Construction of sparse basis by dictionary training for compressive sensing hyperspectral imaging. International Geoscience and Remote Sensing Symposium (IGARSS), pp. 1442–1445.
19. Della Porta, C., Lampe, B., Bekit, A., and Chang, C.-I (2019). Application of universality of random compressed sampling for hyperspectral image processing. Proceedings of the SPIE, Volume 10986, (14 May 2019), pp. 109860P 1–13.
20. Chen, Y., Lin, Z., Zhao, Z. et al. (2014). Deep learning-based classification of hyperspectral data. *IEEE Journal of Selected Topics in Applied Earth Observations and Remote Sensing* 7 (6): 2094–2107.
21. Li, T., Zhang, J. and Zhang, Y. (2014). Classification of hyperspectral image based on deep belief networks. IEEE Internation Conference on Image Processing (ICIP), pp. 5132–5136.
22. Chen, Y., Zhao, X., and Jia, X. (2015). Spectral–spatial classification of hyperspectral data based on deep belief network. *IEEE Journal of Selected Topics in Applied Earth Observations and Remote Sensing* 8 (6): 2381–2392.
23. Camps-Valls, G. and Bruzzone, L. (2005). Kernel-based methods for hyperspectral image classification. *IEEE Transactions on Geoscience and Remote Sensing* 43 (6): 1351–1362.
24. Camps-Valls, G., Gomez-Chova, L., Muñoz-Marí, J. et al. (2006). Composite kernels for hyperspectral image classification. *IEEE Geoscience and Remote Sensing Letters* 3 (1): 93–97.
25. Camps-Valls, G., Tuia, D., Bruzzone, L., and Benediktsson, J.A. (2014). Advances in hyperspectral image classification: Earth monitoring with statistical learning methods. *IEEE Signal Processing Magazine* 31 (1): 45–54.
26. Fauvel, M., Benediktsson, J.A., Chanussot, J. et al. (2008). Spectral and spatial classification of hyperspectral data using SVMs and morphological profiles. *IEEE Transactions on Geoscience and Remote Sensing* 46 (11): 3804–3814.
27. Fauvel, M., Tarabalka, Y., Benediktsson, J.A. et al. (2013). Advances in spectral-spatial classification of hyperspectral images. *Proceedings of IEEE* 101 (3): 652–675.
28. Kang, X., Li, S., and Benediktsson, J.A. (2014). Spectral-spatial hyperspectral image classification with edge-preserving filtering. *IEEE Transactions on Geoscience and Remote Sensing* 52 (5): 2666–2677.
29. Cortes, C. and Vapnik, V. (1995). Support vector machines. *Machine Learning* 20: 273–297.
30. Bishop, C.M. (1995). Neural Networks for Pattern Recognition. New York: Oxford University Press.
31. Schölkopf, B., Herbrich, R., and Smola, A.J. (2001). A Generalized representer theorem. In: Computational Learning Theory. COLT 2001, Lecture Notes in Computer Science, vol. 2111 (ed. D. Helmbold and B. Williamson). Berlin, Heidelber: Springer.
32. Scholkopf, N. and Smola, A. (2001). Learning with Kernels: Support Vector Learning, Regularization, Optimization and Beyond. Cambridge, MA: MIT Press.
33. Xue, B., Yu, C., Wang, Y. et al. (2017). A subpixel target approach to hyperpsectral image classification. *IEEE Transactions on Geoscience and Remote Sensing* 55 (9): 5093–5114.
34. Yu, C., Xue, B., Song, M. et al. (2018). Iterative target–constrained interference-minimized classifier for hyperspectral image classification. *IEEE Journal of Selected Topics in Applied Earth Observation and Remote Sensing* 11 (4): 1095–1117.

35. Vane, G., Green, R.O., Chrien, T.G. et al. (1993). Airborne visible/infrared imaging spectrometer (AVIRIS). *Remote Sensing of Environment* 44 (2-3): 127–143.
36. Gege, P., Beran, D., Mooshuber, W. et al. (1998). System analysis and performances of the new version of the imaging spectrometer ROSIS. In: Proceedings of the First EARSeL Workshop on Imaging Spectroscopy (ed. M. Schaepman, D. Schläpfer and K.I. Itten), 29–35. Paris: EARSeL.
37. Baumgardner, M.F., Biehl, L.L., and Landgrebe, D.A. (2015). 220 Band AVIRIS hyperspectral image data set: June 12, 1992 Indian Pines test site 3. *Purdue University Research Repository* https://doi.org/10.4231/R7RX991C.
38. Universidad del Pais Vasco Grupo de Inteligencia Computacional. http://www.ehu.eus/ccwintco/index.php?title=Hyperspectral_Remote_Sensing_Scenes#Salinas.

# Part IV

## Fusion for Hyperspectral Imaging

# 10

# Hyperspectral and LiDAR Data Fusion

Qian Du[1], Wei Li[2], and Chiru Ge[3]
[1]*Department of Electrical and Computer Engineering, Mississippi State University, Starkville, MS, USA*
[2]*College of Information and Electronics, Beijing Institute of Technology, Beijing, China*
[3]*School of Information Science and Engineering, Shandong Normal University, Jinan, China*

## 10.1 Introduction

Hyperspectral imaging (HSI) sensors acquire hundreds of narrow spectral bands with high spectral resolution, while laser scanning can measure surface topography and 3-D characterization of targets. Extending laser-based intensity into broadband spectral range using an active hyperspectral light detection and ranging (LiDAR) generates additional spectral features for target classification [1–5]. A more flexible strategy is to fuse data from a passive hyperspectral sensor and an active laser scanner [6]. The combined use of HSI and LiDAR results in higher classification accuracies than using each source separately. HSI and LiDAR have been used in combination in many applications successfully, such as biomass estimation [7, 8], vegetation monitoring [9, 10], tree species classification [11, 12], and improved bathymetry estimation [13].

Figure 10.1(a) shows an HSI, whose poor spatial contrast between objects limits its performance in spatial feature presentation. The LiDAR-based digital surface model (DSM) shown in Figure 10.1(b) provides more accurate elevation information and useful spatial contrast. Thus, fusing HSI and LiDAR-based DSM is helpful to synthesize diverse information on the investigated areas. Figure 10.1(c) reflects obvious contrast in the fused image with more information.

Fusion of HSI and LiDAR data has been of great interest for classification purposes. The morphological extinction-profiles [14, 15] were proposed for both HSI and LiDAR feature extraction. In [16], the morphological attribute profiles were demonstrated to produce higher classification performance with two different types of remote sensing data. Graph-based fusion for HSI and LiDAR data classification was presented in [17, 18], and multiple features extracted through different kernels have been fused in a decision phase [19].

Recently, convolution neural network (CNN) has been widely studied in remote sensing community and shown to be more powerful than traditional classifiers, such as support vector

*Advances in Hyperspectral Image Processing Techniques*, First Edition. Edited by Chein-I Chang.

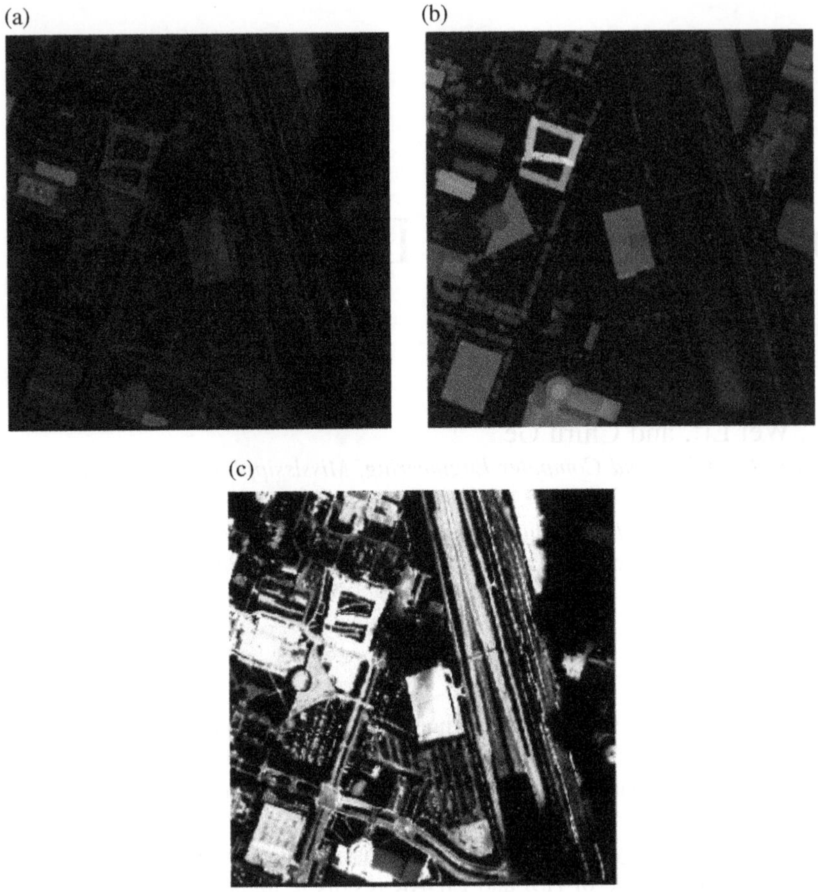

**Figure 10.1** HSI and LiDAR data with their fused image for joint information fusion. (a) HSI, (b) LiDAR, and (c) fused image.

machine (SVM). A CNN-based pixel-pairs feature (CNN-PPF) framework was proposed for HSI classification [20], and a sparse representation combined CNN was further developed [21]. In [22–24], more CNNs were employed for deep feature extraction. In [25], a novel two-branch CNN for multisource remote sensing data classification has proposed, where the CNN architecture is designed to combine features extracted from HSI and LiDAR. For the HSI branch, a dual-tunnel CNN is proposed for spectral-spatial feature extraction on the local HSI patch, where a pure 2-D CNN is utilized to focus spatial information in the local patch windows and a 1-D CNN is employed for spectral feature of the center pixel. Both types of features are subsequently fused in the full-connected layer through concatenation or stacking. Furthermore, a cascade network block, which passes multi-scale features to output, is developed to exploit spatial information in LiDAR. In [26], a postprocessing, i.e. hierarchical random walk layer, has been added to two-branch CNN for better classification, which enforces spatial consistency in the deeper layers of networks. To consider the robustness of small-sample-size situations, an unsupervised patch-to-patch network was designed to extract from both HSI and LiDAR data [27]. The approach of coupled CNNs with a parameter-sharing strategy is developed in [28]. In [29], probability fusion from deep residual networks is proposed for HSI and LiDAR data classification.

In this chapter, three deep learning-based approaches for HSI and LiDAR data fusion are reviewed in detail, and their performance in land cover classification is compared and discussed.

## 10.2 Deep Learning-Based HSI and LiDAR Data Classification

### *10.2.1 Two-Branch CNN for Joint Classification*

The main procedure of the two-branch CNN [25] classification framework is illustrated in Figure 10.2, including a dual-tunnel CNN branch in HSI, a cascade-block CNN branch in LiDAR, followed by a classifier that consists of fully connected layers with softmax loss.

The branch for HSI feature extraction consists of a spectral tunnel and a spatial tunnel. The spectral tunnel concentrates on the center pixel $\mathbf{H}_{ij}^{\text{spec}}$ at the location $p_{ij}$, which only makes up of simple operations including 1-D convolution layer, activation, max-pooling, and batch normalization. Specifically, the leaky rectified linear unit (ReLU) is adopted as the activation which attempts to fix the "dying ReLU" by taking a small coefficient, and the involved batch normalization allows a higher learning rate to accelerate convergence with normalizing data for each training mini-batch. After the convolutional and max-pooling layer, the output spectral features $\mathbf{F}_{ij}^{\text{spec}}$ are then flatten. The spatial tunnel takes a patch centered at the pixel $p_{ij}$ with a radius $r$ (e.g. 4) as input data. The raw data patch $\mathbf{H}_{ij}^{\text{spat}} \in R^{\text{ksize} \times \text{ksize}}(\text{ksize} = 2 \times r + 1)$ are fed into the 2-D CNN tunnel. To ensure consistency of both spectral and spatial tunnels, the architecture of spatial tunnel is the same as the spectral tunnel. The 2-D convolution and batch normalization carry out the spatial features $\mathbf{F}_{ij}^{\text{spat}}$ from the surrounding area. The spatial-spectral features are then concatenated before they are fed into the full-connection layer.

The branch for LiDAR feature extraction included convolution layer, cascade block, and max-pool layer. In this branch, normalized data are first fed into the network following a convolutional operation with a spatial kernel (e.g. a $3 \times 3$ kernel). After follow-up cascade and max-pool operation, features $\mathbf{F}_{lv}$ are extracted. A fully connected layer with dropout takes the flatten $\mathbf{F}_{lv}$ as input, and then the last layer predicts the local patch belonging to the class with the highest probability.

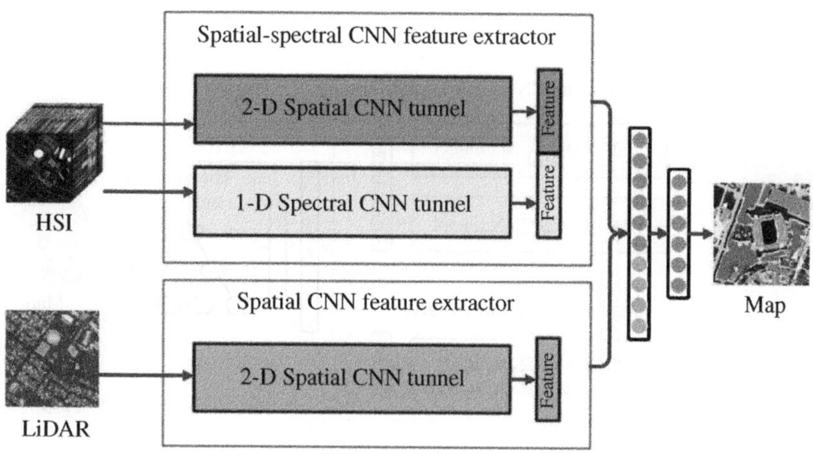


**Figure 10.2** A flowchart of the two-branch CNN for joint classification [25].

All the weights including bias in the proposed network need to be trained. Parameter optimization in the two branches is difficult. In the method, training data pairs are denoted as $\left\{\left[\left(\mathbf{H}_{ij}^{\text{spat}}, \mathbf{H}_{ij}^{\text{spec}}\right), \mathbf{L}_{ij}\right] \sim C_{ij}\right\}$. The $\mathbf{H}_{ij}^{\text{spat}}$ and $\mathbf{H}_{ij}^{\text{spec}}$ are fed into the HSI branch, and the $L_{ij}$ for the LiDAR branch. First, the different branches are trained using separate training data. The fine-tune strategy is employed as a pretrained model. Fine-tune can load the weights from the pretrained model to significantly reduce the computation time. The deep learning methods require a large number of labeled data for training. For remote sensing data, labeling process is expensive and only a few training samples are available usually. To overcome this problem, a random seed is generated for rotating with 90°, flipping left and right, up and down in the training phase. Thus, there are extra samples added to the training set. In order to accelerate the convergence speed, all data are scaled to 0–1. On the other hand, the technique of fine-tune is usually applied to transfer a pretrained model for a large-scale data to a small-scale and similar data. Considering the features extracted by individual branch are suitable to the specific branch, a transferred learning strategy can be used to train the two different branches in a small and very similar training set.

### *10.2.2 Hierarchical Random Walk Network (HRWN)*

The HRWN [26] framework has been designed to classify pixels by fusing multisource remote sensing images in both feature extraction and decision prediction aspects. First of all, random walk process is integrated into deep architecture to encourage coherent classification among pixels that are similar to each other. By doing so, poor localization around boundaries and spatial fragmented classification problems are alleviated. The framework of the proposed HRWN is illustrated in Figure 10.3, which consists of three parts: (i) a dual-tunnel CNN branch that exploits classification potentials (Part I), (ii) the other pixel-level affinity branch that captures local similarities (Part II), and (iii) a novel random walk layer that merges the two branches to obtain a classification map (Part III).

For the input of dual-tunnel CNN branch, both spectral and spatial features are needed, but the original HSI images are limited for low spatial resolution and cloud interference. Therefore, the HSI and LiDAR images with geographical registration are first fused to combine the advantages of both. Classical multi-sensor information fusion methods include color normalized transformation, principal component spectral sharpening, and Gram-Schmidt methods. Specifically, the Gram-Schmidt

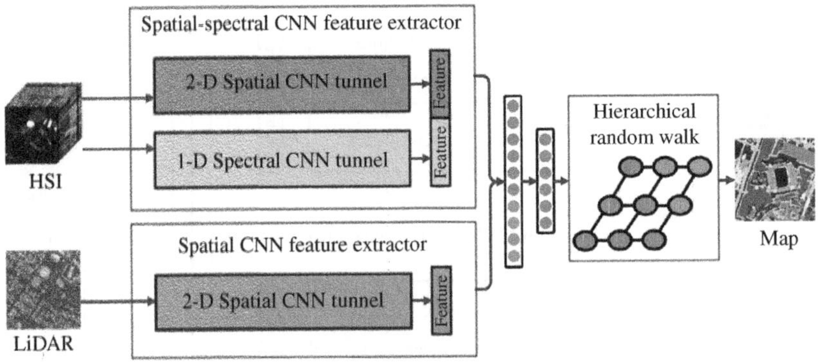


**Figure 10.3** A flowchart of HRWN for joint classification [26]. In the random walk layer, ellipse nodes denote the graph nodes and circle nodes are auxiliary nodes.

method is used for the fusion of HSI and LiDAR after geographical registration. Then, a novel dual-tunnel CNN is designed for the fused image **H**, which consists of a spectral tunnel and a spatial tunnel. For the spatial tunnel, the input is a patch centered at the pixel $p_{ij}$ with a radius $r$. The fused data patch $\mathbf{H}_{ij}^{\text{spa}} \in R^{\text{ksize} \times \text{ksize}}(\text{ksize} = 2 \times r + 1)$ is fed into the 2-D CNN tunnel, which includes simple operations such as 2-D convolution, activation, max-pooling, and batch normalization layers. The spatial feature $\mathbf{F}_{ij}^{\text{spa}}$ around the central target pixel $p_{ij}$ is first extracted by the 2-D convolution and batch normalization operations. The convolution operations are executed with valid padding, and the value of convolution stride *stride* is set to 2, then the range of samples is [(Width × Height)/stride$^2$], where $i = 1, \ldots,$ Width/stride, $j = 1, \ldots,$ Height/stride, Width and Height are the size of the original image. Each convolution process involves certain operations, including 2-D CNN (conv2D) and batch normalization.

The dual-tunnel CNN utilizes both spatial and spectral information, but the large receptive fields in the convolutional layers and the presence of pooling layers lead to low spatial resolution in the deeper CNN layers. To capture the relationship between classes with different elevation information and confirm spatial contrast, a pixel-level affinity branch is employed with its own learning objective as Part II in Figure 10.3. This branch is connected with the LiDAR image with $N$ pixels. The input image is modeled as a weighted, undirected, and connected graph $G = \{V, E\}$, where $V$ is a finite set of vertices with $|V| = N$ and $E$ is a set of edges. Then, **Eu** is a sparse $N \times N$ matrix storing Euclidean distance between each pixel and all its neighbors. Then, the Euclidean distance layer is optimized to predict the pixel similarity matrix **W**, where a weight $\omega_{ij} \in \mathbf{W}$ of edge $e_{ij}$ measures the probability that a random walker will cross this edge. The pixel affinities matrix is used as spatial constraint in the hierarchical random walk layer (Part III) to build up an extended graph.

To integrate the dual-tunnel CNN prediction and pixel affinities, a hierarchical random (Part III) is designed as shown in Figure 10.3. The hierarchical random walk layer is connected to the two branches, including (i) prior distribution $p_i^k \in \mathbf{P}$ obtained by dual-tunnel CNN as a global guide and (ii) affinity matrix **A** reflecting pixel similarity for adjoint and seeds map $s_i^k \in \mathbf{S}$ for the local guide. With the prior information, a graph with priori $G_e$ is constructed. Assume a random walker starts at each unlabeled pixel, the probabilities $r_i^k$ are computed such that random walkers first reach the labeled pixels. For the standard random walker, the probability $r_i^k$ can be computed by minimizing the energy function

$$E_{\text{spatial}}^k\left(\mathbf{r}^k\right) = \left(\mathbf{r}^k\right)^T \mathbf{L}\mathbf{r}^k \tag{10.1}$$

The analytical solution of the energy function can be computed by solving a system of linear equations. Then, the final labeling result can be obtained as

$$\mathbf{R}(i) = \underset{k}{\arg\max}\, r_i^k \tag{10.2}$$

where **R** is the final label that generates the greatest probability.

### *10.2.3 Residual Network-Based Probability Reconstruction Fusion (RNPRF)*

Recently, the spectral-spatial residual network (SSRN) is proposed for HSI classification [30]. The network structure of the SSRN is utilized because it can extract excellent high-level features from 3D data and be more robust and deeper than the traditional 3-D convolutional neural network. Figure 10.4 displays the network of the developed deep feature fusion strategy. In addition to the

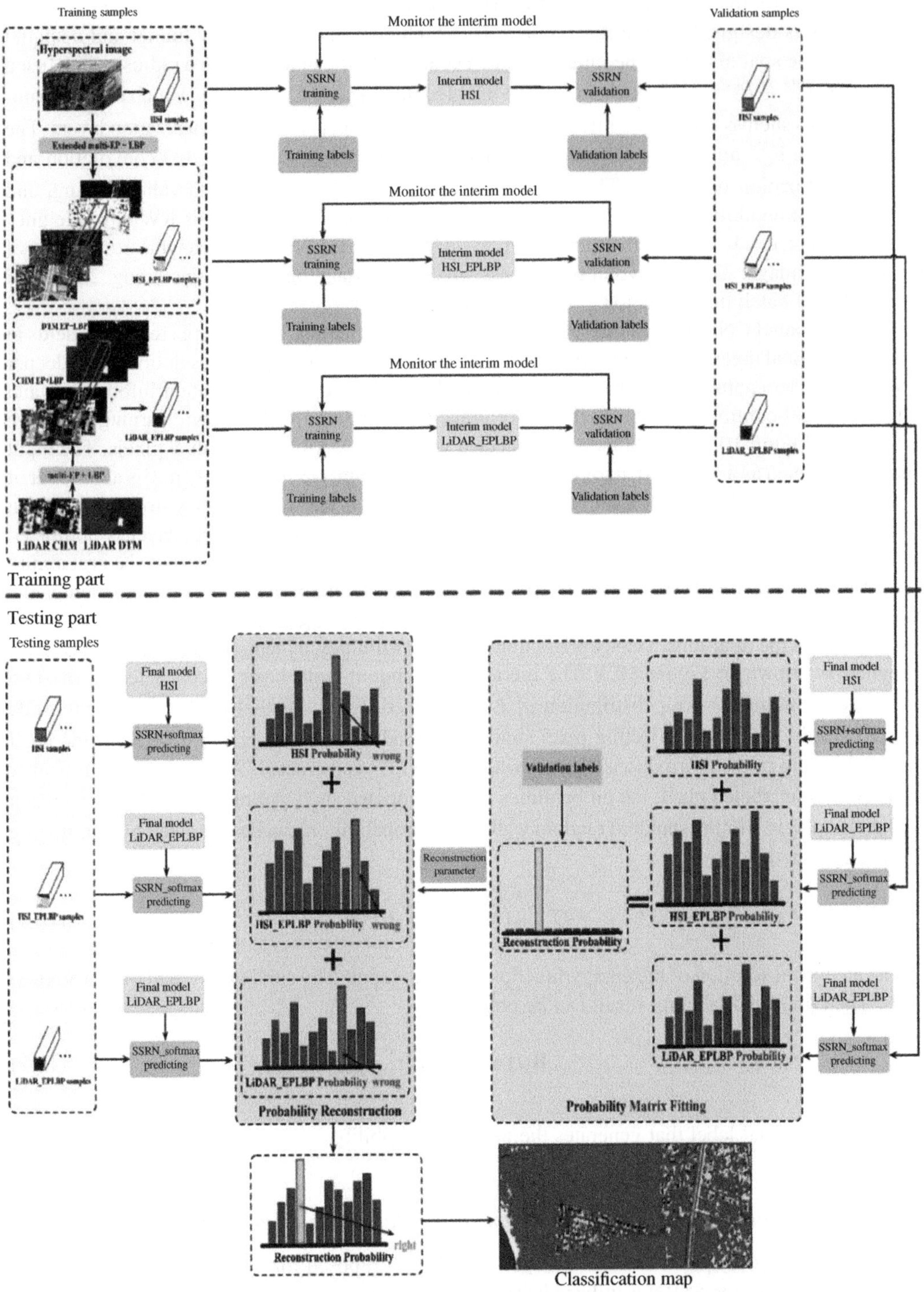


**Figure 10.4** The framework of the RNPRF method for HSI and LiDAR data fusion proposed in Ref. [29], figure 2/IEEE/CC BY 4.0.

original HIS and LiDAR data, extinction profile (EP) and local binary profile (LBP) features are extracted due to their effectiveness [31, 32]. The testing samples of HSI, HSI EPLBP, and LiDAR EPLBP are the input into the SSRN network. Then, deep features of each source are extracted. The size of the deep feature is $1 \times 24$. The three kinds of deep features are stacked to obtain the fusion features whose size is $1 \times 72$. In the end, the fusion features are input into the dense softmax layer to obtain the probability matrix. The probability matrix can be combined by Equation (10.3) and converted to the labels by Equation (10.4). In addition to two weighting parameters, no additional parameters are required for the deep feature fusion strategy.

Let $\mathbf{p}(\mathbf{y}) = [p_1(\mathbf{y})\cdots p_l(\mathbf{y})\cdots p_C(\mathbf{y})]^T$ represent the probability vector for a testing sample $\mathbf{y}$, where $C$ stands for the total number of the classes and $p_l(\mathbf{y})$ is the output of the dense softmax layer and indicates the probability that the sample $\mathbf{y}$ belongs to the $l$-th class. The probability reconstruction is described as

$$\mathbf{p}(\mathbf{y}) = a\mathbf{p}^{\text{HSI}}(\mathbf{y}) + b\mathbf{p}^{\text{HSI\_EPLBP}}(\mathbf{y}) + (1 - a - b)\mathbf{p}^{\text{LiDAR\_EPLBP}}(\mathbf{y}) \tag{10.3}$$

where $\mathbf{p}^{\text{HSI}}(\mathbf{y})$, $\mathbf{p}^{\text{HSI\_EPLBP}}(\mathbf{y})$, and $\mathbf{p}^{\text{LiDAR\_EPLBP}}(\mathbf{y})$ stand for the probability vectors from HSI spectral features, stacked EP and LBP features of HSI, and stacked EP and LBP features of LiDAR, respectively. The parameters $a$ and $b$ control weights of three probability vectors. Finally, the class label is determined by using

$$\text{Class}(\mathbf{y}) = \arg\max_{l} p_l(\mathbf{y}) \tag{10.4}$$

As Figure 10.4 illustrates, a class-4 pixel is misclassified as class-8 (with the largest value of probability) if using the original HSI data alone, it is misclassified as class-10 if using the EPLBP features from the HSI data, and it is mis-assigned with the label of class-9 if using the EPLBP features from the LiDAR data. When using each set of features, the probability for class-4 is not the greatest, but the second greatest. However, when the three probability vectors are appropriately combined, the final class label can be corrected with the rule of winner-take-all.

## 10.3 Experiments

### *10.3.1 Experimental Data*

In this section, the performance of aforementioned methods is evaluated with a standard multisource data set, i.e. the *Houston Data*. The data are composed of HSI and LiDAR images that were introduced for the 2013 GRSS Data Fusion contest. The scene was acquired in 2012 by an airborne sensor over the area of University of Houston campus and neighboring area. The size of the data is $349 \times 1905$ pixels with 2.5 m spatial resolution. The HSI scene consists of 144 spectral bands with wavelength ranging from 0.38 to 1.05 μm including 15 classes. Table 10.1 shows the available training and testing samples.

### *10.3.2 Classification Performance*

To validate the effectiveness, Two-Branch CNN, HRWN, and RNPRF are compared with several other classifiers, such as the standard SVM, recently proposed CNN-PPF [20], Context CNN [33], CRNN [34], and CNN-MRF [35]. All the programs are implemented using Python 3.6 and

**Table 10.1** Number of training and testing samples for the Houston data

| No. | Class name | Train | Test |
|---|---|---|---|
| 1 | Health grass | 198 | 1053 |
| 2 | Stressed grass | 190 | 1064 |
| 3 | Synthetic grass | 192 | 505 |
| 4 | Tress | 188 | 1056 |
| 5 | Soil | 186 | 1056 |
| 6 | Water | 182 | 143 |
| 7 | Residential | 196 | 1072 |
| 8 | Commercial | 191 | 1053 |
| 9 | Road | 193 | 1059 |
| 10 | Highway | 191 | 1036 |
| 11 | Railway | 181 | 1054 |
| 12 | Parking lot 1 | 192 | 1041 |
| 13 | Parking lot 2 | 184 | 285 |
| 14 | Tennis court | 181 | 247 |
| 15 | Running track | 187 | 473 |
| Total | | 2832 | 12 197 |

MATLAB R2018a, and the networks are constructed using Tensorflow with the high-level API Keras. Tensorflow is an open-source software library for numerical computation using data flow graphs, and Keras can be seen as a simplified interface to Tensorflow. All experiments are conducted by a personal computer equipped with Windows 7 and NVIDIA Quadro K2200. Note that SVM is implemented using the LIBSVM toolbox and trained using a Gaussian radial basis function (with $\gamma = 0.5$).

The parameters of all the methods are optimized for fair comparison. Furthermore, several classification frameworks including SVM, CNN-PPF, Two-Branch CNN, Two-Branch CNN(Merge), Context CNN, CRNN, CNN-MRF, and RNPRF are discussed. In Two-Branch CNN(Merge), the input of Two-Branch CNN [25] is changed to the fused data, but all the training and testing samples are exactly the same. In the data preprocessing procedure of the proposed HRWN and other competitive classifiers, data augmentation is used to increase the diversity of training samples, improve the robustness of the model, and avoid overfitting. The used augmentation method applies orientation flipping, injection of noise randomly to the data, which reduces the model's dependence on certain attributes and thereby improving the generalization performance.

Table 10.2 lists the overall accuracy (OA), average accuracy (AA), and Kappa coefficient (Kappa) for the experimental data. As can been seen, HRWN performs obviously better than most of the methods. The Kappa of RNPRF is 93.88%, which is the best among all the classifiers. Due to more robust extraction of global and local features, the hierarchical model for HSI, LiDAR-based DSM, and fused data can provide significant improvement. One of the main challenges in classification is the problem of misclassification. Specifically, it is common to misclassify soil as shrub or grass as crops in natural settings. The HSI data should not be used to differentiate targets consisting of the same material such as road and highway in Houston data. On the other hand, the spatial-spectral information obtained by the dual-tunnel CNN branch can enable the classifier to distinguish targets with similar elevations.

**Table 10.2** Classification accuracy (%) using the Houston data

| | SVM | CNN-PPF | Two-Branch CNN | Two-Branch CNN (Merge) | Context CNN | CNN-MRF | CRNN | HRWN | RNPRFF |
|---|---|---|---|---|---|---|---|---|---|
| 1 | 82.43 | 83.57 | 83.10 | 83.10 | 84.89 | **85.77** | 83.00 | 85.61 | 78.63 |
| 2 | 82.05 | **98.21** | 84.10 | 81.20 | 87.40 | 86.28 | 79.41 | 85.17 | 85.15 |
| 3 | 99.80 | 98.42 | **100.00** | **100.00** | 99.86 | 99.00 | 99.8 | 99.57 | 98.42 |
| 4 | 92.80 | **97.73** | 93.09 | 92.90 | 93.49 | 92.85 | 90.15 | 92.20 | 95.93 |
| 5 | 98.48 | 96.50 | **100.00** | 99.81 | **100.00** | **100.00** | 99.71 | **100.00** | **100.00** |
| 6 | 95.10 | 97.20 | 99.30 | **100.00** | 98.77 | 98.15 | 83.21 | 98.15 | 100.00 |
| 7 | 75.47 | 85.82 | 92.82 | 92.54 | 82.81 | 91.64 | 88.06 | **95.98** | **91.70** |
| 8 | 46.91 | 56.51 | 82.34 | 94.87 | 78.78 | 80.79 | 88.61 | **97.59** | **95.54** |
| 9 | 77.53 | 71.20 | 84.70 | 83.85 | 82.51 | **91.37** | 66.01 | 88.66 | 95.66 |
| 10 | 60.04 | 57.12 | 65.44 | 69.89 | 59.41 | 73.35 | 52.22 | **86.23** | **97.68** |
| 11 | 81.02 | 80.55 | 88.24 | 86.15 | 83.24 | **98.87** | 81.97 | 97.98 | 100.00 |
| 12 | 85.49 | 62.82 | 89.53 | 92.60 | 92.13 | 89.38 | 69.83 | **97.40** | **99.42** |
| 13 | 75.09 | 63.86 | 92.28 | 79.30 | **94.88** | 92.75 | 79.64 | 91.47 | 89.82 |
| 14 | **100.00** | **100.00** | 96.76 | **100.00** | 99.77 | **100.00** | **100.00** | **100.00** | **100.00** |
| 15 | 98.31 | 98.10 | 99.79 | **100.00** | 98.79 | **100.00** | **100.00** | **100.00** | **97.25** |
| OA | 80.49 | 83.33 | 87.98 | 88.91 | 86.90 | 90.61 | 88.55 | **93.61** | **94.36** |
| AA | 83.37 | 83.21 | 90.11 | 90.42 | 89.11 | 92.01 | 90.30 | **94.40** | **95.01** |
| Kappa | 78.98 | 81.88 | 86.98 | 87.96 | 85.89 | 89.87 | 87.56 | **93.09** | **93.88** |

The two largest values in each row are given in bold.

For qualitative evaluation of the classification performance, visual maps are illustrated in Figure 10.5. Also, the ground-truth map and pseudo-color maps of entire image scenes are provided for clarity. HRWN and RNPRF produce the most accurate classification maps. The use of a hierarchical random walk layer enables the HRWN method to produce spatially smooth results with less false boundaries. Interestingly, the simple idea of probability fusion of the RNPRF can result in excellent classification performance. It can be concluded that the visual results are consistent with those in Table 10.2.

## 10.4 Conclusions

Deep learning-based approaches have been investigated for classification of multisource remote sensing data such as HSI and LiDAR. Two-branch CNN model contained two different pipelines. A simple two-tunnel CNN, which consisted of the same architecture of 2-D and 1-D CNNs, was designed for reinforcing the correspondence spatial-spectral information. Besides, a cascade network was devised to combine features at different levels with a shortcut path. On the other hand, HRWN jointly optimized dual-tunnel CNN and pixelwise affinity via a novel random walk layer, which enforced spatial consistency in the deeper layers of the network. In RNPRF, probability fusion from deep residual networks can yield excellent performance, which utilizes EP and LBP spatial features from HSI and LiDAR. Experimental results validated with real Houston data demonstrated that state-of-the-art HRWN and RNPRF can significantly outperform other methods.

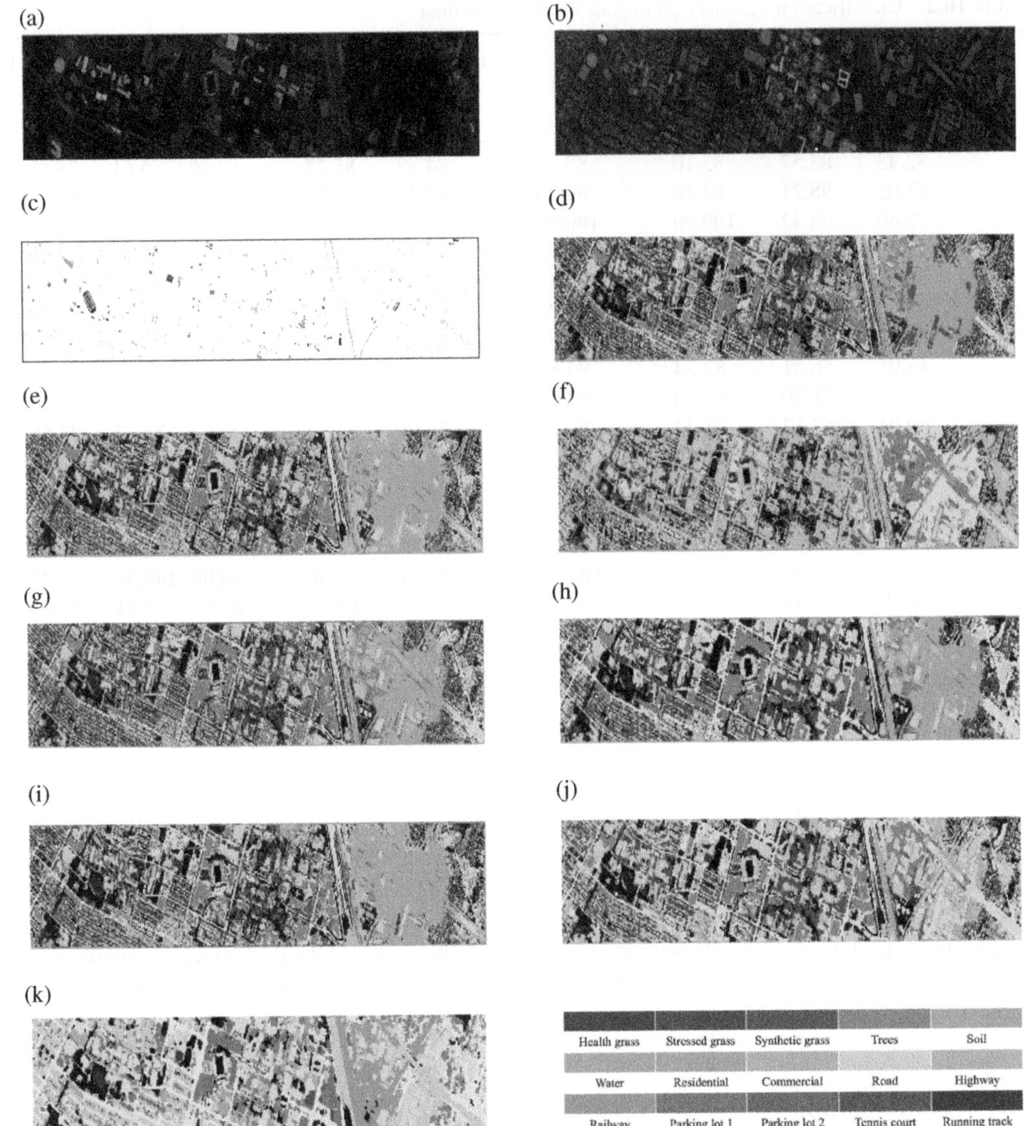


**Figure 10.5** Classification maps for the Houston data obtained with different methods: (a) pseudo-color image for HSI, (b) LiDAR-based DSM, (c) ground-truth map, (d) SVM (80.49%), (e) CNN-PPF (83.33%), (f) Two-Branch CNN (88.91%), (g) Context CNN (86.90%), (h) CNN-MRF(90.61%), (i) CRNN(88.55%), (j) HRWN (93.61%), and (k) RNPRF (94.36%).

## References

1. Kaasalainen, S., Lindroos, T., and Hyyppa, J. (2007). Toward hyperspectral Lidar: measurement of spectral backscatter intensity with a supercontinuum laser source. *IEEE Geoscience and Remote Sensing Letters* 4 (2): 211–215.
2. Vauhkonen, J., Hakala, T., Suomalainen, J. et al. (2013). Classification of spruce and pine trees using active hyperspectral LiDAR. *IEEE Geoscience and Remote Sensing Letters* 10 (5): 1138–1141.

3. Chen, Y., Jiang, C., Hyyppä, J. et al. (2018). Feasibility study of ore classification using active hyperspectral LiDAR. *IEEE Geoscience and Remote Sensing Letters* 15 (11): 1785–1789.
4. Li, W., Jiang, C., Chen, Y. et al. (2019). A liquid crystal tunable filter-based hyperspectral LiDAR system and its application on vegetation red edge detection. *IEEE Geoscience and Remote Sensing Letters* 16 (2): 291–295.
5. Bi, K., Xiao, S., Gao, S. et al. (2020). Estimating vertical chlorophyll concentrations in Maize in different health states using hyperspectral LiDAR. *IEEE Geoscience and Remote Sensing Letters* 58 (11): 8125–8133.
6. Debes, C., Merentitis, A., Heremans, R. et al. (2014). Hyperspectral and LiDAR data fusion: Outcome of the 2013 GRSS data fusion contest. *IEEE Journal of Selected Topics in Applied Earth Observations and Remote Sensing* 7 (6): 2405–2418.
7. Laurin, G.V., Chen, Q., Lindsell, J.A. et al. (2014). Above ground biomass estimation in an African tropical forest with lidar and hyperspectral data. *ISPRS Journal of Photogrammetry and Remote Sensing* 89: 49–58.
8. Du, Y., Wang, J., Liu, Z. et al. (2019). Evaluation on spaceborne multispectral images, airborne hyperspectral, and LiDAR data for extracting spatial distribution and estimating aboveground biomass of wetland vegetation Suaeda salsa. *Journal of Selected Topics in Applied Earth Observations and Remote Sensing* 12 (1): 200–209.
9. Qin, H., Wang, C., Xi, X. et al. (2018). Integration of airborne LiDAR and hyperspectral data for Maize FPAR estimation based on a physical model. *IEEE Geoscience and Remote Sensing Letters* 15 (7): 1120–1124.
10. Melin, M., Hill, R.A., Bellamy, P.E., and Hinsley, S.A. (2019). On bird species diversity and remote sensing—Utilizing Lidar and hyperspectral data to assess the role of vegetation structure and foliage characteristics as drivers of avian diversity. *IEEE Journal of Selected Topics in Applied Earth Observations and Remote Sensing* 12 (7): 2270–2278.
11. Dalponte, M., Bruzzone, L., and Gianelle, D. (2008). Fusion of hyperspectral and LIDAR remote sensing data for classification of complex forest areas. *IEEE Geoscience and Remote Sensing Letters* 46 (5): 1416–1427.
12. Matsuki, T., Yokoya, N., and Iwasaki, A. (2015). Hyperspectral tree species classification of Japanese complex mixed forest with the aid of Lidar data. *IEEE Journal of Selected Topics in Applied Earth Observations and Remote Sensing* 8 (5): 2177–2187.
13. Pan, Z., Glennie, C.L., Fernandez-Diaz, J.C. et al. (2016). Fusion of LiDAR orthowaveforms and hyperspectral imagery for shallow river bathymetry and turbidity estimation. *IEEE Transactions on Geoscience and Remote Sensing* 54 (7): 4165–4177.
14. Ghamisi, P., Souza, R., Benediktsson, J.A. et al. (2016). Extinction profiles for the classification of remote sensing data. *IEEE Transactions on Geoscience and Remote Sensing* 54 (10): 5631–5645.
15. Rasti, B., Ghamisi, P., and Gloaguen, R. (2017). Hyperspectral and LiDAR fusion using extinction profiles and total variation component analysis. *IEEE Geoscience and Remote Sensing Letters* 55 (7): 3997–4007.
16. Khodadadzadeh, M., Li, J., Prasad, S., and Plaza, A. (2015). Fusion of hyperspectral and LiDAR remote sensing data using multiple feature learning. *IEEE Journal of Selected Topics in Applied Earth Observations and Remote Sensing* 8 (6): 2971–2983.
17. Liao, W., Pižurica, A., Bellens, R. et al. (2015). Generalized graph-based fusion of hyperspectral and LiDAR data using morphological features. *IEEE Geoscience and Remote Sensing Letters* 12 (3): 552–556.
18. Xia, J., Liao, W., and Du, P. (2020). Hyperspectral and LiDAR classification with semisupervised graph fusion. *IEEE Geoscience and Remote Sensing Letters* 17 (4): 666–670.
19. Zhao, C., Gao, X., Wang, Y., and Li, J. (2016). Efficient multiple-feature learning-based hyperspectral image classification with limited training samples. *IEEE Transactions on Geoscience and Remote Sensing* 54 (7): 4052–4062.
20. Li, W., Wu, G., Zhang, F., and Du, Q. (2017). Hyperspectral image classification using deep pixel-pair features. *IEEE Transactions on Geoscience and Remote Sensing* 55 (2): 844–853.
21. Liang, H. and Li, Q. (2016). Hyperspectral imagery classification using sparse representations of convolutional neural network features. *Remote Sensing* 8 (2): 99.
22. Chen, Y., Jiang, H., Li, C. et al. (2016). Deep feature extraction and classification of hyperspectral images based on convolutional neural networks. *IEEE Transactions on Geoscience and Remote Sensing* 54 (10): 6232–6251.
23. Makantasis, K., Karantzalos, K., Doulamis, A., and Doulamis, N. (2015). Deep supervised learning for hyperspectral data classification through convolutional neural networks. *Geoscience and Remote Sensing Symposium (IGARSS)*, 2015 IEEE International. IEEE 2015: 4959–4962.

24. Yu, S., Jia, S., and Xu, C. (2017). Convolutional neural networks for hyperspectral image classification. *Neurocomputing* 219: 88–98.
25. Xu, X., Li, W., Ran, Q. et al. (2018). Multisource remote sensing data classification based on convolutional neural network. *IEEE Transactions on Geoscience and Remote Sensing* 56 (2): 937–949.
26. Zhao, X., Tao, R., Li, W. et al. (2020). Joint classification of hyperspectral and LiDAR data using hierarchical random walk and deep CNN architecture. *IEEE Transactions on Geoscience and Remote Sensing* 58 (10): 7355–7370.
27. Zhang, M., Li, W., Du, Q. et al. (2020). Feature extraction for classification of hyperspectral and LiDAR data using patch-to-patch CNN. *IEEE Transactions on Cybernetics* 50 (1): 100–111.
28. Hang, R., Li, Z., Ghamisi, P. et al. (2020). Classification of hyperspectral and LiDAR data using coupled CNNs. *IEEE Geoscience and Remote Sensing Letters* 58 (7): 4939–4950.
29. Ge, C., Du, Q., Sun, W. et al. (2021). Deep residual network-based fusion framework for hyperspectral and LiDAR data. *IEEE Journal of Selected Topics in Applied Earth Observations and Remote Sensing* 14. https://doi.org/10.1109/JSTARS.2021.3054392.
30. Zhong, Z., Li, J., Luo, Z., and Chapman, M. (2018). Spectral-spatial residual network for hyperspectral image classification: a 3-D deep learning framework. *IEEE Transactions on Geoscience and Remote Sensing* 56 (2): 847–858.
31. Ghamisi, P., Höfle, B., and Zhu, X.X. (2017). Hyperspectral and lidar data fusion using extinction profiles and deep convolutional neural network. *IEEE Journal of Selected Topics in Applied Earth Observations and Remote Sensing* 10 (6): 3011–3024.
32. Ge, C., Du, Q., Li, W. et al. (2019). Hyperspectral and LiDAR data classification using kernel collaborative representation based residual fusion. *IEEE Journal of Selected Topics in Applied Earth Observations and Remote Sensing* 12 (6): 1963–1973.
33. Lee, H. and Kwon, H. (2017). Going deeper with contextual CNN for hyperspectral image classification. *IEEE Transactions on Image Processing* 26 (10): 4843–4855.
34. Wu, H. and Prasad, S. (2017). Convolutional recurrent neural networks for hyperspectral data classification. *Remote Sensing* 9 (3): 298.
35. Cao, X., Zhou, F., Xu, L. et al. (2018). Hyperspectral image classification with Markov random fields and a convolutional neural network. *IEEE Transactions on Image Processing* 27 (5): 2354.

# 11

# Hyperspectral Data Fusion Using Multidimensional Information

Lifu Zhang[1], Xia Zhang[1], Mingyuan Peng[2], Xuejian Sun[1], and Xiaoyang Zhao[1,3]
[1]*Hyperspectral Remote Sensing Application Division, Aerospace Information Research Institute, Chinese Academy of Sciences, Beijing, China*
[2]*Land Satellite Remote Sensing Application Center, MNR, Beijing, China*
[3]*College of Natural Resources and Environment, University of Chinese Academy of Sciences, Beijing, China*

## 11.1 Introduction of Remote Sensing Data Fusion

### *11.1.1 Basic Idea of Data Fusion*

Data fusion is the process of synthesizing different indicators or data from different sources through some algorithms to obtain richer information than that from a single data source. According to the type of fusion data source, remote sensing data fusion can be divided into homogeneous remote sensing data fusion (remote sensing data to be fused is obtained by the same imaging method, and the main purpose is to improve the temporal, spatial, and spectral resolution of data); heterogeneous remote sensing data fusion (remote sensing data to be fused is obtained by different imaging means, such as optical-thermal infrared and optical-radar data); remote sensing-site data fusion (fusion of large-area low-precision remote sensing data and high-precision site observation data is to obtain continuous, high-precision, and large area surface data); remote sensing-non-observation data fusion (remote sensing data and non-remote sensing observation data are fused, which is also known as data assimilation) [1]. According to different fusion levels, remote sensing data fusion can be divided into pixel-level fusion (the original data are directly fused), feature-level fusion (the feature is extracted from original data and then is fused), decision-level fusion (the original data are passed through the ground features interpretation and then performed fusion) [2], etc.

### *11.1.2 Why Do We Need Data Fusion?*

Global change, agricultural conditions monitoring, resource and environmental surveys, and other application research urgently need remote sensing data, which has promoted the continuous improvement of satellite sensor performance. Satellite remote sensing has entered the "high-resolution era"

*Advances in Hyperspectral Image Processing Techniques*, First Edition. Edited by Chein-I Chang.

with high temporal, high spatial, and high spectral resolution. For example, China's GF-2 satellite has a spatial resolution of 0.8 m; HJ-1A satellite hyperspectral data has a spectral resolution of 5 nm; the US moderate resolution imaging spectrometer MODIS has a temporal resolution of 1–2 days; China's GF-4 satellite can even monitor the fixed area of the earth in real time, and the temporal resolution can reach the minute level; the GF-5 satellite was launched in May 2018, the spatial resolution of the hyperspectral load is 30 m, the spectral resolution is 5–10 nm, and the width is up to 60 km. A large number of high-precision remote sensing ground observation data at home and abroad provide a rich data guarantee for the research and application of remote sensing.

Although the ability of remote sensing to observe the earth has been greatly improved, it still cannot meet various aspects and needs of people for remote sensing data. Take fine crop monitoring as an example: high spatial resolution data (such as GF-1, GF-2, and ZY-3 satellites) can provide rich texture, hue, geometry, and other characteristics, which can meet the crop planting mode and spatial pattern monitoring needs at field scale. However, constrained by factors such as satellite revisit period and width, accurate identification of large-scale crops on a national scale still faces problems such as missing data, inconsistent radiation, and inapplicable model algorithms; moderate or low-resolution remote sensing data (such as MODIS and NPP series) has a short revisit cycle, which can reflect the phenological changes of crops, which is beneficial to the identification of crop planting types and patterns, but it is difficult to describe the details of the spatial distribution of crops for its spatial resolution [3]; hyperspectral resolution data can capture the subtle spectral differences of ground features, and has huge advantages in crop fine classification and early stress warning [4], but there are certain limitations in terms of the current hyperspectral data sources, data quality, and spatial and temporal resolution [5, 6]. Although various resolution remote sensing data have played an important role in the monitoring of crop planting area nationwide, it is still necessary to make full use of the technical advantages of various sensors, so that the temporal, spatial, and spectral index performance of remote sensing data can be coordinated and complementary; however, no breakthrough has been achieved [5].

Limited by the technical bottleneck of the sensor hardware, any single sensor cannot meet the optimal matching of temporal, spatial, and spectral resolution. At a certain signal-to-noise ratio (SNR) level, the key indexes such as temporal, spatial, and spectral resolution of remote sensing sensors are mutually constrained. In order to tackle this problem, researchers have begun to try to synthesize multisource remote sensing data at different scales through data processing, fuse complementary information from different remote sensing data sources, and expect to obtain temporal-spatial- spectral remote sensing data that all indexes can meet the requirements [1, 7, 8].

## 11.2 Common Methods for Data Fusion

In the process of remote sensing data fusion, according to Schmitt [9], the data to be fused are first registered, rearranged, etc. The purpose is to establish a relationship between the data to be fused, and then use the data fusion algorithm to fuse and obtain the reconstructed image. Among them, the most important part of the fusion process of remote sensing data is the fusion algorithm. In this section, the fusion methods are divided into three categories: fusion methods emphasizing on spatial resolution enhancement, fusion methods emphasizing on spectral resolution enhancement and fusion methods emphasizing on temporal resolution enhancement.

### *11.2.1 Fusion Methods Emphasizing on Spatial Resolution Enhancement*

The fusion methods emphasizing on spatial resolution enhancement are mainly the process of obtaining higher spatial resolution than the original data through multisource data fusion. The main

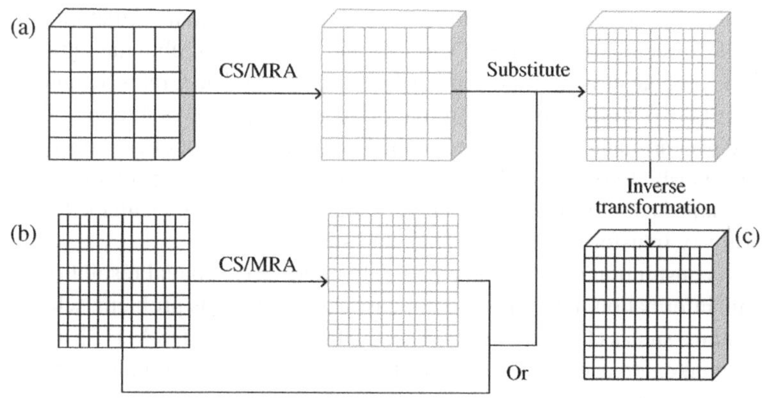


**Figure 11.1** Fusion methods emphasizing on spatial resolution enhancement: (a) low-spatial-resolution multispectral data, (b) panchromatic data, and (c) fused data.

application scenario of spatial dimension enhancement is pansharpening. That is, the spatial dimension information of panchromatic data is used to enhance the spatial resolution of multispectral data, as shown in Figure 11.1. They focus on the improvement of the spatial resolution of the image and may cause distortion in the spectral dimension. Representative methods include component constitution method and multiresolution analysis method.

#### 11.2.1.1 Component Constitution Method

Component Constitution methods refer to the process of projecting remote sensing images into a new transformation space, replacing the components containing spatial information with high-spatial-resolution images and performing inverse transformation to obtain remotely sensed data with improved spatial dimensions [10]. Here multispectral data can be expressed in terms of $M(x_1, \ldots, x_m) \in R^{i\times j\times m}$, and panchromatic images can be expressed in terms of $P(y) \in R^{I\times J}$, in which $i$, $j$, and $m$ can be expressed as the number of rows, columns, and bands of multispectral data; $I$ and $J$ can be expressed as the number of rows and columns. Multispectral data are first subjected to projection transformation to obtain a transformed feature set

$$M'\left(x_1', \ldots, x_m'\right) = F(M(x_1, \ldots, x_m)) \tag{11.1}$$

in which $F()$ is the mapping relationship of the projection transformation. After the transformed feature set is obtained, the component containing spatial information is replaced with a high-spatial-resolution image to obtain $M'\left(x_1', \ldots, x_m'\right)$. In order to reduce spectral distortion, histogram matching is often performed before replacement. Then, inverse transformation is performed on the replaced transform feature set to obtain the fused data set which is

$$M_H\left(x_1', \ldots, x_m' = F'\left(y, \ldots, x_m'\right)\right) \tag{11.2}$$

Commonly used methods include ① intensity-hue-saturation (IHS) transformation [11], which transforms RGB images into three components of intensity, hue, and saturation. The spatial components are mainly concentrated in the intensity component, so the replaced component is the I component; ② the principal component transformation method (PCA) [12], which mainly projects the multispectral data into a new transformation space through linear transformation. Its first principal component is in the direction of maximum variance and has the most abundant information. Therefore, the replaced component is the first principal component. ③ The Gram–Schmidt (GS) transformation method [13] is to simulate the panchromatic data to the spatial resolution of the multispectral, and the simulated panchromatic data are used as the first principal component, together with the multispectral simultaneously, to perform GS transformation,. Then the first principal component of the original panchromatic data is used to perform inverse transformation to obtain the fused data. In addition, there are Brovey transformation (BT) [14], and so on.

The component constitution method can effectively improve the spatial resolution, but it has poor ability to maintain information in the spectral dimension. To solve this problem, scholars have developed many algorithms such as GS adaptive (GSA), GIHS adaptive (GIHSA) [13], and band grouping strategies [15] that take into account the band response function. To a certain extent, the problem of spectral distortion in the process of multispectral fusion is solved.

#### 11.2.1.2 Multiresolution Analysis Method

Multiresolution Analysis Method is to decompose each original data into a series of images of different resolutions, fuse them at different resolutions, and finally perform inverse transformation to obtain the fused image. The method first decomposes the multispectral image band by band into planes to obtain

$$M_m \rightarrow M'_{m \times l}(x_1, x_2, \ldots, x_l) \tag{11.3}$$

Then, perform the same decomposition on the full-color image to get

$$P \rightarrow P'_l \tag{11.4}$$

After obtaining the data at different resolutions, fusion is performed at different scales. Common fusion strategies include ① constitution method, which is to replace the components on the first plane obtained by the decomposition of each band of the multispectral image (MSI) with the corresponding panchromatic image. It is to decompose the components, and then do perform inverse transformation to get the fused data. ② Addition method, that is, only the full-color image is decomposed, and then the decomposed components on the l planes mentioned are added to each band separately. The fused data are obtained after inverse transformation.

Commonly used multiresolution analysis methods are high-pass filtering (HPF) [16], wavelet method [17], curvelet transform [18], smoothing filter-based intensity modulation [19], contourlet transform [20], Laplacian pyramid [9], etc. The multiresolution analysis method can maintain retain the spectral information better than the constitution method, but it requires strict registration of the data, otherwise spatial distortion will occur [21].

### *11.2.2 Fusion Methods Emphasizing on Spectral Resolution Enhancement*

The fusion methods emphasizing on spectral resolution enhancement are mainly a process of obtaining higher spectral resolution through fusion. The main application scenario for the fusion of spectral dimension enhancement is hyperspectral–multispectral fusion. It is to use the hyperspectral data to improve the spectral resolution of multispectral data, as shown in Figure 11.2. Compared with the fusion algorithm for spatial dimension enhancement, it pays more attention to the fidelity of spectral

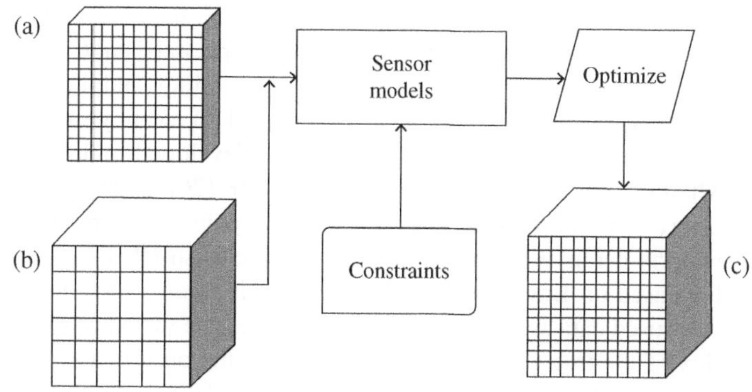


**Figure 11.2** Fusion methods emphasizing on spectral resolution enhancement:: (a) hyperspectral data, (b) multispectral data, and (c) fused data with spectral resolution improved.

dimension information while the spatial dimension is improved. Commonly used methods can be divided into linear optimization method and nonlinear optimization method.

#### 11.2.2.1 Linear Optimization Method

Linear optimization decomposition method assumes that the relationship between observation data and fusion reconstruction data can be expressed linearly, namely

$$X = WZ + n \tag{11.5}$$

Among them, $W$ is a transformation operator. When the measured data are hyperspectral data, W is the spatial fuzzy operator and downsampling operator; when the observed data are multispectral data, $W$ is the spectral response function of the multispectral satellite. Solving $Z$ with known data $X$ can be regarded as an ill-conditioned problem [21], by adding different constraints and solving the optimal solution. According to different principles, it can be divided into spectral-unmixing method, Bayesian method, and sparse representation method.

Spectral unmixing is a powerful method for hyperspectral image (HSI) analysis in recent years. It treats the reflectivity in a limited area as a mixture of different material spectra. It is the linear combination of the endmember and abundances of each pixel in the high and multispectral spectrum [22]. Based on the fusion of spectral unmixing methods, scholars have proposed many methods, such as sub-pixel positioning after unmixing to obtain fusion data [23], and spectral resolution enhancement algorithm (SREM) [24]. Compared to them, the more typical multi-hyperspectral spectral unmixing fusion method [25] can be expressed as a matrix

$$Z = EA + n \approx EA \tag{11.6}$$

in which the $E \in R^{S \times n}$ representing $n$ endmember matrices with $S$ bands, and $A \in R^{n \times N}$ represents the abundance matrices of the n endmembers. Substituting them into formula (11.6), HSI and MSI can be expressed as

$$H = EAW + n = EA_h + n \tag{11.7}$$

$$M = WEA + n = E_m A + n \tag{11.8}$$

in which $H \in R^{n \times S}$ is the observed hyperspectral data, $M \in R^{N \times s}$ is the observed multispectral data, and $E_m$ is the endmember matrix obtained by rebuilding the multispectral data. $E$ is obtained by extracting end elements of hyperspectral data and obtained by downsampling, and then the error n is minimized to obtain $A$, and finally the reconstructed data $Z$ is obtained. In recent years, the fusion methods utilizing non-negative matrix factorization (NMF) have also developed rapidly. For example, Zhi Huang compares NMF and several methods developed by NMF [26]. Naoto Yokoya built a sensor observation model using a combination of spectral response and point spread function (PSF), and used paired non-negative matrix iterative decomposition to obtain reconstructed hyperspectral data [27].

Bayesian method is based on Bayesian model hypothesis. It regards the data to be fused as the observed value, and the fused data as the unobserved true value. The fusion process is solved by calculating the probability of the true value under the premise of the observed value, maximizing the probability, and finally getting the fused data. The representative algorithm is the maximum posterior probability (MAP) [28], namely

$$\hat{Z} = \arg\max_{z} p(Z|M, H) \tag{11.9}$$

in which $\hat{Z}$ is the data after fusion reconstruction, $M$ is the observed multispectral data, $H$ is the observed hyperspectral data, and $p(Z \mid M, H)$ is the maximum probability obtained after observing $M$ and $H$. Using Bayesian criterion and substituting it into formula (11.5), the following formula can be obtained:

$$\hat{Z} = \arg\min_{z} C(Z) \tag{11.10}$$

Among them, the loss function is

$$C(Z) = \frac{1}{2}(M - WZ)^T C_n^{-1}(M - WZ) + \frac{1}{2}\left(Z - \mu_{Z|H}\right)^T C_{Z|X}^{-1}\left(Z - \mu_{Z|H}\right) \tag{11.11}$$

The specific derivation process is described in detail [28]. By solving the gradient of the loss function and making the gradient zero, the final fusion result is obtained through parameter estimation. In addition, representative probabilistic statistical models include Stochastic Mixing Model (SMM) [29, 30], linear Bayesian model [31], Bayesian method combined with wavelet [32, 33], hierarchical Bayesian model [34, 35], etc.

The sparse representation method is to decompose the observation data into a dictionary matrix and a sparse coefficient matrix, and add sparse constraints to solve the sparse coefficients and finally obtain the fused and reconstructed data. Its expression is as follows:

$$X = D\alpha + n \tag{11.12}$$

Among them, $D$ is a sparse dictionary. $\alpha$ is a sparse coefficient requiring to be the sparsest, namely

$$\min_{\alpha}\|\alpha\|_1 \text{ subject to } \|X - D\alpha\|_2^2 < \varepsilon \tag{11.13}$$

Among them, $\|\alpha\|_0$ is L0-norm of $\alpha$ and $\varepsilon$ is the residual. Since this problem is an NP-hard problem [36], it is common to relax its constraints to L1-norm when solving

$$\min_{\alpha}\|\alpha\|_1 \text{ subject to } \|X - D\alpha\|_2^2 < \varepsilon \tag{11.14}$$

Common process is to substitute the linear model of the sensor on the basis of Eq. (11.14), use the sparse unmixing method to obtain the dictionary and sparse matrix, and finally perform fusion data reconstruction. Common methods include compressed sensing [37], sparse expression [35, 38], analytical sparse model [39], etc.

Linear optimization decomposition methods can effectively improve the spectral dimension resolution of multispectral data, and have better fidelity effect on spatial dimension. However, the decomposition optimization method is based on the assumption of the linear observation model. In the actual imaging process, the linear relationship hypothesis often cannot be satisfied.

### 11.2.2.2 Nonlinear Optimization Method

Most of the above methods are based on the assumption that the relationship between sensors is linear, but this assumption has not been rigorously verified. At the same time, the parameters of optimization methods rely more on prior knowledge [21]. The nonlinear optimization method is based on artificial neural networks, convolutional neural networks (CNN), etc., and can better learn the nonlinear relationship between the input and output of the system. Artificial Neural Network (ANN) is to simulate the structure of biological neurons. These neurons have simple structures but are closely connected with each other. The basic structure is as follows:

$$y = f\left(\sum_{k=1}^{K} x_i w_{ij} + b\right) \tag{11.15}$$

For a set of inputs $(x_1, x_2, \ldots, x_K)$, a set of nonlinear functions of weight $w$ is obtained after training and adjustment so that the error between the output $y$ and the expected output value is minimized. By adjusting the input weight $w$, the neural network automatically learns how to obtain the desired result. For the traditional ANN, each neuron must receive the output from all neurons in the previous layer, resulting in a huge number of neurons which causes huge computational loss. CNN solve this problem by abandoning global connectivity. Every neuron in CNN has a receptive field, which only processes feature values produced from neurons in a certain neighborhood, which is more suitable for images with more localized spatial features. The output of the neuron can be regarded as the convolution of the output features of the neuron in the previous layer,

$$y_{i,j} = f\left(\sum_{n=1}^{K}\sum_{m=1}^{K} w_{n,m} x_{i+n,j+m}\right) \tag{11.16}$$

It can be written as a convolution form:

$$y = f(b + w * x) \tag{11.17}$$

in which $*$ represents the convolution operation. In recent years, with the explosive development of deep learning research, scholars have studied the use of deep learning methods for remote sensing image fusion. This method is mostly based on the assumption that the relationship between panchromatic images with high and low spatial resolution and MSI with high and low spatial resolution are the same and non-linear. Huang et al. used deep neural networks (DNN) to fuse multispectral and panchromatic images, added pretraining and fine-tuning learning stages within the DNN framework, and increased the accuracy of fusion [40]. Based on this, scholars have developed many algorithms such as PNN [41] DNN based on the MRA framework [42], DRPNN [43], and MSDCNN [21].

For the research of hyperspectral and multispectral fusion, Frost Palsson uses the 3D convolutional neural networks (3DCNN) model to fuse hyperspectral and multispectral data. In order to overcome the computational complexity caused by the large amount of hyperspectral data, the hyperspectral data are processed of dimensionality reduction, which brings higher accuracy than MAP algorithm [44].

Compared with linear optimization methods, nonlinear methods can better describe the nonlinear relationship between images of different resolutions and has strong transferability. However, the problem is that the premise assumptions have not been strictly verified. Also, the computational complexity of CNN is relatively large and requires a large number of samples. It is often difficult to obtain a very good fusion result with a small number of training samples.

### *11.2.3 Fusion Methods Emphasizing on Temporal Resolution Enhancement*

The fusion methods emphasizing on temporal dimension enhancement are mainly to simulate the remote sensing data with missing phase using data fusion so as to achieve improvement in temporal resolution. The main application scenario of fusion emphasizing on temporal resolution enhancement is multispectral–multispectral spatiotemporal fusion. It is more common to use MODIS data and Landsat data fusion to obtain missing-phase data with Landsat spatial resolution, as shown in Figure 11.3. Typical methods include weight-based method, linear optimization method, and nonlinear optimization method.

#### 11.2.3.1 Weight-Based Method

The weight-based method assumes that there is a linear relationship between the high-spatial and low-time resolution images and the low-spatial and high-resolution images. Based on this, the relationship between the two is established, and sliding window is used to determine the center pixel value. Also, the weight coefficients are used to predict and reconstruct missing time images to obtain high spatial

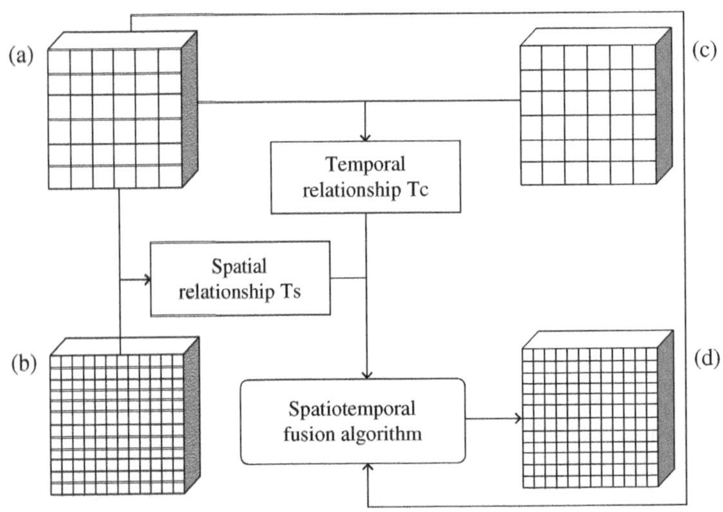


**Figure 11.3** Fusion methods emphasizing improvements of temporal resolution development: (a) data with low spatial and high temporal resolution on date T1, (b) data with low temporal and high spatial resolution on date T1, (c) data with low spatial and high temporal resolution on date T2, and (d) fused data with high spatial and high temporal resolution on date T2.

and temporal resolution images. The method originated from the spatial and temporal adaptive reflectance fusion model (STARFM) [45]. It is one of the most widely used algorithms in spatiotemporal fusion. It assumes that the landcovers will not change from the imaging time to the predicted time, and the basic form is

$$L\left(x_{w/2}, y_{w/2}, t_0\right) = \sum_{i=1}^{w}\sum_{j=1}^{w}\sum_{k=1}^{n} W_{ijk}(M(x_i, y_i, t_0) + L(x_i, y_i, t_k) - M(x_i, y_i, t_k)) \quad (11.18)$$

in which $(x_i, y_i)$ is the location of the pixel. $t_k$ is the imaging time. $L(x_i, y_i, t_k)$ is the Landsat reflectivity data (or high-resolution reflectivity dataset), $M(x_i, y_i, t_k)$ is the MODIS reflectivity data that have been upsampled to Landsat data resolution (or low-resolution dataset upsampled to high-resolution dataset resolution). $w$ is the size of the sliding window. $(x_{w/2}, y_{w/2})$ is the central pixel of the sliding window. $W_{ijk}$ is the weight value related to the calculation of the difference in spectrum, time, and position between the pixel and the central pixel.

Although the STARFM algorithm is easy to understand and widely used, it still has the following problems: (i) although the STARFM algorithm has a good effect in predicting gradual change, it cannot predict instantaneous disturbance events in the short term. (ii) STARFM does not consider the reflectivity directivity. (iii) STARFM relies on the hypothesis of pure, homogeneous pixels of MODIS low-resolution pixels. Scholars have made improvements, such as the STAARCH model [46], Enhanced STARFM algorithm (ESTARFM) [8], mESTARFM [47], and RWSTFM [48]. In addition to the visible-near infrared satellite data fusion, the weight-based method is also widely used in the fusion of surface temperature data such as bilateral filtering method [49], SADFAT [50], STITFM [51], and NDVI data fusion such as STVIFM [52].

The advantage of the weight function method is that the model is intuitive and widely used, but its problems are: (i) objects with changing shapes and short-term transient changes cannot be well predicted, (ii) the correlation between sensors has not been strictly proven, and (iii) the size of the sliding window and other parameters need to be set manually and cannot be automatically obtained.

### 11.2.3.2 Linear Optimization Method

The linear optimization method is also based on linear assumption whose principle is similar to Section 11.2.2.1. The reconstructed image is obtained by adding constraints to obtain the optimal solution. The principle is based on spectral-unmixing method, Bayesian method, and sparse representation method.

The spectral unmixing method is also based on the spectral linear unmixing model (See for section 11.2.2.1). The common steps are as follows [53]: ① unsupervised classification of high-spatial-and-low-temporal-resolution data (HSLT); ② spectral unmixing of low-spatial-and-high-temporal data (LSHT) using classification obtained by ①; the spectral information from the unmixed data is added to the high-space and low-temporal data to obtain the fused high-spatial-and-high-temporal data(HSHT). Commonly used algorithms are MMT spectral unmixing algorithm [54], STDFM algorithm [55], ESTDFM [56], MSTDFA [57], soft clustering [58], OB-STVIUM algorithm (Object-Based Image Analysis, OBIA) [59], etc. The advantage of the spatiotemporal fusion method based on spectral unmixing is that it considers the problem of low-resolution image mixed pixels which improves the accuracy of fusion. However, this method is mostly based on the linear spectral unmixing method, and its premise assumptions have not been strictly verified; during the endmember determination process, when the temporal difference between the landcover type map and the time interval to be predicted is too large, its accuracy cannot be guaranteed. Also, it relies on the basis that the feature type does not change in the predicted time, but this situation sometimes cannot be satisfied.

Bayesian method is using Bayesian probability as a fusion framework to predict high spatial resolution data in the missing phase. Generally, the relationship between the missing image and the observed image is first established, and then the missing image is reconstructed by maximizing the conditional probability of the missing image under the condition of the observed image, that is, the Maximum a posteriori (MAP). According to Xiaolin Zhu, the relationship between missing images and observed images can be divided into temporal model and scale model [60]. The temporal model refers to establishing the relationship between data at different dates and depicting the slow and instantaneous changes of the ground features, such as NDVI-BSFM [61]. The scale model refers to modeling the relationship between the missing image and the observed image at the same date. It is to establish the relationship between the observed image cell and the missing image cell through a PSF. For example, in Xue's work, the joint covariance of the low-resolution image is used as the temporal model, together with the bilateral filtering of low-resolution images and the HPF of high-resolution images to establish a Bayesian fusion method [62]. The advantage of the Bayesian method is that it does not require the band response to be strictly matched.

For the fusion algorithm based on sparse representation, the most representative algorithm is SParse-representation-based SpatioTemporal reflectance Fusion Model (SPSTFM) [63]. The difference image pair of the before and after the missing image is jointly trained to get the dictionary, and the dictionary is used to reconstruct the two difference images, and the final fusion reconstruction result is obtained using the observed image and difference images. Based on the idea of temporal residual image, scholars have also developed many fusion methods based on sparse representation, such as EBSCDL [51], ELM-based method [64], bSBL-SCDL/msSBL-SCDL [65], and CSSF [66].

Although linear optimization method achieve better spatiotemporal fusion reconstruction effect, it is difficult to satisfy the actual imaging scenario due to the assumption that the model is assumed to be linear.

#### 11.2.3.3 Nonlinear Optimization Method

The weight- based method, Bayesian method, and optimization decomposition method are all based on the linear relationship between the sensor observation models, yet it has not been strictly proved. Deep learning methods can well describe the nonlinear relationship. Some scholars have also explored the use of deep learning methods for spatiotemporal fusion. Vahid Moosavi combines wavelet algorithm and ANN, adaptive neuro-fuzzy inference system (ANFIS) and support vector machine (SVM) to propose the prediction-phase WAIFA algorithm for Landsat thermal infrared data [67]. In recent years, the use of CNN for spatiotemporal fusion algorithms such as STFDCNN [53] and DCSTFN [68] have also been gradually proposed.

The nonlinear optimization method can learn and accurately describe the nonlinear relationship between the known and missing time-phase images, and has more mobility and accuracy than the linear optimization decomposition method. However, its ability is related to network architecture design and parameter settings. Poor network structures often fail to obtain good results, and require a large number of training samples and training time.

### *11.2.4 Quality Assessment of Data Fusion*

So far, scholars have proposed many fusion algorithms. However, the fusion effect of the fusion algorithm needs to be evaluated qualitatively and quantitatively. Qualitative evaluation of fused images is usually achieved through visual interpretation. Visual comparison is performed on the fused single-band image or RGB composite image, and the effect of fusion is qualitatively evaluated from the image texture and color changes. Quantitative evaluation usually uses different numerical evaluation indexes for quantitative evaluation, or through information extraction such as target detection, ground classification, and other applications for quantitative evaluation. Among them, according to whether there are reference images, commonly used quality assessment method of data fusion can be divided into two categories [69].

#### 11.2.4.1 Quality Assessment with Reference Image

The fusion of reference images often uses the data simulated by the original data to obtain reference images, which mainly abides by the Wald criterion [70]. The Wald criterion includes two parts, consistency property and synthesis property. Consistency property means that the fused data and the data obtained after downsampling should be the same as the original image. Synthesis property means that the data obtained after downsampling of the original images after the fusion can use the original A0 data as reference data. In most cases, the Wald criterion is effective in evaluating fusion quality, but there are certain restrictions in the following situations [71]: ①when the data contains noise, the criterion is invalid; ②the applicability of the criterion depends on the downsampling factor.

Fusion evaluation indexes with reference images usually achieve fusion quality evaluation by comparing the relevant statistics on the spectral dimension and spatial dimension between the reference image and the fusion image. Common evaluation indexes are spectral angle mapper (SAM) [33], spectral information divergence (SID) [72], universal image quality index (UIQI, Q) [73], root mean square error (RMSE) [35], relative average spectral error (RASE) [74], SNR [33], peak signal noise ratio (PSNR) [75], the correlation coefficient (CC) [76], the structural similarity metric (SSIM) [77], Erreur Relative Globale Adimensionnelle de Synthèse (ERGAS) [78], Mean Bias (MB) [79], etc. The summary, best values and references of each index are shown in Table 11.1.

#### 11.2.4.2 Quality Assessment Without Reference Image

In most fusion scenes, reference images are difficult to obtain. Therefore, it is more meaningful to directly use the original data and fusion reconstruction data to evaluate the indexes on the original resolution scale under the fusion target without reference images. Common fusion quality evaluation indexes include Quality with No Reference (QNR) [76], Mean Value [10], Variance [10], standard deviation [10], Information Entropy [10], Average Gradient [10], etc. The summary, formula, best value, and applicable scope of each index are shown in Table 11.2.

## 11.3 Enhancing Spectral Resolution of Multispectral Data Using Deep Learning Method

The rapid development of remote sensing applications has been promoting the resolution of satellite sensors, especially for the hyperspectral sensors. Yet, due to the restriction of sensors, hyperspectral data suffer from narrow swath and long revisiting time, resulting in the insufficiency of hyperspectral data for remote sensing applications. Here, we proposed a method of data spectral enhancement method frame utilizing the technique of convolution neural networks (CNN) to enhance the spectral resolution of multispectral data (SRECNN) to acquire adequate hyperspectral data.

### *11.3.1 Data Fusion Methods Using Deep Learning*

Deep learning is a subfield of machine learning. It is composed of multiple layers to learn abstractions with multiple data feature levels [80]. Its history and brief introduction can be seen in the review article [81].

**Table 11.1** Fusion result assessment methods with reference image

| Quality indices | Introduction | Common fusion scenario | | |
|---|---|---|---|---|
| | | Spatial enhancement | Spectral enhancement | Temporal enhancement |
| Spectral information divergence (SID) | Calculating the cross entropy of the reference image and the fusion image pixel to reflect the spectral distortion. | | ✓ | |
| Spectral angle mapper (SAM) | The spectral angle between the reference image and the fused image pixel-by-pixel vector is calculated to reflect the spectral distortion of the fused data. | | ✓ | |
| Universal image quality index (UIQI, Q) | Use the covariance, mean, and variance between the reference image and the fused image to evaluate the spatial details of the fused data. | ✓ | | ✓ |
| Root mean square error (RMSE) | Commonly used to measure the error between the reference image and the fused image can be calculated band by band, or add all bands to obtain the total error. | ✓ | | ✓ |
| Relative average spectral error (RASE) | Based on the root mean square, the average performance of the fusion algorithm on all bands of the image is measured. | | ✓ | |
| Signal-to-noise ratio (SNR) | The difference between the reference image and the fused image is regarded as noise, the reference image as a signal, and the ratio is calculated to measure the distortion of the image. | ✓ | | |
| Peak signal noise ratio (PSNR) | It is based on the mean square error (MSE). The quality of the fused reconstructed image is reflected by calculating the ratio of the maximum peak value of the reconstructed image and the mean square error of the two images. | ✓ | | |
| Correlation coefficient (CC) | The correlation coefficient is obtained by calculating the difference between the pixels of the fused image and the reference image and the average pixel. | ✓ | | ✓ |
| Structural similarity metric (SSIM) | The overall fusion quality is measured by calculating the mean, variance, and covariance of the fusion image and the reference impact. | ✓ | | ✓ |
| Erreur Relative Globale Adimensionnelle de Synthèse (ERGAS) | Based on the RMSE, the scale relationship between the fusion image and the observation image is considered. | ✓ | ✓ | |
| Mean Bias (MB) | Calculate the difference between the average of the fusion image and the reference image to measure the fusion effect. | ✓ | | |

**Table 11.2** Fusion result assessment methods without reference image

| Quality indices | Introduction | Common fusion scenario | | |
|---|---|---|---|---|
| | | Spatial enhancement | Spectral enhancement | Temporal enhancement |
| Quality with No Reference (QNR) | There is no need to fuse reference data, and measure and evaluate from the comprehensive indicators of spectral distortion and spatial distortion. Suitable for multispectral-panchromatic image fusion. | ✓ | ✓ | |
| Mean value | The average value of all the pixels of each band image of the fused reconstruction data. | ✓ | | ✓ |
| Variance | The dispersion of each band image relative to the average value of the fused reconstruction data. | ✓ | | ✓ |
| Standard deviation | The dispersion of each band image relative to the average value of the fused reconstruction data. | ✓ | | ✓ |
| Information entropy | The information content of each band image of fusion reconstruction data. | ✓ | | ✓ |
| Average gradient | Single-band image detail contrast change rate, characterizing image clarity | ✓ | | ✓ |

In remote sensing fusion method, there are many deep-learning-based methods proposed. Huang et al. used DNN to fuse multispectral and panchromatic images, added pretraining and fine-tuning learning stages within the DNN framework, and improved the accuracy of the fusion [40]. Based on this, scholars have developed many algorithms, such as PNN [41], DNN based on the MRA framework [42], DRPNN [43], and MSDCN [21]. To study hyperspectral and multispectral fusion, Frost Palsson uses a 3DCNN model to fuse hyperspectral and multispectral data. In order to overcome the computational complexity caused by large amounts of hyperspectral data, dimensionality reduction processing of hyperspectral data brings higher accuracy than MAP algorithm [44]. Vahid Moosavi combined wavelet algorithm and ANN, ANFIS and SVM to propose the WAIFA algorithm for the prediction stage of Landsat thermal infrared data [67]. CNN for spatiotemporal fusion algorithms such as STFDCNN [53] and DCSTFN [68] are the new trend in remote sensing data fusion.

## *11.3.2 Spectral Resolution Enhancement Method via Convolutional Neural Networks (SRECNN)*

In tradition, the relationship between data from different sensors and ground true data is usually represented in the form of linear combinations. It can be represented in a unified formula as

$$Y_i = XB_iS_iR_i + n_i \tag{11.19}$$

in which $Y_i$ represents the data from sensor $i$. $X$ represents the ground truth data(or the quality-improved data), $B_i$ represents the blurring function which is related to PSF of the sensor, $S_i$ represents the downsampling function, $R_i$ represents the spectral response function, and the $n_i$ represents the random bias of the sensor. $B_i$, $S_i$, $R_i$ differs according to the sensor's characteristics. However, the linear hypothesis of sensor is too idealistic for real imaging process. Thus, nonlinear relationship between different sensors can also be inferred.

Here, basic idea of SRECNN is proposed. The relationship, or in other words, mappings F between sensors are nonlinear, which can be represented as

$$H(\alpha, \lambda, \delta, t, \ldots) = F(M(\alpha, \lambda, \delta, t, \ldots)) \tag{11.20}$$

in which $\alpha$ is the landcover types, $\lambda$ is the wavelength, $\delta$ is the observation angle, and $t$ is the imaging time. There are many factors that influence the values for each pixel, and it is difficult to determine the factors, the weights, and powers of different factors. Thus, here we obtain the mappings utilizing the tool of deep convolutional neural network model. The CNN can learn the nonlinear models without determining the parameters. We suppose the mappings between overlapping parts of hyperspectral data and multispectral data can be a good simulation of mappings between sensors, and the relationship of sensors can be applied to the nonoverlapping area of the two datasets. Note hyperspectral data as $H$ with $n_H$ bands, multispectral data as $M$ with $n_M$ bands. The basic idea of the method aims at finding the nonlinear relationship between $H$ and $M$. Suppose these two data are overlapped, and note the overlapped hyperspectral data as $H_{\text{ov}}$, the overlapped multispectral data as $M_{\text{ov}}$. When the landcovers are similar in the overlapped area and none overlapped area, the relationship $F$ between $H$ and $M$ can be viewed as the same as the relationship between $H_{\text{ov}}$ and $M_{\text{ov}}$. It could be written in the following form:

$$F: H_{\text{ov}} = F(M_{\text{ov}}) \rightarrow F: H = F(M) \tag{11.21}$$

Yet, the hyperspectral data possess hundreds of bands resulting in huge volume and redundancy [82], which may cause huge computational complexity. Appropriate data dimensional reduction method can solve this problem. Principal Component Analysis (PCA) is a widely utilized method of dimensional reduction. It can reduce the dimension of hyperspectral data, preserve a maximal amount of variability at the same time [83], and leave out noise information [82]. As PCA is a kind of linear transformation [84], based on the former hypothesis that the relationship of hyperspectral and multispectral is nonlinear, we can arrive at the conclusion that the PCA component of hyperspectral data cannot be linear but nonlinear. So, the relationship changes into:

$$F: H'_{\text{ov}}(1, 2, \ldots, n_M) = F(M_{\text{ov}}) \rightarrow F: H'(1, 2, \ldots, n_M) = F(M) \tag{11.22}$$

in which $H'$ and $H'_{\text{ov}}$ represents PCA components of hyperspectral data and overlapping area of hyperspectral data, respectively. Here, we preserve the same number of PCA components as the number of PCA bands as to make the network more of flexibility and stability.

With the hypothesis that the relationship between PCA components of hyperspectral data and multispectral data is nonlinear and can be mapped via CNN, here we propose the basic frame of SRECNN. The basic workflow is shown in Figure 11.4, in which the cubes represent three-dimensional remote sensing data cube with different samples, columns, and bands. First, the hyperspectral data are transformed into n components of PCA. The multispectral data are cropped according to the overlapped area with hyperspectral data. And then, both cropped multispectral data and PCAs of hyperspectral data are input into the CNN structure for training. In the CNN structure, the multispectral data are the network input while the PCAs of hyperspectral data are the network output. After the training process, we input the whole scene of multispectral data into the trained CNN structure and to get the predicted PCA components. Finally, the predicted components are transformed via inverse PCA to get the final restored spectrally enhanced hyperspectral data.

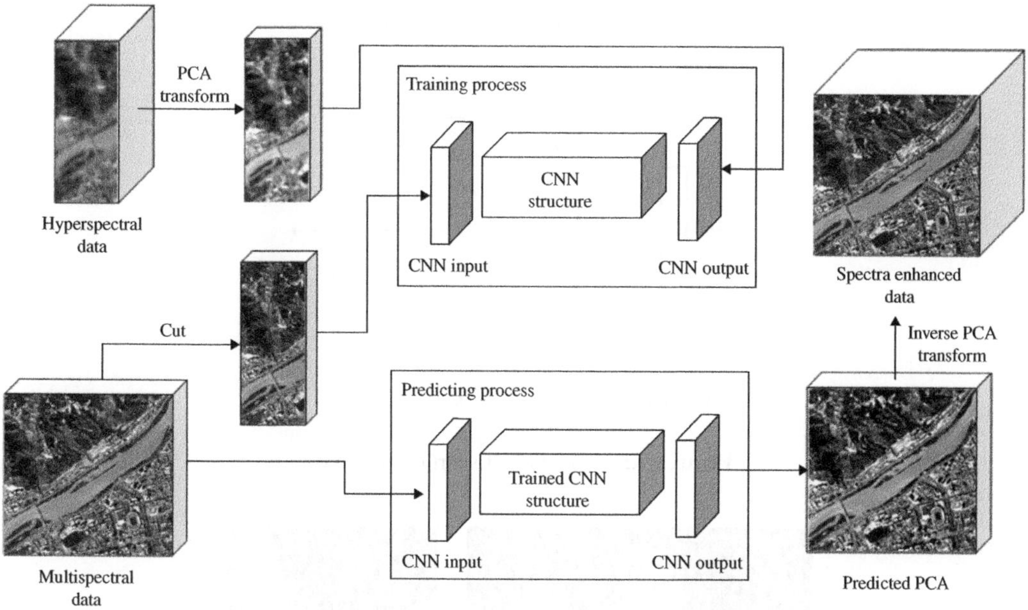


**Figure 11.4** Basic workflow of SRECNN.

Here, CNN architecture can be viewed as an optimal function mapping from multispectral data to PCA of hyperspectral data, and its structure is important in modeling nonlinear relationships. In this article, we compare two architectures: the simple three-layered CNN architecture and the three-layered ResNet [39] architecture. The three-layered CNN architecture is to directly learn the nonlinear mapping between the hyperspectral data and multispectral data. It consists three layers of convolutional neural layers, each followed by a batch normalization layer and a leaky rectified linear unit (ReLu) layer. The ResNet learns the residual part of the nonlinear mappings, which can be represented as

$$y = \mathcal{F}(x, \{W_i\}) + x \tag{11.23}$$

in which x represents the input of a certain layer, y represents the output of this layer, and the $\mathcal{F}(x, \{W_i\})$ is the part where ResNet architecture learns. The residual learning can be realized by the structure of "short connections," which are those skipping some layers [85] shown in Figure 11.5. Three blocks of "short connections" are plied together to compare the results. It is said that residual learning is a useful preconditioning to identity mapping. If the optimal function is close to an identity mapping, it should be easier to learn the perturbations from identity mapping than to learn the optimal function in a new one[39], thus it is easier to optimize than plain networks.

### 11.3.3 Application 1: Extending the Swath of Hyperspectral Data

In order to show the SRECNN's performance in swath extension, we use the real dataset to give an example of extending the swath of hyperspectral data. The real dataset is the Hyperion data and the simultaneously sensed ALI multispectral data, which are shown in Figure 11.6. Despite the fact that there are sensors with better performances compared to the Hyperion and ALI dataset, the reason for

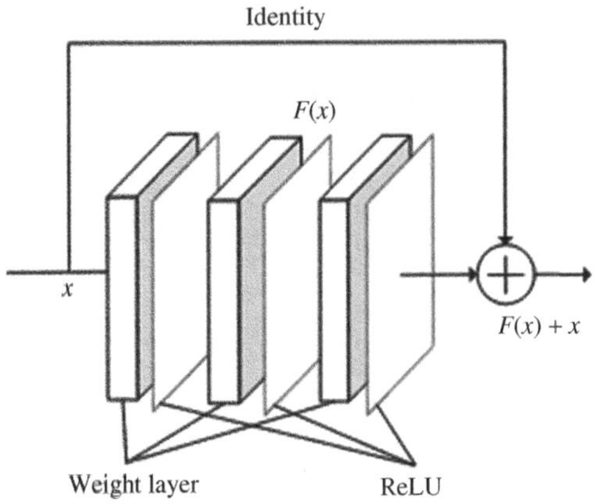


**Figure 11.5** Structure of short connections.

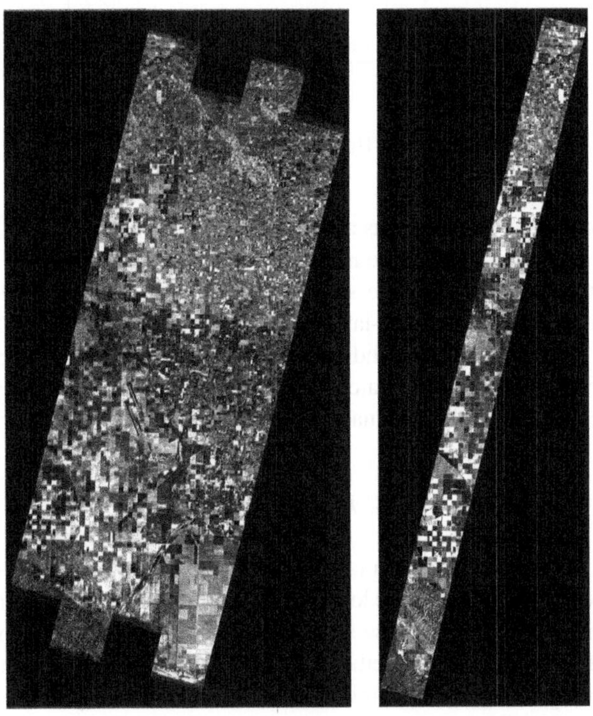

**Figure 11.6** RGB composites of real data. Left scene is the ALI multispectral data, right scene is the Hyperion hyperspectral data.

choosing these two datasets is their strict reregistration. The use of the two sensors can minimize the effect of correction bias. The dataset II is sensed on 30 May 2016. The overlapped area of Hyperion data and ALI data were masked, cropped, and piled to form the training dataset. The predicting dataset is the piled ALI data scene. Necessary preprocessing process includes noise reduction by MNF and radiometric correction by ENVI 11.3.

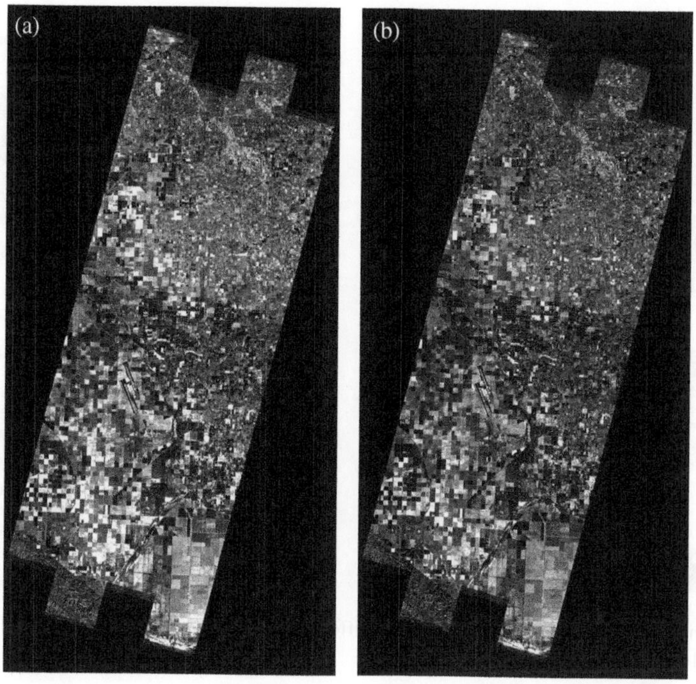


**Figure 11.7** RGB composites of: (a) ALI multispectral data and (b) fused hyperspectral data.

The real dataset is spectrally enhanced and swath extended using SRECNN. The real data experiment represents the case where the data are large and enough for training. The data are divided into patches of 64, and the number of principal components is set to be 20. The RGB composites of ALI mulitspectral data and fused hyperspectral data are shown in Figure 11.7. The RGB composites of fused data by different methods are shown in Figure 11.8. To evaluate the results of spectral resolution improvements, here we perform SRECNN to extend the swath of hyperspectral data, clip and mask to get the overlapped area, and compare its quantitative assessment with the fusion results of CNMF and CRISP, which are shown in Table 11.3. The quantitative results show that SRECNN can outperform CNMF and CRISP in most quality indices, and it can extend the swath of hyperspectral data which the other two fusion methods cannot.

## *11.3.4 Application 2: Cloud Removal of GF-5 Hyperspectral Data*

The main process of the cloud removal of GF-5 hyperspectral data is shown in Figure 11.9. First, the clouded hyperspectral data are transformed using PCA, and are used to generate cloud mask. Here, we use the K-Means method to generate approximate estimation of the cloud mask. Then, the cloud mask, PCA transformed hyperspectral data, and multispectral data are used to generate the training patch pairs. Here, the hyperspectral data and auxiliary multispectral data are divided into small patches as training patches. The patch is discarded if there exists any cloud in the patch. After obtaining training patches, the ResNet model is trained to get the nonlinear mapping $F(\cdot)$ between GF-5 hyperspectral data and the auxiliary multispectral data. Finally, the cloud area was predicted and restored using the trained models.

The test scene of GF-5 hyperspectral data was obtained in Germany, on 29 May 2018. It possesses 1024 pixels in lines and 768 pixels in columns with 180 bands ranging from 390.12 to 1028.98 nm. The test scene is mainly covered by crop fields. The data was processed with QUick atmospheric

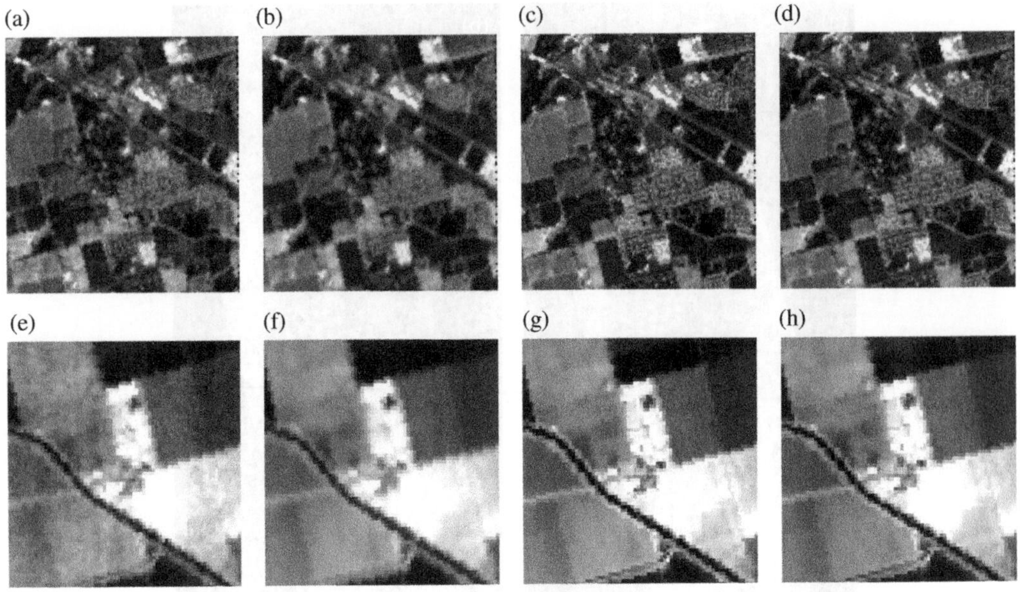


**Figure 11.8** Comparisons of fused data in detail by different methods: (a) and (e) original ALI data, (b) and (f) SRECNN, (c) and (g) CNMF, and (d) and (h) CRISP.

**Table 11.3** Quantitative results of SRECNN(ResNet), CNMF, and CRISP (Dataset I)

| | CC(↑) | SAM(↓) | RMSE(↓) | PSNR(↑) | UIQI (↑) | SSIM(↓) |
|---|---|---|---|---|---|---|
| SRECNN-ResNet | **0.9928** | **1.5073** | **17.2664** | **40.5132** | **0.9924** | **0.0531** |
| CNMF | 0.9416 | 1.9143 | 61.4562 | 30.1956 | 0.9371 | 0.0736 |
| CRISP | 0.9425 | 2.4016 | 66.2334 | 29.5332 | 0.9407 | 0.0764 |

correction and noise reduction using PCA by ENVI 11.2. Then, the test data were added clouds which was derived from other scenes of GF-5 hyperspectral data and cloud cover is approximately 0.249. Figures 11.10 and 11.11 show the RGB composites of the original and synthetic hyperspectral data and Figure 11.12 shows several spectra of the added cloud.

The auxiliary multispectral data is the Sentinel-2a multispectral data obtained on 28 May 2018. We chose the 7 bands of the data with spatial resolution of 20 m ranging from 492.1 to 864.0 nm. Also, the auxiliary data were possessed with QUick atmospheric correction by ENVI 11.2.

To evaluate the effectiveness of this method, the result was assessed by visual interpretation and quantitative assessments. Figure 11.13 shows the overall and detail scene of RGB composites of the cloud removal result in the cloud-blocked scenes. We can see from the result that the clouds have been completely removed, and the spatial details of the data are well preserved. The similarity of the corresponding spectra shows the ability of this method to restore the spectral information of the blocked area (Figure 11.14).

For the quantitative assessment, we used the quality indices of PSNR, CC, RMSE, and SAM to evaluate the results with the original GF-5 hyperspectral data. The restored data were calculated with the original data in the whole scene. Figure 11.15 shows the indices before and after the cloud

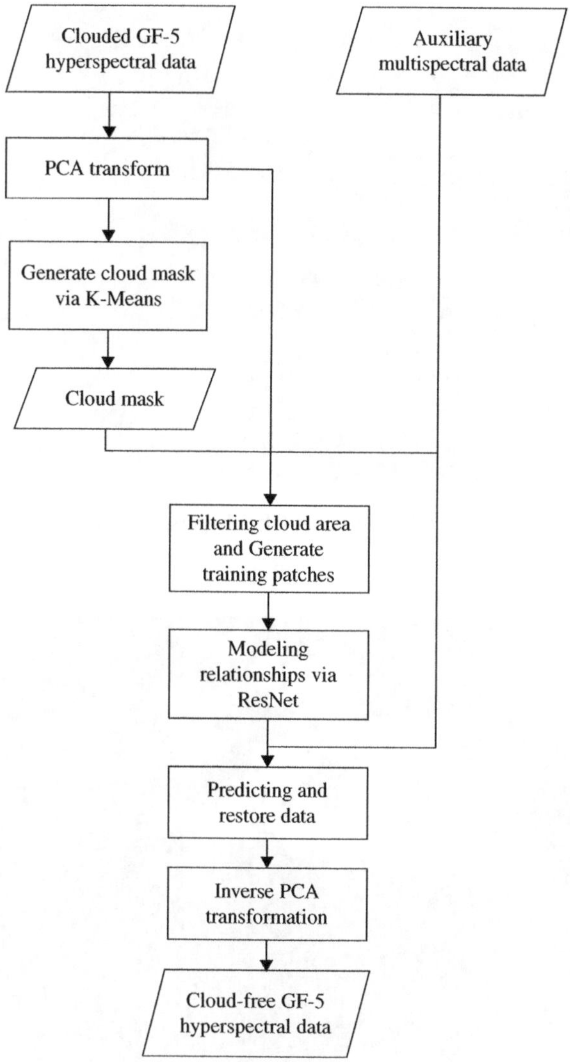


**Figure 11.9** Flowchart of the method.

removal procedures. As we can see in the figure, SAM decreases from 7.09 to 1.93, which shows the spectral feature was restored and close to the original hyperspectral data. PSNR increases from 19.39 to 37.11, RMSE decreases from 15911.1 to 267.62, and CC increases from 0.356 to 0.965 which shows the restoration accuracy has been increased.

## 11.4 Multidimensional Datasets (MDD)

Remote sensing satellites have continuously observed the earth for a long time, and have accumulated a large number of long-term series, multi-resolution, and multi-spectral series data. The existing MODIS and Landsat series of remote sensing satellite observation data has become an important remote sensing data source for global change research. Continuous long-term satellite observation

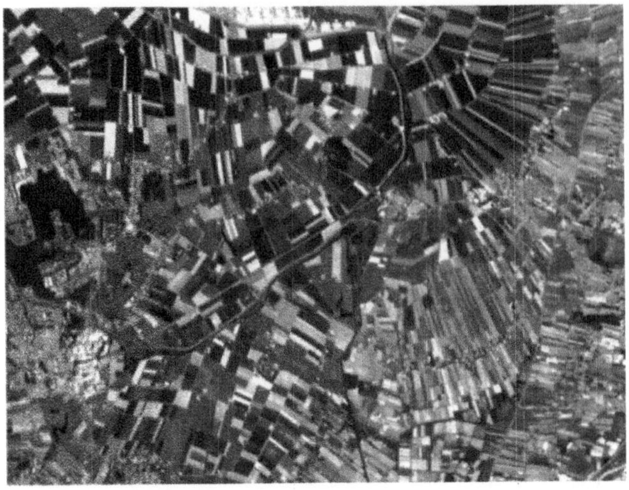

**Figure 11.10** Original GF-5 hyperspectral data.

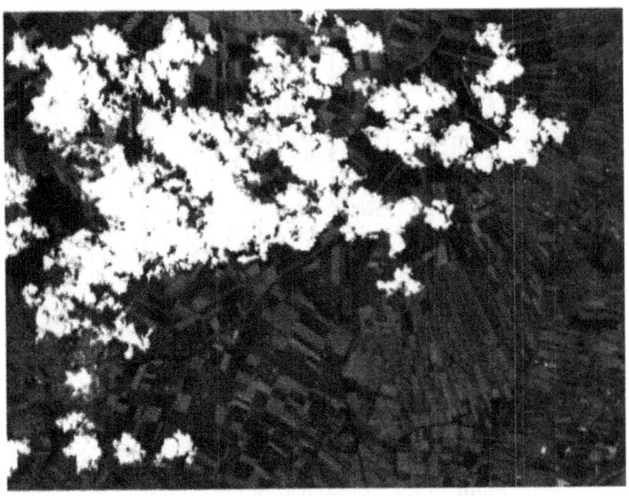

**Figure 11.11** Synthesized GF-5 data with cloud.

data plays an important role in vegetation drought monitoring, but traditional remote sensing data sets usually use a file in a common format as the basic storage unit, such as Geo-TIFF (Tagged Image File Format), HDF (Hierarchical Data Format), as well as the formats that come with remote sensing commercial software (such as ENVI and ERDAS). For these data formats, each file corresponds to one-time data; HDF and Geo-TIFF data formats use more complex data structures (tree structure or linked list structure). Because the commonly used remote sensing data format is basically based on the three-dimensional cube model storage, it is very inconvenient for the long-term sequence analysis of the data. Therefore, Zhang Lifu et al. [86] developed a multi-dimensional data format, namely, MDD (Multi-Dimensional Dataset) data format, based on the data characteristics of multi-temporal phase, multi-space, and multi-spectral data [86].

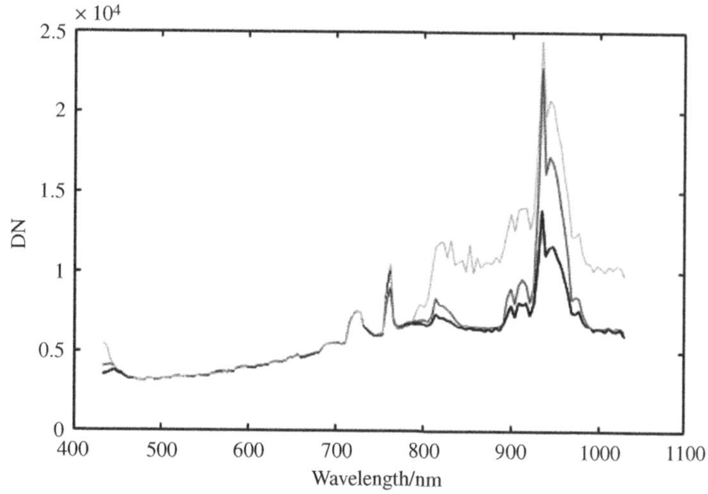


**Figure 11.12** Cloud spectra of the test data.

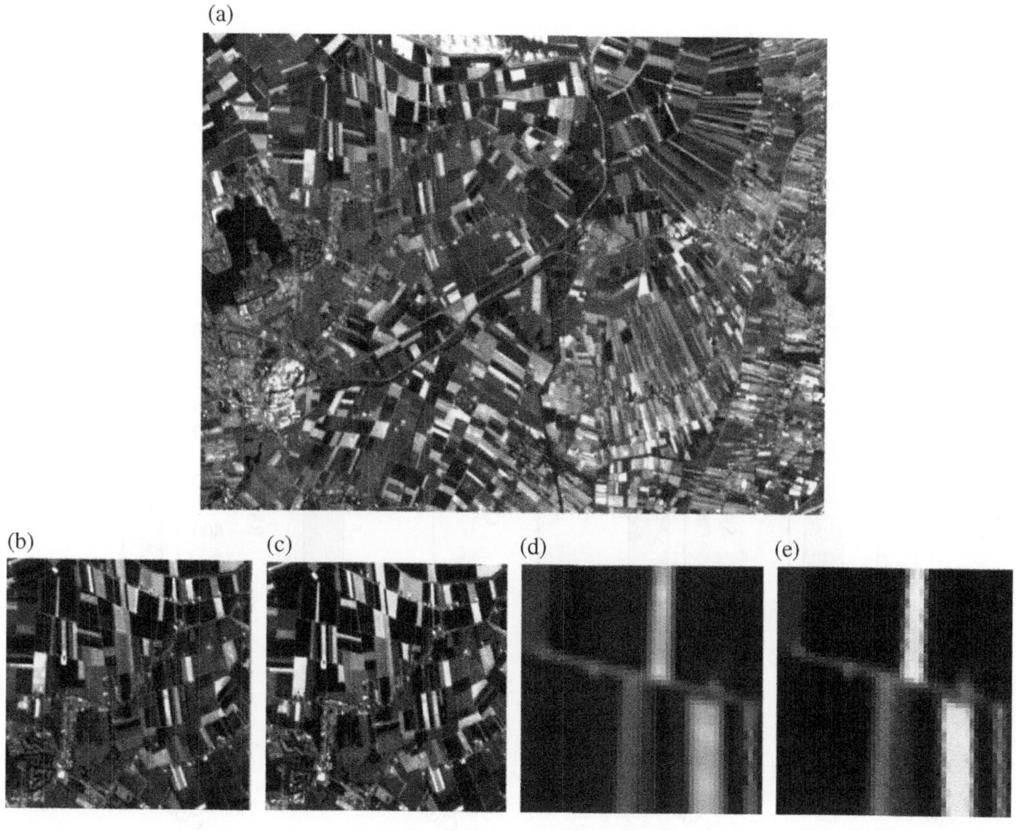


**Figure 11.13** Cloud-removed results in spatial domain (RGB composites): (a) whole scene of the restored area, (b) original data, (c) restored data, (d) original data, and (e) restored data.

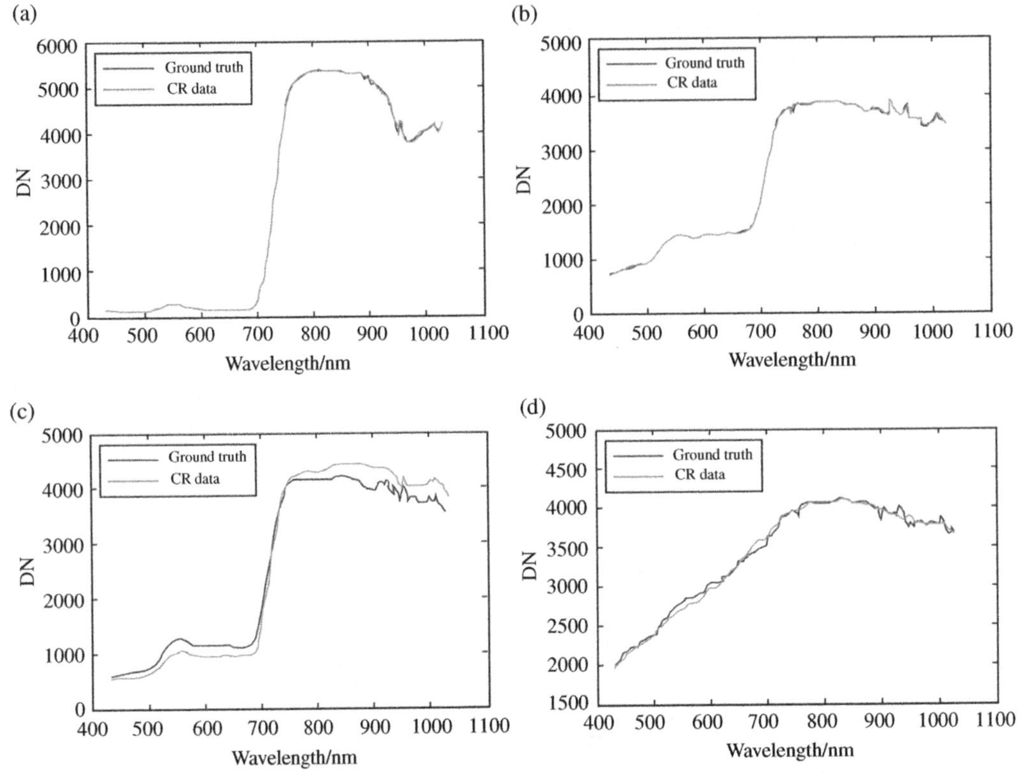


**Figure 11.14** Spectra of typical land covers of original and cloud-removed data. (a) Vegetation A, (b) Vegetation B, (c) Vegetation C, and (d) Soil.

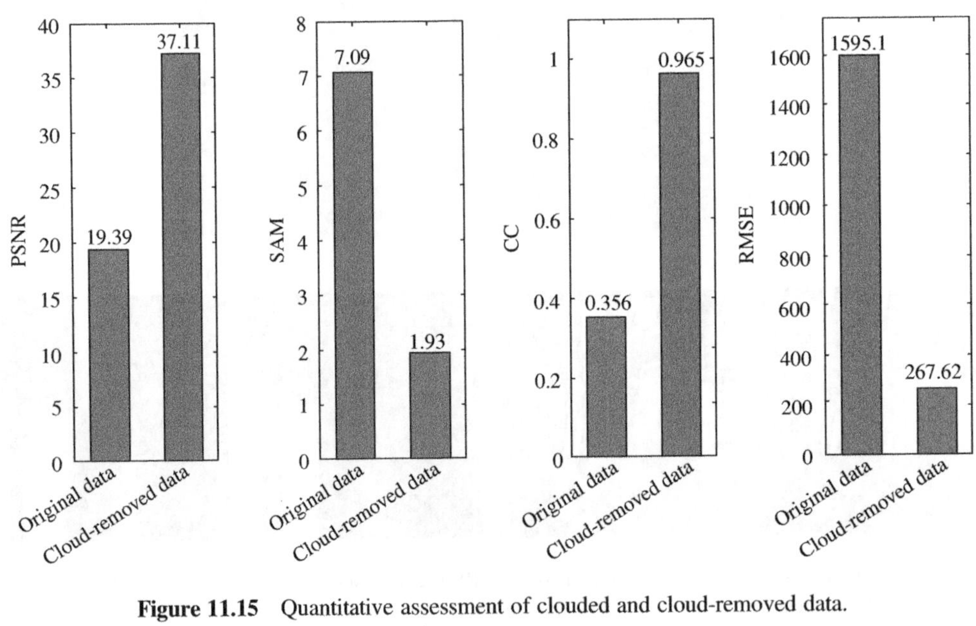


**Figure 11.15** Quantitative assessment of clouded and cloud-removed data.

## 11.4.1 Introduction of Multidimensional Datasets (MDD)

According to the different storage order of multidimensional data, MDD can be divided into five basic data storage formats: Temporal Sequential in Band (TSB), Temporal Sequential in Pixel (TSP), Temporal Interleaved by Band (TIB), Temporal Interleaved by Pixel (TIP), and Temporal Interleaved by Spectrum (TIS).

### 11.4.1.1 Temporal Sequential in Band (TSB)

Suppose the multi-temporal remote sensing dataset contains $T_1$, $T_2$, and $T_3$ images, each time image contains 3 bands, and each band contains 4 pixels (represented by 4 small squares), as shown in Figure 11.16. The order in which the small squares are arranged represents the storage order of all pixels in the TSB format. The dataset contains four dimensions of temporal dimension, spectral dimension, and spatial "row" and "column."

The TSB storage format first stores the pixel data of the first band of time $T_1$ in the order of first row and then column, and then stores the pixel data of each band of $T_1$ in turn, and finally the cube data of all times according to the above rules according to time. The order is stored sequentially. This storage structure can be expressed as

$$P_{\text{TSB}} = F_{\text{TSB}}(t) = F_{\text{TSB}}(1) + F_{\text{TSB}}(2) + \cdots + F_{\text{TSB}}(T), \quad t\epsilon[1, T] \tag{11.24}$$

In the formula, $F_{\text{TSB}}(t)$ represents the cube image dataset at each time. Equation (11.25) indicates that the data storage sequentially stores image cubes at different times in the order of time $T$. In each cube image data, stored in the order of bands, it can be expressed as

$$F_{\text{TSB}}(t) = B_{\text{TSB}}(t,s) = B_{\text{TSB}}(t,1) + B_{\text{TSB}}(t,2) + \cdots + B_{\text{TSB}}(t, S), \quad s\epsilon[1, S] \tag{11.25}$$

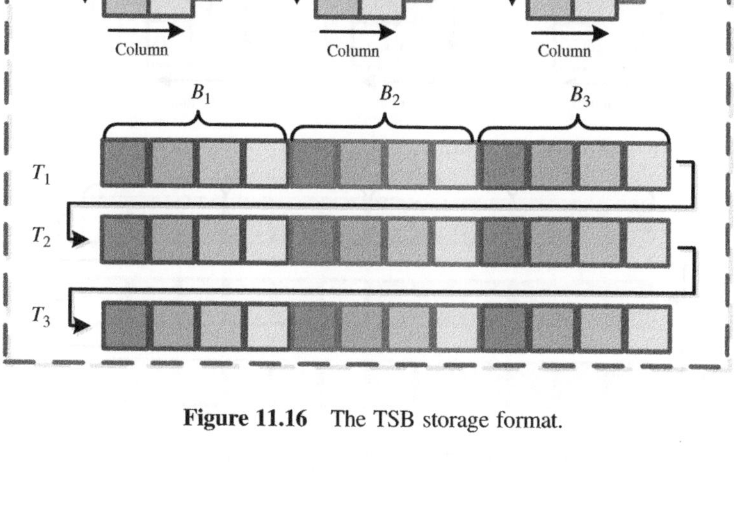


**Figure 11.16** The TSB storage format.

In the storage of each band, it is stored in line-by-line order, which can be expressed as

$$\begin{aligned} B_{\mathrm{TSB}}(t, s) &= A(t, s, c, r) \\ &= A(t, s, [1:C], 1) + A(t, s, [1:C], 2) + \cdots + A(t, s, [1:C], R), \quad r \in [1, R] \end{aligned} \quad (11.26)$$

The TSB format organizes all the band data at each time together, and is suitable for operating on the data of several bands at a time and the processing of the band in the spatial dimension. Applications such as extracting spectral cube data composed of all bands at one or more times, spectral operations, and spatial domain filtering can be organized in the TSB data format.

#### 11.4.1.2 Temporal Sequential in Pixel (TSP)

The TSP storage format adopts the principle of spectral dimension priority. First, the cube data at $T_1$ time is stored in order of the spectral information of the first pixel in sequence, and then the spectral information of each pixel point is stored in the order of "row first, column after." Finally, according to the above rules, the cube data of all times are stored in chronological order, as shown in Figure 11.17; this storage structure can be expressed as

$$P_{\mathrm{TSP}} = F_{\mathrm{TSP}}(t) = F_{\mathrm{TSP}}(1) + F_{\mathrm{TSP}}(2) + \cdots + F_{\mathrm{TSP}}(T), \quad t \in [1, T] \quad (11.27)$$

In the formula, the storage of each cube can be expressed by the following formula:

$$\begin{aligned} F_{\mathrm{TSP}}(t) &= A(t, [1:S], c, r) \\ &= A(t, [1:S], 1, 1) + A(t, [1:S], 2, 1) + \cdots + A(t, [1:S], C, 1) \\ &\quad + A(t, [1:S], 1, 2) + A(t, [1:S], 2, 2) + \cdots + A(t, [1:S], C, 2) \\ &\quad \cdots \\ &\quad + A(t, [1:S], 1, R) + A(t, [1:S], 2, R) + \cdots + A(t, [1:S], C, R) \end{aligned} \quad (11.28)$$

The meaning of the variables in the above formula is the same as Table 11.4.

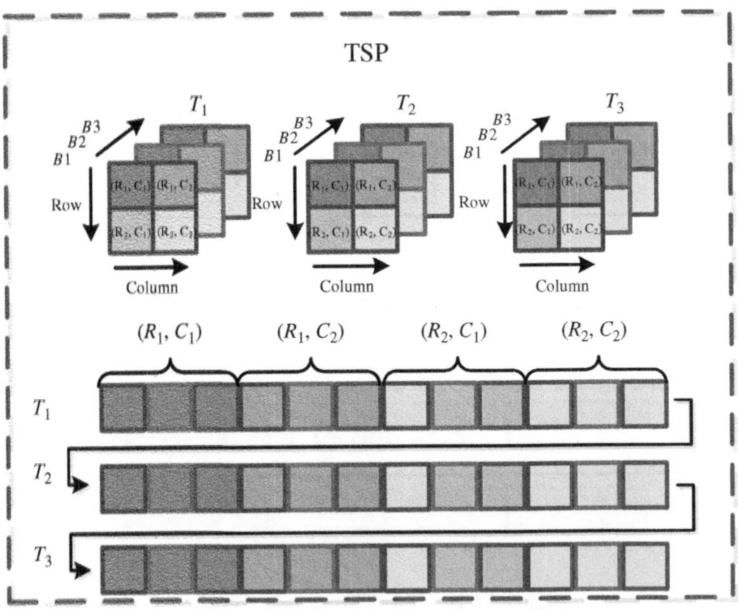


**Figure 11.17** The TSP storage format.

**Table 11.4** The meaning of each expression or variable

| Expression/variable | Description |
|---|---|
| $t$ | A certain time whose value range is [1, $T$] |
| $s$ | A certain spectral band whose value range is [1, $S$] |
| $c$ | A certain column whose value range is [1, $C$] |
| $r$ | A certain row whose value range is [1, $R$] |
| $A(t, s, c, r)$ | Pixels in band $s$, column $c$, and row $r$ in the spectral cube $A$ on time $t$ |
| $B_{XXX}$ | Data representing a band |
| $F_{XXX}$ | Represents cube data composed of multiple bands |
| $P_{XXX}$ | Represents a four-dimensional dataset of space-time spectrum, in which XXX is one of the five formats |

The characteristic of TSP format is to store the spectral data of each time image together, which is suitable for operating on the spectral data of one time. Operations such as extracting the spectral curve of a pixel or an area or performing spectral characterization of images at different times can be organized in TSP data format, such as calculating the spectral slope and aspect, spectral binary code, spectral absorption index, spectral derivative and etc.

### 11.4.1.3 Temporal Interleaved by Band (TIB)

As shown in Figure 11.18, the TIB storage format is based on the principle of band priority. First, all the pixels in the first band $T_1$ time are stored in the order of "row first, column after," and then all the images of the first band at different times are sequentially stored. The metadata is stored together in

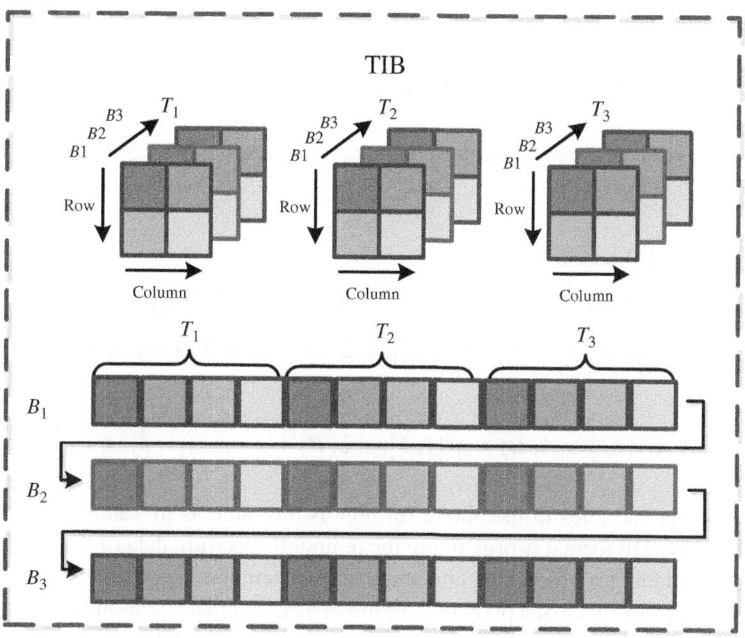


**Figure 11.18** The TIB storage format.

the order of "first in line and second in line," and finally the band data are stored in chronological order according to the above rules. This storage structure can be expressed as

$$P_{TIB} = F_{TIB}(s) = F_{TIB}(1) + F_{TIB}(2) + \cdots + F_{TIB}(S), \quad s\epsilon[1, S] \tag{11.29}$$

In the formula, the cube represented by $F_{TIB}(s)$ is a data cube composed of the same band in different time series. The storage order is to store the first band in time series, and then store the second band until all the bands are stored. $F_{TIB}(s)$ can be expressed as

$$F_{TIB}(s) = B_{TIB}(t, s) = B_{TIB}(1, s) + B_{TIB}(2, s) + \cdots + B_{TIB}(T, S) \tag{11.30}$$

In the formula, the storage order of each band is

$$\begin{aligned} B_{TIB}(t, s) &= A(t, s, c, r) \\ &= A(t, s, [1:C], 1) + A(t, s, [1:C], 2) + \cdots + A(t, s, [1:C], R) \end{aligned} \tag{11.31}$$

The meaning of the variables in the above formula is the same as Table 11.4.

The characteristic of TIB format is to organize all time data of a band together, which is suitable for extracting time series cube data of a band. When the time series cube of one band needs to be extracted or the data of three times need to be selected for a certain band for false color synthesis display, the TIB data format can be adopted for organization.

#### 11.4.1.4 Temporal Interleaved by Pixel (TIP)

As shown in Figure 11.19, TIP first organizes the data of the first pixel point of the first band at all times in chronological order, and then other pixels are stored according to the rule of "row first, column after." The structure can be expressed as

$$P_{TIP} = F_{TIP}(s) = F_{TIP}(1) + F_{TIP}(2) + \cdots + F_{TIP}(S), \quad s\epsilon[1, S] \tag{11.32}$$

In the formula, $F_{TIP}(s)$ represents a cube composed of the same pixel spectrum at different time images, which can be expressed as

$$\begin{aligned} F_{TIP}(s) &= A([1:T], s, c, r) \\ &= A([1:T], s, 1, 1) + A([1:T], s, 2, 1) + \cdots + A([1:T], s, C, 1) \\ &\quad + A([1:T], s, 1, 2) + A([1:T], s, 2, 2) + \cdots + A([1:T], s, C, 2) \\ &\quad \cdots \\ &\quad + A([1:T], s, 1, R) + A([1:T], s, 2, R) + \cdots + A([1:T], s, C, R) \end{aligned} \tag{11.33}$$

The meaning of the variables in the above formula is the same as in Table 11.4.

The highlight of the TIP format is organizing the temporal spectrum data of the pixels in the image together, which is suitable for processing and analyzing the temporal spectrum of the pixels. Extracting the temporal spectrum curve of a pixel or a region in a certain band, smoothing and filtering in the

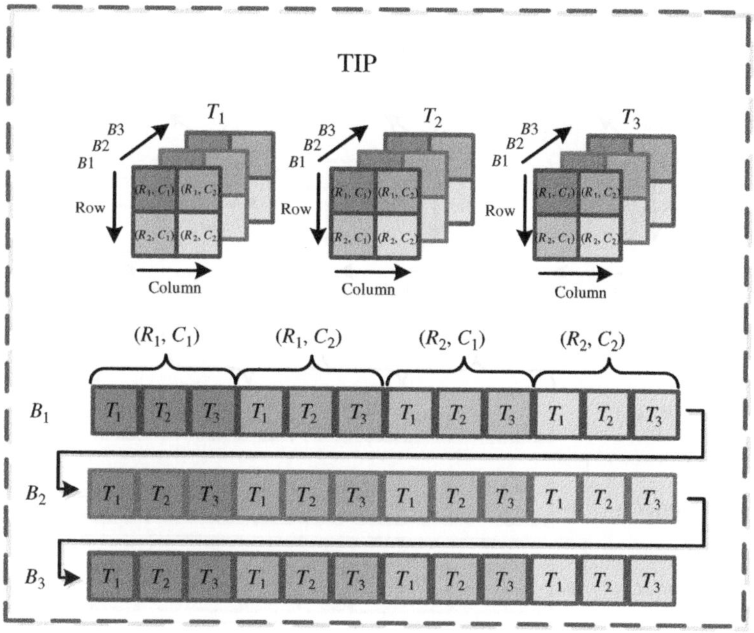


**Figure 11.19** The TIP storage format.

time dimension, fitting the time spectrum curve, and performing predictive analysis can be organized in TIP data format.

### 11.4.1.5 Temporal Interleaved by Spectrum (TIS)

As shown in Figure 11.20, the TIS storage format first stores the first pixel of all bands in chronological order according to the principle of band priority, and then according to this rule, the other pixels are spatially followed by the principle of "row first, column after" for storage. The structure can be expressed as

$$\begin{aligned} P_{\mathrm{TIS}} = B_{\mathrm{TIS}}(c, r) &= B_{\mathrm{TIS}}(1, 1) + B_{\mathrm{TIS}}(2, 1) + \cdots + B_{\mathrm{TIS}}(C, 1) \\ &+ B_{\mathrm{TIS}}(1, 2) + B_{\mathrm{TIS}}(2, 2) + \cdots + B_{\mathrm{TIS}}(C, 2) \\ &\ldots \\ &B_{\mathrm{TIS}}(1, R) + B_{\mathrm{TIS}}(2, R) + \cdots + B_{\mathrm{TIS}}(C, R) \end{aligned} \tag{11.34}$$

In the formula,

$$\begin{aligned} B_{\mathrm{TIS}}(c, r) &= A(t, [1:S], c, r) \\ &= A(1, [1:S], c, r) + A(2, [1:S], c, r) + \cdots + A(T, [1:S], c, r), \ t\epsilon[1:T] \end{aligned} \tag{11.35}$$

The meaning of the variables in the above formula is the same as Table 11.4.

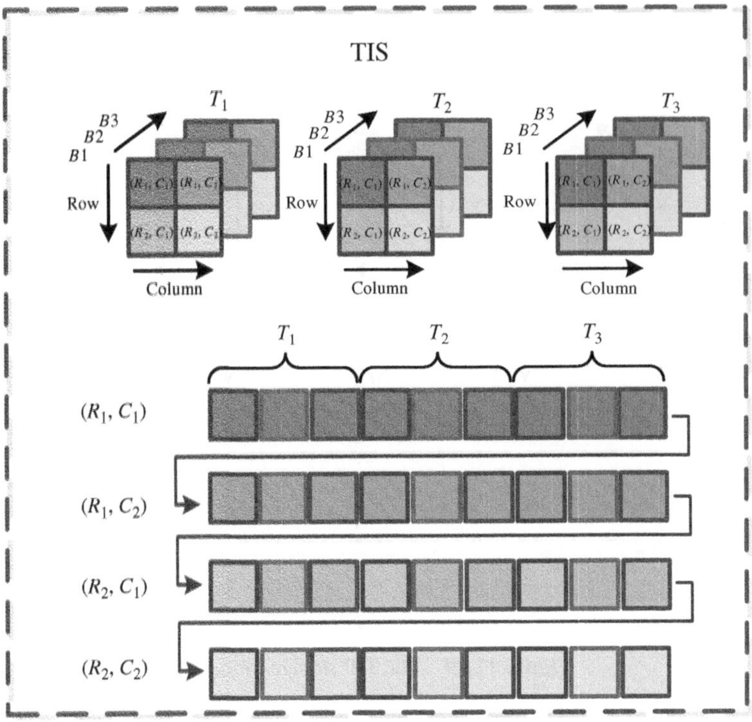


**Figure 11.20** The TIS storage format.

The characteristic of the TIS format is to organize the spectral data of each pixel in the entire time series, which is suitable for extracting the time series data of the spectral curve of the pixel. To extract all the spectral curves of a certain pixel in a time range and visualize these curves in three dimensions, and analyze the change of the spectrum of the pixel with time, it can be organized in TIS format.

In order to facilitate the construction, reading, and analysis of multidimensional time-space-spectrum data, a multidimensional analysis software module, namely Multidimensional Data Analysis (MDA) software module, which is integrated in Multi-dimensional Analysis of Remote Sensing (MARS), a multi-dimensional remote sensing analysis software the download address of MARS software is: http://www.geodoi.ac.cn/WebCn/doi.aspx?Id=702.

### *11.4.2 Spatial-Temporal Data Fusion via Multidimensional Datasets*

To date, huge amount of remote sensing datasets have been obtained by remote sensors, and they provide rich information for land cover change detection, agricultural monitoring, and forest disturbance [87–89]. However, due to the limitation of capacity of satellite sensors, remote sensing data usually suffer from the deficiency of obtaining both high spatial and temporal resolution at the same time. Besides, remote sensor data are also affected by cloud contamination and different imaging conditions. These greatly hinder the applications of remote sensing datasets.

Inspired by the implementation of 3DCNN in video-related tasks, this article proposed a long-time series completion method of utilizing 3D fully CNN. This method piles the long-time remote sensing images in chronological order and constructs four-dimensional datasets to process multi-temporal images synchronously. The model extracts spatiotemporal features of both high spatiotemporal and low spatial resolution data (HTLS) and low spatiotemporal and low spatial resolution data (LTHS) data using two branches of 3D fully CNN. The method has the following advantages. (i) It processes multi-temporal images of a 4D dataset synchronously, thus reducing the fusion time and manual intervention. (ii) It extracts spatial features and temporal features together in four dimensions, which utilizes more information compared to traditional spatiotemporal fusion method. (iii) The architecture of the network is simple and easy to implement. The rest of the article is organized as follows. Section 11.4.2.1 illustrates the model in detail, including its problem formation and architecture. Section 11.4.2.2 shows the experimental results of the proposed method on two datasets, and they are compared with results derived from four spatiotemporal fusion methods.

### 11.4.2.1 Basic Architecture

Note four-dimensional high-temporal-low-spatial data (HTLS) as $M(T_M, W_M, H_M, B) \in R^{T_M \times W_M \times H_M \times B_M}$, in which $T_M$, $W_M$, $H_M$, $B_M$ represents temporal range, width, height, and number of bands of HTLS, respectively. Note four-dimensional low-temporal-high-spatial data (LTHS) as $L(T_L, W_L, H_L, B)R^{T_L \times W_L \times H_L \times B_L}$, in which $T_L$, $W_L$, $H_L$, $B_L$ represents temporal range, width, height, and number of bands of LTHS, respectively. And our goal is to obtain four-dimensional high-temporal-high-spatial data (HTHS) $P(T_M, W_L, H_L, B_L) \in R^{T_L \times W_L \times H_L \times B_L}$. We hypothesize that we can derive data-oriented spatiotemporal features of HTLS and LTHS data, fuse and merge these features together to get the restored HTHS data. The spatiotemporal features can be extracted using 3DCNN. In one layer of 3DCNN, a set of 3D feature extractors (or in other words, kernels) slide and convolve over an input to extract the high-dimensional spatiotemporal features, and a 3D bias was added to it. The operation in one 3DCNN layer can be represented as

$$h_i^l(x, y, z) = \sigma\left(b_i^l + \sum_k \sum_{u, v, w} h_k^{l-1}(x-u, \ y-v, \ z-w)\mathrm{W}_{ki}^l(u, \ v, \ w)\right) \tag{11.36}$$

in which $h_i^l$ and $h_k^{l-1}$ is the $i$th 3D feature data in the $l$th layer and the $k$th 3D feature data in the previous $l-1$th layer, respectively. $W_{ki}^l$ is the 3D convolutional filter and $b_i^l$ is the 3D bias. $\sigma(\cdot)$ is the nonlinear activation function such as ReLu [90]. Here, we use three-layered full convolutional neural networks (FCN) as the spatiotemporal feature extractor.

The basic structure of the method is as shown in Figure 11.21. The whole architecture can be divided into two parts: the 3D feature extraction and 3D feature fusion part. The model has two inputs, the four-dimensional HTLS and four-dimensional LTHS. The output is the four-dimensional HTHS. All 4D data are arranged as single-date images piled along the time axis. The HTLS and LTHS data are first each input into a three full convolutional layered feature extraction model to obtain data-oriented spatiotemporal features. Next, the extracted features are fused and merged into one feature, and it undergoes feature-oriented mapping and finally the HTHS data is obtained.

The 3D feature extraction model consists two branches of 3DCNN. Each subnetwork consists three layers of convolutional layers. To maintain the size of each output feature, the input of each layer undergoes zero-padding on the edge. The stride of convolutional filter is set to 1*1 empirically. The size of the output feature of each convolutional layer is shown in the figure. Each convolutional layer is followed by a ReLU activation function, a commonly used activation function which shows excellent performance.

The 3D feature fusion model first merges the two extracted features into one feature by concatenating the two features along the band axis. After that, the feature goes through three convolutional

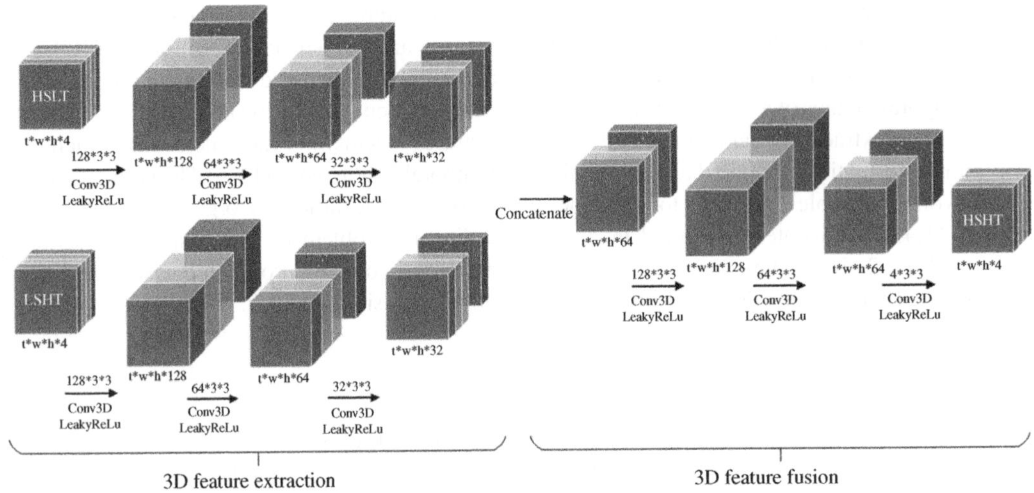


**Figure 11.21** Basic structure of 3D spatial-temporal data fusion via multidimensional datasets.

layers to further fuse into HTHS data. The subnetwork structure is basically the same with the subnetwork of the 3D feature extraction model.

### 11.4.2.2 Dataset Experiment

#### *11.4.2.2.1 Datasets*

There are two datasets to test the efficiency of the proposed method (Figure 11.22). Dataset I is the nine pairs of simulated Landsat-MODIS images. The Landsat OLI-8 data are obtained in California and the time range is from 3 June 2013 to 25 October 2013. The Landsat OLI-8 data L1TP data were used to derive nine MODIS images by downsampling to the spatial resolution of 500 m. The data were resized into size of 800*800 and the number of bands is 4. Dataset II is the Landsat-MODIS images obtained in Coleambally Irrigation Area (CIA) [91], an irrigation system located in southern New South Wales, Australia. The data consist of 17 pairs of cloud-free Landsat-MODIS images obtained during 2001–2002 austral summer growing season. All Landsat images were Landsat-5 TM and were processed with atmospheric correction. MODIS images used in CIA were MODIS Terra MOD09GA Collection 5 data and coregistered to Landsat images.

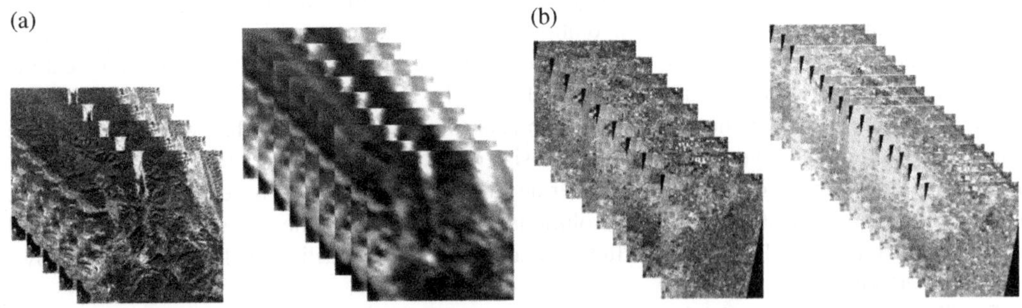


**Figure 11.22** Simulated datasets and CIA datasets: (a) Landsat and simulated MODIS images of Dataset I and (b) Landsat and MODIS images of Dataset II.

The experiments were conducted using KERAS 2.0. The epoch was set to be 1000 originally, and during the training, an early-stop strategy was utilized. When the training loss does not further decrease and remain the same level for 30 epochs, the training will be stopped and the best model was restored. The learning rate was set to be 0.0003 empirically. We used the strategy of reducing learning rate when training loss did not decrease for 10 epochs. The optimizer used here was the ADAM optimizer [92].

For both Dataset I and Dataset II, we all arrange all Landsat-MODIS pairs in chronological order and form the 4D datasets. All 4D datasets are divided into small patches with size of 32, and all the datasets can be represented as $R^{s\times T\times p\times p\times B}$, where s is the number of all 4D patches, $T$ is the number of dates of the datasets, $p$ is the size of the patches, and $B$ is the number of bands of the datasets. To test the effectiveness of our method, we set Landsat images on even dates as the missing images, and test the predicted Landsat images with the images on odd dates. Data training were conducted on odd data pairs. In the prediction stage, all data pairs are input into the trained model to get the predicted missing Landsat images. Figure 11.23 shows the training loss and validation loss for these two datasets and the epoch axis is displayed in logarithmic axis. Both experiments ended early before 1000 epoch and converge quickly in 10 epochs.

To show further details of prediction results, we chose key date of both datasets and the whole imagery and the zoomed imagery of fusion results are shown in false-color composite. For the simulated data, Figure 11.24 shows the fusion results on 7 September 2013 (6th date). The images in the first row are the whole scenes of the reference image and predicted images from different models. The images in the second and the third row are the further zoom images of the corresponding models. All images are displayed in the false-color bands. We can see from the tone and texture that results by the proposed method and ESTARFM are the closest to the reference images. The images derived by STARFM suffer from textual distortion such as "no data" pixel and cannot capture temporal change well. The tone and detail texture of results predicted by FSDAF and DCSTFN are close to reference images, yet they fail to capture the temporal changes of the lake. For the CIA datasets, Figure 11.25 shows the fusion results on 22 February (10th date). Like scenes shown above, the first row is the whole scene and the second and the third rows are the zoomed scenes. It is obvious to see that the prediction derived from the proposed method is the closest to the original Landsat images. The results of STARFM, like simulated datasets, suffer from serious textual distortion. The texture of results of ESTARFM is close to reference image, but its tone is slightly different from the original Landsat image. FSDAF's result is close to original images, but some details are slightly different. The results of DCSTFN are accurate in overall but suffers from slight blur.

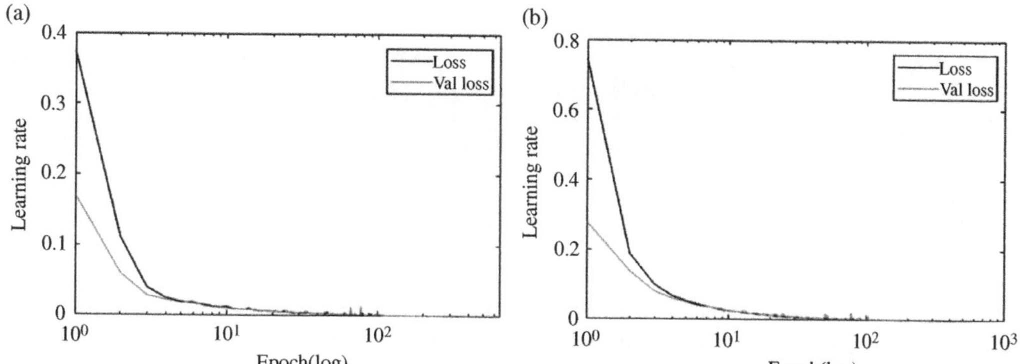


**Figure 11.23** Training loss and validation loss of both experiments: (a) simulated dataset and (b) CIA dataset.

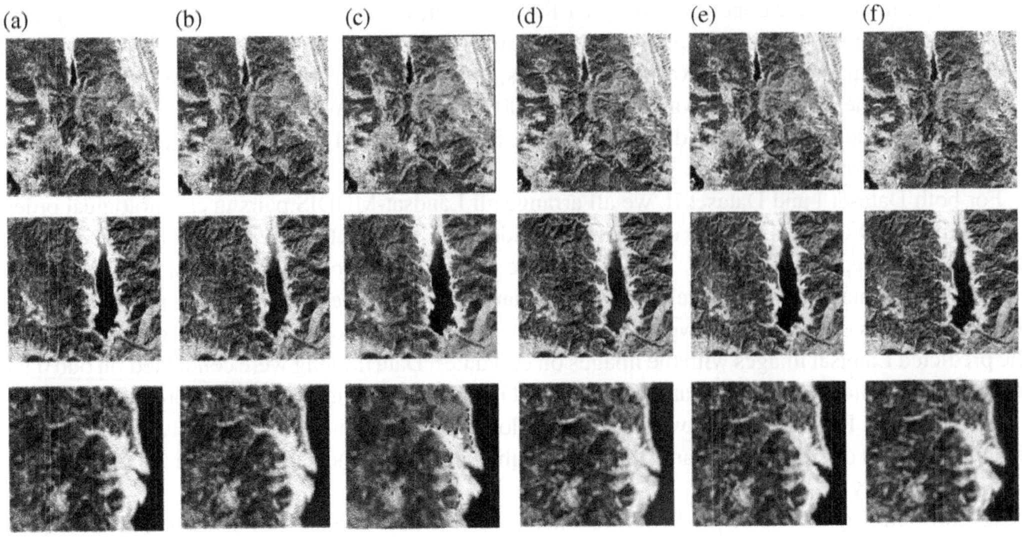


**Figure 11.24** Fusion results of simulated dataset on 6th date (false-color composite): (a) Landsat OLI8, (b) Ours, (c) STARFM, (d) ESTARFM, (e) FSDAF, and (f) DCSTFN.

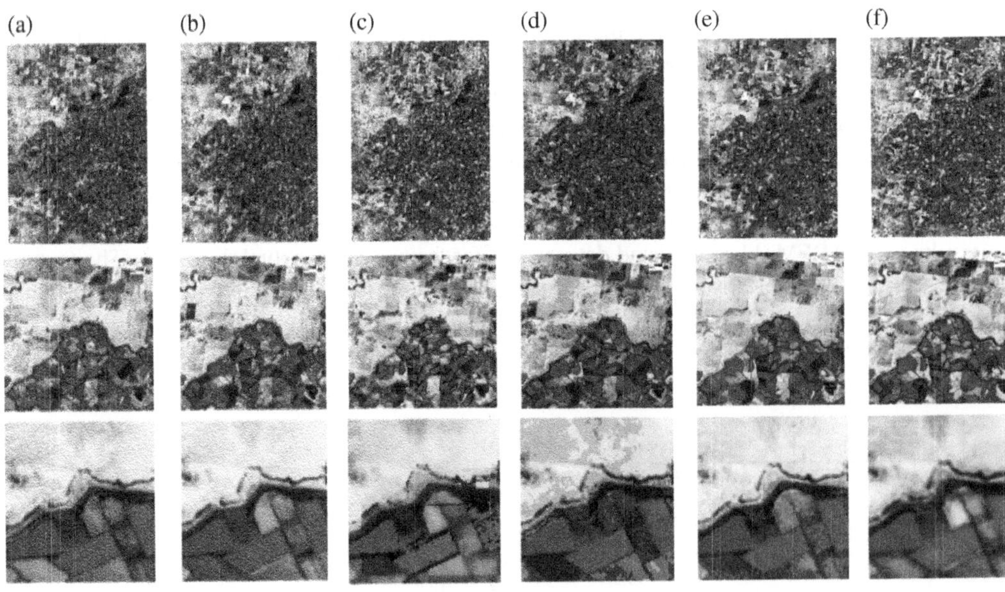


**Figure 11.25** Fusion results of CIA dataset on 10th date (false-color composite): (a) Landsat OLI8, (b) Ours, (c) STARFM, (d) ESTARFM, (e) FSDAF, and (f) DCSTFN.

We also calculated quality indices mentioned in the previous part for prediction results on all predicted dates, and made plots to show the efficiency and accuracy of our method compared to STARFM, ESTARFM, FSDAF, and DCSTFN, as shown in Figures 11.26 and 11.27. For both the simulated dataset and CIA dataset, we can draw the conclusion that: (i) among all the models used in the experiments,

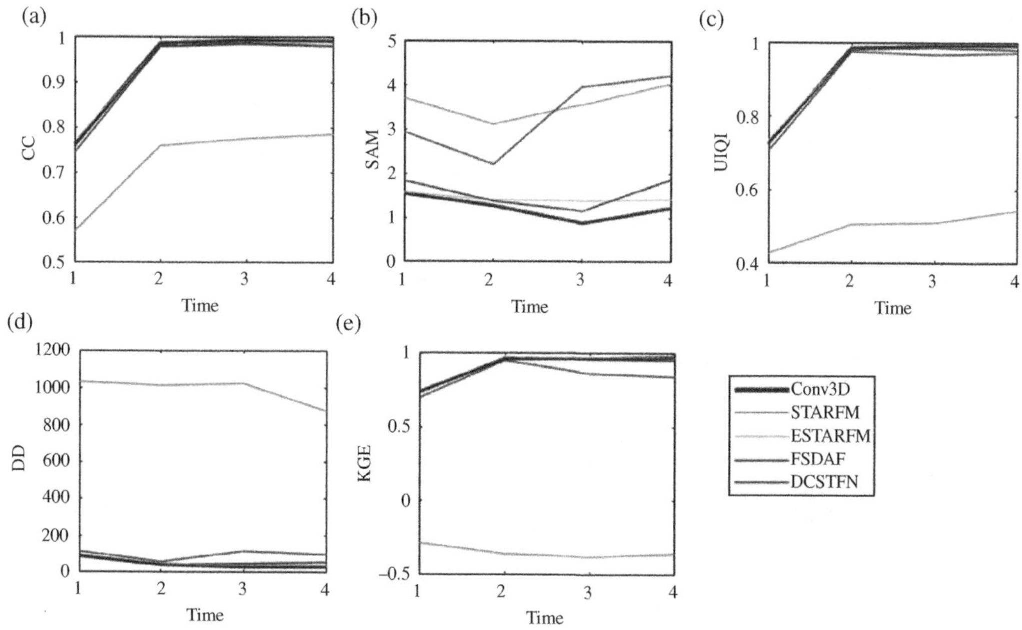


**Figure 11.26** Quantitative indices for four prediction dates in simulated dataset: (a) CC, (b) SAM, (c) UIQI, (d) DD, and (e) KGE.

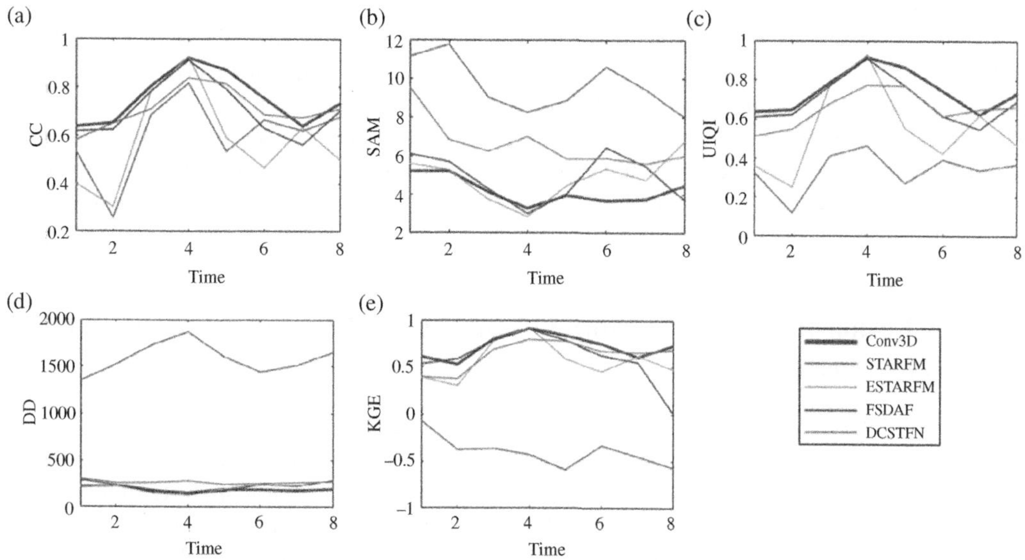


**Figure 11.27** Quantitative indices for eight prediction dates in CIA dataset: (a) CC, (b) SAM, (c) UIQI, (d) DD, and (e) KGE.

**Table 11.5** Quantitative evaluations for the fusion results of simulated data (in average)

| | Conv3D(GPU) | STARFM | ESTARFM | FSDAF | DCSTFN |
|---|---|---|---|---|---|
| CC(↑) | **0.932856** | 0.721424 | 0.931586 | 0.926118 | 0.920663 |
| SAM(↓) | **1.226440** | 3.594455 | 1.439799 | 1.55579 | 3.32276 |
| Q(↑) | **0.922783** | 0.496499 | 0.922384 | 0.915973 | 0.903919 |
| KGE(↑) | **0.913032** | −0.34513 | 0.916931 | 0.90834 | 0.840969 |
| DD(↓) | **47.776330** | 981.5255 | 49.90806 | 57.04557 | 911.30401 |
| Total time(↓) | **45 m 23.14 s** | 1 h 50 m 11.09 s | 2 h 10 m 18.05 s | 46 m 49.21 s | 1 h 45 m 46.27 s |

**Table 11.6** Quantitative evaluations for the fusion results of CIA data (in average)

| | Conv3D (GPU) | STARFM | ESTARFM | FSDAF | DCSTFN |
|---|---|---|---|---|---|
| CC(↑) | **0.750389** | 0.597477 | 0.574164 | 0.701842 | 0.706214 |
| SAM(↓) | **4.221983** | 9.621269 | 4.859649 | 4.849458 | 6.617819 |
| Q(↑) | **0.742306** | 0.33688 | 0.547202 | 0.691564 | 0.649607 |
| DD(↓) | **191.4393** | 1583.97 | 210.3523 | 223.201 | 267.7885 |
| KGE(↑) | **0.718881** | −0.39805 | 0.564562 | 0.593536 | 0.626164 |
| Total time(↓) | **56 m 19.73 s** | 19 h 22 m 6.58 s | 17 h 26 m 28.54 s | 17 h 42 m 6.35 s | 7 h 21 m 31.22 s |

STARFM performs the worst in all the indices. (ii) Although the ESTARFM, FSDAF, and DCSTFN can obtain relatively good result, the performances of different dates are not as stable as the proposed method which we can see from the variance of the plots. (iii) The proposed method outperforms the other four models in most dates in CC, SAM, UIQI, DD, and KGE.

To further show the overall results predicted by the proposed method, STARFM, ESTARFM, FSDAF, and DCSTFN, the averages of the quality indices across all dates are calculated. Also, the overall time for different models is also recorded to show the efficiency of the five methods. The results of both datasets are shown in Tables 11.5 and 11.6. We can see that for both datasets, our method shows the better results compared to other four models. Also, our method used the least total time for predicting multiple temporal Landsat images, longer the temporal range, the more obvious the improvement in efficiency. These proved the accuracy and the efficiency of our method.

## 11.5 A Fusion-Related Case Study: Comparison of Fusion Methods on GF-5 Hyperspectral Data

### *11.5.1 Backgrounds*

The Gaofen-5 (GF-5) satellite was successfully launched on 9 May 2018. It is the first satellite in China to achieve high spectral resolution observation in the Major Project of High Resolution Earth Observation System, and it is the world's first full-spectrum hyperspectral satellite to make comprehensive observations of the atmosphere and land [93]. The spatial resolution of GF-5 hyperspectral data is 30 m [94]. Compared with multispectral satellites such as Gaofen-1 (GF-1) and Gaofen-2 (GF-2), its spatial resolution is still insufficient in many applications [95]. Using spatial-spectral fusion methods to fuse hyperspectral data with panchromatic image or MSI is an effective technical method to improve the spatial resolution of HSI [4]. Therefore, it is of great significance to discuss the fusion effects of different methods on GF-5 hyperspectral data for information mining and promotion application of GF-5 hyperspectral data.

Yokoya et al. studied the fusion performance of the latest 10 fusion methods on hyperspectral and multispectral data [96]. This study provides a good direction for the fusion of hyperspectral and multispectral data. However, this study only used simulated hyperspectral data, so the conclusion was limited in practical applications, such as different hardware design of GF-5 imager, different spatial resolution, and actual acquisition geometry. Ren Kai et al. fused GF-5 hyperspectral data with GF-1, GF-2, Sential-2A multispectral data, and comprehensively evaluated the performance of nine fusion methods [97]. However, it involves fewer domestic multispectral satellite data sources, such as not involving Beijing-2 (BJ-2), GF-1C, and GF-1D.

In this study, based on the principle that the algorithm is easy to use and suitable for generalization, six fusion methods, GS, GSA, Coupled Non-negative Matrix Factorization (CNMF), Color Resolution Improvement Software Package with Butterworth (CRISP-B) or Wavelet transform (CRISP-W), and Generalized Laplacian Pyramid (GLP) are, respectively, used to perform fusion experiments on GF-5 hyperspectral data and multispectral data from BJ-2, GF-2, and GF-1/1C/1D domestic satellites. Visual interpretation, five fusion evaluation indexes, including CC, UIQI, SAM, ERGAS, and PSNR, and classification application as well as time costs are used to comprehensively evaluate the fusion results.

This study provides method decision support for the fusion of GF-5 hyperspectral data and other domestic satellite multispectral data, which is helpful for the application and promotion of GF-5 hyperspectral data.

### *11.5.2 Experiment*

The multispectral data of GF-1/1C/1D, GF-2, and BJ-2 from Poyang Lake, the Yellow River estuary, and Dongting Lake area, as well as GF-5 hyperspectral data of the same area at similar time were fused separately. The data are shown in Table 11.7.

The experiment data include typical features such as bare soil, buildings, roads, water bodies, farmland, breeding areas, paddy field, and farmland. The image data are all Level-1 products, and the pixels are radiance values. ENVI is used to preprocess each group data (A–F) separately. First, RPC orthorectification module is used to orthorectify the multispectral and hyperspectral data, the GF-5 hyperspectral data are resampled to 4m in the A-B group, and 8m in the C-F group. Then, multispectral data are regarded as the reference image and the hyperspectral data as the image to be registered. The homonymy points are selected, and each group of images are registered. Finally, the same area of 400 × 400 pixels is selected for fusion experiments.

**Table 11.7** Experiment data

| ID | Area | Center longitude | Center latitude | Multispectral sensor (Spatial resolution) | Date | Hyperspectral sensor (Spatial resolution) | Date |
|---|---|---|---|---|---|---|---|
| A | Poyang Lake | 116.39E | 28.77N | BJ-2 (4 m) | 20181002 | GF-5 (30 m) | 20181005 |
| B | Dongting Lake | 112.63E | 28.67N | GF-2 (4 m) | 20181005 | | 20181005 |
| C | Poyang Lake | 116.31E | 28.75N | GF-1 (8 m) | 20181006 | | 20181007 |
| D | Poyang Lake | 116.44E | 28.72N | GF-1C (8 m) | 20181008 | | 20181007 |
| E | The Yellow River estuary | 118.94E | 37.60N | GF-1D (8 m) | 20181103 | | 20181101 |
| F | Dongting Lake | 112.63E | 28.67N | GF-1D (8 m) | 20181006 | | 20181005 |

Among them, the filter function of GS is the spectral response function; the preset pure pixel number of CNMF is automatically determined by the virtual dimension method; the CRISP cutoff frequency is 20, and the number of wavelet filter layers of the CRISP-W is 3; GLP uses the least square method to calculate the regression coefficient matrix.

## *11.5.3 Visual Analysis*

In order to facilitate visual interpretation and qualitative evaluation, the fusion results are displayed in true color images with the same band combination (GF-1/1C/1D, R: Band 3 680 nm, G: Band 2 576 nm, B: Band 1 502 nm; GF-2, R: Band 3 660 nm, G: Band 2 555 nm, B: Band 1 485 nm; BJ-2, R: Band 3 635 nm, G: Band2 550 nm, B: Band1 475 nm; GF-5 and fusion image, R: Band59 638.4 nm, G: Band38 548.47 nm, B: Band17 458.59 nm), the stretching method also remains the same (Figure 11.28).

In general, GS cannot effectively improve the spatial resolution, and the fusion result is the fuzziest, especially in A, B, C, D, and F, where the ground features are broken and complex. CNMF has obvious color distortions, which are shown in the six groups of experiments. Both GSA and GLP can effectively improve the spatial resolution. The texture of the ground features is clearer, but the performance of the color fidelity is slightly different. In B, C, E, and F, the color of the GSA fusion result is similar to the MSI. The color of the GLP fusion result is closer to the hyperspectral data, which shows to a certain extent that the spectral fidelity of GLP is better than GSA. The fusion results of CRISP-B and CRISP-W methods are relatively close, and the hue is natural. All six groups of experiments show good spectral retention ability, but the spatial resolution improvement ability is not as good as GLP.

In summary, when the types of ground boundary rules are relatively simple, the six methods are not much different. When the type of ground fragmentation is complex, the spectral retention ability of GLP, CRISP-B, and CRISP-W is better, and the spatial resolution improvement ability of GLP and GSA is better. GS cannot effectively improve the spatial resolution, and there is color distortion in the CNMF fusion results.

## *11.5.4 Index Evaluation*

Table 11.8 shows the quantitative evaluation results generated based on different experimental data and different fusion methods. In longitudinal comparison, it can be seen that groups C, D, E, and F are closer to groups A and B. CC and UIQI are closer to 1; PSNR is larger, and SAM is smaller. That is, the fused image is more related to the reference image, the brightness and contrast are more similar, and the spatial reconstruction quality and spectrum retention ability are better. This is because the two groups of experiments in A and B are all fused with a MSI with a spatial resolution of 4 m and a HSI with a spatial resolution of 30 m, while the four groups of experiments with C, D, E, and F are all 8 m MSI and 30 m HSI fusion. The ERGAS performance is relatively abnormal. This is because the data in Group E have a large range of water, and the average pixel value in the image band is low, which makes the ERGAS value too large. Therefore, when compared between experimental groups, the ERGAS index should not be relied on too much. It is not difficult to find that GF-2 is better than BJ-2 data in the fusion experiment with a spatial resolution of 4 m in multispectral data; and in the 8 m experiment, GF-1D data perform better. In summary, the smaller the difference in the spatial resolution of a group of fused images, the better the fusion effect. When fusing GF-5 hyperspectral data with 4 m multispectral data, GF-2 can be preferentially selected, and when fusing with 8 m multispectral data, GF-1D can be preferentially selected. A comparison between different fusion methods shows that GLP is the best. In terms of spatial reconstruction, GLP is the best. The performance of CRISP-B and CRISP-W are relatively good. Especially, when fusing GF-5 and BJ-2, in these three fusion results, CC have reached 0.84, UIQI have reached 0.81, and PSNR are above 24.11. This shows that the image source has little influence on the spatial reconstruction ability of CRISP-B,

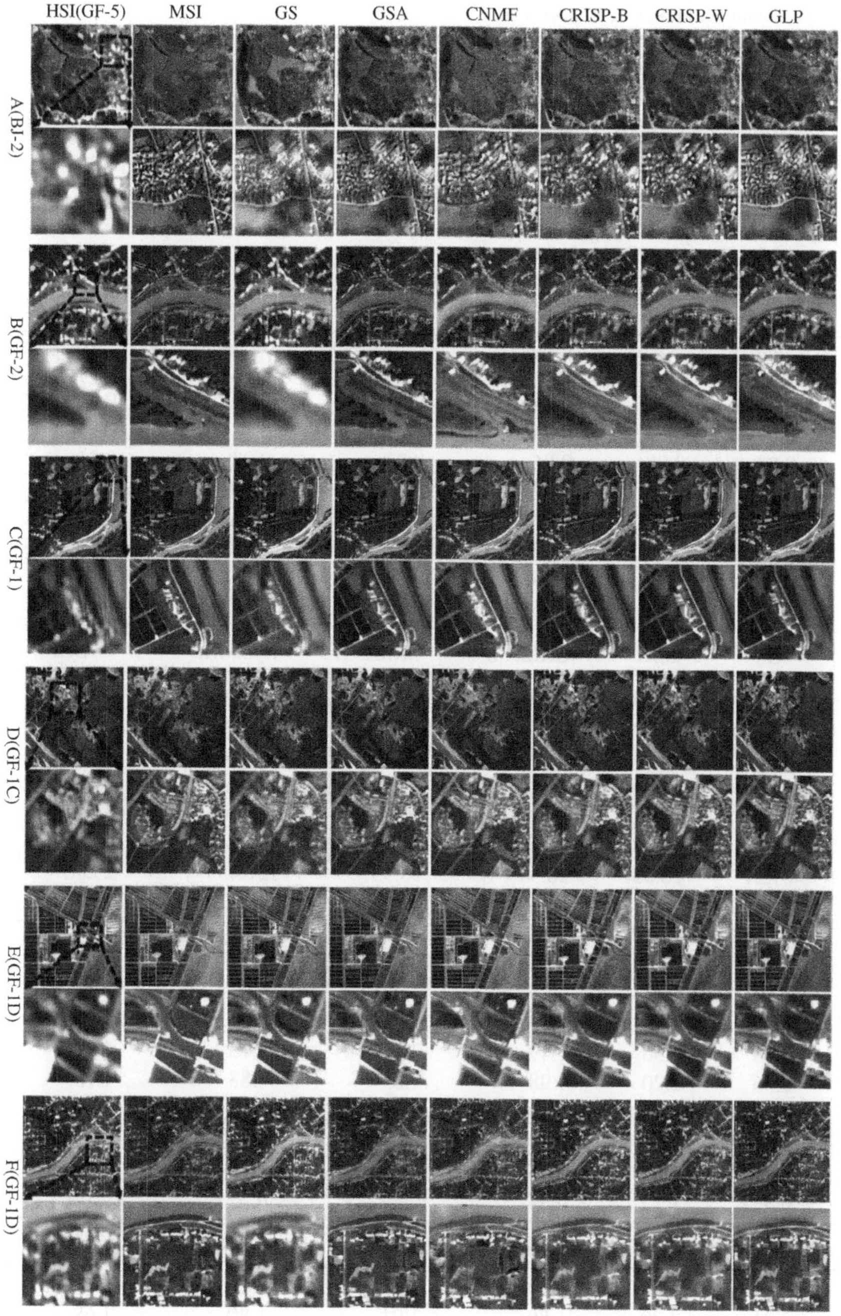


**Figure 11.28** Fusion results of different methods and data. Note: On the left, A–F are the number of experiments and the sensors noted in brackets are the multispectral data sources.

CRISP-W, and GLP. In terms of spectral retention, CNMF, GLP, CRISP-B, and CRISP-W all showed good results. CRISP-W is slightly better than CRISP-B; the average of both SAM are higher, and the standard deviation is very low, thus the performances are stable. GLP and CNMF have the highest SAM average, but the standard deviation is large, which is mainly affected by the A group

**Table 11.8** Evaluation indexes of fusion results

| **CC** | GS | GSA | CNMF | CRISP-B | CRISP-W | GLP | Mean |
|---|---|---|---|---|---|---|---|
| A | 0.59 | 0.74 | 0.75 | 0.84 | 0.84 | 0.87 | 0.77 |
| B | 0.92 | 0.83 | 0.86 | 0.88 | 0.89 | 0.94 | 0.89 |
| F | 0.94 | 0.89 | 0.87 | 0.93 | 0.93 | 0.92 | 0.91 |
| E | 0.92 | 0.92 | 0.92 | 0.96 | 0.96 | 0.95 | 0.94 |
| D | 0.97 | 0.94 | 0.91 | 0.96 | 0.96 | 0.96 | 0.95 |
| C | 0.97 | 0.95 | 0.94 | 0.95 | 0.95 | 0.96 | 0.95 |
| Mean | 0.89 | 0.88 | 0.87 | 0.92 | 0.92 | 0.93 | |
| Std | 0.13 | 0.07 | 0.06 | 0.04 | 0.04 | 0.03 | |

| **UIQI** | GS | GSA | CNMF | CRISP-B | CRISP-W | GLP | Mean |
|---|---|---|---|---|---|---|---|
| A | 0.58 | 0.70 | 0.74 | 0.81 | 0.82 | 0.85 | 0.75 |
| B | 0.91 | 0.81 | 0.85 | 0.87 | 0.88 | 0.93 | 0.88 |
| F | 0.94 | 0.88 | 0.86 | 0.92 | 0.93 | 0.91 | 0.91 |
| E | 0.92 | 0.91 | 0.92 | 0.96 | 0.96 | 0.95 | 0.94 |
| D | 0.97 | 0.94 | 0.91 | 0.95 | 0.96 | 0.96 | 0.95 |
| C | 0.97 | 0.95 | 0.93 | 0.94 | 0.95 | 0.96 | 0.95 |
| Mean | 0.88 | 0.87 | 0.87 | 0.91 | 0.91 | 0.93 | |
| Std | 0.14 | 0.09 | 0.06 | 0.05 | 0.05 | 0.04 | |

| **PSNR** | GS | GSA | CRISP-B | CRISP-W | CNMF | GLP | Mean |
|---|---|---|---|---|---|---|---|
| A | 21.57 | 21.18 | 24.54 | 211.01 | 23.39 | 211.17 | 23.48 |
| B | 32.21 | 23.92 | 211.84 | 26.55 | 26.28 | 29.00 | 27.30 |
| E | 27.61 | 26.79 | 29.74 | 30.06 | 27.58 | 29.03 | 28.47 |
| F | 38.78 | 27.36 | 29.29 | 29.74 | 28.23 | 28.76 | 30.36 |
| C | 34.19 | 29.72 | 29.44 | 30.33 | 29.66 | 31.37 | 30.79 |
| D | 32.91 | 30.23 | 31.16 | 31.85 | 29.54 | 31.55 | 31.21 |
| Mean | 31.21 | 26.53 | 28.34 | 28.93 | 27.45 | 29.15 | |
| Std | 11.42 | 3.17 | 2.34 | 2.36 | 2.15 | 2.11 | |

| **SAM** | GSA | GS | CNMF | GLP | CRISP-B | CRISP-W | Mean |
|---|---|---|---|---|---|---|---|
| A | 4.49 | 4.31 | 2.85 | 2.77 | 2.33 | 2.17 | 3.15 |
| B | 3.20 | 3.74 | 1.79 | 1.78 | 2.36 | 2.11 | 2.50 |
| F | 2.48 | 2.36 | 1.49 | 2.06 | 2.13 | 1.95 | 2.08 |
| E | 2.56 | 2.20 | 1.98 | 1.85 | 1.83 | 1.75 | 2.03 |
| C | 2.20 | 1.67 | 1.94 | 1.72 | 2.13 | 1.88 | 1.92 |
| D | 1.94 | 1.06 | 1.81 | 1.68 | 1.83 | 1.63 | 1.66 |
| Mean | 2.81 | 2.56 | 1.98 | 1.98 | 2.10 | 1.91 | |
| Std | 0.84 | 1.13 | 0.42 | 0.38 | 0.21 | 0.19 | |

| **ERGAS** | GS | GSA | CNMF | GLP | CRISP-B | CRISP-W | Mean |
|---|---|---|---|---|---|---|---|
| E | 4.14 | 4.10 | 3.49 | 3.50 | 3.26 | 3.10 | 3.60 |
| F | 2.31 | 2.29 | 1.80 | 1.91 | 1.80 | 1.68 | 1.97 |
| C | 2.07 | 2.12 | 2.00 | 1.71 | 2.04 | 1.84 | 1.96 |
| A | 2.90 | 2.29 | 1.61 | 1.51 | 1.40 | 1.31 | 1.84 |
| D | 1.63 | 1.88 | 1.85 | 1.66 | 1.68 | 1.54 | 1.71 |
| B | 1.67 | 1.55 | 1.05 | 0.87 | 1.15 | 1.05 | 1.22 |
| Mean | 2.45 | 2.37 | 1.97 | 1.86 | 1.89 | 1.75 | |
| Std | 0.87 | 0.81 | 0.75 | 0.80 | 0.68 | 0.65 | |

**Table 11.9** Runtimes comparison of different fusion methods

| Test (s) | GS | GSA | CRISP-W | GLP | CRISP-B | CNMF |
|---|---|---|---|---|---|---|
| A | ENVI interaction | 11.4 | 8.6 | 13.5 | 16.4 | 417.3 |
| B | | 11.0 | 8.6 | 13.4 | 16.1 | 422.9 |
| C | | 11.0 | 8.7 | 13.3 | 16.4 | 426.0 |
| D | | 11.0 | 8.6 | 13.4 | 16.1 | 474.2 |
| E | | 11.0 | 8.8 | 13.6 | 16.7 | 466.0 |
| F | | 11.0 | 8.7 | 14.7 | 16.6 | 442.4 |
| Mean | | 11.1 | 8.7 | 13.7 | 16.4 | 441.5 |

multisource fusion experiment. Therefore, in order to better maintain the spectral information, GLP can be preferentially selected when fusing between homologous data, and CRISP-W is preferentially used when fusing between multisource data. The spatial reconstruction ability and spectral retention ability of GS and GSA are both poor, especially in Experiment A. Although the GS method performs better in the four group (BCDF) experiments, such as the average of the CC and UIQI of the four groups are about 0.95, the average value of PSNR reaches the highest, but this is because the GS method fails to effectively remove the noise value of the hyperspectral data. The content of this part will be explained in detail in the discussion part. In addition, GSA is more stable than GS in both spectral retention and spatial resolution improvement. ERGAS is a comprehensive index, which has certain advantages when using images with certain registration errors as reference images. In summary, combined with the comprehensive index ERGAS, it can be concluded that the CRISP-W method is the best and the most stable; CRISP-B and GLP are slightly less effective; CNMF is in the middle, and GS and GSA are the worst.

As can be seen from Table 11.9, the running time of GSA, CRISP-W, GLP, CRISP-B, and CNMF increases in sequence, and the running time of CNMF is too long to be desirable. The operating costs of the other five methods are within the acceptable range.

### *11.5.5 Classification Application*

The landscape type of image has a great influence on the evaluation results. The diversity of landscape size and natural features leads to highly variable spatial and spectral information. Table 11.10 shows the label categories in the A, B, C, D, and F study areas and the number of specific samples

**Table 11.10** Evaluation indexes of fusion results

| Test (pixels) | Bearing land | | Building | | Roads | | Water body | | Farmland | | Breeding area | | Paddy field | |
|---|---|---|---|---|---|---|---|---|---|---|---|---|---|---|
| | Tr | Te | Tr | Te | Tr | Te | Tr | Te | Tr | Te | Tr | Te | Tr | Te |
| A | 30 | 970 | 30 | 970 | 30 | 970 | 30 | 970 | 30 | 970 | / | / | / | / |
| B | 30 | 970 | 30 | 970 | 30 | 970 | 30 | 970 | 30 | 970 | / | / | / | / |
| C | 30 | 970 | 30 | 970 | 30 | 970 | 30 | 970 | 30 | 970 | 30 | 970 | 30 | 970 |
| D | 30 | 970 | 30 | 970 | 30 | 970 | 30 | 970 | 30 | 970 | 30 | 970 | 30 | 970 |
| F | 30 | 970 | 30 | 970 | 30 | 970 | 30 | 970 | 30 | 970 | 30 | 970 | / | / |

Note: Tr means train; Te means test.

**Table 11.11** Overall Accuracy (OA) and Kappa coefficient

| Criterion | Test | HSI | MSI | GS | CNMF | GLP | CRISP-W | GSA | CRISP-B |
|---|---|---|---|---|---|---|---|---|---|
| OA (%) | A | 81.32 | 87.96 | 87.84 | 87.46 | 811.57 | 86.04 | 88.72* | 88.39* |
| | B | 81.09 | 82.08 | 78.14 | 79.98 | 81.24 | 81.05 | 88.29*** | 83.59* |
| | C | 88.26 | 86.41 | 86.52 | 87.78 | 88.06 | 88.92* | 84.36 | 90.06*** |
| | D | 88.47 | 90.66 | 90.03 | 92.64* | 92.53* | 92.75* | 90.24 | 93.43** |
| | F | 88.76 | 90.55 | 88.47 | 91.56 | 90.86 | 92.11* | 94.06*** | 93.25** |
| Kappa | A | 0.767 | 0.850 | 0.848 | 0.843 | 0.820 | 0.826 | 0.859* | 0.855* |
| | B | 0.764 | 0.776 | 0.727 | 0.750 | 0.766 | 0.763 | 0.854*** | 0.795* |
| | C | 0.863 | 0.841 | 0.843 | 0.857 | 0.861 | 0.871* | 0.818 | 0.884** |
| | D | 0.866 | 0.891 | 0.884 | 0.914* | 0.913* | 0.916* | 0.886 | 0.923** |
| | F | 0.865 | 0.887 | 0.862 | 0.899 | 0.890 | 0.905* | 0.929*** | 0.919** |

Note: the more *, the better the classification effect.

used for training and testing. The number of samples in each category is equal, which avoids the different proportion of ground features to affect the evaluation results. There are few types of features in the E study area. In order to ensure the effectiveness of the method comparison, E is not classified. In addition, the classification results of fusion images can be identified more effectively when the ground truth is fine or the main ground features are boundaries.

Table 11.11 shows the original MSI and HSI in different study areas, the overall classification accuracy of each fusion result, and Kappa coefficient. The classification result of GS is the worst, because GS has the worst spatial resolution improvement ability, which affects classification, especially fine classification. The effects of CNMF, GLP, and CRISP-W are centered, among which, groups C, D, and F are better than groups A and B. This may be due to the small difference in spatial resolution of the images to be fused in groups C, D, and F, and the fusion results are relatively good, so the classification results are relatively better. In the two groups A and B, GSA showed strong superiority. Although the spectral retention of the GSA fusion result is poor, the component substitution increases the spectral divergence of the ground features. It is also found in visual analysis that the spatial information of GSA is closer to the MSI, and the spatial resolution has been effectively improved, thereby improving the precision of fine classification. CRISP-B has shown good results in the classification of all fusion results. When the number of samples is small, the overall accuracy of the original GF-5 classification results is improved by about 4% points. Figure 11.29 shows the ground truth, classification results of HSI and MSI, and CRISP-W, GSA, CRISP-B fusion image classification results with relatively good results. It can be seen that in the two groups of A and B, the road division by GSA is very clear, and the roads separated by CRISP-B and CRISP-W are thicker, and some are in the form of sheets. The green land and bare soil separated by CRISP-B are basically lumpy. In the two groups C and D, the classification results of GSA are relatively broken. In group F, there is high classification accuracy of buildings and roads by GSA fusion, making the overall classification effect good.

In summary, the classification accuracy of different fusion methods is not significantly different for water bodies with a large degree of discrimination. For buildings and roads with broken ground features and borders that are difficult to identify, the classification effect of GSA fusion results is the best, but GSA will also partially break up flaky ground objects such as bare soil and farmland. The result of CRISP-B fusion results is stable, and the overall classification accuracy is high.

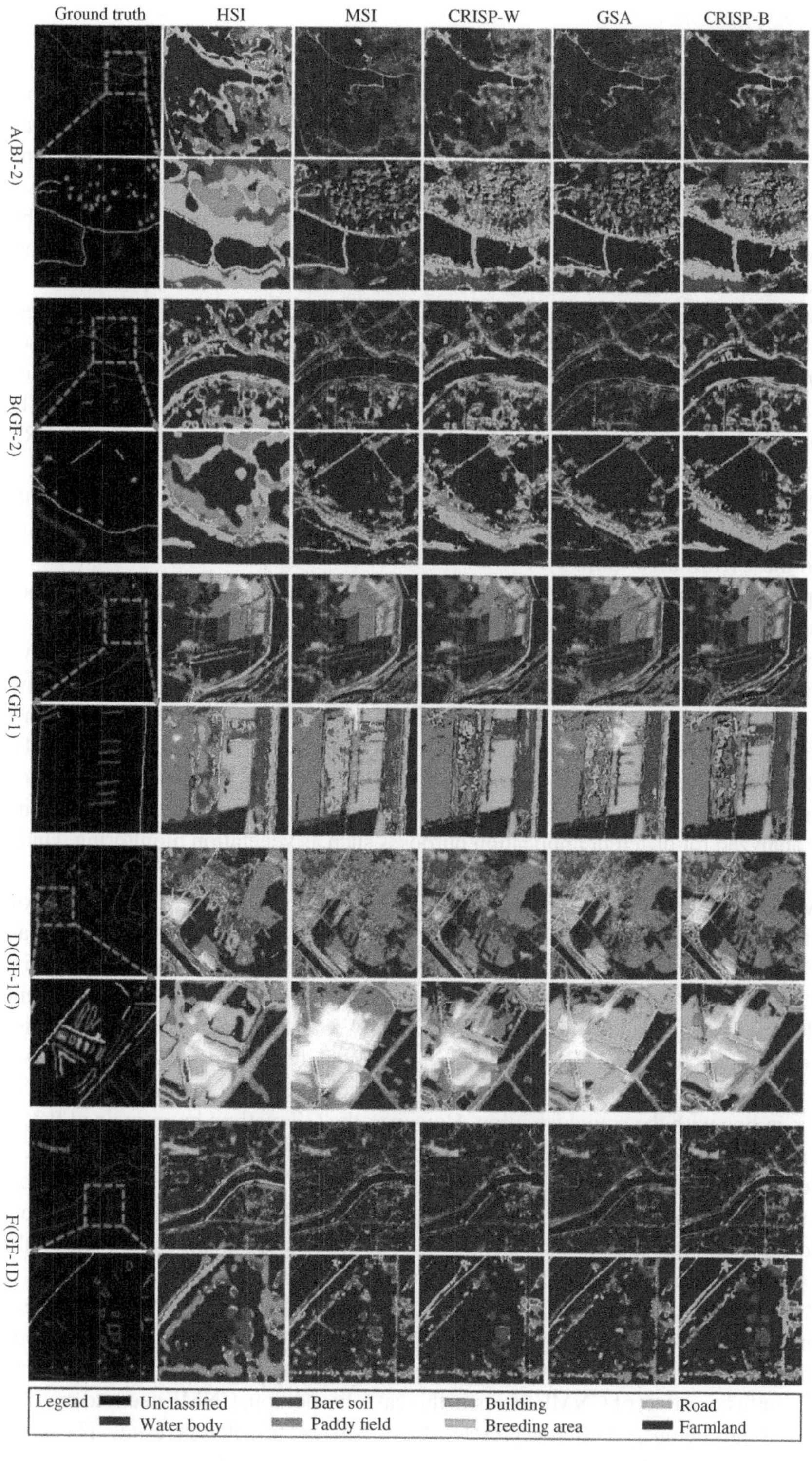


**Figure 11.29** Classification result. Note: A–D and F are the number of the experiments. The sensors noted in brackets are the multispectral data sources.

### *11.5.6 Discussion*

1. The CRISP model uses Wavelet and Butterworth filters to perform the frequency domain filtering process. The advantage of wavelet filter is that the processing speed is relatively fast, but when the image registration accuracy is low, there will be obvious "patches" in the fused image. Butterworth filter is relatively less affected by the accuracy of registration, showing strong stability. In this case, the experimental data registration accuracy is higher, CRISP-W effect is better than CRISP-B, which shows the necessity of strict registration before fusion. However, whether it is a Wavelet filter or a Butterworth filter, different fusion images need different filter parameters to achieve the best fusion effect, thereby making it difficult to choose the basis function, filter cutoff frequency, and number of decomposition layers. The selected parameters are not ideal, and will cause further information loss in the fusion image space and spectrum. The main reason for these problems is that the image reconstruction process in the CRISP model is entirely based on mathematical statistics theory, and statistical optimization is performed mathematically. The lack of a clear physical meaning makes it easy to cause unstable results. In addition, in the CRISP model, the simulation process of the high spatial resolution HSI and the subsequent filtering process are completely independent two parts. If the hyperspectral data with high spatial resolution can be directly reconstructed without the need for filter fusion process, it will be an important improvement to the CRISP algorithm.
2. In the process of band transformation, the GSA fusion algorithm will cause loss of spectral information and produce spectral distortion. A good fusion algorithm can not only increase the spatial texture information of the image, but also keep the original information of the spectrum. That is, no or little information distortion occurs. However, there is no absolute correlation between the fusion quality of various fusion methods, that is, spectrum, spatial fidelity, and remote sensing applications. The results of this experiment show that the GSA spectrum is seriously distorted, but at the same time increases the spectral separation of ground features, and is particularly suitable for the classification of roads and buildings. In the next step, the fusion algorithm can be further analyzed for different application purposes such as vegetation monitoring, land productivity inversion, environmental assessment, mineral exploration, and target detection.
3. Observing the original HSI, it is found that except for group E, bands in the front part of other five group images have different levels of noise and stripes, and the quality is poor. It can also be seen from Figure 11.30 (methods are discussed here of which the spectrum holding effect is better and most stable) that the CC trend of Group E is the steadiest. In the other five sets of experiments, the CC values of the front bands of CRISP-B and CRISP-W all have increasing trends. This is because based on the fusion of CRISP-B and CRISP-W, image noise and stripes are greatly improved, which is quite different from the original image (Figure 11.31). Therefore, the CC value is lower, which is in line with the actual situation. However, the GS method has remained flat; that is, it is very similar to the original noisy image, and does not have the effect of CRISP-B and CRISP-W to remove noise. In the A, B, C, and F experiments, the CC of the GS method has a steep downward trend after about the 80th band (center wavelength 728.06 nm), while other methods have an upward trend and maintain a relatively stable. The state, to a certain extent, shows that CRISP-B and CRISP-W have a retention advantage over the GS in the infrared band.
4. The CNMF model meets the non-negative constraints of the linear mixed model. Because the objective function has obvious non-convexity, the model has a large number of convergence points; it may converge to a local minimum, resulting in a nonunique decomposition matrix. This may be the reason for the poor fusion effect of CNMF fusion in this case. In addition, CNMF needs to preset the number of unmixed end-members. Different preset methods will lead to different results. The influence of the preset end-member methods on the fusion result can be further explored.

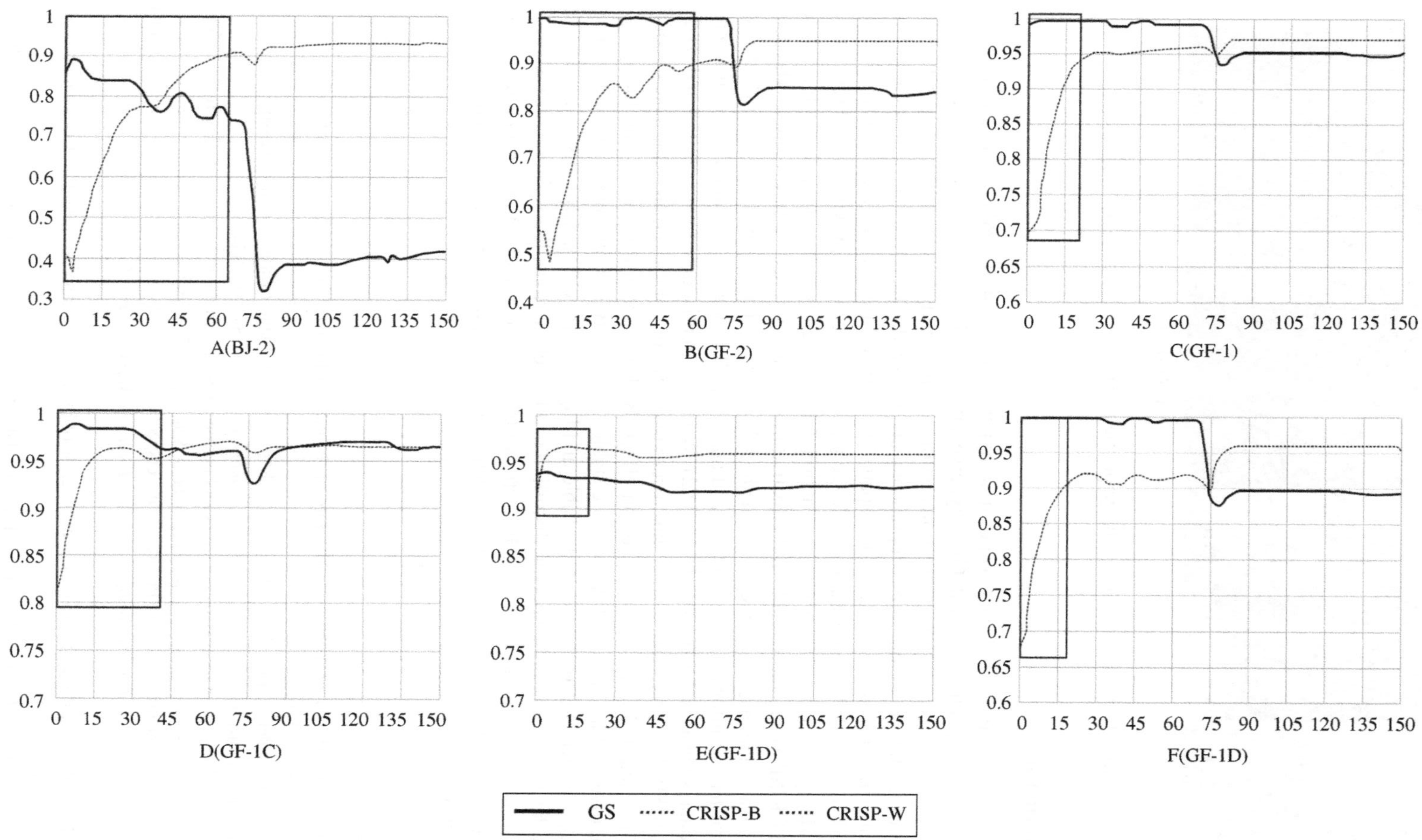


**Figure 11.30** CC trend of different bands.

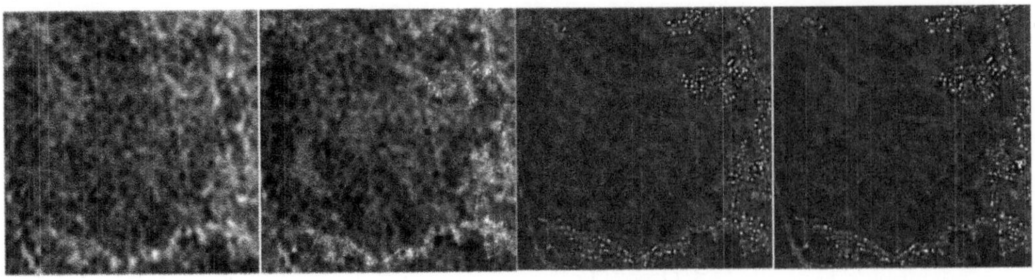


**Figure 11.31** The first band image of origin and fusion result of test A.

## 11.6 Conclusion

This study selects six simple and easy-to-promote hyperspectral fusion methods to fuse multiple domestic satellite multispectral data with GF-5 hyperspectral data. Through visual analysis, five classic evaluation indexes, classification application accuracy, and running time cost, the advantages and disadvantages of the fusion method are comprehensively analyzed. The results show that the fusion image series are the same and the smaller the spatial resolution difference is, the better the fusion result is. CRISP-B, CRISP-W, and GLP can achieve a good balance in improving spatial resolution and spectral fidelity. In terms of spatial reconstruction, GLP is slightly better and more stable, while CRISP-B and CRISP-W are more stable and effective in maintaining spectral information. The data source will have a certain impact on the fusion method. In the tasks that require high spectral fidelity, such as spectral feature information extraction and analysis, GLP is more suitable for the fusion of homologous data (such as GF-5 and GF-1/1C/1D/2). When the multisource images (GF-5 and BJ-2) are merged, CRISP-W is preferred. CNMF has a certain degree of color distortion and takes a long time to run. GSA and GS have the worst fusion effect. Both the spectral retention and the spatial resolution improvement ability of GSA are more stable than GS. Based on small sample, the classification effect of CRISP-B fusion result is stable and high-accuracy. The GSA fusion results are rich in spatial details. Although the spectral distortion is relatively serious, it also increases the spectral distinction of the ground objects, which is still suitable for accurately drawing buildings and roads. This study provides a basis for method selection for the application research of GF-5 hyperspectral data in fusion.

## References


1. Zhang, L. and Shen, H. (2016). Progress and future of remote sensing data fusion. *Yaogan Xuebao/Journal of Remote Sensing* 20: 1050–1061.
2. Pohl, C. and Van Genderen, J.L. (1998). Review article multisensor image fusion in remote sensing: concepts, methods and applications. *International Journal of Remote Sensing* 19 (5): 823–854.
3. Chen, Z., Ren, J., Tang, H. et al. (2016). Progress and perspectives on agricultural remote sensing research and applications in China. *Yaogan Xuebao/Journal of Remote Sensing* 20: 748–767.
4. Tong, Q., Xue, Y., and Zhang, L. (2014). Progress in hyperspectral remote sensing science and technology in China over the past three decades. *IEEE Journal of Selected Topics in Applied Earth Observations & Remote Sensing* 7 (1): 70–91.
5. Qingxi, T., Bing, Z., and Lifu, Z. (2016). Current progress of hyperspectral remote sensing in China. *Journal of Remote Sensing* 20: 689–707.
6. Wang, J. and Wang, Y. (2010). Noise model of hyperspectral imaging system and influence on radiation sensitivity. *Journal of Remote Sensing* 14 (4): 607–620.

7. Huang, B., Zhang, H., Song, H. et al. (2013). Unified fusion of remote-sensing imagery: generating simultaneously high-resolution synthetic spatial-temporal-spectral earth observations. *Remote Sensing Letters* 4 (4-6): 561–569.
8. Zhu, X., Jin, C., Feng, G. et al. (2010). An enhanced spatial and temporal adaptive reflectance fusion model for complex heterogeneous regions. *Remote Sensing of Environment* 114 (11): 2610–2623.
9. Schmitt, M. and Zhu, X.X. (2016). Data fusion and remote sensing: an ever-growing relationship. *IEEE Geoscience & Remote Sensing Magazine* 4 (4): 6–23.
10. Ghassemian, H. (2016). A review of remote sensing image fusion methods. *Information Fusion* 32: 75–89. S1566253516300173.
11. Tu, T.M., Su, S.C., Shyu, H.C. et al. (2001). A new look at IHS-like image fusion methods. *Information Fusion* 2 (3): 177–186.
12. Shettigara, V.K. (1992). A generalized component substitution technique for spatial enhancement of multispectral images using higher resolution data. *Photogrammetric Engineering & Remote Sensing* 58 (5): 561–567.
13. Aiazzi, B., Baronti, S., and Selva, M. (2007). Improving component substitution pansharpening through multivariate regression of MS + Pan data. *IEEE Transactions on Geoscience and Remote Sensing* 45 (10): 3230–3239.
14. Tu, T., Lee, Y.-C., Chang, C.-P., and Huang, P. (2005). Adjustable intensity-hue-saturation and Brovey transform fusion technique for IKONOS/QuickBird imagery. *Optical Engineering* 44 (11): 116201. https://doi.org/10.1117/1.2124871.
15. Chen, Z., Pu, H., Wang, B. et al. (2014). Fusion of hyperspectral and multispectral images: a novel framework based on generalization of Pan-sharpening methods. *IEEE Geoscience and Remote Sensing Letters* 11 (8): 1–1.
16. Chavez, P.S., Sides, S.C., and Anderson, J.A. (1991). Comparison of three different methods to merge multiresolution and multispectral data: Landsat TM and SPOT panchromatic. *Photogrammetric Engineering & Remote Sensing* 57 (3): 265–303.
17. Nunez, J., Otazu, X., Fors, O. et al. (1999). Multiresolution-based image fusion with additive wavelet decomposition. *IEEE Transactions on Geoscience and Remote Sensing* 37 (3): 1204–1211.
18. Nencini, F., Garzelli, A., Baronti, S. et al. (2007). Remote sensing image fusion using the curvelet transform. *Information Fusion* 8 (2): 143–156.
19. Liu, J.G. (2000). Smoothing filter-based intensity modulation: a spectral preserve image fusion technique for improving spatial details. *International Journal of Remote Sensing* 21: 3461–3472.
20. Do, M.N. and Vetterli, M. (2006). The contourlet transform: an efficient directional multiresolution image representation. *IEEE Transactions on Image Processing* 14 (12): 2091–2106.
21. Yuan, Q., Wei, Y., Meng, X. et al. (2018). A multiscale and multidepth convolutional neural network for remote sensing imagery pan-sharpening. *IEEE Journal of Selected Topics in Applied Earth Observations & Remote Sensing* 11 (3): 978–989.
22. Adams, J. and Smith, M. (1987). Spectral mixture modeling: a new analysis of rock and soil types at the Viking Lander 1 Site. *Journal of Geophysical Research* 91 (B8): 8098. https://doi.org/10.1029/jb091ib08p08098.
23. Gross, H.N. and Schott, J.R. (1998). Application of spectral mixture analysis and image fusion techniques for image sharpening. *Remote Sensing of Environment* 63 (2): 85–94.
24. Sun, X., Zhang, L., Yang, H. et al. (2015). Enhancement of spectral resolution for remotely sensed multispectral image. *IEEE Journal of Selected Topics in Applied Earth Observations and Remote Sensing* 8 (5): 2198–2211.
25. Bendoumi, M.A., He, M., and Mei, S. (2014). Hyperspectral image resolution enhancement using high-resolution multispectral image based on spectral unmixing. *IEEE Transactions on Geoscience and Remote Sensing* 52 (10): 6574–6583.
26. Zhi, H., Yu, X., Wang, G. et al. (2008). Application of several non-negative matrix factorization-based methods in remote sensing image fusion. Fifth International Conference on Fuzzy Systems & Knowledge Discovery, FSKD 2008, Jinan, Shandong, China (18–20 October 2008), Proceedings, vol. 4.
27. Yokoya, N., Yairi, T., Iwasaki, A. et al. (2012). Coupled nonnegative matrix factorization unmixing for hyperspectral and multispectral data fusion. *IEEE Transactions on Geoscience and Remote Sensing* 50 (2): 528–537.
28. Hardie, R.C., Eismann, M.T., and Wilson, G.L. (2004). MAP estimation for hyperspectral image resolution enhancement using an auxiliary sensor. *IEEE Transactions on Image Processing* 13 (9): 1174–1184.

29. Eismann, M.T. and Hardie, R.C. (2003). Resolution enhancement of hyperspectral imagery using coincident panchromatic imagery and a stochastic mixing model. IEEE Workshop on Advances in Techniques for Analysis of Remotely Sensed Data.
30. Eismann, M.T. and Hardie, R.C. (2005). Hyperspectral resolution enhancement using high-resolution multispectral imagery with arbitrary response functions. *IEEE Transactions on Geoscience and Remote Sensing* 43 (3): 455–4611.
31. ZhiRong, G., Bin, W., and LiMing, Z. (2007). Remote sensing image fusion based on Bayesian linear estimation. *Science in China. Series F, Information Sciences* 50 (2): 227–240.
32. Zhang, Y. (2012). Wavelet-based Bayesian fusion of multispectral and hyperspectral images using Gaussian scale mixture model. *International Journal of Image and Data Fusion* 3 (1): 23–37.
33. Zhang, Y., De Backer, S., and Scheunders, P. (2009). Noise-resistant wavelet-based bayesian fusion of multispectral and hyperspectral images. *IEEE Transactions on Geoscience and Remote Sensing* 47 (11): 3834–3843.
34. Wei, Q., Dobigeon, N., and Tourneret, J.-Y. (2014). Bayesian fusion of hyperspectral and multispectral images. Paper presented at the 2014 IEEE International Conference on Acoustics, Speech and Signal Processing (ICASSP).
35. Wei, Q., Dobigeon, N., and Tourneret, J.-Y. (2015). Bayesian fusion of multispectral and hyperspectral images with unknown sensor spectral response. *2014 IEEE International Conference on Image Processing, ICIP* 2014: 698–702.
36. Elad, M., Figueiredo, M.A.T., Ma, Y. et al. (2010). On the role of sparse and redundant representations in image processing. *Proceedings of the IEEE* 98 (6): 972–982.
37. Li, S. and Yang, B. (2011). A new pan-sharpening method using a compressed sensing technique. *IEEE Transactions on Geoscience and Remote Sensing* 49 (2): 738–746.
38. Huang, B., Song, H., Hengbin, C. et al. (2014). Spatial and spectral image fusion using sparse matrix factorization. *IEEE Transactions on Geoscience and Remote Sensing* 52 (3): 1693–1704.
39. Han, C., Zhang, H., Gao, C. et al. (2016). A remote sensing image fusion method based on the analysis sparse model. *IEEE Journal of Selected Topics in Applied Earth Observations & Remote Sensing* 9 (1): 439–453.
40. Huang, W., Xiao, L., Liu, H. et al. (2015). A new pan-sharpening method with deep neural networks. *IEEE Geoscience and Remote Sensing Letters* 12 (5): 1037–1041.
41. Giuseppe, M., Davide, C., Luisa, V. et al. (2016). Pansharpening by convolutional neural networks. *Remote Sensing* 8 (7): 594.
42. Azarang, A. and Ghassemian, H. (2017). A new pansharpening method using multi resolution analysis framework and deep neural networks. Paper presented at the 2017 3rd International Conference on Pattern Recognition and Image Analysis (IPRIA).
43. Wei, Y., Yuan, Q., Shen, H. et al. (2017). Boosting the accuracy of multispectral image pansharpening by learning a deep residual network. *IEEE Geoscience and Remote Sensing Letters* 14 (10): 1795–1799.
44. Palsson, F., Sveinsson, J., and Ulfarsson, M. (2017). Multispectral and hyperspectral image fusion using a 3-d-convolutional neural network. *IEEE Geoscience and Remote Sensing Letters* 14: 639–643.
45. Gao, F., Masek, J., Schwaller, M. et al. (2006). On the blending of the Landsat and MODIS surface reflectance: predicting daily Landsat surface reflectance. *IEEE Transactions on Geoscience and Remote Sensing* 44 (8): 2207–2218.
46. Hilker, T., Wulder, M.A., Coops, N.C. et al. (2009). A new data fusion model for high spatial- and temporal-resolution mapping of forest disturbance based on Landsat and MODIS. *Remote Sensing of Environment* 113 (8): 1613–1627.
47. Fu, D., Chen, B., Wang, J. et al. (2013). An Improved image fusion approach based on enhanced spatial and temporal the adaptive reflectance fusion model. *Remote Sensing* 5: 6346–6360.
48. Huang, B., Wang, J., Song, H. et al. (2013). A rigorously-weighted spatiotemporal fusion model with uncertainty analysis. *Remote Sensing* 9: 990.
49. Huang, B. et al. (2013). Generating high spatiotemporal resolution Land surface temperature for urban heat Island Monitoring. *IEEE Geoscience and Remote Sensing Letters* 10 (5): 1011–10111.
50. Weng, Q., Fu, P., and Gao, F. (2014). Generating daily land surface temperature at Landsat resolution by fusing Landsat and MODIS data. *Remote Sensing of Environment* 145: 55–67.

51. Wu, B., Huang, B., Zhang, L. et al. (2015). An error-bound-regularized sparse coding for spatiotemporal reflectance fusion. *IEEE Transactions on Geoscience and Remote Sensing* 53 (12): 1–13.
52. Liao, C., Wang, J., Pritchard, I. et al. (2017). A spatio-temporal data fusion model for generating NDVI time series in heterogeneous regions. *Remote Sensing* 9: 1125.
53. Song, H., Liu, Q., Wang, G. et al. (2018). Spatiotemporal satellite image fusion using deep convolutional neural networks. *IEEE Journal of Selected Topics in Applied Earth Observations & Remote Sensing* 11 (3): 821–829.
54. Zhukov, B., Oertel, D., Lanzl, F. et al. (1999). Unmixing-based multisensor multiresolution image fusion. *IEEE Transactions on Geoscience and Remote Sensing* 37 (3): 1212–1226.
55. Mingquan, W., Zheng, N., Changyao, W. et al. (2012). Use of MODIS and Landsat time series data to generate high-resolution temporal synthetic Landsat data using a spatial and temporal reflectance fusion model. *Journal of Applied Remote Sensing* 6 (1): 1–14.
56. Zhang, W., Li, A., Jin, H. et al. (2013). An enhanced spatial and temporal data fusion model for fusing Landsat and MODIS surface reflectance to generate high temporal Landsat-like data. *Remote Sensing* 5: 5346–5368.
57. Wu, P., Shen, H., Zhang, L. et al. (2015). Integrated fusion of multi-scale polar-orbiting and geostationary satellite observations for the mapping of high spatial and temporal resolution land surface temperature. *Remote Sensing of Environment* 156: 169–181.
58. Amorós-López, J., Gómez-Chova, L., Alonso, L. et al. (2013). Multitemporal fusion of Landsat/TM and ENVISAT/MERIS for crop monitoring. *International Journal of Applied Earth Observation and Geoinformation* 23 (Complete): 132–141.
59. Lu, M., Chen, J., Tang, H. et al. (2016). Land cover change detection by integrating object-based data blending model of Landsat and MODIS. *Remote Sensing of Environment* 184: 374–386.
60. Zhu, X., Cai, F., Tian, J. et al. (2018). Spatiotemporal fusion of multisource remote sensing data: literature survey, Taxonomy, principles, applications, and future directions. *Remote Sensing* 10 (4): 527.
61. Liao, L., Song, J., Wang, J. et al. (2016). Bayesian method for building frequent Landsat-like NDVI datasets by integrating MODIS and Landsat NDVI. *Remote Sensing* 8 (6): 452.
62. Xue, J., Leung, Y., and Fung, T. (2017). A bayesian data fusion approach to spatio-temporal fusion of remotely sensed images. *Remote Sensing* 9 (12): 1310.
63. Huang, B. and Song, H. (2012). Spatiotemporal reflectance fusion via sparse representation. *IEEE Transactions on Geoscience and Remote Sensing* 50 (10): 3707–3716.
64. Liu, X., Deng, C., Wang, S. et al. (2016). Fast and accurate spatiotemporal fusion based upon extreme learning machine. *IEEE Geoscience and Remote Sensing Letters* 13 (12): 2039–2043. https://doi.org/10.1109/LGRS.2016.2622726.
65. Wei, J., Wang, L., Liu, P. et al. (2016). Spatiotemporal fusion of remote sensing images with structural sparsity and semi-coupled dictionary learning. *Remote Sensing* 9: 21.
66. Wei, J., Wang, L., Liu, P. et al. (2017). Spatiotemporal fusion of MODIS and Landsat-7 reflectance images via compressed sensing. *IEEE Transactions on Geoscience and Remote Sensing* 55 (12): 7126–7139.
67. Moosavi, V., Talebi, A., Mokhtari, M.H. et al. (2015). A wavelet-artificial intelligence fusion approach (WAIFA) for blending Landsat and MODIS surface temperature. *Remote Sensing of Environment* 169: 243–254.
68. Tan, Z., Peng, Y., Di, L. et al. (2018). Deriving high spatiotemporal remote sensing images using deep convolutional network. *Remote Sensing* 10 (7): 1066.
69. Jagalingam, P. and Hegde, A.V. (2015). A review of quality metrics for fused image. *Aquatic Procedia* 4: 133–142.
70. Wald, L., Ranchin, T., and Mangolini, M. (1997). Fusion of satellite images of different spatial resolutions: assessing the quality of resulting images. *Photogrammetric Engineering & Remote Sensing* 63: 691–699.
71. Thomas, C. and Wald, L. (2007). Analysis of changes in quality assessment with scale. Paper presented at the 2006 9th International Conference on Information Fusion (10–13 July 2006).
72. Chang, C.-I (2002). Spectral information divergence for hyperspectral image analysis. IEEE 1999 International Geoscience and Remote Sensing Symposium. IGARSS'99 (Cat. No.99CH36293).
73. Li, D., Hao, M., Zhang, J., Hu, B., and Lu, Q.(2012). A universal hypercomplex color image quality index. Paper presented at the 2012 IEEE International Instrumentation and Measurement Technology Conference Proceedings (13–16 May 2012).

74. Choi, M. (2006). A new intensity-hue-saturation fusion approach to image fusion with a tradeoff parameter. *IEEE Transactions on Geoscience and Remote Sensing* 44 (6): 1672–1682.
75. Huynh-Thu, Q. and Ghanbari, M. (2008). Scope of validity of PSNR in image/video quality assessment. *Electronics Letters* 44 (13): 800–801.
76. Alparone, L., Wald, L., Chanussot, J. et al. (2007). Comparison of pansharpening algorithms: outcome of the 2006 GRS-S data fusion contest. *IEEE Transactions on Geoscience and Remote Sensing* 45: 3012–3021.
77. Wang, Z. (2004). Image quality assessment: from error visibility to structural similarity. *IEEE Transactions on Image Processing* 13 (4): 600–612.
78. Zurita-Milla, R., Clevers J.G.P.W., and Schaepman, M.E. (2008). Unmixing-based landsat TM and MERIS FR data fusion. *IEEE Geoscience and Remote Sensing Letters* 5: 453–457.
79. Yusuf, Y., Sri Sumantyo, J.T., and Kuze, H. (2013). Spectral information analysis of image fusion data for remote sensing applications. *Geocarto International* 28 (4): 291–310.
80. Lecun, Y., Bengio, Y., and Hinton, G. (2015). Deep learning. *Nature* 521 (7553): 436.
81. Minar, M. R. and J. Naher (2018). Recent advances in deep learning: an overview. https://doi.org/10.13140/RG.2.2.24831.10403
82. Farrell, M.D. and Mersereau, R.M. (2005). On the impact of PCA dimension reduction for hyperspectral detection of difficult targets. *IEEE Geoscience and Remote Sensing Letters* 2 (2): 192–1911.
83. Richards, J.A. (1999). Remote Sensing Digital Image Analysis: An Introduction, 1ee. Berlin, Heidelberg: Springer Berlin Heidelberg.
84. Linting, M., Meulman, J., Groenen, P., and Koojj, A. (2009). Nonlinear principal components analysis: introduction and application. *Psychological Methods* 12: 336–358. https://doi.org/10.1037/1082-989X.12.3.336.
85. Chen, C. and Qi, F. (2018). Single image super-resolution using deep CNN with dense skip connections and inception-ResNet. 2018 9th International Conference on Information Technology in Medicine and Education (ITME), pp. 999–1003. doi: 10.1109/ITME.2018.00222.
86. Zhang, L., Chen, H., Sun, X. et al. (2017). Designing spatial-temporal-spectral integrated storage structure of multidimensional remote sensing images. *Yaogan Xuebao/Journal of Remote Sensing* 21: 62–73.
87. Brink, A.B. and Eva, H.D. (2009). Monitoring 25 years of land cover change dynamics in Africa: A sample based remote sensing approach. *Applied Geography* 29 (4): 501–512.
88. Eberhardt, I., Schultz, B., Rizzi, R. et al. (2016). Cloud cover assessment for operational crop monitoring systems in tropical areas. *Remote Sensing* 8: 1.
89. Pflugmacher, D., Cohen, W.B., Kennedy, R.E. et al. (2014). Using Landsat-derived disturbance and recovery history and lidar to map forest biomass dynamics. *Remote Sensing of Environment* 151: 124–137.
90. Nair, V. and Hinton, G.E. (2010). Rectified linear units improve restricted boltzmann machines vinod nair. Proceedings of the 27th International Conference on Machine Learning (ICML-10), (21–24 June 2010), Haifa, Israel.
91. Emelyanova, I.V., Mcvicar, T.R., Van Niel, T.G. et al. (2013). Assessing the accuracy of blending Landsat–MODIS surface reflectances in two landscapes with contrasting spatial and temporal dynamics: a framework for algorithm selection. *Remote Sensing of Environment* 133: 193–209.
92. Kingma, D.P. and Ba, J. (2014). Adam: a method for stochastic optimization. *Computer Science*. abs/1412.6980.
93. Sun, Y., Jiang, G., Yunduan, L.I. et al. (2018). GF-5 satellite: overview and application prospects. *Spacecraft Recovery & Remote Sensing* 39 (3): 1–13.
94. Liu, Y. (2018). Visible-shortwave infrared hyperspectral imager of GF-5 satellite. *Spacecraft Recovery & Remote Sensing* 39 (3): 25–28.
95. Li, X.T., Lu, J.X., Song, X.N. et al. (2018). Application of the GF satellite data in flood disaster monitoring. *International Geoscience and Remote Sensing Symposium (IGARSS)* 2018: 7293–7296.
96. Yokoya, N., Grohnfeldt, C., and Chanussot, J. (2017). Hyperspectral and multispectral data fusion: a comparative review of the recent literature. *IEEE Geoscience & Remote Sensing Magazine* 5 (2): 29–56.
97. Ren, K., Sun, W., Meng, X. et al. (2020). Fusing China GF-5 hyperspectral data with GF-1, GF-2 and Sentinel-2A multispectral data: which methods should be used? *Remote Sensing* 12: 882.

# 12

# Fusion of Band Selection Methods for Hyperspectral Imaging

Yulei Wang[1], Lin Wang[2], and Chein-I Chang[1,3]
[1]*Center for Hyperspectral Imaging in Remote Sensing (CHIRS), School of Information and Technology, Dalian Maritime University, Dalian, China*
[2]*School of Physics and Optoelectronic Engineering, Xidian University, Xi'an, China*
[3]*Remote Sensing Signal and Image Processing Laboratory, Department of Computer Science and Electrical Engineering, University of Maryland Baltimore County, Baltimore, MD, USA*

## 12.1 Introduction

Hyperspectral imaging has emerged as a promising technique in remote sensing [1] due to its use of hundreds of contiguous spectral bands. However, it is also traded for an issue of how to effectively utilize such a wealth of spectral information. In various applications, such as classification, target detection, spectral unmixing, endmember finding/extraction, and so on, the spectral properties of material substances of interest may respond to different ranges of wavelengths. In this case, not all spectral bands are useful. Consequently, it is crucial to find appropriate wavelength ranges for particular applications of interest. This leads to a need of BS, which has become increasingly important in hyperspectral data exploitation.

Over the past years, many BS methods have been developed and reported in the literature [1–32]. In general, it can be categorized into two groups based on application-independent and application-dependent approaches. One is made up of BS methods designed based on data structures and characteristics. This type of BS methods generally relies on band prioritization (BP) criteria specified by data statistics such as variance, signal-to-noise ratio (SNR), entropy, information divergence (ID) to rank spectral bands [4] so that bands can be selected according to their ranked orders. As a result, such BP-based methods are generally unsupervised and completely determined by data itself not applications but have two major issues. The first one is inter-band correlation. That is, when a band is selected by its high priority, it is very likely that its adjacent bands will be also selected because of their close correlation with the selected band. In this case, band de-correlation is generally required to be implemented in conjunction with BP. However, how to appropriately choose a threshold to determine how close is sufficient correlation is challenging. A second issue is that a BP criterion may be good for one application but may not be good for another application. The other type of

*Advances in Hyperspectral Image Processing Techniques*, First Edition. Edited by Chein-I Chang.

BS methods is designed based on particular applications, for example classification [4, 8–17], detection [18]. Therefore, they are generally supervised and most likely requires training sample data sets. Since such application-based BS methods are heavily determined by applications, one BS method works for one application may not be applicable to another.

This chapter takes a quite different approach by looking into a set of BS methods, each of which can come from either group described above, and fusing bands selected by these BS methods. Because each BS method produces its own band subset that is different one from another, it is advantageous if we can fuse these band subsets to select bands that work for different applications. Consequently, the fused band subset should be able to produce a better band subset that is more robust to various applications.

More specifically, assume that there are $p$ BS methods. Let $\Omega_{n_{\text{BS}}}^{(j)} = \left\{\mathbf{b}_{l_1}^{j}, \mathbf{b}_{l_2}^{j}, \ldots, \mathbf{b}_{l_{n_{\text{BS}}(j)}}^{j}\right\}$ be the band subset generated by a particular $j$th BS method where $n_{\text{BS}}$ is the number of bands in the $j$th band subset $\Omega_{n_{\text{BS}}}^{(j)}$. In order to take advantage of these $p$ band subsets, $\left\{\Omega_{n_{\text{BS}}}^{(j)}\right\}_{j=1}^{p}$ the idea is to first find their union to form a JBS, i.e. $\Omega_{\text{JBS}}^{p} = \cup_{j=1}^{p}\Omega_{n_{\text{BS}}}^{(j)}$. Since it is very likely that one band selected by one BS method may be also selected by another BS method, we calculate the frequency of each band in the JBS, $\Omega_{\text{JBS}}^{p}$, which appear in individual band subsets, $\Omega_{n_{\text{BS}}}^{(1)}, \Omega_{n_{\text{BS}}}^{(2)}, \ldots, \Omega_{n_{\text{BS}}}^{(p)}$. By virtue of such frequencies we can calculate the probability of each band being selected in these $p$ band subsets, $\left\{\Omega_{n_{\text{BS}}}^{(j)}\right\}_{j=1}^{p}$ which can be further used as the priority probability assigned to this particular band. Accordingly, the higher the priority probability of a band is, the more significant the band is. Then a band fusion method can be designed by ranking bands according to their assigned priority probabilities. The resulting technique is called BSF. Because JBS is obtained by taking the union of all band subsets, the number of bands in JBS is at least equal to or greater than $n_{\text{BS}}$. So, there may be more than $n_{\text{BS}}$ bands in which case, BSF uses the frequency-calculated priority probabilities assigned to bands in JBS to select $n_{\text{BS}}$ bands from JBS.

Interestingly, according to how fusion takes place, two versions of BSF can be further developed, referred to as progressive BSF (PBSF), which fuse a smaller number of band subsets one at a time progressively, as well as simultaneous BSF (SBSF), which fuse all band subsets at once in one shot operation. There are several benefits that can be gained from BSF.

1. It improves individual BS methods.
2. There is no need of band de-correlation for BSF, which has been a major issue in many BP-based BS methods due to their use of BP as a criterion to select bands.
3. It can adapt to different data structures characterized by statistics and also can be applicable to various applications. This is because bands selected by BSF come from different band subsets that are obtained by various BP criteria or application-based BS methods.
4. Although BSF does not implement any BP criterion, it can actually prioritize bands according to their appearing frequencies in different band subsets.
5. BSF is flexible and can be implemented in various forms, specifically progressive fusion that can be carried out by different numbers of BS methods progressively.

## 12.2 Band Selection Fusion

Generally speaking, there are two logical approaches to fusing bands. One is to find overlapped bands by taking the intersection of $\left\{\Omega_{n_{\text{BS}}}^{(j)}\right\}_{j=1}^{p}$, $\tilde{\Omega}_{\text{JBS}}^{p} = \cap_{j=1}^{p}\Omega_{n_{\text{BS}}}^{(j)}$. The main problem with this approach is that on many occasions $\tilde{\Omega}_{\text{JBS}}^{p}$ may end up with an empty set, i.e. $\tilde{\Omega}_{\text{JBS}}^{p} = \emptyset$ or too small if it is not empty. The other way is to find the joint bands by taking the union of $\left\{\Omega_{n_{\text{BS}}}^{(j)}\right\}_{j=1}^{p}$, i.e. $\Omega_{\text{JBS}}^{p} = \cup_{j=1}^{p}\Omega_{n_{\text{BS}}}^{(j)}$. The main issue arising from this approach is that $\Omega_{\text{JBS}}^{p}$ may turn out to be too large in most cases. In this case, BS becomes meaningless. In order to resolve both dilemmas, we develop a new approach to fusing $p$ band subsets, $\left\{\Omega_{n_{\mathrm{BS}}}^{(j)}\right\}_{j=1}^{p}$, called BSF method in accordance with frequency of a band appearing in these $p$ band subsets. Its idea is similar to finding a gray-level histogram of an image where each BS method corresponds to a gray level.

Suppose that there are $p$ BS methods to be fused. Two versions of BSF can be developed.

### 12.2.1 Simultaneous Band Selection Fusion

The first BSF method is to fuse all band subsets, $\left\{\Omega_{n_{\mathrm{BS}}}^{(j)}\right\}_{j=1}^{p}$, produced by $p$ BS methods.

---

**Simultaneous Band Selection Fusion (SBSF)**

---

1. Assume that $n_{\mathrm{BS}}$ is given a priori or estimated.
2. Let $\Omega_{n_{\mathrm{BS}}(j)} = \left\{\mathbf{b}_{l_1}^j, \mathbf{b}_{l_2}^j, \ldots, \mathbf{b}_{l_{n_{\mathrm{BS}}(j)}}^j\right\}$ be generated by the $j$th BS method.
3. Find $\Omega_{\mathrm{JBS}}^p = \cup_{j=1}^p \Omega_{n_{\mathrm{BS}}(j)}$. Let $\mathbf{b}_{l_m}^j$ be a band in $\Omega_{n_{\mathrm{BS}}(j)}$ with $1 \le l_m \le l_{n_j}$, where $l_m$ is the band number of the $j$th BS method, and $l_{n_j}$ is the maximal band number of the $j$th BS method.
4. Calculate the frequency of each band $\mathbf{b}_{l_m}^j$ appearing in $\Omega_{\mathrm{JBS}}^p$ by
$$n\left(\mathbf{b}_{l_m}^j\right) = \sum_{k=1}^{p} I_{\Omega_{n_k}^k}\left(\mathbf{b}_{l_m}^j\right) \tag{12.1}$$
where $I_{\Omega_{n_{\mathrm{BS}}}^k}\left(\mathbf{b}_{l_m}^j\right)$ is an indicator function defined by
$$I_{\Omega_{n_{\mathrm{BS}}(k)}}\left(\mathbf{b}_{l_m}^j\right) = \begin{cases} 1; & \text{if } \mathbf{b}_{l_m}^j \in \Omega_{n_{\mathrm{BS}}(k)} \\ 0; & \text{if } \mathbf{b}_{l_m}^j \notin \Omega_{n_{\mathrm{BS}}(k)} \end{cases} \tag{12.2}$$
5. For $1 \le j \le p$, $1 \le l_m \le n_j$, we calculate the band fusion probability (BFP) of $\mathbf{b}_{l_m}^j$ as
$$p\left(\mathbf{b}_{l_m}^j\right) = \frac{n(\mathbf{b}_{l_m}^j)}{\sum_{\mathbf{b}_l^k \in \Omega_{\mathrm{JBS}}^p} n(\mathbf{b}_l^k)} \tag{12.3}$$
6. Rank all the bands in $\Omega_{\mathrm{JBS}}^p$ according to their BFP $\left\{p\left(\mathbf{b}_{l_n}^k\right)\right\}_{k=1,n=1}^{p,n_{\mathrm{BS}}(k)}$ calculated by (12.3), that is
$$\mathbf{b}_{l_m}^j \succ \mathbf{b}_{l_n}^k \Leftrightarrow p\left(\mathbf{b}_{l_m}^j\right) > p\left(\mathbf{b}_{l_n}^k\right) \tag{12.4}$$
7. where "$A \succ B$" indicates "A has a higher priority than B." On some occasions, when $\mathbf{b}_{l_m}^j \in \Omega_{\mathrm{JBS}}^j$ and $\mathbf{b}_{l_n}^k \in \Omega_{n_{\mathrm{BS}}(k)}$ with $j<k$ may have the same priority, i.e. $p\left(\mathbf{b}_{l_m}^j\right) = p\left(\mathbf{b}_{l_n}^k\right)$. It should be noted that when $p\left(\mathbf{b}_{l_m}^j\right) = p\left(\mathbf{b}_{l_n}^k\right)$, $\mathbf{b}_{l_m}^j$ and $\mathbf{b}_{l_n}^k$ have the same priority. In this case, if $\mathbf{b}_{l_m}^j$ has a higher priority in $\Omega_{n_j}^j$ than $\mathbf{b}_{l_n}^k$ in $\Omega_{n_k}^k$, then $\mathbf{b}_{l_m}^j \succ \mathbf{b}_{l_n}^k$.
8. Finally, select $n_{\mathrm{BS}}$ bands from $\Omega_{\mathrm{JBS}}^p$.

---

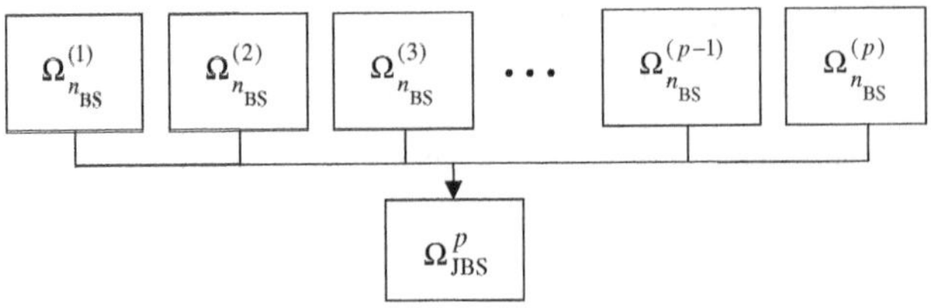


**Figure 12.1** Diagram of SBSF.

Figure 12.1 describes a diagram of fusing $p$ band subsets, $\left\{\Omega^{(j)}_{n_{BS}}\right\}^{p}_{j=1}$ simultaneously, denoted by $\{BS1, BS2, ..., BSp\}$, where BS$j$ produces the band subset $\Omega^{(j)}_{n_{BS}}$.

## *12.2.2 Progressive Band Selection Fusion*

A second BSF method expands SBSF in multiple stages in a progressive manner where each stage essentially fuses two-band subsets at a time by SBSF.

---

**Progressive Band Selection Fusion (PBSF)**

1. Assume that $n_{BS}$ is given a priori or estimated.
2. Randomly pick two BS methods that produced two band subsets, denoted by $\Omega_{n_{BS}(1)}$ and $\Omega_{n_{BS}(2)}$.
3. Find $\Omega^2_{JBS} = \Omega_{n_{BS}(1)} \cup \Omega_{n_{BS}(2)}$. Let $\mathbf{b}^1_{l_m}$ be a band in $\Omega_{n_{BS}(1)}$ with $1 \le l_m \le l_{n_1}$ and Let $\mathbf{b}^2_{l_k}$ be a band in $\Omega_{n_{BS}(2)}$ with $1 \le l_k \le l_{n_2}$.
4. If a band $\mathbf{b}^2_l$ in $\Omega^2_{JBS}$ is found in $\Omega_{n_{BS}(1)} \cap \Omega_{n_{BS}(2)}$, then $n(\mathbf{b}^2_l) = 2$. Otherwise, $n(\mathbf{b}^2_l) = 1$ if $\mathbf{b}^2_l \in \Omega^2_{JBS} - \left(\Omega_{n_{BS}(1)} \cap \Omega_{n_{BS}(2)}\right)$.
5. For $3 \le j \le p$ with $1 \le l_m \le n_j$, pick any $\Omega_{n_{BS}(j)}$ to form $\Omega^j_{JBS} = \Omega^{j-1}_{JBS} \cup \Omega_{n_{BS}(j)}$.
   (i) If a band $\mathbf{b}^j_l$ in $\Omega^j_{JBS}$ is also found in $\Omega^{j-1}_{JBS} \cap \Omega_{n_{BS}(j)}$ with its corresponding band $\mathbf{b}^{j-1}_{l_{j-1}} \in \Omega^{j-1}_{JBS}$, then $n(\mathbf{b}^j_l) = n\left(\mathbf{b}^{j-1}_{l_{j-1}}\right) + 1$.
   (ii) Else
   (a) If $\mathbf{b}^j_l \in \Omega^{j-1}_{JBS} - \Omega_{n_{BS}(j)}$, then $n(\mathbf{b}^j_l) = n\left(\mathbf{b}^{j-1}_{l_{j-1}}\right)$
   (b) Else, if $\mathbf{b}^j_l \in \Omega_{n_{BS}(j)} - \Omega^{j-1}_{JBS}$, then $n(\mathbf{b}^j_l) = 1$.
   (iii) Let $\Omega^j_{JBS} \leftarrow \Omega^j_{JBS,n_{BS}}$ comprised of bands with the first $n_{BS}$ priorities.
6. Rank all the bands in $\Omega^j_{JBS}$ according to $\{n(\mathbf{b}^j_l)\}_{\mathbf{b}^j_l \in \Omega^j_{JBS}}$, that is

$$\mathbf{b}^j_l \succ \mathbf{b}^j_k \Leftrightarrow n(\mathbf{b}^j_l) > n(\mathbf{b}^j_k) \tag{12.5}$$

where "$A \succ B$" indicates "$A$ has a higher priority than $B$." It should be noted that when $p\left(\mathbf{b}^j_{l_m}\right) = p\left(\mathbf{b}^k_{l_n}\right)$, $\mathbf{b}^j_{l_m}$ and $\mathbf{b}^k_{l_n}$ have the same priority. In this case, if $\mathbf{b}^j_{l_m}$ has a higher priority in $\Omega_{n_{BS}(j)}$ than $\mathbf{b}^k_{l_n}$ in $\Omega_{n_{BS}(k)}$, then $\mathbf{b}^j_{l_m} \succ \mathbf{b}^k_{l_n}$.

---

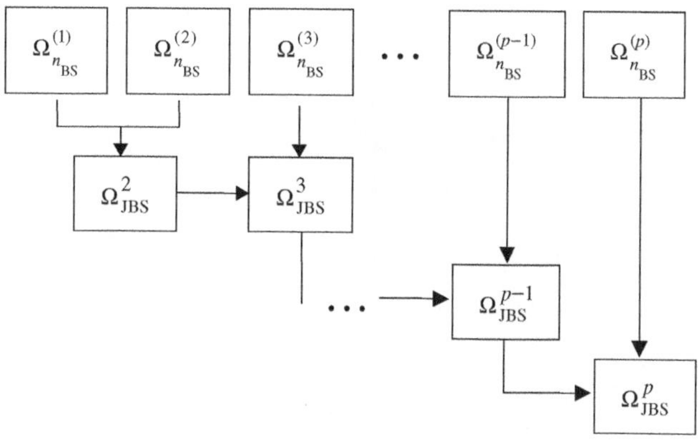


**Figure 12.2** Diagram of PBSF.

Figure 12.2 depicts a diagram to illustrate how to implement PBSF progressively, denoted by BS1 – BS2 – ⋯ – BS$p$, where BS$j$ produces the band subset $\Omega^{(j)}_{n_{BS}}$.

It is worth noting that the above PBSF can be also implemented in a more general fashion. It does not have to fuse two band subsets at a time, but rather small varying numbers, for example three band subsets in the first stage, then four band subsets in the second stage, and so on.

Once the number of bands is determined for BS such as VD or $n_{BS} = \max_{1 \leq j \leq p}\{n_j\}$ where $n_j$ is the number of bands selected by $j$th BS method, we can select $n_{BS}$ bands from $\left\{\Omega^{(j)}_{n_{BS}}\right\}_{j=1}^{p}$ according to (12.5). However, there is a key difference between SBSF and PBSF. That is, SBSF waits until the final $\Omega^{p}_{JBS}$ is generated to select $n_{BS}$ bands, while PBSF selects $n_{BS}$ bands from $\Omega^{j}_{JBS}$ after each fusion.

As noted, one major issue arising in selection of prioritized bands is that once a band with high priority is selected, its adjacent bands may be also selected due to their close inter-band correlation with the selected band. With the use of BSF, this issue can be resolved because bands are selected according to their frequencies appearing in different sets of band subsets not the priority orders.

## 12.3 Experiments

In this section, two applications are studied to demonstrate the utility of fusing various BS methods using real hyperspectral images.

### *12.3.1 Linear Spectral Unmixing*

The image scene shown in Figure 12.3 was acquired by the airborne hyperspectral digital imagery collection experiment (HYDICE) sensor in August 1995 from a flight altitude of 10 000 ft. This scene has been studied extensively by many reports such as [1, 21]. There are 15 square panels with three different sizes, 3 m × 3 m, 2 m × 2 m, and 1 m × 1 m, respectively.

Due to the approximately 1.56 m ground sampling distance, each of panels in the first column except the first row contains two panel pixels highlighted by dark gray, $p_{211}$, $p_{221}$ in row 2, $p_{311}$, $p_{312}$ in row 3, $p_{411}$, $p_{412}$ in row 4, $p_{511}$, $p_{521}$ in row 5 as shown in Figure 12.3. All the 11 remaining panels in Figure 12.3 contain one single panel pixel for each panel also highlighted by dark gray, $p_{11}$, $p_{12}$, $p_{13}$ in row 1, $p_{22}$, $p_{23}$ in row 2, $p_{32}$, $p_{33}$ in row 3, $p_{42}$, $p_{43}$ in row 4, $p_{52}$, $p_{53}$ in row 5. Therefore, there are a total of 19 panel pixels. Figure 12.3(b) shows their precise spatial locations with the pixels in yellow (Y pixels), indicating panel pixels mixed with the BKG. The virtual dimensionality (VD) of

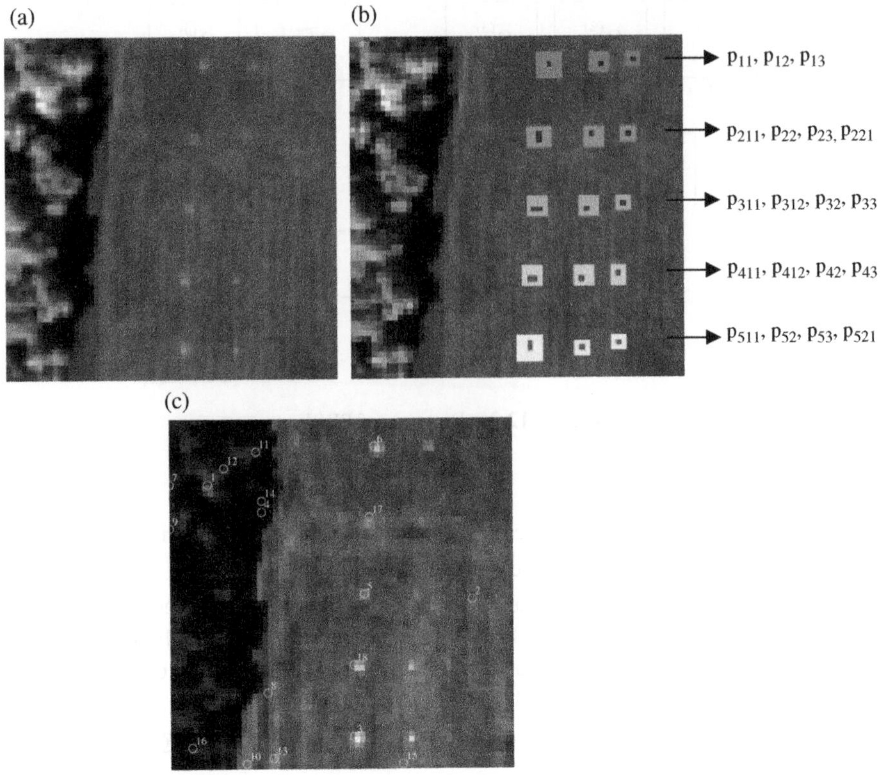


**Figure 12.3** (a) A HYDICE panel scene that contains 15 panels, (b) ground truth map of spatial locations of the 15 panels, (c) 18 target pixels found by automatic target generation process (ATGP), where the color maps of (a) and (b) can be found in Ref. [1], figure 1.15/with permission of John Wiley & Sons.

this scene has been estimated by the HFC method in [21–23] as 9. However, according to [24, 25], $n_{BS} = 9$ seemed to be insufficient because when the ATGP developed in [26] was used to find 9 target pixels in Figure 12.3(c) where only three panel pixels could be found among 9 ATGP-found target pixels. In order for ATGP to find five panel pixels with each panel pixel corresponding to one individual row, it requires 18 pixels to do so as also shown in Figure 12.3(c).

It should be noted that ATGP has been shown in [27] to essentially the same as vertex component analysis (VCA) [28] and simplex growing algorithm (SGA) [29] as long as their initial conditions are chosen the same. Accordingly, ATGP can be used for a general purpose for target detection and endmember finding. So, in the following experiments the value of VD, $n_{VD}$, used for BS was set to $n_{BS} = 18$ [30, 31]. Over the past years, many BS methods were developed [1–20]. It is impossible to cover all such methods. Instead, we have selected some representatives for our experiments, for example second order statistics-based BP criteria: variance, constrained band selection (CBS), signal-to-noise ratio (SNR); and higher order statistics BP criteria: entropy (*E*), information divergence (ID). SBSF and PBSF were then used to fuse these BS methods. Tables 12.2–12.3 list 18 bands selected by various SBSF and PBSF methods. Figures 12.4–12.6 show 18 target pixels found by ATGP using 18 bands selected in Tables 12.1–12.3, respectively. Table 12.4 lists dark gray panel pixels in Figure 12.3(b) found by ATGP in Figures 12.4–12.6 where ATGP using only 18 bands selected by S, UBS and E-ID-CBS (PBSF) could find panel pixels in each of five different rows. The last column in Table 12.4 shows whether all the five categories of targets (dark gray panels

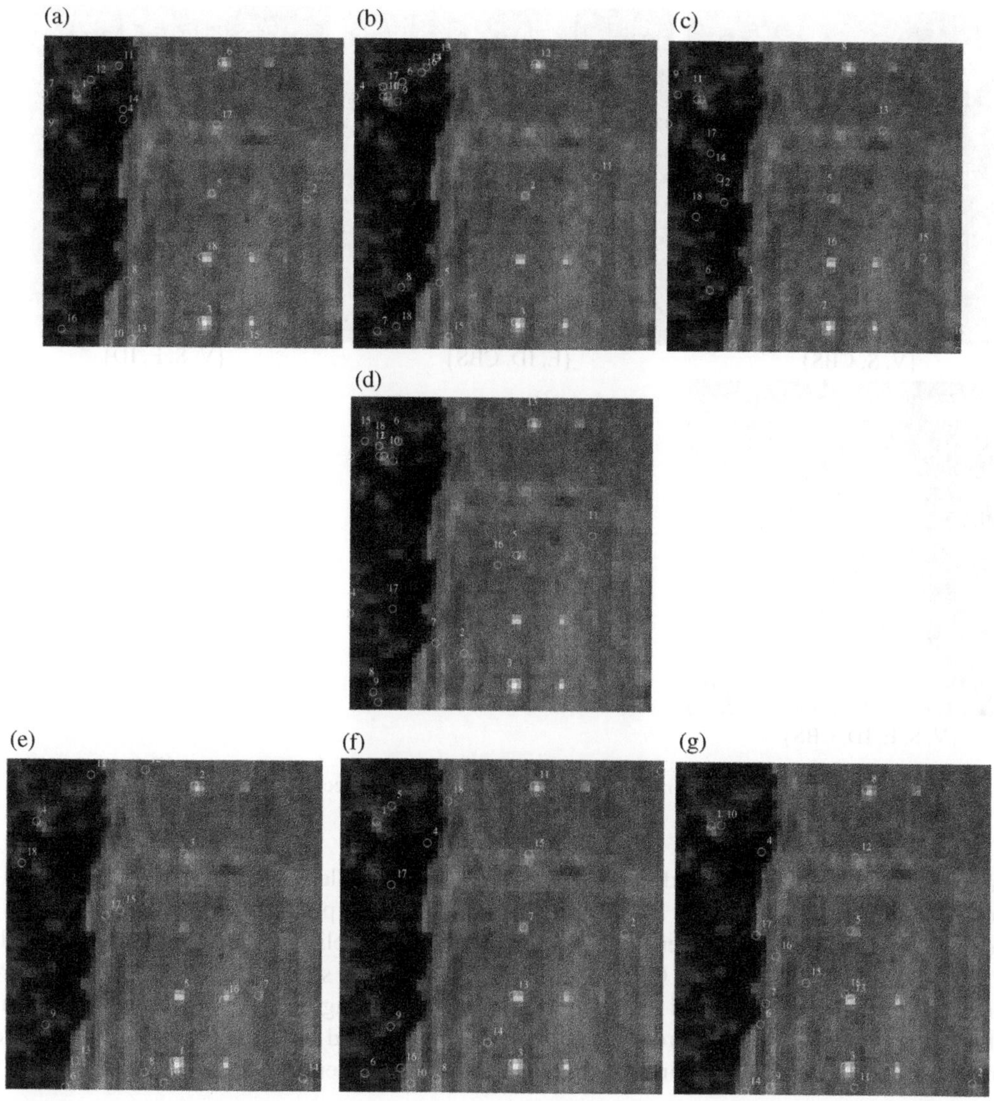


**Figure 12.4** 18 pixels found by ATGP using full bands and selected bands in Table 12.1. (a) Fullbands, (b) V, (c) S, (d) E, (e) ID, (f) CBS, (g) UBS.

in Figure 12.3(b) were found using $p = 18$, if yes, this column would give the order and the exact endmember where the last pixel was found as the fifth dark panel pixel in Figure 12.3(b).

Now, we used the 18 ATGP-found target pixels in Figures 12.4–12.6 as image endmembers for fully constrained least squares (FCLS) developed in [32] to perform linear spectral unmixing. Table 12.5 tabulates their total FCLS-unmixed errors. For comparison, we also included the results using full bands and 18 bands selected by uniform band selection (UBS) in Table 12.2 where the smallest unmixed errors produced by BS ad BSF methods are boldfaced.

As we can see from Table 12.5, using five pure panel pixels found by S, UBS, and E–ID–CBS (PBSF) did not necessarily produce best unmixing results. As a matter of fact, the best results

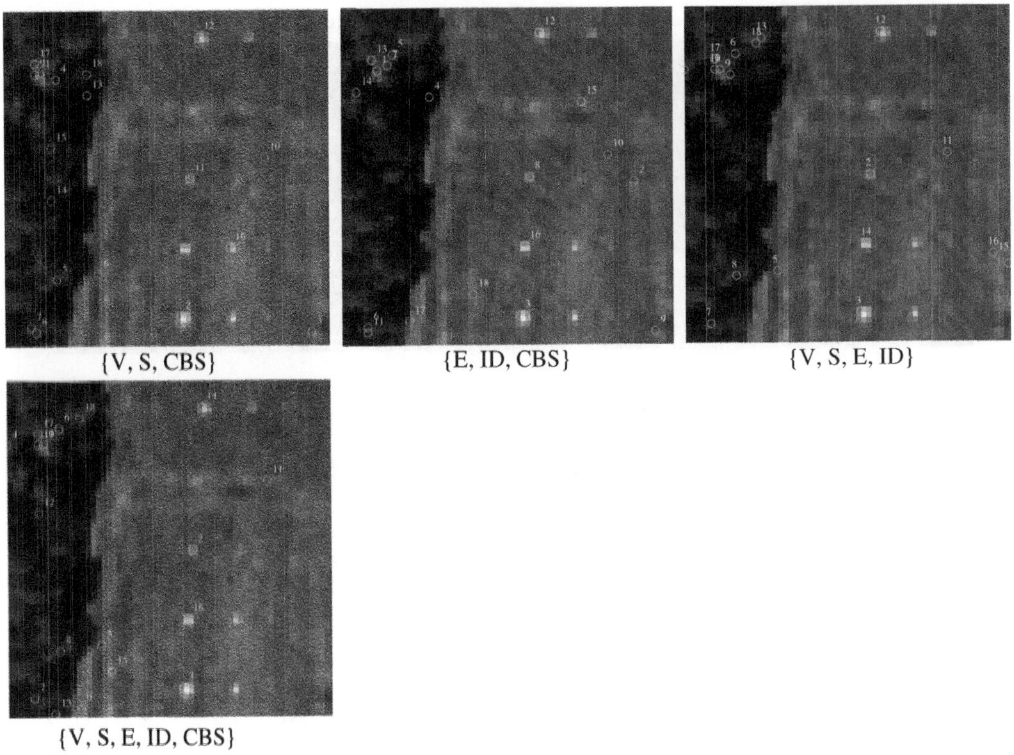


**Figure 12.5** 18 pixels found by ATGP using SBSF and selected bands in Table 12.2.

was FCLS using the bands selected by ID, which produced the smallest unmixed errors. These experiments demonstrated that in order for linear spectral unmixing to perform effectively, finding endmembers is not critical. But rather finding appropriate target pixels is more important and crucial. This was also confirmed in [33]. On the other hand, using bands selected by S and CBS also performed better than using full band for spectral unmixing. Interestingly, if we further used BSF methods, then the FCLS-unmixed errors produced by bands selected by PBSF-based methods were smaller than that produced by using full bands and single BP criteria except ID.

## *12.3.2 Hyperspectral Image Classification*

Three popular hyperspectral images available on the website http://www.ehu.eus/ccwintco/index.php?title=Hyperspectral_Remote_Sensing_Scenes which have been studied extensively for hyperspectral image classification were used for experiments.

### 12.3.2.1 AVIRIS Data

The first one is a well-known airborne visible/infrared imaging spectrometer (AVIRIS) image scene, Purdue Indiana Indian Pines test site shown in Figure 12.7(a) with its ground truth of 17 class maps in Figure 12.7(b). It has a size of 145 × 145 × 220 pixel vectors including water absorption bands (bands 104–108 and 150–163, 220).

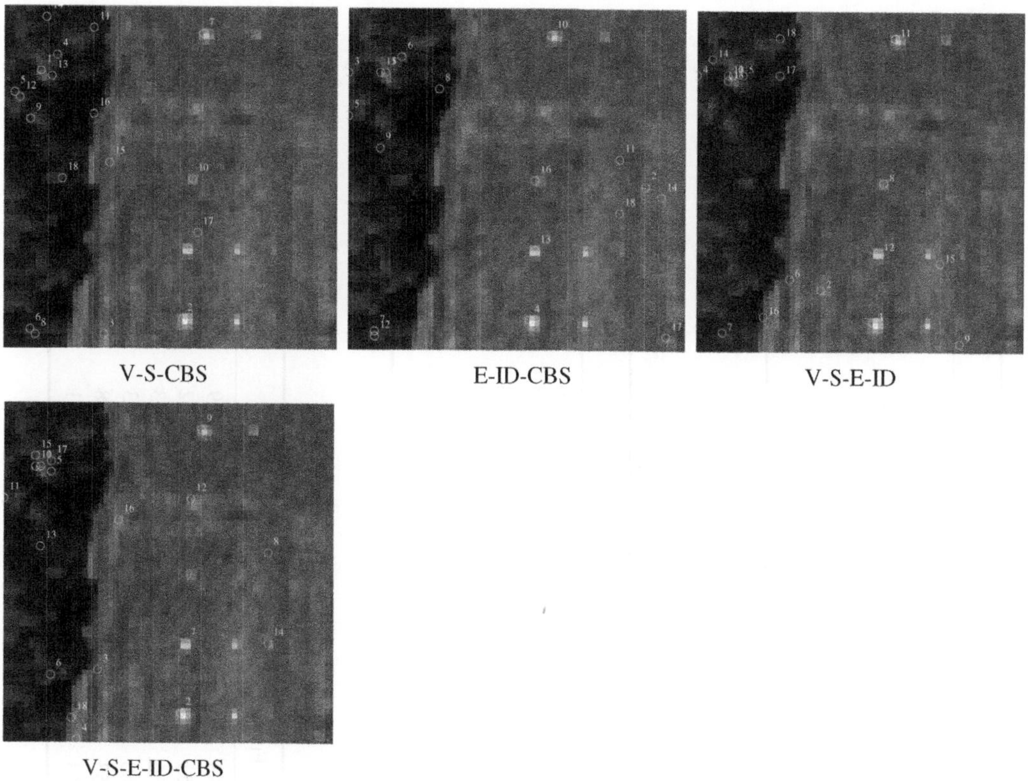


**Figure 12.6** 18 pixels found by ATGP using PBSF and selected bands in Table 12.3.

A second AVIRIS data used for experiments was Salinas scene shown in Figure 12.8(a), which was captured by the AVIRIS sensor over Salinas Valley, California, and with a spatial resolution of 3.7-m per pixel with spectral resolution of 10 nm. It has size of 512 × 217 × 224 including 20 water absorption bands, 108–112, 154–167, 224. Figure 12.8(b, c) shows the color composite of the Salinas image along with the corresponding ground truth class labels.

#### 12.3.2.2 ROSIS Data

The last hyperspectral image data used for experiments was the University of Pavia image shown in Figure 12.9, which is an urban area surrounding the University of Pavia, Italy. It was recorded by the ROSIS-03 satellite sensor. It has size of 610 × 340 × 115 with a spatial resolution of 1.3-m per pixel and a spectral coverage ranging from 0.43 to 0.86 μm with spectral resolution of 4 nm (12 most noisy channels were removed before experiments). Nine classes of interest plus BKG class, class 0 are considered for this image.

According to the recent work [1, 32, 33], VD was used to estimate $n_{BS}$ for the three hyperspectral images using the Harsanyi–Farrand–Chang (HFC) method in [21] and noise-whitened HFC method [22, 23]. Table 12.6 tabulates the values of $n_{BS}$, $n_{BS} = 18$ for the Purdue data, $n_{BS} = 21$ for Salinas, and $n_{BS} = 14$ for the University of Pavia with the false alarm probability $10^{-4}$ where a/b indicates the values of VD estimated by HFC/NWHFC methods.

**Table 12.1** 18 bands selected by variance, SNR, entropy, ID and CBS

| BS methods | Selected bands ($p = 18$) | | | | | | | | | | | | | | | | | |
|---|---|---|---|---|---|---|---|---|---|---|---|---|---|---|---|---|---|---|
| V | 60 | 61 | 67 | 66 | 65 | 59 | 57 | 68 | 62 | 64 | 56 | 78 | 77 | 76 | 79 | 63 | 53 | 80 |
| S | 78 | 80 | 93 | 91 | 92 | 95 | 89 | 94 | 90 | 88 | 102 | 96 | 79 | 82 | 105 | 62 | 107 | 108 |
| E | 65 | 60 | 67 | 53 | 66 | 61 | 52 | 68 | 59 | 64 | 62 | 78 | 77 | 57 | 79 | 49 | 76 | 56 |
| ID | 154 | 157 | 156 | 153 | 150 | 158 | 145 | 164 | 163 | 160 | 142 | 144 | 148 | 143 | 141 | 152 | 155 | 135 |
| CBS | 62 | 77 | 63 | 61 | 13 | 91 | 30 | 69 | 76 | 56 | 38 | 45 | 16 | 20 | 39 | 34 | 24 | 47 |
| UBS | 1 | 10 | 19 | 28 | 37 | 46 | 55 | 64 | 73 | 82 | 91 | 100 | 109 | 118 | 127 | 136 | 145 | 154 |

**Table 12.2** 18 bands selected by various SBSF methods

| SBSF methods | Fused bands ($p = 18$) | | | | | | | | | | | | | | | | | |
|---|---|---|---|---|---|---|---|---|---|---|---|---|---|---|---|---|---|---|
| {V, S} | 78 | 80 | 62 | 79 | 60 | 61 | 67 | 93 | 66 | 91 | 65 | 92 | 59 | 95 | 57 | 89 | 68 | 94 |
| {E, ID} | 65 | 154 | 60 | 157 | 67 | 156 | 53 | 153 | 66 | 150 | 61 | 158 | 52 | 145 | 68 | 164 | 59 | 163 |
| {V, S, CBS} | 62 | 78 | 61 | 77 | 80 | 63 | 91 | 76 | 56 | 79 | 60 | 67 | 93 | 66 | 13 | 65 | 92 | 59 |
| {E, ID, CBS} | 62 | 77 | 61 | 76 | 56 | 65 | 154 | 60 | 157 | 63 | 67 | 156 | 53 | 153 | 13 | 66 | 150 | 91 |
| {V, S, E, ID} | 78 | 62 | 79 | 60 | 65 | 61 | 80 | 67 | 53 | 66 | 59 | 57 | 68 | 64 | 56 | 77 | 76 | 154 |
| {V, S, E, ID, CBS} | 62 | 78 | 61 | 77 | 76 | 56 | 79 | 60 | 65 | 80 | 63 | 67 | 53 | 66 | 91 | 59 | 57 | 68 |

**Table 12.3** 18 bands selected by various PBSF methods

| PBSF methods | Fused bands ($p = 18$) | | | | | | | | | | | | | | | | | |
|---|---|---|---|---|---|---|---|---|---|---|---|---|---|---|---|---|---|---|
| V–S | 78 | 80 | 62 | 79 | 60 | 61 | 67 | 93 | 66 | 91 | 65 | 92 | 59 | 95 | 57 | 89 | 68 | 94 |
| E–ID | 65 | 154 | 60 | 157 | 67 | 156 | 53 | 153 | 66 | 150 | 61 | 158 | 52 | 145 | 68 | 164 | 59 | 163 |
| V–S–CBS | 62 | 78 | 61 | 77 | 80 | 63 | 91 | 76 | 56 | 79 | 60 | 67 | 93 | 66 | 13 | 65 | 92 | 59 |
| E–ID–CBS | 62 | 77 | 61 | 76 | 56 | 65 | 154 | 60 | 157 | 63 | 67 | 156 | 53 | 153 | 13 | 66 | 150 | 91 |
| V–S–E–ID | 62 | 65 | 60 | 67 | 78 | 66 | 61 | 53 | 80 | 154 | 52 | 93 | 68 | 91 | 59 | 157 | 64 | 92 |
| V–S–E–ID–CBS | 62 | 154 | 78 | 157 | 65 | 156 | 80 | 153 | 60 | 150 | 93 | 158 | 61 | 145 | 91 | 164 | 67 | 163 |

**Table 12.4** R panel pixels found by ATGP in Figures 12.1–12.3

| Various BS and BSF methods | $p = 18$ | $p = 9$ | Last pixel found as the fifth R panel pixel using $p = 18$ |
|---|---|---|---|
| Full bands | $p_{11}$, $p_{312}$, $p_{411}$, $p_{521}$ | $p_{11}$, $p_{312}$, $p_{521}$ | No |
| V | $p_{11}$, $p_{312}$, $p_{521}$ | $p_{312}$, $p_{521}$ | No |
| S | $p_{11}$, $p_{22}$, $p_{311}$, $p_{412}$, $p_{521}$ | $p_{11}$, $p_{311}$, $p_{521}$ | 16th pixel, $p_{412}$ |
| E | $p_{11}$, $p_{311}$, $p_{521}$ | $p_{311}$, $p_{521}$ | No |
| ID | $p_{11}$, $p_{211}$, $p_{412}$, $p_{521}$ | $p_{11}$, $p_{211}$, $p_{412}$, $p_{521}$ | No |
| CBS | $p_{11}$, $p_{312}$, $p_{411}$, $p_{521}$ | $p_{312}$, $p_{521}$ | No |
| UBS | $p_{11}$, $p_{211}$, $p_{311}$, $p_{412}$, $p_{521}$ | $p_{11}$, $p_{311}$, $p_{521}$ | 13th pixel, $p_{412}$ |
| V–S–CBS (PBSF) | $p_{11}$, $p_{312}$, $p_{42}$, $p_{521}$ | $P_{412}$, $p_{521}$ | No |
| E–ID–CBS (PBSF) | $p_{11}$, $p_{22}$, $p_{312}$, $p_{412}$, $p_{521}$ | $p_{11}$, $p_{412}$, $p_{521}$ | 16th pixel, $p_{412}$ |
| V–S–E–ID (PBSF) | $p_{11}$, $p_{312}$, $p_{412}$, $p_{521}$ | $p_{521}$ | No |
| V–S–E–ID–CBS (PBSF) | $p_{11}$, $p_{312}$, $p_{412}$, $p_{521}$ | $p_{11}$, $p_{521}$, $p_{53}$ | No |
| {V, S, CBS} (SBSF) | $p_{11}$, $p_{312}$, $p_{521}$ | $p_{412}$ | No |
| {E, ID, CBS} (SBSF) | $p_{11}$, $p_{311}$, $p_{412}$, $p_{521}$ | $p_{22}$, $p_{521}$ | No |
| {V, S, E, ID} (SBSF) | $p_{11}$, $p_{312}$, $p_{412}$, $p_{521}$ | $p_{412}$, $p_{521}$ | No |
| {V, S, E, ID, CBS} (SBSF) | $p_{11}$, $p_{211}$, $p_{412}$, $p_{521}$ | $p_{11}$, $p_{312}$, $p_{521}$ | No |

**Table 12.5** Total FCLS-unmixed errors produced by using 18 target pixels in Table 12.4 and full bands, UBS, OSPR-BP. B-BP, SFFS, SBBS, BF-SFFS and BF-SBBS using bands selected in Tables 12.2–12.3

| BS methods | Unmixed error |
|---|---|
| Full bands | 222.09 |
| UBS | 245.58 |
| V | 268.71 |
| S | 211.27 |
| E | 296.72 |
| ID | **22.104** |
| CBS | 207.32 |
| V–S–CBS (PBSF) | 209.22 |
| E–ID–CBS (PBSF) | 195.27 |
| V–S–E–ID (PBSF) | 201.51 |
| V–S–E–ID–CBS (PBSF) | **96.203** |
| {V, S, CBS} (SBSF) | 181.70 |
| {E, ID, CBS} (SBSF) | 228.63 |
| {V, S, E, ID} (SBSF) | 263.62 |
| {V, S, E, ID, CBS} (SBSF) | 249.34 |

Boldfaced entries indicate best results.

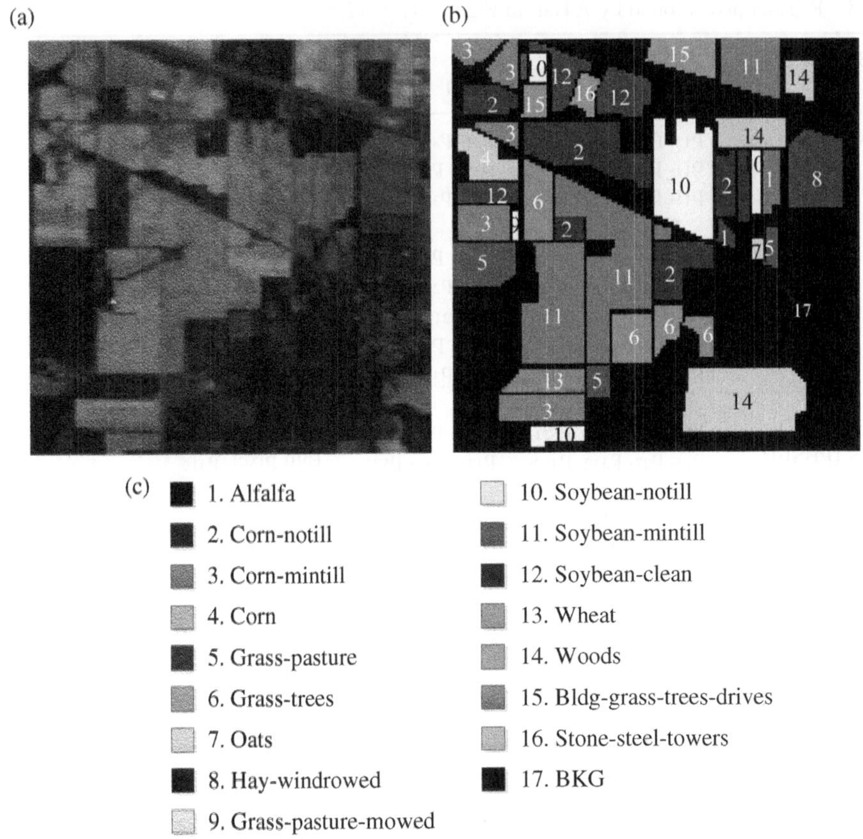


**Figure 12.7** Purdue's Indiana Indian Pines scene with band 186 (2162.56 nm) shown in (a) and 16 classes identified in (b) and its provided ground truth labels in (c), where its color maps can be found in Ref. [35], figure 4/with permission of IEEE.

Then UBS, variance (V), SNR (S), entropy (E), ID, CBS, and various versions of SBSF and PBSF were implemented to find desired bands listed in Table 12.7.

Once the bands were selected in Table 12.7, two types of classification techniques were implemented for performance evaluation. One is a commonly used edge preserving filtering (EPF)-based spectral–spatial approach developed in [34], which referred to as EPF-B-c, EPF-G-c, EPF-B-g, and EPF-G-g were shown to be among best classification techniques where "B" and "G" are used to specify bilateral filter and guided filter, respectively, and "g" and "c" indicate that the first principal component and color composite of three principal components are used as reference images. Therefore, in the following experiments the performance of various BSF techniques will be evaluated and compared to these four EPF-based techniques because of two main reasons. One is that these four techniques are available on website and we could re-implement them for comparison. Another is that these four techniques were compared to other existing spectral–spatial classification methods in [34] to show their superiority.

While EPF-based methods are pure pixel-based classification techniques, the other type of classification techniques is a subpixel detection-based method, which was recently developed in [35], called iterative CEM (ICEM). In order to make fair comparison, ICEM was modified without nonlinear expansion. In addition, the ICEM implemented here is little different from the ICEM with band selection nonlinear expansion (BSNE) in [35] in the sense that the ground truth has been used to

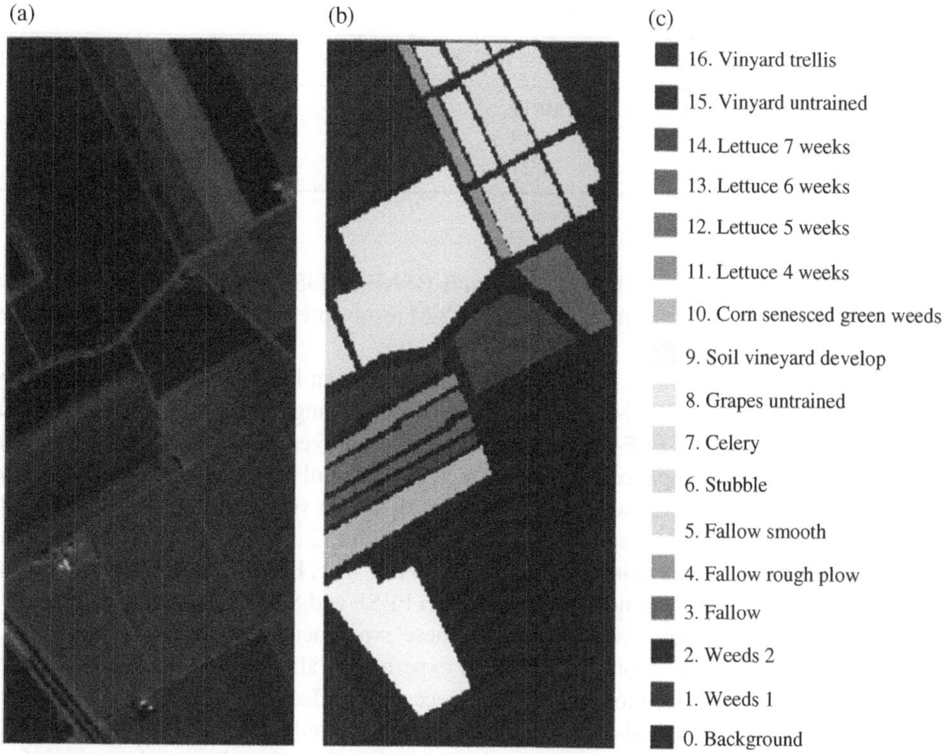


**Figure 12.8** Salinas scene shown in (a) and ground truth of Salinas scene with 16 classes identified in (b) and its provided ground truth labels in (c), where its color maps can be found in Ref. [35], figure 8/with permission of IEEE.

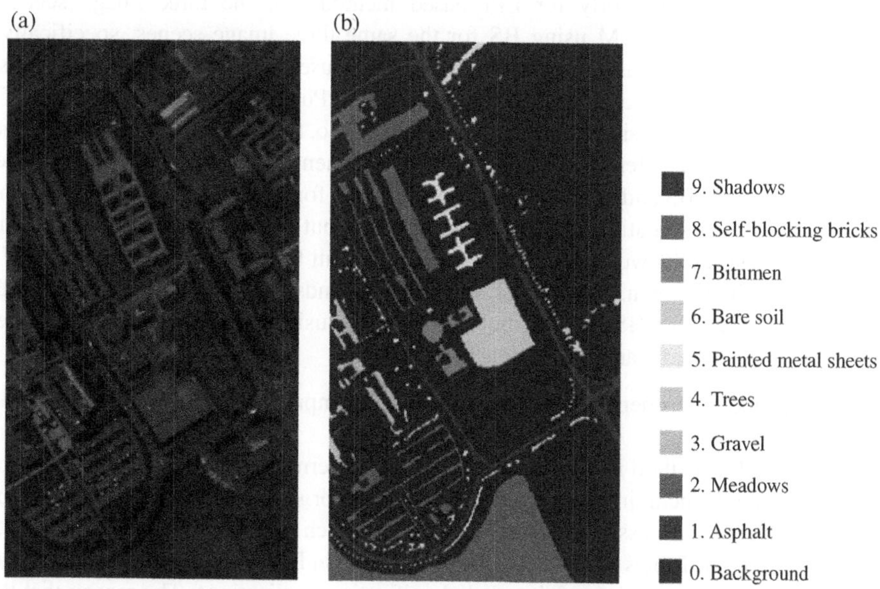


**Figure 12.9** Ground truth of University of Pavia scene with nine classes. (a) University of Pavia scene, (b) color ground-truth image, where its color maps can be found in Ref. [35], figure 12/with permission of IEEE.

**Table 12.6** $n_{BS}$ estimated by HFC//NWHFC

| | $P_F = 10^{-1}$ | $P_F = 10^{-2}$ | $P_F = 10^{-3}$ | $P_F = 10^{-4}$ | $P_F = 10^{-5}$ |
|---|---|---|---|---|---|
| Purdue | 73/21 | 49/19 | 35/18 | 27/18 | 25/17 |
| Salinas | 32/33 | 28/24 | 25/21 | 21/21 | 20/20 |
| University of Pavia | 25/34 | 21/27 | 16/17 | 14/14 | 13/12 |

update new image cubes instead of the ICEM with BSNE in [35], which used classified results to update new image cubes. As a consequence, the ICEM results presented in this chapter were better than ICEM with BSNE [35].

Tables 12.8–12.10 calculate $P_{OA}$, $P_{AA}$, $P_{PR}$ for Purdue's Indian Pines, Salinas, and University of Pavia produced by EPF-B-c, EPF-G-c, EPF-B-g, and EPF-G-g using full bands and bands selected in Table 12.7 respectively, where EPF-methods could not be improved by BS in classification. This is mainly due to their use of principal components, which contains full band information compared with BS methods that only retain selected band information. It is also very interesting to note that compared to experiments conducted for spectral unmixing in Section 12.3.1 where ID was shown to be the best BS method, ID here was the worst BS for all of the 4 EPF-B-c, EPF-G-c, EPF-B-g, and EPF-G-g methods. What is more is that whenever the BSF (both PBSF and SBSF) included ID as one of BS methods, the classification results were also worst. These experiments demonstrated that ID was not suitable to be used for classification. Furthermore, experiments also showed that one BS method, which is effective for one application, may not be necessarily effective for another application.

In contrast to EPF-based methods, which could not be improved by BS, ICEM coupled with BS behaved completely oppositely. Tables 12.11–12.13 calculate $P_{OA}$, $P_{AA}$, $P_{PR}$, and number of iterations for Purdue's Indian Pines, Salinas, and University of Pavia produced by ICEM using full bands and bands selected in Table 12.7, respectively, where the best results are boldfaced. Apparently, all the $P_{OA}$ and $P_{AA}$ results produced by ICEM in Tables 12.11–12.13 were much better than that produced by EPF-based methods in Tables 12.8–12.10, but PPR results were reversed. Most interestingly, ID, which performed poorly for EPF-based methods for the three image scenes, now worked very effectively for ICEM using BS for the same three image scenes, specifically using BSF methods that included ID as one of BS methods to be fused. Compared with the results using full bands, all BS and BSF methods performed better for both Purdue's data and University of Pavia, and slightly worse than ICEM using full bands for Salinas. Also, the experimental results showed that three image scenes do have different image characteristics when ICEM was used as a classifier with bands selected by various BS and BSF methods. For example, for the Purdue data all the BS and BSF methods did improve classification results in $P_{OA}$ and $P_{AA}$ but not PPR. But this was not true for Salinas where the best results were still produced by using full bands. And for University of Pavia, the results were right in between. That is, ICEM using full bands generally performed better than its using bands selected by single BS methods but worse than its using bands selected by BSF methods.

Several concluding remarks are worthwhile.

1. From our extensive experiments, BSF methods generally improve and perform better than a single BS method.
2. Whether or not a BS method is effective is completely determined by applications. For example, ID and BSF methods including ID performed very well for spectral unmixing and also for classification provided that a classifier is judiciously chosen such as ICEM in Tables 12.11–12.13 not EPF-based methods in Tables 12.8–12.10. This suggests that BSF methods fusing ID may actually perform well for both spectral unmixing and classification applications. The reason that they did

**Table 12.7** Bands selected by various BS methods and various versions of SBSF, PBSF

| BS methods | Purdue Indian Pines (18 bands) | Salinas (21 bands) | University of Pavia (14 bands) |
|---|---|---|---|
| UBS | 1/13/25/37/49/61/73/85/97/109/121/133/145/157/169/181/193/205 | 1/11/21/31/41/51/61/71/81/91/101/111/121/131/141/151/161/171/181/191/201 | 1/8/15/22/29/36/43/50/57/64/71/78/85/92 |
| V | 29/28/27/26/25/30/42/32/41/24/33/23/31/43/22/44/39/21 | 45/46/42/47/44/48/52/51/53/41/54/55/50/56/49/57/43/58/40/32/34 | 91/88/90/89/87/92/93/95/94/96/82/86/83/97 |
| S | 28/27/26/29/30/123/121/122/25/120/124/119/24/129/131/127/130/125 | 46/45/74/52/55/71/72/56/53/73/54/76/75/57/48/70/50/77/44/51/47 | 63/62/64/61/65/60/59/66/67/58/68/69/48/57 |
| E | 41/42/43/44/39/29/28/48/49/25/51/50/52/27/45/31/24/38 | 42/47/46/45/44/51/41/55/53/52/48/54/56/49/50/57/35/40/58/36/37 | 91/90/88/92/89/87/95/93/94/96/82/83/86/97 |
| ID | 156/157/158/220/155/159/161/160/162/95/4/219/154/2/190/32/153/1 | 107/108/109/110/111/112/113/114/115/116/152/153/154/155/156/157/158/159/160/161/162 | 8/10/9/11/7/12/13/14/15/6/16/17/18/19 |
| CBS (LCMV–BCC) | 9/114/153/198/191/159/152/163/161/130/167/150/219/108/160/180/215/213 | 153/154/113/152/167/114/223/222/224/166/115/107/112/168/116/165/221/109/174/151/218 | 37/38/39/40/36/32/41/33/42/31/34/30/43/35 |
| V–S–CBS (PBSF) | 9/28/29/114/27/153/26/198/25/191/30/159/24/152/123/163/42/161 | 45/153/46/154/47/113/52/152/44/167/55/114/48/223/51/222/56/224/53/166/54 | 37/63/38/91/39/62/40/88/36/64/32/90/41/61 |
| E–ID–CBS (PBSF) | 153/159/161/9/156/41/157/42/158/114/220/43/155/44/160/198/39/162 | 42/107/47/153/46/154/45/109/44/113/51/152/41/112/55/114/53/115/52/116/48 | 8/91/37/90/10/88/38/92/9/89/39/87/11/95 |
| V–S–E–ID (PBSF) | 28/29/27/41/26/25/30/42/43/44/39/156/32/157/24/33/123/23 | 45/46/42/52/47/55/44/53/48/54/56/51/41/107/108/50/74/49/109/57/43 | 91/88/90/89/87/92/95/8/63/10/62/93/9/94 |
| V–S–E–ID–CBS (PBSF) | 28/156/29/157/27/158/26/220/30/155/25/159/24/161/41/160/42/162 | 113/107/112/109/45/153/47/154/46/44/152/51/167/52/114/55/223/53/222/48/224 | 37/91/38/90/39/88/8/40/36/63/10/32/41/92 |
| {V, S, CBS} (SBSF) | 28/29/27/26/25/30/24/130/9/114/153/198/191/123/159/42/121/152 | 45/46/47/52/44/55/48/51/56/53/54/50/57/153/154/42/74/113/152/167/71 | 37/63/91/38/62/88/39/64/90/40/61/89/36/65 |
| {E, ID, CBS} (SBSF) | 153/159/161/160/219/9/41/156/42/114/157/43/158/44/198/220/39/155 | 107/153/154/109/113/152/112/114/115/116/42/47/108/46/45/110/44/111/167/51/41 | 8/37/91/10/38/90/9/39/88/11/40/92/7/36 |
| {V, S, E, ID} (SBSF) | 28/29/27/25/24/41/42/26/43/44/30/39/32/31/156/157/158/220 | 45/46/47/52/44/55/48/51/56/53/54/50/57/42/41/49/40/58/107/108/74 | 91/88/90/89/92/87/93/95/94/96/82/83/86/97 |
| {V, S, E, ID, CBS} (SBSF) | 28/29/27/25/24/41/42/26/43/153/44/30/39/159/161/32/160/130 | 45/46/47/52/44/55/48/51/56/53/54/50/57/42/107/153/154/109/113/152/112 | 91/88/90/89/92/87/93/95/94/96/82/83/86/97 |

**Table 12.8** $P_{OA}$, $P_{AA}$, $P_{PR}$ for Purdue's Indian Pines produced by EPF-B-c, EPF-G-c, EPF-B-g, and EPF-G-g using full bands and bands selected in Table 12.7

| BS and BSF methods | EPF-B-c | | | EPF-G-c | | | EPF-B-g | | | EPF-G-g | | |
|---|---|---|---|---|---|---|---|---|---|---|---|---|
| | $P_{OA}$ | $P_{AA}$ | $P_{PR}$ | $P_{OA}$ | $P_{AA}$ | $P_{PR}$ | $P_{OA}$ | $P_{AA}$ | $P_{PR}$ | $P_{OA}$ | $P_{AA}$ | $P_{PR}$ |
| Full bands | 0.9584 | 0.9829 | 0.9773 | 0.9679 | 0.9875 | 0.9826 | 0.9603 | 0.9840 | 0.9784 | 0.9616 | 0.9844 | 0.9789 |
| V | 0.9239 | 0.9665 | 0.9588 | 0.9257 | 0.9660 | 0.9613 | 0.9252 | 0.9667 | 0.9591 | 0.9180 | 0.9627 | 0.9557 |
| S | 0.9328 | 0.9687 | 0.9516 | 0.9351 | 0.9692 | 0.9556 | 0.9342 | 0.9696 | 0.9531 | 0.9280 | 0.9648 | 0.9475 |
| E | 0.9322 | 0.9707 | 0.9570 | 0.9358 | 0.9711 | 0.9605 | 0.9336 | 0.9710 | 0.9570 | 0.9250 | 0.9667 | 0.9530 |
| ID | 0.8236 | 0.8817 | 0.8682 | 0.8406 | 0.8979 | 0.8892 | 0.8245 | 0.8838 | 0.8697 | 0.8304 | 0.8884 | 0.8767 |
| CBS | 0.8593 | 0.9402 | 0.9337 | 0.8660 | 0.9483 | 0.9419 | 0.8604 | 0.9416 | 0.9353 | 0.8632 | 0.9442 | 0.9372 |
| V–S–CBS (PBSF) | 0.9558 | 0.9804 | 0.9754 | 0.9660 | 0.9857 | 0.9812 | 0.9583 | 0.9816 | 0.9765 | 0.9601 | 0.9826 | 0.9776 |
| E–ID–CBS (PBSF) | 0.9039 | 0.9372 | 0.9403 | 0.9208 | 0.9469 | 0.9506 | 0.9071 | 0.9388 | 0.9420 | 0.9102 | 0.9407 | 0.9437 |
| V–S–E–ID (PBSF) | 0.9332 | 0.9693 | 0.9610 | 0.9490 | 0.9775 | 0.9733 | 0.9366 | 0.9709 | 0.9634 | 0.9407 | 0.9733 | 0.9668 |
| V–S–E–ID–CBS (PBSF) | 0.9342 | 0.9610 | 0.9513 | 0.9466 | 0.9766 | 0.9640 | 0.9360 | 0.9706 | 0.9530 | 0.9389 | 0.9721 | 0.9552 |
| {V, S, CBS} (SBSF) | 0.9146 | 0.9448 | 0.9457 | 0.9301 | 0.9555 | 0.9582 | 0.9186 | 0.9473 | 0.9484 | 0.9218 | 0.9496 | 0.9511 |
| {E, ID, CBS} (SBSF) | 0.9477 | 0.9775 | 0.9731 | 0.9590 | 0.9836 | 0.9806 | 0.9496 | 0.9784 | 0.9744 | 0.9523 | 0.9799 | 0.9760 |
| {V, S, E, ID} (SBSF) | 0.9220 | 0.9609 | 0.9585 | 0.9380 | 0.9717 | 0.9692 | 0.9261 | 0.9636 | 0.9610 | 0.9291 | 0.9656 | 0.9631 |
| {V, S, E, ID, CBS} (SBSF) | 0.9337 | 0.9675 | 0.9516 | 0.9448 | 0.9741 | 0.9641 | 0.9357 | 0.9681 | 0.9531 | 0.9359 | 0.9680 | 0.9537 |

**Table 12.9** $P_{OA}$, $P_{AA}$, $P_{PR}$ for Salinas produced by EPF-B-c, EPF-G-c, EPF-B-g, and EPF-G-g using full bands and bands selected in Table 12.7

| BS and BSF methods | EPF-B-c | | | EPF-G-c | | | EPF-B-g | | | EPF-G-g | | |
|---|---|---|---|---|---|---|---|---|---|---|---|---|
| | $P_{OA}$ | $P_{AA}$ | $P_{PR}$ | $P_{OA}$ | $P_{AA}$ | $P_{PR}$ | $P_{OA}$ | $P_{AA}$ | $P_{PR}$ | $P_{OA}$ | $P_{AA}$ | $P_{PR}$ |
| Full bands | 0.8973 | 0.9282 | 0.9177 | 0.8896 | 0.9313 | 0.9186 | 0.8938 | 0.9269 | 0.9146 | 0.8932 | 0.9389 | 0.9121 |
| V | 0.8264 | 0.9011 | 0.8816 | 0.8297 | 0.9029 | 0.8805 | 0.8276 | 0.8938 | 0.8814 | 0.8255 | 0.9040 | 0.8782 |
| S | 0.8054 | 0.8161 | 0.7873 | 0.8096 | 0.8409 | 0.8573 | 0.8051 | 0.8126 | 0.7834 | 0.8000 | 0.8406 | 0.8223 |
| E | 0.8361 | 0.9062 | 0.8901 | 0.8296 | 0.9107 | 0.8826 | 0.8371 | 0.9027 | 0.8878 | 0.8352 | 0.9129 | 0.8859 |
| ID | 0.6525 | 0.5372 | 0.5232 | 0.6441 | 0.5289 | 0.5123 | 0.6532 | 0.5360 | 0.5282 | 0.6499 | 0.5360 | 0.5357 |
| CBS | 0.8119 | 0.8630 | 0.8630 | 0.8000 | 0.8010 | 0.8519 | 0.8116 | 0.8231 | 0.8628 | 0.8024 | 0.8407 | 0.8572 |
| V–S–CBS (PBSF) | 0.8336 | 0.8587 | 0.8687 | 0.8278 | 0.8255 | 0.8009 | 0.8363 | 0.8577 | 0.8698 | 0.8315 | 0.8526 | 0.8693 |
| E–ID–CBS (PBSF) | 0.7698 | 0.8091 | 0.7945 | 0.7473 | 0.7879 | 0.7728 | 0.7658 | 0.8076 | 0.7926 | 0.7650 | 0.8064 | 0.7949 |
| V–S–E–ID (PBSF) | 0.8315 | 0.9153 | 0.8836 | 0.8249 | 0.9078 | 0.8725 | 0.8321 | 0.9130 | 0.8830 | 0.8239 | 0.9045 | 0.8734 |
| V–S–E–ID–CBS (PBSF) | 0.6825 | 0.7197 | 0.7449 | 0.6623 | 0.6884 | 0.7266 | 0.6794 | 0.6865 | 0.7408 | 0.6791 | 0.7237 | 0.7403 |
| {V, S, CBS} (SBSF) | 0.8540 | 0.9036 | 0.8964 | 0.8529 | 0.8969 | 0.8942 | 0.8506 | 0.8881 | 0.8923 | 0.8531 | 0.9139 | 0.8935 |
| {E, ID, CBS} (SBSF) | 0.7793 | 0.7980 | 0.8099 | 0.7541 | 0.7733 | 0.7330 | 0.7710 | 0.7916 | 0.7991 | 0.7760 | 0.8059 | 0.8102 |
| {V, S, E, ID} (SBSF) | 0.7771 | 0.8484 | 0.8495 | 0.7704 | 0.8409 | 0.8247 | 0.7740 | 0.8491 | 0.8510 | 0.7761 | 0.8461 | 0.8441 |
| {V, S, E, ID, CBS} (SBSF) | 0.8300 | 0.8884 | 0.8677 | 0.8134 | 0.8680 | 0.8451 | 0.8279 | 0.8827 | 0.8483 | 0.8205 | 0.8814 | 0.8654 |

**Table 12.10** $P_{OA}$, $P_{AA}$, $P_{PR}$ for University of Pavia produced by EPF-B-c, EPF-G-c, EPF-B-g, and EPF-G-g using full bands and bands selected in Table 12.7

| BS and BSF methods | EPF-B-c | | | EPF-G-c | | | EPF-B-g | | | EPF-G-g | | |
|---|---|---|---|---|---|---|---|---|---|---|---|---|
| | $P_{OA}$ | $P_{AA}$ | $P_{PR}$ | $P_{OA}$ | $P_{AA}$ | $P_{PR}$ | $P_{OA}$ | $P_{AA}$ | $P_{PR}$ | $P_{OA}$ | $P_{AA}$ | $P_{PR}$ |
| Full bands | 0.9862 | 0.9848 | 0.9818 | 0.9894 | 0.9901 | 0.9863 | 0.9866 | 0.9852 | 0.9829 | 0.9853 | 0.9837 | 0.9820 |
| V | 0.9055 | 0.9332 | 0.8676 | 0.9139 | 0.9408 | 0.8776 | 0.9055 | 0.9344 | 0.8672 | 0.9067 | 0.9365 | 0.8697 |
| S | 0.8852 | 0.9205 | 0.8487 | 0.8910 | 0.9222 | 0.8556 | 0.8852 | 0.9190 | 0.8487 | 0.8850 | 0.9197 | 0.8489 |
| E | 0.9055 | 0.9332 | 0.8676 | 0.9139 | 0.9408 | 0.8776 | 0.9055 | 0.9344 | 0.8672 | 0.9067 | 0.9365 | 0.8697 |
| ID | 0.6543 | 0.7688 | 0.6659 | 0.6657 | 0.7814 | 0.6707 | 0.6543 | 0.7695 | 0.6662 | 0.6523 | 0.7676 | 0.6593 |
| CBS | 0.7227 | 0.8507 | 0.7320 | 0.7402 | 0.8567 | 0.7398 | 0.7218 | 0.8488 | 0.7292 | 0.7177 | 0.8439 | 0.7226 |
| V–S–CBS (PBSF) | 0.9811 | 0.9829 | 0.9731 | 0.9859 | 0.9865 | 0.9808 | 0.9820 | 0.9833 | 0.9750 | 0.9806 | 0.9810 | 0.9735 |
| E–ID–CBS (PBSF) | 0.9088 | 0.9482 | 0.8867 | 0.9245 | 0.9566 | 0.9017 | 0.9126 | 0.9497 | 0.8900 | 0.9109 | 0.9474 | 0.8888 |
| V–S–E–ID (PBSF) | 0.8831 | 0.9268 | 0.8537 | 0.8990 | 0.9375 | 0.8653 | 0.8838 | 0.9290 | 0.8531 | 0.8822 | 0.9277 | 0.8520 |
| V–S–E–ID–CBS (PBSF) | 0.8662 | 0.9301 | 0.8441 | 0.8757 | 0.9365 | 0.8499 | 0.8695 | 0.9318 | 0.8456 | 0.8628 | 0.9298 | 0.8412 |
| {V, S, CBS} (SBSF) | 0.9178 | 0.9497 | 0.8947 | 0.9260 | 0.9567 | 0.9030 | 0.9186 | 0.9498 | 0.8954 | 0.9175 | 0.9472 | 0.8941 |
| {E, ID, CBS} (SBSF) | 0.8848 | 0.9135 | 0.8481 | 0.9014 | 0.9256 | 0.8623 | 0.8880 | 0.9152 | 0.8505 | 0.8882 | 0.9138 | 0.8497 |
| {V, S, E, ID} (SBSF) | 0.8504 | 0.9101 | 0.8368 | 0.8615 | 0.9210 | 0.8488 | 0.8508 | 0.9092 | 0.8370 | 0.8517 | 0.9088 | 0.8391 |
| {V,S,E,ID,CBS} (SBSF) | 0.9055 | 0.9332 | 0.8676 | 0.9139 | 0.9408 | 0.8776 | 0.9055 | 0.9344 | 0.8672 | 0.9067 | 0.9365 | 0.8697 |

**Table 12.11** $P_{OA}$, $P_{AA}$, $P_{PR}$, and number of iterations calculated by ICEM using full bands and bands selected in Table 12.7 for Purdue's data

| BS and BSF methods | $P_{OA}$ | $P_{AA}$ | $P_{PR}$ | Iteration times |
|---|---|---|---|---|
| Full bands | 0.9650 | 0.9673 | **0.9018** | 24 |
| V | 0.9715 | 0.9767 | 0.8909 | 30 |
| S | 0.9717 | 0.9750 | 0.8826 | 29 |
| E | 0.9700 | 0.9736 | 0.8940 | 30 |
| ID | 0.9728 | 0.9762 | 0.8940 | 31 |
| CBS | 0.9729 | 0.9791 | 0.8871 | 30 |
| UBS | 0.9699 | 0.9753 | 0.8852 | 29 |
| V–S–CBS (PBSF) | 0.9738 | 0.9745 | 0.8822 | 30 |
| E–ID–CBS (PBSF) | 0.9720 | 0.9760 | 0.08873 | 30 |
| V–S–E–ID (PBSF) | 0.9742 | **0.9773** | 0.8833 | 30 |
| V–S–E–ID–CBS (PBSF) | **0.9751** | 0.9760 | 0.8925 | 33 |
| {V, S, CBS} (SBSF) | 0.9701 | 0.9729 | 0.8781 | 27 |
| {E, ID, CBS} (SBSF) | 0.9737 | 0.9762 | 0.8912 | 31 |
| {V, S, E, ID} (SBSF) | 0.9723 | 0.9750 | 0.8953 | 31 |
| {V, S, E, ID, CBS} (SBSF) | 0.9701 | 0.9767 | 0.8867 | 31 |

Boldfaced entries indicate best results.

**Table 12.12** $P_{OA}$, $P_{AA}$, $P_{PR}$, and number of iterations calculated by ICEM using full bands and bands selected in Table 12.7 for Salinas

| BS and BSF methods | $P_{OA}$ | $P_{AA}$ | $P_{PR}$ | Iteration times |
|---|---|---|---|---|
| Full bands | **0.9697** | **0.9662** | 0.9446 | 13 |
| V | 0.9621 | 0.9587 | **0.9467** | 19 |
| S | 0.9622 | 0.9573 | 0.9392 | 19 |
| E | 0.9622 | 0.9584 | 0.9445 | 18 |
| ID | 0.9588 | 0.9569 | 0.9432 | 20 |
| CBS | 0.9608 | 0.9581 | 0.9382 | 17 |
| UBS | 0.9609 | 0.9609 | 0.9418 | 15 |
| V–S–CBS (PBSF) | 0.9595 | 0.9530 | 0.9331 | 17 |
| E–ID–CBS (PBSF) | 0.9640 | 0.9597 | 0.9417 | 19 |
| V–S–E–ID (PBSF) | 0.9595 | 0.9520 | 0.9330 | 17 |
| V–S–E–ID–CBS (PBSF) | 0.9601 | 0.9548 | 0.9385 | 17 |
| {V, S, CBS} (SBSF) | 0.9577 | 0.09525 | 0.9423 | 16 |
| {E, ID, CBS} (SBSF) | 0.9615 | 0.9601 | 0.9439 | 19 |
| {V, S, E, ID} (SBSF) | 0.9645 | 0.9570 | 0.9423 | 19 |
| {V, S, E, ID, CBS} (SBSF) | 0.9659 | 0.9603 | 0.9457 | 19 |

Boldfaced entries indicate best results.

not work well in Tables 12.8–12.10 because the chosen EPF-based classification methods were not effective for BS.

3. The number of bands, $n_{BS}$, to be selected has also significant impact on the results. The $n_{BS}$ used in the experiments was selected based on VD, which is completely determined by image data characteristics regardless of applications. It is our belief that in order for BS to perform effectively, the value of $n_{BS}$ should be determined by a specific application.

**Table 12.13** $P_{OA}$, $P_{AA}$, $P_{PR}$, and number of iterations calculated by ICEM using full bands and bands selected in Table 12.7 for University of Pavia

| BS and BSF methods | $P_{OA}$ | $P_{AA}$ | $P_{PR}$ | Iteration times |
|---|---|---|---|---|
| Full bands | 0.8853 | 0.8868 | 0.6878 | 75 |
| V | 0.8764 | 0.8731 | 0.6898 | 77 |
| S | 0.8722 | 0.8736 | 0.6868 | 92 |
| E | 0.8763 | 0.8730 | 0.6898 | 77 |
| ID | 0.8690 | 0.8553 | 0.6870 | 100 |
| CBS | 0.8842 | 0.8844 | 0.6906 | 100 |
| UBS | 0.8836 | 0.8857 | 0.6876 | 82 |
| V–S–CBS (PBSF) | 0.8886 | 0.8817 | 0.6962 | 92 |
| E–ID–CBS (PBSF) | 0.8965 | 0.8881 | **0.7078** | 99 |
| V–S–E–ID (PBSF) | 0.8904 | 0.8783 | 0.6974 | 87 |
| V–S–E–ID–CBS (PBSF) | **0.8993** | 0.8893 | 0.6998 | 100 |
| {V, S, CBS} (SBSF) | 0.8900 | 0.8878 | 0.6966 | 99 |
| {E, ID, CBS} (SBSF) | 0.8917 | **0.8906** | 0.6816 | 84 |
| {V, S, E, ID} (SBSF) | 0.8764 | 0.8731 | 0.6898 | 77 |
| {V, S, E, ID, CBS} (SBSF) | 0.8764 | 0.8731 | 0.6898 | 77 |

Boldfaced entries indicate best results.

## 12.4 Conclusion

In general, a BS method is developed for a particular purpose. So, different BS methods produce different band subsets. Consequently, as a BS method is developed for one application, it may not work for another. It is a great advantage if different BS methods designed for various applications can be fused so that the fused band set not only can work for one application but also work for other applications as well. The BSF presented in this chapter fits this need. It develops two versions of BSF called simultaneous BSF (SBSF) and progressive BSF (PBSF). The main idea of BSF is to fuse bands by prioritizing fused bands according to their frequencies appearing in different band subsets. As a result, the fused band subset is more robust to various applications than a band subset produced by a single BS method. Additionally, such fused band subset generally takes care of the band de-correlation issue. Several contributions are worth noting. First and foremost is its idea of BSF, which has never been explored in the past. Second, fusion of different BS methods allows users to select most effective and significant bands among different band subsets produced by different BS methods. Third, since fused bands are selected from different band subsets, their inter-band correlation is largely reduced to avoid selection of high correlated bands. Fourth, one bad band selected by a BS method will not have much effect on BSF performance because it may be filtered out by fusion. Lastly and most importantly, bands can be fused according to practical applications, simultaneously or progressively. For example, PBSF has potential in future hyperspectral data exploitation space communication in which case BSF can take place during hyperspectral data transmission [31].

## References


1. Chang, C.-I (2013). *Hyperspectral Data Processing: Signal Processing Algorithm Design and Analysis*. New Jersey: Wiley.
2. Mausel, P.W., Kramber, W.J., and Lee, J.K. (1990). Optimum band selection for supervised classification of multispectral data. *Photogrammetric Engineering and Remote Sensing* 56 (1): 55–60.

3. Conese, C. and Maselli, F. (1993). Selection of optimum bands from TM scenes through mutual information analysis. *ISPRS Journal of Photogrammetry and Remote Sensing* 48 (3): 2–11.
4. Chang, C.-I, Du, Q., Sun, T.S., and Althouse, M.L.G. (1999). A joint band prioritization and band decorrelation approach to band selection for hyperspectral image classification. *IEEE Transactions on Geoscience and Remote Sensing* 37 (6): 2631–2641.
5. Keshava, N. (2004). Distance metrics and band selection in hyperspectral processing with applications to material identification and spectral libraries. *IEEE Transactions on Geoscience and Remote Sensing* 42 (7): 1552–1565.
6. Martínez-Usó, A., Pla, F., Sotoca, J.M., and García-Sevilla, P. (2007). Clustering-based hyperspectral band selection using information measures. *IEEE Transactions on Geoscience and Remote Sensing* 45 (12): 4158–4171.
7. Zare, A. and Gader, P. (2008). Hyperspectral band selection and endmember detection using sparisty promoting priors. *IEEE Geoscience and Remote Sensing Letters* 5 (2): 256–260.
8. Du, Q. and Yang, H. (2008). Similarity-based unsupervised band selection for hyperspectral image analysis. *IEEE Transactions on Geoscience and Remote Sensing Letters* 5 (4): 564–568.
9. Xia, W., Wang, B., and Zhang, L. (2013). Band selection for hyperspectral imagery: a new approach based on complex networks. *IEEE Geoscience and Remote Sensing Letters* 10 (5): 1229–1233.
10. Huang, R. and He, M. (2005). Band selection based feature weighting for classification of hyperspectral data. *IEEE Geoscience and Remote Sensing Letters* 2 (2): 156–159.
11. Koonsanit, K., Jaruskulchai, C., and Eiumnoh, A. (2012). Band selection for dimension reduction in hyper spectral image using integrated information gain and principal components analysis technique. *International Journal of Machine Learning and Computing* 2 (3): 248–251.
12. Yang, H., Du, Q., Su, H., and Sheng, Y. (2011). An efficient method for supervised hyperspectral band selection. *IEEE Geoscience and Remote Sensing Letters* 8: 138–142.
13. Su, H., Du, Q., Chen, G., and Du, P. (2014). Optimized hyperspectral band selection using particle swam optimization. *IEEE Journal of Selected Topics in Applied Earth Observations and Remote Sensing* 7 (6): 2659–2670.
14. Su, H., Yong, B., and Du, Q. (2016). Hyperspectral band selection using improved firefly algorithm. *IEEE Geoscience and Remote Sensing Letters* 13 (1): 68–72.
15. Yuan, Y., Zhu, G., and Wang, Q. (2015). Hyperspectral band selection using multitask sparisty pursuit. *IEEE Transactions on Geoscience and Remote Sensing* 53 (2): 631–644.
16. Yuan, Y., Zheng, X., and Lu, X. (2017). Discovering diverse subset for unsupervised hyperspectral band selection. *IEEE Transactions on Image Processing* 26 (1): 51–64.
17. Chen, P. and Jiao, L. (2017). Band selection for hyperspectral image classification with spatial-spectral regularized sparse graph. *Journal of Applied Remote Sensing* 11 (1): 010501-1–010501-8.
18. Geng, X., Sun, K., and Ji, L. (2014). Band selection for target detection in hyperspectral imagery using sparse CEM. *Remote Sensing Letters* 5 (3): 1022–1031.
19. Chang, C.-I and Wang, S. (2006). Constrained band selection for hyperspectral imagery. *IEEE Transactions on Geoscience and Remote Sensing* 44 (6): 1575–1585.
20. Chang, C.-I and Liu, K.-H. (2014). Progressive band selection for hyperspectral imagery. *IEEE Transactions on Geoscience and Remote Sensing* 52 (4): 2002–2017.
21. Harsanyi, J.C., Farrand, W., and Chang, C.-I (1994). Detection of subpixel spectral signatures in hyperspectral image sequences. *Annual Meeting, Proceedings of American Society of Photogrammetry & Remote Sensing*, Reno (1994), 236–247.
22. Chang, C.-I (2003). *Hyperspectral Imaging: Techniques for Spectral Detection and Classification*. New York, NY: Kluwer Academic/Plenum Publishers.
23. Chang, C.-I and Du, Q. (2004). Estimation of number of spectrally distinct signal sources in hyperspectral imagery. *IEEE Transactions on Geoscience and Remote Sensing* 42 (3): 608–619.
24. Chang, C.-I, Jiao, X., Du, Y., and Chang, M.L. (2010). A review of unsupervised hyperspectral target analysis. *EURASIP Journal on Advanced in Signal Processing* 2010: 503752.

25. Chang, C.-I, Jiao, X., Du, Y., and Chen, H.M. (2011). Component-based unsupervised linear spectral mixture analysis for hyperspectral imagery. *IEEE Transactions on Geoscience and Remote Sensing* 49 (11): 4123–4137.
26. Ren, H. and Chang, C.-I (2003). Automatic spectral target recognition in hyperspectral imagery. *IEEE Transactions on Aerospace and Electronic Systems* 39 (4): 1232–1249.
27. Chang, C.-I, Chen, S.Y., Li, H.C., and Wen, C.-H. (2016). A comparative analysis among ATGP, VCA and SGA for finding endmembers in hyperspectral imagery. *IEEE Journal of Selected Topics in Applied Earth Observation and Remote Sensing* 9: 4280–4306.
28. Nascimento, J.M.P. and Bioucas-Dias, J.M. (2005). Vertex component analysis: a fast algorithm to unmix hyperspectral data. *IEEE Transactions on Geoscience and Remote Sensing* 43 (4): 898–910.
29. Chang, C.-I, Wu, C., Liu, W., and Ouyang, Y.C. (2006). A growing method for simplex-based endmember extraction algorithms. *IEEE Transactions on Geoscience and Remote Sensing* 44: 2804–2819.
30. Heinz, D. and Chang, C.-I (2001). Fully constrained least squares linear mixture analysis for material quantification in hyperspectral imagery. *IEEE Transactions on Geoscience and Remote Sensing* 39 (3): 529–545.
31. Chang, C.-I (2016). *Real-Time Progressive Image Processing: Endmember Finding and Anomaly Detection*. New York: Springer.
32. Chang, C.-I (2018). A review of virtual dimensionality for hyperspectral imagery. *IEEE Journal of Selected Topics in Applied Earth Observation and Remote Sensing* 11 (4): 1285–1305.
33. Chang, C.-I (2017). *Real Time Recusive Hyperspectral Sample and Band Processing*. New York: Springer Chapter 4.
34. Kang, X., Li, S., and Benediktsson, J.A. (2014). Spectral-spatial hyperspectral image classification with edge-preserving filtering. *IEEE Transactions on Geoscience and Remote Sensing* 52: 2666–2677.
35. Xue, B., Yu, C., Wang, Y. et al. (2017). A subpixel target approach to hyperpsectral image classification. *IEEE Transactions on Geoscience and Remote Sensing* 55 (9): 5093–5114.

# Part V

## Hyperspectral Data Unmixing

# 13

# Model-Inspired Deep Neural Networks for Hyperspectral Unmixing

Yuntao Qian[1], Fengchao Xiong[2], Minchao Ye[3], and Jun Zhou[4]
[1]*College of Computer Science, Zhejiang University, Hangzhou, China*
[2]*School of Computer Science and Engineering, Nanjing University of Science and Technology, Nanjing, China*
[3]*College of Information Engineering, China Jiliang University, Hangzhou, China*
[4]*School of Information and Communication Technology, Griffith University, Nathan, QLD, Australia*

Hyperspectral unmixing is a significant tool to learn the material constitution and distribution of a scene. The traditional model-based unmixing approaches rely on the definition of physical mechanism, introduce mathematical descriptions of the underlying mixing process, including various linear or nonlinear spectral mixture models, and then use well-designed iterative optimization algorithms to solve these models. Unfortunately, defining an accurate model is always difficult due to the complicated reflection interaction mechanism and imaging environment. Inferencing an optimal solution is also difficult due to the exhaustive hyperparameter selection and time-consuming optimization procedure. Data-driven learning-based approaches motivated by the great success of deep learning are becoming a new trend in hyperspectral unmixing. Various supervised and unsupervised learning-based unmixing networks have been proposed, typically including convolutional neural networks (NNs) and encoder–decoder networks. These approaches learn the spectral mixture models from data in an end-to-end manner instead of hand-crafted definitions in model-based methods. However, the network architectures of purely data-driven learning-based methods are generic rather than specified for the unmixing task and heavily rely on the quality and quantity of training samples due to the lack of physical interpretability. In this chapter, leveraging model-based and learning-based approaches, we propose model-inspired learning methods, which incorporate the prior knowledge of the unmixing model and optimization algorithm into the network architecture by the unfolding technique. Specifically, based on the linear mixture model (LMM) with various constraints/regularizers such as nonnegativity, sparsity, and low-rankness, several model-inspired supervised and unsupervised learning-based deep NNs are derived by unrolling their corresponding iterative optimization algorithms.

*Advances in Hyperspectral Image Processing Techniques*, First Edition. Edited by Chein-I Chang.

## 13.1 Model-Based and Learning-Based Spectral Unmixing

Hyperspectral imaging has been widely adopted in many applications thanks to the spectral information contained in hundreds of continuous narrow bands. Due to the limited spatial resolution of imaging sensors, many pixels in a hyperspectral image (HSI) contain reflections from more than one type of material in their receptive fields. Spectral unmixing aims to decompose a mixed pixel into a collection of constituent spectra, or endmembers, and their corresponding fractional abundances. This has become a critical process to improve the utility of HSIs.

### *13.1.1 Mode-Based Spectral Unmixing*

The spectral unmixing contains two closely related tasks: endmember identification and abundance estimation. Some methods tackle them separately in a sequential pipeline [1]. They firstly determine the endmembers in an HSI using an endmember identification method, e.g. vertex component analysis (VCA) [2], or directly obtain all possible endmembers from a spectral library [3]. Then the corresponding abundances can be considered as a regression problem with various constraints related to unmixing such as nonnegativity and sparsity. To solve the constrained regression problems, a number of iterative optimization algorithms have been proposed, which not only allow easy incorporation of constraints but also allow better regularization for an ill-conditioned model. A fully constrained least squares (FCLS) algorithm was proposed for linear unmixing, which imposes both nonnegative and sum-to-one constraints on abundances [4]. Sparsity constraint is based on the fact that a mixed pixel always only contains a few constituted endmembers, especially if the spectral library contains a very large number of spectral signatures. Many sparsity constrained linear regression algorithms can be used for abundance estimation [3], such as matching pursuit (MP), orthogonal matching pursuit (OMP), and iterative shrinkage thresholding algorithm (ISTA) [5, 6].

Some methods estimate both endmembers and abundances simultaneously, which is also called blind unmixing. Nonnegative matrix factorization (NMF) is a widely used tool for blind unmixing, thanks to its inherent advantages in learning a part-based representation of data [7]. A number of variations of NMF have been proposed, most of which embed a range of constraints into the abundance and/or endmember estimation to incorporate spatial, spectral, and joint spectral–spatial information into NMF model [8, 9]. In order to address information loss when converting a 3-D HSI into a 2-D matrix, nonnegative tensor factorization (NTF) was proposed for blind unmixing [10, 11]. The solutions for NMF and NTF are mainly based on iterative optimization algorithms, such as Lagrange multipliers, alternating direction method of multipliers [12], and adaptive half-thresholding algorithm [13].

The above conventional unmixing approaches are based on the knowledge of spectral mixing mechanism, i.e. spectral mixture models must be constructed. Unfortunately, defining an accurate model is always difficult due to the complicated reflection interaction mechanism and environment. In practice, LMM is more frequently used than the nonlinear model due to its simplicity and clearness, but it is only a very rough approximation of real mixing process. Furthermore, inferencing an optimal solution is also hard and time-consuming.

### *13.1.2 Learning-Based Spectral Unmixing*

Motivated by the great success of deep learning (DL), recently learning-based unmixing has attracted many attentions. Learning-based methods are data-driven, which directly perform unmixing by learning from a large number of training samples [14, 15]. Learning-based unmixing methods can be categorized into supervised and unsupervised. For a supervised setting, the network is trained with the spectra paired with their corresponding abundances at pixels in order to produce a function to predict the abundances. Zhang et al. [16] proposed a two-stage convolutional NN for estimating the abundances from the extracted convolutional features. The abundances of training samples can be obtained by using other abundance estimation methods or site investigation.

Unsupervised learning-based unmixing methods do not require any supervised information, and the blind unmixing task can be cast as an unsupervised learning problem that simultaneously estimates the endmembers and abundances from data. The unmixing task can be regarded as finding the low-dimensional representation in the subspace spanned by endmembers. Autoencoders (AEs) first map the input data into a low-dimensional representation with an encoder and then transform the representation back to the original spectral space with a decoder. The low-dimensional representation can be regarded as abundances, and the endmembers are embedded as learnable weights in the decoder. For this reason, AEs have been widely accepted as a useful tool for the blind unmixing. Specifically, Su et al. employed variational autoencoder (VAE) to perform unmixing and stacked denoising AEs to generate favorable initialization for the subsequent VAE [17]. Instead of Euclidian distance, EndNet used spectral angle distances (SADs) in the loss function to consider the nonlinearity of endmembers [18]. Min et al. [19] further introduced the Wasserstein distance as a regularization term to better consider the distribution similarity between the observation and the reconstruction. Under the observation that the abundance maps of different endmembers are nearly orthogonal, Dou et al. embedded the orthogonal sparse prior into the AE to better consider the relationship among abundance maps [20]. Qu et al. introduced an untied denoising AE (uDAS), which decouples the decoder from the encoder weights for more accurate endmember extraction [21]. Convolutional AE was adopted in [22] to include the spatial and the spectral structure of HSIs for endmember and abundance map estimation. Hong et al. [23] proposed a two-stream Siamese-based network where the unmixing network has an AE-like architecture, and an additional network learns from the pure or nearly pure endmembers to guide the learning of the encoding part of the unmixing network. Gao et al. [24] cascaded two AEs for hyperspectral unmixing and introduced a self-perception loss to ensure the cycle consistency between spectral reconstruction and abundance estimation.

Moreover, AEs have also been introduced for nonlinear unmixing. Su et al. [25] developed two AEs to, respectively, consider the linear and bilinear mixing processes between materials. Zhao et al. [26] further decomposed the decoder into the linear mixture part and an additive nonlinear mixture part to perform nonlinear unmixing. In [27], the long short-term memory network (LSTM) and attention mechanism are embedded into the encoder to capture the spectral correlation information and focus on parts of the input during learning selectively. In addition to AE-based unmixing [28, 29] learned the deep priors of the abundance maps rather than hand-crafted priors and integrated them in the unmixing framework to boost the unmixing quality.

In general, most of the above methods fail to consider the spectral physical mixing process and introduce a general network that requires exhausting trials rather than a specified network for the unmixing task. The absence of the spectral mixture model makes it not easy to understand why data-driven NNs can perform unmixing, i.e. they are "black box" models. Moreover, their overparameterized architectures always demand many training samples.

## 13.2 Model-Inspired Learning for Spectral Unmixing

### *13.2.1 Why Should Model-Driven and Data-Driven Techniques Be Combined?*

We are surrounded by massive data, such as videos, images, audio, and more. These data have underlying structures rather than randomly organized. Models are mathematical descriptions of the generative mechanisms of data and lay a solid foundation for data analysis. Taking spectral unmixing task as an example, it is an ill-posed problem that seeks to decompose an acquired HSI **X** into an endmember matrix **A** and an abundance matrix **S**. To depart from trivial and unstable solutions, we must characterize the underlying formulation process of **X**. A spectral mixture model is a mathematical character of **X** that relates the light with the acquired spectra. If we assume that the observed light interacts with only one material, a linear spectral mixture model can be built that represents each spectrum with a linear combination of endmembers, so that the unmixing task becomes possible. As model-based methods are based on the prior knowledge of the underlying physics, they are interpretable and do not heavily rely on the data. The interpretability also makes it convenient to embed additional domain knowledge in a form of regularization to further narrow the solutions. For example, the nonnegativity constraint that requires the spectral values and fractional contributions to be never negative, and the abundance sum-to-one (ASC) constraint that ensures the total contributions of a pixel equal one, are commonly adopted.

However, model-based unmixing methods rely on well-designed optimization, which usually requires a high number of numerical iterations and tedious hyperparameter tuning to obtain the endmembers and abundances. This makes it challenging to deploy model-based methods, especially in time-sensitive and low-resource scenarios. In practice, an accurate mathematical model describing the true spectrum mixing process is too difficult to express analytically. To make the unmixing problem tractable, we typically simplify the underlying material interaction with linear or nonlinear modeling, etc. However, the subjectively simplified models cannot capture the dynamic variance and complexity of real scenarios, limiting the real-world practicality of model-based unmixing.

In contrast to model-driven methods using prior knowledge and intuition to reason at the problem level in devising inference algorithms, data-driven methods use data and universal approximators to solve the problem at the computational level in learning inference functions. Attributing to their strong representation ability, learning-based methods provide competitive performance. However, complicated network architectures hinder deep analysis of the functionality of different modules and a comprehensive understanding of the rationality behind their unmixing mechanisms. Therefore, it is preferable to design spectral unmixing methods that combine the advantages of both model-based and learning-based methods, i.e. of high interpretability while supporting powerful learning.

### *13.2.2 How Are Model-Inspired Unmixing Network Architectures Designed?*

Model-inspired NN benefits from model-based and learning-based methods by embedding prior knowledge of problems into deep network architecture in the form of established mathematical models. Deep unfolding technique is commonly used for transferring a model and its associated iterative inference algorithm to a network architecture [30, 31]. Deep unfolding starts from the iterative algorithm deduced by the model-based methods and maps each iteration into a typical layer of a deep NN. By stacking a predefined number of layers, a hierarchical deep network architecture can be obtained. Deep unfolding provides an attractive way to inherit the advantages of model-based methods, for example, high interpretability and good generalization capability and absorb the merits of learning-based methods such as strong learning ability and computational efficiency. Attributing to this, deep unfolding networks have been successfully applied in many tasks, such as hyperspectral fusion [32–34], pansharpening [35], rain removal [36], image super-resolution [37], and deblurring [38]. Our work falls in the line of deep unfolding and converts the constrained LMM into an end-to-end network to achieve more effective spectral unmixing.

Given a specific task, a model-based approach that requires an iterative inference algorithm is first constructed based on the task background. Secondly unfold the iterations into a layer-wise structure and then untie the model parameters across layers to obtain network architectures that can easily be trained discriminatively using gradient-based methods [30, 39]. Different from the traditional model-based approaches requiring accurate modeling that is always a very difficult task, the model used for deep unfolding only needs a very rough and broad definition of the task because the model-based approach is only used to construct the network architecture, and the learning-based solution can be learned by training data.

Assume a general formulation of a model, determined by parameters Θ, which specifies the relationships between the output of interest **y** and the observed variables **x**. In most cases, the close-form solution of model does not exist, so numerical iterative methods are always used. Given an iterative algorithm, **y** can be iteratively optimized by the following update rule:

$$\mathbf{y}^{(k+1)} = f_{\Theta}\left(\mathbf{x}, \mathbf{y}^{(k)}\right) \tag{13.1}$$

where $k$ is the iteration index. After the updating converges after $K$ iterations, the final **y** is obtained by $\mathbf{y}^{(K)}$. The model-inspired network, rather than considering the iterations as an algorithm, unfolds iterations as a sequence of layers in a network, and the parameters Θ are untied across layers as the transformation and activation function in each layer. Different from the model-based methods, Θ will be learned as the parameters of NN through training samples, so that the learned $\Theta'$ is not the same as in the original model. This learning ability can make the actually required number of layers $L$ much less than the number of iterations $K$ required in the iterative algorithm. In the following sections, we introduce several model-inspired deep NNs for the spectral unmixing task.

## 13.3 Unfolded Iterative Shrinkage-Thresholding Model for Supervised Abundance Estimation

In Section 13.1.2, we introduce the basic idea of model-based NNs. Here, we assume the endmembers are known and present an unfolded iterative shrinkage-thresholding algorithm (ISTA) for supervised abundance estimation.

### *13.3.1 Linear Mixture Model*

The LMM represents each pixel as a linear combination of $R$ endmembers weighted by their corresponding abundances. Given an HSI with $N$ pixels and $L$ bands, LMM can be formally formulated as

$$\mathbf{X} = \mathbf{AS} + \mathbf{E} \tag{13.2}$$

where $\mathbf{A} \in \mathbb{R}^{L \times R}$ is the endmember matrix, $\mathbf{S} \in \mathbb{R}^{R \times N}$ denotes abundances, and **E** accounts for measurement errors.

Generally speaking, the endmembers can be extracted ahead via many methods such as VCA [2]. Only a few endmembers exist in a pixel, so the abundance estimation problem can be cast as a sparse linear coding problem with sparsity and nonnegativity constraints, i.e.

$$E_{\mathrm{S}}(\mathbf{X}, \mathbf{A}) = \frac{1}{2}\|\mathbf{X} - \mathbf{AS}\|_2^2 + \lambda\|\mathbf{S}\|_1 \quad \texttt{s.t.} \ \mathbf{S} \succeq 0 \tag{13.3}$$

### 13.3.2 *Model Optimization*

The $L_1$-regularized $L_2$ problem defined in Eq. (13.3) has no closed-form solution. Iterative optimization that uses an iterative sequence of subproblems with each solved efficiently is required to solve this problem. ISTA is such an iterative algorithm that iteratively calculates the following recursive equation to convergence:

$$\mathbf{S}^{(k+1)} = h_\theta\left(\mathbf{WX} + \mathbf{DS}^{(k)}\right) \tag{13.4}$$

**W** and **S** are closely related to the endmember matrix **A**:

$$\mathbf{W} = \frac{1}{\beta}\mathbf{A}^T, \quad \mathbf{D} = \mathbf{I} - \frac{1}{\beta}\mathbf{A}^T\mathbf{A} \tag{13.5}$$

where $\beta$ is the step size. $h_\theta$ is the shrinkage function. Note that when $h$ has a nonnegativity constraint, the function $h$ is a component-wise vector one-sided shrinkage function with a vector of thresholds $\theta$:

$$h_\theta(\mathbf{y}) = \max(\mathbf{y} - \theta, 0) \tag{13.6}$$

In the standard ISTA, all the thresholds are set to $\theta = \dfrac{\lambda}{\beta}$.

### 13.3.3 *Unfolded ISTA for Abundance Estimation*

We unfold LMM-based ISTA into a NN-like architecture that can be trained to infer a mixed pixel's abundance vector directly. Figure 13.1(a) illustrates two unfolded network architectures of the iterations of ISTA in Eq. (13.4). Both are nonlinear, parameterized, feed-forward networks that untie the parameters of ISTA across a fixed number of layers. Each layer represents an iteration in ISTA and running the network fulfills the optimization of ISTA.

Provided with a supervised data set $\{\mathbf{x}_n, \mathbf{s}_n\}_{n=1:N}$ with $N$ pixels, all the parameters can be discriminatively learned from the data with the loss function defined as:

$$\mathcal{L}_\Theta = \frac{1}{2N}\sum_{n=1}^{N}\|\hat{\mathbf{s}}_n - \mathbf{s}_n\|_2^2 = \frac{1}{2N}\sum_{n=1}^{N}\|f(\mathbf{x}_n, \Theta) - \mathbf{s}_n\|_2^2 \tag{13.7}$$

where $\hat{\mathbf{s}}$ is the model output, i.e. estimated abundance vector. $\Theta = \{\mathbf{W}', \mathbf{D}', \theta'\}$ is the parameter set of the network, originated from $\{\mathbf{W}, \mathbf{D}, \theta\}$ in Eq. (13.4) of ISTA. These parameters are learned during the training procedure in our NN, but they are required to be known in the ISTA. The fixed parameters in ISTA are changed to be the learned parameters in model-inspired NN, so that the number of layers in our network is significantly less than the number of iterations required in the ISTA.

In Figure 13.1(a), all the parameters are shared in every layer. In order to increase representation capacity and the flexibility of architecture, we can let $\Theta = \{\mathbf{W}', \mathbf{S}', \theta'\}$ unshared in each layer by setting $\Theta_l = \{\mathbf{W}'_l, \mathbf{S}'_l, \theta'_l\}_{l=1:L}$, where $l$ is the layer index and $L$ is the number of layers. As a sequence, the untied NN architecture is shown in Figure 13.1(b). The total parameter set is $\Theta = \{\Theta_0, \ldots, \Theta_l, \ldots, \Theta_L\}$, in which $\Theta_0 = \{\mathbf{W}'_0, \theta'_0\}$. These two NNs are abbreviated as MNN-AE-1 and MNN-AE-2, respectively, in which MNN represents model and algorithm inspired NN, AE represents abundance estimation, and 1 and 2 represent parameters shared and unshared across layers. Both MNN-AE-1 and MNN-AE-2 have their own merits, e.g. coupled architecture

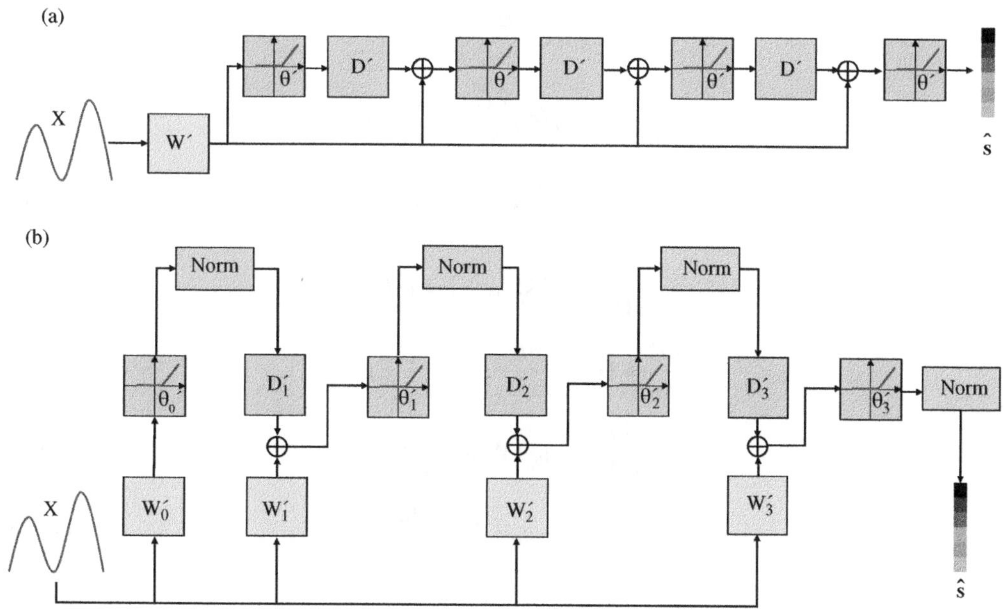


**Figure 13.1** Two abundance estimation NN architectures inspired from LMM and ISTA. (a) MNN-AE-1, (b) MNN-AE-2.

in MNN-AE-1 is closer to ISTA, while decoupled one in MNN-AE-2 has better approximation when their network sizes are the same.

Stochastic gradient descent algorithm can be used to train MNN-AE-1 and MNN-AE-2, which minimizes the loss function:

$$\Theta^{(k+1)} = \Theta^{(k)} - \eta^{(k)} \frac{\partial \mathcal{L}\Theta^{(k)}}{d\Theta^{(k)}} \tag{13.8}$$

where $\eta^{(k)}$ is the learning rate. We can compute the derivatives of $\mathcal{L}$ with respect to each parameter recursively. Since the updating rules of MNN-AE-1 can be easily obtained from those of MNN-AE-2, here we only consider MNN-AE-2. With the help of back-propagation algorithm, we can calculate the gradients for all network parameters starting from the output down to the input in a sequence of $\Theta_L \to \cdots \to \Theta_l \to \cdots \to \Theta_0$ for each training iteration.

At first, we define some intermediate variables in MNN-AE-2 for the convenience of derivation:

$$\mathbf{c}_l = \mathbf{W}_l\mathbf{x} + \mathbf{D}_l\hat{\mathbf{a}}_{l-1}, \quad l = 1:L \tag{13.9}$$

$$\hat{\mathbf{s}}_l = \begin{cases} h_{\theta'_l}(\mathbf{c}_l), & l = 1:L \\ h_{\theta'_l}\left(\mathbf{W}'_0\mathbf{x}\right), & l = 0 \end{cases} \tag{13.10}$$

Then, the gradient of loss with respect to parameters can be compute d as:

$$\frac{\partial \mathcal{L}}{\partial \hat{\mathbf{a}}_L} = \hat{\mathbf{s}}_L - \mathbf{s} \tag{13.11}$$

$$\frac{\partial \mathcal{L}}{\partial \mathbf{c}_l} = g_{\theta'_l}(\mathbf{c}_l)\frac{\partial \mathcal{L}}{\partial \hat{\mathbf{s}}_l}, \quad l = 1:L \tag{13.12}$$

$$\frac{\partial \mathcal{L}}{\partial \theta'_l} = -g_{\theta'_l}(\mathbf{c}_l)\frac{\partial \mathcal{L}}{\hat{\mathbf{s}}_l}, \quad l = 1:L \tag{13.13}$$

$$\frac{\partial \mathcal{L}}{\partial \mathbf{W}'_l} = \frac{\partial \mathcal{L}}{\partial \mathbf{c}_l}\mathbf{x}^T, \quad l = 1:L \tag{13.14}$$

$$\frac{\partial \mathcal{L}}{\partial \mathbf{D}_l} = \frac{\partial \mathcal{L}}{\partial \mathbf{c}_l}\hat{\mathbf{s}}_{l-1}^T, \quad l = 1:L \tag{13.15}$$

$$\frac{\partial \mathcal{L}}{\partial \hat{\mathbf{a}}_l} = \mathbf{D}_{l+1}^T\frac{\partial \mathcal{L}}{\partial \mathbf{c}_{l+1}}, \quad l = 1:L-1 \tag{13.16}$$

$$\frac{\partial \mathcal{L}}{\partial \mathbf{W}_0} = g_{\theta'_0}\left(\mathbf{W}'_0\mathbf{x}\right)\frac{\partial \mathcal{L}}{\partial \hat{\mathbf{s}}_0}\mathbf{x}^T \tag{13.17}$$

$$\frac{\partial \mathcal{L}}{\partial \theta'_0} = -g_{\theta'_0}\left(\mathbf{W}'_0\mathbf{x}\right)\frac{\partial \mathcal{L}}{\partial \hat{\mathbf{s}}_0} \tag{13.18}$$

where

$$g_\theta(\mathbf{y}) = \begin{cases} 1 & \mathbf{y} \geq \theta \\ 0 & \mathbf{y} < \theta \end{cases} \tag{13.19}$$

As $l$ starts from $L$ and ends at 0, one iteration of back-propagation is finished.

The overall abundance estimation procedure includes the following steps:

1. Sample several pixels and estimate their corresponding abundance vector by any traditional unmixing method to construct a supervised data set $\{\mathbf{x}_n, \mathbf{s}_n\}_{n=1:N}$.
2. Train MNN-AE-1 or MNN-AE-2 with stochastic gradient descent algorithm. As MNN-AE-1 and MNN-AE-2 are obtained from the LMM and ISTA algorithm, the model parameters can be initialized with the prior known endmember set $\mathbf{A}$ and ISTA parameter $\theta$.
3. Input any new mixed pixel, the learned MNN-AE-1 or MNN-AE-2 outputs its abundance vector. The speed of this step is very fast, thanks to that it is only a feed-forward computation. It must be kept in mind that input samples should be drawn from the same distribution as the training samples, for example all mixed pixels are collected by the same hyperspectral imager and on a similar geographic scene.

### *13.3.4 Experimental Results on Synthetic Data*

Here, we conduct a series of experiments to illustrate the performances of MNN-AE-1 and MNN-AE-2 in abundance estimation. Learning-based CNN [16] and model-based ISTA [5] are selected for comparison.

#### 13.3.4.1 Data Generation

We follow two steps to produce the synthetic data. Firstly, six pure signatures (Carnallite, Ammonio-jarosite, Almandine, Brucite, Axinite, and Chlonte) are selected from the USGS library to generate

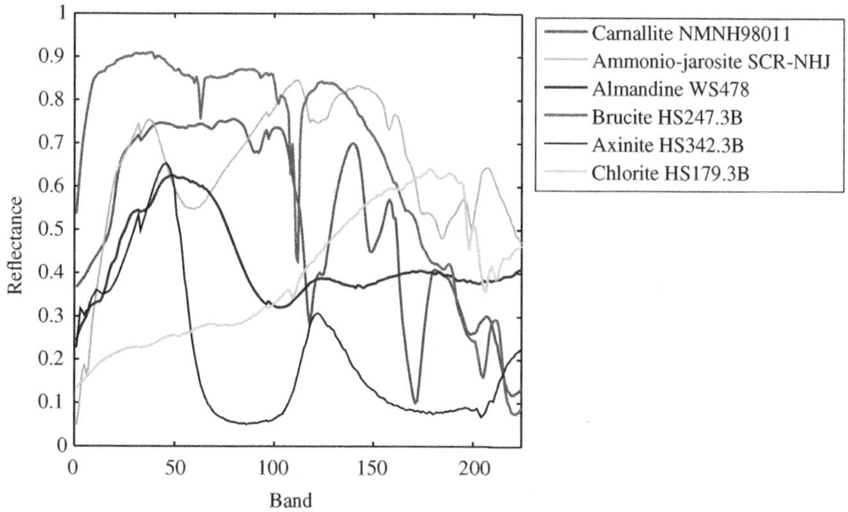


**Figure 13.2** The spectral signatures of six endmembers used in the synthetic data.

endmembers. They contain 224 bands covering wavelengths from 0.38 to 2.5 μm. Figure 13.2 demonstrates the spectral signatures of these endmembers.

The abundances are generated by the following steps:

1. $z^2$ nonoverlapping patches with the same size of $z \times z$ pixels are extracted from a $z^2 \times z^2$ synthetic image.
2. Each pixel of a region is assigned with two kinds of randomly selected endmembers whose fractions are initialized with $\beta$ and $1 - \beta$, respectively.
3. Each abundance map is convolved with a Gaussian filter whose variance is set as 2 to make the resulted map highly mixed.
4. The fractions of all the endmembers in each pixel are rescaled, by which the sum-to-one constraint of fractions is met.

Moreover, zero-mean white Gaussian noise is added to the clean HSI to test the robustness of unmixing methods. The signal-to-noise ratio (SNR) is given by

$$\text{SNR} = 10 \log_{10} \frac{E[\mathbf{y}^T\mathbf{y}]}{E[\mathbf{e}^T\mathbf{e}]} \tag{13.20}$$

where $\mathbf{y}$ and $\mathbf{e}$ are the clean signal and the noise at a pixel, respectively, and $E[\cdot]$ denotes the expectation operator. In experiments on synthetic data, we set the number of endmembers as $R = 6$ and the patch size as $z = 8$, so the generated image contains $64 \times 64 = 4096$ pixels.

#### 13.3.4.2 Evaluation Index

Root-mean-square error (RMSE) is utilized to quantitatively evaluate the performance of unmixing. RMSE compares the error between the reference abundance map $\mathbf{A}_r$ of the $r$th endmember and its estimated map $\hat{\mathbf{A}}_r$, which is defined as:

$$\text{RMSE}_r = \left(\frac{1}{N} \mid \mathbf{A}_r - \hat{\mathbf{A}}_r \mid^2\right)^{\frac{1}{2}} \tag{13.21}$$

where $N$ is the number of pixels used for test.

#### 13.3.4.3 Experiment Setting

The training pixels of MNN-AE-1, MNN-AE-2, and CNN are randomly selected from a synthetic HSI while the remaining pixels are used as testing samples. The experiments are done under two settings. In the first setting, for MNN-AE-1, MNN-AE-2, and CNN, the abundances of the training samples come from the ground truth data, while for ISTA the endmembers are also fully precise, and all of them can be obtained in synthetic generation procedure. This experiment is to evaluate the performance when the known information is completely right. For the second setting, the above endmembers and abundances are obtained by VCA-FCLS, i.e. the endmembers are extracted by VCA [2], and then the abundances are estimated by FCLS [4]. This experiment is to evaluate the performance when the known information is not perfect. This case is very common in the real applications.

#### 13.3.4.4 Impact of Sampling Strategy

In this experiment, we set SNR = ∞. Random and fixed samplings, i.e. pixel selection in a fixed region, are used to generate training samples. We varied the sample size with {256, 500, 1000}. As shown in Table 13.1, the more training samples, the smaller the RMSE, which is consistent with the fact that more training samples contribute to better performance. The results are much better with random sampling, because the abundance distribution is better covered. In addition, unfolding ISTA also produced competitive RMSE at a smaller training scale, demonstrating the effectiveness of learning-based unmixing.

#### 13.3.4.5 Impact of Number of Layers

This experiment mainly analyzes the influence of network layers on the unmixing performance. The synthetic data in this experiment were generated by setting $\beta = 0.8$, SNR = 15 dB, the number of training samples was set as 1000, and the number of layers was set from 1 to 7. Generally, networks with more layers lead to better approximation but introduce more learning parameters, leading to overfitting problems. Figure 13.3 shows the results of MNN-AE-1 and MNN-AE-2 under the two experimental settings: absolutely correct training sample and VCA-FCLS-based training sample. As we can see, they both require very few layers (2 or 3) to achieve good performance in two different settings. This shows that our networks have high learning capability even with a very small network size. We set $L = 2$ for our abundance estimation networks in the next experiments.

**Table 13.1** RMSEs of two sampling strategies

| Sampling strategy | 256 | 500 | 1000 |
|---|---|---|---|
| Random sampling | 0.0010 | 3.783e–04 | 7.049e–05 |
| Fixed sampling | 0.0271 | 0.0153 | 0.0020 |

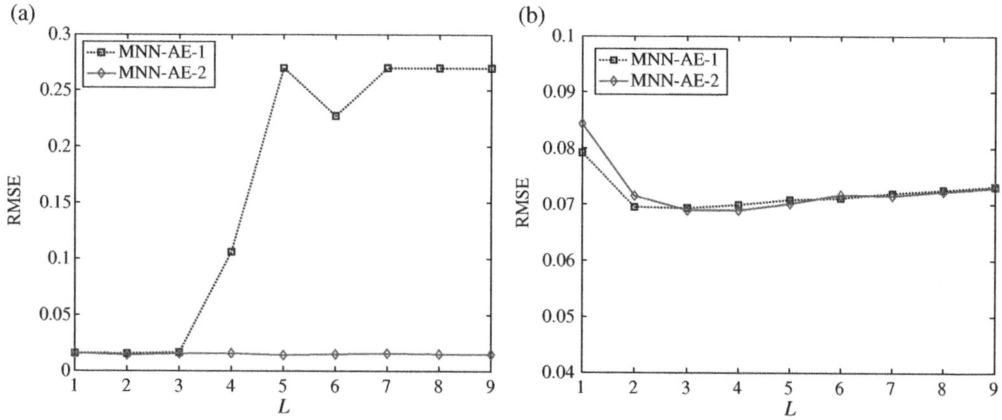


**Figure 13.3** RMSE of MNN-AE-1 and MNN-AE-2 with respect to different numbers of layers under two training conditions, respectively: (a) ground-truth, (b) VCA-FCLS.

#### 13.3.4.6 Impact of Number of Training Samples

This experiment aims to study the influence of training samples on unmixing performance. The data of this experiment is also generated with $\beta = 0.8$, SNR = 15 dB. The training set size is set to {256, 500, 1000, 2000, 4096}. The results are shown in Figure 13.4, indicating that the performance of our two networks is superior to CNN, especially in the case of small training sample size and unsatisfactory training sample quality. Experimental results show that the model-based NN has obvious advantages over the ordinary NN.

#### 13.3.4.7 Robustness to Noise

This experiment was designed to evaluate the robustness of the abundance estimation method against noise. The resulting clean HSIs is contaminated by different levels of noise with SNR = {15 dB, 20 dB, 25 dB, 30 dB, $\infty$}. The results are shown in Figure 13.5. From the figure, we can find that when

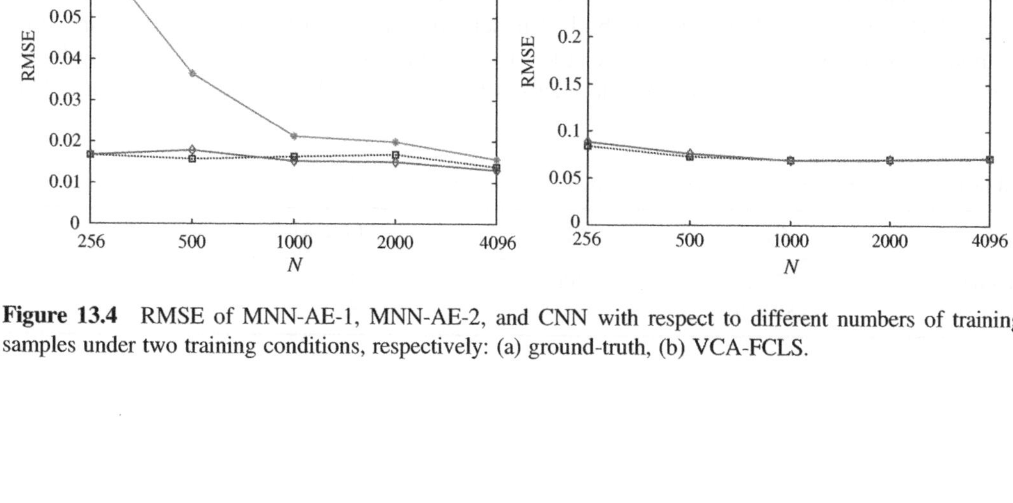


**Figure 13.4** RMSE of MNN-AE-1, MNN-AE-2, and CNN with respect to different numbers of training samples under two training conditions, respectively: (a) ground-truth, (b) VCA-FCLS.

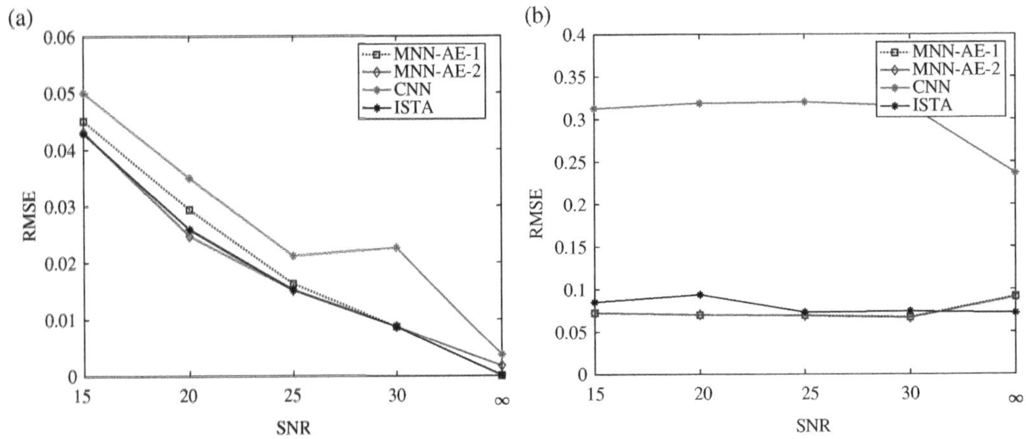


**Figure 13.5** RMSE of all four abundance estimation methods with respect to different noise levels under two training conditions, respectively: (a) ground-truth, (b) VCA-FCLS.

the ground truth data is used as the training sample and endmembers are known, the performance of all the methods becomes better with the reduction of the noise level. However, when the training sample or the known endmember is not perfect, the noise level has no obvious influence on the four methods. This may be because more training samples do not play a positive role when their quality is not good. ISTA behaves much like our method because our NNs are unfolded from ISTA.

#### 13.3.4.8 Running Time Comparison

The aim of this experiment is to compare the running time of all methods. All the methods are tested on a Windows machine equipped with Intel(R) Core(TM) i5-4440 CPU@3.10 GHz and 16 GB RAM. All algorithms are implemented in Python and Matlab. For an HSI containing $p$ bands and $R$ endmembers, MNN-AE-1 and MNN-AE-2 with $L$ layers, respectively, has $O((p + R + 1)R)$ and MnN-AE-2 requires $O((p + R + 1)RL)$ parameters. For CNN, according to its network architecture [16], the number of parameters is $O(3600p + 151R)$, which is larger than those of MNN-AE-1 and MNN-AE-2.

To show the efficiency of the proposed method on large data sets, we add another synthetic HSI with a size 640 × 640 using the same generation method. For both synthetic data sets with different sizes, the same number of training pixels, 1000, is used for MNN-AE-1, MNN-AE-2, and CNN, and all pixels are used for the test. Table 13.2 reports the running times of ISTA, CNN, MNN-AE-1, and MNN-AE-2 in both data sets. In contrast to ISTA, whose running time is approximately linear with respect to the size of the data set, the three learning-based approaches are almost unaffected by the size of the data set and have good scalability. This

**Table 13.2** Running time(s) comparison on two synthetic data sets with different sizes

| Method | ISTA | CNN | MNN-AE-1 | MNN-AE-2 |
|---|---|---|---|---|
| 64 × 64 | 878.2 | 3065.0 | 5.4 | 7.5 |
| 640 × 640 | 89 839 | 3115.3 | 6.8 | 8.2 |

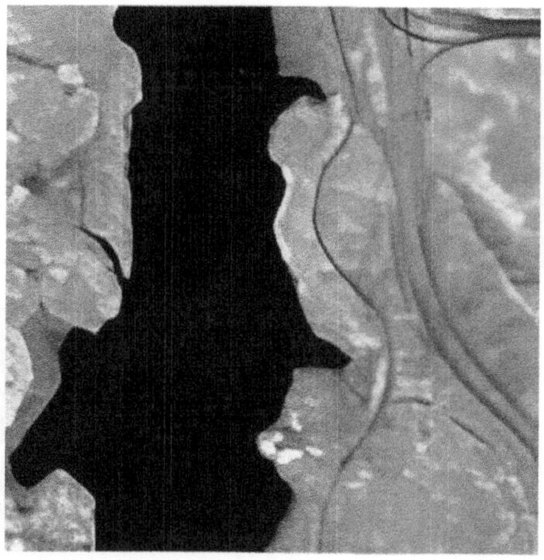

**Figure 13.6** Band image of Jasper Ridge.

is because the training time for MNN-AE-1, MNN-AE-2, and CNN depends on the size of the training set. After the training step, the abundance estimation problem is the network's feedforward process, which is computationally very fast. CNN is also a learning-based method but takes more time to train the network, indicating that our network architecture has a smaller scale and better convergence performance.

### *13.3.5 Experimental Results on Real-World Data*

Now we turn to experiment with real-world HSI, the Jasper Ridge data set. The reference endmember and abundances were obtained by the method of Qian et al. [40]. Firstly, several highly likely pure pixels were selected from an HSI, and then their average signatures were used as reference endmembers. After all the endmembers were ready, CVX MATLAB software was used to obtain their corresponding abundance maps.

The Jasper Ridge data set was acquired by the AVIRIS (Airborne Visible/Infrared Imaging Spectrometer) sensor over Jasper Ridge, Central California, USA. The original image contains $512 \times 614$ pixels and 224 electromagnetic bands covering the electromagnetic spectrum from 380 to 2500 nm with a resolution of nearly 10 nm. A sub-scene of $100 \times 100$ pixels was cropped for the experiment, and its 80th band image was shown in Figure 13.6(a). It is assumed that there are four types of endmembers, namely soil, water, tree, and road. 1000 pixels were randomly selected as training samples for MNN-AE-1, MNN-AE-2, and CNN, and their abundances were the same as the reference sample and ISTA uses reference endmembers. Figure 13.7 shows the abundance obtained by four methods. We can see that they all get similar abundance maps close to the reference one.

## 13.4 Model-Inspired Network Architectures for Blind Unmixing

In Section 13.3, we assume the endmembers are known. In general, an accurate estimation of abundance is difficult. In this section, under the framework of LMM, we further introduce two unfolded blind unmixing NNs that simultaneously extract endmembers and estimate abundances.

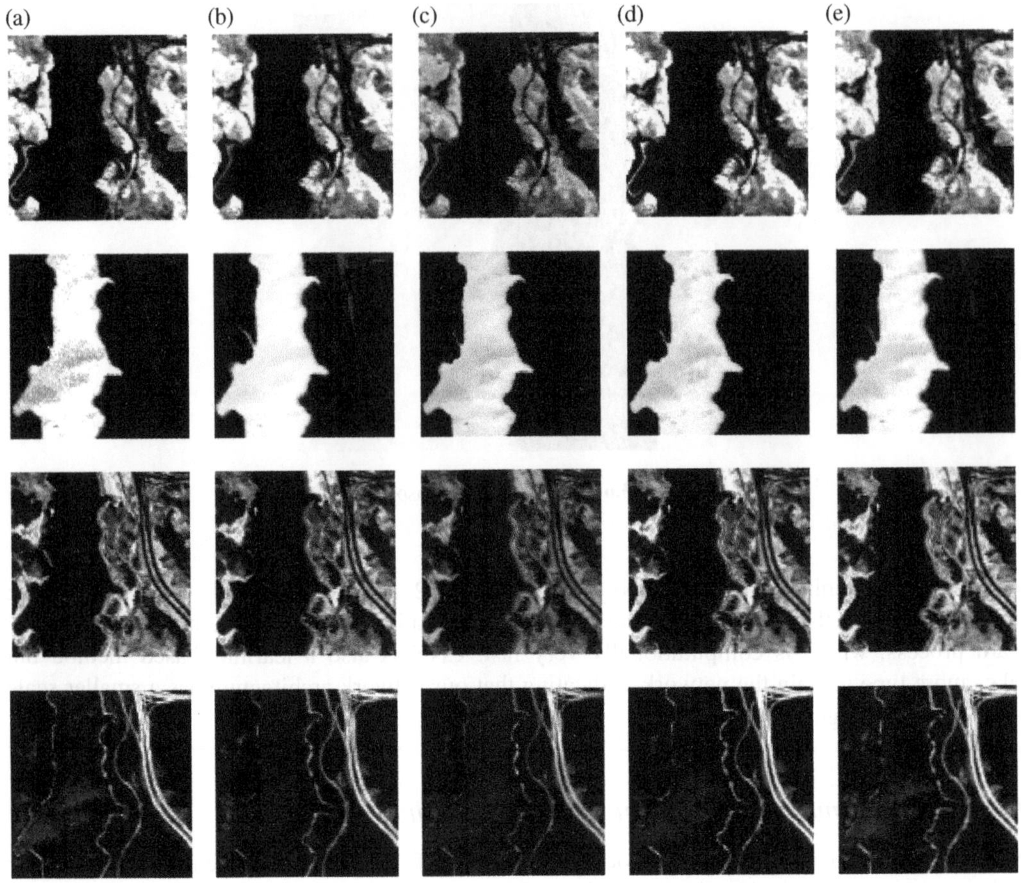

**Figure 13.7** Estimated abundance maps by four methods on the Jasper Ridge data set. From top to bottom, the rows are the abundance maps of tree, water, soil, and road, respectively. (a) CNN, (b) ISTA, (c) MNN-AE-1, (d) MNN-AE-2, (e) reference.

### *13.4.1 Unsupervised Model-Inspired NN for Blind Unmixing*

Based on MNN-AE-1 and MNN-AE-2, we further add the endmember matrix **A** as the last layer and change their output from $\hat{\mathbf{s}}$ to reconstructed input $\hat{\mathbf{x}}$ to yield the blind unmixing networks as Figure 13.8. They are called MNN-BU-1 and MNN-BU-2, respectively, where BU represents blind unmixing. The first has shared parameters in different layers as MNN-AE-1, while the second is similar to MNN-AE-2 without shared parameters.

If the endmember matrix **A** is known and fixed in MNN-BU-1 and MNN-BU-2, they are equivalent to MNN-AE-1 and MNN-AE-2. If the endmembers are unknown, **A** in MNN-BU-1 and MNN-BU-2 can be set as variable/parameter of NNs that must be learned during training. The training problem is completely unsupervised, that is it not only does not require any labeled training pixels but it also estimates endmembers and abundance. Unsupervised MNN-BU-1 and MNN-BU-2 can be considered as an encoder–decoder architecture. The part before the last layer is an encoder that transforms the pixel's original spectrum into an encoded vector (abundance vector). The last layer is a decoder that recovers the spectrum of pixels from the encoded data.

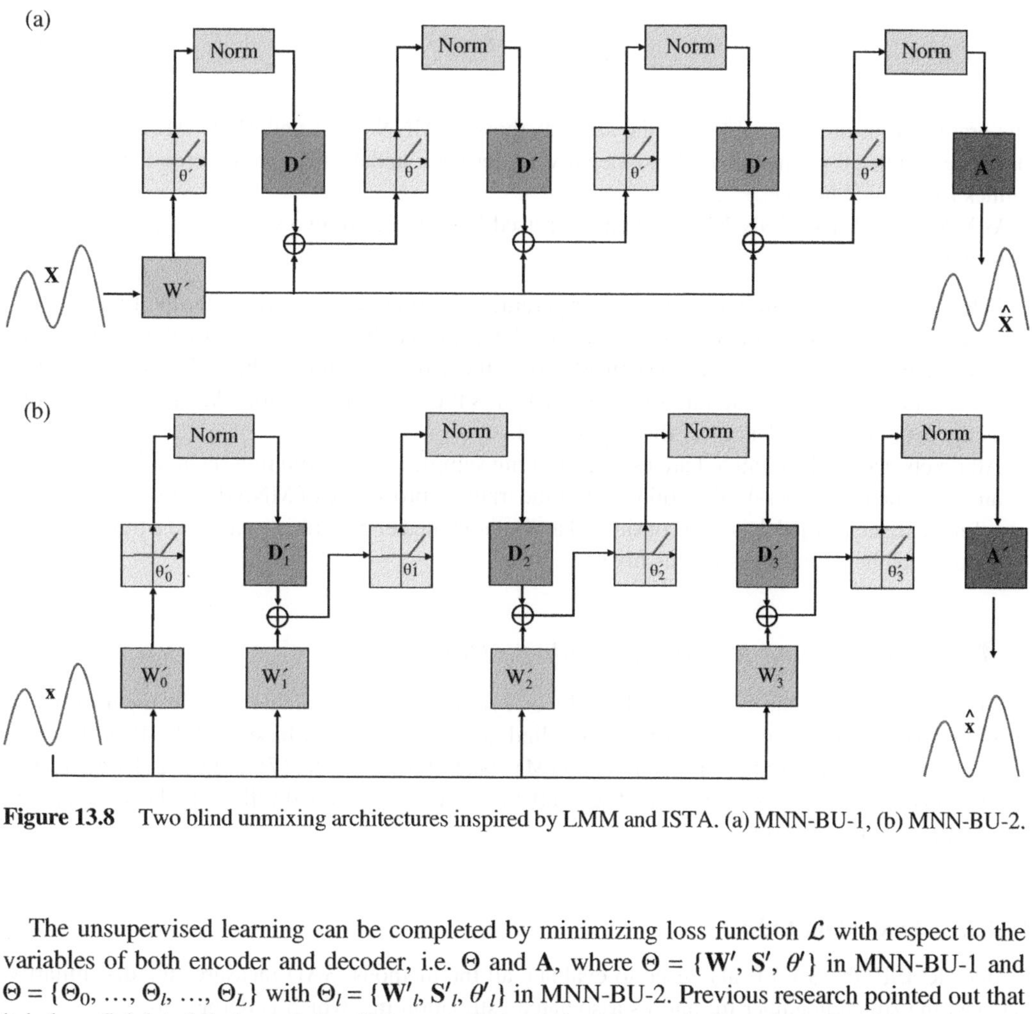


**Figure 13.8** Two blind unmixing architectures inspired by LMM and ISTA. (a) MNN-BU-1, (b) MNN-BU-2.

The unsupervised learning can be completed by minimizing loss function $\mathcal{L}$ with respect to the variables of both encoder and decoder, i.e. $\Theta$ and $\mathbf{A}$, where $\Theta = \{\mathbf{W}', \mathbf{S}', \theta'\}$ in MNN-BU-1 and $\Theta = \{\Theta_0, \ldots, \Theta_l, \ldots, \Theta_L\}$ with $\Theta_l = \{\mathbf{W}'_l, \mathbf{S}'_l, \theta'_l\}$ in MNN-BU-2. Previous research pointed out that it is beneficial to blind unmixing if the encoder and decoder are independent. For this reason, we assume these two sets of variables are untied, i.e. $\Theta = \{\mathbf{W}', \mathbf{S}', \theta'\}$ and $\mathbf{A}$ are not enforced by any correlated constraint. The loss function is defined as:

$$\mathcal{L}_{\Theta,\mathbf{A}} = \frac{1}{2N}\sum_{n=1}^{N} \|\hat{\mathbf{x}}_n - \mathbf{x}_n\|_2^2 = \frac{1}{2N}\sum_{n=1}^{N} \|f(\mathbf{x}_n, \Theta, \mathbf{A}) - \mathbf{x}_n\|_2^2 \tag{13.22}$$

where $\hat{\mathbf{x}}$ is the model output, i.e. the reconstructed input pixel's spectrum.

Like the training of MNN-AE-1 and MNN-AE-2, the gradient of loss with respect to parameters of MNN-BU-1 and MNN-BU-2 also can be computed. The gradients of loss with respect to the last two variables are as follows:

$$\frac{\partial \mathcal{L}}{\partial \mathbf{A}} = (\mathbf{A}\hat{\mathbf{s}}_L - \mathbf{x})\hat{\mathbf{s}}_L^T \tag{13.23}$$

$$\frac{\partial \mathcal{L}}{\partial \hat{\mathbf{s}}_L} = \mathbf{A}^T(\mathbf{A}\hat{\mathbf{s}}_L - \mathbf{x}) \tag{13.24}$$

The gradients for other variables are the same as those in MNN-AE-1 and MNN-AE-2. In order to make the estimated endmembers **A** nonnegative, after each update step, we enforce their negative entities to a tiny value, i.e. $e^{-6}$.

As MNN-BU-1 and MNN-BU-2 are unsupervised NNs, there are two schemes of accomplishing blind unmixing:

1. Select some pixels as training samples for parameter learning of network. After the training procedure of MNN-BU-1 or MNN-BU-2 is finished, all variables/parameters of the network are determined. The learned **A** can be considered as the estimated endmembers. The spectrum of a pixel is input into the trained network, the $\hat{\mathbf{s}}_L$ that is the output item before **A**, can be considered as the estimated abundances of this input pixel.
2. All pixels need to be unmixed are used as training samples, i.e. the training set and the testing set are the same in unsupervised learning. After the training procedure of MNN-BU-1 or MNN-BU-2 is finished, the learned **A** is also considered as the estimated endmembers, and $\hat{\mathbf{s}}_L$ of a training pixel is taken as the abundances of this pixel.

## *13.4.2 Experimental Results on Synthetic Data*

In this part, we present the experimental results on synthetic data to evaluate the effectiveness of the proposed MNN-BU-1 and MNN-BU-2 for blind unmixing. Region-based NMF (R-NMF) [8], matrix-vector nonnegative tensor factorization (MV-NTF) [10], DAEN [17], uDAS [21] are selected as alternative methods for comparison. The synthetic data was generated following the same procedure in the supervised cases.

### 13.4.2.1 Evaluation Index

Besides RMSE, the SAD is also used to evaluate all the methods. SAD measures the dissimilarity between the $r$th endmember $\mathbf{m}_r$ and its associated estimation $\hat{\mathbf{m}}_r$, which is defined as:

$$\text{SAD}_r = \arccos\left(\frac{\mathbf{m}_r^{\text{T}}\hat{\mathbf{m}}_r}{\|\mathbf{m}_r\|\|\hat{\mathbf{m}}_r\|}\right) \tag{13.25}$$

The average SADs and RMSEs of all the endmembers are calculated to compare the performance of unmixing methods.

### 13.4.2.2 Impact of Number of Layers

The synthetic HSI was generated with $\beta = 0.8$ and SNR = 25 dB; 1000 randomly selected pixels are used for training and the number of layers $L$ is set from 1 to 9. Figure 13.9 shows SAD and RMSE of MNN-BU-1 and MNN-BU-2 with respect to the number of layers. With the number of layers changing, MNN-BU-2 is more stable than MNN-BU-1. Still, both of them have good performance when

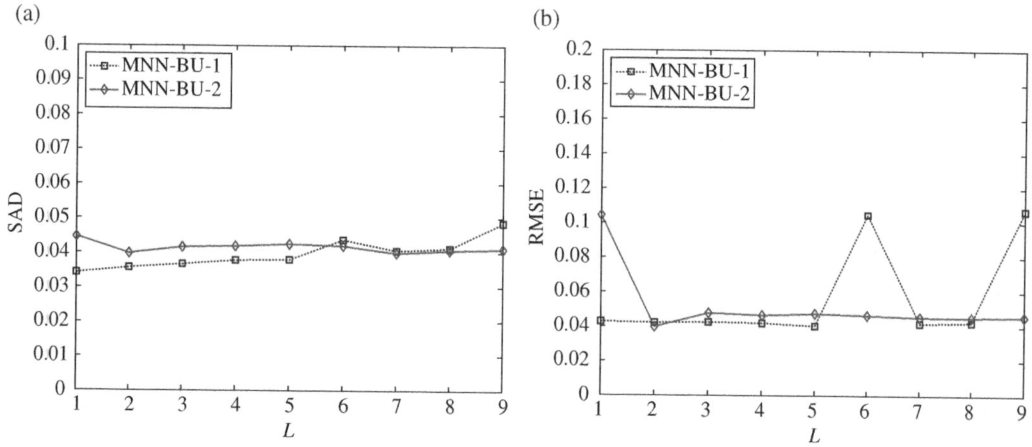


**Figure 13.9** Performance of MNN-BU-1 and MNN-BU-2 with respect to the numbers of layers in terms of (a) SAD, (b) RMSE.

the number of layers is small, implying the strong approximation capability of our networks even with very little size. In the following experiments we set $L = 2$ for blind unmixing.

#### 13.4.2.3 Impact of Number of Training Samples

We set $\beta = 0.8$ and SNR = 25 dB to generate the synthetic data. The training samples with the size $N = \{256, 500, 1000, 2000, 4096\}$ are selected from the data set, respectively, and all pixels are used for testing. The unmixing results of our network are shown in Figure 13.10. With the increase of the number of training samples, MNN-BU-1 and MNN-BU-2 show better unmixing performance, especially the SAD gets more rewards than RMSE with the increase of the size of the training set, possibly

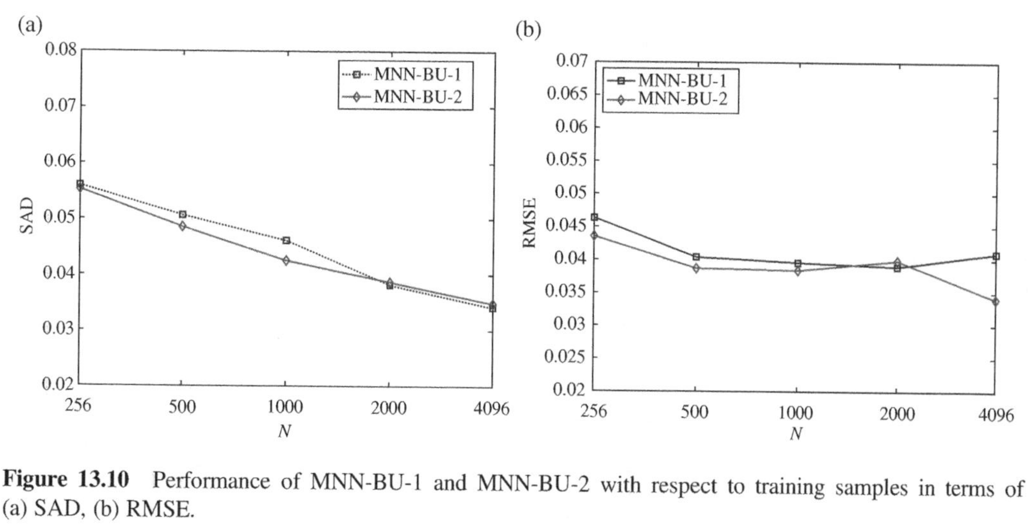


**Figure 13.10** Performance of MNN-BU-1 and MNN-BU-2 with respect to training samples in terms of (a) SAD, (b) RMSE.

because endmember extraction fully relies on all training samples. At the same time, RMSE is more dependent on each pixel. Since MNN-BU-2 is slightly better than MNN-BU-1, we only use MNN-BU-2 in the following experiments.

#### 13.4.2.4 Robustness to Noise

This experiment aims to test the performance of various methods in noisy HSIs. We set $\beta = 0.8$ to produce clean HSI, which is then contaminated by different levels of noise, with SNR = {15 dB, 20 dB, 25 dB, 30 dB, $\infty$}. We used all 4096 pixels for training, the same as the experimental setup for DAEN and uDAS in their original papers. Figure 13.11 shows the performance of all the compared methods, among which MNN-BU-2 has the best performance. MNN-BU-2 has a similar RMSE as DAEN but has better SAD. The NN methods, including MNN-BU-2, DAEN, and uDAS perform better than those based on matrix decomposition and tensor decomposition.

#### 13.4.2.5 Running Time Comparison

The synthetic image was generated by setting $z = 8$, $\beta = 0.8$, and SNR = 25 dB. We trained with all 4096 training samples. Table 13.3 gives the running times of all methods. Because DAEN has two AEs for initialization and blind-source separation, it costs the longest convergence time. MNN-BU-2 and uDAS require less time.

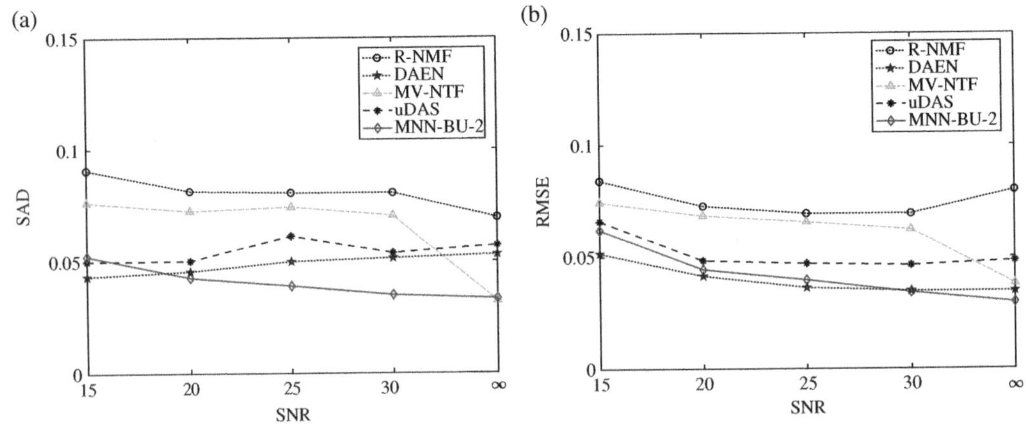


**Figure 13.11** Performance of all five blind unmixing methods with respect to different noise levels in terms of (a) SAD, (b) RMSE.

**Table 13.3** Running time comparison

| | DAEN | R-NMF | MV-NTF | uDAS | MNN-BU-2 |
|---|---|---|---|---|---|
| Time (s) | 6276.4 | 43.6 | 14.4 | 3.1 | 9.3 |

**Table 13.4** Means and standard deviations of the SAD on Jasper Ridge data set

| | R-NMF | MV-NTF | DAEN | uDAS | MNN-BU-2 |
|---|---|---|---|---|---|
| Tree | **0.0407 ± 0.80%** | 0.2209 ± 2.17% | 0.1774 ± 0.82% | 0.1511 ± 1.37% | 0.0468 ± 0.32% |
| Water | 0.1596 ± 0.69% | 0.2352 ± 1.65% | 0.3237 ± 1.42% | 0.1742 ± 11.39% | **0.1179 ± 0.23%** |
| Soil | 0.0867 ± 1.12% | 0.1752 ± 4.67% | 0.1123 ± 2.03% | 0.1639 ± 6.09% | **0.0357 ± 0.43%** |
| Road | 0.7402 ± 0.28% | 0.1741 ± 4.35% | **0.0588 ± 1.59%** | 0.0597 ± 0.27% | 0.0901 ± 0.37% |
| Mean | 0.2568 ± 0.27% | 0.2015 ± 1.77% | 0.1680 ± 0.43% | 0.1373 ± 3.27% | **0.0726 ± 0.11%** |

The bold value means the best performance.

### *13.4.3 Experimental Results on Real-World Data*

In the experiment, 1000 pixels were randomly selected from an HSI to learn the parameters of MNN-BU-2 and all pixels were used for testing. We sampled five times for experiments. Table 13.4 illustrates SAD results of all. MNN-BU-2, DAEN, and uDAS have better performance than R-NMF and MV-NTF. It is further proved that the learning-based blind unmixing methods have obvious advantages over the traditional unmixing methods. Model-inspired MNN-BU-2 achieves the best results among all the methods. According to the estimated endmember and reference endmember in Figure 13.12, it can be obviously found that the endmember estimated by MNN-BU-2 is closest to the reference ones. The abundance maps corresponding to all four endmembers are shown in Figure 13.13.

## 13.5 NMF-Inspired Sparse Autoencoder for Hyperspectral Unmixing

In Section 13.4, a blind unmixing method, MNN-BU, based on ISTA and LMM is introduced. MNN-BU achieves better results than both model-based and DL-based methods. However, the decoder module is interpreted as an additional endmember matrix directly connected with the encoder (abundances), which fails to consider unfolding the optimization of endmembers. Here, we propose an NMF-inspired sparse autoencoder (NMF-SAE) for hyperspectral unmixing [41]. By introducing some learnable weights and uniting them across the layers as parameters, NMF-SAE further simultaneously unfolds the update rules of endmembers and abundances, and achieves improved unmixing performance.

### *13.5.1 Model Optimization*

The optimization problem in Eq. (13.3) can be decomposed into two subproblems, respectively, for estimating **A** and **S**, i.e.

$$\min_{\mathbf{S}} \frac{1}{2}\|\mathbf{X}-\mathbf{A}_0\mathbf{S}\|_F^2 + \lambda\|\mathbf{S}\|_1 \quad \text{s.t. } \mathbf{1}_R^T\mathbf{S} = \mathbf{1}_N^T,\ \mathbf{S} \succeq 0 \tag{13.26}$$

$$\min_{\mathbf{A}} \frac{1}{2}\|\mathbf{X}-\mathbf{A}\mathbf{S}_0\|_F^2 \quad \text{s.t. } \mathbf{A} \succeq 0 \tag{13.27}$$

where $\mathbf{A}_0$ and $\mathbf{S}_0$ are the initial values of **A** and **S**, respectively. Both problems can be solved by ISTA. Taking **A** for example, under the framework of ISTA, **A** can be obtained by recursively executing the following equation until convergence:

$$\mathbf{A}_k = \mathbf{prox}\left(\mathbf{A}_{k-1} - \alpha\mathbf{S}_0^T(\mathbf{A}_{k-1}\mathbf{S}_0 - \mathbf{X})\right) \tag{13.28}$$

where $\alpha > 0$ is the step size usually set as $\frac{1}{\|\mathbf{S}_0^T\mathbf{S}_0\|_2}$. **prox** is an operator guarantying the nonnegativity of **A** and defined as $\mathbf{prox}(\mathbf{a}) = \max(\mathbf{a}, 0)$.

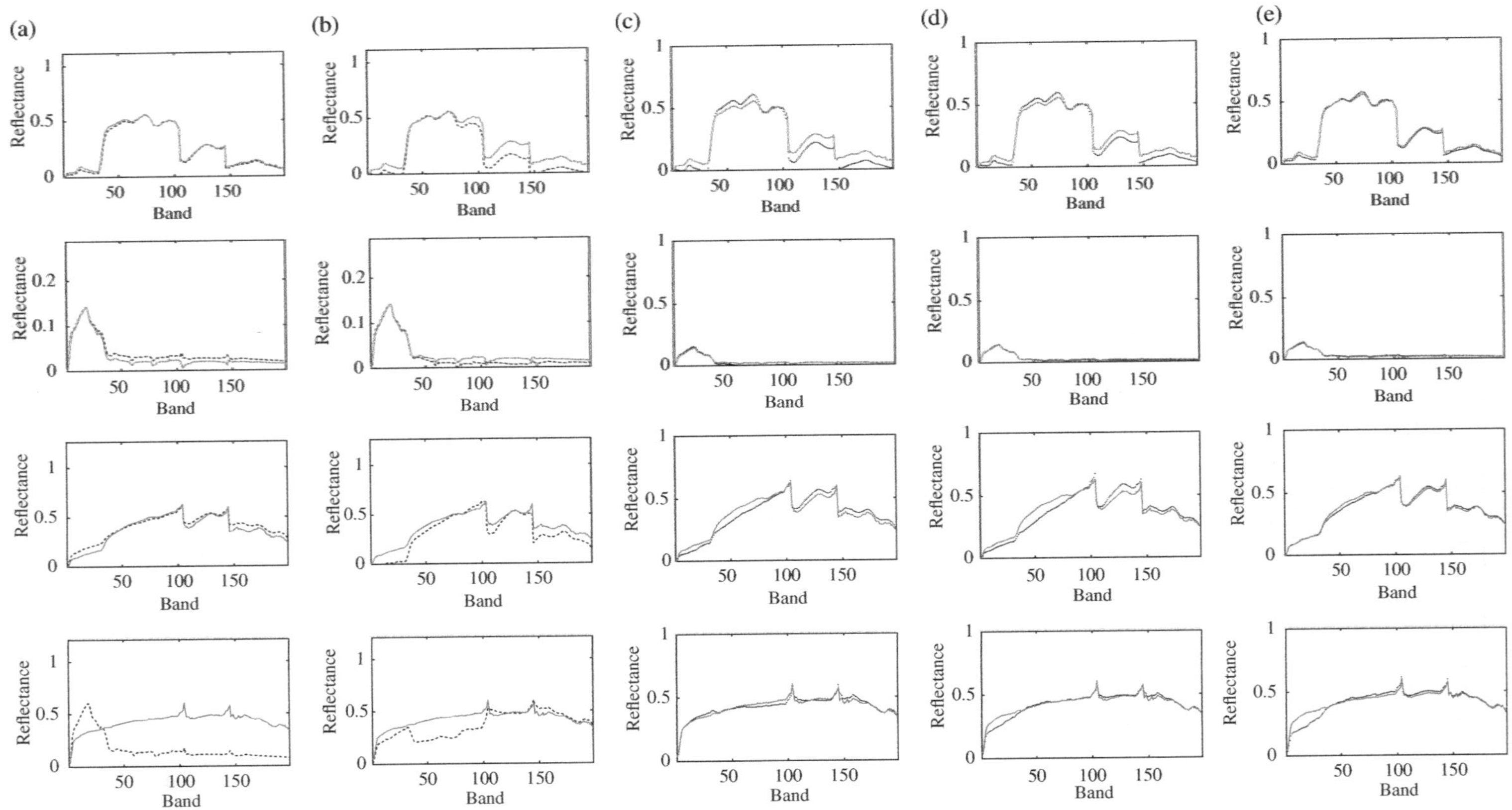


**Figure 13.12** Estimated endmembers by all five blind unmixing methods on the Jasper Ridge data set. From top to bottom, the rows are the spectral signatures of tree, water, soil, and road, respectively. Solid lines denote the reference endmembers and dashed lines denote the estimated endmembers. (a) R-NMF, (b) MV-NTF, (c) DAEN, (d) uDAS, (e) MNN-BU-2.

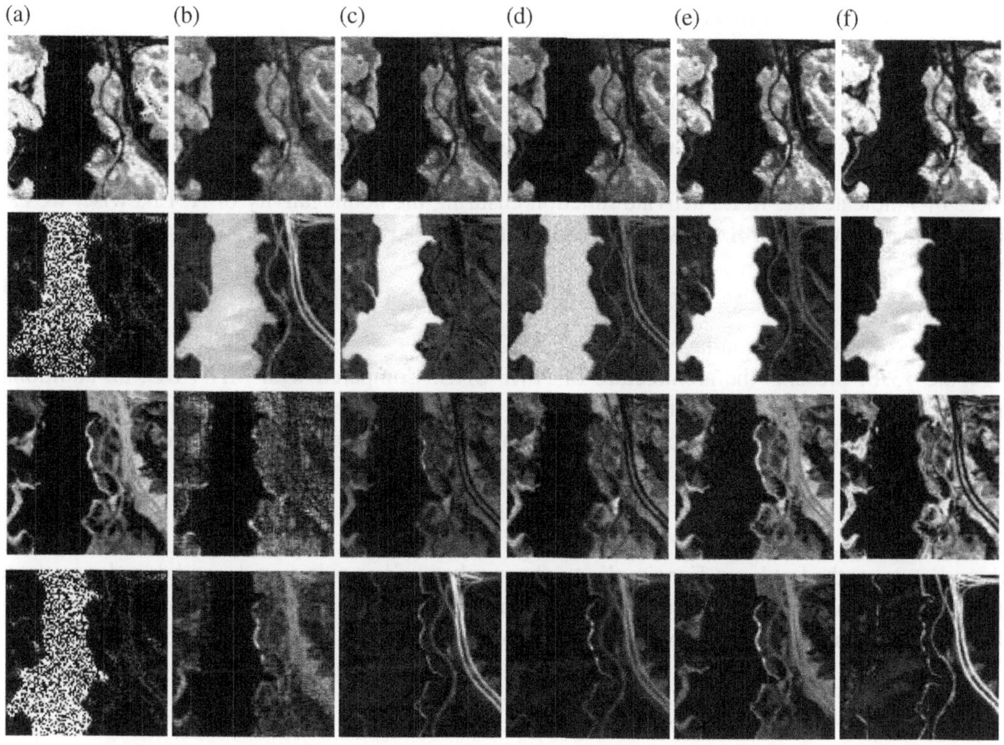


**Figure 13.13** Estimated abundance maps by all five blind unmixing methods on the Jasper Ridge data set. From top to bottom, the rows are the abundance maps of tree, water, soil, and road, respectively. (a) R-NMF, (b) MV-NTF, (c) DAEN, (d) uDAS, (e) MNN-BU-2, (f) reference.

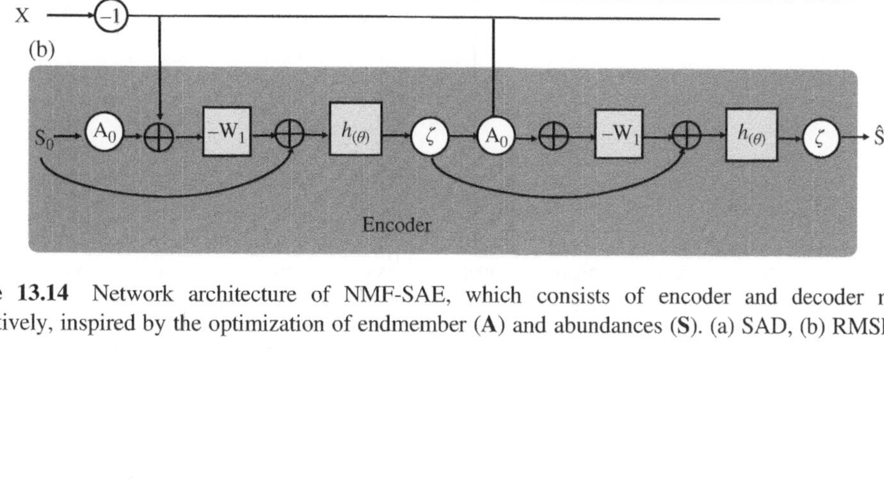


**Figure 13.14** Network architecture of NMF-SAE, which consists of encoder and decoder modules, respectively, inspired by the optimization of endmember (**A**) and abundances (**S**). (a) SAD, (b) RMSE.

### 13.5.2 Network Architecture

By introducing a learnable parameter $\mathbf{W}_2 = \alpha \mathbf{S}_0^T$, Eq. (13.28) can be unfolded into a network structure with a fixed number of layers ($K$), whose consecutive layers are connected by

$$\mathbf{A}_k = \sigma(\mathbf{A}_{k-1} - (\mathbf{A}_{k-1}\mathbf{S}_0 - \mathbf{X})\mathbf{W}_2) \tag{13.29}$$

where $\sigma(\cdot)$ indicates the ReLu activate function ensuring the nonnegativity of $\mathbf{A}$. Likewise, we can obtain the subnetwork for $\mathbf{S}$ whose consecutive layers are given by

$$\mathbf{S}_k = \zeta\left(h_{(\theta)}(\mathbf{S}_{k-1} - \mathbf{W}_1(\mathbf{A}_0\mathbf{S}_{k-1} - \mathbf{X}))\right) \tag{13.30}$$

where $\zeta(\cdot)$ is a normalization layer satisfying ASC.

Figure 13.14 presents the architecture of NMF-SAE. Compared with MNN-BU, an additional unfolded decoder is connected to the encoder. All the parameters can be learned by the training data through backpropagation. The reconstruction error of $\mathbf{X}$ is taken as the loss function, i.e.

$$\mathcal{L}_\Theta = \frac{1}{2}\|\mathbf{A}_K\mathbf{S}_K - \mathbf{X}\|_F^2 \tag{13.31}$$

where $K$ indexes the last layer. For simplicity, the numbers of layers in encoder and decoder are set to the same. Once the network is trained, $\mathbf{A}$ can be obtained by the decoder sub-network. For new samples, their abundance $\mathbf{S}$ can be obtained by the forward process of encoder sub-network.

### 13.5.3 Experimental Results on Synthetic Data

Here we investigate the unmixing performance of NMF-SAE by analyzing the influence of training samples and comparing with the state-of-the-art methods. The synthetic data is generated the same as that in MNN-BU.

#### 13.5.3.1 Influence of the Number of Training Samples

Setting SNR = 25 dB, Figure 13.15 presents the effect of training samples on SAD and RMSE. NMF-SAE can provide very competitive unmixing even with only 256 pixels (6.25% of all pixels), which is

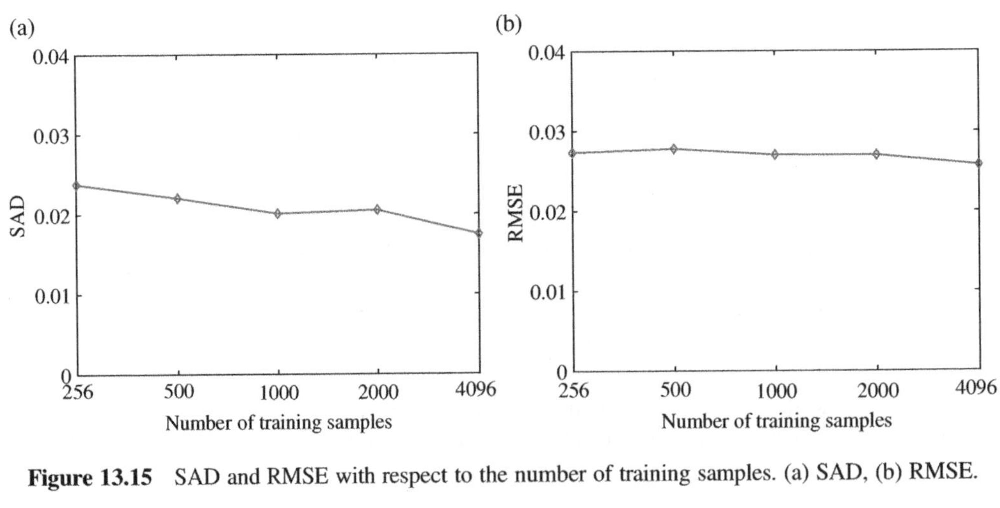


**Figure 13.15** SAD and RMSE with respect to the number of training samples. (a) SAD, (b) RMSE.

a clear indication of the strong learning ability of NMF-SAE. On the other hand, more training time is required as the training sample increases. Considering the balance between training time and unmixing accuracy, we set the number of training samples as 1000 in the rest experiments.

#### 13.5.3.2 Comparison with the State-of-the-Arts

Here, we show the unmixing performance of all competing methods under different SNRs of Gaussian noise. Except for our method, all other learning-based methods use all pixels for training. Figure 13.15 shows the experimental results. In general, the learning-based approach is superior to TV-RSNMF and MV-NTF in most cases, which means that the learning-based approach makes more use of available training data to enhance representation capabilities. Compared with DAEN, uDAS, and MNN-BU, NMF-SAE achieved better SAD and RMSE by using about 25% of pixels for training. The main reason is that NMF-SAE incorporates the linear spectral mixture model into a deep NN and combines the advantages of good interpretation of model-based methods and the strong learning ability of learning-based methods. Additional consideration of unfolding the optimization of endmembers and more residual connections contribute to improved performance over MNN-BU.

### *13.5.4 Experimental Results on Real-World Data*

Table 13.5 compares the unmixing accuracy on Jasper Ridge HSI. Similar to the observation on the synthetic data, the learning-based methods are superior to TV-RSNMF and MV-NTF. Compared with the other three learning-based methods, NMF-SAE obtains relatively more accurate endmembers by providing an average SAD of 0.0671. Figure 13.16 shows a qualitative comparison between abundance obtained by NMF-SAE and corresponding references. As can be seen, the estimated abundances are a perceptual and visual match to the references. In general, the promising unmixing performance of the NMF-SAE real-world data set evidently confirms the unmixing ability of NMF-SAE again.

**Table 13.5** Means and standard deviations of the SAD on Jasper Ridge data set

| Algorithm | TV-RSNMF | MV-NTF | DAEN | uDAS | MNN-BU | NMF-SAE |
|---|---|---|---|---|---|---|
| Tree | 0.0776 ± 4.91% | 0.2209 ± 2.17% | 0.1774 ± 0.82% | 0.1511 ± 1.37% | **0.0468 ± 0.32%** | 0.0494 ± 1.90% |
| Water | 0.1337 ± 1.31% | 0.2352 ± 1.65% | 0.3237 ± 1.42% | 0.1742 ± 11.39% | 0.1179 ± 0.23% | **0.0729 ± 0.50%** |
| Soil | 0.1047 ± 10.11% | 0.1752 ± 4.67% | 0.1123 ± 2.03% | 0.1639 ± 6.09% | **0.0357 ± 0.43%** | 0.0527 ± 4.17% |
| Road | 0.6515 ± 27.46% | 0.1741 ± 4.35% | **0.0588 ± 1.59%** | 0.0597 ± 0.27% | 0.0901 ± 0.37% | 0.0932 ± 0.57% |
| Mean | 0.2419 ± 3.06% | 0.2015 ± 1.77% | 0.1680 ± 0.43% | 0.1373 ± 3.27% | 0.0726 ± 0.11% | **0.0671 ± 1.47%** |

The bold value means the best performance.

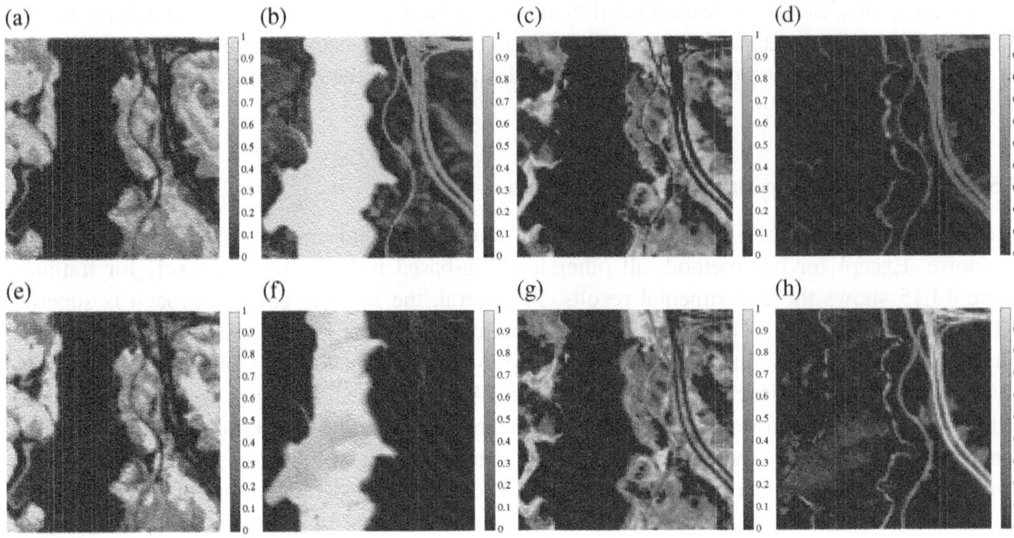


**Figure 13.16** Abundance maps comparison. Estimated abundance map of each endmember: (a) tree, (b) water, (c) soil, (d) road, (e)–(h) are associated reference abundance maps.

## 13.6 Learning a Deep Alternating Neural Network for Hyperspectral Unmixing

The endmembers of MNN-BU and NMF-SAE are hard-coded into the network via a learnable endmember matrix during training. This means once the network is trained, the endmembers are fixed at the testing stage, possibly causing incompatibility between endmembers and abundances, especially in complex scenarios with perturbations. This section presents a novel interpretable deep alternating NN (SNMF-Net) for hyperspectral unmixing to address this problem [44]. It not only equips with a more robust physical model ($L_p$-NMF) but also conforms to the alternating minimization of $L_p$-NMF.

### *13.6.1 Model Formulation*

MNN-BU adopts the $L_1$ norm of $\mathbf{S}$ to produce sparse solution, which tends to punish more on large component. An alternative approach is to use $L_p$ norm ($0 < p \leq 1$), resulting the following $L_p$-NMF-based unmixing model:

$$\mathcal{L}(\mathbf{A}, \mathbf{S}) = \frac{1}{2}\|\mathbf{X} - \mathbf{AS}\|_F^2 + \lambda\|\mathbf{S}\|_p \quad \text{s.t. } \mathbf{1}_R^T\mathbf{S} = \mathbf{1}_N^T, \ \ \mathbf{A}, \mathbf{S} \succeq 0 \tag{13.32}$$

where

$$\|\mathbf{S}\|_p = \sum_{r=1}^{R}\sum_{n=1}^{N}(s_{rn})^p \tag{13.33}$$

### 13.6.2 Model Optimization

Generally, alternating least square (ALS) algorithm [45] is a powerful tool to solve Eq. (13.32). ALS alternately optimizes the endmembers and abundances with the other variable fixed. Specifically, fixing **S**, the subproblem for optimizing **A** corresponds to

$$\mathbf{A}_{k+1} = \min_{\mathbf{A}} \frac{1}{2} \|\mathbf{X} - \mathbf{A}\mathbf{S}_k\|_F^2, \quad \text{s.t.,} \quad \mathbf{A} \succeq 0 \tag{13.34}$$

where $k$ refers to the iteration number. The update rule of **A** is the same as that in NMF-SAE, i.e.

$$\mathbf{A}_{k+1} = \left[\mathbf{A}_k - t_{1_k}(\mathbf{A}_k\mathbf{S}_k - \mathbf{X})\mathbf{S}_k^T\right]_+ \tag{13.35}$$

where $t_{1_k}$ is the step size and $[\cdot]_+ = \max(\cdot, 0)$ guarantees the nonnegativity of **A**.

Similarly, when **A** is fixed, we arrive at the optimization problem for **S**:

$$\mathbf{S}_{k+1} = \min_{\mathbf{S}} \frac{1}{2} \|\mathbf{X} - \mathbf{A}_{k+1}\mathbf{S}\|_F^2 + \lambda\|\mathbf{S}\|_p, \quad \text{s.t.,} \quad \mathbf{S} \succeq 0, \mathbf{1}_R^T\mathbf{S} = \mathbf{1}_N^T \tag{13.36}$$

Equation (13.36) can be decomposed into two parts, respectively, accounting for nonnegative sparsity constraint and sum-to-one constraint. The solution of **S** concerning nonnegative sparsity constraint is

$$\mathbf{S}_{k+1} = \left[\text{GST}_{(\lambda,p)}\left(\mathbf{S}_k - t_{2_k}\mathbf{A}_{k+1}^T(\mathbf{A}_{k+1}\mathbf{S}_k - \mathbf{X})\right)\right]_+ \tag{13.37}$$

where $t_{2_k}$ is also the step size. Parameterized by $\lambda$ and $p$, $\text{GST}_{(\lambda, p)}(\cdot)$ is the generalized shrinkage thresholding (GST) operator for solving the following $L_p$ norm regularized sparse minimization problem:

$$\min_s \frac{1}{2}(y - s)^2 + \lambda|s|^p \tag{13.38}$$

The close-form definition of $\text{GST}_{(\lambda, p)}(\cdot)$ is

$$s = \begin{cases} \text{sign}(y)S_p^{\text{GST}}(|y|; \lambda), & \text{if } |y| > \Delta_p^{\text{GST}}(\lambda) \\ 0, & \text{otherwise} \end{cases} \tag{13.39}$$

where the threshold $\Delta_p^{\text{GST}}(\lambda)$ is given by

$$\Delta_p^{\text{GST}}(\lambda) = (2\lambda(1-p))^{\frac{1}{2-p}} + \lambda p(2\lambda(1-p))^{\frac{p-1}{2-p}} \tag{13.40}$$

Zuo et al. [46] have given the solution of GST($|y|$;$\lambda$) by performing the following iteration steps:

$$x_{t+1} = |\, y \,| - \lambda p x_t^{p-1} \tag{13.41}$$

where $x_0 = |y|$. After every iteration, the columns of $\mathbf{S}_{k+1}$ are normalized to satisfy the ASC constrain.

## *13.6.3 SNMF-NET*

In this section, we first establish the connection between the proximal gradient method and deep NN, then introduce the details of SNMF-NET and address its learning and implementation issues.

### 13.6.3.1 Connection Between Proximal Gradient Method and DNN

As shown in Figure 13.17, the proposed SNMF-Net is an LMM-based network and alternately optimizes endmembers and abundances by End-Net and Abun-Net. Both End-Net and Abun-Net are obtained by unfolding the iterative optimization of **A** and **S**. In essence, the optimization problems of **A** and **S** share the following unconstrained problem:

$$\min_{\mathbf{u}} \frac{1}{2}\|\mathbf{B}\mathbf{u}-\mathbf{v}\|_2^2 + f(\mathbf{u}) \tag{13.42}$$

where the first part is convex and differentiable, representing the reconstruction term and the second part is neither necessarily convex nor differentiable but has a fast proximal operator, representing the regularization term.

Proximal gradient algorithm is a widely used framework for such problem with the following update rule comprising of two fundamental steps, i.e.

$$\mathbf{u}' = \mathbf{u}_k - t\mathbf{B}^T(\mathbf{B}\mathbf{u}_k - \mathbf{v}) \tag{13.43}$$

$$\mathbf{u}_{k+1} = \mathbf{prox}_{tf}(\mathbf{u}') \tag{13.44}$$

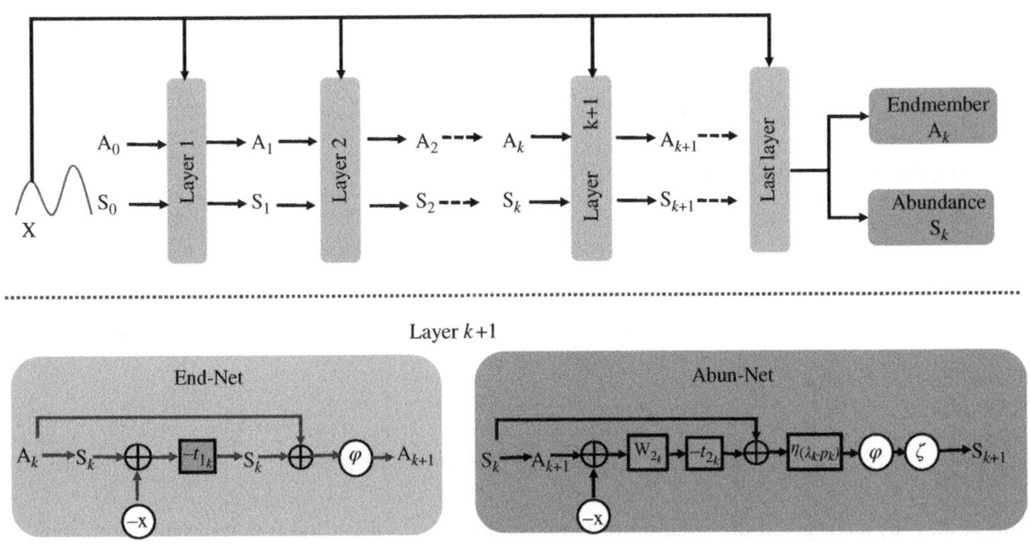


**Figure 13.17** The framework of proposed SNMF-Net. Based on $L_p$ norm-based sparse unmixing model, SNMF-Net takes the observed spectrum as the input and alternately predicts the endmembers and abundances using End-Net and Abun-Net submodules. All the parameters (boxed) can be linked with the optimization of $L_p$-NMF and the unmixing problem and support data-driven training.

where **prox** is the proximal operator, for example soft-thresholding operator, and $t > 0$ is the step size. Running Eqs. (13.43) and (13.44), a number of iterations till convergence yield the solution of **u**.

From another perspective, the two steps of updating **u** can be associated with a typical local block of a recurrent NN (RNN). Specifically, Eq. (13.43) accounts for a linear operation to compute an update in the direction of the gradient, while Eq. (13.44) mimics an element-wise nonlinearity with a proximal operator **prox**. Soft threshold operator, for example, can be defined as $\mathcal{S}_\lambda(u) = \text{ReLu}(u - \lambda) - \text{ReLu}(-u - \lambda)$. When the number of iterations is truncated to $K$, and the RNN is unfolded $K$ times, the network becomes a $K$-layer feedforward NN and supports discriminative learning. Introducing two differentiable variables, respectively, responsible for step size $t$ and $\mathbf{W} = \mathbf{B}^T$ as network parameters, Eqs. (13.43) and (13.44) can be unfolded into a network with consecutive layers being a function parameterized by $\{\mathbf{W}, t\}$ such that

$$\mathbf{u}_{k+1} = \varphi(\mathbf{u}_k - t\mathbf{W}(\mathbf{B}\mathbf{u}_k - \mathbf{v})) \tag{13.45}$$

Here, $\varphi(\cdot)$ is a nonlinear activation function and $k$ indexes the layer corresponding to $k$th iteration in the iterative update procedure.

### 13.6.3.2 $L_p$-NMF Inspired Deep Alternating Neural Network

Following the above perspective, we introduce the detailed network architecture for endmembers and abundances.

**End-Net for Endmembers:** Based on the iterative update in Eq. (13.35), the unfolded subnetwork for endmember can be defined by recursion:

$$\mathbf{A}_{k+1} = \sigma(\mathbf{A}_k - t_{1_k}(\mathbf{A}_k\mathbf{S}_k - \mathbf{X})\mathbf{W}_{1_k})) \tag{13.46}$$

where $\sigma(\cdot) = \max(.,0)$ is the ReLu activation function allowing for the nonnegativity of **A**.

From Eq. (13.46), we can find that $W_{1_k}$ is the same size as the input sample, indicating that the network can only process a fixed number of pixels at a time to satisfy the matrix multiplication. In practice, we always expect the network to accept any number of pixels for testing. In addition, End-Net produces a set of endmembers for a given pixel, which may change when different sets of pixels are provided. This does not make sense for the unmixing task because all pixels in the scene should share the same set of endmembers. To meet the above requirements, we replace $\mathbf{W}_{1_k}$ with $\mathbf{S}_k^T$, so that the network can accept any number of pixels, and all pixels share the same set of endmembers. Accordingly, the network defined in Eq. (13.46) follows the iteration

$$\mathbf{A}_{k+1} = \varphi\left(\mathbf{A}_k - t_{1_k}(\mathbf{A}_k\mathbf{S}_k - \mathbf{X})\mathbf{S}_k^T\right) \tag{13.47}$$

Unlike the original network defined in Eq. (13.46), only the step sizes $\{t_{1_k}\}$ are required to learn in the network defined by Eq. (13.47), significantly simplifying network training.

**Abun-Net for Abundances:** Different from **A**, the update rule of **S** contains an additional iterative optimization of $S_p^{\text{GST}}$, which makes it very difficult to analyze the relationship between $\mathbf{S}_k$ and $(\lambda, p)$. For simplicity, we only iterate Eq. (13.41) for one time, yielding the following solution for GST:

$$\eta_{(\lambda,p)}(y) = \begin{cases} \text{sign}(y)\left(|y| - \lambda p(|y|)^{p-1}\right), & \text{if } |y| > \Delta_p^{\text{GST}}(\lambda) \\ 0, & \text{otherwise} \end{cases} \tag{13.48}$$

As MNN-BU, we also choose different parameters $\{\lambda_k, p_k\}$ across the iteration of $\mathbf{S}$. Consequently, the updating rule for $\mathbf{S}_{k+1}$ in Eq. (13.37) becomes:

$$\mathbf{S}_{k+1} = \left[\eta_{(\lambda_k,p_k)}\left(\mathbf{S}_k - t_{2_k}\mathbf{A}_{k+1}^T(\mathbf{A}_{k+1}\mathbf{S}_k - \mathbf{X})\right)\right]_+ \tag{13.49}$$

Accordingly, the succeeding layers in Abun-Net are given by

$$\mathbf{S}_{k+1} = \zeta\left(\sigma\left(\eta_{(\lambda_k,p_k)}\left(\mathbf{S}_k - t_{2_k}\mathbf{W}_{2_k}(\mathbf{A}_{k+1}\mathbf{S}_k - \mathbf{X})\right)\right)\right) \tag{13.50}$$

where $\zeta(\cdot)$ is a normalization layer satisfying ASC. The parameter set of Abun-Net is $\{t_{2_k}, \mathbf{W}_{2_k}, p_k, \lambda_k\}$. Instead of inefficiently tuning $p$ to obtain the desired sparsity level, $p$ is also layer-wise learned from data, offering better adaptation to the underlying properties of the data.

**Network Training:** SNMF-Net is trained by minimizing the Euclidean distance between the reconstructed and input spectrum. Since adding an auxiliary classifier to the middle layer helps propagate the gradient back to all layers [47], we put the output of all layers into the loss function. Given training data set $\mathbf{X} = \{\mathbf{x}_1, \ldots, \mathbf{x}_N\}$ including $N$ spectrum, the loss function can be defined as

$$\mathcal{L}_\Theta = \frac{1}{2N}\sum_{k=1}^{K} \|\mathbf{X} - \mathbf{A}_k\mathbf{S}_k\|_F^2 \tag{13.51}$$

where $\Theta = \{t_{1_k}, t_{2_k}, \mathbf{W}_{2_k}, p_k, \lambda_k\}$ are the parameters to be learned and $K$ is the number of layers. To learn the network, the partial derivative of $\mathcal{L}_\Theta$ concerning each variable should be computed first. For simplicity, we consider an individual training sample $\mathbf{x}$ and define the following intermediate variables:

$$\begin{aligned}
\widetilde{\mathbf{A}}_k &= \mathbf{A}_{k-1} - t_{1_{k-1}}(\mathbf{A}_{k-1}\mathbf{s}_{k-1} - \mathbf{x})\mathbf{s}_{k-1}^T \\
\widetilde{\mathbf{s}}_k &= \mathbf{s}_{k-1} - t_{2_{k-1}}\mathbf{W}_{2_{k-1}}(\mathbf{A}_k\mathbf{s}_{k-1} - \mathbf{x}) \\
\hat{\mathbf{s}}_k &= \varphi\left(\eta_{(\lambda_{k-1},p_{k-1})}(\widetilde{\mathbf{s}}_k)\right)
\end{aligned} \tag{13.52}$$

The partial differentials of $\mathcal{L}_\Theta$ with respect to each variable are as follows:

$$\begin{aligned}
&\frac{\partial \mathcal{L}}{\partial t_{2_k}} = \frac{\partial \widetilde{\mathbf{s}}_{k+1}}{\partial t_{2_k}}\frac{\partial \mathcal{L}}{\partial \widetilde{\mathbf{s}}_{k+1}} \qquad \frac{\partial \mathcal{L}}{\partial p_k} = \frac{\partial \widetilde{\mathbf{s}}_{k+1}}{\partial p_k}\frac{\partial \mathcal{L}}{\partial \widetilde{\mathbf{s}}_{k+1}} \\
&\frac{\partial \mathcal{L}}{\partial \lambda_k} = \frac{\partial \widetilde{\mathbf{s}}_{k+1}}{\partial \lambda_k}\frac{\partial \mathcal{L}}{\partial \widetilde{\mathbf{s}}_{k+1}} \qquad \frac{\partial \mathcal{L}}{\partial \mathbf{W}_{2_k}} = \frac{\partial \mathcal{L}}{\partial \widetilde{\mathbf{s}}_{k+1}}\frac{\partial \widetilde{\mathbf{s}}_{k+1}}{\partial \mathbf{W}_{2_k}} \\
&\frac{\partial \mathcal{L}}{\partial \widetilde{\mathbf{s}}_{k+1}} = \frac{\partial \hat{\mathbf{s}}_{k+1}}{\partial \widetilde{\mathbf{s}}_{k+1}}\frac{\partial \mathbf{s}_{k+1}}{\partial \hat{\mathbf{s}}_{k+1}}\frac{\partial \mathcal{L}}{\partial \mathbf{s}_{k+1}} \qquad \frac{\partial \mathcal{L}}{\partial \widetilde{\mathbf{A}}_{k+1}} = \frac{\partial \mathcal{L}}{\partial \mathbf{A}_{k+1}}\frac{\partial \mathbf{A}_{k+1}}{\partial \widetilde{\mathbf{A}}_{k+1}} \\
&\frac{\partial \mathcal{L}}{\partial t_{1_k}} = \frac{\partial \widetilde{\mathbf{A}}_{k+1}}{\partial t_{1_k}}\frac{\partial \mathcal{L}}{\partial \widetilde{\mathbf{A}}_{k+1}}
\end{aligned} \tag{13.53}$$

$\frac{\partial \mathbf{A}_{k+1}}{\partial \widetilde{\mathbf{A}}_{k+1}}$ and $\frac{\partial \hat{\mathbf{s}}_{k+1}}{\partial \widetilde{\mathbf{s}}_{k+1}}$ are both related to corresponding activation functions and can be respectively obtained by

$$\frac{\partial \mathbf{A}_{k+1}}{\partial \widetilde{\mathbf{A}}_{k+1}} = \mathcal{I}_{\widetilde{\mathbf{A}}_{k+1} > 0} \qquad \frac{\partial \hat{\mathbf{s}}_{k+1}}{\partial \widetilde{\mathbf{s}}_{k+1}} = \mathcal{I}_{\eta_{(\lambda_k,p_k)}(\widetilde{\mathbf{s}}_{k+1}) > 0} \tag{13.54}$$

where $\mathcal{I}$ is an indicator function. According to Eq. (13.52), we can get

$$\frac{\partial \widetilde{\mathbf{s}}_{k+1}}{\partial t_{2_k}} = \mathbf{W}_{2_k}(\mathbf{x} - \mathbf{A}_{k+1}\mathbf{s}_{k+1}) \quad \frac{\partial \widetilde{\mathbf{A}}_{k+1}}{\partial t_{1_k}} = (\mathbf{x} - \mathbf{A}_k\mathbf{s}_k)\mathbf{s}_k^T$$
$$\frac{\partial \widetilde{\mathbf{s}}_{k+1}}{\partial \mathbf{W}_{2_k}} = t_{2_k}(\mathbf{x} - \mathbf{A}_{k+1}\mathbf{s}_k) \tag{13.55}$$

Based on Eq. (13.48), each element of $\frac{\partial \widetilde{\mathbf{S}}_{k+1}}{\partial p_{k+1}}$ and $\frac{\partial \widetilde{\mathbf{S}}_{k+1}}{\partial \lambda_{k+1}}$ can, respectively, be derived by

$$\frac{\partial \widetilde{s}_{k+1}}{\partial p_k} = \begin{cases} 0, & \widetilde{s}_{k+1} \leq \Delta_{p_k}^{\mathrm{GST}}(\lambda_k) \\ -\lambda \mathrm{sign}(\widetilde{s}_{k+1})|\widetilde{s}_{k+1}|^{p-1}(p\ln(|\widetilde{s}_{k+1}|) + 1), & \text{otherwise} \end{cases} \tag{13.56}$$

$$\frac{\partial \widetilde{s}_{k+1}}{\partial \lambda_k} = \begin{cases} 0, & \widetilde{s}_{k+1} \leq \Delta_{p_k}^{\mathrm{GST}}(\lambda_k) \\ -\mathrm{sign}(\widetilde{s}_{k+1})p(|\widetilde{s}_{k+1}|)^{p-1} & \text{otherwise} \end{cases} \tag{13.57}$$

Based on ASC, $\frac{\partial \mathbf{s}_{k+1}}{\partial \hat{\mathbf{s}}_{k+1}}$ is

$$\frac{\partial \mathbf{s}_{k+1}}{\partial \hat{\mathbf{s}}_{k+1}} = \frac{1}{T}\mathbf{I} - \frac{1}{T^2}\hat{\mathbf{s}}_{k+1}\mathbf{1}_R^T \tag{13.58}$$

where $T = \sum \hat{\mathbf{s}}_{k+1}$.

$\frac{\partial \mathcal{L}}{\partial \mathbf{A}_k}$ and $\frac{\partial \mathcal{L}}{\partial \mathbf{s}_k}$ can be respectively computed as

$$\begin{aligned}
\frac{\partial \mathcal{L}}{\partial \mathbf{s}_K} &= \mathbf{A}_K^T(\mathbf{A}_K\mathbf{s}_K - \mathbf{x}) \\
\frac{\partial \mathcal{L}}{\partial \mathbf{A}_K} &= (\mathbf{A}_K\mathbf{s}_K - \mathbf{x})\mathbf{s}_K^T - t_{2_{K-1}}\mathbf{W}_{2_{K-1}}^T\frac{\partial \mathcal{L}}{\partial \widetilde{\mathbf{s}}_K}\mathbf{s}_{K-1}^T \\
\frac{\partial \mathcal{L}}{\partial \mathbf{s}_k} &= \mathbf{A}_k^T(\mathbf{A}_k\mathbf{s}_k - \mathbf{x}) + t_{1_k}(\mathbf{x} - 2\mathbf{A}_k\mathbf{s}_k)\frac{\partial \mathcal{L}}{\partial \widetilde{\mathbf{A}}_{k+1}} \\
&\quad + (\mathbf{I} - t_{2_k}\mathbf{W}_{2_k}\mathbf{A}_{k+1})\frac{\partial \mathcal{L}}{\partial \widetilde{\mathbf{s}}_{k+1}}, \quad k = 1 : K-1 \\
\frac{\partial \mathcal{L}}{\partial \mathbf{A}_k} &= (\mathbf{A}_k\mathbf{s}_k - \mathbf{x})\mathbf{s}_k^T + \frac{\partial \mathcal{L}}{\partial \widetilde{\mathbf{A}}_{k+1}}(\mathbf{I} - t_{1_k}\mathbf{s}_k\mathbf{s}_k^T) \\
&\quad - t_{2_{k-1}}\mathbf{W}_{2_{k-1}}^T\frac{\partial \mathcal{L}}{\partial \widetilde{\mathbf{s}}_k}\mathbf{s}_{k-1}^T, \quad k = 1 : K-1
\end{aligned} \tag{13.59}$$

Substituting Eqs. (13.54), (13.55), (13.56), (13.57), (13.59) into Eq. (13.53), we can get the derivations of all the variables. Once they are obtained, all the parameters in Θ can be learned by end-to-end training via any gradient descent method. The training process can be summarized as follows:

1. Estimate the endmember matrix $\mathbf{A}_0$ using any endmember extraction algorithm, e.g. VCA.

2. Randomly sample a number of pixels to construct the training set $\mathbf{X} = \{\mathbf{x}_1, \ldots, \mathbf{x}_N\}$ and calculate their initial abundances $\mathbf{S}_0$ by any traditional unmixing methods, for example FCLS.
3. Initialize the network with $\mathbf{A}_0$ and $\mathbf{S}_0$. In contrast to alternative deep learning based methods, the embedded models make it easy to integrate the prior knowledge on unmixing into the network for enhanced learning.
4. Train SNMF-Net with stochastic gradient descent algorithm till convergence to learn the network paprameters.

**Network Implementation:** Due to the high interpretability of SNMF-Net, prior knowledge can be reasonably integrated into the network to achieve satisfactory unmixing. NMF-based unmixing methods usually use endmembers extracted from VCA [2] and abundance estimates from FCLS reference [4] to initialize factor matrices. For this reason, we also adopt this strategy to initialize $\mathbf{A}_0$ and $\mathbf{S}_0$. After that, $t_{1_k}$ and $t_{2_k}$ are accordingly set as $\frac{1}{\|\mathbf{S}_0\mathbf{S}_0^T\|_2 + \epsilon}$ and $\frac{1}{\|\mathbf{A}_0^T\mathbf{A}_0\|_2 + \epsilon}$ where $\epsilon$ is a tiny number given by 0.01. $p$ is initially assigned as 0.5 based on the observation in [40].

We initialize $\mathbf{W}_{2_k}$ following mutual-coherence minimization in sparse representation theory. In sparse representation, mutual coherence is an important index to describe the sparse recovery quality of dictionary defined by

$$\mu(\mathbf{D}) = \max_{1 \le i \ne j \le L} \frac{\mathbf{D}_i^T\mathbf{D}_j}{\|\mathbf{D}_i\|_2\|\mathbf{D}_j\|_2} \tag{13.60}$$

A dictionary with low coherence implies better sparse recovery ability. Based on above definition, we define a generalized mutual coherence as follows:

**Definition 13.1** For a dictionary $\mathbf{D} \in \mathbb{R}^{L \times R}$, its generalized mutual coherence (GMC) is given by

$$\hat{\mu}(\mathbf{D}) = \inf_{\substack{\mathbf{W} \in \mathbb{R}^{L \times R} \\ \mathbf{W}_i^T\mathbf{D}_i = 1,\, 1 \le i \le R}} \left\{ \max_{1 \le i \ne j \le R} \mathbf{W}_i^T\mathbf{D}_j \right\} \tag{13.61}$$

As pointed out in [48], if $\mathbf{W} \in \mathcal{W}(\mathbf{D})$ where $\mathcal{W}(\mathbf{D}) = \{\mathbf{W} \in \mathbb{R}^{L \times R} : \text{attains the infium in Eq.(61)}\}$, the sparsity recovery error is bounded and $\mathbf{W}$ can be precomputed by solving the following problem:

$$\arg\min_{\mathbf{W} \in \mathbb{R}^{L \times R}} \|\mathbf{W}^T\mathbf{D}\|_F^2 \quad \text{s.t.} \mathbf{W}^\mathbf{T}\mathbf{D} = \mathbf{I} \tag{13.62}$$

In this way, there is no need to learn $\mathbf{W}$ in a data-driven manner, significantly reducing the training complexity and improving the robustness of a network.

When it comes to unmixing, the endmembers can be considered as dictionaries. Intuitively, we can calculate $\mathbf{W}_{2_k}$ using Eq. (13.62) when the endmembers are ready. However, the endmembers are also unknown in the blind unmixing task, meaning that it is unrealistic to obtain an exact solution of $\mathbf{W}_{2_k}$ in advance. Fortunately, as mentioned above, VCA can provide a good initial estimation of endmembers. For this regard, we alternatively still keep $\mathbf{W}_{2_k}$ learnable and initial them as the values computed by Eq. (13.62). By this means, the prior knowledge on spare representation theory is integrated into the network, enhancing network learning as shown in the experiment section.

### 13.6.4 Experimental Results on Synthetic Data

In this section, we comprehensively evaluate the unmixing ability of the proposed SNMF-Net on both synthetic data. The synthetic data is generated in the same way mentioned above. Seven methods are selected as alternative methods to be compared with our method, including Arctan-NMF [49], TV-RSNMF [42], MV-NTF [10], uDAS [21], UnDIP [29], DAEN [17], and MNN-BU [43].

#### 13.6.4.1 The Influence of Layers

We first analyze the influence of layers on unmixing accuracy. The clean HSI was degraded by Gaussian noise with SNR = 30 dB and the training size was set to 500. Figure 13.18 shows the changes of SAD and RMSE when altering $K$ from 3 to 30 with an interval of 3. It can be seen that when $K$ increases from 3 to 9, both SAD and RMSE decrease rapidly, indicating the positive effect of deep models. After reaching the bottom at $K = 9$, SAD gradually increases, and RMSE remains almost stable. This may be because network parameters increase with the number of layers, which requires more training data. Since the best SAD and RMSE are obtained when $K = 9$, we set the number of layers to 9 in the following experiments. This experiment shows very few layers of SNMF-Net can produce promising unmixing, confirming its strong representing ability.

When $K = 9$, we demonstrate the learned $p$ concerning the number of layers. As shown in Figure 13.19, in the first few layers, $L_p$ norm is close to $L_0$ norm, indicating that the network tends to obtain a very sparse estimation of abundance maps, which helps to suppress the side-effect of perturbation. When $K$ is greater than 7, the $p$ value increases to 0.2636, which means the network is learning to generate less sparse solutions. This is also in line with that as the number of iterations increases, the optimization algorithm gradually converges, and smaller penalties should be added. Moreover, compared with the fixed $L_1$ norm, our network is more adaptive to the data as $p$ is learned. Unlike the $L_1$ norm, the proposed can yield sparser and unbiased abundances because the gradient of learned $L_p$ norm decreases as $x$ increases, which means large magnitude components are less likely to be shrunk to 0. In short, this experiment verifies the interpretability of the SNMF-Net.

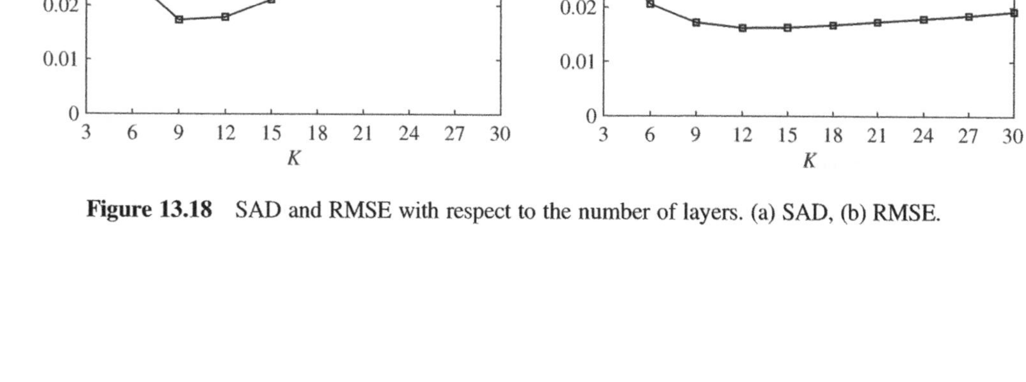


**Figure 13.18** SAD and RMSE with respect to the number of layers. (a) SAD, (b) RMSE.

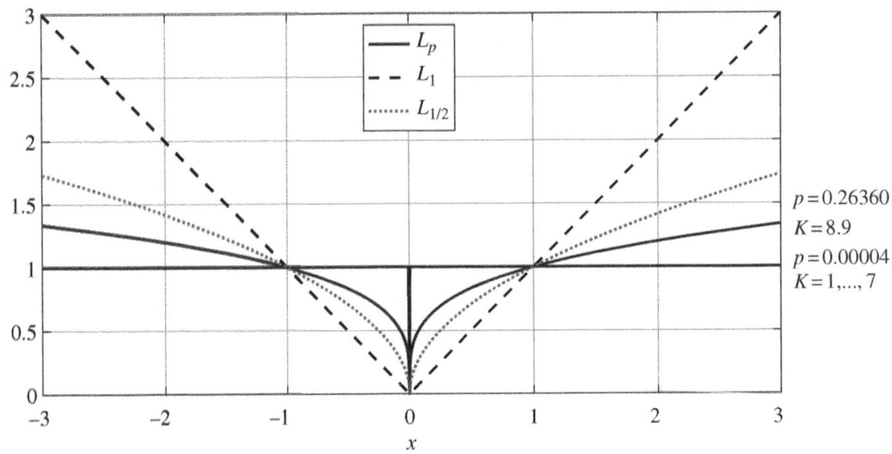


**Figure 13.19** Learned $L_p$ sparse penalty with respect to $K$ values.

#### 13.6.4.2 Influence of the Number of Training Samples

In this experiment, we evaluate the effect of training samples on unmixing. We set SNR = 30 dB to generate the noisy HSI. Figure 13.20 plots the SAD and RMSE changes with the number of training samples. As can be seen from Figure 13.20, without using all the pixels for training, i.e. only 500 pixels (12.21% of all the pixels), the network can achieve as satisfying a performance as unmixing with all pixels. This phenomenon clearly shows the strong learning and generalization ability of SNMF-Net. The main reason is that SNMF-NET considers the spectrum mixing model and $L_p$-NMF alternate optimization to construct the network architecture. All the key components of the SNMF-Net are interpretable, helping to integrate prior knowledge of unmixing into the network to enhance learning.

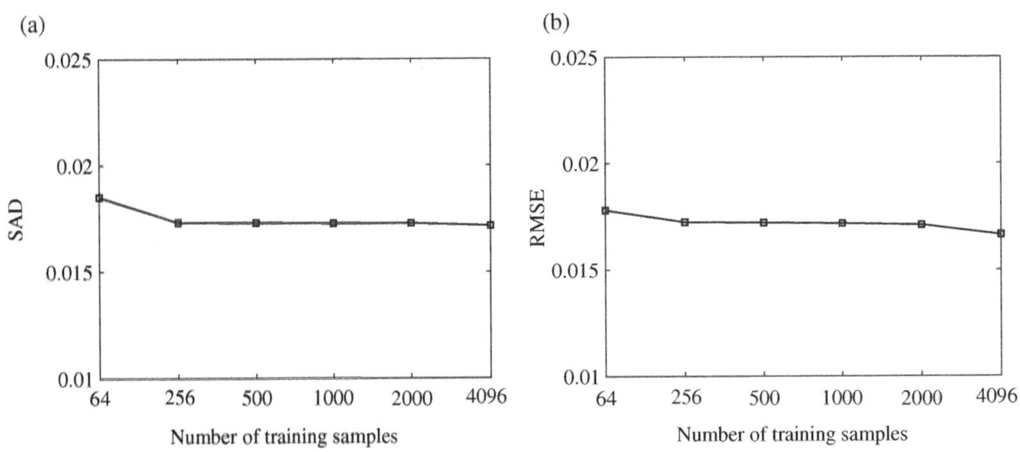


**Figure 13.20** SAD and RMSE with respect to the number of training samples. (a) SAD, (b) RMSE.

#### 13.6.4.3 Influence of Initialization

As described in the network implementation, we use Eq. (13.62) to initialize $\mathbf{W}_{2_k}$, which is induced by GMC theory to enhance learning. Here, we examine the effects of this initialization. The number of training samples was fixed at 500, and the SNR was changed from 15, 20, 25, and 30 dB to ∞ (no noise). Figure 13.21 reports the SAD and RMSE changes with or without GMC-based initialization. Here, "SNMF-Net without GMC" refers to setting $\mathbf{W}_{2_k} = \mathbf{A}^T$. It can be clearly observed that GMC-based initialization helps SNMF-Net achieve significantly better and more robust unmixing performance. This experiment shows GMC-based initialization's effectiveness and the benefit of interpretable network architecture.

Attributing to the embedded physical model, the prior knowledge learned or based on domain knowledge concerning unmixing can be flexibly incorporated into the network to enhance network learning. We further initialized the network by NMF whose results are denoted as SNMF-Net-NMF. Figure 13.22 compares SAD and RMSE with respect to different initializations. As can be seen, using NMF to initialize the network yields lower SAD and RMSE because NMF can provide more accurate

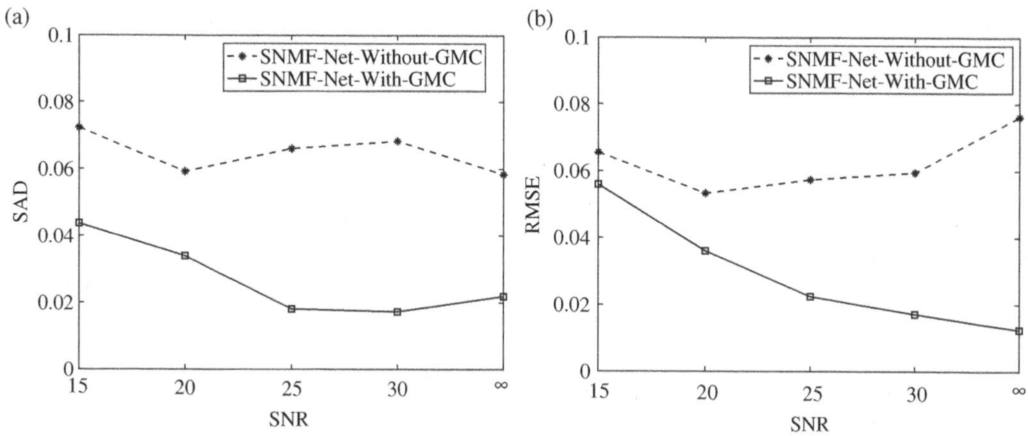


**Figure 13.21** The impact of GMC-based initialization. (a) SAD, (b) RMSE.

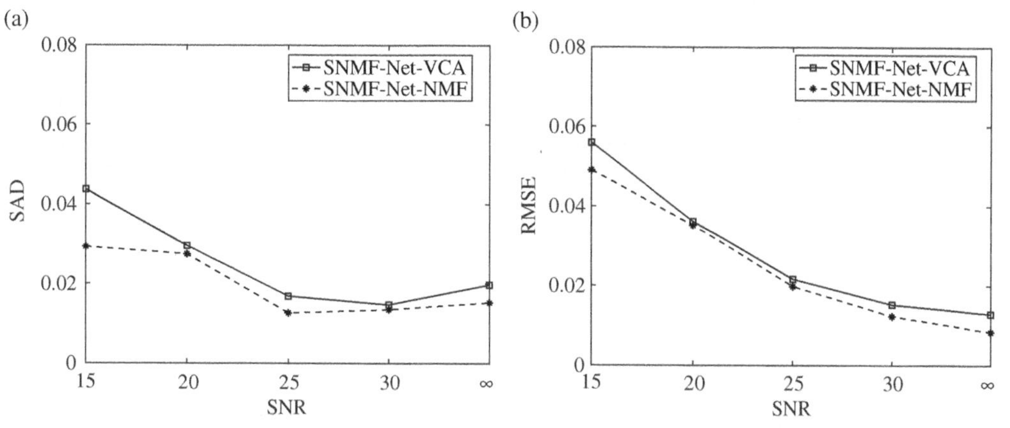


**Figure 13.22** The impact of different initializations. (a) SAD, (b) RMSE.

endmembers and abundances, especially in lower SNR cases. This experiment shows the benefits of the model-based NN, i.e. easy incorporation of prior knowledge.

#### 13.6.4.4 Robustness to Noise

We further compared SNMF-Net with alternative methods to demonstrate its superiority in unmixing. The SNRs were selected as 15 dB, 20 dB, 25 dB, 30 dB, and ∞ (noise-free). Based on the above experimental observations, SNMF-Net was trained with 500 pixels. The other compared methods utilized all the pixels to conduct unmixing. Figure 13.23 presents the unmixing performance of all the methods regarding SNR settings. In general, most of the unmixing methods show a decreasing trend with the increase of SNR. UnDIP provides inferior SAD and RMSE because it estimates endmembers and abundances in sequence rather than simultaneously. The failure of endmember extraction inevitably leads to poor estimation of abundance. Moreover, it can be seen from the figure that MV-NTF outperforms TV-RSNMF, especially in high SNR scenarios, thanks to its inherent advantages of preserving the spatial-spectral structure of HSIs. Compared with three model-based methods, DAEN, uDAS, and SNMF-Net provide more desirable unmixing. Among them, the most promising unmixing results belong to the proposed SNMF-Net, even using only 12.21% of pixels. This attributes to that SNMF-Net is equipped with the hybrid advantages of model-based methods and learning-based methods.

#### 13.6.4.5 Experiments on Real-world Data

Table 13.6 provides a quantitative comparison of all the competing methods on real-world Jasper Ridge HSI with respect to the SAD index. From Table 13.6, it can be seen that MV-NTF performs better than TV-RSNMF, thanks to the consideration of the spectral-spatial structure of HSIs powered by tensor algebra. The standard deviations of UnDIP are zero because UnDIP adopts the simplex volume maximization (SiVM) [50] method to extract endmembers, which is insensitive to initializations. Compared with DAEN, learning-based methods with sparsity constraints, i.e. uDAS, MNN-BU, and SNMF-Net perform better. MNN-BU and SNMF-Net surpass uDAS because they

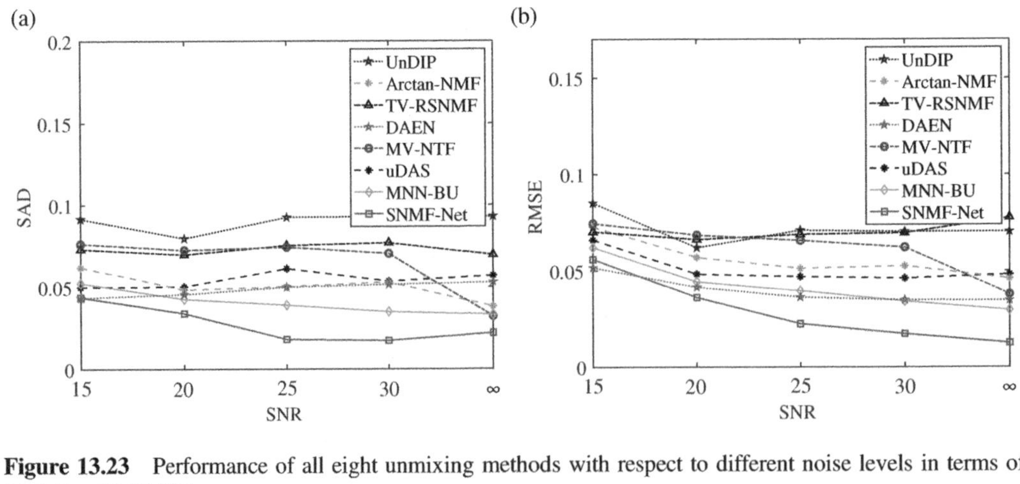


**Figure 13.23** Performance of all eight unmixing methods with respect to different noise levels in terms of (a) SAD, (b) RMSE.

**Table 13.6** Means and standard deviations of the SAD on Jasper Ridge data set

| Algorithm | TV-RSNMF | MV-NTF | DAEN | uDAS | MNN-BU | UnDIP | SNMF-Net |
|---|---|---|---|---|---|---|---|
| Tree | 0.0776 ± 4.91% | 0.2209 ± 2.17% | 0.1774 ± 0.82% | 0.1511 ± 1.37% | **0.0468 ± 0.32%** | 0.1493 ± 0% | 0.0478 ± 1.62% |
| Water | 0.1337 ± 1.31% | 0.2352 ± 1.65% | 0.3237 ± 1.42% | 0.1742 ± 11.39% | 0.1179 ± 0.23% | 0.2529 ± 0% | **0.0855 ± 1.77%** |
| Soil | 0.1047 ± 10.11% | 0.1752 ± 4.67% | 0.1123 ± 2.03% | 0.1639 ± 6.09% | **0.0357 ± 0.43%** | 0.1144 ± 0% | 0.0443 ± 3.09% |
| Road | 0.6515 ± 27.46% | 0.1741 ± 4.35% | **0.0588 ± 1.59%** | 0.0597 ± 0.27% | 0.0901 ± 0.37% | 0.2757 ± 0% | 0.0875 ± 0.05% |
| Mean | 0.2419 ± 3.06% | 0.2015 ± 1.77% | 0.1680 ± 0.43% | 0.1373 ± 3.27% | 0.0726 ± 0.11% | 0.1981 ± 0% | **0.0663 ± 1.04%** |

The bold value means the best performance.

simultaneously take the advantages of model-based and learning-based methods for unmixing. Though provided with better performance than two model-based methods, UnDIP provides unsatisfactory SAD than alternative learning-based methods. The main reason is that UnDIP precomputes the endmembers by SiVM and keeps endmembers fixed for abundance estimation. Proposed SNMF-Net stands out among all the methods. On one hand, SNMF-Net adopts alternating network architecture to estimate endmembers and abundances, which fully accords with the alternating optimization of $L_p$-NMF. On the other hand, SNMF-Net absorbs the spectral mixture model for network constructing and incorporates the prior knowledge on the unmixing problem and sparse representation theory for enhanced learning.

Figures 13.24 and 13.25, respectively, demonstrate the endmembers and abundance maps generated by all the methods. The higher brightness (black to white) indicates larger abundance values (0–1). Learning-based approaches produce more attractive abundances because of higher learning capacity. SNMF-Net matches best with reference targets and ranks first among all methods, especially the soil and road target, whereas other methods such as TV-RSNMF and MV-NTF have some difficulties in accurate estimation. The experiment further proves the superior unmixing ability of SNMF-Net.

## 13.7 Conclusion

In this chapter, we introduce four model-based NNs for hyperspectral unmixing. Experimental results on both synthetic and real-world HSIs show the advantages of model-based NNs over ordinary NNs and model-based methods, i.e. higher interpretability, stronger learning ability, less requirement of training samples, and easy integration of domain knowledge. Model-inspired NNs applied to spectral unmixing are new topics that will motivate more research. As our work in this chapter is an original and pioneering attempt, it provides plenty of room for future work in the following directions: (i) adopting other sparse coding algorithms such as AMP [51] and Fast ISTA [6] instead of ISTA to construct unfolded architectures; (ii) using a patch-based processing scheme or convolutional sparse coding model instead of pixel-based network architecture to incorporate spatial structure in HSIs into the networks; (iii) unfolding kernel-based sparse coding algorithms to extend the proposed methods to nonlinear mixture model.

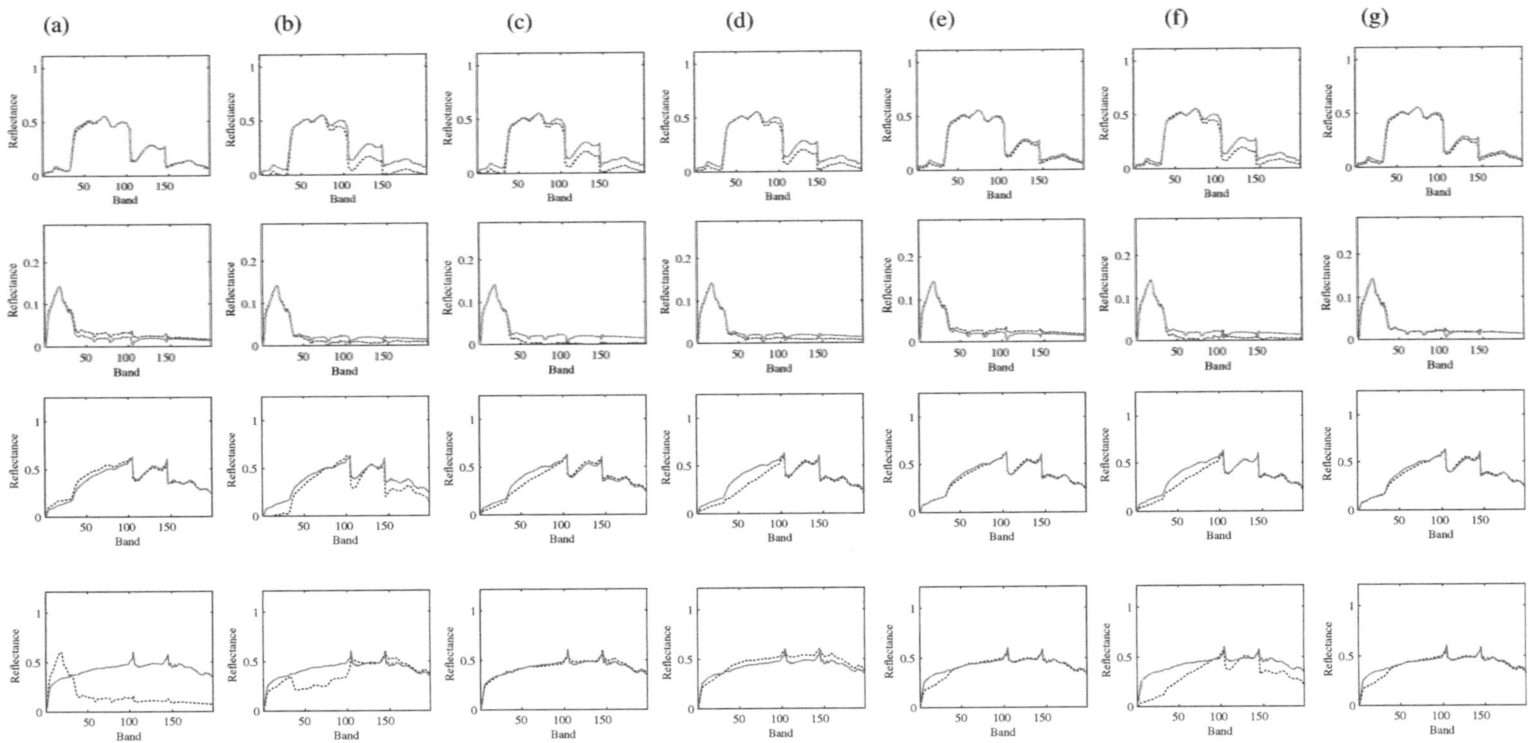


**Figure 13.24** Estimated endmembers by seven unmixing methods on the Jasper Ridge data set. From top to bottom, the rows are the spectral signatures of tree, water, soil, and road, respectively. Solid lines denote the reference endmembers and dashed lines denote the estimated endmembers. (a) TV-RSNMF, (b) MV-NTF, (c) DAEN, (d) uDAS, (e) MNN-BU, (f) UnDIP, (g) SNMF-Net.

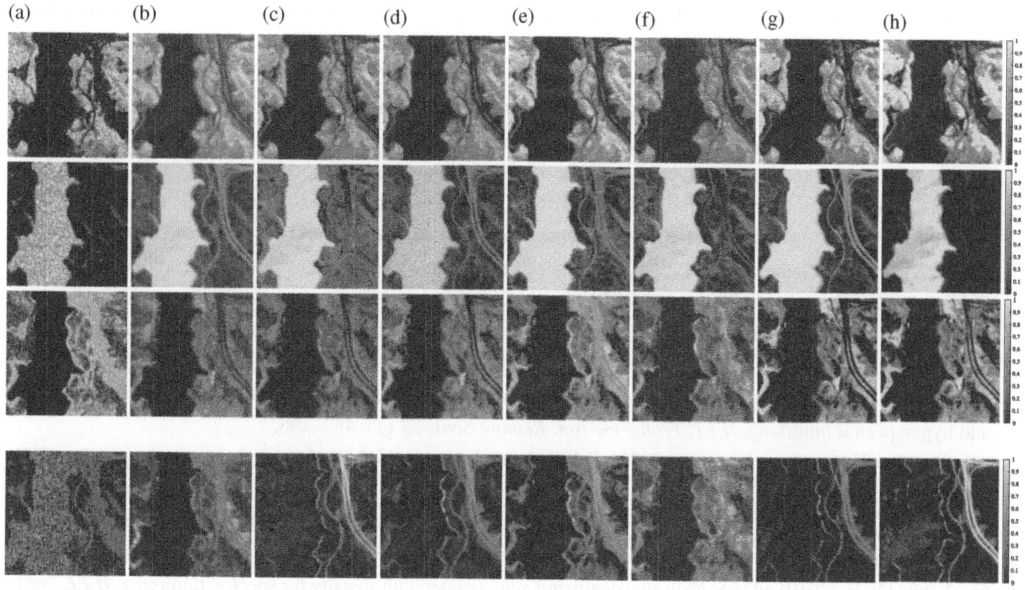


**Figure 13.25** Estimated abundance maps by seven unmixing methods on the Jasper Ridge data set. From top to bottom, the rows are the abundance maps of tree, water, soil, and road, respectively. (a) TV-RSNMF, (b) MV-NTF, (c) DAEN, (d) uDAS, (e) MNN-BU, (f) UnDIP, (g) SNMF-Net, (h) reference.

## References


1. Bioucas-Dias, J.M., Plaza, A., Dobigeon, N. et al. (2012). Hyperspectral unmixing overview: geometrical, statistical, and sparse regression-based approaches. *IEEE J. Sel. Topics Appl. Earth Observ. Remote Sens.* 5 (2): 354–379.
2. Nascimento, J. and Bioucas-Dias, J. (2005). Vertex component analysis: a fast algorithm to unmix hyperspectral data. *IEEE Trans. Geosci. Remote Sens.* 43 (4): 898–910.
3. Iordache, M.-D., Bioucas-Dias, J., and Plaza, A. (June 2011). Sparse unmixing of hyperspectral data. *IEEE Trans. Geosci. Remote Sens.* 49 (6): 2014–2039.
4. Heinz, D.C. and Chang, C.-I (2001). Fully constrained least squares linear spectral mixture analysis method for material quantification in hyperspectral imagery. *IEEE Trans. Geosci. Remote Sens.* 39 (3): 529–545.
5. Daubechies, I., Defrise, M., and De Mol, C. (2004). An iterative thresholding algorithm for linear inverse problems with a sparsity constraint. *Commun. Pure Appl. Math.* 57 (11): 1413–1457.
6. Beck, A. and Teboulle, M. (2009). A fast iterative shrinkage-thresholding algorithm for linear inverse problems. *SIAM J. Imag. Sci.* 2 (1): 183–202.
7. Jia, S. and Qian, Y. (Jan. 2009). Constrained nonnegative matrix factorization for hyperspectral unmixing. *IEEE Trans. Geosci. Remote Sens.* 47 (1): 161–173.
8. Tong, L., Zhou, J., Li, X. et al. (Apr. 2017). Region-based structure preserving nonnegative matrix factorization for hyperspectral unmixing. *IEEE J. Sel. Topics Appl. Earth Observ. Remote Sens.* 10 (4): 1575–1588.
9. Huang, R., Li, X., and Zhao, L. (Oct. 2019). Spectral-spatial robust nonnegative matrix factorization for hyperspectral unmixing. *IEEE Trans. Geosci. Remote Sens.* 57 (10): 8235–8254.
10. Qian, Y., Xiong, F., Zeng, S. et al. (2017). Matrix-vector nonnegative tensor factorization for blind unmixing of hyperspectral imagery. *IEEE Trans. Geosci. Remote Sens.* 55 (3): 1776–1792.
11. Xiong, F., Qian, Y., Zhou, J., and Tang, Y.Y. (2019). Hyperspectral unmixing via total variation regularized nonnegative tensor factorization. *IEEE Trans. Geosci. Remote Sens.* 57 (4): 2341–2357.

12. Boyd, S., Parikh, N., Chu, E. et al. (2011). Distributed optimization and statistical learning via the alternating direction method of multipliers. *Found. Trends Mach. Learn.* 3 (1): 1–122.
13. Wang, W. and Qian, Y. (2015). Adaptive $L_{1/2}$ sparsity-constrained NMF with half-thresholding algorithm for hyperspectral unmixing. *IEEE J. Sel. Topics Appl. Earth Observ. Remote Sens.* 8 (6): 2618–2631.
14. Khajehrayeni, F. and Ghassemian, H. (2020). Hyperspectral unmixing using deep convolutional autoencoders in a supervised scenario. *IEEE J. Sel. Topics Appl. Earth Observ. Remote Sens.* 13: 567–576.
15. Zhou, L., Xiao, B., Liu, X. et al. (2019). Latent distribution preserving deep subspace clustering. *Proceedings of the International Joint Conference on Artificial Intelligence (IJCAI)*, Macao, China (10–16 August 2019), 4440–4446.
16. Zhang, X., Sun, Y., Zhang, J. et al. (2018). Hyperspectral unmixing via deep convolutional neural networks. *IEEE Geosci. Remote Sens. Lett.* 15 (11): 1755–1759.
17. Su, Y., Li, J., Plaza, A. et al. (Jul. 2019). DAEN: deep autoencoder networks for hyperspectral unmixing. *IEEE Trans. Geosci. Remote Sens.* 57 (7): 4309–4321.
18. Ozkan, S., Kaya, B., and Akar, G.B. (Jan. 2019). EndNet: sparse autoencoder network for endmember extraction and hyperspectral unmixing. *IEEE Trans. Geosci. Remote Sens.* 57 (1): 482–496.
19. Min, A., Guo, Z., Li, H., and Peng, J. (2022). JMnet: joint metric neural network for hyperspectral unmixing. *IEEE Trans. Geosci. Remote Sens.* 60: 1–12.
20. Dou, Z., Gao, K., Zhang, X. et al. (2020). Hyperspectral unmixing using orthogonal sparse prior-based autoencoder with hyper-laplacian loss and data-driven outlier detection. *IEEE Trans. Geosci. Remote Sens.* 58 (9): 6550–6564.
21. Qu, Y. and Qi, H. (2019). uDAS: an untied denoising autoencoder with sparsity for spectral unmixing. *IEEE Trans. Geosci. Remote Sens.* 57 (3): 1698–1712.
22. Palsson, B., Ulfarsson, M.O., and Sveinsson, J.R. (2021). Convolutional autoencoder for spectralspatial hyperspectral unmixing. *IEEE Trans. Geosci. Remote Sens.* 59 (1): 535–549.
23. Hong, D., Gao, L., Yao, J. et al. (2021). Endmember-guided unmixing network (EGU-Net): a general deep learning framework for self-supervised hyperspectral unmixing. *IEEE Trans. Neural Netw. Learn. Syst.* 1–14.
24. Gao, L., Han, Z., Hong, D. et al. (2022). CyCU-Net: cycle-consistency unmixing network by learning cascaded autoencoders. *IEEE Trans. Geosci. Remote Sens.* 60: 1–14.
25. Su, Y., Xu, X., Li, J. et al. (2021). Deep autoencoders with multitask learning for bilinear hyperspectral unmixing. *IEEE Trans. Geosci. Remote Sens.* 59 (10): 8615–8629.
26. Zhao, M., Wang, M., Chen, J., and Rahardja, S. (2022). Hyperspectral unmixing for additive nonlinear models with a 3-D-CNN autoencoder network. *IEEE Trans. Geosci. Remote Sens.* 60: 1–15.
27. Zhao, M., Yan, L., and Chen, J. (2021). LSTM-DNN based autoencoder network for nonlinear hyperspectral image unmixing. *IEEE J. Sel. Topics Signal Process.* 15 (2): 295–309.
28. Zhao, M., Wang, X., Chen, J., and Chen, W. (2022). A plug-and-play priors framework for hyperspectral unmixing. *IEEE Trans. Geosci. Remote Sens.* 60: 1–13.
29. Rasti, B., Koirala, B., Scheunders, P., and Ghamisi, P. (2022). UnDIP: hyperspectral unmixing using deep image prior. *IEEE Trans. Geosci. Remote Sens.* 60: 1–15.
30. Hershey, J.R., Roux, J.L., and Weninger, F. (2014). Deep unfolding: model-based inspiration of novel deep architectures, arXiv:1409.2574 (arXiv preprint).
31. Wisdom, S., Powers, T., Pitton, J., and Atlas, L. (2017). Building recurrent networks by unfolding iterative thresholding for sequential sparse recovery. *Proceedings of the IEEE International Conference on Acoustics, Speech, and Signal Processing*, 4346–4350, New Orleans, LA, USA (5–8 March 2017).
32. Dong, W., Zhou, C., Wu, F. et al. (2021). Model-guided deep hyperspectral image super-resolution. *IEEE Trans. Image Process.* 30: 5754–5768.
33. He, J., Li, J., Yuan, Q. et al. (2021). Spectral response function-guided deep optimization-driven network for spectral super-resolution. *IEEE Trans. Neural Netw. Learn. Syst.* 1–15.
34. Shen, D., Liu, J., Wu, Z. et al. (2022). ADMM-HFNet: a matrix decomposition-based deep approach for hyperspectral image fusion. *IEEE Trans. Geosci. Remote Sens.* 60: 1–17.

35. Cao, X., Fu, X., Hong, D. et al. (2022). PanCSC-Net: a model-driven deep unfolding method for pansharpening. *IEEE Trans. Geosci. Remote Sens.* 60: 1–13.
36. Wang, H., Xie, Q., Zhao, Q., and Meng, D. (2020). A model-driven deep neural network for single image rain removal. *Proceedings of the IEEE Conference on Computer Vision and Pattern Recognition (CVPR)* (14–19 June 2020). Virtual.
37. Liu, D., Wang, Z., Wen, B. et al. (2016). Robust single image super-resolution via deep networks with sparse prior. *IEEE Trans. Image Process.* 25 (7): 3194–3207.
38. Li, Y., Tofighi, M., Geng, J. et al. (2020). Efficient and interpretable deep blind image deblurring via algorithm unrolling. *IEEE Trans. Comput. Imaging* 6: 666–681.
39. Gregor, K. and LeCun, Y. (2010). Learning fast approximations of sparse coding. *Proceedings of the International Conference on Machine Learning (ICML)*, Haifa, Israel (21–25 June 2010), 399–406.
40. Qian, Y., Jia, S., Zhou, J., and Robles-Kelly, A. (2011). Hyperspectral unmixing via $L_{1/2}$ sparsity-constrained nonnegative matrix factorization. *IEEE Trans. Geosci. Remote Sens.* 49 (11): 4282–4297.
41. Xiong, F., Zhou, J., Ye, M. et al. (2021). Nmf-sae: an interpretable sparse autoencoder for hyperspectral unmixing. *Proceedings of the IEEE International Conference on Acoustics, Speech, and Signal Processing (ICASSP)*, Toronto, Canada (6–11 June 2021).
42. He, W., Zhang, H., and Zhang, L. (2017). Total variation regularized reweighted sparse nonnegative matrix factorization for hyperspectral unmixing. *IEEE Trans. Geosci. Remote Sens.* 55 (7): 3909–3921.
43. Qian, Y., Xiong, F., Qian, Q., and Zhou, J. (2020). Spectral mixture model inspired network architectures for hyperspectral unmixing. *IEEE Trans. Geosci. Remote Sens.* 58 (10): 7418–7434.
44. Xiong, F., Zhou, J., Tao, S. et al. (2022). Snmf-net: learning a deep alternating neural network for hyperspectral unmixing. *IEEE Trans. Geosci. Remote Sens.* 60: 1–16.
45. Lin, C.-J. (2007). Projected gradient methods for nonnegative matrix factorization. *Neural Comput.* 19 (10): 2756–2779.
46. Zuo, W., Meng, D., Zhang, L. et al. (2013). A generalized iterated shrinkage algorithm for non-convex sparse coding. *Proceedings of the International Conference on Computer Vision (ICCV)*, Sydney, Australia (8–12 April 2013), 217–224.
47. Szegedy, C., Liu, W., Jia, Y. et al. (2015). Going deeper with convolutions. *Proceedings of the IEEE Conference on the Computer Vision and Pattern Recognition (CVPR)*, Boston, USA (7–12 June 2015), 1–9.
48. Liu, J. and Chen, X. (2019). ALISTA: analytic weights are as good as learned weights in LISTA. *Proceedings of the International Conference on Learning Representations (ICLR)*, New Orleans, USA (6–9 May 2019).
49. Salehani, Y.E. and Gazor, S. (2017). Smooth and sparse regularization for NMF hyperspectral unmixing. *IEEE J. Sel. Topics Appl. Earth Observ. Remote Sens.* 10 (8): 3677–3692.
50. Heylen, R., Burazerovic, D., and Scheunders, P. (2011). Fully constrained least squares spectral unmixing by simplex projection. *IEEE Trans. Geosci. Remote Sens.* 49 (11): 4112–4122.
51. Borgerding, M., Schniter, P., and Rangan, S. (2017). AMP-inspired deep networks for sparse linear inverse problems. *IEEE Trans. Signal Process.* 65 (16): 4293–4308.

# 14

# Analytical Fully Constrained Least Squares Linear Spectral Mixture Analysis

Chein-I Chang[1] and Hsiao-Chi Li[2]
[1]*Remote Sensing Signal and Image Processing Laboratory, Department of Computer Science and Electrical Engineering, University of Maryland Baltimore County, Baltimore, MD, USA*
[2]*Department of Electrical Engineering, National Taipei University of Technology, Taipei, Taiwan, Republic of China*

## 14.1 Introduction

Linear spectral mixture analysis (LSMA) is one of the earliest applications in hyperspectral imaging. It has found its great potential and promise in many applications [1], specifically in linear spectral unmixing (LSU), which assumes that a data sample vector can be linearly mixed by a set of so-called endmembers via a linear mixing model and then unmixes the data sample vectors in terms of abundance fractions of these endmembers. More specifically, assume that $L$ is the number of spectral bands and $\mathbf{r}$ is an $L$-dimensional image pixel vector. Let $\mathbf{m}_1, \mathbf{m}_2, \ldots, \mathbf{m}_p$ denote $p$ endmembers. A linear mixing model of $\mathbf{r}$ is given by

$$\mathbf{r} = \mathbf{M}\boldsymbol{\alpha} + \mathbf{n} \tag{14.1}$$

where $\mathbf{M}$ is an $L \times p$ endmember matrix, denoted by $\mathbf{M} = [\mathbf{m}_1\mathbf{m}_2 \cdots \mathbf{m}_p]$, $\boldsymbol{\alpha} = (\alpha_1, \alpha_2, \ldots, \alpha_p)^T$ is a $p \times 1$ abundance column vector with $\alpha_j$ being the abundance fraction of the $j$th endmember $m_j$, and $\mathbf{n}$ is noise or can be interpreted as a measurement or model error.

Two challenging issues in making LSMA effective need to be solved. One is endmember finding as a preprocessing prior to LSMA. Many works have been reported along this line over the past years, such as N-finder algorithm (N-FINDR) discussed in Chapter 8. Once endmembers are found, a follow-up second issue is how to take care of two physical constraints imposed on LSMA, viz. abundance sum-to-one constraint (ASC), $\sum_{j=1}^{p}\alpha_j = 1$, and abundance non-negativity constraint (ANC), $\alpha_j \geq 0$, for all $1 \leq j \leq p$ [2]. Since the first issue of endmember finding is an important subject in

*Advances in Hyperspectral Image Processing Techniques*, First Edition. Edited by Chein-I Chang.

hyperspectral imaging and is studied in detail in Chapter 8 separately, this chapter is devoted to the second issue, abundance-constrained LSMA (AC-LSMA).

In order to solve AC-LSMA, we first find a partially ASC-constrained LSMA to derive sum-to-one-constrained least square (SCLS) solution, then find a non-negativity-constrained least square (NCLS) solution for the ASC-constrained LSMA to yield fully AC-LSMA, FCLS. While finding the SCLS solution is straightforward, finding NCLS is not trivial at all. Its difficulty lies on solving an inequality-constrained optimization problem which does not have analytical solutions and must rely on numerical algorithms to find approximate optimal solutions. So, NCLS utilizes the Kuhn–Tucker conditions to develop an NCLS numerical algorithm to find approximate NCLS solutions. As a result, NCLS uses the SCLS solution as the initial condition to finally generate FCLS solution. It is important to note that due to the inequality constraint, i.e. ANC, the Lagrange multiplier method is not applicable. To resolve this issue, Ren and Chang [3] developed an elegant and simple fully AC-LSMA method, called a modified FCLS (MFCLS) which converts the $p$-inequality ANC to an equality constraint, called absolute abundance sum-to-one constraint (AASC) defined by $\sum_{j=1}^{p} |\alpha_j| = 1$ so that the Lagrange multiplier method can be used to find numerical NCLS solution. However, although AASC is an equality constraint, its absolute value is continuous but not differentiable. Accordingly, MFCLS derives two equations which can be iterated numerically to find a numerical NCLS solution. So, a key difference between FCLS and MFCLS is that FCLS appeals for the Kuhn–Tucker condition to find the fully $\left(\sum_{j=1}^{p}\alpha_j = 1, \{\alpha_j \geq 0\}_{j=1}^{p}\right)$-constrained LSMA solution, whereas MFCLS makes use of the Lagrange multiplier method to iteratively generate $\left(\sum_{j=1}^{p}\alpha_j = 1, \sum_{j=1}^{p} |\alpha_j| = 1\right)$-constrained LSMA solution. Nevertheless, both still have to rely on numerical algorithms to find approximate optimal solutions. This chapter looks into AASC and further develops an analytical modified non-negativity-constrained least squares (AMNCLS) method that can solve the AASC-constrained optimization problem analytically. Therefore, combining SCLS and AMNCLS yields a set of equations which can be solved by Cramer's rule to derive analytical FCLS (AFCLS) in closed forms. As a result, the AFCLS-unmixed results using analytical solutions are more accurate than FCLS-unmixed and MFCLS-unmixed results resulting from numerical solutions.

## 14.2 Linear Spectral Mixture Analysis

Suppose that $L$ is the number of spectral bands and $\mathbf{r}$ is an $L$-dimensional image pixel vector. Assume that there are $p$ material substances, $\mathbf{t}_1, \mathbf{t}_2, \ldots, \mathbf{t}_p$, present in an image scene. Let $\mathbf{m}_1, \mathbf{m}_2, \ldots, \mathbf{m}_p$ denote their corresponding substance signatures, which are generally referred to as digital numbers (DN). A linear mixture model of $\mathbf{r}$ in (14.1) models the spectral signature of $\mathbf{r}$ as a linear combination of $\mathbf{m}_1, \mathbf{m}_2, \ldots, \mathbf{m}_p$ with appropriate abundance fractions specified by $\alpha_1, \alpha_2, \ldots, \alpha_p$. More precisely, $\mathbf{r}$ is an $L \times 1$ column vector and $\mathbf{M}$ be an $L \times p$ substance spectral signature matrix, denoted by $\mathbf{M} = [\mathbf{m}_1\mathbf{m}_2 \cdots \mathbf{m}_p]$, where $\mathbf{m}_j$ is an $L \times 1$ column vector represented by the spectral signature of the $j$-th substance $\mathbf{t}_j$ resident in the pixel vector $\mathbf{r}$. Let $\boldsymbol{\alpha} = (\alpha_1, \alpha_2, \ldots, \alpha_p)^T$ be a $p \times 1$ abundance column vector associated with $\mathbf{r}$ where $\alpha_j$ denotes the abundance fraction of the $j$th substance signature $\mathbf{m}_j$ present in the pixel vector $\mathbf{r}$. In order to restore the pixel vector $\mathbf{r}$, we assume that the spectral signature of the pixel vector $\mathbf{r}$ is linearly mixed by $\mathbf{m}_1, \mathbf{m}_2, \ldots, \mathbf{m}_p$, the spectral signatures of the $p$ substances, $\mathbf{t}_1, \mathbf{t}_2, \ldots, \mathbf{t}_p$, as follows.

By inspecting (14.1), the linear model used by LSMA is similar to a linear model used by Wiener filter except that (14.1) explores correlation among $p$ substance signatures compared to the latter which uses a linear model to account for $p$ the past observed data samples to perform prediction.

Furthermore, by virtue of (14.1), a hyperspectral image viewed as an image cube can be restored by a set of $p$ abundance fraction maps.

A classical approach to solving (14.1) is the least squares estimation given by

$$\hat{\boldsymbol{\alpha}}^{\text{LS}}(\mathbf{r}) = \left(\mathbf{M}^T\mathbf{M}\right)^{-1}\mathbf{M}^T\mathbf{r} \tag{14.2}$$

where $\hat{\boldsymbol{\alpha}}^{\text{LS}}(\mathbf{r}) = \left(\hat{\alpha}_1^{\text{LS}}(\mathbf{r}), \hat{\alpha}_2^{\text{LS}}(\mathbf{r}), \ldots, \hat{\alpha}_p^{\text{LS}}(\mathbf{r})\right)$ and $\hat{\alpha}_j^{\text{LS}}(\mathbf{r})$ is the abundance fraction of the $j$th substance signature $\mathbf{m}_j$ estimated from the data sample vector $\mathbf{r}$. Here, the data sample vector is included to emphasize that the abundance estimate is determined by $\mathbf{r}$.

In a completely different approach, Harsanyi and Chang developed an SNR approach to solving (14.1) [4], which represents (14.1) as a standard signal detection model where $\mathbf{M}\boldsymbol{\alpha}$ is a desired signal vector to be detected and $\mathbf{n}$ is a corrupted noise. Since we are interested in detecting one substance at a time, we can divide the set of the $p$ substances, $\mathbf{t}_1, \mathbf{t}_2, \ldots, \mathbf{t}_p$, into a desired substance, say $\mathbf{t}_p$ and a class of undesired substances, $\mathbf{t}_1, \mathbf{t}_2, \ldots, \mathbf{t}_{p-1}$. In this case, a logical approach is to eliminate the effects caused by the undesired substances $\mathbf{t}_1, \mathbf{t}_2, \ldots, \mathbf{t}_{p-1}$ that are considered as interferers to $\mathbf{t}_p$ before the detection of $\mathbf{t}_p$ takes place. With annihilation of the undesired substance signatures, the detectability of $\mathbf{t}_p$ can be therefore enhanced. In doing so, we first separate $\mathbf{m}_p$ from $\mathbf{m}_1, \mathbf{m}_2, \ldots, \mathbf{m}_p$ in $\mathbf{M}$ and rewrite (14.1) as

$$\mathbf{r} = \alpha_p\mathbf{d} + \boldsymbol{\gamma}\mathbf{U} + \mathbf{n} \tag{14.3}$$

where $\mathbf{d} = \mathbf{m}_p$ is the desired spectral signature of $\mathbf{t}_p$ and $\mathbf{U} = [\mathbf{m}_1\mathbf{m}_2\cdots\mathbf{m}_{p-1}]$ is the undesired substance spectral signature matrix made up of $\mathbf{m}_1, \mathbf{m}_2, \ldots, \mathbf{m}_{p-1}$ which are the spectral signatures of the remaining $p-1$ undesired substances, $\mathbf{t}_1, \mathbf{t}_2, \ldots, \mathbf{t}_{p-1}$. Here, without loss of generality, we assume that the desired substance is a single substance $\mathbf{t}_p$. According to the derivation provided in [4], the solution can be obtained by

$$\hat{\alpha}_p^{\text{OSP}}(\mathbf{r}) = \mathbf{d}^T P_{\mathbf{U}}^{\perp}\mathbf{r} \tag{14.4}$$

where $P_{\mathbf{U}}^{\perp} = \mathbf{I} - \mathbf{U}\mathbf{U}^{\#} = \mathbf{I} - \mathbf{U}\left(\mathbf{U}^T\mathbf{U}\right)^{-1}\mathbf{U}^T$ and $\mathbf{U}^{\#}$ is the pseudo inverse of $\mathbf{U}$ given by $(\mathbf{U}^T\mathbf{U})^{-1}\mathbf{U}^T$. It has been shown in [5–8] that

$$\hat{\alpha}_p^{\text{LS}}(\mathbf{r}) = \left(\mathbf{d}^T P_{\mathbf{U}}^{\perp}\mathbf{d}\right)^{-1}\hat{\alpha}_p^{\text{OSP}}(\mathbf{r}). \tag{14.5}$$

According to (14.4), $\hat{\alpha}_p^{\text{LS}}(\mathbf{r})$ and $\hat{\alpha}_p^{\text{OSP}}(\mathbf{r})$ only differ by a constant $\left(\mathbf{d}^T P_{\mathbf{U}}^{\perp}\mathbf{d}\right)^{-1}$ but implication is quite different because $\hat{\boldsymbol{\alpha}}^{\text{LS}}(\mathbf{r})$ is an estimator compared to $\hat{\alpha}_p^{\text{LS}}(\mathbf{r})$ is a detector. In addition, $\hat{\boldsymbol{\alpha}}^{\text{LS}}(\mathbf{r})$ estimates all abundance fractions of $p$ substance signatures, $\mathbf{m}_1, \mathbf{m}_2, \ldots, \mathbf{m}_p$, while $\hat{\alpha}_p^{\text{LS}}(\mathbf{r})$ detects the presence of the substance signature $\mathbf{m}_p$ via its detected abundance fraction in which case it can be performed one substance signature at a time. So, $\hat{\alpha}_p^{\text{LS}}(\mathbf{r})$ can be considered as normalized $\hat{\alpha}_p^{\text{OSP}}(\mathbf{r})$ by $\left(\mathbf{d}^T P_{\mathbf{U}}^{\perp}\mathbf{d}\right)$, which is actually SNR [1].

In order to faithfully restore the spectral information contained in data sample vectors via LSMA, three issues need to be addressed. One is determination of $p$, i.e. number of substance signatures to be used to form a linear mixing model in (14.1). Another is to find a desired set of $p$ substance signatures, $\mathbf{m}_1, \mathbf{m}_2, \ldots, \mathbf{m}_p$, to form a signature matrix $\mathbf{M}$ in an unsupervised manner. The last issue is to impose two physical constraints on model (14.1), which are ASC, $\sum_{j=1}^{p}\alpha_j = 1$, and (14.2) ANC, $\alpha_j \geq 0$ for all

$1 \leq j \leq p$. While the first issue can be determined by a recently developed concept, virtual dimensionality (VD) in [1, 9–12], the second issue is discussed in Chapter 8. So, in what follows, we will only deal with the last issue, AL-LSMA with abundance vector $\boldsymbol{\alpha}$ imposed by ASC and ANC.

## 14.3 Fully Constrained Least Squares Method

For mathematical tractability, the solutions given by (14.2) and (14.4) are abundance-unconstrained in the sense that no physical constraints are imposed on the model (14.1). However, in order to restore data integrity to satisfy reality, abundance constraints must be imposed on the substance signatures in (14.1) by either ASC or ANC on the abundance vector $\boldsymbol{\alpha}$ in (14.1). In this section, three AC-LSMA problems can be derived from (14.1) as follows.

### *14.3.1 Abundance Sum-to-one-Constrained LSMA*

One simplest way to impose a constraint on (14.1) is ASC. Since there is no physical constraint imposed on the solution $\hat{\boldsymbol{\alpha}}^{\text{LS}}(\mathbf{r})$. In other words, a SCLS problem is

$$\min_{\boldsymbol{\alpha}}\left\{(\mathbf{r}-\mathbf{M}\boldsymbol{\alpha})^T(\mathbf{r}-\mathbf{M}\boldsymbol{\alpha})\right\} \text{ subject to } \sum_{j=1}^{p}\alpha_j = 1. \tag{14.6}$$

To solve (14.6), we use the Lagrange multiplier $\lambda_1$ to constrain the ASC, $\mathbf{1}^T\boldsymbol{\alpha} = 1$ by

$$J(\boldsymbol{\alpha}, \lambda_1) = \frac{1}{2}(\mathbf{M}\boldsymbol{\alpha}-\mathbf{r})^T(\mathbf{M}\boldsymbol{\alpha}-\mathbf{r}) + \lambda_1\left(\mathbf{1}^T\boldsymbol{\alpha}-1\right) \tag{14.7}$$

where $\mathbf{1}$ is the unity vector given by $\mathbf{1} = \underbrace{(1, 1, \ldots, 1)}_{p}^T$. Differentiating (14.8) with respect to $\boldsymbol{\alpha}$ and $\lambda_1$ yields

$$\begin{aligned}
&\left.\frac{\partial J(\boldsymbol{\alpha},\lambda)}{\partial \boldsymbol{\alpha}}\right|_{\hat{\boldsymbol{\alpha}}^{\text{SCLS}}(\mathbf{r})} = \left(\mathbf{M}^T\mathbf{M}\right)\hat{\boldsymbol{\alpha}}^{\text{SCLS}}(\mathbf{r}) - \mathbf{M}^T\mathbf{r} + \lambda_1^*\mathbf{1} = 0 \\
&\Rightarrow \hat{\boldsymbol{\alpha}}^{\text{SCLS}}(\mathbf{r}) = \left(\mathbf{M}^T\mathbf{M}\right)^{-1}\mathbf{M}^T\mathbf{r} - \lambda_1^*\left(\mathbf{M}^T\mathbf{M}\right)^{-1}\mathbf{1} \\
&\Rightarrow \hat{\boldsymbol{\alpha}}^{\text{SCLS}}(\mathbf{r}) = \hat{\boldsymbol{\alpha}}^{\text{LS}}(\mathbf{r}) - \lambda_1^*\left(\mathbf{M}^T\mathbf{M}\right)^{-1}\mathbf{1}
\end{aligned} \tag{14.8}$$

and

$$\left.\frac{\partial J(\boldsymbol{\alpha}, \lambda_1)}{\partial \lambda_1}\right|_{\lambda_1^*} = \mathbf{1}^T\hat{\boldsymbol{\alpha}}^{\text{LS}}(\mathbf{r}) - \mathbf{1} = 0. \tag{14.9}$$

It should be noted that (14.8) and (14.9) must be solved simultaneously in such a way that both the optimal solutions $\hat{\boldsymbol{\alpha}}_{\text{SCLS}}(\mathbf{r})$ and $\lambda_1^*$ appear in (14.8). Using (14.9) further implies

$$\mathbf{1}^T\hat{\boldsymbol{\alpha}}^{\text{LS}}(\mathbf{r}) - \lambda_1^*\mathbf{1}^T\left(\mathbf{M}^T\mathbf{M}\right)^{-1}\mathbf{1} = 1 \Rightarrow \lambda_1^* = -\left[\mathbf{1}^T\left(\mathbf{M}^T\mathbf{M}\right)^{-1}\mathbf{1}\right]^{-1}\left(1 - \mathbf{1}^T\hat{\boldsymbol{\alpha}}^{\text{LS}}(\mathbf{r})\right). \tag{14.10}$$

Substituting (14.10) into (14.9) yields

$$\begin{aligned}\hat{\boldsymbol{\alpha}}^{\mathrm{SCLS}}(\mathbf{r}) &= \hat{\boldsymbol{\alpha}}^{\mathrm{LS}}(\mathbf{r}) + \left(\mathbf{M}^T\mathbf{M}\right)^{-1}\mathbf{1}\left[\mathbf{1}^T\left(\mathbf{M}^T\mathbf{M}\right)^{-1}\mathbf{1}\right]^{-1}\left(1-\mathbf{1}^T\hat{\boldsymbol{\alpha}}^{\mathrm{LS}}(\mathbf{r})\right)\\ &= P_{\mathbf{M},\mathbf{1}}^{\perp}\hat{\boldsymbol{\alpha}}^{\mathrm{LS}}(\mathbf{r}) + \left(\mathbf{M}^T\mathbf{M}\right)^{-1}\mathbf{1}\left[\mathbf{1}^T\left(\mathbf{M}^T\mathbf{M}\right)^{-1}\mathbf{1}\right]^{-1}\end{aligned} \tag{14.11}$$

where

$$P_{\mathbf{M},\mathbf{1}}^{\perp} = \mathbf{I} - \left(\mathbf{M}^T\mathbf{M}\right)^{-1}\mathbf{1}\left[\mathbf{1}^T\left(\mathbf{M}^T\mathbf{M}\right)^{-1}\mathbf{1}\right]^{-1}\mathbf{1}^T. \tag{14.12}$$

### 14.3.2 Abundance Non-negativity-Constrained LSMA

Another crucial physical constraint imposed on the solution $\hat{\boldsymbol{\alpha}}^{\mathrm{LS}}(\mathbf{r})$ is that their generated abundance fractions $\hat{\alpha}_1^{\mathrm{LS}}(\mathbf{r}), \hat{\alpha}_2^{\mathrm{LS}}(\mathbf{r}), \ldots, \hat{\alpha}_p^{\mathrm{LS}}(\mathbf{r})$ must be non-negative, i.e. ANC must be imposed on (14.1). A general approach to solving such an ANC-constrained problem is to introduce an LSE-based objective function imposed by the ANC for $j \in \{1, 2, \ldots, p\}$ defined as

$$J = \frac{1}{2}(\mathbf{M}\boldsymbol{\alpha} - \mathbf{r})^T(\mathbf{M}\boldsymbol{\alpha} - \mathbf{r}) + \boldsymbol{\lambda}(\boldsymbol{\alpha} - \mathbf{c}) \tag{14.13}$$

where $\boldsymbol{\lambda} = (\lambda_1, \lambda_2, \ldots, \lambda_p)^T$ is a Lagrange multiplier vector and a constraint vector $\mathbf{c} = (c_1, c_2, \ldots, c_p)^T$ with $c_j > 0$ for $1 \leq j \leq p$. In analogy with (14.9), we obtain

$$\begin{aligned}&\left.\frac{\partial J(\boldsymbol{\alpha},\lambda)}{\partial\boldsymbol{\alpha}}\right|_{\hat{\boldsymbol{\alpha}}^{\mathrm{NCLS}}(\mathbf{r})} = \left(\mathbf{M}^T\mathbf{M}\right)\hat{\boldsymbol{\alpha}}^{\mathrm{NCLS}}(\mathbf{r}) - \mathbf{M}^T\mathbf{r} + \boldsymbol{\lambda} = 0\\ &\Rightarrow \hat{\boldsymbol{\alpha}}^{\mathrm{NCLS}}(\mathbf{r}) = \left(\mathbf{M}^T\mathbf{M}\right)^{-1}\mathbf{M}^T\mathbf{r} - \left(\mathbf{M}^T\mathbf{M}\right)^{-1}\boldsymbol{\lambda}\\ &\Rightarrow \hat{\boldsymbol{\alpha}}^{\mathrm{NCLS}}(\mathbf{r}) = \hat{\boldsymbol{\alpha}}^{\mathrm{LS}}(\mathbf{r}) - \left(\mathbf{M}^T\mathbf{M}\right)^{-1}\boldsymbol{\lambda}\end{aligned} \tag{14.14}$$

which implies

$$\begin{aligned}&\left(\mathbf{M}^T\mathbf{M}\right)^{-1}\boldsymbol{\lambda}^* = \hat{\boldsymbol{\alpha}}^{\mathrm{LS}}(\mathbf{r}) - \hat{\boldsymbol{\alpha}}^{\mathrm{NCLS}}(\mathbf{r})\\ &\Rightarrow \boldsymbol{\lambda}^* = \left(\mathbf{M}^T\mathbf{M}\right)\hat{\boldsymbol{\alpha}}^{\mathrm{LS}}(\mathbf{r}) - \left(\mathbf{M}^T\mathbf{M}\right)\hat{\boldsymbol{\alpha}}^{\mathrm{NCLS}}(\mathbf{r})\\ &\Rightarrow \boldsymbol{\lambda}^* = \left(\mathbf{M}^T\mathbf{M}\right)\left[\left(\mathbf{M}^T\mathbf{M}\right)^{-1}\mathbf{M}^T\mathbf{r}\right] - \left(\mathbf{M}^T\mathbf{M}\right)\hat{\boldsymbol{\alpha}}^{\mathrm{NCLS}}(\mathbf{r})\\ &\Rightarrow \boldsymbol{\lambda}^* = \mathbf{M}^T\mathbf{r} - \mathbf{M}^T\mathbf{M}\hat{\boldsymbol{\alpha}}^{\mathrm{NCLS}}(\mathbf{r}).\end{aligned} \tag{14.15}$$

To further satisfy ANC, the following Kuhn–Tucker conditions must be implemented

$$\begin{aligned}\lambda_i &= 0, \quad i \in P\\ \lambda_i &< 0, \quad i \in R\end{aligned} \tag{14.16}$$

where $P$ and $R$ represent passive and active sets that contain indices representing negative or positive abundances, respectively. By virtue of (14.14)–(14.16), a numerical algorithm, referred to as NCLS, can be designed to start off with the initial estimate given by $\hat{\boldsymbol{\alpha}}^{\text{LS}}(\mathbf{r})$ in (14.3). If all abundance fractions are positive, the NCLS stops. Otherwise, all indices correspond to negative and zero abundance fractions are moved to passive set $P$ and all positive abundance indices are moved to $R$. According to Kuhn–Tucker conditions (14.16), any $\lambda_i$ with index $i \in P$ is set to zero, and other indices are calculated based on (14.15). If all $\lambda_i$'s are negative, the NCLS stops. If not, the corresponding most negative index is moved from $R$ to $P$. A new vector $\boldsymbol{\lambda}$ is then recalculated based on the modified index sets and a new Lagrange multiplier vector is further implemented in order to find a new set of abundance fractions. By comparing with index set $S$, any negative abundance indices are shuffled from $P$ to $R$. By iteratively implementing (14.14) and (14.15) with using $\hat{\boldsymbol{\alpha}}^{\text{LS}}(\mathbf{r})$ in (14.3) as an initial abundance vector, the NCLS can be derived to find an optimal solution, $\hat{\boldsymbol{\alpha}}^{\text{NCLS}}(\mathbf{r})$ [13]. A detailed step-by-step algorithmic implementation of NCLS is given in the following:

---

***NCLS Algorithm***

---

1. Initialization: Set the passive set $P^{(0)} = \{1, 2, \ldots, p\}$ and $R^{(0)} = \emptyset$ and $k = 0$.
2. Compute $\hat{\boldsymbol{\alpha}}^{LS}(\mathbf{r})$ by (14.3) and let $\hat{\boldsymbol{\alpha}}^{\text{NCLS},(k)}(\mathbf{r}) = \boldsymbol{\alpha}^{\text{LS}}(\mathbf{r})$.
3. At the $k$th iteration, if all components in $\hat{\boldsymbol{\alpha}}^{\text{NCLS},(k)}(\mathbf{r})$ are non-negative, the algorithm is terminated. Otherwise, continue.
4. Let $k = k + 1$.
5. Move all indices in $P^{(k-1)}$ that correspond to negative components of $\hat{\boldsymbol{\alpha}}^{\text{NCLS},(k-1)}(\mathbf{r})$ to $R^{(k-1)}$, respectively. Create a new index $S^{(k)}$ and set it equal to $R^{(k)}$.
6. Let $\hat{\boldsymbol{\alpha}}_{R^{(k)}}$ denote the vector consisting of all components of $\hat{\boldsymbol{\alpha}}^{\text{NCLS},(k-1)}(\mathbf{r})$ in $R^{(k)}$.
7. Form a matrix $\boldsymbol{\Phi}_{\boldsymbol{\alpha}}^{(k)}$ by deleting all rows and columns in the matrix $(\mathbf{M}^T\mathbf{M})^{-1}$ that are specified by $P^{(k)}$.
8. Calculate $\boldsymbol{\lambda}^{(k)} = \left(\boldsymbol{\Phi}_{\boldsymbol{\alpha}}^{(k)}\right)^{-1}\hat{\boldsymbol{\alpha}}_{R^{(k)}}$. If all components in $\lambda^{(k)}$ are negative, go to step 13. Otherwise, continue.
9. Calculate $\lambda_{\max}^{(k)} = \arg\left\{\max_j \lambda_j^{(k)}\right\}$ and move the index in $R^{(k)}$ that corresponds to $\lambda_{\max}^{(k)}$ to $P^{(k)}$.
10. Form another matrix $\Psi_{\lambda}^{(k)}$ by deleting every column of $(\mathbf{M}^T\mathbf{M})^{-1}$ that are specified by $P^{(k)}$.
11. Set $\hat{\boldsymbol{\alpha}}_{S^{(k)}} = \hat{\boldsymbol{\alpha}}^{\text{NCLS},(k)} - \Psi_{\lambda}^{(k)}\boldsymbol{\lambda}^{(k)}$.
12. If any component of $\hat{\boldsymbol{\alpha}}_{S^{(k)}}$ in $S^{(k)}$ are negative, then move these components from $P^{(k)}$ to $R^{(k)}$. Go to step 6.
13. Form another matrix $\Psi_{\lambda}^{(k)}$ by deleting every column of $(\mathbf{M}^T\mathbf{M})^{-1}$ that are specified by $P^{(k)}$.
14. Set $\hat{\boldsymbol{\alpha}}^{\text{NCLS},(k)} = \hat{\boldsymbol{\alpha}}^{\text{LS}} - \Psi_{\lambda}^{(k)}\boldsymbol{\lambda}^{(k)}$. Go to step 3.

---

### *14.3.3 Abundance Fully Constrained LSMA*

Since SCLS in Section 14.3.1 only satisfies ASC but not ANC, its generated abundance fraction may be negative. On the other hand, because NCLS in Section 14.3.2 does not impose ASC, its generated abundance fractions do not necessarily sum up to one. Accordingly, a fully AC-LSMA must be imposed by both ASC and AN to solve the following constrained optimization problem:

$$\min_{\boldsymbol{\alpha}\in\Delta}\left\{(\mathbf{r}-\mathbf{M}\boldsymbol{\alpha})^T(\mathbf{r}-\mathbf{M}\boldsymbol{\alpha})\right\} \text{ subject to } \Delta=\left\{\alpha \mid \alpha_j\geq 0 \quad \text{for } \forall j, \ \sum_{j=1}^{p}\alpha_j=1\right\}. \tag{14.17}$$

The optimal solution to (14.18) first takes advantage of the LS solution, $\hat{\boldsymbol{\alpha}}^{\text{LS}}(\mathbf{r})$, as an initial estimate to derive

$$\hat{\boldsymbol{\alpha}}^{\text{FCLS}}(\mathbf{r})=P_{\mathbf{M},\mathbf{1}}^{\perp}\hat{\boldsymbol{\alpha}}^{\text{LS}}(\mathbf{r})+\left(\mathbf{M}^T\mathbf{M}\right)^{-1}\mathbf{1}\left[\mathbf{1}^T\left(\mathbf{M}^T\mathbf{M}\right)\mathbf{1}\right]^{-1} \tag{14.18}$$

and

$$P_{\mathbf{M},\mathbf{1}}^{\perp}=\mathbf{I}_{L\times L}-\left(\mathbf{M}^T\mathbf{M}\right)^{-1}\mathbf{1}\left[\mathbf{1}^T\left(\mathbf{M}^T\mathbf{M}\right)\mathbf{1}\right]^{-1}\mathbf{1}^T. \tag{14.19}$$

It then uses SCLS together with ANC by introducing a new signature matrix **N** and an observation vector **s** into the NCLS specified by

$$\mathbf{N}=\begin{bmatrix}\eta\mathbf{M}\\ \mathbf{1}^T\end{bmatrix} \text{ and } \mathbf{s}=\begin{bmatrix}\eta\mathbf{r}\\ 1\end{bmatrix} \tag{14.20}$$

where $\eta$ is a parameter to control the effect of ASC on the NCLS and defined as the reciprocal of the maximal element in the matrix $\mathbf{M}=[m_{ij}]$, i.e. $\eta=1/\max_{ij}\{m_{ij}\}$. The utilization of $\eta$ in (14.20) controls the impact of ASC. Using (14.20), an FCLS algorithm can be derived directly from the NCLS algorithm described in the previous section by replacing **M**, **r**, and $\hat{\boldsymbol{\alpha}}^{\text{LS}}(\mathbf{r})$ used in the NCLS algorithm with **N**, **s**, and $\hat{\boldsymbol{\alpha}}^{\text{SCLS}}(\mathbf{r})$.

---

***FCLS Algorithm***

---

1. Initialization: Set $\mathbf{N}=\begin{bmatrix}\eta\mathbf{M}\\ \mathbf{1}^T\end{bmatrix}$ and $\mathbf{s}=\begin{bmatrix}\eta\mathbf{r}\\ 1\end{bmatrix}$, where $\delta\approx 1/\max(\max(\mathbf{M}))$.
2. Compute $\hat{\boldsymbol{\alpha}}^{\text{SCLS}}$ using (14.11) with **N** and **s**. Set $\hat{\boldsymbol{\alpha}}^{\text{FCLS},(0)}=\hat{\boldsymbol{\alpha}}^{\text{SCLS}}$.
3. At the $k$th iteration. If all components in $\hat{\boldsymbol{\alpha}}^{\text{FCLS},(k)}$ are positive, the algorithm is terminated. Otherwise, continue.
4. Let $k=k+1$.
5. Move all indices in $P^{(k-1)}$ that correspond to negative components of $\hat{\boldsymbol{\alpha}}^{\text{FCLS},(k-1)}$ to $R^{(k-1)}$ and the resulting index sets are denoted by $P^{(k)}$ and $R^{(k)}$, respectively.
6. Let $\hat{\boldsymbol{\alpha}}_{R^{(k)}}$ denote the vector consisting of all components $\hat{\boldsymbol{\alpha}}_{\text{SCLS}}$ in $R^{(k)}$.
7. Form a steering matrix $\Phi_{\alpha}^{(k)}$ by deleting all rows and columns in the matrix $(\mathbf{M}^T\mathbf{M})^{-1}$ that are specified by $P^{(k)}$. Then, append $\left(\left(\mathbf{M}^T\mathbf{M}\right)_{[:,R^{(k)}]}^{-1}\right)^T\mathbf{1}$ after the last column, $\left(\left(\mathbf{M}^T\mathbf{M}\right)^{-1}\mathbf{1}\right)_{R^{(k)}}$ after the last row, and $\mathbf{1}^T(\mathbf{M}^T\mathbf{M})^{-1}\mathbf{1}$ to the left-most corner to $\Phi_{\alpha}^{(k)}$.
8. Append a zero to the last row of $\hat{\boldsymbol{\alpha}}_{R(k)}$. Calculate $\boldsymbol{\lambda}^{(k)}=\left(\boldsymbol{\Phi}_{\alpha}^{(k)}\right)^{-1}\hat{\boldsymbol{\alpha}}_{R(k)}$.
9. If all components expect the last row in $\lambda^{(k)}$ are negative, go to step 11. Otherwise, continue.

10. Find $\lambda_{\max}^{(k)} = \max_j \lambda_j^{(k)}$ and move the index in $R^{(k)}$ that corresponds to $\lambda_{\max}^{(k)}$ to $P^{(k)}$. Go to step 7.
11. Form another matrix $\Psi_\lambda^{(k)}$ by deleting every column of $(\mathbf{M}^T\mathbf{M})^{-1}$ specified by $P^{(k)}$.
12. Let $\lambda_1^{(k)} = \boldsymbol{\lambda}^{(k)}(l)$ and $\boldsymbol{\lambda}_2^{(k)} = \boldsymbol{\lambda}^{(k)}[1 : l-1]$, where $l$ equals to the last row of $\lambda^{(k)}$.
13. Set $\hat{\boldsymbol{\alpha}}^{\text{FCLS},(k)} = \hat{\boldsymbol{\alpha}}^{\text{SCLS}} - \left(\left(\mathbf{M}^T\mathbf{M}\right)^{-1}\mathbf{1}\right)\lambda_1^{(k)}\Psi_\lambda^{(k)}\boldsymbol{\lambda}_2^{(k)}$. Go to step 3.

## 14.4 Modified Fully Constrained Linear Squares Method

Despite that FCLS can use the Kuhn–Tucker conditions to find optimal solution numerically, it unfortunately cannot use the Lagrange multiplier method to solve ANC. In this section, we propose an alternative approach to modify this constraint. Instead of directly dealing with the inequalities, $\alpha_j \geq 0$ for each $1 \leq j \leq p$ as abundance non-negativity-constrained least squares (NCLS) does numerically, here we replace ANC with an AASC, $\sum_{j=1}^{p} |\alpha_j| = 1$, an idea which was originally envisioned in Ren and Chang [3] with more details in [1, 14] and has been overlooked in the past. The advantage of AASC is that the Lagrange multiplier method is now applicable and can be used to derive an analytical solution that leads to a desired optimal NCLS solution. With this in mind, a modified least squares linear mixing problem using AASC, $\sum_{j=1}^{p} |\alpha_j| = 1$, can be cast as follows:

$$\min_{\boldsymbol{\alpha}\in\Delta}\left\{(\mathbf{r}-\mathbf{M}\boldsymbol{\alpha})^T(\mathbf{r}-\mathbf{M}\boldsymbol{\alpha})\right\} \tag{14.21}$$

subject to

$$\boldsymbol{\Delta} = \left\{\boldsymbol{\alpha}\middle|\sum_{j=1}^{p} |\alpha_j| = 1\right\}. \tag{14.22}$$

As mentioned previously, the main difficulty with solving constrained linear mixing problems is the constraint of abundance non-negativity that prohibits us from using the Lagrange multiplier method to find solutions analytically. In this section, we propose an alternative approach to modify this constraint. Instead of directly dealing with the inequalities, $\alpha_j \geq 0$ for each $1 \leq j \leq p$, we replace them with an AASC, $\sum_{j=1}^{p} |\alpha_j| = 1$. The advantage of AASC is that the Lagrange multiplier method is now applicable and can be used to derive an iterative algorithm that leads to a desired optimal constrained least squares solution. Additionally, AASC also allows us to exclude negative abundances from solutions. In other words, the only possibility to satisfy both constraints, sum-to-one ($\sum_{j=1}^{p}\alpha_j = 1$) and AASC ($\sum_{j=1}^{p}|\alpha_j| = 1$), is that all the abundances $\{\alpha_j\}_{j=1}^{p}$ must be non-negative. So, a modified least squares linear mixing problem with constraints, ASC, $\sum_{j=1}^{p}\alpha_j = 1$ and AASC, $\sum_{j=1}^{p} |\alpha_j| = 1$ can be cast as follows:

$$\min_{\boldsymbol{\alpha}\in\Delta}\left\{(\mathbf{r}-\mathbf{M}\boldsymbol{\alpha})^T(\mathbf{r}-\mathbf{M}\boldsymbol{\alpha})\right\} \tag{14.23}$$

subject to

$$\boldsymbol{\Delta} = \left\{\boldsymbol{\alpha}\middle|\sum_{j+1}^{p}\alpha_j = 1 \text{ and } \sum_{j=1}^{p} |\alpha_j| = 1\right\}. \tag{14.24}$$

Using a similar argument that derived the SCLS solution, we write

$$J(\boldsymbol{\alpha}) = \left(\frac{1}{2}\right)(\mathbf{r}-\mathbf{M}\boldsymbol{\alpha})^T(\mathbf{r}-\mathbf{M}\boldsymbol{\alpha}) - \lambda_1\left(\sum_{j=1}^{p}\alpha_j - 1\right) - \lambda_2\left(\sum_{j=1}^{p}|\alpha_j| - 1\right). \tag{14.25}$$

Differentiating (14.25) with respect to $\boldsymbol{\alpha}$ and setting to zero yields

$$\left.\frac{\partial J(\boldsymbol{\alpha})}{\partial \boldsymbol{\alpha}}\right|_{\hat{\boldsymbol{\alpha}}_{\text{MFCLS}}} = 0$$

$$\Rightarrow \hat{\boldsymbol{\alpha}}_{\text{MFCLS}} = \left(\mathbf{M}^T\mathbf{M}\right)^{-1}\left[\mathbf{M}^T\mathbf{r} - \lambda_1\mathbf{1} - \lambda_2\text{sign}(\boldsymbol{\alpha})\right] \tag{14.26}$$

$$\Rightarrow \hat{\boldsymbol{\alpha}}_{\text{MFCLS}} = \boldsymbol{\alpha}_{\text{LS}} - \left(\mathbf{M}^T\mathbf{M}\right)^{-1}\left[\lambda_1\mathbf{1} + \lambda_2\text{sign}(\boldsymbol{\alpha}_{\text{LS}})\right]$$

where $\hat{\boldsymbol{\alpha}}_{\text{LS}} = \left(\mathbf{M}^T\mathbf{M}\right)^{-1}\mathbf{M}^T\mathbf{r}$ is the unconstrained least squares estimate of $\boldsymbol{\alpha}$ given by (14.3). We then substitute $\hat{\boldsymbol{\alpha}}_{\text{LS}}$ for $\boldsymbol{\alpha}$ in the following constraints:

$$\sum_{j=1}^{p}\alpha_j = \mathbf{1}^T\boldsymbol{\alpha} = 1 \tag{14.27}$$

$$\sum_{j=1}^{p}|\alpha_j| = \text{sign}(\boldsymbol{\alpha})^T\boldsymbol{\alpha} = 1 \tag{14.28}$$

to compute $\lambda_1$ and $\lambda_2$ where sign($\boldsymbol{\alpha}$) is a vector given by the sign function of $\boldsymbol{\alpha}$ with the $j$th component being the sign of $\alpha_j$, namely, $\text{sign}(\boldsymbol{\alpha}) = (\beta_1, \beta_2, \ldots, \beta_p)^T$ where $\beta_j$ is defined by

$$\beta_j = \frac{\alpha_j}{|\alpha_j|}; \quad \text{if } \alpha_j \neq 0 \text{ and } 0; \quad \text{if } \alpha_j = 0 \tag{14.29}$$

The solution to (14.24)–(14.25) can then be obtained by iteratively computing $\lambda_1$, $\lambda_2$, and $\hat{\boldsymbol{\alpha}}_{\text{MFCLS}}$ using (14.27)–(14.29). The details of implementing this algorithm are as follows:

---

***MFCLS Iterative Algorithm***

---

1. Initialization: Set $\hat{\boldsymbol{\alpha}}^{\text{MFCLS}} = \hat{\boldsymbol{\alpha}}^{\text{SCLS}}$ using (14.11)
2. Compute $\lambda_1$, $\lambda_2$ using (14.26)
3. Compute $\hat{\boldsymbol{\alpha}}^{\text{MFCLS}} = \hat{\boldsymbol{\alpha}}^{\text{SCLS}} - \left(\mathbf{M}^T\mathbf{M}\right)^{-1}\left[\lambda_1\mathbf{1} + \lambda_2\text{sign}\left(\hat{\boldsymbol{\alpha}}^{\text{SCLS}}\right)\right]$
4. If there exists one component in $\hat{\boldsymbol{\alpha}}^{\text{MFCLS}}$, which is negative, go to step 2. Otherwise, stop.

---

It should be noted that in step 1, the algorithm is initialized by taking advantage of the SCLS solution $\hat{\boldsymbol{\alpha}}^{\text{SCLS}}$ obtained by (14.11). The stopping criterion described in step 4 provides a rule of when the algorithm will be terminated in which case all components must be non-negative. In general, it requires a fair amount of computation time to reach this requirement. So, in real implementation, we suggest a simpler rule by checking $\sum_{j=1}^{p} | \alpha_j | - 1 < \varepsilon$ for a preselected threshold $\varepsilon$ that will guarantee a quick termination.

There are two main differences between FCLS and MFCLS. One is their used constraints. FCLS uses ASC and ANC to impose LSMA, whereas MFCLS imposes ASC and AASC on LSMA. The other is that FCKS utilizes the Kuhn–Tucker condition to find numerical solutions for fully AC-LSMA as opposed to MFCLS which takes advantage of the Lagrange multiplier method to find numerical solutions for fully AC-LSMA.

## 14.5 Analytical Non-negativity-Constrained Linear Squares Method

The advantage of AASC used by MFCLS is that the Lagrange multiplier method is now applicable and can be used to derive an numerical solution that leads to a desired optimal NCLS solution. Additionally, AASC also allows us to exclude inequality constraints from finding solutions. In other words, the only possibility to satisfy both constraints, sum-to-one ($\sum_{j=1}^{p}\alpha_j = 1$) and AASC ($\sum_{j=1}^{p}|\alpha_j| = 1$), is that all the abundances $\{\alpha_j\}_{j=1}^{p}$ must be non-negative. With this in mind, we can also adopt a similar treatment to derive the analytical SCLS solution in (14.11) to derive an analytical AASC-based NCLS as follows:

$$\min_{\boldsymbol{\alpha}\in\boldsymbol{\Delta}}\left\{(\mathbf{r}-\mathbf{M}\boldsymbol{\alpha})^T(\mathbf{r}-\mathbf{M}\boldsymbol{\alpha})\right\} \quad \text{subject to } \boldsymbol{\Delta}=\left\{\boldsymbol{\alpha}\left|\sum_{j=1}^{p} | \alpha_j | = 1\right.\right\}. \tag{14.30}$$

Following a similar argument that derived the SCLS solution, we re-express an analytical abundance non-negativity least squares-constrained (ANCLS) problem via a Lagrange multiplier $\lambda_2$ as

$$J(\boldsymbol{\alpha},\lambda_2) = (1/2)(\mathbf{r}-\mathbf{M}\boldsymbol{\alpha})^T(\mathbf{r}-\mathbf{M}\boldsymbol{\alpha}) + \lambda_2\left(\sum_{j=1}^{p} | \alpha_j | - 1\right). \tag{14.31}$$

Now, if we define sign($\boldsymbol{\alpha}$) as a sign vector of $\boldsymbol{\alpha}$ given by $\text{sign}(\boldsymbol{\alpha}) = (\text{sign}(\alpha_1), \ldots, \text{sign}(\alpha_p))^T$ with the $j$th component being the sign of $\alpha_j$, then for $1 \leq j \leq p$,

$$\text{sign}(\alpha_j) = \begin{cases} +1; & \text{if } \alpha_j > 0 \\ -1; & \text{if } \alpha_j > 0. \\ 0; & \text{if } \alpha_j > 0 \end{cases} \tag{14.32}$$

Using a unit step $u(t)$ as $u(t) = \begin{cases} 1; & \text{if } t > 0 \\ 0; & \text{if } t < 0 \\ \text{not defined}; & \text{if } t = 0 \end{cases}$, then (14.32) can be re-expressed as

$$\text{sign}(\alpha_j) = \begin{cases} u(\alpha_j) - u(-\alpha_j); & \text{if } \alpha_j \neq 0 \\ 0; & \text{if } \alpha_j = 0 \end{cases}. \tag{14.33}$$

Using (14.32) and 14.33), we can further express (14.31) as

$$J(\boldsymbol{\alpha},\lambda_2) = (1/2)(\mathbf{r}-\mathbf{M}\boldsymbol{\alpha})^T(\mathbf{r}-\mathbf{M}\boldsymbol{\alpha}) + \lambda_2\left(\text{sign}(\boldsymbol{\alpha})^T\boldsymbol{\alpha}-1\right). \tag{14.34}$$

Since the derivative of the unit step function $u(t)$ is the delta function $\delta(t)$ defined by $\delta(0) = 1$ and zero, otherwise, the derivative of $\text{sign}(\hat{\boldsymbol{\alpha}})$,

$$\frac{d\left[\text{sign}\left(\alpha_j\right)\right]}{\mathrm{d}\alpha_j} = \delta\left(\alpha_j\right) + \delta\left(-\alpha_j\right) + \delta(0) \tag{14.35}$$

with $\delta(\boldsymbol{\alpha}) = (\delta(\alpha_1), \ldots, \delta(\alpha_p))^T$ and $\delta\left(\alpha_j\right) = \begin{cases} 0; & \text{if } \alpha_j \neq 0 \\ 1; & \text{if } \alpha_j = 0 \end{cases}$. In this case, $\frac{d(\text{sign}(\boldsymbol{\alpha}))}{\mathrm{d}\boldsymbol{\alpha}} = \left(\frac{\mathrm{d}(\text{sign}(\alpha_1)}{\alpha_1}, \ldots, \frac{\mathrm{d}\left(\text{sign}\left(\alpha_p\right)\right)}{\alpha_p}\right)^T$ can be found as

$$\begin{aligned}\left[\frac{\mathrm{d}\left(\text{sign}\left(\alpha_j\right)\right)}{\mathrm{d}\alpha_j}\right]\alpha_j &= \left[\delta\left(\alpha_j\right) + \delta\left(-\alpha_j\right) + \delta(0)\right]\alpha_j \\ &= \left[\delta\left(\alpha_j\right) + \delta\left(-\alpha_j\right)\right]\left(\alpha_j \neq 0\right) + \delta(0)\left(\alpha_j = 0\right) = 0\end{aligned} \tag{14.36}$$

which results in $\left[\frac{\mathrm{d}(\text{sign}(\boldsymbol{\alpha}))}{\mathrm{d}\boldsymbol{\alpha}}\right]^T\boldsymbol{\alpha} = 0$. Differentiating (14.34) with respect to $\boldsymbol{\alpha}$ and $\lambda_2$, using (14.35) and setting to zero yields

$$\begin{aligned}&\left.\frac{\partial J(\boldsymbol{\alpha},\lambda_2)}{\partial\boldsymbol{\alpha}}\right|_{\hat{\boldsymbol{\alpha}}^{\text{ANCLS}}(\mathbf{r})} = 0 \\ &\Rightarrow \left(\mathbf{M}^T\mathbf{M}\right)\boldsymbol{\alpha} - \mathbf{M}^T\mathbf{r} + \lambda_2^*\left[\left(\mathrm{d}(\text{sign}(\boldsymbol{\alpha})/\mathrm{d}\boldsymbol{\alpha}\right)^T\boldsymbol{\alpha} + (\text{sign}(\boldsymbol{\alpha}))\right]\Big|_{\boldsymbol{\alpha}=\hat{\boldsymbol{\alpha}}^{\text{ANCLS}}(\mathbf{r})} = 0 \\ &\Rightarrow \hat{\boldsymbol{\alpha}}^{\text{ANCLS}}(\mathbf{r}) = \left(\mathbf{M}^T\mathbf{M}\right)^{-1}\mathbf{M}^T\mathbf{r} - \lambda_2^*\left(\mathbf{M}^T\mathbf{M}\right)^{-1}\left[\left(\text{sign}\left(\hat{\boldsymbol{\alpha}}^{\text{ANCLS}}(\mathbf{r})\right)\right)\right] \\ &\Rightarrow \hat{\boldsymbol{\alpha}}^{\text{ANCLS}}(\mathbf{r}) = \hat{\boldsymbol{\alpha}}_{\text{LS}}(\mathbf{r}) - \lambda_2^*\left(\mathbf{M}^T\mathbf{M}\right)^{-1}\left[\left(\text{sign}\left(\hat{\boldsymbol{\alpha}}^{\text{ANCLS}}(\mathbf{r})\right)\right)\right]\end{aligned} \tag{14.37}$$

$$\left.\frac{\partial J(\boldsymbol{\alpha},\lambda_2)}{\partial\lambda_2}\right|_{\lambda_2^*} = \text{sign}(\boldsymbol{\alpha})^T\boldsymbol{\alpha} - \mathbf{1}\Big|_{\boldsymbol{\alpha}=\hat{\boldsymbol{\alpha}}^{\text{ANCLS}}(\mathbf{r})} = \left(\text{sign}\left(\hat{\boldsymbol{\alpha}}^{\text{ANCLS}}(\mathbf{r})\right)\right)^T\boldsymbol{\alpha}^{\text{ANCLS}}(\mathbf{r}) - 1 = 0 \tag{14.38}$$

where $\hat{\boldsymbol{\alpha}}_{\text{LS}}(\mathbf{r}) = \left(\mathbf{M}^T\mathbf{M}\right)^{-1}\mathbf{M}^T\mathbf{r}$ is the unconstrained least squares estimate of $\boldsymbol{\alpha}$ given by (14.2). It should be noted that $\mathrm{d}\left(\text{sign}\left(\hat{\boldsymbol{\alpha}}^{\text{ANCLS}}(\mathbf{r})\right)\right)^T\hat{\boldsymbol{\alpha}}^{\text{ANCLS}}(\mathbf{r})$ in the second equality of (14.37) is zero since $\mathrm{d}\text{sign}(\hat{\boldsymbol{\alpha}})/\mathrm{d}\boldsymbol{\alpha} = 0$ due to (14.36).

Similar to (14.24) and (14.25) the optimal solutions, $\hat{\boldsymbol{\alpha}}^{\text{MNCLS}}(\mathbf{r})$ and $\lambda_2^*$ must be solved by (14.37, 14.38) simultaneously. For simplicity, we let $\mathbf{b} = \text{sign}\left(\hat{\boldsymbol{\alpha}}^{\text{MNCLS}}(\mathbf{r})\right)$. It is worth noting that $\mathbf{b}$ is not necessarily to equal to unity vector $\mathbf{1}$ unless $\alpha_j^{\text{MNCLS}} > 0$ for all $1 \leq j \leq p$. Then, the AASC constraint of $\left(\text{sign}\left(\hat{\boldsymbol{\alpha}}^{\text{ANCLS}}(\mathbf{r})\right)\right)^T\hat{\boldsymbol{\alpha}}^{\text{ANCLS}}(\mathbf{r}) = \mathbf{b}^T\hat{\boldsymbol{\alpha}}^{\text{ANCLS}}(\mathbf{r}) = 1$ in (14.37) yields

$$\begin{aligned}&\mathbf{b}^T\hat{\boldsymbol{\alpha}}_{\text{ANCLS}}(\mathbf{r})=1\\&\Rightarrow \mathbf{b}^T\hat{\boldsymbol{\alpha}}^{\text{LS}}(\mathbf{r})-\lambda_2^*\mathbf{b}^T\left(\mathbf{M}^T\mathbf{M}\right)^{-1}\mathbf{b}=1\\&\Rightarrow -\lambda_2^*\mathbf{b}^T\left(\mathbf{M}^T\mathbf{M}\right)^{-1}\mathbf{b}=1-\mathbf{b}^T\hat{\boldsymbol{\alpha}}^{\text{LS}}(\mathbf{r})\\&\Rightarrow \lambda_2^*=-\left[\mathbf{b}^T\left(\mathbf{M}^T\mathbf{M}\right)^{-1}\mathbf{b}\right]^{-1}\left(1-\mathbf{b}^T\hat{\boldsymbol{\alpha}}^{\text{LS}}(\mathbf{r})\right)\end{aligned}\quad . \tag{14.39}$$

So, we can derive an analytical AASC-constrained least squares-constrained (ANCLS) solution which has the following analytical forms:

$$\begin{aligned}\hat{\boldsymbol{\alpha}}^{\text{ANCLS}}(\mathbf{r})&=\hat{\boldsymbol{\alpha}}^{\text{LS}}(\mathbf{r})+\left(\mathbf{M}^T\mathbf{M}\right)^{-1}\left[\mathbf{b}^T\left(\mathbf{M}^T\mathbf{M}\right)^{-1}\mathbf{b}\right]^{-1}\left(1-\mathbf{b}^T\hat{\boldsymbol{\alpha}}^{\text{LS}}(\mathbf{r})\right)\\&=P_{\mathbf{M},\mathbf{b}}^{\perp}\hat{\boldsymbol{\alpha}}^{\text{LS}}(\mathbf{r})+\left(\mathbf{M}^T\mathbf{M}\right)^{-1}\left[\mathbf{b}^T\left(\mathbf{M}^T\mathbf{M}\right)^{-1}\mathbf{b}\right]^{-1}\end{aligned} \tag{14.40}$$

where

$$P_{\mathbf{M},\mathbf{b}}^{\perp}=\mathbf{I}-\left(\mathbf{M}^T\mathbf{M}\right)^{-1}\left[\mathbf{b}^T\left(\mathbf{M}^T\mathbf{M}\right)^{-1}\mathbf{b}\right]^{-1}\mathbf{b}^T. \tag{14.41}$$

## 14.6 Analytical Fully Constrained Least Squares Method

With AASC coupled with ASC, an analytical fully AC-LSMA can be cast as follows:

$$\min_{\boldsymbol{\alpha}\in\Delta}\left\{(\mathbf{r}-\mathbf{M}\boldsymbol{\alpha})^T(\mathbf{r}-\mathbf{M}\boldsymbol{\alpha})\right\} \tag{14.42}$$

subject to

$$\Delta=\left\{\boldsymbol{\alpha}\middle|\sum_{j=1}^{p}\alpha_j=1 \text{ and } \middle|\sum_{j=1}^{p}|\alpha_j|=1\right\}. \tag{14.43}$$

By introducing two Lagrange multipliers, $\lambda_1$ and $\lambda_2$, we derive the following objective function:

$$\begin{aligned}J(\boldsymbol{\alpha},\lambda_1,\lambda_2)&=\left(\frac{1}{2}\right)(\mathbf{r}-\mathbf{M}\boldsymbol{\alpha})^T(\mathbf{r}-\mathbf{M}\boldsymbol{\alpha})+\lambda_1\left(\sum_{j=1}^{p}\alpha_j-1\right)+\lambda_2\left(\sum_{j=1}^{p}|\alpha_j|-1\right)\\&=\left(\frac{1}{2}\right)(\mathbf{r}-\mathbf{M}\boldsymbol{\alpha})^T(\mathbf{r}-\mathbf{M}\boldsymbol{\alpha})+\lambda_1\left(\sum_{j=1}^{p}\alpha_j-1\right)+\lambda_2\left(\text{sign}(\boldsymbol{\alpha})^T\boldsymbol{\alpha}-1\right).\end{aligned} \tag{14.44}$$

Differentiating (14.44) with respect to $\boldsymbol{\alpha}$, $\lambda_1$, and $\lambda_2$, respectively, and setting to zero yields

$$\begin{aligned}&\left.\frac{\partial J(\boldsymbol{\alpha},\lambda_1,\lambda_2)}{\partial\boldsymbol{\alpha}}\right|_{\hat{\boldsymbol{\alpha}}^{\text{AFCLS}}(\mathbf{r})}=0\\&\Rightarrow \left(\mathbf{M}^T\mathbf{M}\right)\boldsymbol{\alpha}-\mathbf{M}^T\mathbf{r}+\lambda_1^*\mathbf{1}+\lambda_2^*\left[(\partial\text{sign}(\boldsymbol{\alpha})/d\boldsymbol{\alpha})^T\boldsymbol{\alpha}+(\text{sign}(\boldsymbol{\alpha}))^T\mathbf{1}\right]\Big|_{\boldsymbol{\alpha}=\hat{\boldsymbol{\alpha}}^{\text{AFCLS}}(\mathbf{r})}=0\\&\Rightarrow \hat{\boldsymbol{\alpha}}^{\text{MFCLS}}(\mathbf{r})=\left(\mathbf{M}^T\mathbf{M}\right)^{-1}\left[\mathbf{M}^T\mathbf{r}-\lambda_1^*\mathbf{1}-\lambda_2^*\left(\left(\text{sign}\left(\hat{\boldsymbol{\alpha}}^{\text{AFCLS}}(\mathbf{r})\right)\right)^T\mathbf{1}\right)\right]=0\\&\Rightarrow \hat{\boldsymbol{\alpha}}^{\text{AFCLS}}(\mathbf{r})=\hat{\boldsymbol{\alpha}}^{\text{LS}}(\mathbf{r})-\left(\mathbf{M}^T\mathbf{M}\right)^{-1}\left(\lambda_1^*\mathbf{1}+\lambda_2^*\left(\text{sign}\left(\hat{\boldsymbol{\alpha}}^{\text{AFCLS}}(\mathbf{r})\right)\right)^T\mathbf{1}\right)\end{aligned} \tag{14.45}$$

$$\left.\frac{\partial J(\boldsymbol{\alpha}, \lambda_1, \lambda_2)}{\partial \lambda_1}\right|_{\lambda_1^*} = \mathbf{1}^T \hat{\boldsymbol{\alpha}}^{\text{AFCLS}}(\mathbf{r}) - \mathbf{1} = 0 \tag{14.46}$$

$$\left.\frac{\partial J(\boldsymbol{\alpha}, \lambda_1, \lambda_2)}{\partial \lambda_2}\right|_{\lambda_2^*} = \left(\text{sign}\left(\hat{\boldsymbol{\alpha}}^{\text{AFCLS}}(\mathbf{r})\right)\right)^T \hat{\boldsymbol{\alpha}}^{\text{AFCLS}}(\mathbf{r}) - \mathbf{1} = 0. \tag{14.47}$$

More specifically, the constraint of $\mathbf{1}^T \hat{\boldsymbol{\alpha}}^{\text{AFCLS}}(\mathbf{r}) = 1$ in (14.46) derives

$$\begin{aligned} &\mathbf{1}^T \hat{\boldsymbol{\alpha}}^{\text{LS}}(\mathbf{r}) - \left(\lambda_1^* \mathbf{1}^T \left(\mathbf{M}^T\mathbf{M}\right)^{-1}\mathbf{1} + \lambda_2^* \mathbf{1}^T \left(\mathbf{M}^T\mathbf{M}\right)^{-1}\mathbf{b}\right) = 1 \\ &\Rightarrow \lambda_1^* \mathbf{1}^T \left(\mathbf{M}^T\mathbf{M}\right)^{-1}\mathbf{1} + \lambda_2^* \mathbf{1}^T \left(\mathbf{M}^T\mathbf{M}\right)^{-1}\mathbf{b} = \left(\mathbf{1}^T \hat{\boldsymbol{\alpha}}^{\text{LS}}(\mathbf{r}) - 1\right) \end{aligned} \tag{14.48}$$

and the constraint of $\mathbf{b}^T \hat{\boldsymbol{\alpha}}^{\text{MFCLS}}(\mathbf{r}) = 1$ in (14.46) further implies

$$\begin{aligned} &\mathbf{b}^T \hat{\boldsymbol{\alpha}}^{\text{LS}}(\mathbf{r}) + \mathbf{b}^T \left(\mathbf{M}^T\mathbf{M}\right)^{-1}\left(\lambda_1^* \mathbf{1} + \lambda_2^* \mathbf{b}\right) = 1 \\ &\Rightarrow \lambda_1^* \mathbf{b}^T \left(\mathbf{M}^T\mathbf{M}\right)^{-1}\mathbf{1} + \lambda_2^* \mathbf{b}^T \left(\mathbf{M}^T\mathbf{M}\right)^{-1}\mathbf{b} = \left(\mathbf{b}^T \hat{\boldsymbol{\alpha}}^{\text{LS}}(\mathbf{r}) - 1\right) \end{aligned}. \tag{14.49}$$

Let $\mathbf{a} = (\mathbf{M}^T\mathbf{M})^{-1}\mathbf{1}$ and $\mathbf{c} = (\mathbf{M}^T\mathbf{M})^{-1}\mathbf{b}$. Then, (14.48) and (14.49) become

$$\lambda_1^* \mathbf{1}^T \mathbf{a} + \lambda_2^* \mathbf{1}^T \mathbf{c} = \left(\mathbf{1}^T \hat{\boldsymbol{\alpha}}^{\text{LS}}(\mathbf{r}) - 1\right) \tag{14.50}$$

$$\lambda_1^* \mathbf{c}^T \mathbf{a} + \lambda_2^* \mathbf{c}^T \mathbf{b} = \left(\mathbf{b}^T \hat{\boldsymbol{\alpha}}^{\text{LS}}(\mathbf{r}) - 1\right). \tag{14.51}$$

The solution to solving (14.45) and (14.48) simultaneously for $\lambda_1$ and $\lambda_2$ can be found by Cramer's rule and given by

$$\lambda_1^* = \frac{\begin{vmatrix} \left(\mathbf{1}^T \hat{\boldsymbol{\alpha}}^{\text{LS}}(\mathbf{r}) - 1\right) & \mathbf{1}^T \mathbf{b} \\ \left(\mathbf{b}^T \hat{\boldsymbol{\alpha}}^{\text{LS}}(\mathbf{r}) - 1\right) & \mathbf{c}^T \mathbf{a} \end{vmatrix}}{\begin{vmatrix} \mathbf{1}^T \mathbf{a} & \mathbf{1}^T \mathbf{c} \\ \mathbf{c}^T \mathbf{a} & \mathbf{c}^T \mathbf{b} \end{vmatrix}}, \quad \lambda_2^* = \frac{\begin{vmatrix} \mathbf{1}^T \mathbf{a} & \left(\mathbf{1}^T \hat{\boldsymbol{\alpha}}^{\text{LS}}(\mathbf{r}) - 1\right) \\ \mathbf{c}^T \mathbf{a} & \left(\mathbf{b}^T \hat{\boldsymbol{\alpha}}^{\text{LS}}(\mathbf{r}) - 1\right) \end{vmatrix}}{\begin{vmatrix} \mathbf{1}^T \mathbf{a} & \mathbf{1}^T \mathbf{c} \\ \mathbf{c}^T \mathbf{a} & \mathbf{c}^T \mathbf{b} \end{vmatrix}}. \tag{14.52}$$

Accordingly, $\hat{\boldsymbol{\alpha}}^{\text{AFCLS}}(\mathbf{r})$ specified by (14.45) can be obtained analytically by finding $\hat{\boldsymbol{\alpha}}^{\text{LS}}(\mathbf{r})$ via (14.2) and $\lambda_1^*$ and $\lambda_2^*$ via (14.52).

$$\lambda_1^* = \frac{\begin{vmatrix} \left(\mathbf{1}^T \hat{\boldsymbol{\alpha}}^{\text{LS}}(\mathbf{r}) - 1\right) & \mathbf{1}^T \mathbf{c} \\ \left(\mathbf{b}^T \hat{\boldsymbol{\alpha}}^{\text{LS}}(\mathbf{r}) - 1\right) & \mathbf{b}^T \mathbf{c} \end{vmatrix}}{\begin{vmatrix} \mathbf{1}^T \mathbf{a} & \mathbf{1}^T \mathbf{c} \\ \mathbf{b}^T \mathbf{a} & \mathbf{b}^T \mathbf{c} \end{vmatrix}}, \quad \lambda_2^* = \frac{\begin{vmatrix} \mathbf{1}^T \mathbf{a} & \left(\mathbf{1}^T \hat{\boldsymbol{\alpha}}^{\text{LS}}(\mathbf{r}) - 1\right) \\ \mathbf{b}^T \mathbf{a} & \left(\mathbf{b}^T \hat{\boldsymbol{\alpha}}^{\text{LS}}(\mathbf{r}) - 1\right) \end{vmatrix}}{\begin{vmatrix} \mathbf{1}^T \mathbf{a} & \mathbf{1}^T \mathbf{c} \\ \mathbf{b}^T \mathbf{a} & \mathbf{b}^T \mathbf{c} \end{vmatrix}} \tag{14.53}$$

which are not derived in any literature. Accordingly, $\hat{\boldsymbol{\alpha}}^{\text{AFCLS}}(\mathbf{r})$ specified by (14.45) can be obtained analytically by finding $\hat{\boldsymbol{\alpha}}^{\text{LS}}(\mathbf{r})$ via (14.2) and $\lambda_1^*$ and $\lambda_2^*$ via (14.53) simultaneously to solve an optimal set of the $p$-dimensional abundance vector $\boldsymbol{\alpha}$, $\lambda_1$, and $\lambda_2$ altogether using the following simultaneously algebraic equations:

$$\begin{bmatrix} (\mathbf{M}^T\mathbf{M})\boldsymbol{\alpha} - \mathbf{M}^T\mathbf{r} + \lambda_1\mathbf{1} + \lambda_2\text{sign}(\boldsymbol{\alpha})|(\text{sign}(\boldsymbol{\alpha}(\mathbf{r}))^T\boldsymbol{\alpha}(\mathbf{r})\mathbf{1}^T\boldsymbol{\alpha}(\mathbf{r}) \end{bmatrix} = \begin{bmatrix} \mathbf{0}_{p\times 1} \\ 1 \\ 1 \end{bmatrix} \quad (14.54)$$

which leads to AFCLS. The key difference between MFCLS and AFCLS is that the former assumes its initial abundance estimator already satisfies ASC and then finds iterative solutions via (14.26–14.28) compared to the latter which directly solves all the three equations specified by (14.54) analytically. Consequently, AFCLS is in fact the one which produces the optimal solutions as opposed to suboptimal numerical solutions produced by FCLS and MFCLS.

## 14.7 Experiments

The image scene shown in Figure 14.1 was acquired by the airborne Hyperspectral Digital Imagery Collection Experiment (HYDICE) sensor in August 1995 from a flight altitude of 10 000 ft with the ground sampling distance approximately 1.56 m.

It has 210 spectral channels ranging from 0.4 to 2.5 μm with spectral resolution of 10 nm. The low signal/high noise bands: bands 1–3 and bands 202–210; and water vapor absorption bands: bands

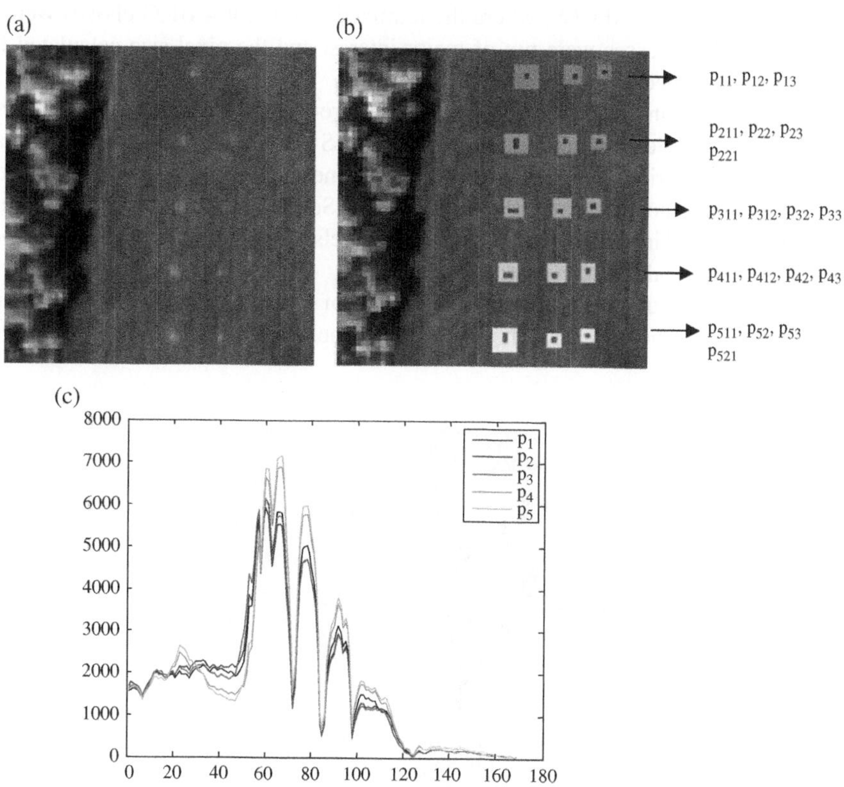


**Figure 14.1** (a) A HYDICE panel scene which contains 15 panels, (b) ground-truth map of spatial locations of the 15 panels. (c) Spectra of $\mathbf{p}_1$, $\mathbf{p}_2$, $\mathbf{p}_3$, $\mathbf{p}_4$ and $\mathbf{p}_5$, where its color map can be found in Ref. [11], figure 1.15/with permission of John Wiley & Sons.

101–112 and bands 137–153, were removed. So, a total of 169 bands were used for the experiments. It has size of 64 × 64 pixel vectors shown in Figure 14.1(a) along with its ground truth provided in Figure 14.1(b) used for experiments where the center and boundary pixels of objects are highlighted by dark gray and light gray, respectively. Each element in Figure 14.1(b) is a square panel and denoted by $p_{ij}$ with rows indexed by $i = 1, 2, \ldots, 5$ and columns indexed by $j = 1, 2, 3$. The 1.56 m-spatial resolution of the image scene suggests that most of the 15 panels are one pixel in size except that the panels in the first column with the second, third, fourth, and fifth rows which are two-pixel panels, denoted by $p_{211}$, $p_{221}$, $p_{311}$, $p_{312}$, $p_{411}$, $p_{412}$, $p_{511}$, and $p_{521}$. As a result, there are a total 19 panel pixels. Figure 14.1(b) shows the precise spatial locations of these 19 panel pixels where the dark gray pixels (DG pixels) are the panel center pixels and the pixels in light gray pixels (LG pixels) are panel pixels mixed with the background. Figure 14.1(c) plots the five panel spectral signatures obtained from Figure 14.1(b), where the $i$th panel signature, denoted by $\mathbf{p}_i$ was generated by averaging dark gray panel center pixels in row $i$. These panel signatures will be used to represent target knowledge of the panels in each row.

According to the ground truth there are five panel signatures that were obtained from Figure 14.1(c) and the other four BKG signatures were obtained by prior knowledge as the areas marked in Figure 14.2 as interferer, tree, grass, and road to make up five target classes representing five panel signatures and four BKG signatures. These signatures represent exactly nine signatures estimated by the VD. This fact provides further evidence that VD is an effective estimation method to estimate the number of image endmembers for spectral unmixing.

Table 14.1 tabulates the unmixed abundance fractions of 19 panel pixels by the three unmixing methods, FCLS, MFCLS and AFCLS, where the unmixed results for 4 BKG classes were not displayed because the BKG classes were not of major interest and the pixel size calculated from the ground truth is also provided in the second column.

Interestingly, the results in Table 14.1 show that FCLS produced the minimum least squares error, while MFCLS required the least computing time and AFCLS was the worst one. But this was mainly due to the fact that the prior knowledge which used nine signatures to produce the results in Table 14.1 was generally not sufficient as shown in [15]. So, we used automatic target generation process (ATGP) developed in [16] to generate 34 target pixels shown in Figure 14.3 which were used to form a linear mixing model to perform LSMA.

Table 14.2 tabulates the unmixed abundance fractions of 19 panel pixels by the three unmixing methods, FCLS, MFCLS and AFCLS, where the same conclusions made for Table 14.1 were true.

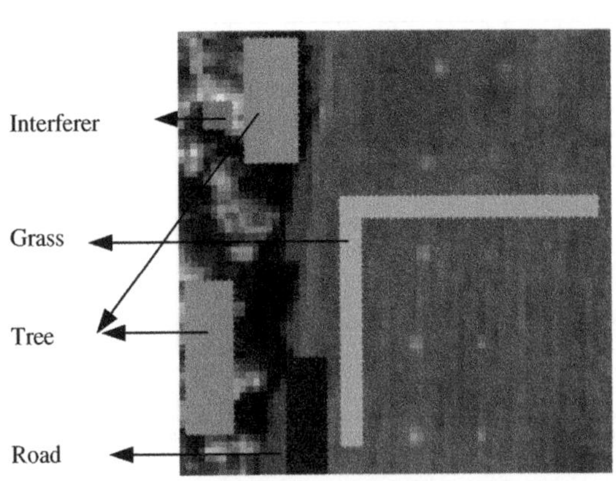


**Figure 14.2** Four BKG classes obtained by marked areas, where its color map can be found in Ref. [11], figure 1.17/with permission of John Wiley & Sons.

**Table 14.1** Unmixed abundance fractions of 19R pixels by FCLS, MFCLS, and AFCLS

| Computing time | | FCLS | MFCLS | AFCLS |
|---|---|---|---|---|
| Panel pixels | Ground truth | 0.0217 | 0.0154 | 3.0554 |
| $p_{11}$ | 1 | 0.7948 | 0.7617 | 0.6488 |
| $p_{12}$ | 1 | 0.5541 | 0.5549 | 0.5126 |
| $p_{13}$ | 0.4 | 0.0367 | 0.0355 | 0.0009 |
| $p_{211}$ | 1 | 0.5734 | 0.5436 | 0.5714 |
| $p_{212}$ | 1 | 0.9163 | 0.9156 | 0.7865 |
| $p_{22}$ | 1 | 0.8827 | 0.8828 | 0.8660 |
| $p_{23}$ | 0.4 | 0.4666 | 0.4670 | 0.4012 |
| $p_{311}$ | 1 | 0.9142 | 0.9136 | 0.9002 |
| $p_{312}$ | 1 | 0.9068 | 0.9025 | 0.8814 |
| $p_{32}$ | 1 | 0.6521 | 0.6647 | 0.5802 |
| $p_{33}$ | 0.4 | 0.2103 | 0.2105 | 0.2111 |
| $p_{411}$ | 1 | 0.5481 | 0.5094 | 0.4456 |
| $p_{412}$ | 1 | 0.4375 | 0.4378 | 0.4294 |
| $p_{42}$ | 1 | 0.6956 | 0.6910 | 0.6288 |
| $p_{43}$ | 0.4 | 0.1577 | 0.1572 | 0.1036 |
| $p_{511}$ | 1 | 0.8304 | 0.8304 | 0.8201 |
| $p_{521}$ | 1 | 1.0119 | 1.0000 | 1.0000 |
| $p_{52}$ | 1 | 0.9354 | 0.9353 | 0.9360 |
| $p_{53}$ | 0.4 | 0.1377 | 0.1374 | 0.1137 |
| Least squares errors | | 7.1121E+02 | 7.4680E+02 | 7.5832E+02 |

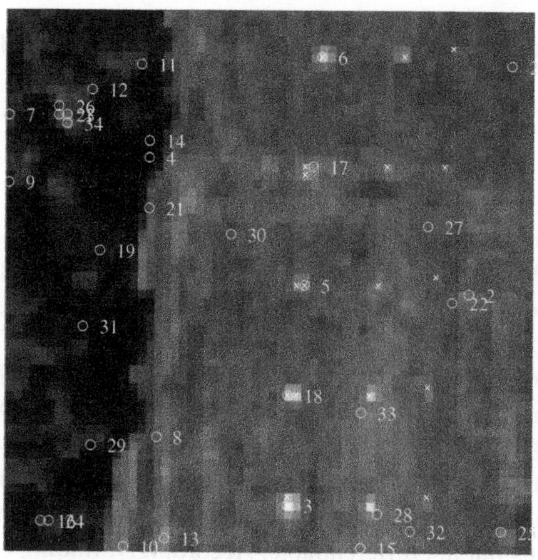


**Figure 14.3** 34 ATGP-generated target pixels.

**Table 14.2** Unmixed abundance fractions of 19R pixels by FCLS, MFCLS, and AFCLS using 34 target signatures generated by ATGP in Figure 14.3

| | | FCLS | MFCLS | AFCLS |
|---|---|---|---|---|
| Computing time | | 0.0233 | 0.0187 | 3.1359 |
| Panel pixels | Ground truth | | | |
| $p_{11}$ | 1 | 1.0000 | 1.0000 | 1.0000 |
| $p_{12}$ | 1 | 0.3748 | 0.3747 | 0.3828 |
| $p_{13}$ | 0.4 | 0.0320 | 0.0319 | 0.0091 |
| $p_{211}$ | 1 | 0.5061 | 0.5053 | 0.5078 |
| $p_{212}$ | 1 | 1.0000 | 1.0000 | 1.0000 |
| $p_{22}$ | 1 | 0.6646 | 0.6646 | 0.6655 |
| $p_{23}$ | 0.4 | 0.3457 | 0.3457 | 0.3558 |
| $p_{311}$ | 1 | 0.8631 | 0.8631 | 0.8654 |
| $p_{312}$ | 1 | 1.0000 | 1.0000 | 1.0000 |
| $p_{32}$ | 1 | 0.5319 | 0.5317 | 0.5263 |
| $p_{33}$ | 0.4 | 0.3461 | 0.3461 | 0.3520 |
| $p_{411}$ | 1 | 1.0000 | 1.0000 | 1.0000 |
| $p_{412}$ | 1 | 0.3578 | 0.3578 | 0.3872 |
| $p_{42}$ | 1 | 0.7095 | 0.7095 | 0.7102 |
| $p_{43}$ | 0.4 | 0.2345 | 0.2345 | 0.1825 |
| $p_{511}$ | 1 | 0.7245 | 0.7245 | 0.7242 |
| $p_{521}$ | 1 | 1.0000 | 1.0000 | 1.0000 |
| $p_{52}$ | 1 | 0.7773 | 0.7773 | 0.7767 |
| $p_{53}$ | 0.4 | 0.1508 | 0.1508 | 0.1519 |
| Least squares errors | | 2.0153E+02 | 2.0160E+02 | 2.0510E+02 |

However, if we look into individual R panel pixels, AFCLS actually produced better accuracy for most panel pixels.

## 14.8 Conclusion

Fully AC-LSMA implements two physical constraints, ASC and ANC. The major difficulty with implementing fully AC-LSMA is ANC which is an inequality constraint. As a result, commonly used constrained optimization technique, Lagrange multiplier method, cannot be applied. To resolve this dilemma, the FCLS developed by Heinz and Chang appealed for the Kuhn–Tucker conditions to find numerical solutions to approximate the optimal solutions. Later, Ren and Chang further introduced a novel idea which converts an inequality-based ANC to an equality-based AASC to make the equality–inequality (ASC, ANC)-constrained LSMA a two-equality (ASC, AASC)-constrained optimization problem so that the Lagrange multiplier method is applicable to finding fully AC-LSMA. The resulting (ASC, AASC)-constrained LSMA is referred to as MFCLS. Despite that AASC is an equality constraint, it is continuous but not differentiable. Accordingly, finding closed forms for solutions is not available. In this case, MFCLS derives two equations which can be iterated to find numerical solutions. Although FCLS and MFCLS can find solutions to fully AC-LSMA by using different constraints, they both still have to rely on numerical algorithms to find approximate optimal solutions. To further resolve this issue, the (ASC, AASC)-constrained optimization problem is solved by an analytic approach, to be called the AMNCLS method. Combining SCLS and AMNCLS yields AFCLS which can find FCLS solutions in closed forms via Cramer's rule analytically.

## References


1. Chang, C.-I (2003). *Hyperspectral Imaging: Techniques for Spectral Detection and Classification*. New York: Kluwer Academic/Plenum Publishers.
2. Heinz, D. and Chang, C.-I (2001). Fully constrained least squares linear mixture analysis for material quantification in hyperspectral imagery. *IEEE Transactions on Geoscience and Remote Sensing* 39: 529–545.
3. Ren, H. and Chang, C.-I (1999). A constrained least squares approach to hyperspectral image classification. In: *Conference on Information Science and Systems*, 17–19. Baltimore, MD: Johns Hopkins University.
4. Wong, E. and Chang, C.-I (2012). Modified full abundance-constrained spectral unmixing. *Proceedings of Conference High-Performance Computing in Remote Sensing*, SPIE 8539, Edinburgh, United Kingdom (24–27 September 2012).
5. Harsanyi, J.C. and Chang, C.-I (1994). Hyperspectral image classification and dimensionality reduction: an orthogonal subspace projection approach. *IEEE Transactions on Geoscience and Remote Sensing* 32 (4): 779–785.
6. Tu, T.M., Chen, C.-H., and Chang, C.-I (1997). *A posteriori* least squares orthogonal subspace projection approach to weak signature extraction and detection. *IEEE Transactions on Geoscience and Remote Sensing* 35 (1): 127–139.
7. Chang, C.-I, Zhao, X., Althouse, M.L.G., and Pan, J.-J. (1998). Least squares subspace projection approach to mixed pixel classification in hyperspectral images. *IEEE Transactions on Geoscience and Remote Sensing* 36 (3): 898–912.
8. Chang, C.-I (2005). Orthogonal subspace projection revisited: a comprehensive study and analysis. *IEEE Transactions on Geoscience and Remote Sensing* 43 (3): 502–518.
9. Chang, C.-I (ed.) (2007). Chapter 3: information-processed matched filters for hyperspectral target detection and classification. In: *Hyperspectral Data Exploitation: Theory and Applications*, 47–74. Wiley.
10. Chang, C.-I and Du, Q. (2004). Estimation of number of spectrally distinct signal sources in hyperspectral imagery. *IEEE Transactions on Geoscience and Remote Sensing* 42 (3): 608–619.
11. Chang, C.-I (2013). *Hyperpscetral Data Processing*, Chapter 4. Wiley.
12. Chang, C.-I (2017). *Real-Time Recursive Hyperspectral Sample and Band Processing: Algorithm Architecture and Implementation*. New York: Springer.
13. Chang, C.-I (2018). A review of virtual dimensionality for hyperspectral imagery. *IEEE Journal of Selected Topics in Applied Earth Observation and Remote Sensing* 11 (4): 1285–1305.
14. Chang, C.-I and Heinz, D. (2000). Constrained subpixel detection for remotely sensed images. *IEEE Transactions on Geoscience and Remote Sensing* 38 (3): 1144–1159.
15. Chang, C.-I, Jiao, X., Du, Y., and Chen, H.M. (2011). Component-based unsupervised linear spectral mixture analysis for hyperspectral imagery. *IEEE Transactions on Geoscience and Remote Sensing* 49 (11): 4123–4137.
16. Ren, H. and Chang, C.-I (2003). Automatic spectral target recognition in hyperspectral imagery. *IEEE Transactions on Aerospace and Electronic Systems* 39 (4): 1232–1249.

# 15

# Swarm Intelligence Optimization-Based Spectral Unmixing

Lianru Gao[1], Xu Sun[1], Zhu Han[1,2], Lina Zhuang[3], Wenfei Luo[4], and Bing Zhang[1,2]
[1]*Key Laboratory of Digital Earth Science, Aerospace Information Research Institute, Chinese Academy of Sciences, Beijing, China*
[2]*College of Resources and Environment, University of Chinese Academy of Sciences, Beijing, China*
[3]*Department of Mathematics, Hong Kong Baptist University, Hong Kong, China*
[4]*School of Geographical Science, South China Normal University, Guangzhou, China*

## 15.1 Introduction

Mixed pixels are common in hyperspectral image due to relatively low spatial resolution. The main purpose of spectral unmixing is to analyze the materials (called endmembers) contained in the mixed pixel and their proportions (called abundances). The basic models normally adopted in spectral unmixing are LMM, NLMM, and NCM. Among them, LMM is widely used in mixed pixel problems, because it can satisfy the physical principles of the spectral mixing process under certain conditions, and the form of LMM is simple. The endmember extraction algorithm based on LMM can be further divided into two main categories: geometrical and statistical. Of these approaches, geometrical methods have the longest history of study, and these algorithms are the most abundant. Geometrical methods are based on convex geometry concepts. According to different assumptions, the geometrical methods can be classified in two types: pure pixel-based methods and minimum volume-based methods. Pure pixel-based methods seek endmembers under the assumption that there are pure pixels in the scene. Typical pure pixel-based methods are the vertex component analysis (VCA) [1], NFINDR [2], and pixel purity index (PPI) [3]. Minimum volume-based methods aim to find the simplex of endmembers with minimum volume, such as minimum volume-constrained nonnegative matrix factorization (MVCNMF) [4], minimum volume simplex analysis (MVSA) [5], and robust minimum volume enclosing simplex (RMVES) [6]. Compared to LMM, NLMM is used to describe multiple scattering effects. Therefore, NLMM is more advantageous for dealing with mixed pixel problems in high resolution and complex scenes. As an extension of LMM, NCM uses the probability to describe the uncertainty of the spectrum. It regards the endmember as a random

*Advances in Hyperspectral Image Processing Techniques*, First Edition. Edited by Chein-I Chang.

variable with a given probability distribution and uses specific parameters to represent the endmember variability.

SI algorithms are an important branch of artificial intelligence. Over the past years, these methods have developed rapidly and have demonstrated great potential for solving the complicated optimization problem. Typical SI algorithms include particle swarm optimization (PSO), ant colony optimization (ACO), and artificial bee colony (ABC) algorithm. The mixed pixel decomposition problem can be transformed into optimization problems with different objective functions. Therefore, it is very suitable for searching the optimal solution by using the SI algorithms. In recent years, many unmixing methods based on SI have been proposed. Zhang et al. [7] used ACO and discrete particle swarm algorithm (DPSO) to extract the endmember by solving LMM problem. In 2013, Zhang et al. [8] adopted the elite ant strategy to improve the endmember extraction algorithm based on ACO. Gao et al. [9] compared the effects of different preprocessing methods on the endmember extraction results of ACO and proposed a multi-algorithm fusion endmember extraction strategy based on ACO. Sun et al. [10] proposed the ABC unmixing algorithm to complete the endmember extraction without relying on the assumption of pure pixels. Zhuang et al. [11] proposed a new Bayesian method for improved parameter estimation in the NCM. Gao et al. [12] took the quantitative and comparative analysis of different preprocessing implementations for DPSO and concluded that VCA is the most reliable preprocessing implementation. Su et al. [13] improved the objective function of ACO and DPSO and proposed an endmember extraction algorithm for discrete ABC optimization. Gao et al. [14] proposed the region-based stochastic expectation maximization (EM) to learn endmember variances from spatial information and the proposed approach achieved better endmember variance estimates. Luo et al. [15] used dual population PSO algorithm to solve the bilinear spectral mixing model.

In the remainder of this chapter, the common spectral mixing models are first reviewed. Then, typical SI optimization-based approaches are introduced. Third, the proposed methods are compared using the real hyperspectral image to validate the effectiveness. Finally, the conclusions and discussion for future work are provided.

## 15.2 Pixel Mixing Models

1. *Linear Mixing Model (LMM)*

There are two assumptions in LMM: (i) no interaction between photons; (ii) a mixed pixel's spectrum is a linear superposition of the spectrum of each material represented in the pixel based on its corresponding proportion. The LMM approach is the basis for most spectral unmixing algorithms. Consider a $B$-band hyperspectral image $\{\mathbf{x}_i\}_{i=1}^{N}$ which contains $N$ pixels. There are $M$ endmembers $\{\mathbf{e}_j\}_{j=1}^{M}$ in each pixel $\mathbf{x}_i \in \mathbb{R}^B$. Then, the LMM is

$$\mathbf{x}_i = \sum_{j=1}^{M} \alpha_{ij}\mathbf{e}_j + \boldsymbol{\varepsilon}_i \tag{15.1}$$

where $\alpha_{ij}$ is the abundance, the proportion of the $j$th endmember in the $i$th pixel, and $\boldsymbol{\varepsilon}_i$ represents the random error.

Since $\alpha_{ij}$ is the abundance, theoretically, it should be nonnegative and the sum of which for each pixel should equal to one, i.e.

$$\alpha_{ij} \geq 0, \quad \forall i, \ \forall j \tag{15.2}$$

$$\sum_{j=1}^{M} \alpha_{ij} = 1, \quad \forall i \tag{15.3}$$

Equations (15.2) and (15.3) are the constraints for Eq. (15.1), where the former is the "nonnegative" constraint, and the latter is the "sum-to-one" constraint.

For the estimation of the values $\{\hat{\alpha}_{ij}\}_{N\times M}$ of $\{\alpha_{ij}\}_{N\times M}$ after the endmember extraction process, $\{\mathbf{x}_i\}_{i=1}^{N}$ and $\{\mathbf{e}_j\}_{j=1}^{M}$ are required. If Eqs. (15.2) and (15.3) are not taken into consideration, a least-squares analysis can be applied. While obtaining solution, considering only Eq. (15.3) is called "sum-to-one"-constraint least-squares (SCLS) analysis and considering Eqs. (15.2) and (15.3) is called "fully constrained" least-squares (FCLS) analysis [16, 17]. Mature algorithms for both approaches are available at present.

LMM-based endmember extraction methods can be categorized into geometric, statistical, sparsity-based, and spatial contextual information-based methods [18]. Among these approaches, the geometric methods have the longest history and are the most abundant. Typical geometric methods are NFINDR, VCA, the sequential maximum angle convex cone (SMACC) model [19], alternating volume maximization (AVMAX) [20], MVCNMF, etc.

2. *Normal Compositional Model (NCM)*

Let $\mathbf{X} = \{\mathbf{x}_1, \ldots, \mathbf{x}_N\} \in \mathbb{R}^{B\times N}$ denote a hyperspectral image (HSI) with $N$ pixel vectors (the columns of $\mathbf{X}$) of size $B$ (the number of bands of the sensor). In hyperspectral unmixing problem, NCM represents each pixel $\mathbf{x} \in \mathbb{R}^{B}$ as:

$$\mathbf{x} = \sum_{m=1}^{M} \alpha_m \mathbf{e}_m \tag{15.4}$$

where $\mathbf{e} \in \mathbb{R}^{B\times M}$, $\boldsymbol{\alpha} \in \mathbb{R}^{M}$, and $M$ denote Gaussian distributed endmembers, abundance vector, and number of endmembers, respectively. Herein abundance vector is constrained as nonnegativity and sum-to-one. In order to consider inherent spectral uncertainty lying in materials, endmembers are modeled as Gaussian random variables in NCM. We may represent $m$th Gaussian endmember as $\mathbf{e}_m \sim \mathcal{N}(\boldsymbol{\mu}_m, \mathbf{C}_m)$, where $\boldsymbol{\mu}_m$ is mean and $\mathbf{C}_m$ is covariance matrix. Mean and covariance matrix are considered to be sufficient to capture endmember spectral variability.

NCM can be considered as a two-stage hierarchical model for observation $\mathbf{x}$ [21]. We have the conditional probability density function:

$$f(\mathbf{x} \mid \boldsymbol{\theta}_1, \boldsymbol{\theta}_2) = \int f(\mathbf{x} \mid \boldsymbol{\alpha}, \boldsymbol{\theta}_1) g(\boldsymbol{\alpha} \mid \boldsymbol{\theta}_2) \mathrm{d}\boldsymbol{\alpha} \tag{15.5}$$

It refers to two unknown parameters: $\boldsymbol{\theta} = \{\boldsymbol{\theta}_1, \boldsymbol{\theta}_2\}$. Parameter $\boldsymbol{\theta}_1 = \{\boldsymbol{\mu}_1, \mathbf{C}_1, \boldsymbol{\mu}_2, \mathbf{C}_2, \ldots, \boldsymbol{\mu}_M, \mathbf{C}_M\}$ represents endmember statistics. Parameter $\boldsymbol{\theta}_2$ describes the prior distribution of abundances. In this work, the abundance of each pixel is assumed to be a normal distributed vector, so we have

$$\boldsymbol{\theta}_2 = \boldsymbol{\mu}', \mathbf{C}' \tag{15.6}$$

With these considerations, the functions in the right-hand side of Eq. (15.5) can be represented in detail as follows.

The conditional probability density function is normal,

$$f(\mathbf{x} \mid \boldsymbol{\alpha}, \boldsymbol{\theta}_1) = \frac{1}{(2\pi)^{1/2}} \frac{1}{|\mathbf{C}(\boldsymbol{\alpha})|^{1/2}} \exp\left\{ -\frac{1}{2}[\mathbf{x} - \boldsymbol{\mu}(\boldsymbol{\alpha})]^T \mathbf{C}^{-1}(\boldsymbol{\alpha})[\mathbf{x} - \boldsymbol{\mu}(\boldsymbol{\alpha})] \right\} \tag{15.7}$$

with a mean vector

$$\boldsymbol{\mu}(\boldsymbol{\alpha}) = \sum_{m=1}^{M} \alpha_m \boldsymbol{\mu}_m, \tag{15.8}$$

and the covariance matrix

$$\mathbf{C}(\boldsymbol{\alpha}) = \sum_{m=1}^{M} \alpha_m^2 \mathbf{C}_m. \tag{15.9}$$

The prior probability density function $g(\boldsymbol{\alpha}, \boldsymbol{\theta}_2)$ is also assumed to be normal as

$$g(\boldsymbol{\alpha}, \boldsymbol{\theta}_2) \sim \mathcal{N}(\boldsymbol{\mu}', \mathbf{C}'). \tag{15.10}$$

Assume that the observation of each pixel is independent, then the joint density function for the entire image,

$$\mathbf{X} = (\mathbf{x}_1, \mathbf{x}_2, \ldots, \mathbf{x}_N) \tag{15.11}$$

is given by

$$f(\mathbf{X} \mid \boldsymbol{\theta}_1, \boldsymbol{\theta}_2) = \prod_{i=1}^{N} \int f(\mathbf{x}_i \mid \boldsymbol{\alpha}_i, \boldsymbol{\theta}_1) g(\boldsymbol{\alpha}_i, \boldsymbol{\theta}_2) \mathrm{d}\boldsymbol{\alpha}_i. \tag{15.12}$$

3. *Non-Linear Mixture Model (NLMM)*

Given a hyperspectral image with $N$ pixels, an NLMM that does not consider intimate mixture can be summarized as

$$\mathbf{x}[n] = nl(\mathbf{E}, \mathbf{a}[n], \boldsymbol{\theta}[n]) + \boldsymbol{\varepsilon}, \quad n = 1, \ldots, N \tag{15.13}$$

where $\mathbf{x}[n] = [x_1[n], \ldots, x_B[n]]^{\mathbf{T}}$, $n = 1, \ldots, N$ is the $n$th pixel vector comprising $B$ bands, $nl(\mathbf{E}, \mathbf{a}[n], \boldsymbol{\theta}[n])$ is a nonlinear mixing function, $\mathbf{E} = [\mathbf{e}_1, \mathbf{e}_2, \ldots, \mathbf{e}_M] \in \mathbb{R}^{B \times M}$ is the endmember matrix whose $i$th column $\mathbf{e}_i = [e_{1,i}, \ldots, e_{B,i}]^{\mathbf{T}} \in \mathbb{R}^B$ represents the $i$th endmember signature vector, $M$ is the total number of endmembers, $\mathbf{a}[n] = [a_1[n], a_2[n], \ldots, a_M[n]]^{\mathbf{T}} \in \mathbb{R}^M$ denotes the abundance fraction for the $n$th pixel, $\boldsymbol{\theta}[n]$ is the vector of other model parameters including nonlinear parameters, and $\boldsymbol{\varepsilon}$ is a noise vector.

For example, the nonlinear mixing function $nl$ of Fan's model (FM) [22] can be written by

$$nl_{\mathrm{FM}}(\mathbf{E}, \mathbf{a}[n], \boldsymbol{\theta}[n]) = \mathrm{FM}(\mathbf{E}, \mathbf{a}[n]) = \mathbf{Ea}[n] + \sum_{i=1}^{M-1} \sum_{j=i+1}^{M} a_i[n] a_j[n] \mathbf{e}_i \odot \mathbf{e}_j \tag{15.14}$$

and that of generalized bilinear model (GBM) [23] can be written by

$$nl_{\text{GBM}}(\mathbf{E}, \mathbf{a}[n], \boldsymbol{\theta}[n]) = \mathbf{E}\mathbf{a}[n] + \sum_{i=1}^{M-1} \sum_{j=i+1}^{M} \theta_{i,j}[n] a_i[n] a_j[n] \mathbf{e}_i \odot \mathbf{e}_j \tag{15.15}$$

## 15.3 Swarm Intelligence Optimization-Based Approaches

### 15.3.1 Approach Based on LMM

1. *Pure Pixel-Based Algorithm*

Assume that the image consists of pure pixels, SI algorithms target at searching the best image endmembers that would have the minimum root mean square error (RMSE) after hyperspectral scene reconstruction by using LMM:

$$\text{RMSE}\left(\{\mathbf{x}_i\}_{i=1}^{N}, \mathbf{E}\right) = \frac{1}{N}\sum_{i=1}^{N} \sqrt{\frac{1}{B}\|\mathbf{x}_i - \hat{\mathbf{x}}_i\|_2^2} \tag{15.16}$$

where $\{\mathbf{x}_i\}_{i=1}^{N}$ is the pixel spectra and $\mathbf{E} = \left\{\mathbf{e}_j\right\}_{j=1}^{M}$ is the endmember set. Based on the pure pixel assumption, $\mathbf{E} = \left\{\mathbf{e}_j\right\}_{j=1}^{M} \subset \{\mathbf{x}_i\}_{i=1}^{N}$. Based on LMM, variable $\hat{\mathbf{x}}_i$ is the $i$-th remixed pixel, i.e. $\hat{\mathbf{x}}_i = \sum_{j=1}^{M} a_{ij}\mathbf{e}_j, \quad i = 1, 2, \ldots, N$. Therefore, the endmember extraction process can be defined as a combinatorial optimization problem. Its objective function is:

$$\min \text{RMSE}\left(\{\mathbf{x}_i\}_{i=1}^{N}, \mathbf{E}\right) \tag{15.17}$$

$$\text{s.t.} \quad \mathbf{E} \in C\left(\{\mathbf{x}_i\}_{i=1}^{N}, M\right) \tag{15.18}$$

where $C\left(\{\mathbf{x}_i\}_{i=1}^{N}, M\right)$ is the subsets of $\{\mathbf{x}_i\}_{i=1}^{N}$, which contain $M$ elements, and also the feasible solution space. Specifically, the solution space is discrete. Consider image $\{\mathbf{x}_i\}_{i=1}^{N}$ and endmember number $M$, then $\mathbf{R}_{N,M} = \left\{(r_1, r_2, \ldots, r_N) \mid r_i \in \{0, 1\}, \ \sum_{i=1}^{N} r_i = M\right\}$ is defined as the solution space where $\mathbf{r} \in \mathbf{R}_{N,\ M}$ represents an n-dimensional potential solution. Thus, the mapping relationship, represented by $G$, between the image and the feasible solution space is:

$$\begin{aligned} G &: C\left(\{\mathbf{x}_i\}_{i=1}^{N}, M\right) \to \mathbf{R}_{N,M} \\ \mathbf{E} &\to \mathbf{r} = (r_1, r_2, \ldots, r_N) \end{aligned} \tag{15.19}$$

where

$$r_i = \begin{cases} 1 & \mathbf{x}_i \in \mathbf{E} \\ 0 & \mathbf{x}_i \notin \mathbf{E} \end{cases} \tag{15.20}$$

Suppose $G^{-1}$ represents the inverse mapping of $G$ (defined by Eq. (15.19)), the SI-based methods' adaptability function would become:

$$f(\mathbf{r}) = \text{RMSE}\left(\{\mathbf{x}_i\}_{i=1}^{N}, G^{-1}(\mathbf{r})\right) \tag{15.21}$$

Hence, the endmember extraction has become a problem of searching for the optimal **r**, which indeed is a discrete solution that SI-based algorithms can search for.

PSO, first proposed by Eberhart and Kennedy [24], simulates the behavior in a bird flock or fish school searching their habitat (or food), so as to find the overall optimal solution for optimization problems. In nature, when a bird flock flies through an area and searches for habitats, each bird evaluates the surrounding environment and decide whether the place is suitable for habitation individually, and exchange information within a certain range by tweeting. Each bird will adjust the flight direction and speed based on the information from others and their own experience, and continue to seek its habitat. Eventually, with this strategy, the flock can find the most suitable habitat in a region.

In an optimization problem, particles are used in PSO for searching in the feasible solution space, where the particles move continuously, and the value of the adaptability function is calculated. Each particle has two attributes (position and velocity) and remembers its own and global optimum position through the whole searching process. Since the feasible solution space $X_{N,M}$ is a discrete set, the PSO used on such domain is called discrete particle swarm optimization (DPSO) [25].

In the feasible solution space, as each particle is searched, the position of particle is the feasible solution. In other words, the $k$th particle's position at moment $t$ is $\mathbf{r}_k(t) \in R_{N,M}$. Apparently, $\mathbf{r}_{k,\text{best}}(t)$, $\mathbf{r}_{\text{gbest}}(t) \in R_{N,M}$. If set $V_{n,q} = \left\{ (v_1, v_2, \ldots, v_n) \mid v_i \in \{-1,0,1\}, \sum_{i=1}^{n} v_i = 0, \sum_{i=1}^{n} |v_i| = 2q \right\}$ is defined, then $\mathbf{v} \in V_{n,q}$ expresses a numeral string with $n$-digit composed of 0, 1 and −1 (or a $n$-dimensional vector). The number of 1 and −1 is $q$. Assume $\mathbf{E}_1, \mathbf{E}_2 \in C\left(\{\mathbf{r}_i\}_{i=1}^{N}, M\right)$ and $\mathbf{x}_1 = G(\mathbf{E}_1)$, $\mathbf{x}_2 = G(\mathbf{E}_2)$, then $\mathbf{x}_1, \mathbf{x}_2 \in X_{N,M}$. The velocity from $\mathbf{x}_2$ to $\mathbf{x}_1$ is $\mathbf{v} = \mathbf{x}_1 - \mathbf{x}_2 = (v_1, v_2, \ldots, v_n) \in V_{n,q}$, which satisfies:

$$v_i = \begin{cases} 1 & \mathbf{x}_i \in \mathbf{E}_1, \quad \mathbf{x}_i \notin \mathbf{E}_2 \\ -1 & \mathbf{x}_i \notin \mathbf{E}_1, \quad \mathbf{x}_i \in \mathbf{E}_2 \\ 0 & \text{other} \end{cases} \tag{15.22}$$

where **v** describes the difference between $\mathbf{E}_1$ and $\mathbf{E}_2$, whereas $q$ denotes the extent of such difference. Therefore, $q$ is referred to as the magnitude of velocity. $\mathbf{r}_{k,\text{best}}(t) - \mathbf{r}_k(t) \in V_{n,q_1}$, $\mathbf{r}_{\text{gbest}}(t) - \mathbf{r}_k(t) \in V_{n,q_2}$ is obtained based on the definitions of position and velocity.

In $R_{N,M}$, velocity is updated with the following equation:

$$\mathbf{v}_k(t+1) = \begin{cases} P\left(\left(\mathbf{r}_{k,\text{best}}(t) - \mathbf{r}_k(t)\right) + \left(\mathbf{r}_{\text{gbest}}(t) - \mathbf{r}_k(t)\right)\right) \\ Q\left(\mathbf{r}_k(t)\right) \end{cases} \tag{15.23}$$

where $P$ and $Q$ denote the random selection function. $P(\mathbf{r})$ represents a random selection of a position with the value of 1 from the $\mathbf{r}'$s positive component or a position with the value of −1 from the $\mathbf{r}'$s negative component, the remaining position values are 0, then $\mathbf{v}_k(t+1) \in V_{n,1}$ or $\mathbf{r}_k(t) \in R_{N,M}$ is obtained. $Q(\mathbf{r})$ represents a random selection of a position with the value of −1 is randomly selected from the $\mathbf{r}'$s positive component or a position with the value of 1 is randomly selected from the $\mathbf{r}'$s negative component, the remaining position values are 0, then $\mathbf{v}_k(t+1) \in V_{n,1}$ is obtained. $P(\mathbf{r})$, referred to as "directional movement," represents velocity obtained according to self and social experience. $Q(\mathbf{r})$, referred to as "random movement," represents the selection of a velocity randomly without experience. Within every iteration, particles will choose directional or random movement randomly. In other words, if a random selection probability is set to $p \in (0, 1)$, then the probability of particles that select directional movement is $p$ while the probability of the selection of random movement is $1 - p$. Moreover, a particle can only undergo random movement, if it is at the global optimal position, then $(\mathbf{r}_{k,\text{ best}}(t) - \mathbf{r}_k(t)) + (\mathbf{r}_{\text{gbest}}(t) - \mathbf{r}_k(t)) = \mathbf{0}$.

The huge computational cost poses a great challenge to SI due to the extremely large feasible solution space, which is filled with numerous vectors corresponding to the image pixels. A solution to this issue is preprocessing, which provides a small number of candidate endmembers. The PPI is used to extract candidate endmembers, and these candidate endmembers retrieved are used to obtain the final results by DPSO.

2. *Minimum Volume-Based Algorithm*

Corresponding pure pixels of some materials are difficult to find in hyperspectral images with low signal-to-noise ratio (SNR) or low spatial resolution. The pure pixel assumption is not acceptable in this scenario. Therefore, endmember extraction no longer has a feasible solution as shown in Eq. (15.19). Instead, it should be modified to:

$$\mathbf{E} \in \mathbb{R}_{+}^{B \times M} \tag{15.24}$$

Additionally, it should be reminded that Eq. (15.17) is no longer a suitable objective function in this situation. The reason is that the RMSE becomes smaller as the estimated endmembers form a larger simplex that contains all pixels. These endmembers extracted no longer have a true physical meaning corresponding to the pure materials in the image. It also does not meet the minimum volume requirements. In order to deal with this issue, a volume penalty term is added in the objective function shown in Eq. (15.17), that is:

$$f(\mathbf{E}) = \text{RMSE}\left(\{\mathbf{x}_i\}_{i=1}^{N}, \mathbf{E}\right) + \mu \cdot V(\mathbf{E}) \tag{15.25}$$

where $V(\mathbf{E})$ denotes the endmember simplex volume (the volume term of the objective function). $\text{RMSE}\left(\{\mathbf{x}_i\}_{i=1}^{N}, \mathbf{E}\right)$ is the error of the objective function. $\mu$ represents the penalty coefficient, which is adopted to adjust the weight of error and volume.

Equation (15.25) is used as the objective function, while Eq. (15.24) is the feasible solution, in the endmember extraction process based on SI and minimum volume. Rather than a discrete optimization problem, it is continuous, which means that DPSO is not a suitable method. However, an effective SI algorithm, ABC, showed good performance when dealing with continuous optimization problems [10]. ABC can be categorized as a minimum volume-based algorithm because the objective Eq. (15.25) solved by ABC has the intention of volume minimization.

ABC algorithms simulate the bees' foraging behavior in nature, and they are often adopted to solve optimization problems. In ABC, the colony's search space is the feasible solution space and a food source is a feasible solution for the problem to be solved. Fitness is the amount of nectar in each food source. It is related to the value of the objective function that is generated by corresponding solution. The higher the fitness and the more bees the food source attracts, the better solution it achieves. Bees can be categorized into three types: employed bees, onlooker bees, and scout bees [26].

Each employed bee corresponds to a source of food (and the corresponding fitness value). It searches the neighborhood near the food source to find a new one. If the fitness of the new source of food is better than that of the original one, the new food source (and corresponding fitness value) will be recorded. Else, the employed bee will discard the new food source and continue searching near the original one. Assume that $X_i = (x_{i1}, x_{i2}, \ldots, x_{i,B\times N})^T \in \mathbb{R}_+^{B\times(B-1)}$ denotes the $i$th source of food (the $i$th employed bee's location), then the search in its neighborhood is

$$x'_{ij} = x_{ij} + \Phi(x_{ij} - x_{kj}) \tag{15.26}$$

where $k$, different from $i$, is another source of food selected randomly; $j$ is an integer chosen from $\{1, 2, \ldots, B\times(B-1)\}$ randomly, and $\Phi \in [0, 1]$ is a random number with uniform distribution.

Each onlooker bee chooses a food source with a specific probability (referred to as the follow probability) and looks for that kind of food source throughout the neighborhood according to the fitness values of all food sources found by the employed bees. When a new source of food with better fitness is located, only the corresponding employed bee's food source will be updated. If not, the new food source will be discarded. Onlooker bees' probability of following the $j$th food is

$$P_j = \frac{\text{fit}_j}{\sum_{i=1}^{N_e} \text{fit}_i} \tag{15.27}$$

where $N_e$ is the total number of food sources, which could also be referred to as the total number of employed bees. $\text{fit}_i$ is the fitness of the $i$th food source. If the value of the objective function of the optimization problem is nonnegative, then the fitness of food source $X_i$ is

$$\text{fit}_i = \frac{1}{f(X_i)} \tag{15.28}$$

where $f(x)$ is the objective function of the optimization problem. $f(X_i)$ with a smaller value implies larger $\text{fit}_i$ and $P_i$. Therefore, better food sources attract more onlooker bees who will search in their vicinity and eventually will obtain more solutions with high quality. For endmember extraction, the objective function is Eq. (15.25) and the feasible solution (food source) is Eq. (15.24).

An employed bee will discard its current source of food and do a "role change," transforming into a scout bee, if it cannot seek a better source of food in its current neighborhood for a long time. The scout bee search for sources of food in the feasible solution space randomly and obtains their fitness, and then it turns itself into an employed bee. This role change strategy enables the ABC algorithm to easily leave the local minima, thereby search for a global optimal solution effectively.

Among all sources of food (including the abandoned ones), the best one will be logged during the search process. This solution will be regarded as the optimal solution if it is not updated in a long period of time. According to Xiao [27], while solving the proposed endmember extraction optimization model, a standard and universal optimization model, it can certainly be converged by increasing the number of iterations.

### *15.3.2 Approach Based on NCM*

There is not a directly methodology to estimate NCM parameters because of the intractable integrals [28]. This section studies a new algorithm based on the framework of "winner-take-all" version of EM to deal with the integral computation [29]. The method PSO-EM also considers the spatial information into the prior probability density function of abundance [30]. Finally, PSO would be introduced as an optimization technique to find the "winner" abundance in the E step of EM.

#### 15.3.2.1 "Winner-Take-All" Version of the EM

The EM algorithm finds maximum likelihood parameter estimates in unmixing problems [31] where abundance is considered as the hidden variable [32–34]. It estimates the unknown parameters of NCM iteratively, starting from some initial guess [35]. Each iteration consists of an expectation step (E step) and a maximization step (M step). And the EM and its incremental versions have been proven mathematically to be convergent in Neal and Hinton [29].

1. *Initialization Strategy*

The EM seeks to maximize the log-likelihood function

$$L(\mathbf{X},\boldsymbol{\theta}_1,\boldsymbol{\theta}_2) = \log f(\mathbf{X} \mid \boldsymbol{\theta}_1,\boldsymbol{\theta}_2) = \sum_{i=1}^{N} \log \int f(\mathbf{x}_i \mid \boldsymbol{\alpha}_i,\boldsymbol{\theta}_1)\, g(\boldsymbol{\alpha}_i,\boldsymbol{\theta}_2)\mathrm{d}\boldsymbol{\alpha}_i \tag{15.29}$$

where refers to the unknown parameters $\{\boldsymbol{\theta}_1,\boldsymbol{\theta}_2\}$ needed to be initialized. If a good initialization can be provided, then the convergence of EM would be speed up significantly since the initialized parameters are close to the true values in the solution space. Here, we suggest some strategies for initialization as follows.

Here, parameter $\boldsymbol{\theta}_1$ is used to describe the information of endmember classes. The means of endmember classes are suggested being initialized by the automatic endmember extraction techniques such as the famous VCA or N-FINDR algorithm [36]. Then an adequate approach for endmember covariance initialization has been found to work very well in our experiments. That is to use the covariance of neighbor pixels those near the endmember extracted previously in the image.

Then, the distribution of abundance, described by parameter $\boldsymbol{\theta}_2$, is used to be assumed as a uniform prior probability density in the early literature. In this section, however, spatial information would be considered into the prior distribution of abundances based on the assumption that the neighbor spectra are more likely to be similar. Here, the rough abundances of neighbor pixels are suggested initializing the parameter $\boldsymbol{\theta}_2$. More precisely, a rough abundance map can be received after the conventional unmixing process with the help of endmember extracted and least-squares unmixing technique [37, 38]. The mean value and covariance matrix of neighbor abundances are calculated for initializing the prior distribution of central abundance. And different size of structuring element can be used to define neighborhood around the central pixel. Because of the similar spectra assumption, the size of structuring element should not be too large and that would be discussed in detail in the following experiments. In our experiment, the structuring element of size $7 \times 7$ pixels are usually used.

2. *E Step*

The original E step is trying to find the distribution for hidden abundance variables in Eq. (15.27) and obtain the expectation value according to the distribution. Estimating the distribution on abundance variable would be really difficult because the abundance vector is a continuous as well as high-dimensional variable. In order to solve this problem, an incremental version of EM: "winner-take-all" **version** is employed in this paper, in which, for the iterations, the distributions over hidden abundance variables are restricted to those in which a single value has probability one. Assuming the particular distribution, the expectation value of hidden abundance turns to a single value that has the maximum probability.

E step: finds the $\boldsymbol{\alpha}^{(t)}$ that maximizes the log-likelihood function given the known values for the observed variables $\mathbf{X}$ and the current estimated of the parameters $\left\{\boldsymbol{\theta}_1^{(t)},\boldsymbol{\theta}_2^{(t)}\right\}$

$$\boldsymbol{\alpha}^{(t)} = \arg\max_{\boldsymbol{\alpha}} \sum_{i=1}^{N} \log \int f\left(\mathbf{x}_i \mid \boldsymbol{\alpha}_i, \boldsymbol{\theta}_1^{(t)}\right) g\left(\boldsymbol{\alpha}_i, \boldsymbol{\theta}_2^{(t)}\right) d\boldsymbol{\alpha}_i \tag{15.30}$$

3. *M Step*

M step: re-estimates the parameters $\left\{\boldsymbol{\theta}_1^{(t+1)}, \boldsymbol{\theta}_2^{(t+1)}\right\}$ to those with maximum likelihood value

$$\boldsymbol{\theta}^{(t+1)} = \arg\max_{\boldsymbol{\theta}} \sum_{i=1}^{N} \log f\left(\mathbf{x}_i \mid \boldsymbol{\alpha}_i^{(t)}, \boldsymbol{\theta}_1\right) g\left(\boldsymbol{\alpha}_i^{(t)}, \boldsymbol{\theta}_2\right) \tag{15.31}$$

More precisely, parameter update function can be derived from Eq. (15.31) as follows [39]:

$$\boldsymbol{\mu}_m^{(t+1)} = \frac{\sum_{i=1}^{N} \alpha_{\text{im}}^{(t)} \mathbf{x}_i}{\sum_{i=1}^{N} \alpha_{\text{im}}^{(t)}} \tag{15.32}$$

$$\mathbf{C}_m^{(t+1)} = \frac{\sum_{i=1}^{N} \alpha_{\text{im}}^{(t)} (\mathbf{x}_i - \boldsymbol{\mu}(\boldsymbol{\alpha}))(\mathbf{x}_i - \boldsymbol{\mu}(\boldsymbol{\alpha}))^T}{\sum_{i=1}^{N} \alpha_{\text{im}}^{(t)}} \tag{15.33}$$

$$\boldsymbol{\mu}_i'^{(t+1)} = \frac{1}{N_{\text{nb}}} \sum_{\substack{k=1 \\ k \in \Omega_i}}^{N_{\text{nb}}} \boldsymbol{\alpha}_k^{(t)} \tag{15.34}$$

$$\mathbf{C}_i'^{(t+1)} = \frac{1}{N_{\text{nb}} - 1} \sum_{\substack{k=1 \\ k \in \Omega_i}}^{N_{\text{nb}}} \left(\boldsymbol{\alpha}_k^{(t)} - \boldsymbol{\mu}_i'^{(t+1)}\right)\left(\boldsymbol{\alpha}_k^{(t)} - \boldsymbol{\mu}_i'^{(t+1)}\right)^T \tag{15.35}$$

where $N_{\text{nb}}$ is the number of neighbor pixels those locate in the defined structuring element and $\Omega_i$ refers to the set of indices $k$ for all the neighbor pixels of $i$th pixel.

4. *Convergence Criterion*

After enough iterations of EM, the likelihood would converge to a stationary point [40]. Thus, the algorithm can stop iterating if $\log L\left(\mathbf{X}, \boldsymbol{\theta}_1^{(t)}, \boldsymbol{\theta}_2^{(t)}\right)$ and $\log L\left(\mathbf{X}, \boldsymbol{\theta}_1^{(t-1)}, \boldsymbol{\theta}_2^{(t-1)}\right)$ are close enough (below a small threshold).

The whole framework of EM used for unmixing has been afore presented, but there is still a problem has not been solved. That is how to find the $\boldsymbol{\alpha}^{(t)}$ that maximizes the log-likelihood function in the E step. Actually, it is an optimization problem as well as a NP complete problem [41] when we are trying to find the best abundance point in a high-dimensional solution space [42]. Luckily, there are some powerful intelligent optimization techniques that have been proposed for that problem. Next part of this section would studies PSO method to deal with this problem.

#### 15.3.2.2 PSO for Abundance Optimization

PSO is a global optimal searching algorithm based on SI theory, originally attributed to Kennedy and Eberhart [43] and Shi [44], and was first intended for simulating social behavior, as a stylized representation of the movement of organisms in a bird flock or fish school. As discussed in the previous part, PSO algorithm is desired to find the abundances that maximize the log-likelihood function in each E-step. This paper just employs the classical version of PSO [45, 46] where the log-likelihood function is regarded as an optimization fitness function.

In the original PSO, each abundance sample of a potential solution is treated as a particle with its own position and velocity on $M$ dimensions in the search space, which flies through the solution space by a certain velocity. Furthermore, the flying velocity is dynamically fine-tuned based on the flying experience of the individual and the group. More precisely, each particle updates its flying velocity and position by two extreme values: one is the best solution it has achieved so far, called local extreme value $\mathbf{p}_i$, and the other is the best value obtained so far by any particle in the entire population, referred to as global extreme value $\mathbf{p}_g$. Given the two extreme values, each particle can therefore update its flying velocity $\mathbf{v}$ and position $\mathbf{s}$ as

$$\mathbf{v}_i \leftarrow w_i \mathbf{v}_i + c_1 \times \text{rand}() \times (\mathbf{p}_i - \mathbf{s}_i) + c_2 \times \text{rand}() \times (\mathbf{p}_g - \mathbf{s}_i) \tag{15.36}$$

$$\mathbf{s}_i \leftarrow \mathbf{s}_i + \mathbf{v}_i \tag{15.37}$$

where $\{\mathbf{v}_i,\ \ i = 1, 2, \ldots, P\}$ denotes the velocity of the $i$th particle, $P$ is the total number of particles, $w$ means inertia weight, $c_1$ and $c_2$ are both acceleration constants, usually chosen in the interval [0, 2], then rand()~$U(0,\ 1)$ means independent uniform random variable. In our experiment, these parameters are usually be set as follows: $w = 0.1$, $c_1 = 0.3$, $c_2 = 0.6$. In the PSO process, the velocity of each particle is iteratively adjusted so that the particle stochastically oscillates around $\mathbf{p}_i$ and $\mathbf{p}_g$ location.

There are still some questions to be discussed before using PSO. The first one is how many particles should be used in the unmixing problem. Actually, it depends on the situation. More particles mean more possibility to find the best solution as well as more computing times. The second one is how to initialize the particle position. A good initialization of iteration process can shorten the time to find the best solution since the initial position may be closed to the best point. Here, we suggest generating particle samples according to the prior distribution of abundance discussed in this section, and then a small random disturbing value is added into each initial position to keep the particles from locating too close. The third one is when does the iteration process converge [47]. In practice, we usually stop iterating if all the local best positions $\mathbf{p}_i(i = 1, \ldots, P)$ in the swarm are close enough (below a small threshold) [48]. More precisely, in our experiment, if the distance between any two local best positions is smaller than 0.001, the iteration process of PSO would be stopped. It means that if the difference between any two abundance values estimated by the particles is smaller than 0.001, the swarm is considered to reach a steady state.

#### 15.3.2.3 Unmixing Based on PSO-EM Algorithm

The overall flow chart of PSO-EM method is described in Figure 15.1. Based on the formulation of NCM, the unmixing problem refers to two unknown parameters $\{\boldsymbol{\theta}_1, \boldsymbol{\theta}_2\}$ and a hidden variable $\boldsymbol{\alpha}$. They are self-estimated in the iteration process. Finally, the outputs of PSO-EM are abundance map and the information of endmembers (i.e. means and covariance matrices). Note that the dimension reduction step [49–51] is not necessary, but it is strongly suggested reducing the dimension since the PSO-EM is a time-consuming algorithm. As shown in Figure 15.1, PSO-EM is also a friendly method for hyperspectral unmixing problem since there is just two parameters need to be provided by user, i.e. the number of endmembers and the number of particles used in PSO. If there is not any prior knowledge, the number of endmembers can be given by some automatic estimation algorithm [52, 53], such as virtual dimensionality (VD) [54] and HySime [55].

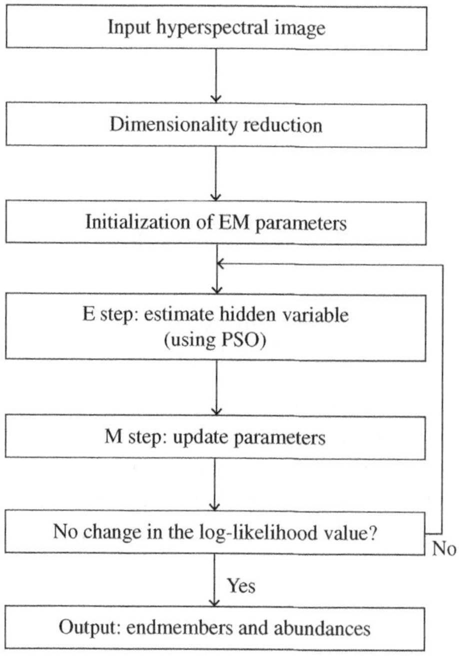


**Figure 15.1** PSO-EM method flowchart.

### *15.3.3 Approach Based on NLMM*

In this section, we introduce an alternating PSO framework for nonlinear unmixing, denoted by APSO. For clarity, we consider a simple nonlinear model, i.e. the FM, which only includes two groups of model parameters, namely **E** and **a**[*n*]. It can be easy to extend to more complicated models, such as the GBM and the polynomial post-nonlinear mixing model (PPNMM) [56], which have more than two groups of model parameters. The details can be referred to our works [57, 58]. Furthermore, when we consider constrains for the optimization problem, e.g. simplex volume, a strategy of multi-objective optimization can be employed [15]. We only present the core part of the alternating PSO framework herein.

Considering Eq. (15.14), the unmixing problem can be written as

$$\min_{\mathbf{E}\geq\mathbf{0},\,\mathbf{a}[n]\in Z}\sum_{n=1}^{N}\|\mathbf{x}[n]-\mathrm{FM}(\mathbf{E},\mathbf{a}[n])\|_2^2 \tag{15.38}$$

where $Z=\{\mathbf{a}\in\mathbb{R}^M|\mathbf{a}\geq\mathbf{0},\quad \mathbf{a}^{\mathrm{T}}\cdot\mathbf{1}=1\}$.

In the alternating PSO framework, we set up several swarms for the estimation of each group of parameters. The algorithm alternately updates the swarms by minimizing the objective function until a terminal condition is met. Considering problem (15.38), two swarms, namely SE and SA, are established to estimate **E** and **a**[*n*], respectively, where $n = 1, \ldots, N$. The whole procedure is described as follows.

*APSO framework for unmixing FM model*

1. Initial $\mathbf{E}^{(0)}$ for spectral endmember and $\mathbf{a}[n]^{(0)}$, $n = 1, \ldots, N$, for spectral abundance. Set $i := 0$.
2. (For SE) Update spectral endmembers by minimizing $\min_{\mathbf{E} \geq \mathbf{0},} \sum_{n=1}^{N} \left\| \mathbf{x}[\mathrm{n}] - \mathrm{FM}\left(\mathbf{E}, \mathbf{a}[\mathrm{n}]^{(i)}\right) \right\|_2^2$. Obtain $\mathbf{E}^{(i+1)}$ by each iteration.
3. (For SA) Update spectral abundances by minimizing $\min_{\mathbf{a}[n] \in Z} \sum_{n=1}^{N} \|\mathbf{x}[n] - \mathrm{FM}(\mathbf{E}^{(i)}, \mathbf{a}[n])\|_2^2$. Obtain $\mathbf{a}[n]^{(i+1)}$ by each iteration.
4. Set $i := i + 1$. If not satisfy stopping criterion (measuring whether the log-likelihood value change), then go to step 2. Otherwise, output $\mathbf{E}^{(i-1)}$ and $\mathbf{a}[n]^{(i-1)}$.

For each swarm, the updating operation follows the standard PSO operation. There are several particles in a swarm. Each particle $i$ has its own current position $\mathbf{X}_i$, current velocity $\mathbf{V}_i$, and a personal best position $\mathbf{P}_i$ in the search space, where $1 \leq i \leq s$ and $s$ is the size of the swarm. The swarm has a global best position $\mathbf{P}_G$ by selecting the best position from the particles.

Specifically, the updating operation of each particle at time step $j$ can be performed the following operations.

$$\mathbf{V}_i^{(j+1)} = w\mathbf{V}_i^{(j)} + c_1 r_1^{(j)}\left(\mathbf{P}_i^{(j)} - \mathbf{X}_i^{(j)}\right) + c_2 r_2^{(j)}\left(\mathbf{P}_G{}^{(j)} - \mathbf{X}_i^{(j)}\right) \tag{15.39}$$

for all $i \in 1, \ldots, s$, where $w$ is an inertial weight; $c_1$ and $c_2$ are acceleration coefficients; $r_1$ and $r_2$ are random numbers following a uniform distribution in the range (0, 1).

Then we can obtain their current position

$$\mathbf{X}_i^{(j+1)} = \mathbf{X}_i^{(j)} + \mathbf{V}_i^{(j+1)} \tag{15.40}$$

The personal best position of each particle is updated by

$$\mathbf{P}_i^{(j+1)} \begin{cases} \mathbf{P}_i^{(j)}, & \text{if } f\left(\mathbf{X}_i^{(j+1)}\right) \geq f\left(\mathbf{P}_i^{(j)}\right) \\ \mathbf{X}_i^{(j+1)}, & \text{if } f\left(\mathbf{X}_i^{(j+1)}\right) < f\left(\mathbf{P}_i^{(j)}\right) \end{cases} \tag{15.41}$$

where $f(\bullet)$ represents the objective function.

Finally, we can obtain the global best position for the swarm

$$\mathbf{P}_G{}^{(j+1)} = \underset{\mathbf{P}_i^{(j+1)}}{\arg\min}\left\{ f\left(\mathbf{P}_i^{(j+1)}, \quad \text{for all } i \in 1, \ldots, s\right)\right\} \tag{15.42}$$

Another issue of PSO algorithms is a high-dimensional problem. In problem (15.38), $B$ and $N$ can be very large so that the performance of PSO may be greatly degenerate. A strategy can improve the high-dimensional searching capability [59]. Considering (15.38) in the $k$th dimension for SE, the optimization subproblem is written as

$$\min_{\mathbf{E}_k \geq \mathbf{0}, \mathbf{a}[n] \in Z} \sum_{n=1}^{N} \|\mathbf{x}_k[n] - \mathrm{FM}(\mathbf{E}_k, \mathbf{a}[n])\|_2^2 \tag{15.43}$$

where symbol $\mathbf{E}_k$ represents the $k$th row of $\mathbf{E}$ and

$$\mathrm{FM}(\mathbf{E}_k, \mathbf{a}[n]) = \sum_{i=1}^{M-1} \sum_{j=i+1}^{M} a_i[n] a_j[n] e_{k,i} e_{k,j} \tag{15.44}$$

The hypothesis of SE can be separated and rewritten in the $B$-dimensional vector form:

$$\mathrm{FM}(\mathbf{E}, \mathbf{a}[n]) = [\mathrm{FM}(\mathbf{E}_1, \mathbf{a}[n]), \ldots, \mathrm{FM}(\mathbf{E}_B, \mathbf{a}[n])]^T \tag{15.45}$$

Therefore, we calculate (15.45) separately by evaluating the sub-objective function (15.44) using the operations of (15.41) and (15.42):

$$\mathbf{P}_i^{(j+1)} = \left[\mathbf{p}_{i,1}^{(j+1)}, \mathbf{p}_{i,2}^{(j+1)}, \ldots, \mathbf{p}_{i,B}^{(j+1)}\right]^T \tag{15.46}$$

where each component $k$ is

$$\mathbf{p}_{i,k}^{(j+1)} = \begin{cases} \left(\mathbf{P}_i^{(j+1)}\right)_{k,:}, & \text{if } F_k\left(\mathbf{X}_i^{(j+1)}\right) \geq F_k\left(\mathbf{P}_i^{(j)}\right) \\ \left(\mathbf{X}_i^{(j+1)}\right)_{k,:}, & \text{if } F_k\left(\mathbf{P}_i^{(j+1)}\right) \geq F_k\left(\mathbf{X}_i^{(j)}\right) \end{cases} \tag{15.47}$$

where $(\bullet)_{k,:}$ represents the $k$th row of the matrix $\bullet$ and

$$F_k(\bullet_k) = \sum_{n=1}^{N} \|\mathbf{x}_k[n] - \mathrm{FM}(\bullet_k, \mathbf{a}[n])\|_2^2 \tag{15.48}$$

Then, the global best particle can be updated by:

$$\mathbf{P}_{\mathbf{G}}^{(j+1)} = \left[\mathbf{p}_{\mathbf{G},1}^{(j+1)}, \mathbf{p}_{\mathbf{G},2}^{(j+1)}, \ldots, \mathbf{p}_{\mathbf{G},B}^{(j+1)}\right]^T \tag{15.49}$$

where each component is obtained by:

$$\mathbf{p}_{\mathbf{G},k}^{(j+1)} = \left(\underset{\mathbf{P}_j^{(j+1)} \in S}{\arg\min} \left\{F_k\left(\mathbf{P}_j^{(j+1)}\right)\right\}\right)_{k,:} \tag{15.50}$$

Similarly, we can separate the optimization problem into $N$ subproblem as follows for SA:

$$\min_{\mathbf{a}[n] \in Z} \left\|\mathbf{x}[n] - \mathrm{FM}\left(\mathbf{E}^{(j)}, \mathbf{a}[n]\right)\right\|_2^2, \quad n \in \{1, \ldots, N\} \tag{15.51}$$

So the particles in SA can perform the operations of (15.41) and (15.42) by

$$\mathbf{P}_i^{(j+1)} = \left[\mathbf{p}_{i,1}^{(j+1)}, \mathbf{p}_{i,2}^{(j+1)}, \ldots, \mathbf{p}_{i,N}^{(j+1)}\right] \tag{15.52}$$

where each component $k$ is

$$\mathbf{P}_{i,k}^{(j+1)} = \begin{cases} \left(\mathbf{P}_i^{(j+1)}\right)_{:,k}, & \text{if } F_k^{(SA)}\left(\mathbf{X}_i^{(j+1)}\right) \geq F_k^{(SA)}\left(\mathbf{P}_i^{(j)}\right) \\ \left(\mathbf{X}_i^{(j+1)}\right)_{:,k}, & \text{if } F_k^{(SA)}\left(\mathbf{P}_i^{(j+1)}\right) \geq F_k^{(SA)}\left(\mathbf{X}_i^{(j)}\right) \end{cases} \tag{15.53}$$

where $(\bullet)_{:,k}$ represents the $k$th column of the matrix $\bullet$ and

$$F_k^{(SA)}(\bullet_k) = \left\| \mathbf{x}[k] - \text{FM}\left(\mathbf{E}^{(j)}, \mathbf{a}[k]\right) \right\|_2^2 \tag{15.54}$$

Then, the global best particle of SA can be updated by:

$$\mathbf{P}_{\mathbf{G}}^{(j+1)} = \left[\mathbf{p}_{\mathbf{G},1}^{(j+1)}, \mathbf{p}_{\mathbf{G},2}^{(j+1)}, \ldots, \mathbf{p}_{\mathbf{G},B}^{(j+1)}\right]^T \tag{15.55}$$

where each component is obtained by:

$$\mathbf{p}_{\mathbf{G},k}^{(j+1)} = \left( \underset{\mathbf{P}_j^{(j+1)} \in S}{\arg\min} \left\{ F_k^{(SA)}\left(\mathbf{P}_j^{(j+1)}\right) \right\} \right)_{:,k} \tag{15.56}$$

## 15.4 Experiments

This section first provides brief descriptions of the two data sets used in the experiment, which were collected by the airborne visible infrared imaging spectrometer (AVIRIS). In order to verify the performance of different algorithms, the observed scenes ensure diversity (e.g. vegetation and mineral). Next, the experimental evaluation is detailed, including experimental results and performance assessments.

### 15.4.1 *Experimental Data*

Two AVIRIS data sets presented in this section are Moffett Field data and Cuprite data, which can be generally used for endmember extraction and abundance inversion.

#### 15.4.1.1 Moffett Field Data

The considered hyperspectral image was acquired over Moffett Field in California, USA, in 1997 by JPL spectro-imager AVIRIS. The data is available on the web site http://aviris.jpl.nasa.gov/html/aviris.freedata.html. The image has been used in many fields to illustrate and compare hyperspectral information processing algorithms [60–62]. This section considers a subset image of size $50 \times 50$, extracted from a large image, as depicted in Figure 15.2, and it was the same region in Dobigeon et al. [63] and Eches et al. [64]. The subset image contains three substances: vegetation, water, and soil. The VCA method can be used to extract the endmember spectrum, and the locations of the estimated endmembers are shown in Figure 15.3. Due to water absorption, the affected bands are removed and 189 bands are finally used.

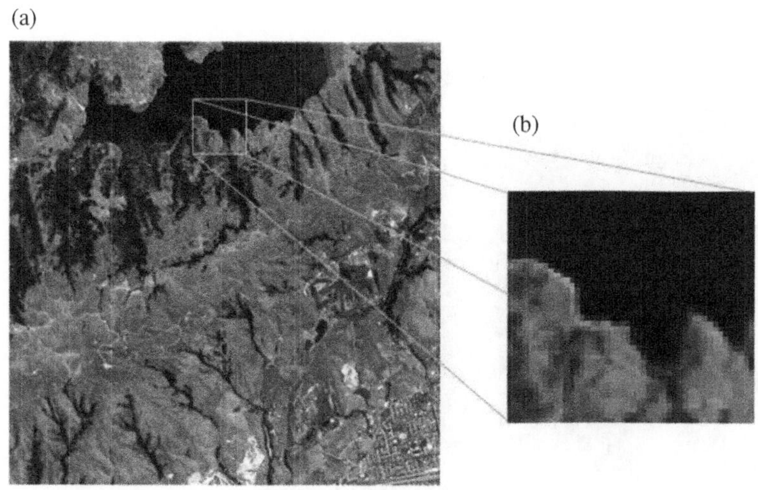


**Figure 15.2** Moffett Field scene, (a) original Moffett Field image, (b) the region of subset image.

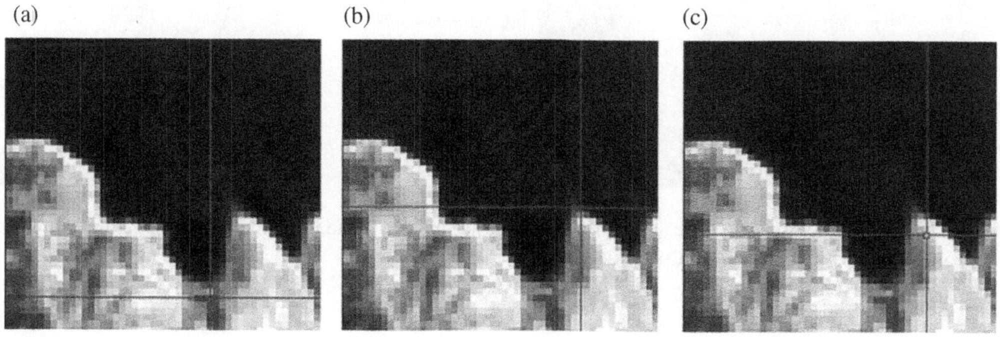


**Figure 15.3** Locations of endmembers extracted by the VCA method. (a) vegetation, (b) water, (c) soil.

#### 15.4.1.2 Cuprite Data

One of the most widely used hyperspectral images available in public domain is Cuprite mining site, Nevada, in 1995, as shown in Figure 15.4(a). We select a subset image (see Figure 15.4(b)) containing different minerals. The captured image size is 400 × 350 pixel, and the spatial resolution is 20 m. There are 224 bands, ranging from 370 to 2480 nm. It is recommended that bands 1–2, 104–113, 148–167, and 221–224 can be removed prior to data processing due to water absorption and noise. Since the identification of minerals in hyperspectral images mainly depends on the absorption characteristics of mineral components and is easily affected by iron in the visible spectrum, the range of 1.9908–2.4790 μm in 50 bands is usually used to identify minerals.

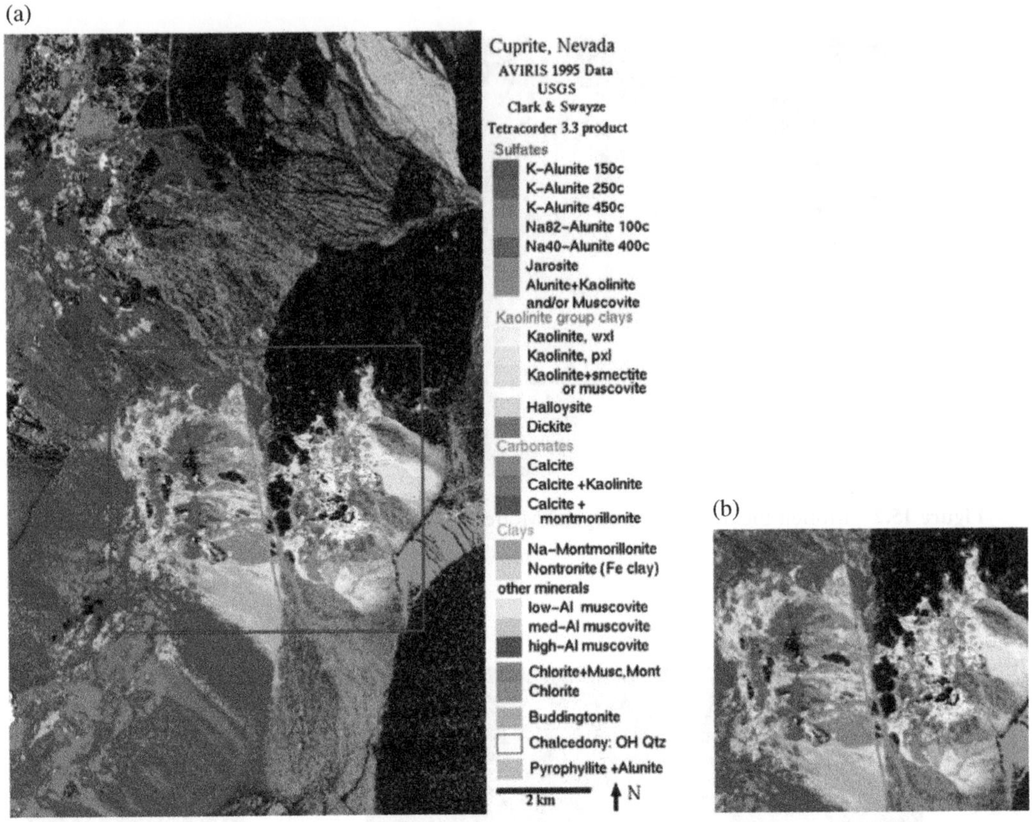


**Figure 15.4** Cuprite image scene. (a) Original Cuprite image scene, (b) the region of subset image. *Source:* Chang [65]/with permission of John Wiley & Sons.

## *15.4.2 Results of Approaches Based on LMM*

### 15.4.2.1 Pure Pixel-Based Algorithm

In this section, we will carry out experiments of pure pixel-based algorithms in the Cuprite scene data. Here, the application of DPSO for endmember extraction is performed. In addition, the experiments of VCA, NFINDR, and PPI are used for comparison. According to the HySime algorithm, the number of extracted endmembers is 6. As for the parameter setting of DPSO, the random probability is set to 0.2. DPSO is a method based on PSO to find feasible solutions. If the feasible solution space is too large, it will affect the search rate and reduce the calculation efficiency. Therefore, the maximum noise fraction (MNF) algorithm is used to perform the dimensionality reduction processing on the original image, and the PPI algorithm is used to obtain 80 candidate endmembers.

In our performance comparison, two evaluation criteria are used: the spectral angle distance (SAD) and reconstruction error (RE). The SAD is computed to evaluate the accuracy of endmember identification. The smaller the SAD value is, the smaller the spectral difference is. The RE is used to reflect the difference between the original image and the reconstructed image. The smaller the RE value is, the better the endmember and abundance results of the algorithms are. The SAD and RE are, respectively, defined as:

$$\mathrm{SAD}(\mathbf{e}_{\mathrm{true}}, \mathbf{e}_{\mathrm{est}}) = \arccos\left(\frac{\mathbf{e}_{\mathrm{true}}^T \mathbf{e}_{\mathrm{est}}}{\|\mathbf{e}_{\mathrm{true}}\|_2 \cdot \|\mathbf{e}_{\mathrm{est}}\|_2}\right) \tag{15.57}$$

$$\mathrm{RE} = \sqrt{\frac{1}{\mathrm{BN}} \sum_{i=1}^{N} \|\mathbf{x}_i - \hat{\mathbf{x}}_i\|_2^2} \tag{15.58}$$

where $\mathbf{e}_{\mathrm{est}}$ and $\mathbf{e}_{\mathrm{true}}$ represent the estimated and true endmember, respectively. $\mathbf{x}_i$ and $\hat{\mathbf{x}}_i$ represent the original image vector and reconstructed image vector.

There are a lot of minerals in the Cuprite scene. According to the endmember number result of HySime and mineral spectrum extracted by all algorithms, we use some minerals for experimental comparison, including alunite, calcite, desert varnish, halloysite, jarosite, kaolinite, montmorillonite, muscovite, chalcedony, and nontronite. The SAD results between the endmember extracted by the pure pixel-based algorithms and the USGS spectral library tabulate in Table 15.1. The corresponding endmembers extracted by each algorithm are shown in Figure 15.5. The smaller the SAD value is, the closer the extracted endmembers are to the USGS mineral spectrum. As seen from the results, the extracted endmembers corresponding to different algorithms are duplicate, e.g. NFINDR extracts two montmorillonite and muscovite endmembers. In the term of quality, DPSO can extract different minerals and the SAD results are the lowest compared to other algorithms, indicating that the extracted endmembers of DPSO are closer to the true mineral spectrum.

Figure 15.6 shows the FCLS results for the endmembers extracted by different algorithms and the reconstruction maps. It can be seen that the abundance maps of DPSO are similar to reference map in Figure 15.4(b). Moreover, the reconstruction result reveals that the DPSO can obtain better endmember and abundance results compared with other algorithms.

#### 15.4.2.2 Minimum Volume-Based Algorithm

Here, the experiment is performed by using MVCNMF, MVSA, and RMVES for comparison. The extracted endmember results are shown in Table 15.2 and Figure 15.7. Compared with the pure

**Table 15.1** The SAD results between the endmember extracted by the pure pixel-based algorithms and the USGS spectral library

| | VCA | NFINDR | PPI | DPSO |
|---|---|---|---|---|
| Alunite | 4.324 | 4.431 | 4.358 | 4.291 |
| Calcite | 4.458 | — | 4.729 | 3.546 |
| Desert Varnish | 3.114 | 3.112 | 1.902 | 1.565 |
| Jarosite | — | — | 4.140 | — |
| Kaolinite | 3.514 | — | — | 3.234 |
| Montmorillonite | 6.194 | 4.925 | 7.611 | 5.132 |
| Montmorillonite | 4.621 | 5.949 | — | — |
| Muscovite | — | 5.167 | 4.202 | 4.125 |
| Muscovite | — | 5.865 | — | — |
| Average | 4.371 | 4.908 | 4.490 | 3.649 |

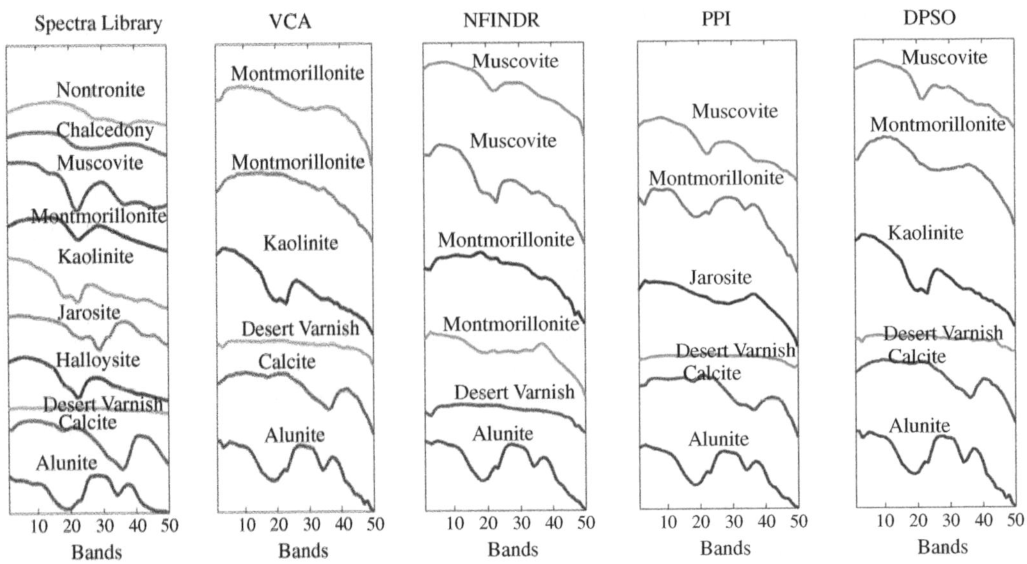


**Figure 15.5** Endmember extraction curve for each pure pixel-based algorithm.

pixel-based algorithms, the SAD results of the minimum volume hypothesis algorithms are higher, while the RE results are lower. These results demonstrate that algorithms based on pure pixel assumptions are more suitable to solve the unmixing problem in the Cuprite scene.

As seen from the result, MVCNMF, RMVES, and ABC can extract six different endmembers, but only the extracted endmembers of MVSA are duplicate. These results indicate that most minimum volume-based algorithms can extract different endmembers. Although the SAD results between six endmembers extracted by ABC and the actual minerals are not always optimal, the average SAD result is the smallest, which shows the effectiveness and robustness of ABC. The abundance maps estimated by the minimum volume-based algorithms and reconstruction results are shown in Figure 15.8. In Figure 15.8, the abundance maps estimated by ABC, such as Alunite and Calcite, are closer to the actual abundance map. Correspondingly, the distribution of RE result in ABC is more uniform than other algorithms. Therefore, the experimental results validate that the endmembers extracted by ABC can better contain the information of the Cuprite scene.

As the model based on the external minimum volume, ABC method does not require the simplex formed by the endmembers to completely contain the convex hull formed by the pixels. This enhances the algorithm's robustness and allows it to produce a better endmember extraction result.

## *15.4.3 Results of Approaches Based on NCM*

### 15.4.3.1 Experiment Using Moffett Field Data

Two previously developed methods are presented in order to compare the performance of the PSO-EM method. The first is the fully constrained least-squares (FCLS) method based on LMM assumption, the other is the Markov chain Monte Carlo (MCMC) method based on NCM assumption [66]. The abundance maps estimated by the FCLS, the MCMC, and the PSO-EM method are shown in Figure 15.9. Visually, the abundance maps estimated by the MCMC method seem similar to those estimated by the FCLS method rather than PSO-EM. This is because the MCMC and PSO-EM

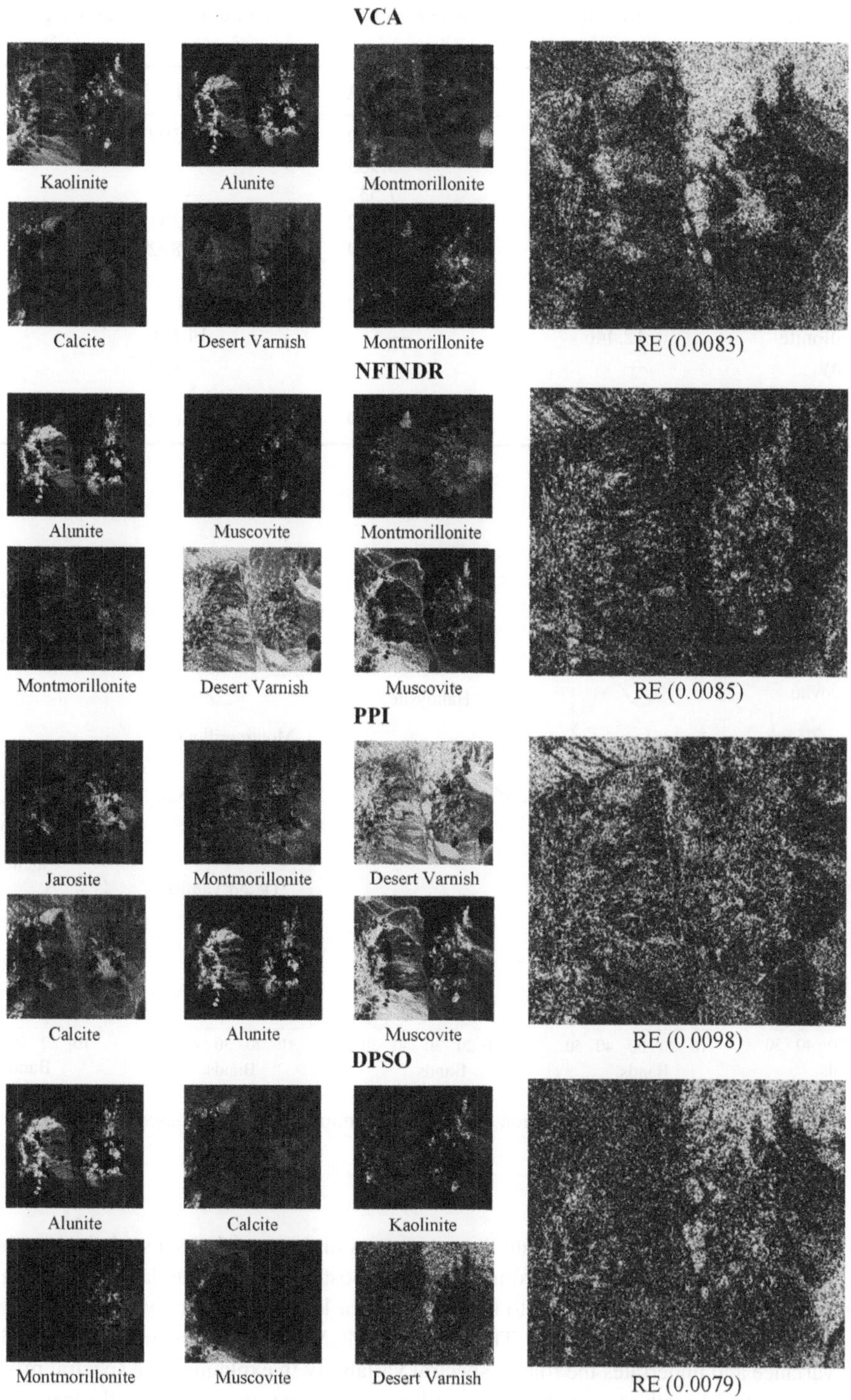


**Figure 15.6** Abundance maps estimated by the pure pixel-based algorithms and the reconstruction error maps.

**Table 15.2** The SAD results between the endmember extracted by the minimum volume-based algorithms and the USGS spectral library

| | MVCNMF | MVSA | RMVES | ABC |
|---|---|---|---|---|
| Alunite | 12.707 | 13.059 | 6.597 | 6.936 |
| Calcite | 9.085 | 5.963 | — | 5.775 |
| Calcite | — | 7.532 | — | — |
| Desert Varnish | 3.714 | 3.672 | 3.341 | 2.556 |
| Halloysite | 12.651 | 14.529 | 11.882 | 13.360 |
| Jarosite | 9.216 | — | — | — |
| Kaolinite | — | 7.017 | — | — |
| Montmorillonite | 12.446 | — | 14.176 | 3.929 |
| Chalcedony | — | — | 4.714 | — |
| Nontronite | — | — | 5.382 | 6.747 |
| Average | 9.970 | 8.629 | 7.682 | 6.551 |

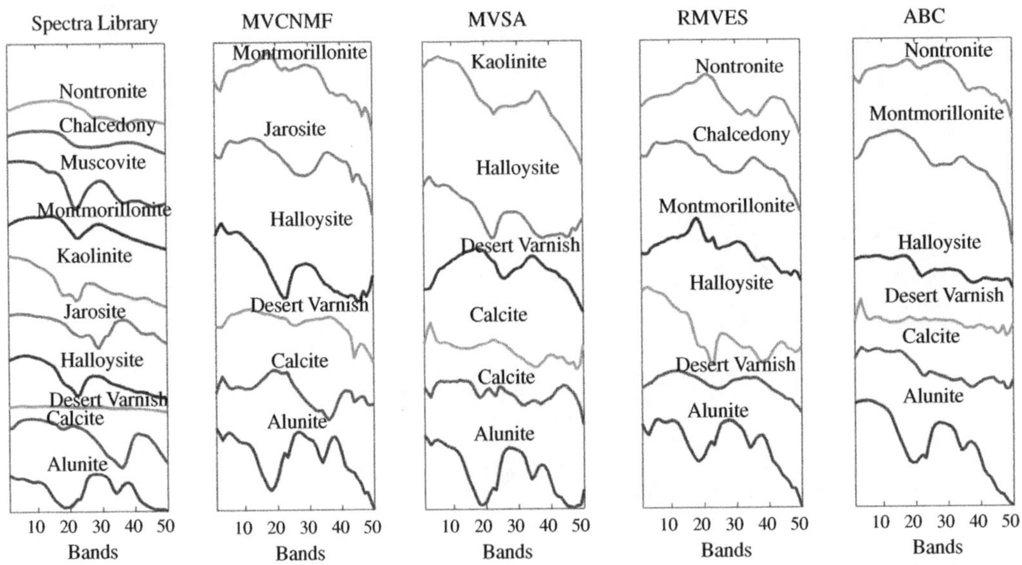


**Figure 15.7** Endmember extraction curve for each minimum volume-based algorithm.

employ entirely different ways to estimate the parameters of the model. As a statistical method, the MCMC method generated samples according to the joint posterior of the abundances, the endmember variance, and one hyperparameter. Due to the lack of prior knowledge of hyperspectral images, the endmember variance cannot be obtained. Therefore, the MCMC method may underestimate the endmember variance and underrates the role of spectral variability in explaining the spectral mixture. In contrast, the PSO-EM method uses the machine learning strategy to automatically learn the parameters of endmember distributions. As a result, it can avoid the influence of prior knowledge in hyperspectral images and obtain better unmixing accuracy.

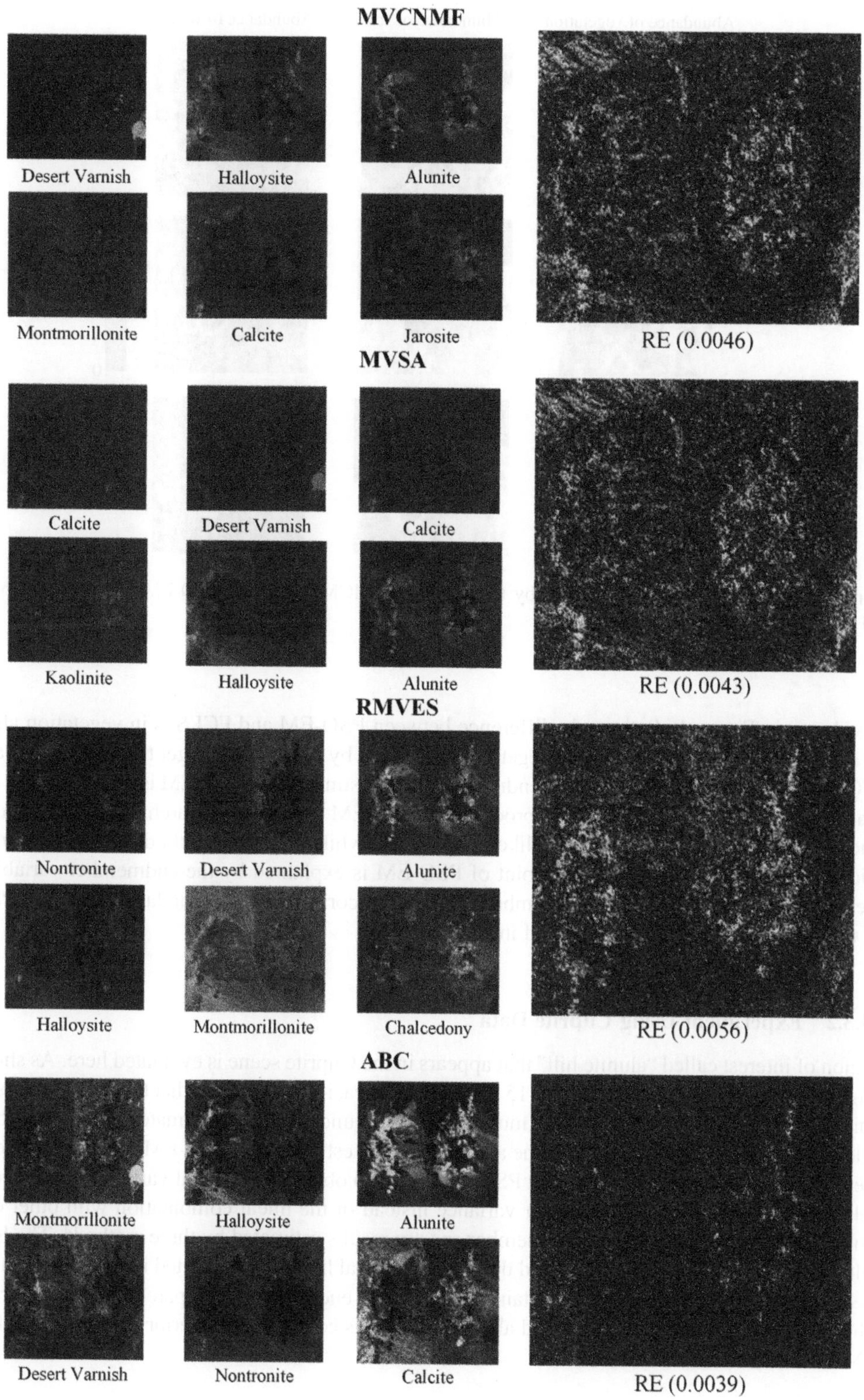


**Figure 15.8** Abundance maps estimated by the minimum volume-based algorithms and the reconstruction error maps.

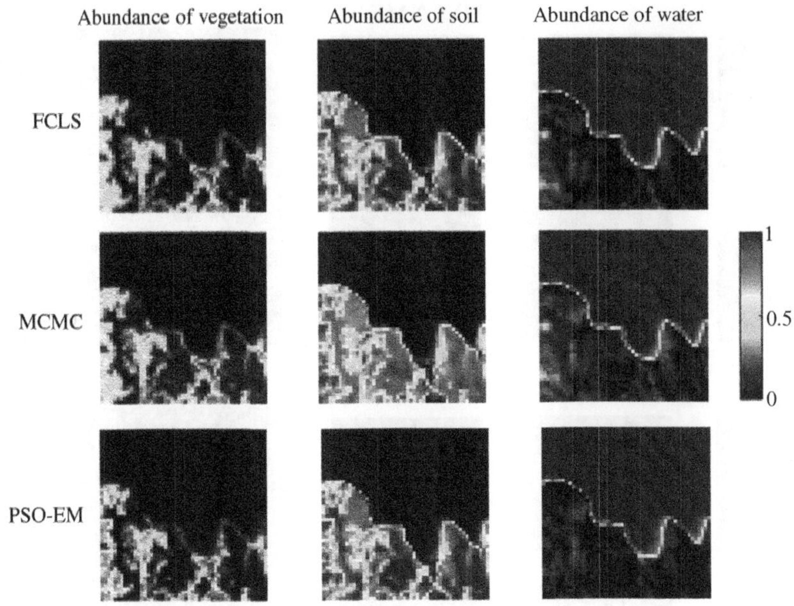


**Figure 15.9** Abundance maps estimated by the FCLS, the MCMC, and the PSO-EM methods in Moffett Field scene.

As shown in Figure 15.9, the only difference between PSO-EM and FCLS is in vegetation abundance maps. The abundance values of vegetation estimated by PSO-EM is larger than those estimated by FCLS. It can be clearly seen that the endmember point estimated by PSO-EM is different from that extracted by VCA. During the iteration process, the PSO-EM method will search for the endmember in the direction with the maximum log-likelihood value, while the VCA method will search for the vertices. For example, the vegetation point of PSO-EM is explained by the endmember variability rather than the mixed by other endmembers. Thus, the corresponding abundance value of PSO-EM would be larger, as it has happened in Figure 15.9.

#### 15.4.3.2 Experiment Using Cuprite Data

A region of interest called "alunite hill" that appears in the Cuprite scene is evaluated here. As shown in Figure 15.10, the size of alunite hill is $15 \times 20$. In this area, it is considered that the endmembers are dominated by two materials: alunite and muscovite. The abundance maps estimated by three methods are depicted in Figure 15.11. Visually, the abundance map estimated by PSO-EM obviously covers a larger area. The result demonstrates that PSO-EM tends to obtain the spectral variation so that some pixels are considered as the endmember variance instead of the linear combination with other endmembers. Figure 15.12 shows the endmember spectra results estimated by three methods. The SAD results between the estimated spectra and the USGS spectral library are reported in Table 15.3. It can be seen that the PSO-EM method can obtain more accurate endmembers compared with the other two methods. Therefore, the endmember and abundance results confirm the superior performance of the PSO-EM method.

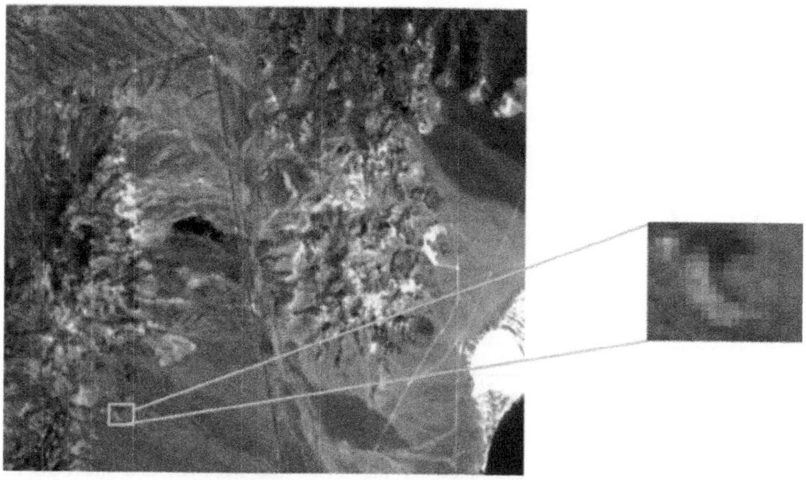

**Figure 15.10** Alunite hill image.

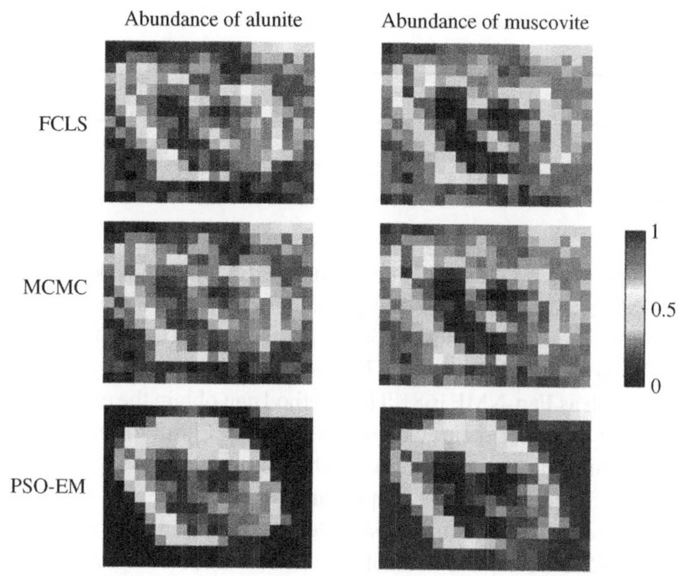


**Figure 15.11** Abundance maps estimated by the FCLS, the MCMC, and the PSO-EM methods in Cuprite scene.

## *15.4.4 Results of Approaches Based on NLMM*

### 15.4.4.1 Experiment Using Moffett Field Data

The proposed alternating PSO method is tested using two real hyperspectral data. The results are compared with those obtained using two groups of unmixing methods. The first group is based on classical supervised algorithms, where the prior knowledge of endmembers is determined by VCA. The

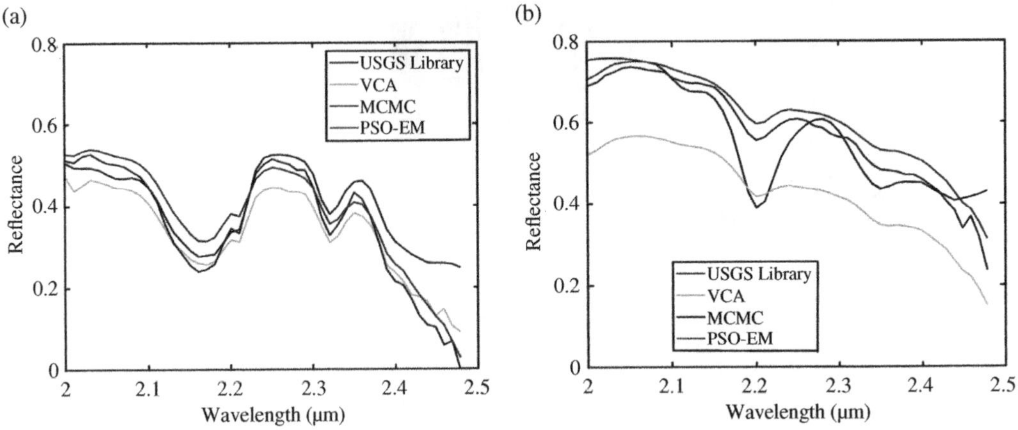


**Figure 15.12** The endmember spectra estimated by the VCA, the MCMC, and the PSO-EM methods, compared with the USGS spectral library. (a) Alunite, (b) muscovite.

**Table 15.3** The SAD results between the estimated spectra and USGS spectral library for the VCA, MCMC, and PSO-EM

| Mineral | VCA | MCMC | PSO-EM |
|---|---|---|---|
| Alunite | 4.324 | 4.111 | 4.067 |
| Muscovite | 6.309 | 6.339 | 6.179 |

first supervised algorithm is based on gradient descent in GBM, which is denoted by GBA. It should be noted that the GBM includes the FM as the particular case, where the nonlinear parameters are fixed to one. In addition, the constrained nonlinear least squares (CNLS) algorithm is used as the classic supervised bilinear unmixing method [67]. The second group of algorithms is a NMF method for spectral unmixing, such as Fan-NMF [68]. This method can obtain the endmember and abundance results simultaneously. In the experiment, we use the VCA algorithm for initialization. To obtain the robust results for the algorithms, the maximum number of iterations is set to 500 for all algorithms. The convergence tolerance threshold is set empirically to $10^{-6}$.

Figure 15.13 shows the abundance maps estimated by the compared algorithms. Here, the RE results showed in Table 15.4 are used to evaluate the performance of algorithms. Table 15.4 reveals that APSO can achieve smaller RE values than those provided by other algorithms. In the abundance maps of vegetation and water, it can be seen that APSO is more robust than other algorithms. However, for the soil abundance map, VCA-GBA is better than APSO. Generally speaking, APSO can get better unmixing result than other algorithms.

#### 15.4.4.2 Experiment Using Cuprite Data

In the alunite hill experiment, the abundance maps estimated by the VCA-GBA, VCA-CNLS, Fan-NMF, and APSO method are displayed in Figure 15.14. The corresponding RE results are tabulated in Table 15.5. Due to the lack of ground truth, precise verification of inversion results with real data is

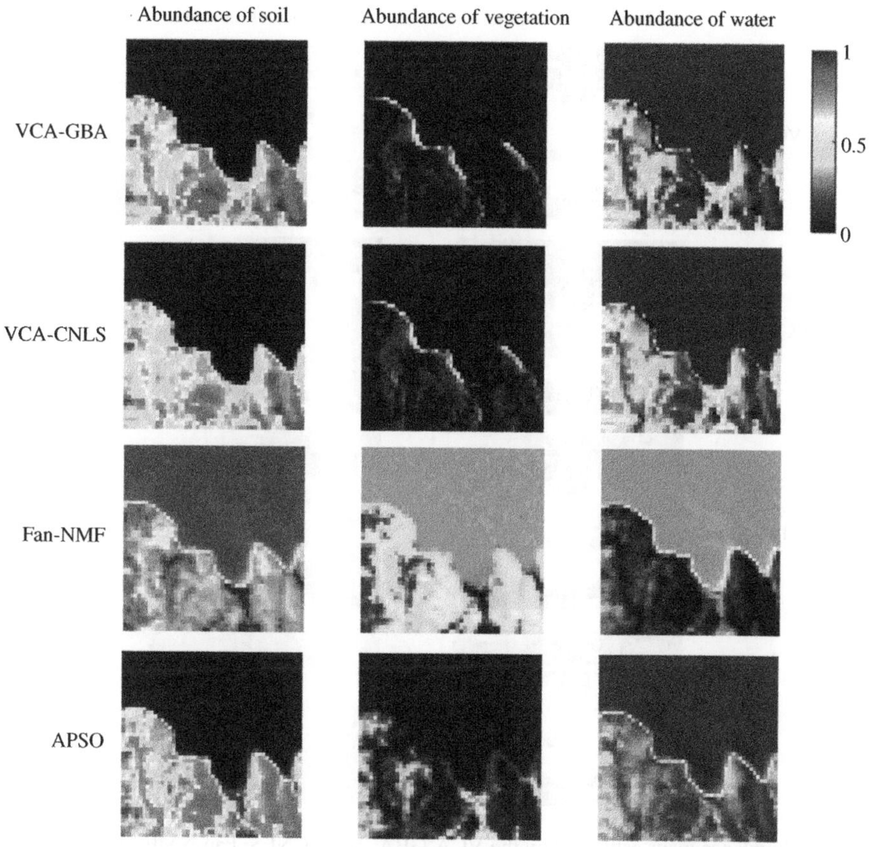


**Figure 15.13** Abundance maps estimated by the VCA-GBA, VCA-CNLS, Fan-NMF, and APSO method in Moffett Field scene.

**Table 15.4** The RE results in Moffett Field scene

| | VCA-GBA | VCA-CNLS | Fan-NMF | APSO |
|---|---|---|---|---|
| RE | 0.1221 | 0.1234 | 0.0262 | 0.0213 |

elusive. Most of experiments use the RE between the original and the reconstructed hyperspectral image as the evaluation criteria. It can be seen that the VCA-CNLS method has the poor abundance map and high RE result, which illustrates that the unmixing performance of this method is not effective compared with other methods. The APSO method can obtain the lowest RE result and slight accurate abundance map, but it has high computational cost mainly due to the optimization procedure using PSO. In order to evaluate the endmember result, the SAD results are tabulated in Table 15.6 and Figure 15.15 shows the corresponding endmember spectra estimated by comparative methods. In the alunite result, APSO is better than the other two methods, but it is not as good as the VCA method in the muscovite result. However, in general, the proposed APSO method can get more accurate endmember and abundance results.

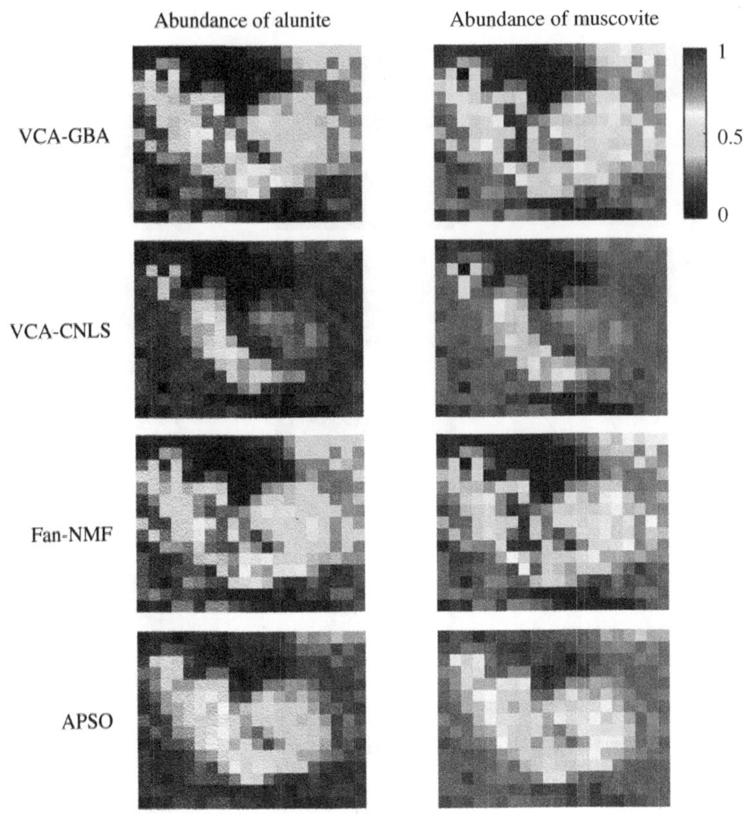


**Figure 15.14** Abundance maps estimated by the VCA-GBA, VCA-CNLS, Fan-NMF, and APSO methods in Cuprite scene.

**Table 15.5** The RE results in Cuprite scene

| | VCA-GBA | VCA-CNLS | Fan-NMF | APSO |
|---|---|---|---|---|
| RE | 0.0183 | 0.0209 | 0.0166 | 0.0161 |

**Table 15.6** The SAD results between the estimated spectra and USGS spectral library for the VCA, Fan-NMF, and APSO

| Mineral | VCA | Fan-NMF | APSO |
|---|---|---|---|
| Alunite | 4.324 | 4.852 | 4.167 |
| Muscovite | 6.309 | 6.678 | 6.491 |

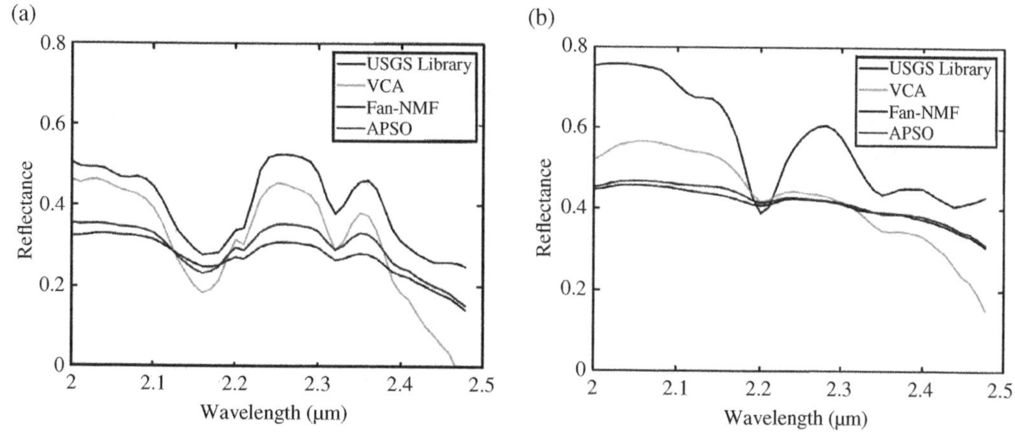


**Figure 15.15** The endmember spectra estimated by the VCA, the MCMC, and the PSO-EM methods, compared with the USGS spectral library. (a) Alunite spectrum, (b) muscovite spectrum.

## 15.5 Conclusion

SI algorithm has unique advantages in solving some difficult optimization problems and has achieved good results in the application of unmixing of hyperspectral images. This chapter develops four SI algorithms to improve the unmixing results, called DPSO, ABC, PSO-EM, and APSO. These algorithms are based on LMM, NCM, and NLMM, respectively. The idea of DPSO is to expand PSO in a discrete feasible solution space to improve the accuracy of unmixing. Its advantage is that it does not rely on LMM and can adjust objective function automatically. Unlike DPSO, ABC can search for endmembers in a continuous feasible solution space. In addition, the volume penalty is considered to enhance the robustness of ABC. In the NCM, PSO is employed to find the best abundance estimation by maximizing the likelihood function. Moreover, the proposed PSO-EM method applies the EM technique to update the parameters in NCM, thereby obtaining good unmixing results. By combining PSO with multiobjective optimization, the complicated NLMM problem is separated into several subproblems, which improves the high-dimensional searching capability. Therefore, APSO can have good performance in unmixing. Experiments in the AVIRIS data are conducted to demonstrate the effectiveness and robustness of these four SI algorithms compared with tradition unmixing algorithms.

It is worth noting that there are still some nonnegligible difficulties in the application of SI algorithms for specific problems. An obvious problem that hinders the application of SI algorithms is the contradiction between search space, population size, and computational efficiency. These problems can be easily solved by parallel computing platforms, such as general-purpose graphics processing units (GP-GPUs). In the future works, developing the parallel version of unmixing algorithms to improve the computation speed will be the focus of future research.

## Acknowledgments

The authors acknowledge the Jet Propulsion Laboratory for generously providing the AVIRIS data. This work was supported by the National Natural Science Foundation of China under Grant 42030111 and Grant 41722108.

## References


1. Nascimento, J.M.P. and Dias, J.M.B. (2005). Vertex component analysis: a fast algorithm to unmix hyperspectral data. *IEEE Transactions on Geoscience and Remote Sensing* 43 (4): 898–910.
2. Winter, M.E. (1999). N-Findr: an algorithm for fast autonomous spectral end-member determination in hyperspectral data. *SPIE Proceedings Series* 3753: 266–275.
3. Boardman, J.W., Kruse, F.A., and Green, R.O. (1995). Mapping target signatures via partial unmixing of aviris data. *Proceedings of the Fifth Annual JPL Airborne Earth Science Workshop*, Pasadena, California (23–26 January 1995), 23–26.
4. Miao, L. and Qi, H. (2007). Endmember extraction from highly mixed data using minimum volume constrained nonnegative matrix factorization. *IEEE Transactions on Geoscience and Remote Sensing* 45: 765–777.
5. Jun, L. and Dias, J.M.B. (2008). Minimum volume simplex analysis: a fast algorithm to unmix hyperspectral data. *IEEE Transactions on Geoscience and Remote Sensing* 3: 250–253.
6. Ambikapathi, A., Chan, T., and Ma, W. (2011). Chance constrained robust minimum-volume enclosing simplex algorithm for hyperspectral unmixing. *IEEE Transactions on Geoscience and Remote Sensing* 49: 4194–4209.
7. Zhang, B., Sun, X., Gao, L., and Yang, L. (2011). Endmember extraction of hyperspectral remote sensing images based on the ant colony optimization (ACO) algorithm. *IEEE Transactions on Geoscience and Remote Sensing* 49 (7): 2635–2646.
8. Zhang, B., Gao, J., Gao, L., and Sun, X. (2013). Improvements in the ant colony optimization algorithm for endmember extraction from hyperspectral images. *IEEE Journal of Selected Topics in Applied Earth Observations and Remote Sensing* 6 (2): 522–530.
9. Gao, L., Gao, J., Li, J. et al. (2015). Multiple algorithm integration based on ant colony optimization for endmember extraction from hyperspectral imagery. *IEEE Journal of Selected Topics in Applied Earth Observations and Remote Sensing* 8 (6): 2569–2582.
10. Sun, X., Yang, L., Zhang, B. et al. (2015). An endmember extraction method based on artificial Bee Colony Algorithms for hyperspectral remote sensing images. *Remote Sensing* 7 (12): 16363–16383.
11. Zhuang, L., Zhang, B., Gao, L. et al. (2015). Normal endmember spectral unmixing method for hyperspectral imagery. *IEEE Journal of Selected Topics in Applied Earth Observations and Remote Sensing* 8 (6): 2598–2606.
12. Gao, L., Zhuang, L., and Zhang, B. (2016). Region-based estimate of endmember variances for hyperspectral image unmixing. *IEEE Geoscience and Remote Sensing Letters* 13 (12): 1807–1811.
13. Su, Y., Sun, X., Gao, L. et al. (2016). Improved discrete swarm intelligence algorithms for endmember extraction from hyperspectral remote sensing images. *Journal of Applied Remote Sensing* 10 (4): 045018.
14. Gao, L., Zhuang, L., Wu, Y. et al. (2016). A quantitative and comparative analysis of different preprocessing implementations of DPSO: a robust endmember extraction algorithm. *Soft Computing* 20 (12): 4669–4683.
15. Luo, W., Gao, L., Plaza, A. et al. (2016). A new algorithm for bilinear spectral unmixing of hyperspectral images using particle swarm optimization. *IEEE Journal of Selected Topics in Applied Earth Observations and Remote Sensing* 9 (12): 5776–5790.
16. Chang, C.-I (2007). *Hyperspectral Data Exploitation: Theory and Applications*. Hoboken, NJ: Wiley-Interscience.
17. Heinz, D.C. (2001). Fully constrained least squares linear spectral mixture analysis method for material quantification in hyperspectral imagery. *IEEE Transactions on Geoscience and Remote Sensing* 39 (3): 529–545.
18. Bioucas-Dias, J.M., Plaza, A., and Dobigeon, N. (2012). Hyperspectral unmixing overview: geometrical, statistical, and sparse regression-based approaches. *IEEE Journal of Selected Topics in Applied Earth Observations and Remote Sensing* 5 (2): 354–379.
19. Gruninger, J.H., Ratkowski, A.J., and Hoke, M.L. (2004). The sequential maximum angle convex cone (SMACC) endmember model. *Proceedings of SPIE* 5425: 1–14.
20. Chan, T., Ma, W., and Ambikapathi, A. (2011). A simplex volume maximization framework for hyperspectral endmember extraction. *IEEE Transactions on Geoscience and Remote Sensing* 49 (11): 4177–4193.
21. Eismann, M.T. and Stein, D.W.J. (2007). Stochastic mixture modeling. In: *Hyperspectral Data Exploitation: Theory and Applications*, vol. 148 (ed. C.-I Chang). Hoboken, NJ: Wiley.

22. Fan, W., Hu, B., Miller, J., and Li, M. (2009). Comparative study between a new nonlinear model and common linear model for analyzing laboratory simulated-forest hyperspectral data. *International Journal of Remote Sensing* 30 (11): 2951–2962.
23. Halimi, A., Altmann, Y., Dobigeon, N., and Tourneret, J. (2011). Nonlinear unmixing of hyperspectral images using a generalized bilinear model. *IEEE Transactions on Geoscience and Remote Sensing* 49 (11): 4153–4162.
24. Eberhart, R. and Kennedy, J. (1995). A new optimizer using particle swarm theory. *Proceedings of the Sixth International Symposium on Micro Machine and Human Science*, Nagoya, Japan (4–6 October 1995), 39–43.
25. Zhang, B., Sun, X., Gao, L., and Yang, L. (2011). Endmember extraction of hyperspectral remote sensing images based on the discrete particle swarm optimization algorithm. *IEEE Transactions on Geoscience and Remote Sensing* 49 (11): 4173–4176.
26. Karaboga, D. and Basturk, B. (2007). A powerful and efficient algorithm for numerical function optimization: artificial bee colony (ABC) algorithm. *Journal of Global Optimization* 39 (3): 459–471.
27. Xiao, Y. (2011). Research on Bee Colony Algorithm and its applications in image processing. Ph.D. Thesis. South China University of Technology, Guangzhou, China (in Chinese).
28. Bishop, C.M. and Nasrabadi, N.M. (2006). *Pattern Recognition and Machine Learning*. New York: Springer.
29. Neal, R.M. and Hinton, G.E. (1998). A view of the EM algorithm that justifies incremental, sparse, and other variants. *Learning in Graphical Models* 89: 355–368.
30. Zhang, B., Zhuang, L., Gao, L. et al. (2014). PSO-EM: a hyperspectral unmixing algorithm based on normal compositional model. *IEEE Transactions on Geoscience and Remote Sensing* 52 (12): 7782–7792.
31. Boyd, S.P. and Vandenberghe, L. (2004). *Convex Optimization*. Cambridge University Press.
32. Bilmes, J.A. (1998). A gentle tutorial of the EM algorithm and its application to parameter estimation for Gaussian mixture and hidden Markov models. *International Computer Science Institute* 4 (510): 126.
33. Dempster, A.P., Laird, N.M., and Rubin, D.B. (1977). Maximum likelihood from incomplete data via the EM algorithm. *Journal of the Royal Statistical Society: Series B (Methodological)* 39 (1): 1–38.
34. Redner, R.A. and Walker, H.F. (1984). Mixture densities, maximum likelihood and the EM algorithm. *SIAM Review* 26 (2): 195–239.
35. Laird, N., Lange, N., and Stram, D. (1987). Maximum likelihood computations with repeated measures: application of the EM algorithm. *Journal of the American Statistical Association* 82 (397): 97–105.
36. Plaza, A., Martinez, P., Pérez, R., and Plaza, J. (2004). A quantitative and comparative analysis of endmember extraction algorithms from hyperspectral data. *IEEE Transactions on Geoscience and Remote Sensing* 42 (3): 650–663.
37. Bro, R. and De Jong, S. (1997). A fast non-negativity-constrained least squares algorithm. *Journal of Chemometrics* 11 (5): 393–401.
38. Lawson, C.L. and Hanson, R.J. (1974). *Solving Least Squares Problems*. Englewood Cliffs, NJ: SIAM.
39. McLachlan, G. and Krishnan, T. (2007). *The EM Algorithm and Extensions*. Hoboken, NJ: Wiley.
40. Wu, C. (1983). On the convergence properties of the EM algorithm. *The Annals of Statistics* 11 (1): 95–103.
41. Kann, V. (1992). *On the Approximability of NP-Complete Optimization Problems*. Stockholm: Royal Institute of Technology.
42. Ausiello, G. (1999). *Complexity and Approximability Properties: Combinatorial Optimization Problems and Their Approximability Properties*. Germany: Springer.
43. Kennedy, J. and Eberhart, R. (1995). Particle swarm optimization. *Proceedings of the IEEE International Conference on Neural Networks*, Perth, Western Australia, Australia (27 November–1 December 1995),1942–1948.
44. Shi, Y. (1998). A modified particle swarm optimizer. *Proceedings of IEEE ICEC Conference*, Anchorage (4–9 May 1998).
45. Kennedy, J. (2010). Particle swarm optimization. In: *Encyclopedia of Machine Learning* (ed. C. Sammut and G.I. Webb), 760–766. New York: Springer.
46. Poli, R., Kennedy, J., and Blackwell, T. (2007). Particle swarm optimization. *Swarm Intelligence* 1 (1): 33–57.
47. Clerc, M. and Kennedy, J. (2002). The particle swarm – explosion, stability, and convergence in a multidimensional complex space. *IEEE Transactions on Evolutionary Computation* 6 (1): 58–73.

48. Trelea, I.C. (2003). The particle swarm optimization algorithm: convergence analysis and parameter selection. *Information Processing Letters* 85 (6): 317–325.
49. Chang, C.-I and Du, Q. (1999). Interference and noise-adjusted principal components analysis. *IEEE Transactions on Geoscience and Remote Sensing* 37 (5): 2387–2396.
50. Green, A.A., Berman, M., Switzer, P., and Craig, M.D. (1988). A transformation for ordering multispectral data in terms of image quality with implications for noise removal. *IEEE Transactions on Geoscience and Remote Sensing* 26 (1): 65–74.
51. Lee, J.B., Woodyatt, A.S., and Berman, M. (1990). Enhancement of high spectral resolution remote-sensing data by a noise-adjusted principal components transform. *IEEE Transactions on Geoscience and Remote Sensing* 28 (3): 295–304.
52. Chang, C.-I and Du, Q. (2004). Estimation of number of spectrally distinct signal sources in hyperspectral imagery. *IEEE Transactions on Geoscience and Remote Sensing* 42 (3): 608–619.
53. Luo, B., Chanussot, J., Douté, S., and Zhang, L. (2013). Empirical automatic estimation of the number of endmembers in hyperspectral images. *IEEE Geoscience and Remote Sensing Letters* 10 (1): 24–28.
54. Chang, C.-I (2003). *Hyperspectral Imaging: Techniques for Spectral Detection and Classification*. New York: Springer.
55. Nascimento, J.M.P. and Dias, J.M.B. (2007). Hyperspectral signal subspace estimation. *IEEE International Geoscience and Remote Sensing Symposium (IGARSS)* 2007: 3225–3228.
56. Altmann, Y., Halimi, A., Dobigeon, N., and Tourneret, J.-Y. (2012). Supervised nonlinear spectral unmixing using a postnonlinear mixing model for hyperspectral imagery. *IEEE Transactions on Image Processing* 21 (6): 3017–3025.
57. Zhong, L., Luo, W., and Gao, L. (2016). A particle swarm optimization algorithm for unmixing the polynomial post-nonlinear mixing model. *2016 9th International Congress on Image and Signal Processing, BioMedical Engineering and Informatics (CISP-BMEI) IEEE*, Datong, China (15–17 October 2016), 15–17.
58. Zhong, L., Luo, W., and Gao, L. (2016). Particle swarm optimization for nonlinear spectral unmixing: a case study of generalized bilinear model. *2016 12th International Conference on Natural Computation and 13th Fuzzy Systems and Knowledge Discovery (ICNC-FSKD) IEEE*, Changsha, China (13–15 August 2016), 13–15.
59. Yang, B., Luo, W., and Wang, B. (2017). Constrained nonnegative matrix factorization based on particle swarm optimization for hyperspectral unmixing. *IEEE Journal of Selected Topics in Applied Earth Observations and Remote Sensing* 10 (8): 3693–3710.
60. Akgun, T., Altunbasak, Y., and Mersereau, R.M. (2005). Super-resolution reconstruction of hyperspectral images. *IEEE Transactions on Image Processing* 14: 1860–1875.
61. Christophe, E., Léger, D., and Mailhes, C. (2005). Quality criteria benchmark for hyperspectral imagery. *IEEE Transactions on Geoscience and Remote Sensing* 43 (9): 2103–2114.
62. Yokoya, N., Grohnfeldt, C., and Chanussot, J. (2017). Hyperspectral and multispectral data fusion: a comparative review of the recent literature. *IEEE Geoscience and Remote Sensing Magazine* 5 (2): 29–56.
63. Dobigeon, N., Moussaoui, S., Coulon, M. et al. (2009). Joint Bayesian endmember extraction and linear unmixing for hyperspectral imagery. *IEEE Transactions on Signal Processing* 57 (11): 4355–4368.
64. Eches, O., Dobigeon, N., Mailhes, C., and Tourneret, J.Y. (2009). Unmixing hyperspectral images using a normal compositional model and MCMC methods. *Proceedings of the IEEE/SP 15th Workshop on Statistical Signal Processing*, Cardiff, United Kingdom (31 August–3 September 2009), 646–649.
65. Chang, C.-I (2013). *Hyperspectral Data Processing: Algorithm Design and Analysis*. Hoboken, NJ: Wiley.
66. Eches, O., Dobigeon, N., Mailhes, C., and Tourneret, J.Y. (2010). Bayesian estimation of linear mixtures using the normal compositional model. Application to hyperspectral imagery. *IEEE Transactions on Image Processing* 19 (6): 1403–1413.
67. Pu, H., Chen, Z., Wang, B. et al. (2015). Constrained least squares algorithms for nonlinear unmixing of hyperspectral imagery. *IEEE Transactions on Geoscience and Remote Sensing* 53 (3): 1287–1303.
68. Eches, O. and Guillaume, M. (2014). A bilinear–bilinear nonnegative matrix factorization method for hyperspectral unmixing. *IEEE Geoscience and Remote Sensing Letters* 11 (4): 778–782.

# 16

# Spectral-Spatial Robust Nonnegative Matrix Factorization for Hyperspectral Unmixing

Risheng Huang[1], Xiaorun Li[1], and Liaoying Zhao[2]
[1]*College of Electrical Engineering, Zhejiang University, Hangzhou, Zhejiang, China*
[2]*School of Computer Science, Hangzhou Dianzi University, Hangzhou, Zhejiang, China*

## 16.1 Introduction

Hyperspectral data are imageries collected by hypersepctral sensors, which consist of hundreds of contiguous narrow spectral band images. Due to the limitation of hypersepctral sensors' spatial resolution and mixing effects of ground materials, the observed spectrum for a pixel is commonly mixed by several materials' spectra. Therefore, hyperspectral unmixing (HU) is an important problem for the exploiting of remotely sensed hyperspectral data. The task of HU can be divided into two main steps, endmember extraction and abundance estimating. For endmember extraction, a number of algorithms have been proposed and widely used to find endmembers, including pixel purity index (PPI) [1], N-FINDR [2], vertex component analysis (VCA) [3], and the simplex growing algorithm (SGA) [4]. The relationships among these methods have been investigated in Chang et al. [5], Li and Chang [6], and Chang et al. [7]. When the endmembers have been extracted, the abundances can be estimated by solving the constrained optimization problem under constraints including abundance nonnegativity constraint (ANC), abundance sum-to-one constraint (ASC), and so on [8].

Since the pure pixels may be absent in the hyperspectral data, a group of algorithms have been developed to consider the unsupervised linear HU as a blind source separation (BSS) problem [9–11]. Among these methods, nonnegative matrix factorization (NMF) has shown to be attractive for the promising in extracting sparse and interpretable representations from a data matrix. NMF aims to decompose a data matrix into two low-rank matrices with nonnegative constraint: a basis matrix and a coefficient matrix, which provide an intuitive and interpretable representation of data [12]. As an unsupervised method, NMF has been applied to HU and shown its advantages in many situations [13, 14]. Unfortunately, due to the nonconvexity of NMF's objection function, the solution space of NMF is quite large and make the algorithm easy to fall into local minima. To reduce the solution space, additional constraints have been introduced into NMF guided by the inherent

*Advances in Hyperspectral Image Processing Techniques*, First Edition. Edited by Chein-I Chang.

characteristics of hyperspectral applications [10, 15, 16]. Miao and Qi [17] proposed a minimum volume constrained NMF (MVCNMF). Wang et al. [18] proposed an endmember dissimilarity constrained NMF method for HU. Constraints on abundances have also been exploited and used for HU [19–21].

In real applications, noises commonly exist in hyperspectral data and seriously degrade the unmixing performance of HU methods. A bunch of robust HU methods have been proposed to alleviate the influence of noises [22–24]. In Jiang et al. [23], the trail-and-error method is used to estimate noise variance and determine the regularization parameter, which helps to weaken the noise effects on endmember and abundance estimation. He et al. [25] adopted the TV regularizer to denoise the abundance maps, which improves the robustness of the proposed method to noise. Ma et al. [26] proposed a robust sparse unmixing method with $\ell_{2,1}$ norm to achieve robustness for noises and outliers. $\ell_{2,1}$ norm is a rotational invariant norm firstly proposed by Ding et al. [27], which can effectively handle noises and outliers [28, 29]. Kong et al. [30] proposed a robust formulation of NMF applying $\ell_{2,1}$ norm loss function ($\ell_{2,1}$-RNMF) and provided efficient and elegant updating rules, which can accommodate outliers and noises in a better way than standard NMF. Yang et al. [31] proposed a robust NMF via joint sparse and graph regularization model, which also applies $\ell_{2,1}$ norm into objective function to handle noisy and sparse data effectively. The above algorithms using $\ell_{2,1}$ norm all consider that there exist noises and outliers in the samples (pixels). Nevertheless, for real-world HSIs, there also commonly exist band noises, and the noise intensities in different bands are various.

In the most real HU experiments, the noisy bands are manually selected and eliminated, which requires the prior knowledge of noisy bands and may miss important information provided by these bands [32]. For this consideration, Wang et al. [33] proposed a robust HU method with correntropy-based metric, which automatically assigns adaptive weights to noisy bands and noise-free bands to suppress the contribution of noisy bands. As a symmetric form of $\ell_{2,1}$ norm, $\ell_{1,2}$ norm is firstly used in the Group Lasso as a sparsity constraint to drive as many feature weights to zero as possible [34]. Interestingly, $\ell_{1,2}$ norm can also be applied to HU to achieve robustness to band noise. He et al. [32] proposed a new coupled blind HU to unmix the low-noisy and noisy bands of HSIs simultaneously. When applied as objective function, $\ell_{1,2}$ norm also possesses the ability to automatically allocate adaptive weight in the feature dimension (spectral dimension). Therefore, $\ell_{1,2}$ norm can also be introduced into NMF to derive a robust NMF formulation [35], which has good robustness to noisy bands by rows.

Although the hyperspectral data are often corrupted by noisy bands and noisy pixels, few of existing methods consider the two formulations of noise (noise by pixel and noise by band) simultaneously. For this consideration, this chapter presents a spectral-spatial robust NMF model incorporating both $\ell_{2,1}$ and $\ell_{1,2}$ norms, which is insensitive to both noisy bands and noisy pixels and is able to automatically allocate adaptive weights for each band and pixel.

Furthermore, in $\ell_{2,1}$-RNMF or $\ell_{1,2}$-RNMF, the contribution of each pixel or band in the objective function is assigned with a weight directly decided by the residue of corresponding pixel or band. When meeting the data points with nero-zero residues, $\ell_{2,1}$-RNMF or $\ell_{1,2}$-RNMF may run into singularity problem. More seriously, due to the weights are directly proportional to the inverses of residues, the objective function may be dominated by those data points with small residues, which omits the important information of other data points and may cause the algorithms perform unsatisfactorily, even worse than standard NMF. For hyperspectral data, since the intensity levels of noise are various, bands or pixels with small residues exist commonly and should also be aligned with appropriate weights to avoid overwhelming the contribution of other bands or pixels. Therefore, the Huber's M-estimator with a "cutoff" parameter is introduced to our robust model to achieve better allocation strategy of weights.

The main contributions of this chapter can be summarized as three aspects. Firstly, the problems of robust NMF methods based on $\ell_{2,1}$ norm or $\ell_{1,2}$ norm are analyzed and the idea of considering band noise and pixel noise of two formulations are put forward in the HU model. Secondly, the Huber's M-estimator [36] is introduced to solve the singularity problem and weight allocating problem of robust NMF methods based on $\ell_{2,1}$ norm or $\ell_{1,2}$ norm, which helps to achieve more reasonable weight allocating and make full use of the information of low-noisy and noisy bands or pixels. Thirdly, a spectral-spatial robust NMF effectively combining the robustness of $\ell_{2,1}$ and $\ell_{1,2}$ norms is utilized for HU, which is robust to noise by band and noise by pixel simultaneously. The superiorities of our method have been validated via both synthetic experiments and real data experiments.

The remainder of this chapter is organized as follows. Section 16.2 gives a brief introduction of the NMF algorithm and then introduces $\ell_{2,1}$-RNMF and $\ell_{1,2}$-RNMF. Section 16.3 proposes the spectral-spatial robust NMF model and provides the corresponding updating rules for the proposed model. Sections 16.4–16.6 validate the effectiveness of the proposed method on synthetic data and real hyperspectral images. Finally, Section 16.7 concludes this chapter.

## 16.2 Robust NMF

In the classical linear mixing model (LMM), the observed pixels in HSI are assumed as linear mixtures of several spectral signatures called endmembers. Consider hyperspectral image data $X = [x_1, x_2, \ldots, x_N]$, where $X \in \mathbb{R}_+^{L \times N}$ and $N$ is the number of pixels. Based on LMM, the hyperspectral data $X$ could be represented as

$$X = WH + E, \tag{16.1}$$

where $W = [\mathbf{w}_1, \mathbf{w}_2, \ldots \mathbf{w}_P] \in \mathbb{R}_+^{L \times P}$ denotes the endmember matrix, $H \in \mathbb{R}_+^{P \times N}$ denotes the abundances of respective endmembers, and $E$ is a residual term. The NMF algorithm is designed to find an approximate factorization of $X$ such that $X \approx WH$, where $W \geq 0$ and $H \geq 0$. To quantify the quality of the approximate factorization, the Euclidean distance is commonly used to measure the distance between $X$ and $WH$. The loss function of NMF using Frobenius (i.e. quadratic) norm is defined as follows:

$$f(W, H) = \frac{1}{2}\|X - WH\|_F^2 = \sum_{l=1}^{L}\sum_{n=1}^{N}(X - WH)_{ln}^2. \tag{16.2}$$

The multiplication update rules for standard NMF algorithm are presented in Lee and Seung [37] to locally minimize the cost function in (16.2).

$$W_{lp} = W_{lp}\frac{(XH^T)_{lp}}{(WHH^T)_{lp}}, \tag{16.3}$$

$$H_{pn} = H_{pn}\frac{(W^TX)_{pn}}{(W^TWH)_{pn}}. \tag{16.4}$$

### *16.2.1 RNMF Using $\ell_{2,1}$ Norm*

One of the main drawbacks of NMF is that it is sensitive to the outliers and noises. In hypersepctral data, the pixels are commonly corrupted by various types of noise including Gaussian noise, impulse noise, and dead pixel of lines. To alleviate this problem, $\ell_{1,2}$ norm is introduced to NMF to derive a robust formulation of the cost function [30] as follows:

$$\|X - WH\|_{2,1} = \sum_{n=1}^{N} \sqrt{\sum_{l=1}^{L} (X - WH)_{ln}^2}, \tag{16.5}$$

which can also be formulated as

$$\|X - WH\|_{2,1} = \sum_{n=1}^{N} \sum_{l=1}^{L} \frac{(X - WH)_{ln}^2}{\sqrt{\sum_{l=1}^{L} (X - WH)_{ln}^2}}. \tag{16.6}$$

Comparing (16.6) with the formula form of (16.2), it could be seen that each element $(X–WH)_{ln}$ in $\ell_{2,1}$ norm enjoys a weight $1/\sqrt{\sum_{l=1}^{L} (X - WH)_{ln}^2}$, corresponding to its located column's error. By this robust formulation, the influence of those columns with large errors due to high noise intensities is reduced respectively. Therefore, Robust NMF using $\ell_{2,1}$ norm ($\ell_{2,1}$-RNMF) has good robustness to noisy samples (pixels) by columns. $\ell_{2,1}$-RNMF is formulated as follows:

$$\begin{aligned} &\min_{W,H} f(W, H) = \|X - WH\|_{2,1}, \\ &\text{s.t.} \quad W \geq 0, \ \ H \geq 0. \end{aligned} \tag{16.7}$$

The multiplicative update rules for (16.7) are given by

$$W_{lp} \leftarrow W_{lp} \frac{(XD_aH^T)_{lp}}{(WHD_aH^T)_{lp}}, \tag{16.8}$$

$$H_{pn} \leftarrow H_{pn} \frac{(W^TXD_a)_{pn}}{(W^TWHD_a)_{pn}}, \tag{16.9}$$

where $D_a$ is a diagonal matrix with the diagonal element

$$(D_a)_{nn} = \frac{1}{\sqrt{\sum_{l=1}^{L} (X - WH)_{ln}^2}} = \frac{1}{\|x_n - Wh_n\|_2}, \tag{16.10}$$

where $x_n$, $h_n$ denote the $n$th column of $X$, $H$, respectively. The adaptive weight for each sample (pixel) is contained in the matrix $D_a$.

### 16.2.2 *RNMF Using* $\ell_{1,2}$ Norm

As demonstrated in Section 16.2.1, $\ell_{2,1}$-RNMF is robust to samples (pixels). As a symmetric form of $\ell_{2,1}$ norm, $\ell_{1,2}$ norm can also be introduced to NMF to derive a formulation robust to band noises by row, which is formulated as follows:

$$\begin{aligned} \|X-WH\|_{1,2} &= \sum_{l=1}^{L}\sqrt{\sum_{n=1}^{N}(X-WH)_{ln}^2} \\ &= \sum_{l=1}^{L}\sum_{n=1}^{N}\frac{(X-WH)_{ln}^2}{\sqrt{\sum_{n=1}^{N}(X-WH)_{ln}^2}}. \end{aligned} \tag{16.11}$$

Therefore, each element $(X-WH)_{ln}$ in $\ell_{1,2}$ norm also enjoys a weight $1/\sqrt{\sum_{n=1}^{N}(X-WH)_{ln}^2}$ corresponding to its located row's error. Robust NMF using $\ell_{1,2}$ norm ($\ell_{1,2}$-RNMF) has good robustness to noises by rows. When applied to hyperspectral data, $\ell_{1,2}$-RNMF aligns different weights for low-noise and noisy bands accordingly to alleviate the influence of noisy bands. $\ell_{1,2}$-RNMF is formulated as follows:

$$\begin{aligned} &\min_{W,H} f(W,H) = \|X-WH\|_{1,2}, \\ &\text{s.t.} \quad W \geq 0, \ H \geq 0. \end{aligned} \tag{16.12}$$

The multiplicative update rules for (16.12) are given by

$$W_{lp} \leftarrow W_{lp}\frac{(D_bXH^T)_{lp}}{(D_bWHH^T)_{lp}}, \tag{16.13}$$

$$H_{pn} \leftarrow H_{pn}\frac{(W^TD_bX)_{pn}}{(W^TD_bWH)_{pn}}, \tag{16.14}$$

where $D_b$ is a diagonal matrix with the diagonal element

$$(D_b)_{ll} = \frac{1}{\sqrt{\sum_{n=1}^{N}(X-WH)_{ln}^2}} = \frac{1}{\|x_l - w_lH\|_2}, \tag{16.15}$$

where $x_l$, $w_l$ denote the $l$th row of $X$, $W$, respectively. The adaptive weight for each band is contained in the matrix $D_b$.

## 16.3 Spectral-Spatial Robust Nonnegative Matrix Factorization (SSRNMF)

For the previous methods using Frobenius norm or $\ell_1$-norm is that they treat the two indexes $n$ and $l$ with no difference, which can be seen from (16.2). Nevertheless, these two indexes have different physical meaning: $n$ runs through the spatial dimension (different pixels), and $l$ run through the spectral dimension (different bands). As shown in (16.5) and (16.11), for $\ell_{2,1}$ and $\ell_{1,2}$ norms, the two indexes are run in different orders, respectively, which help to reveal and handle the abnormal data points in different dimensions. Accordingly, we are motivated to develop a robust unmixing model achieving robustness in both spectral dimension and spatial dimension. Nevertheless, simple combination of $\ell_{2,1}$ and $\ell_{1,2}$ norms may not be feasible since $\ell_{2,1}$-RNMF and $\ell_{1,2}$-RNMF suffer from some problems. Both $\ell_{2,1}$-RNMF and $\ell_{1,2}$-RNMF can be seen as a weighted version of NMF. In $\ell_{2,1}$-RNMF and $\ell_{1,2}$-RNMF, the adaptive weight for a pixel or a band in the objective is automatically decided by the residue from last update. However, when meet those pixels or bands with near-zero residues, $\ell_{2,1}$-RNMF and $\ell_{1,2}$-RNMF may come across the singularity problem (blow-up when the denominator is near 0), which leads to the failure of matrix decomposition. This problem can be temporarily solved by adding a small quantity number to the denominator. More seriously, these two methods exist evident defects in weight allocation. Although $\ell_{2,1}$-RNMF and $\ell_{1,2}$-RNMF can efficiently suppress the contribution of noisy part through adaptive weight proportional to the reciprocal of residue, the objective functions of these two methods may also be dominated by the data points with small residues and lead to the degrading of performance. A small number of data points with small residues may occupy an extremely large proportion in objective function and overwhelm the contribution of other data points that are in majority. Therefore, as robust methods, $\ell_{2,1}$-RNMF and $\ell_{1,2}$-RNMF are robust to data points with high noise intensities but are easy to be seriously disturbed by those data points with near-zero residues. The above problems can be solved via the following proposed SSRNMF model.

### *16.3.1 SSRNMF Model*

In hyperspectral data, there exist noisy bands with low SNRs due to water vapor and atmospheric effects [38]. Simultaneously, there often exist some pixels with large DN variations caused by sudden and unexpected change of the spectral range of the illumination source or sensor's corruption by the electronic charge, which can be seen as noisy pixels [39]. Since band noises and pixel noises commonly exist in hyperspectral data simultaneously, we employ both $\ell_{2,1}$ and $\ell_{1,2}$ norms into objective function to achieve robustness in band and pixel dimensions simultaneously. $\ell_{2,1}$ norm enables our model to effectively handle pixels with noises or outliers by column and $\ell_{1,2}$ norm enables our model to effectively handle bands with noises or outliers by rows. Meanwhile, to solve the problems of $\ell_{2,1}$-RNMF and $\ell_{1,2}$-RNMF discussed in Section 16.2.2, a popular robust estimation called Huber's M-estimator $\rho(\cdot)$ is introduced to formulate our model to achieve better weight assignations according to the reconstruction errors. The Huber's M-estimator is defined as

$$\rho(t) = \begin{cases} t^2 & \text{if } |t| \leq c \\ 2c|t| - c^2 & \text{if } |t| > c \end{cases} \tag{16.16}$$

with a cutoff parameter $c$ for the regularization effect of the residues' weights. Apparently, the "small" residues ($|t| \leq c$) are measured by their squares (the Euclidean metric). And the "large" residues ($|t| > c$) are measured by the $\ell_1$ form. It is worth to point out that our model is different from the traditional robust estimation methods, which consider that there are only one kind of noise formulation (by band or by pixel). Our model considers that there exist "outlier" bands or "outlier" pixels simultaneously, which is more physically consistent with the practical situations in the hyperspectral images. Combining the $\ell_{2,1}$ norm, $\ell_{1,2}$ norm, and the Huber's M-estimator, our proposed spectral-spatial robust unmixing model is defined as follows:

$$\begin{gathered} \min_{W,H} f(W,H) = f_1(W,H) + f_2(W,H) \\ = \sum_{n=1}^{N} \rho_a\left(\sqrt{\sum_{l=1}^{L}(X - WH)_{ln}^2}\right) + \sum_{l=1}^{L} \rho_b\left(\sqrt{\sum_{n=1}^{N}(X - WH)_{ln}^2}\right), \\ \text{s.t.} \quad W \geq 0, \ H \geq 0. \end{gathered} \tag{16.17}$$

where $\rho_a(\cdot)$ and $\rho_b(\cdot)$ are the Huber's M-estimators for pixel and band, respectively.

There are two cutoff parameters $c_a$ and $c_b$ for the two parts of (16.17), respectively. In the proposed spectral-spatial robust NMF, the residues by columns and by rows are estimated through $W$ and $H$ achieved by a previous NMF process

$$r_n = \|x_n - Wh_n\|_2, \tag{16.18}$$

$$r_l = \|x_l - w_l H\|_2, \tag{16.19}$$

where $r_n$ and $r_l$ are the residues of $n$th pixel and $l$th band, respectively. The indexes of columns with residues $r_n \leq c_a$ and $r_n > c_a$ are written as $\text{index}_1^{(\text{pixel})}$, $\text{index}_2^{(\text{pixel})}$, respectively. And the indexes of rows with residues $r_l \leq c_b$ and $r_l > c_b$ are written as $\text{index}_1^{(\text{pixel})}$, $\text{index}_2^{(\text{pixel})}$.

The data matrix $X$ can be rearranged as

$$X_r = \begin{bmatrix} X_{11}, X_{12} \\ X_{21}, X_{22} \end{bmatrix}, \tag{16.20}$$

where

$$X_{11} = X\left(\text{index}_1^{(\text{band})}, \text{index}_1^{(\text{pixel})}\right),$$

$$X_{12} = X\left(\text{index}_1^{(\text{band})}, \text{index}_2^{(\text{pixel})}\right),$$

$$X_{21} = X\left(\text{index}_2^{(\text{band})}, \text{index}_1^{(\text{pixel})}\right),$$

$$X_{22} = X\left(\text{index}_2^{(\text{band})}, \text{index}_2^{(\text{pixel})}\right).$$

In $X_r$, $\begin{bmatrix} X_{12} \\ X_{22} \end{bmatrix} \cdot \in \mathbb{R}_+^{(L_1+L_2)\times N_2}$ contains the columns whose residue $r_n$ is larger than the cutoff $c_a$, and $[X_{21}, X_{22}] \in \mathbb{R}_+^{L_2 \times (N_1+N_2)}$ contains the rows whose residue $r_l$ is larger than the cutoff $c_b$. The aim of proposed method is to find $W_1 \in \mathbb{R}_+^{L_1 \times P}$, $W_2 \in \mathbb{R}_+^{L_2 \times P}$, $H_1 \in \mathbb{R}_+^{P \times N_1}$, and $H_2 \in \mathbb{R}_+^{P \times N_2}$ such that

$$\begin{bmatrix} X_{11}, X_{12} \\ X_{21}, X_{22} \end{bmatrix} \simeq \begin{bmatrix} W_1 \\ W_2 \end{bmatrix} \cdot [H_1, H_2], \tag{16.21}$$

where $W_2$ is related the noisy bands and $H_2$ is related to the noisy pixels.

Therefore, the cost function in (16.17) can be further expressed as (16.22). It is worth noting that $c_a$, $c_b$, and the dividing of $X_r$ in (16.20) are fixed to ensure the monotonically decreasing of objective function in (16.22).

$$\begin{aligned} f(W,H) &= f_1(W,H) + f_2(W,H) \\ &= \left\| \begin{bmatrix} X_{11} \\ X_{21} \end{bmatrix} - \begin{bmatrix} W_1 H_1 \\ W_2 H_1 \end{bmatrix} \right\|_F^2 + 2c_a \left\| \begin{bmatrix} X_{12} \\ X_{22} \end{bmatrix} - \begin{bmatrix} W_1 H_2 \\ W_2 H_2 \end{bmatrix} \right\|_{2,1} \\ &\quad + \|[X_{11}, X_{12}] - [W_1 H_1, W_1 H_2]\|_F^2 \\ &\quad + 2c_b \|[X_{21}, X_{22}] - [W_2 H_1, W_2 H_2]\|_{1,2}. \end{aligned} \tag{16.22}$$

### 16.3.2 Update Rules for SSRNMF

In this section, we present the multiplicative updating rules for SSRNMF. The objective function in (16.22) can be optimized in an alternative updating way. Let $\Theta_1 \in \mathbb{R}^{L_1 \times P}$, $\Theta_2 \in \mathbb{R}^{L_2 \times P}$, $\Xi_1 \in \mathbb{R}^{P \times N_1}$, and $\Xi_2 \in \mathbb{R}^{P \times N_2}$ be the corresponding Lagrange multipliers for the constraint $W_1 \geq 0$, $W_2 \geq 0$, $H_1 \geq 0$, and $H_2 \geq 0$, respectively. Consider the Lagrange function as

$$\begin{aligned}\mathcal{L} = &\left\| \begin{bmatrix} X_{11} \\ X_{21} \end{bmatrix} - \begin{bmatrix} W_1 H_1 \\ W_2 H_1 \end{bmatrix} \right\|_F^2 + 2c_a \left\| \begin{bmatrix} X_{12} \\ X_{22} \end{bmatrix} - \begin{bmatrix} W_1 H_2 \\ W_2 H_2 \end{bmatrix} \right\|_{2,1} \\ &+ \| [X_{11}, X_{12}] - [W_1 H_1, W_1 H_2] \|_F^2 \\ &+ 2c_b \| [X_{21}, X_{22}] - [W_2 H_1, W_2 H_2] \|_{1,2} \\ &+ \mathrm{Tr}(\Theta_1 W_1^T) + \mathrm{Tr}(\Theta_2 W_2^T) + \mathrm{Tr}(\Xi_1 H_1^T) + \mathrm{Tr}(\Xi_2 H_2^T).\end{aligned} \tag{16.23}$$

Take the partial derivatives of $\mathcal{L}$ with respect to $W_1$, $W_2$, $H_1$, and $H_2$.

$$\begin{aligned}\frac{\partial \mathcal{L}}{\partial W_1} = &\, 4W_1 H_1 H_1^T + 2W_1 H_2 H_2^T + 2c_a W_1 H_2 D_a H_2^T \\ &- 4X_{11} H_1^T - 2X_{12} H_2^T - 2c_b X_{12} D_a H_2^T + \Theta_1,\end{aligned} \tag{16.24}$$

where $D_a$ is a diagonal matrix with diagonal element

$$\begin{aligned}D_a(n_2 n_2) &= \frac{1}{\sqrt{\sum_{l=1}^{L} \left( \begin{bmatrix} X_{12} \\ X_{22} \end{bmatrix} - \begin{bmatrix} W_1 H_2 \\ W_2 H_2 \end{bmatrix} \right)_{l n_2}^2}} \\ &= 1 \Big/ \left\| \left( \begin{bmatrix} X_{12} \\ X_{22} \end{bmatrix} - \begin{bmatrix} W_1 H_2 \\ W_2 H_2 \end{bmatrix} \right)_{n_2} \right\|_2 .\end{aligned} \tag{16.25}$$

$$\begin{aligned}\frac{\partial \mathcal{L}}{\partial W_2} = &\, 2W_2 H_1 H_1^T + 2c_a W_2 H_2 D_a H_2^T + 2c_b D_b W_2 [H_1, H_2] \begin{bmatrix} H_1^T \\ H_2^T \end{bmatrix} \\ &- 2X_{21} H_1^T - 2c_a X_{22} D_a H_2^T - 2c_b D_b [X_{21}, X_{22}] \begin{bmatrix} H_1^T \\ H_2^T \end{bmatrix} + \Theta_2,\end{aligned} \tag{16.26}$$

where $D_b$ is a diagonal matrix with diagonal element

$$D_b(l_2 l_2) = \frac{1}{\| ([X_{21}, X_{22}] - [W_2 H_1, W_2 H_2])_{l_2} \|_2}. \tag{16.27}$$

$$\begin{aligned}\frac{\partial \mathcal{L}}{\partial H_1} = &\, 4W_1^T W_1 H_1 + 2W_2^T W_2 H_1 + 2c_b W_2^T D_b W_2^T H_1 \\ &- 4W_1^T X_{11} - 2W_2^T X_{21} - 2c_b W_2^T D_b X_{21} + \Xi_1,\end{aligned} \tag{16.28}$$

$$\frac{\partial \mathcal{L}}{\partial H_2} = 2c_a\left[W_1^T, W_2^T\right]\begin{bmatrix} W_1 \\ W_2 \end{bmatrix} H_2 D_a + 2W_1^T W_1 H_2 + 2c_b W_2^T D_b W_2^T H_2 \\ -2c_a\left[W_1^T, W_2^T\right]\begin{bmatrix} X_{12} \\ X_{22} \end{bmatrix} D_a - 2W_1^T X_{12} - 2c_b W_2^T D_b X_{22} + \Xi_2. \tag{16.29}$$

According to the Karush–Kuhn–Tucker conditions $(\Theta_1)_{l_1p}(W_1)_{l_1p} = 0$, $(\Theta_2)_{l_2p}(W_2)_{l_2p} = 0$, $(\Xi_1)_{pn_1}(H_1)_{pn_1} = 0$, and $(\Xi_2)_{pn_2}(H_2)_{pn_2} = 0$, we have

$$\left(2W_1H_1H_1^T + W_1H_2H_2^T + c_aW_1H_2D_aH_2^T - 2X_{11}H_1^T - X_{12}H_2^T - c_bX_{12}D_aH_2^T\right).^*W_1 = 0, \tag{16.30}$$

$$\left(W_2H_1H_1^T + c_a W_2 H_2 D_a H_2^T + c_b D_b W_2\left[H_1, H_2\right]\begin{bmatrix} H_1^T \\ H_2^T \end{bmatrix} \right. \\ \left. - X_{21}H_1^T - c_a X_{22} D_a H_2^T - c_b D_b\left[X_{21}, X_{22}\right]\begin{bmatrix} H_1^T \\ H_2^T \end{bmatrix}\right).^*W_2 = 0, \tag{16.31}$$

$$\left(2W_1^TW_1H_1 + W_2^TW_2H_1 + c_bW_2^TD_bW_2^TH_1 - 2W_1^TX_{11} - W_2^TX_{21} - c_bW_2^TD_bX_{21}\right).^*H_1 = 0, \tag{16.32}$$

$$\left(c_a\left[W_1^T, W_2^T\right]\begin{bmatrix} W_1 \\ W_2 \end{bmatrix} H_2 D_a + W_1^T W_1 H_2 + c_b W_2^T D_b W_2^T H_2 \right. \\ \left. - c_a\left[W_1^T, W_2^T\right]\begin{bmatrix} X_{12} \\ X_{22} \end{bmatrix} D_a - W_1^T X_{12} - c_b W_2^T D_b X_{22}\right).^*H_2 = 0. \tag{16.33}$$

Therefore, the corresponding multiplicative update rules of $W_1$ and $W_2$ are given by (16.34), (16.35).

$$(W_1)_{l_1p} \leftarrow (W_1)_{l_1p} \frac{\left(2X_{11}H_1^T + X_{12}H_2^T + c_aX_{12}D_aH_2^T\right)_{l_1p}}{\left(2W_1H_1H_1^T + W_1H_2H_2^T + c_aW_1H_2D_aH_2^T\right)_{l_1p}}, \tag{16.34}$$

$$(W_2)_{l_2p} \leftarrow (W_2)_{l_2p} \frac{\left(X_{21}H_1^T + c_a X_{22} D_a H_2^T + c_b D_b\left[X_{21}, X_{22}\right]\begin{bmatrix} H_1^T \\ H_2^T \end{bmatrix}\right)_{l_2p}}{\left(W_2H_1H_1^T + c_a W_2 H_2 D_a H_2^T + c_b D_b W_2\left[H_1, H_2\right]\begin{bmatrix} H_1^T \\ H_2^T \end{bmatrix}\right)_{l_2p}}. \tag{16.35}$$

The corresponding multiplicative update rules of $H_1$ and $H_2$ are given by (16.36), (16.37).

$$(H_1)_{pn_1} \leftarrow (H_1)_{pn_1} \frac{\left(2W_1^TX_{11} + W_2^TX_{21} + c_bW_2^TD_bX_{21}\right)_{pn_1}}{\left(2W_1^TW_1H_1 + W_2^TW_2H_1 + c_bW_2^TD_bW_2H_1\right)_{pn_1}}, \tag{16.36}$$

$$(H_2)_{pn_2} \leftarrow (H_2)_{pn_2} \frac{\left(c_a\left[W_1^T, W_2^T\right]\begin{bmatrix} X_{12} \\ X_{22} \end{bmatrix} D_a + W_1^T X_{12} + c_b W_2^T D_b X_{22}\right)_{pn_2}}{\left(c_a\left[W_1^T, W_2^T\right]\begin{bmatrix} W_1 \\ W_2 \end{bmatrix} H_2 D_a + W_1^T W_1 H_2 + c_b W_2^T D_b W_2 H_2\right)_{pn_2}}. \tag{16.37}$$

We have proved the objective (16.22) is nonincreasing under the update rules (16.34), (16.35), (16.36), and (16.37).

The final unmixing results $W$ and $H$ are formed by

$$\begin{aligned} W\left(\text{index}_1^{(\text{band})}, :\right) &= W_1, \\ W\left(\text{index}_2^{(\text{band})}, :\right) &= W_2, \\ H\left(:, \text{index}_1^{(\text{pixel})}\right) &= H_1, \\ H\left(:, \text{index}_2^{(\text{pixel})}\right) &= H_2. \end{aligned} \tag{16.38}$$

The procedure of the SSRNMF is summarized in Algorithm 16.1.

---

**Algorithm 16.1: SSRNMF Algorithm**

---

**Input:** The observed mixture data $X$; the number of endmembers $P$.
Output: Endmembers matrix $W$ and abundance matrix $H$.

1: Initialize endmember matrix $W$ by endmember extraction method and abundance matrix $H$ by FCLS.
2: Conduct standard NMF by (16.3, 16.4) and update $W$ and $H$.
3: Compute residues $r_n$, $r_l$ of each pixel and each band via (16.18) and (16.19).
4: Compute $c_a$ = median $(r_n)$ and $c_b$ = median $(r_l)$.
5: Obtain the indexes of pixels with residues $r_n \leq c_a$ and $r_n > c_a$ : $\text{index}_1^{(\text{pixel})}, \text{index}_2^{(\text{pixel})}$.
6: Obtain the indexes of bands with residues $r_l \leq c_b$ and $r_l > c_b$ : $\text{index}_1^{(\text{band})}, \text{index}_2^{(\text{band})}$.
7: Rearrange $X$ as $X_r = \begin{bmatrix} X_{11}, X_{12} \\ X_{21}, X_{22} \end{bmatrix}$ according to (16.20).
8: Initialize $W_1$, $W_2$, $H_1$, and $H_2$.
9: Repeat.
10: Compute $D_a$ according to Eq. (16.25).
11: Compute $D_b$ according to Eq. (16.27).
12: Update $W_1$ by Eq. (16.34).
13: Update $W_2$ by Eq. (16.35).
14: Update $H_1$ by Eq. (16.36).
15: Update $H_2$ by Eq. (16.37).
16: Until convergence.
17: Form the final unmixing results $W$ and $H$ by (16.38).

---

### 16.3.3 Implementation Issues

Both $\ell_{2,1}$-RNMF and $\ell_{1,2}$-RNMF can be seen as a weighted version of NMF, which can suppress the weights of the pixels or bands with high noise intensities effectively. However, the weights for pixels or bands are directly proportional to the inverses of residues by pixel or band in $\ell_{2,1}$-RNMF and $\ell_{1,2}$-RNMF. Therefore, the objective function is heavily dominated by the pixels or bands with nero-zero residues ($r_l = \|x_l - w_l H\|_2 \approx 0$, $r_n = \|x_n - Wh_n\|_2 \approx 0$). The contributions of those pixels or bands with larger residues will be omitted, leading to unsatisfactory unmixing results. And $\ell_{2,1}$-RNMF may come across the singularity problem when the denominator $\|x_n - Wh_n\|_2$ is almost 0. To address above problems, our method introduces Huber's M-estimator with a cutoff parameter to achieve better assignations of weights. For the pixel dimension, when the residue of a pixel is larger than the cutoff $c_a$, the weight of $\ell_{2,1}$ norm is used. Otherwise, the weight of Frobenius norm is used. It is similar for the band dimension. In the proposed method, each pixel and each band are aligned with different weights according to their reconstruction errors. The weight of the $n$th pixel is

$$w_n^{(\text{pixel})} = \begin{cases} 1 & \text{if } \|(x_n - Wh_n)\|_2 \le c_a \\ c_a D_a(n_2 n_2) & \text{if } \|(x_n - Wh_n)\|_2 > c_a \end{cases} \tag{16.39}$$

and the weight of the $l$th band is

$$w_l^{(\text{band})} = \begin{cases} 1 & \text{if } \|(x_l - w_l H)\|_2 \le c_b \\ c_b D_b(l_2 l_2) & \text{if } \|(x_l - w_l H)\|_2 > c_b \end{cases}. \tag{16.40}$$

The effect of the cutoff parameter has been investigated in Ding et al. [27]. As shown in Figure 16.1, the weight curve is not sensitive to the specific value of $c$ since the singularity caused by data points with nero-zero residues is cutoff. The weight curve also shows the two main functions

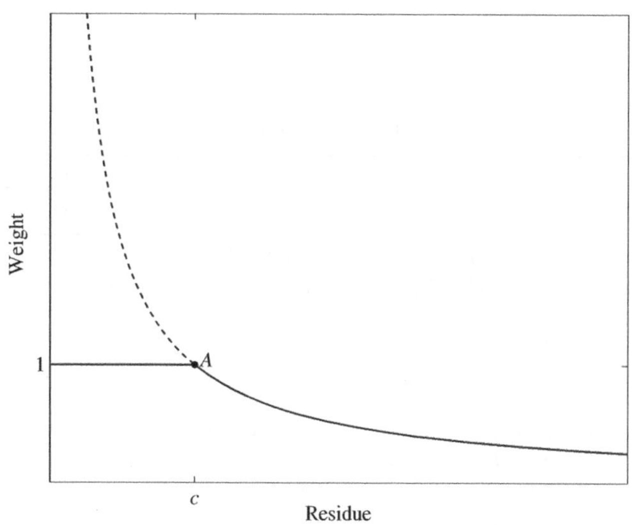


**Figure 16.1** The weight assignation by Huber's M-estimator.

of Huber's M-estimator. The first part of the curve (before the Point A) is to avoid singularity problem and prevent the overwhelming superiority of data points with small residues (the weight is fixed as 1). And the second part of the curve is to suppress the noises and outliers with a weight smaller than 1, which is decided by the residues. The data points with larger residues are aligned with smaller weight. Therefore, we set $c_a$, $c_b$ as the median value of $r_n$, $r_l$ as suggested in Ding et al. [27] to achieve stable and robust performance of the algorithm.

Another issue we concerned about is the initialization approach. To eliminate the interference of noise and outliers, we propose an effective initialization approach by generating initial values from pixels and bands with small residues. The initial $\begin{bmatrix} W_1 \\ W_2 \end{bmatrix}$ is generated by applying endmember extraction on $\begin{bmatrix} X_{11} \\ X_{21} \end{bmatrix}$. And $W_{1(0)}$ is used to estimate the initial abundance matrix by $[H_1, H_2]_{(0)} = \text{FCLS}([X_{11}, X_{12}], W_{1(0)})$.

## 16.3.4 Computation Complexity Analysis

In this section, we analyze the computational complexity of the proposed method. The computational complexity at each iteration in NMF and SSRNMF is presented in Table 16.1. Comparing the update rules of SSRNMF with the standard NMFs, we can find the additional computational cost mainly comes from the computations related to $D_a$ and $D_b$. It is worth noting that $D_a$ and $D_b$ are diagonal matrices, which effectively reduces the computational cost. For the updating of $W$, it takes additional $LN_2 + L_2N$ times of floating-point addition and multiplication. And for the updating of $H$, it takes additional $L_2P + N_2P$ times of floating-point addition and multiplication. Additionally, the computing of $D_a$ costs $LN_2$ additions, $LN_2$ multiplication, and $N_2$ divisions, while $D_b$ costs $L_2N$ additions, $L_2N$ multiplication, and $L_2$ divisions. Furthermore, the cost for the previous NMF step in SSRNMF is $O(LNP)$. Therefore, considering that $L$, $P$ is usually much smaller than $N$, the overall cost of SSRNMF is $O(LNP)$, which is of the same order of magnitude as the computational complexity of NMF.

**Table 16.1** Computational complexity at each iteration in NMF and SSRNMF

| Methods | | Update $W$ | Update $H$ | Update $D_a$ | Update $D_b$ | Overall |
|---|---|---|---|---|---|---|
| NMF | Addition | $LNP + (L + N)P^2$ | $LNP + (L + N)P^2$ | — | — | $O(LNP)$ |
| | Multiplication | $LNP + (L + N)P^2 + LP$ | $LNP + (L + N)P^2 + NP$ | — | — | |
| | Division | $LP$ | $NP$ | — | — | |
| SSRNMF | Addition | $LNP + (L + N)P^2 + LN_2 + L_2N$ | $LNP + (L + N)P^2 + L_2P + N_2P$ | $LN_2$ | $L_2N$ | $O(LNP)$ |
| | Multiplication | $LNP + (L + N)P^2 + LP + LN_2 + L_2N$ | $LNP + (L + N)P^2 + NP + L_2P + N_2P$ | $LN_2$ | $L_2N$ | |
| | Division | $LP$ | $NP$ | $N_2$ | $L_2$ | |
| | Previous NMF step | $O(LNP)$ | | | | |

## 16.4 Experiments on Synthetic Data

In this section, the SSRNMF is tested on synthetic data to evaluate its performance. Five methods, including CENMF [33], RNMF [32], RRLbS [35], $\ell_{2,1}$-RNMF [30], and $\ell_{1,2}$-RNMF [35], are compared with the SSRNMF in the experiments. It is worth noting that sparsity regularizers such as $\ell_1$ regularizer or $\ell_{1/2}$ regularizer are not added to the proposed RRNMF and the compared methods except for RRLbS. In this work, we concentrate on investigating the robustness to band noise and pixel noise. Of course, sparsity constraints can be easily added to the proposed SSRNMF for the hyperspectral scenes requiring for sparsity. The synthetic images are generated following the Gaussian field method [40]. The adopted type of Gaussian field is spheric. The spectral signatures randomly selected from the United States Geological Survey (USGS) digital spectral library are used as endmembers to generate synthetic data, which are displayed in Figure 16.2. In the real hyperspectral images, the noise levels of different bands or different pixels are various. To achieve better simulation of hyperspectral data in accordance with the real-world situation, both band noise and pixel noise are added to the synthetic images, and the noise intensity for different band or pixel is different, which means the SNR value for each band or pixel is different. As shown in Figure 16.3, both band noise and pixel noise are divided into noisy part (low SNR values) and low-noisy part (high SNR values). In each part, the SNR value also varies. Specifically, in our simulated experiments, the two kinds of noise are added as follows:

Band noise: We added zero-mean Gaussian noise to each band of the HSI. For different bands, the levels of noise intensities are different. A percentage of bands are randomly chosen and simulated as noisy bands and other bands are simulated as low-noisy bands. The signal-to-noise ratio (SNR) values of noisy bands are sampled from a Gaussian distribution $\mathrm{SNR}_{\mathrm{band1}} \sim \mathcal{N}(\overline{\mathrm{SNR}}_{\mathrm{band1}}, \delta^2)$, and the SNR

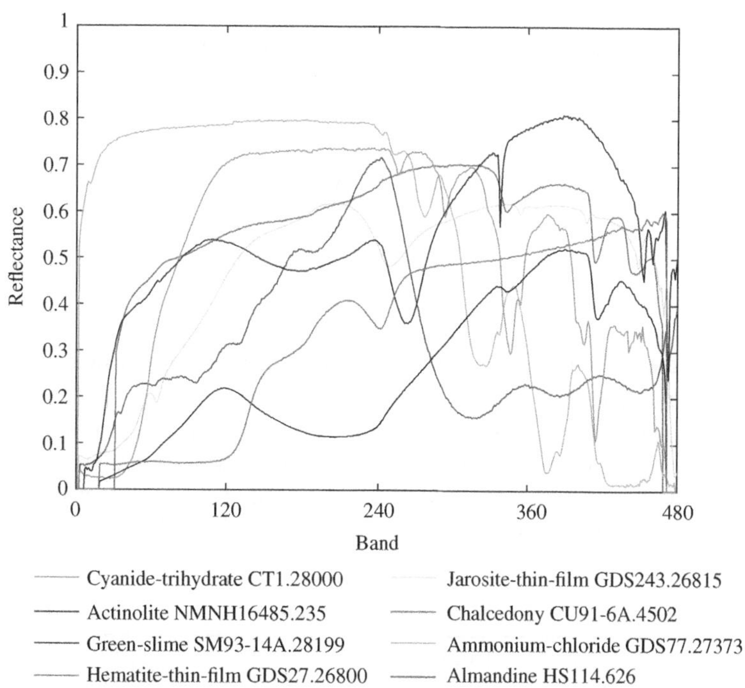


**Figure 16.2** The USGS spectral signatures used in the synthetic data experiments.

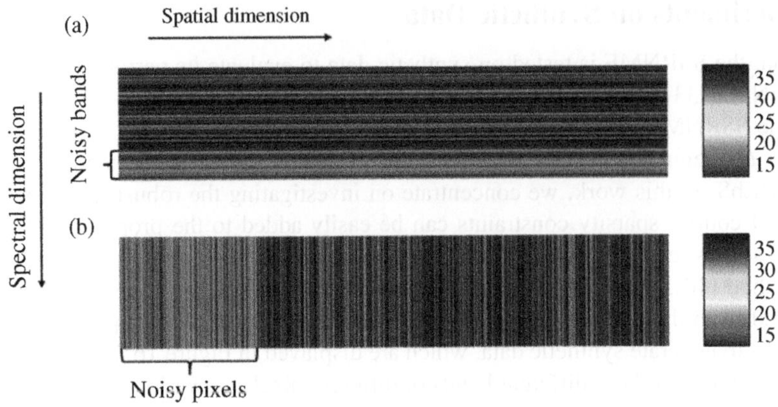


**Figure 16.3** SNR values for noisy part and low-noisy part. (a) SNR for each band, (b) SNR for each pixel.

values of low-noisy bands are sampled from $\mathrm{SNR}_{\mathrm{band2}} \sim \mathcal{N}\left(\overline{\mathrm{SNR}}_{\mathrm{band2}}, \delta^2\right)$. $\delta$ is fixed as 5 in this work. Figure 16.3(a) presents the SNR values for noisy bands and low-noisy bands when $\overline{\mathrm{SNR}}_{\mathrm{band1}} = 10$, $\overline{\mathrm{SNR}}_{\mathrm{band2}} = 30$.

Pixel noise: Similarly, we also added zero-mean Gaussian noise to each pixel of the HSI. For different pixels, the levels of noise intensities are different. A percentage of pixels are simulated as noisy pixels and other pixels are simulated as low-noisy pixels. The SNR values of noisy pixels are sampled from $\mathrm{SNR}_{\mathrm{pixel1}} \sim \mathcal{N}\left(\overline{\mathrm{SNR}}_{\mathrm{pixel1}}, \delta^2\right)$, and the SNR values of low-noisy pixels are sampled from $\mathrm{SNR}_{\mathrm{pixel2}} \sim \mathcal{N}\left(\overline{\mathrm{SNR}}_{\mathrm{pixel2}}, \delta^2\right)$. Figure 16.3(b) presents the SNR values for noisy pixels and low-noisy pixels when $\overline{\mathrm{SNR}}_{\mathrm{pixel1}} = 10$, $\overline{\mathrm{SNR}}_{\mathrm{pixel2}} = 30$.

Two criteria, spectral angle distance (SAD) and root-mean-square error (RMSE), are adopted to evaluate the algorithms' unmixing performance, which are defined as follows:

$$\mathrm{SAD}_p = \cos^{-1}\left(\frac{\mathbf{w}_i^T \hat{\mathbf{w}}_i}{\|\mathbf{w}_i\|\|\hat{\mathbf{w}}_i\|}\right), \tag{16.41}$$

$$\mathrm{RMSE}_p = \sqrt{\frac{1}{N}\sum_{n=1}^{N}\left(h_{pn} - \hat{h}_{pn}\right)^2}, \tag{16.42}$$

where $\mathbf{w}_i$ and $\hat{\mathbf{w}}_i$ are the reference endmember signatures and the estimated signatures, respectively. And $h_{pn}$ and $\hat{h}_{pn}$ are the reference abundances and their estimates.

To ensure reliable comparisons, 20 independent runs are carried out and the results are averaged for the SSRNMF and compared methods in the experiments. All experiments are performed using a PC with an Intel Core I7 CPU and 16 GB of RAM. When the iteration number is set as 200, the time cost of SSRNMF and other compared methods are presented in Table 16.2.

**Table 16.2** Comparison of the time cost of different algorithms

| NMF (s) | CENMF (s) | RNMF (s) | RRLbS (s) | $L_{1,2}$-RNMF (s) | $L_{2,1}$-RNMF (s) | SSRNMF (s) |
|---|---|---|---|---|---|---|
| 1.17 | 25.12 | 21.43 | 12.66 | 4.75 | 5.27 | 19.84 |

### 16.4.1 Robustness Analysis to Noise Composition

In this experiment, we investigate the performance of algorithms under different noise compositions. The number of endmembers is fixed as 5. Three kinds of situations are considered: only noisy bands, only noisy pixels, both noisy bands and noisy pixels.

For the first situation, a number of bands are randomly chosen as noisy bands and are added with high intensities of Gaussian noise. The SNR values of noisy bands are sampled from the Gaussian distribution $\mathrm{SNR}_{\mathrm{band1}} \sim \mathcal{N}\left(\overline{\mathrm{SNR}}_{\mathrm{band1}}, \delta^2\right)$. Other bands' SNR values are sampled from $\mathrm{SNR}_{\mathrm{band2}} \sim \mathcal{N}\left(\overline{\mathrm{SNR}}_{\mathrm{band2}}, \delta^2\right)$. $\overline{\mathrm{SNR}}_{\mathrm{band1}}$ is set to 10 and $\overline{\mathrm{SNR}}_{\mathrm{band2}}$ is set to 30. The ratio of noisy bands varies from 10 to 50% to test the algorithms' robustness to different noise percentages. Figure 16.4 shows the SAD and RMSE values of algorithms in the cases of different ratios. The results show that the SSRNMF performs the best and have good robustness to band noises. CENMF, RNMF, RRLbS, and $\ell_{1,2}$-RNMF also show some robustness to band noise as expected, but the effect is not significant. The SAD and RMSE values of $\ell_{2,1}$-RNMF increase as the number of noisy bands increases, which indicates that the algorithm is easy to be corrupted by band noise.

For the second situation with pixel noise, a number of pixels are randomly chosen as noisy pixels and are added with high intensities of Gaussian noise. $\overline{\mathrm{SNR}}_{\mathrm{pixel1}}$ is set to 10 and $\overline{\mathrm{SNR}}_{\mathrm{pixel2}}$ is set to 30. The ratio of noisy pixels is set to 10, 20, 30, 40, and 50%, respectively. The results are presented in Figure 16.5. We can see that the SSRNMF achieves best results for either SAD or RMSE when the ratio of noise pixels varies, which indicates that SSRNMF also has good robustness to pixel noise. As designed, $\ell_{2,1}$-RNMF should has robustness to noises and outliers in the sample dimension (by pixel). However, the results of $\ell_{2,1}$-RNMF show no obvious superiority to the other methods' results. Especially for the RMSE results, $\ell_{2,1}$-RNMF performs even worse than other algorithms. This phenomenon validates the problem we have pointed out in Section 16.2.2, which is mainly caused by unreasonable weight assignations.

Although $\ell_{2,1}$-RNMF can be seen as a symmetry method of $\ell_{1,2}$-RNMF, the problem of $\ell_{2,1}$-RNMF is more serious than $\ell_{1,2}$-RNMF. In hyperspectral data, the spatial dimension is usually much larger than the spectral dimension. In the unmixing process, the residues of different bands usually won't be very small and almost not possible to be near-zero. Therefore, $\ell_{1,2}$-RNMF seldom meets the singularity problem and shows better performance than NMF when imposed with band noise.

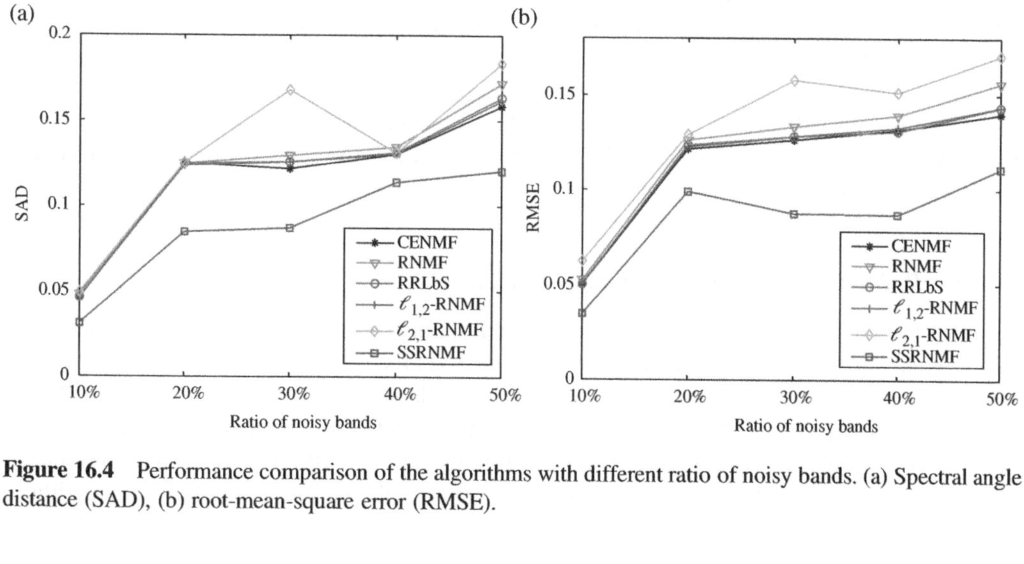


**Figure 16.4** Performance comparison of the algorithms with different ratio of noisy bands. (a) Spectral angle distance (SAD), (b) root-mean-square error (RMSE).

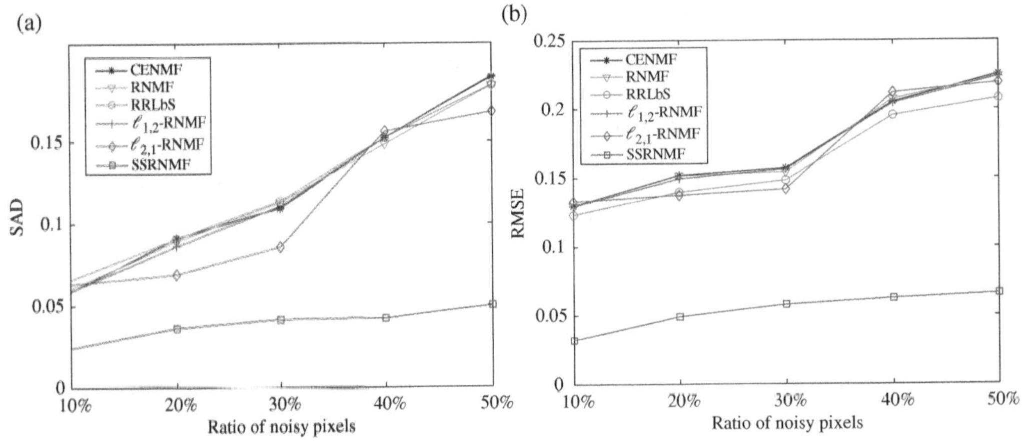


**Figure 16.5** Performance comparison of the algorithms with different ratio of noisy pixels. (a) SAD, (b) RMSE.

However, it's common to have pixels with small residues or near-zero residues, which cause that $\ell_{2,\,1}$-RNMF often run into singular problem and performs unsatisfactory in HU. Comparing the results in Figures 16.4 and 16.5, we can find that SSRNMF shows more obvious advantage over $\ell_{1,\,2}$-RNMF and $\ell_{2,\,1}$-RNMF in Figure 16.5 where only pixel noise is added. This phenomenon is mainly caused because the variety of the residues by pixel is much more obvious than that of the residues by band. Therefore, there is more need for reasonable weight assignations in the pixel dimension, and SSRNMF with Huber estimator has more obvious effect in the case of adding only pixel noise than that of band noise. We further analyze the behaviors of algorithms' weight allocating in the real data experiment.

For the situation with both band noise and pixel noise, a same ratio of bands and pixels are randomly selected as noisy band and noisy pixels. We set $\overline{\text{SNR}}_{\text{band1}} = \overline{\text{SNR}}_{\text{pixel1}} = 10$ and $\overline{\text{SNR}}_{\text{pixel2}} = \overline{\text{SNR}}_{\text{pixel2}} = 30$. The ratio of noisy bands and pixels is set to 10, 20, 30, 40, and 50%, respectively. From Figure 16.6, we can see that the SSRNMF also performs much better than other methods under all cases of ratios, which demonstrates that SSRNMF is not sensitive to band noise and pixel noise simultaneously and has good robustness in both spectral dimension and spatial dimension. With the increase in the number of noisy bands and pixels, the performance of other five methods degrades apparently, which indicates that those methods are more easy to be corrupted by noisy bands and noisy pixels compared with SSRNMF. This further confirms the superiority of SSRNMF in the situation with the presence of both band noise and pixel noise.

### *16.4.2 Robustness Analysis to Noise Intensity*

To further test the robustness of the proposed method, we also test the algorithms on synthetic data with different noise intensity. The number of endmembers is fixed as 5. The ratio of noisy bands and noisy pixels is set to 0.2. The noisy bands and noisy pixels are added with different intensities of noise, namely $\overline{\text{SNR}}_{\text{band1}} = \overline{\text{SNR}}_{\text{pixel1}} = 10, 15, 20, 25, 30$, respectively. The SNRs of other low-noisy bands and pixels are fixed as 30. Figure 16.7 presents the SAD and RMSE results of algorithms under various noise intensities. With the increase of noise intensity, the performance of all six algorithms

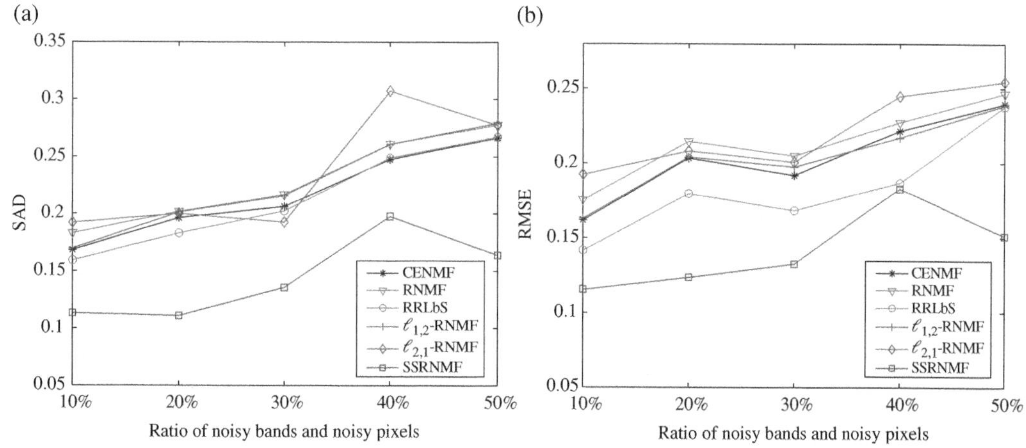


**Figure 16.6** Performance comparison of the algorithms with different ratio of noisy bands and pixels. (a) SAD, (b) RMSE.

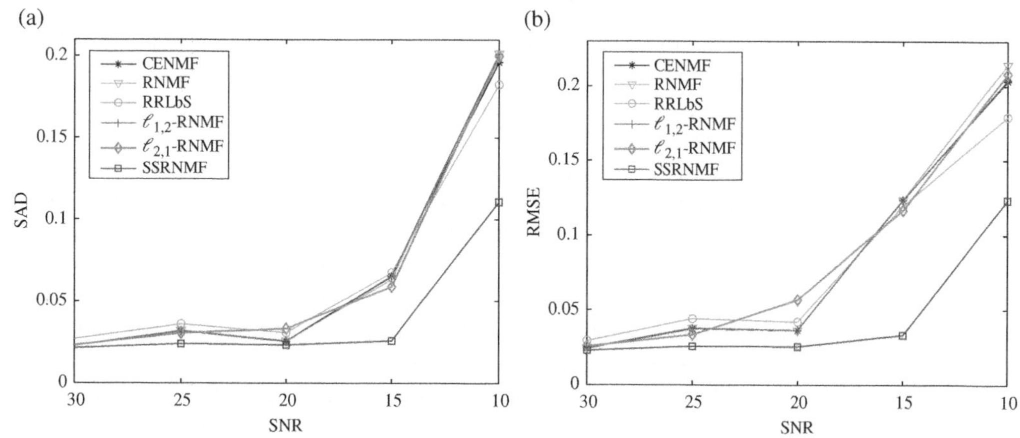


**Figure 16.7** Performance comparison of the algorithms with different ratio of noisy bands and pixels. (a) SAD, (b) RMSE.

degrades as expected. It is obvious that SSRNMF achieves much lower SAD and RMSE values than other methods and shows better robustness to noise intensity.

### *16.4.3 Performance Comparison When the Number of Endmembers Varies*

In this experiment, synthetic data made up of different number of endmembers are used to examine the performance of algorithms. The number of endmembers varies from 3 to 8. The ratio of noisy bands and noisy pixels is set to 0.2. We set $\overline{\text{SNR}}_{\text{band1}} = \overline{\text{SNR}}_{\text{pixel1}} = 10$ and $\overline{\text{SNR}}_{\text{pixel2}} = \overline{\text{SNR}}_{\text{pixel2}} = 30$. The results are as shown in Figure 16.8. We can see from the figure that the performance of algorithms has the trend of degrading when the number of endmembers increases. SSRNMF also shows the best performance in SAD and RMSE.

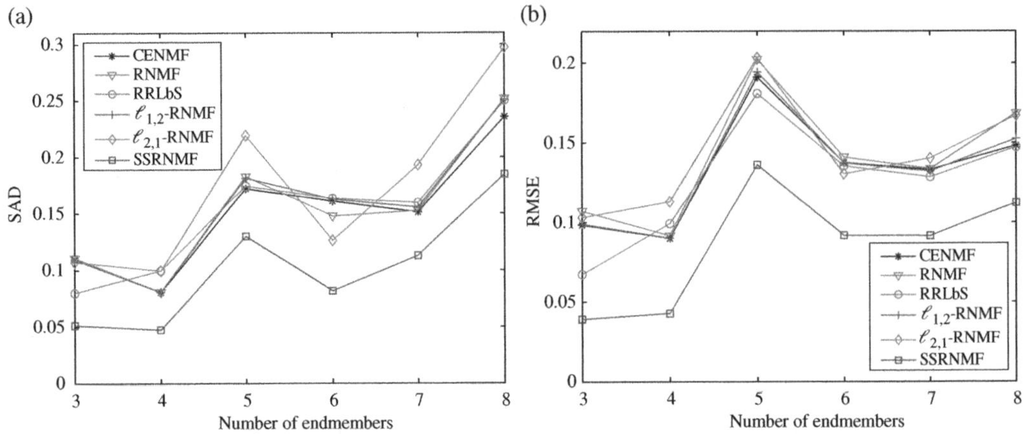


**Figure 16.8** Performance comparison of the algorithms with different ratio of noisy bands and pixels. (a) SAD, (b) RMSE.

## 16.5 Experiments on First Real Data

In this section, we present the unmixing results of proposed method on real-world data. To validate the performance of SSRNMF, the results of five compared algorithms, CENMF, RNMF [32], RRLbS, $\ell_{2,\ 1}$-RNMF and $\ell_{1,\ 2}$-RNMF, are also presented. Each experiment is repeated 10 times and the averaged result is reported.

The first hyperspectral scene used in the experiment is the Urban data set collected by the Hyperspectral Digital Imagery Collection Experiment (HYDICE) sensor over an area located at Copperas Cove near Fort Hood, TX, USA, in October 1995. In Urban data set, the spectral and spatial resolutions are 10 nm and 2 m, respectively. The image consists of a suburban residential area as shown in Figure 16.9, with a size of $307 \times 307$. There are four targets, including of asphalt, grass, roofs, and trees, exist in this area. The data set is composed of 210 spectral bands. Due to water absorption and atmosphere effects, there exists some noisy bands with low-SNRs, including the bands 1–4, 76, 87, 101–111, 136–153, and 198–210. Figure 16.10 displays some band images of Urban data set. As shown in the figure, band 94 is a low-noisy band, while bands 104, 206 are noisy bands seriously corrupted by noise. Meanwhile, it is worth mentioning that the data are also corrupted by noisy pixels such as stripes, which can be seen in Figure 16.10(b) and (c). For the traditional unmixing methods, these bands with low-SNRs are usually manually removed and the noisy pixels are not separately considered. The existence of noisy bands and noisy pixels seriously degrades the performance of most existing unmixing methods. Nevertheless, in real-world applications, the prior knowledge of noisy bands may not be available and the noisy band images may also contains important information. In this experiment, we examine the performance of algorithms on both data removed of noisy band images (i.e. $L = 162$) and data containing noisy band images (i.e. $L = 210$), which can further demonstrate the robustness of SSRNMF to band noise and pixel noise.

### 16.5.1 Results Without the Low-SNR Bands

We first present the unmixing results of algorithms on Urban data set removed of the low-SNR band images. Table 16.3 represents the mean SAD values of different methods on the Urban data. The rows show the results of four targets of interest, i.e. asphalt, grass, trees, and roofs, along with the mean

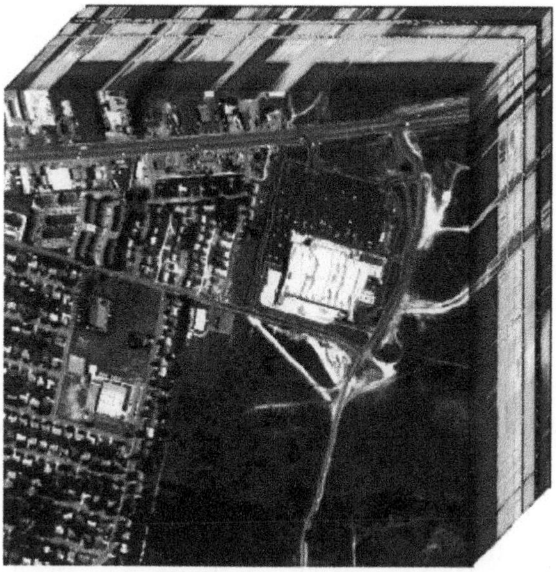

**Figure 16.9** The Hyperspectral Digital Imagery Collection Experiment (HYDICE) Urban data set.

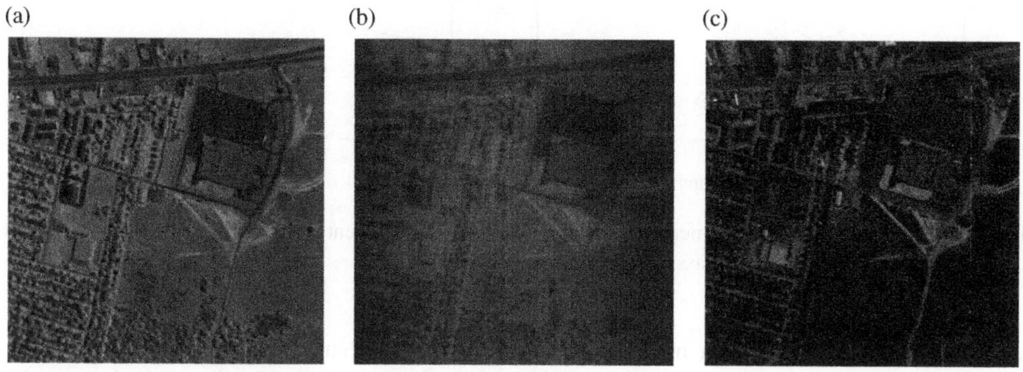


**Figure 16.10** The images of some bands in Urban data set. (a) Band 94, (b) band 104, (c) band 206.

**Table 16.3** The SAD of algorithms on the urban data set removed of noisy bands

| | Spectral angle distance | | | | | |
|---|---|---|---|---|---|---|
| Endmember | CENMF | RNMF | RRLbS | $\ell_{1,2}$-RNMF | $\ell_{2,1}$-RNMF | SSRNMF |
| Asphalt | 0.2172 | 0.1774 | 0.3512 | 0.2109 | 0.1664 | **0.1094** |
| Grass | 0.2587 | 0.2475 | **0.1316** | 0.2779 | 0.2102 | 0.1699 |
| Tree | 0.0913 | 0.1117 | 0.1206 | 0.0936 | **0.0681** | 0.1453 |
| Roof | 0.4255 | 0.4517 | 0.5548 | 0.4559 | 0.4668 | **0.2505** |
| Mean | 0.2482 | 0.2471 | 0.2896 | 0.2596 | 0.2279 | **0.1688** |

The bold values indicate that the method corresponding to the value is the best in the row.

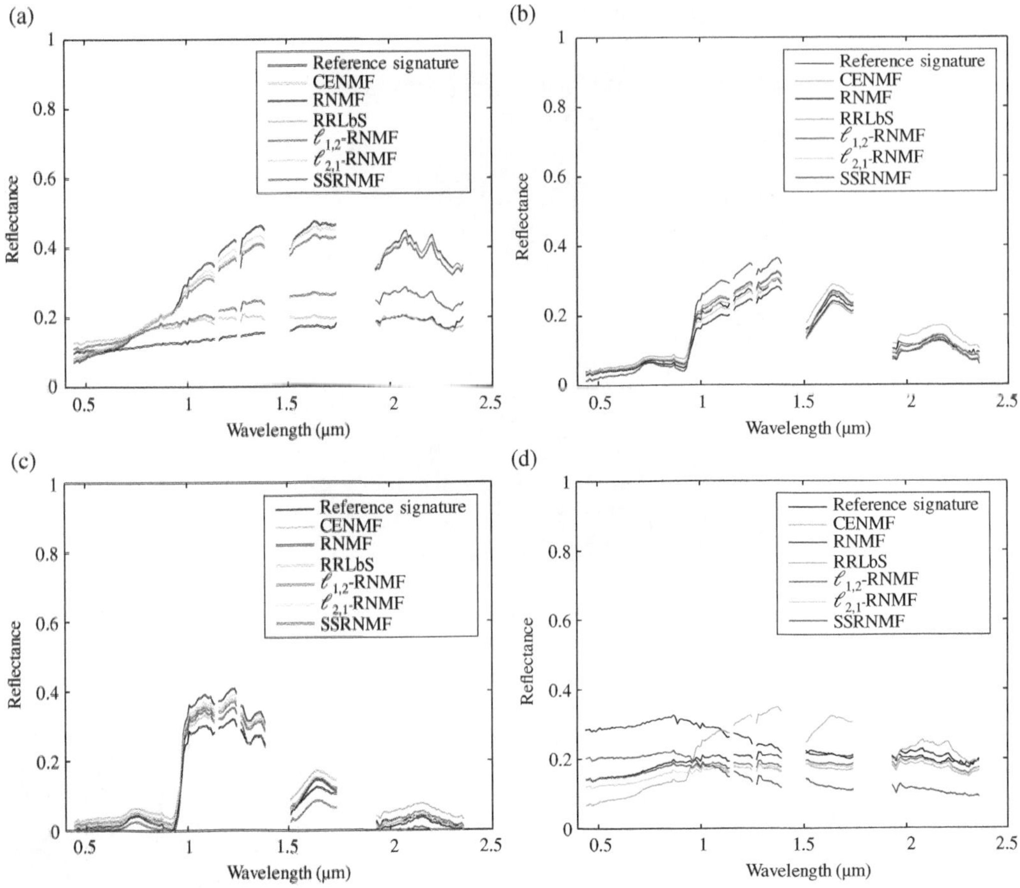


**Figure 16.11** Comparison of endmember signatures estimated by different methods over Urban data removed of noisy bands. (a) Asphalt, (b) Grass, (c) Tree, (d) Roof.

values, respectively. As shown in Table 16.3, it can be seen that the SAD results achieved by SSRNMF are better than those of other methods in general. The SSRNMF achieves the best results in two targets, including Asphalt and Roof, as well as the mean value. Figure 16.11 displays the endmember signatures obtained by different methods with the reference signatures collected the spectral library. The figure demonstrates the results of SSRNMF are in good accordance with the reference endmembers. In Figure 16.12, we present the abundance maps obtained by different methods on the Urban data set removed of noisy bands. As can be observed from Figure 16.12, SSRNMF performs better than the compared methods.

### *16.5.2 Results with the Low-SNR Bands*

We also test the algorithms on Urban data set containing the noisy bands, which includes 210 bands in total. In the iterations of $\ell_{1,\,2}$-RNMF, $\ell_{2,\,1}$-RNMF, and SSRNMF, we can obtain the weight assignations of different bands and pixels. Figure 16.13 presents the weights of all the bands in the last iteration of $\ell_{1,\,2}$-RNMF and SSRNMF, and Figure 16.14 presents the weights of all the pixels in the last

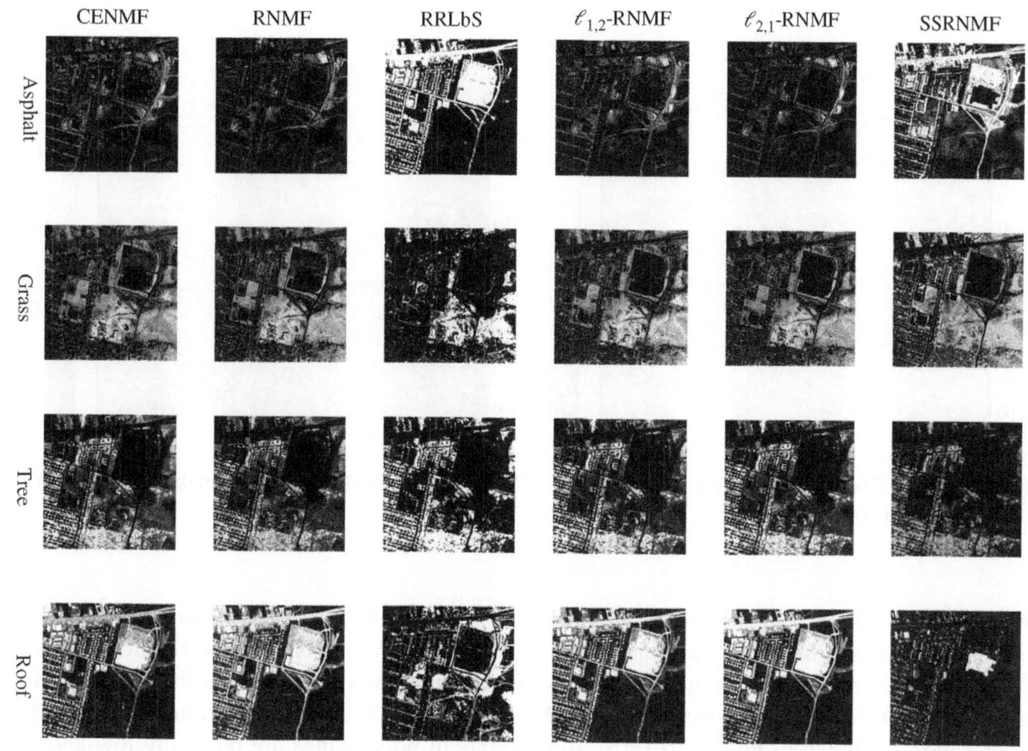


**Figure 16.12** Abundance maps of Urban data removed of noisy bands by CENMF, RNMF, RRLbS, $\ell_{1,2}$-RNMF, $\ell_{2,1}$-RNMF, SSRNMF, respectively, from right column to left column. Each row shows the corresponding abundance maps of a same endmember.

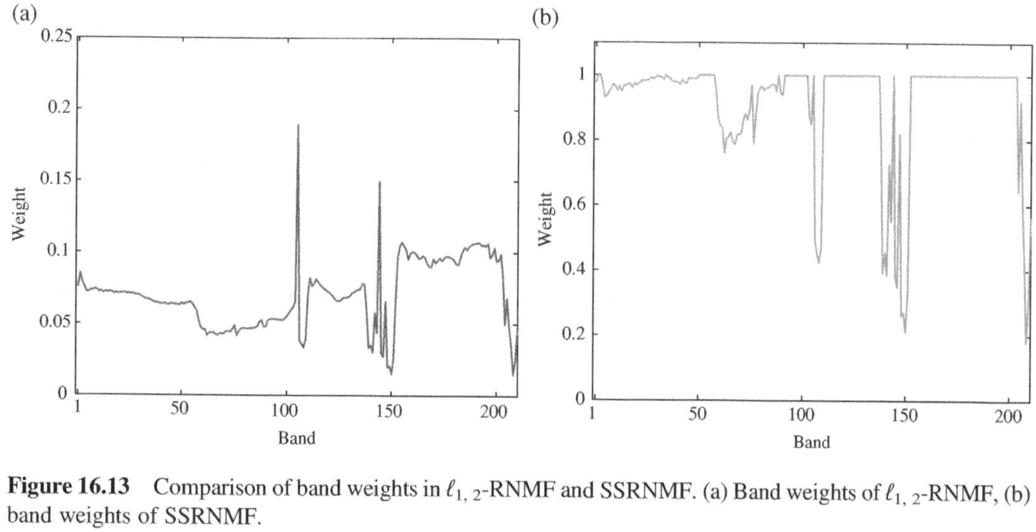


**Figure 16.13** Comparison of band weights in $\ell_{1,2}$-RNMF and SSRNMF. (a) Band weights of $\ell_{1,2}$-RNMF, (b) band weights of SSRNMF.

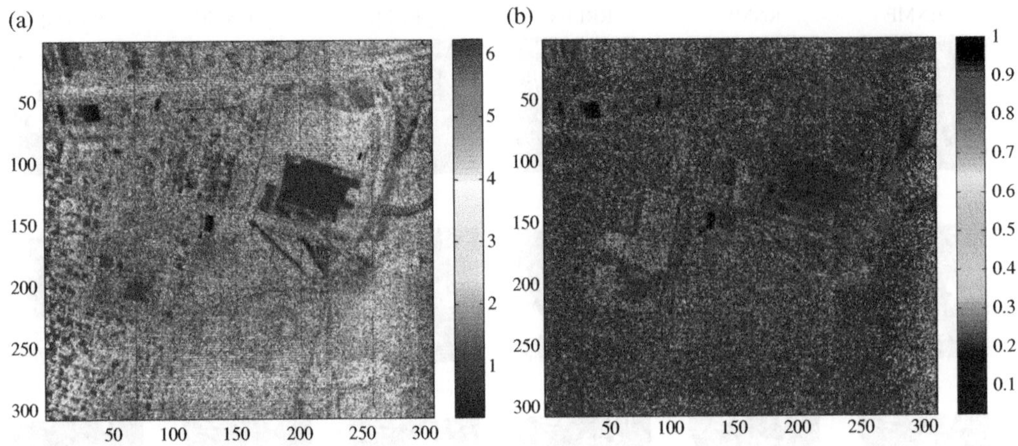


**Figure 16.14** Comparison of pixel weights in $\ell_{2,1}$-RNMF and SSRNMF. (a) Pixel weights of $\ell_{2,1}$-RNMF, (b) pixel weights of SSRNMF.

iteration of $\ell_{2,1}$-RNMF and SSRNMF. Figure 16.13 shows that both $\ell_{1,2}$-RNMF and SSRNMF can assign adaptive weights to different bands with various noise levels. For $\ell_{1,2}$-RNMF, the weights of noisy bands are relatively low and the weights of some low-noisy bands are much higher than the noisy bands' weights. Compared with $\ell_{1,2}$-RNMF, SSRNMF achieves better weight assignation for each band. As shown in Figure 16.13(b), the weights of noisy bands are relatively low and the weights of low-noisy bands are cut off by 1. Therefore, SSRNMF can effectively suppress the contribution of noisy bands and can also avoid the overwhelming superiority of low-noisy bands, which helps to make full use of all bands' information. Figure 16.14 shows the comparison of each pixel's weight in $\ell_{2,1}$-RNMF and SSRNMF. As shown in the figure, both $\ell_{2,1}$-RNMF and SSRNMF assign low weights to noisy pixels, such as the pixels located in the stripe region. However, $\ell_{2,1}$-RNMF assigns higher weights to pixels with small residues, which may cause the overwhelming superiority of these pixels. For SSRNMF, the weights for pixels with small residues are limited and cut off by 1. In general, SSRNMF can automatically assign adaptive weights for both bands and pixels and make better use of all data points' information compared with $\ell_{1,2}$-RNMF and $\ell_{2,1}$-RNMF. Since the reference signatures are also corrupted by noisy bands, the noise corrupted bands are removed and only the low-noisy bands are adopted when computing the SAD values as suggested in He et al. [32]. Table 16.4 represents the mean SAD values of different methods. For four categories, i.e. Asphalt, Tree, Roof, and the mean value, the proposed method achieves the best results. Figure 16.15 displays the endmember signatures obtained by different methods with the reference to signatures collected from the spectral library. The figure demonstrates that SSRNMF extracts all the four endmember signatures successfully, which are in good accordance with the reference endmembers. In Figure 16.16, we present the abundance maps obtained by different methods. When comparing Figures 16.12 and 16.16, it can be observed that all show some robustness as designed. However, the abundance results of the three methods are still inevitably corrupted by the noisy bands. For example, RRLbS performs worse in the noisy case than in the low-noisy case, especially for the Asphalt and the Grass. In general, we can observe that SSRNMF is less affected by the noisy band images with phenomenon of blurring. The abundance results in Figure 16.16 demonstrate that SSRNMF has good robustness to noise compared with other five methods.

**Table 16.4** The SAD of algorithms on the urban data set containing the noisy bands

| | Spectral angle distance | | | | | |
|---|---|---|---|---|---|---|
| Endmember | CENMF | RNMF | RRLbS | $\ell_{1,2}$-RNMF | $\ell_{2,1}$-RNMF | SSRNMF |
| Asphalt | 0.4372 | 0.1654 | 0.3958 | 0.2869 | 0.2284 | **0.1412** |
| Grass | 0.4098 | 0.3791 | **0.3324** | 0.3925 | 0.4038 | 0.4538 |
| Tree | 0.1168 | 0.1181 | 0.1510 | 0.1409 | 0.1241 | **0.1054** |
| Roof | 0.5722 | 0.4457 | 0.4463 | 0.5029 | 0.4539 | **0.3161** |
| Mean | 0.3840 | 0.2771 | 0.3314 | 0.3308 | 0.3026 | **0.2541** |

The bold values indicate that the method corresponding to the value is the best in the row.
*Source:* Based on He et al. [32].

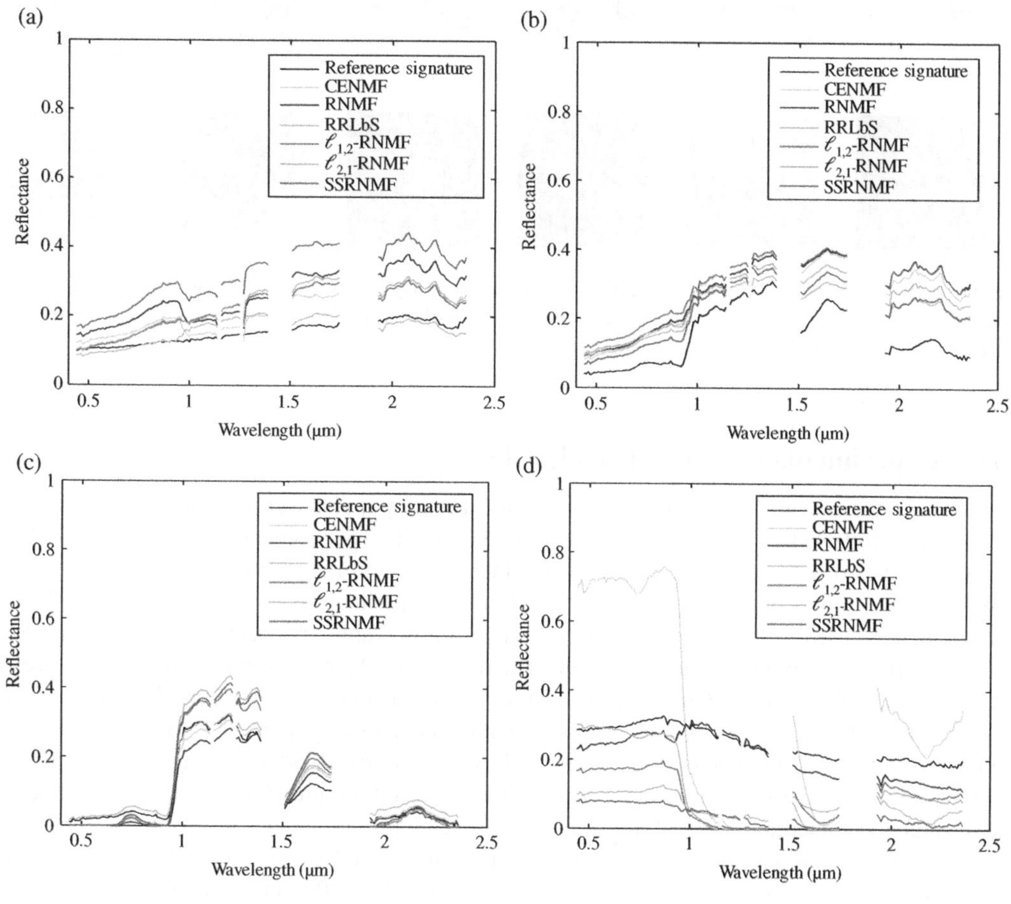


**Figure 16.15** Comparison of endmember signatures estimated by different methods over Urban data containing the noisy bands. (a) Asphalt, (b) Grass, (c) Tree, (d) Roof.

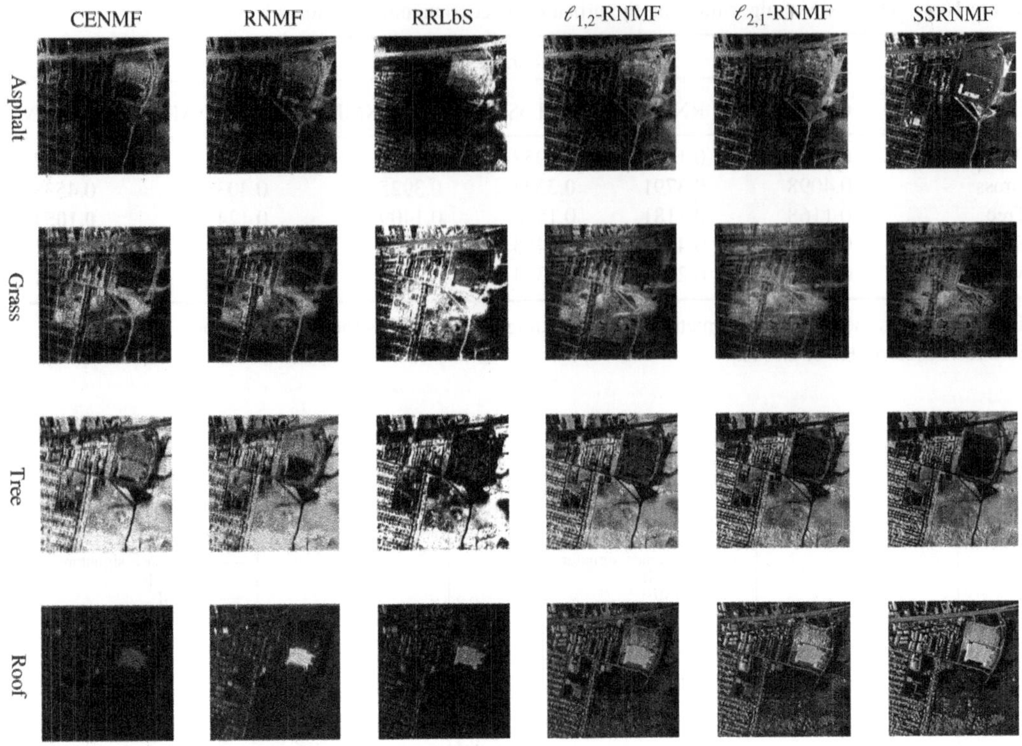


**Figure 16.16** Abundance maps of Urban data containing the noisy bands by CENMF, RNMF, RRLbS, $\ell_{1,2}$-RNMF, $\ell_{2,1}$-RNMF, SSRNMF, respectively, from right column to left column. Each row shows the corresponding abundance maps of a same endmember.

## 16.6 Experiments on Second Real Data

The second hyperspectral scene used in the experiment is the popularly used Jasper data set [33, 35, 41], which is collected by an airborne visible/infrared imaging spectrometer (AVIRIS) sensor. The data set consists of 224 spectral bands of which the wavelength ranges from 0.38 to 2.5 μm. The $100 \times 100$ pixel subset of the Jasper data set used in our experiment is shown in Figure 16.17. All 224 bands were used to test the performance of algorithms. Table 16.5 represents the mean SAD values of different methods on the Jasper data. The rows show the results of four targets of interest, i.e. tree, water, soil, and road, along with the mean values, respectively. As shown in Table 16.5, it can be seen SSRNMF achieves better SAD results than other methods in general. The SSRNMF achieves the lowest SAD values in three targets, including Tree, Water, and Road, as well as the mean value. For Water, SSRNMF also performs quite close to the best result. Figure 16.18 displays the endmember signatures obtained by different methods with the reference to signatures collected the spectral library. The figure demonstrates the endmembers extracted by SSRNMF are in good accordance with the reference signatures. The abundance maps obtained by different methods are shown in Figure 16.19. It can be observed that the SSRNMF achieves comparatively satisfactory abundance results.

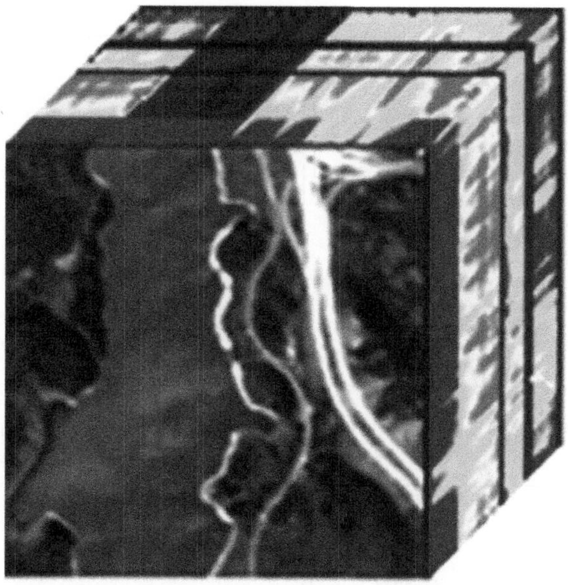

**Figure 16.17** The airborne visible/infrared imaging spectrometer (AVIRIS) Jasper data set.

**Table 16.5** The SAD of algorithms on the jasper data set

| | Spectral angle distance | | | | | |
|---|---|---|---|---|---|---|
| Endmember | CENMF | RNMF | RRLbS | $\ell_{1,2}$-RNMF | $\ell_{2,1}$-RNMF | SSRNMF |
| Tree | 0.1811 | 0.1931 | 0.1791 | 0.1816 | 0.1990 | **0.1753** |
| Water | 0.3034 | 0.2845 | 0.0674 | 0.2547 | 0.1562 | **0.0657** |
| Soil | 0.0744 | 0.0749 | 0.0710 | 0.0711 | **0.0372** | 0.0378 |
| Road | 0.5599 | 0.8461 | 1.2602 | 0.5734 | 0.6696 | **0.2604** |
| Mean | 0.2797 | 0.3496 | 0.3944 | 0.2702 | 0.2655 | **0.1348** |

The bold values indicate that the method corresponding to the value is the best in the row.

## 16.7 Conclusion

In this chapter, we present a robust NMF using $l_{1,2}$ norm ($\ell_{1,2}$-RNMF) and further propose a spectral-spatial robust NMF model (SSRNMF) by incorporating $\ell_{2,1}$ and $\ell_{1,2}$ norms. The main advantage of SSRNMF is that it takes both spatial dimension and spectral dimension into consideration and achieves robustness to band noise and pixel noise simultaneously. In addition, we integrate the Huber's M-estimator into the SSRNMF model to achieve better strategy of weight assignations, which helps to avoid the singularity problem of $\ell_{2,1}$-RNMF and $\ell_{1,2}$-RNMF and improves the unmixing performance effectively. The updating rules of the derived model are learnt and provided. We also give the convergence analysis of the proposed algorithm. To validate the effectiveness and practicability of the algorithm, both experiments on synthetic data and real hyperspectral images are

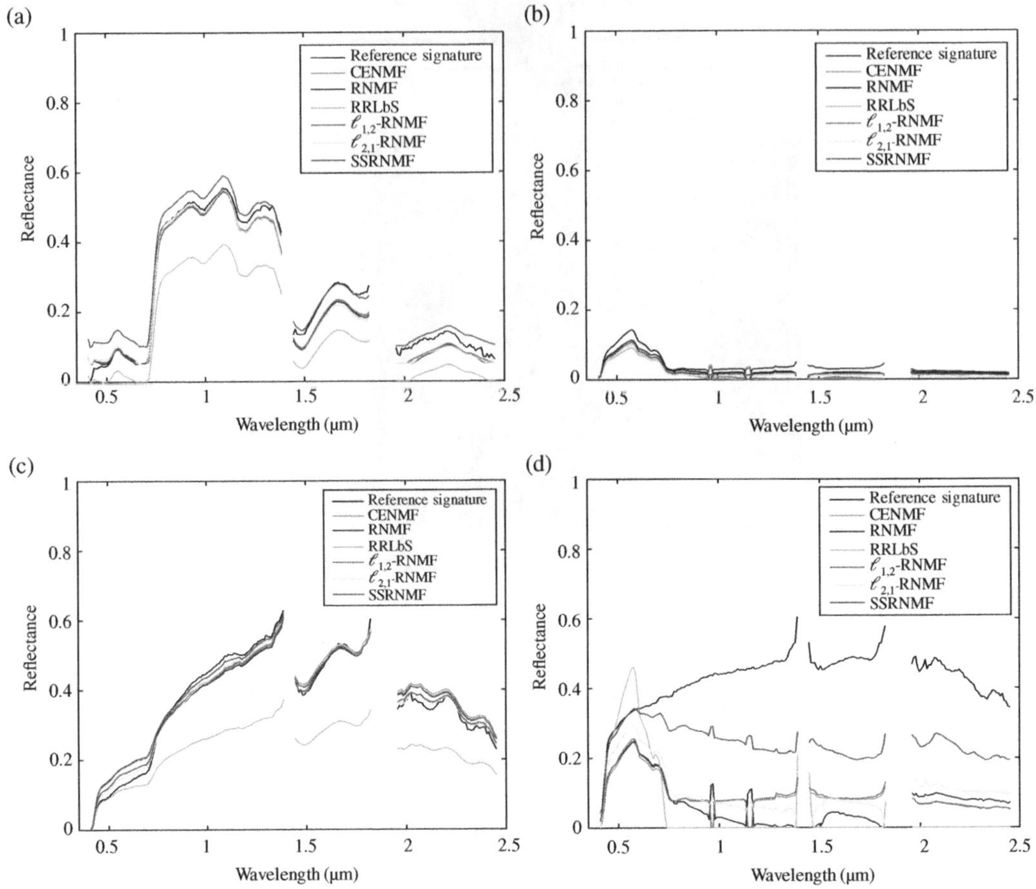


**Figure 16.18** Comparison of endmember signatures estimated by different methods on the Jasper data set. (a) Tree, (b) Water, (c) Soil, (d) Road.

conducted. The results show that the SSRNMF model has good robustness in both spatial dimension and spectral dimension and achieves better performance than the state-of-the-art methods. In addition, we investigate the performance of the algorithm on Urban data sets with and without noise bands, respectively. The experiment results validate that the proposed method can automatically assign adaptive weights to bands and pixels with various noise intensities and can efficiently exploit the useful information latent in pixels and bands.

For the future work, the performance of our method could be further improved. For instance, there is no sparsity constraint imposed on the abundances of pixels in this chapter, which may lead to unsatisfactory performance when applied to HSI composed of areas with various sparsity levels. It would be interesting to extend the proposed model by introducing adaptive sparsity constraint on abundances according to the sparsity level of data. More methods based on mining and making full use of information latent in data itself would also be worthy of further study.

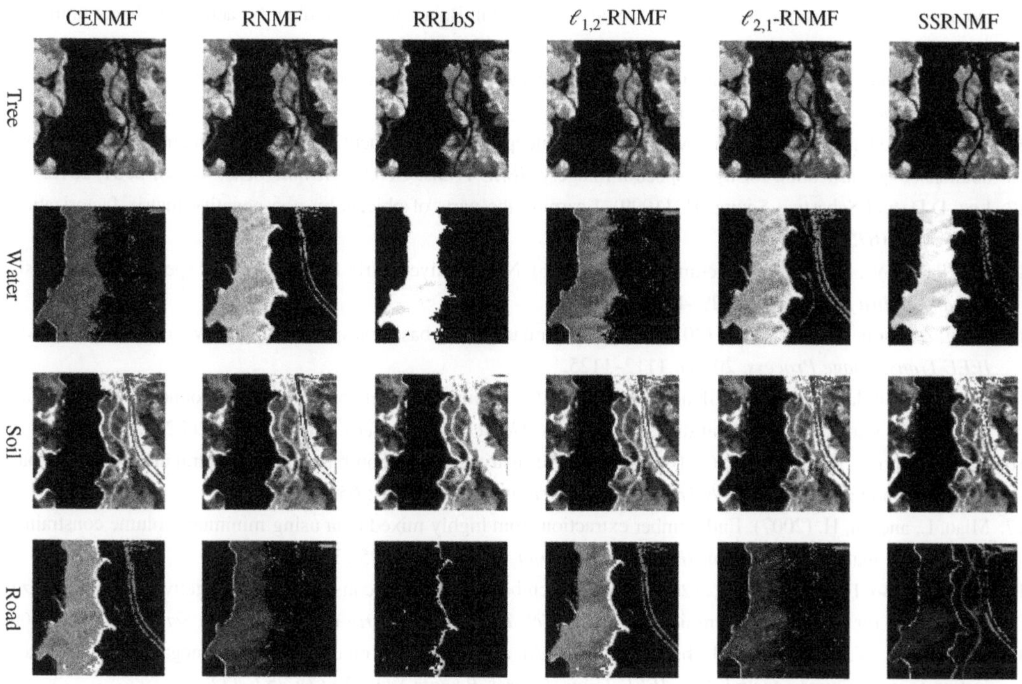


**Figure 16.19** Abundance maps of Jasper data containing the noisy bands by CENMF, RNMF, RRLbS, $\ell_{1,2}$-RNMF, $\ell_{2,1}$-RNMF, SSRNMF, respectively, from right column to left column. Each row shows the corresponding abundance maps of a same endmember.

## References

1. Boardman, J.W. (1994). Geometric mixture analysis of imaging spectrometry data. In: *Proceedings of the International Geoscience and Remote Sensing Symposium*, vol. 4, 2369–2371. IEEE.
2. Winter, M.E. (1999). N-findr: an algorithm for fast autonomous spectral end-member determination in hyperspectral data. In: *Proceedings of the SPIE's International Symposium on Optical Science, Engineering, and Instrumentation*, 266–275. International Society for Optics and Photonics.
3. Nascimento, J.M.P. and Dias, J.M.B. (2005). Vertex component analysis: a fast algorithm to unmix hyperspectral data. *IEEE Trans. Geosci. Remote Sens.* 43 (4): 898–910.
4. Chang, C.-I, Wu, C.-C., Liu , W., and Ouyang, Y.-C. (2006). A new growing method for simplex-based endmember extraction algorithm. *IEEE Trans. Geosci. Remote Sens.* 44 (10): 2804–2819.
5. Chang, C.-I, Wen, C.-H., and Chao-Cheng, W. (2013). Relationship exploration among PPI, ATGP and VCA via theoretical analysis. *Int. J. Comput. Sci. Eng.* 8 (4): 361–367.
6. Li, H.C. and Chang, C.-I (2015). Linear spectral unmixing using least squares error, orthogonal projection and simplex volume for hyperspectral images. Proceedings of the 7th Workshop Hyperspectral Image Signal Processing: Evolution in Remote Sensing (WHISPERS), Tokyo, Japan (2–5 June), 2–5.
7. Chang, C.-I, Chen, S.-Y., Li, H.-C. et al. (2016). Comparative study and analysis among ATGP, VCA and SGA for finding endmembers in hyperspectral imagery. *IEEE J. Sel. Topics Appl. Earth Obser. Remote Sens.* 9 (9): 4280–4306.
8. Heinz, D.C. and Chang, C.-I (2001). Fully constrained least squares linear spectral mixture analysis method for material quantification in hyperspectral imagery. *IEEE Trans. Geosci. Remote Sens.* 39 (3): 529–545.

9. Dobigeon, N., Moussaoui, S., Coulon, M. et al. (2009). Joint Bayesian endmember extraction and linear unmixing for hyperspectral imagery. *IEEE Trans. Signal Process.* 57 (11): 4355–4368.
10. Jia, S. and Qian, Y. (2009). Constrained nonnegative matrix factorization for hyperspectral unmixing. *IEEE Trans. Geosci. Remote Sens.* 47 (1): 161–173.
11. Wang, J. and Chang, C.-I (2006). Applications of independent component analysis in endmember extraction and abundance quantification for hyperspectral imagery. *IEEE Trans. Geosci. Remote Sens.* 44 (9): 2601–2616.
12. Lee, D.D. and Sebastian Seung, H. (1999). Learning the parts of objects by non-negative matrix factorization. *Nature* 401 (6755): 788–791.
13. Paul Pauca, V., Piper, J., and Plemmons, R.J. (2006). Nonnegative matrix factorization for spectral data analysis. *Linear Algebra Appl.* 416 (1): 29–47.
14. Yang, Z., Zhou, G., Xie, S. et al. (2011). Blind spectral unmixing based on sparse nonnegative matrix factorization. *IEEE Trans. Image Process.* 20 (4): 1112–1125.
15. Huck, A., Guillaume, M., and Blanc-Talon, J. (2010). Minimum dispersion constrained nonnegative matrix factorization to unmix hyperspectral data. *IEEE Trans. Geosci. Remote Sens.* 48 (6): 2590–2602.
16. Lei, T., Zhou, J., Qian, Y. et al. (2016). Nonnegative-matrix-factorization-based hyperspectral unmixing with partially known endmembers. *IEEE Trans. Geosci. Remote Sens.* 54 (11): 6531–6544.
17. Miao, L. and Qi, H. (2007). Endmember extraction from highly mixed data using minimum volume constrained nonnegative matrix factorization. *IEEE Trans. Geosci. Remote Sens.* 45 (3): 765–777.
18. Wang, N., Bo, D., and Zhang, L. (2013). An endmember dissimilarity constrained non-negative matrix factorization method for hyperspectral unmixing. *IEEE J. Sel. Topics Appl. Earth Observ. Remote Sens.* 6 (2): 554–569.
19. Liu, X., Xia, W., Wang, B., and Zhang, L. (2011). An approach based on constrained nonnegative matrix factorization to unmix hyperspectral data. *IEEE Trans. Geosci. Remote Sens.* 49 (2): 757–772.
20. Lu, X., Wu, H., Yuan, Y. et al. (2013). Manifold regularized sparse nmf for hyperspectral unmixing. *IEEE Trans. Geosci. Remote Sens.* 51 (5): 2815–2826.
21. Qian, Y., Jia, S., Zhou, J., and Robles-Kelly, A. (2011). Hyperspectral unmixing via $L_{1/2}$ sparsity-constrained nonnegative matrix factorization. *IEEE Trans. Geosci. Remote Sens.* 49 (11): 4282–4297.
22. Févotte, C. and Dobigeon, N. (2015). Nonlinear hyperspectral unmixing with robust nonnegative matrix factorization. *IEEE Trans. Image Process.* 24 (12): 4810–4819.
23. Jiang, X., Gong, M., Zhan, T., and Tang, Z. (2018). Multiobjective sparse unmixing approach with noise removal. In: *Proceedings of the Genetic and Evolutionary Computation Conference (GECCO 2018)*, 681–688. ACM.
24. Li, C., Ma, Y., Mei, X. et al. (2016). Hyperspectral unmixing with robust collaborative sparse regression. *Remote Sens.* 8 (7): 588.
25. He, W., Zhang, H., and Zhang, L. (2017). Total variation regularized reweighted sparse nonnegative matrix factorization for hyperspectral unmixing. *IEEE Trans. Geosci. Remote Sens.* 55 (7): 3909–3921.
26. Ma, Y., Li, C., Mei, X. et al. (2017). Robust sparse hyperspectral unmixing with $\ell_{2,1}$ norm. *IEEE Trans. Geosci. Remote Sens.* 55 (3): 1227–1239.
27. Ding, C., Zhou, D., He, X., and Zha, H. (2006). $R_1$-PCA: rotational invariant $L_1$-norm principal component analysis for robust subspace factorization. In: *Proceedings of the 23rd International Conference on Machine Learning*, 281–288. ACM.
28. Huang, H. and Ding, C. (2008). Robust tensor factorization using $R_1$ norm. In: *Proceedings of the IEEE Conference on Computer Vision and Pattern Recognition (CVPR)*, 1–8. IEEE.
29. Ren, C.-X., Dai, D.-Q., and Yan, H. (2012). Robust classification using $\ell_{2,1}$-norm based regression model. *Pattern Recogn.* 45 (7): 2708–2718.
30. Kong, D., Ding, C., and Huang, H. (2011). Robust nonnegative matrix factorization using L21-norm. In: *Proceedings of the 20th ACM International Conference on Information and Knowledgement Management*, 673–682. ACM.
31. Yang, S., Hou, C., Zhang, C., and Yi, W. (2013). Robust non-negative matrix factorization via joint sparse and graph regularization for transfer learning. *Neural Comput. Appl.* 23 (2): 541–559.

32. He, W., Zhang, H., and Zhang, L. (2016). Sparsity-regularized robust non-negative matrix factorization for hyperspectral unmixing. *IEEE J. Sel. Topics Appl. Earth Observ. Remote Sens.* 9 (9): 4267–4279.
33. Wang, Y., Pan, C., Xiang, S., and Zhu, F. (2015). Robust hyperspectral unmixing with correntropy-based metric. *IEEE Trans. Image Process.* 24 (11): 4027–4040.
34. Yuan, M. and Lin, Y. (2006). Model selection and estimation in regression with grouped variables. *J. Royal Stat. Soc. Ser. B (Stat. Meth.)* 68 (1): 49–67.
35. Zhu, F., Wang, Y., Fan, B. et al. (2014). Effective spectral unmixing via robust representation and learning-based sparsity. arXiv preprint. arXiv:1409.0685.
36. Pinar, M.C. (1997). Duality in robust linear regression using Huber's M-estimator. *Appl. Math. Lett.* 10 (4): 65–70.
37. Lee, D.D. and Seung, H.S. (2001). Algorithms for non-negative matrix factorization. Proceedings of the Advances in Neural Information Processing Systems, Vancouver, British Columbia, Canada (3–8 December)556–562.
38. Shaw, G.A. and Burke, H.-h.K. (2003). Spectral imaging for remote sensing. *Lincoln Lab. J.* 14 (1): 3–28.
39. Zhang, H., He, W., Zhang, L. et al. (2014). Hyperspectral image restoration using low-rank matrix recovery. *IEEE Trans. Geosci. Remote Sens.* 52 (8): 4729–4743.
40. Kozintsev, B. (1999). Computations with Gaussian random fields. Ph.D. thesis. University of Maryland.
41. Aggarwal, H.K. and Majumdar, A. (2016). Hyperspectral unmixing in the presence of mixed noise using joint-sparsity and total variation. *IEEE J. Sel. Topics Appl. Earth Observ. Remote Sens.* 9 (9): 4257–4266.

# Part VI

## Hyperspectral Image Classification

# 17

# Sparse Representation-Based Hyperspectral Image Classification

Haoyang Yu[1], Jun Li[2], Wei Li[3], and Bing Zhang[4]
[1]*Center for Hyperspectral Imaging in Remote Sensing (CHIRS), Information Science and Technology College, Dalian Maritime University, Dalian, China*
[2]*Guangdong Provincial Key Laboratory of Urbanization and Geo-Simulation, School of Geography and Planning, Sun Yat-sen University, Guangzhou, China*
[3]*School of Information and Electronics, Beijing Institute of Technology, Beijing, China*
[4]*Key Laboratory of Digital Earth Science, Aerospace Information Research Institute, Chinese Academy of Sciences, Beijing, China*

## 17.1 Introduction

Remote sensing (RS) is a technology of Earth observation, which has experienced a process from panchromatic image, multispectral image to hyperspectral image (HSI). HSI can be viewed as a data cube, which contains diagnostic continuous spectrum [1]. The syncretic form of image and spectrum brings HSI more details of terrestrial objects and benefits the precision land-use and land-cover mapping [2–4]. HSI classification (HSIC) uses classifier to determine the class of each pixel based on its features. Several classic models, such as support vector machine (SVM) and sparse representation (SR), have shown their availability for HSIC tasks [5–7].

One of the main problems of HSIC is the high correlation between adjacent bands and redundancy in both the spectral and spatial domains. Therefore, it has been demonstrated that HSI is essentially low rank and can be represented sparsely, though the original image is not sparse [8, 9]. In this context, SR-based models have been introduced to HSI for classification [10–12]. The main principle of representation-based classifier is that a testing pixel can be linearly represented by labeled samples [13, 14]. The weight coefficients are solved by an $l_0$-norm or $l_1$-norm penalty, denoted as SR, or an $l_2$-norm penalty, named as collaborative representation (CR) [15]. In SR-based classification, the class of a testing pixel is usually determined by the class that can provide the minimum error with an $l_1$-norm constraint, while collaborative representation classification emphasizes more on the collaboration of all samples and is solved by an $l_2$-norm regularized least square formulation. In [16], SR-based classifier (SRC) is applied to process HSI and proven to offer rather plausible

*Advances in Hyperspectral Image Processing Techniques*, First Edition. Edited by Chein-I Chang.

results compared with SVM and other traditional classifiers. In [17], collaborative representation-based classifier (CRC) is combined with SRC via multiple regularization terms and provides more balanced performance in comparison with its original counterparts for HSIC.

Although SR-based models have shown their capability for HSIC tasks, it should be noted that the inherent mechanism lies on the mutual response of features between the testing pixel and the labeled samples in the dictionary. Especially for HSI, it has been demonstrated that the original spectral information might not be the best effective feature for separation. As mentioned above, there are high redundancy and correlation among the spectral bands [18, 19]. Besides, the number of the original spectral features may be too high for classification purposes. In this context, several feature extraction methods like manifold learning models have been introduced to SR-based models for HSIC [20]. Examples include locality preserving projection (LPP) and locality sensitive discriminant analysis (LSDA) [21]. LPP uses the Laplacian of the graph to compute a transformation matrix that maps the samples to a subspace [22]. With a different objective criterion from classic linear techniques like PCA and linear discriminant analysis (LDA), LPP aims to optimally preserve the neighborhood structure of the data and makes particular sense when data lies on a nonlinear manifold [23, 24]. The linearity of LPP makes it fast and suitable for practical application, and the locality preserving character makes it less sensitive to outliers and noise. LSDA can optimally preserve the local manifold structure, as well as discriminant information. It relies on the construction of a nearest neighbor graph, which can be split into within-class graph and between-class graph to separately characterize the geometrical and discriminant structure of data manifold such that the margin between samples with different class labels is maximized at each local neighborhood.

Another main problem of HSIC is the underutilization of spatial information. Since HSI is not just a combination of disordered signals, the underutilization of spatial coherence may lead to the misclassification. With regard to this problem, second generations of SRC have appeared in the literatures [25–27]. These models are basically under a common sparsity shared framework, with spatial consistency in consideration [28, 29]. In Chen et al. [30], joint SRC (JSRC) is proposed by means of representing a local area around the testing pixel instead of the individual one. In Li et al. [31, 32], joint CRC (JCRC) is put forward as an extension of CRC, with integration of spatial information similar to JSRC. Among these, group SRC (GSRC) is a typical and mature model. Based on the inherent group-structured property, GSRC is designed to reconstruct the dictionary composed of several class-dependent sub-dictionaries such that a testing sample is represented by group atoms rather than individual ones [33]. With the homogeneity assumption, GSRC incorporates the contextual information by simultaneously representing the testing pixel and its neighborhood, and the class label is determined by the class with the minimum error [34].

No matter in the classic SR-based models or in the JR-based frameworks, the sparse coefficient is usually considered with the dictionary together. Specifically, it is generally processed in conjunction with the dictionary to calculate the approximate value of testing pixel, and the classification is implemented according to the class-dependent minimum residual error in the representation process [20, 35]. However, this determination mechanism ignores the role of sparse coefficient to a certain extent. In our research, it has been shown that there is latent discriminant information hidden behind the coefficient, defined as the activity degree (AD) and the neighborhood activity degree (NAD), which can be further exploited for the classification [36–40]. Specifically, AD can be understood as the contribution of labeled samples from different classes in the dictionary to the testing pixel, and NAD can be considered as the contribution of different class-dependent sub-dictionaries to the neighborhood centered on the testing pixel during the representation. Through the estimation of AD, it can establish an equivalent and simplified model to the classic SRC. Based on the evaluation of NAD, it can achieve an effective spectral-spatial classification framework, with spatial coherence in consideration.

The reminder of this chapter is organized as follows: Section 17.2 reviews the classic SR-based and JR-based models. Section 17.3 introduces the proposed SR-based HSI classification models from two aspects. One is approaches in the spectral domain, including SR based on the feature extraction toward samples and SR based on activity analysis toward coefficient. The other one is approaches in the spectral-spatial domains, including the joint processing framework based on group sparsity, and post-processing framework based on neighborhood activity analysis. Section 17.4 evaluates the proposed methods with two widely used real hyperspectral data sets from different sensors. Section 17.5 concludes this paper with some remarks and plausible follow-up research line.

## 17.2 Classic Representation-Based Models

### *17.2.1 Sparse Representation-Based Framework*

Compared with the traditional processing techniques for HSI such as band fitting and SVM, representation-based methods have gained great interest due to their no assumption of data density distribution [41]. By representing testing pixels as a linear combination of a small subset of labeled pixels, they have been proven to provide rather plausible results. The classic models cover SRC and CRC, which have been widely exploited.

1. *SRC*

SRC is a classic SR-based model. It constructs the dictionary through training the labeled samples and then realizes the approximately representation of the testing pixel. The category of the testing pixel is decided by analyzing which class yields the least residual error values between the approximation and the original testing pixel. Concretely, it implements the classification based on several steps as follows.

Firstly, we acquire the labeled samples from different classes to construct the dictionary **D**, of which the **D** covers the same amount of the labeled samples per class. The testing samples can be approximately represented via a sparse linear combination $\mathbf{x}_{i,\,j} \approx \mathbf{D}\boldsymbol{\alpha}$. Moreover, in order to use as few samples as possible to better represent the testing pixel, the weight coefficients $\boldsymbol{\alpha}$ used in representation are sparsely constrained [42]. Finally, the classification is conducted by a residual error-driven decision mechanism, which classifies the testing pixel as the class with the minimum class-dependent residual error using the following formula:

$$\begin{cases} \hat{\boldsymbol{\alpha}}_{i,j} = \arg\min\limits_{\boldsymbol{\alpha}_{i,j}} \left\{ \left\|\mathbf{x}_{i,j} - \mathbf{D}\boldsymbol{\alpha}_{i,j}\right\|_2^2 + \lambda \left\|\boldsymbol{\alpha}_{i,j}\right\|_1 \right\} \\ \text{class}\left(\mathbf{x}_{i,j}\right) = \arg\min\limits_{k} \left\{ \left\|\mathbf{x}_{i,j} - \mathbf{D}\delta_k\left(\hat{\boldsymbol{\alpha}}_{i,j}\right)\right\|_2^2 \right\} \end{cases} \tag{17.1}$$

where $\left\|\boldsymbol{\alpha}_{i,j}\right\|_1 = \sum_{m=1}^{n} |\alpha_m|$ denotes the $l_1$-norm and $\|\cdot\|_2$ is the $l_2$-norm. The $l_0$-norm is to calculate the number of the nonzero elements; the optimization of $l_0$-norm often suffers a combinatorial NP-hard problem. Therefore, the $\boldsymbol{\alpha}$ applies $l_1$-norm to substitute $l_0$-norm, where $l_1$-norm is the closest convex function to the $l_0$-norm. Transforming the original optimization problem into a convex optimization problem to solve. And $\lambda$ is a scalar regularization parameter, $k \in [1, K]$ denotes the class index. As an indicator function, $\delta_k\left(\hat{\boldsymbol{\alpha}}_{i,j}\right)$ defines an indicator operation, zeroing out the elements in $\hat{\boldsymbol{\alpha}}_{i,j}$ that are not corresponding to the labeled class $k$ in **D**.

2. *CRC*

Although compared with the $l_2$-norm minimization, the $l_1$-norm minimization yields more sparse resolution [43]. Some researchers also exploit the property of $l_2$-norm minimization from perspective of collaborative mechanism and establish CRC model based on the regularized least square function. The approximation optimal solution of $\boldsymbol{\alpha}_{i,\,j}$ is firstly obtained by the $l_1$-norm or $l_2$-norm minimization. Then, the second step of the CRC model is to represent the testing pixel by combination of the dictionary and weight vector, such that the final class label is determined based on the class-dependent minimum residual error between the original signal and its approximation. The classification process can be presented as follows:

$$\begin{cases} \hat{\boldsymbol{\alpha}}_{i,j} = \arg\min\limits_{\boldsymbol{\alpha}_{i,j}} \left\{ \left\| \mathbf{x}_{i,j} - \mathbf{D}\boldsymbol{\alpha}_{i,j} \right\|_2^2 + \lambda \left\| \boldsymbol{\alpha}_{i,j} \right\|_2^2 \right\} \\ \text{class}\left(\mathbf{x}_{i,j}\right) = \arg\min\limits_{k} \left\{ \left\| \mathbf{x}_{i,j} - \mathbf{D}\delta_k\left(\hat{\boldsymbol{\alpha}}_{i,j}\right) \right\|_2^2 \right\} \end{cases} \tag{17.2}$$

## 17.2.2 *Joint Representation-Based Framework*

HSI is not just a combination of disordered signals, pixels within a small neighborhood usually consist of similar materials. The underutilization of spatial coherence may lead to the misclassification. In order to incorporate the spatial information, the second generation of SRCs including JSRC, GSRC, and sparse GSRC (SGSRC) has appeared by integrating additional structure sparsity priors.

1. *JSRC*

As the most typical JR-based model, JSRC adopts the dictionary and the sparse coefficient to represent the testing pixel and its neighboring pixels together. With the consideration of the spatial consistency, the basic assumption of JSRC is that the sparse weight vectors associated with the neighborhood centered on the testing pixel could share a common sparsity support. Therefore, the general weight matrix is optimized toward the direction of row sparsity under a mixed norm constraint which can be solved upon the Lasso problem as follows:

$$\hat{\mathbf{A}} = \arg\min_{\mathbf{A}} \left\| \mathbf{X}_{i,j} - \mathbf{D}\mathbf{A} \right\|_F^2 + \lambda \left\| \mathbf{A} \right\|_{1,2} \tag{17.3}$$

Here, $\|\mathbf{A}\|_{1,2} = \sum_{s=1}^{n} \|\boldsymbol{\alpha}^s\|_2$ defines the $l_{1,2}$-norm, where $\boldsymbol{\alpha}^s$ is the $s$th row of $\mathbf{A}$, and $\|\cdot\|_F$ represents the Frobenius-norm constraint. The class label of $\mathbf{x}_{i,\,j}$ in the center of $\mathbf{X}_{i,\,j}$ is decided based on the minimum total residual errors as follows:

$$\text{class}\left(\mathbf{x}_{i,j}\right) = \arg\min_{k} \left\| \mathbf{X}_{i,j} - \mathbf{D}\delta_k\left(\hat{\mathbf{A}}\right) \right\|_F^2 \tag{17.4}$$

where $\delta_k\left(\hat{\mathbf{A}}\right)$ is the indicator operation zeroing out all the elements of $\hat{\mathbf{A}}$ that do not belong to the class $k$.

2. *GSRC*

GSRC has been considered as a relatively complete model due to its joint representation of pixels in neighborhood and activation of group atoms instead of the individual ones in the dictionary and sparse coefficients. In comparison to the original counterparts, GSRC considers the spatial consistency of a group of pixels surrounding a target pixel. It circumvents the instability of sparse coefficients and achieves quite persuasive classification results for HSI.

The GSRC model is designed to exploit the inherent structure-based priors of coefficients and dictionary and incorporated with spatial dependences of pixels in neighborhood. The main idea of GSRC is to encourage sparse coefficients of only certain groups to be active and the remaining groups inactive so that the pixels are represented by a group of atoms belonging to the same class, the matrix representing dictionary may be rearranged as $\mathbf{D} = (\mathbf{D}_1, \ldots, \mathbf{D}_K)$, where $\mathbf{D}_K$ denotes a sub-dictionary with each column vector representing a labeled sample from class $K$. Therefore, the objective function of GSRC can be expressed as

$$\hat{\mathbf{A}} = \arg\min_{\mathbf{A}} \left\|\mathbf{X}_{i,j} - \mathbf{D}\mathbf{A}\right\|_F^2 + \lambda \sum_{g \in G} \omega_g \left\|\mathbf{A}_g\right\|_2 \tag{17.5}$$

where $g \in \{G_1, \ldots, G_K\}$ indexes the $K$ groups corresponding to the sub-dictionaries in $\mathbf{D}$, $\mathbf{A}_g$ presents the weight matrix of each group, and $\omega_g$ defines a compensation parameter for different sizes of groups. Here, $\sum_{g \in G} \left\|\mathbf{A}_g\right\|_2$ defines a group-level $l_{1,2}$-norm regularization and can be optimized by the Group Lasso algorithm. After obtaining the coefficient matrix $\hat{\mathbf{A}}$, the class label of $\mathbf{x}_{i,j}$ can be also determined by minimum total residual errors:

$$\text{class}\left(\mathbf{x}_{i,j}\right) = \arg\min_{k} \left\{ \left\|\mathbf{X}_{i,j} - \mathbf{D}\delta_k\left(\hat{\mathbf{A}}\right)\right\|_F^2 + \lambda \sum_{g \in G} \omega_g \left\|\delta_k\left(\hat{\mathbf{A}}_g\right)\right\|_2 \right\} \tag{17.6}$$

3. *SGSRC*

In the formulations (17.5) and (17.6), the coefficients within each group are not sparse, and all the atoms in the selected groups could be active. If the sub-dictionary is over complete, then enforcing sparsity within each group is necessary. To achieve sparsity within the groups, an $l_1$-norm regularizer can be added to the group Lasso (17.5), which can be written as

$$\hat{\mathbf{A}} = \arg\min_{\mathbf{A}} \frac{1}{2}\left\|\mathbf{X}_{i,j} - \mathbf{D}\mathbf{A}\right\|_F^2 + \lambda_1 \sum_{g \in G} \omega_g \left\|\mathbf{A}_g\right\|_2 + \lambda_2 \sum_{g \in G} \omega_g \left\|\mathbf{A}_g\right\|_1 \tag{17.7}$$

The above optimization problem is referred as the collaborative hierarchical Lasso [44] and the class label of test pixel can be decided by minimum total residual errors.

## 17.3 Sparse Representation-Based Hyperspectral Image Classification

### *17.3.1 Approach in the Spectral Domain*

1. *Feature extraction toward samples (LPSRC)*

Locality-preserving SR-based classification (LPSRC) combining LPP and SR is proposed for HSI classification to address the imbalance between high dimensionality of hyperspectral data and the limited number of training samples. First, LPP is used to reduce the dimensionality of all the training and testing data by finding the optimal linear approximations to the eigenfunctions of the Laplace Beltrami operator on the manifold where the high-dimensional data lies on [21]. Second, SR codes

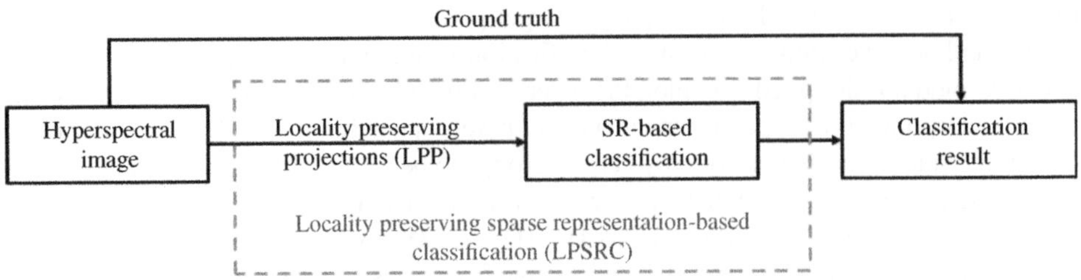


**Figure 17.1** General framework of LPSRC.

the projected testing pixels as sparse linear combinations of all the training samples so as to classify the testing pixels by evaluating which class leads to the minimum approximation error. The framework of LPSRC is shown in Figure 17.1.

1. LPP

Given a set $\mathbf{X} = (\mathbf{x}_1, \mathbf{x}_2, \ldots, \mathbf{x}_n)$ in $\mathbf{R}^d$, a linear dimensionality reduction process is to find a transformation matrix $\mathbf{A}$ that maps these $n$ points to a set $\mathbf{Z} = (\mathbf{z}_1, \mathbf{z}_2, \ldots, \mathbf{z}_n)$ in $\mathbf{R}^m$ ($m \ll d$), such that $\mathbf{z}_i$ can be represented by $\mathbf{x}_i$, where $\mathbf{z}_i = \mathbf{P}\mathbf{x}_i$. Under the assumption that $\mathbf{X} \in \mathcal{M}$ ($\mathcal{M}$ is a manifold embedded in $\mathbf{R}^d$), LPP needs three steps to project the samples:

**Step 1:** Constructing the adjacency graph, where nodes $i$ and $j$ are connected by an edge if $\mathbf{x}_i$ and $\mathbf{x}_j$ are "close" in following two situations:

(a) $\|\mathbf{x}_i - \mathbf{x}_j\|^2 < \varepsilon$, where the norm is the usual Euclidean norm in $\mathbf{R}^l$.
(b) $i$ is among $n$ nearest neighbors of $j$ or $j$ is among $n$ nearest neighbors of $i$.

**Step 2:** Choosing the weights, where the edges are weighted by two variations:

(a) Heat kernel, where $W_{ij} = e^{-\frac{\|\mathbf{x}_i - \mathbf{x}_j\|^2}{t}}$ $(t \in R)$ if node $i$ and $j$ are connected. Here, the weight matrix $W$ of the graph models the manifold structure by preserving local structure and $t$ is a suitable constant. $W$ has an intrinsic connection to the Laplace Beltrami operator on differentiable functions on a manifold, and the choice of $t$ has a great impact on $W$ [45, 46].
(b) Simple-minded, where $W_{ij} = 1$ if and only if nodes $i$ and $j$ are connect by an edge.

**Step 3**: Eigenmaps, where the eigenvectors and eigenvalues for the generalized eigenvectors problem are computed by

$$\mathbf{XLX}^T\mathbf{p} = \lambda\mathbf{XDX}^T\mathbf{p} \tag{17.8}$$

where the $i$th column of matrix $\mathbf{X}$ is $\mathbf{x}_i$, $\mathbf{D}$ is a diagonal matrix with $D_{ii} = \sum_j W_{ji} = \sum_j W_{ij}$ and $\mathbf{L} = \mathbf{D} - \mathbf{W}$ is the Laplacian matrix. Let $\mathbf{P} = (\mathbf{p}_1, \mathbf{p}_2, \ldots, \mathbf{p}_k)$ be the solutions of (17.8) with the increasing eigenvalues, $\lambda_1 < \lambda_2 < \cdots < \lambda_n$.

Thus, the result of LPP is given by

$$\mathbf{z}_i = \mathbf{P}_m{}^T\mathbf{x}_i \tag{17.9}$$

where $\mathbf{P}_m{}^T = (\mathbf{p}_1, \mathbf{p}_2, \ldots, \mathbf{p}_m)$ is a $d \times m$ matrix.

2. LPSRC

Derived from SRC, the implementation of LPSRC has the following two steps: (1) dimensionality reduction, to reduce the dimensionality of training samples **D** and a testing sample set **y** using LPP; (2) classification, to infer the class labels from SRC based on the results of LPP. Thus, the final objective function of LPSRC is given by

$$\hat{\boldsymbol{\alpha}}^{LPSR} = \arg\min_{\boldsymbol{\alpha}^{LPSR}} \left\|\phi(\mathbf{y}) - \phi(\mathbf{D})\boldsymbol{\alpha}^{LPSR}\right\|_2^2 + \lambda\left\|\boldsymbol{\alpha}^{LPSR}\right\|_1 \tag{17.10}$$

where $\phi(\mathbf{D})$ and $\phi(\mathbf{y})$ are computed by LPP with **D** and **y**. After obtaining $\hat{\boldsymbol{\alpha}}^{LPSR}$, the class label of **y** is determined by the minimum residual $r_l^{LPSRC}(\phi(\mathbf{y}))$. The procedures of LPSRC is summarized as follows.

---

**Algorithm LPSRC**

---

**Input:** Available training data **D** and a testing sample set **y**.
**Step 1:** Normalize **D** to have unit $l_2$-norm.
**Step 2:** Obtain $\phi(\mathbf{D})$ and $\phi(\mathbf{y})$ by LPP using **D** and **y**.
**Step 3:** Code $\phi(\mathbf{y})$ over $\phi(\mathbf{D})$ via $l_{1,2}$-minimization according to (17.10).
**Step 4:** Compute the residuals:

$$r_k^{LPSRC}(\phi(\mathbf{y})) = \left\|\phi(\mathbf{y}) - \phi(\mathbf{D}_k)\hat{\boldsymbol{\alpha}}_k^{LPSR}\right\|_2 \tag{17.11}$$

**Step 5:** Identify the class label of **y**.

$$\text{class}(\mathbf{y}) = \arg\min_{k = 1, 2, \ldots, K} r_k^{LPSRC}(\mathbf{y}) \tag{17.12}$$

**Output:** class(**y**).

---

2. *Activity analysis toward coefficient (ADRC)*

Representation-based model represents the testing pixel via a sparse linear combination of the labeled samples. The testing pixel is represented by a single element with nonzero coefficient from some certain classes as follows:

$$\mathbf{x}_{i,j} \approx \alpha_p \mathbf{x}_t^k + \alpha_q \mathbf{x}_{t_1}^{k'} + \alpha_m \mathbf{x}_{t_2}^{k''} + \cdots + \alpha_n \mathbf{x}_{t_n}^{k^*} \tag{17.13}$$

where $\|\alpha_p\|_d \geq \|\alpha_q\|_d \geq \|\alpha_m\|_d \geq \cdots \geq \|\alpha_n\|_d$ denoted the weight atoms with decreasing order of measurement $l_d$-norm in $\hat{\boldsymbol{\alpha}}_{i,j}$. Since $\hat{\boldsymbol{\alpha}}_{i,j}$ is sparsely constrained, most atoms in it are zero. It means that the samples participated in the representation of current testing pixel are actually the ones whose weight values are not zero. Moreover, the samples with larger values of measurement play decisive roles to the representation, which means that the attribution of these samples with higher degree of activity can largely determine the class of the testing pixel.

Based on the above analysis, here a new concept is defined from the sparse coefficient to sub-dictionary, which is called activity degree (AD), estimates the contribution of the labeled samples from different classes in representing the testing pixel $\mathbf{x}_{i,j}$. The AD of a certain class $k$ is calculated by the measurement of $l_d$-norm ($d = 1$ or 2) to its corresponding weight vector $\boldsymbol{\alpha}_{i,j}^{k}$ as follows:

$$\mathrm{AD}_k = \left\| \boldsymbol{\alpha}_{i,j}^{k} \right\|_d . \tag{17.14}$$

With the consideration of AD, a classifier named AD-driven representation-based classifier (ADRC) is established, of which the main principle is to divide the testing pixel into the class with the maximum AD. Specifically, Figure 17.2 shows a schematic diagram of the ADRC in a simplified situation. Assuming that only three atoms, i.e. $\alpha_p$, $\alpha_q$, $\alpha_m$ ($\|\alpha_p\|_d > \|\alpha_q\|_d > \|\alpha_m\|_d$), that are not zero after the optimization of $\hat{\boldsymbol{\alpha}}_{i,j}$, and the samples in $\mathbf{D}$ corresponding to these three atoms belong to class $k$, $k'$, and $k''$, respectively. According to (17.13) and (17.14), the class label of $\mathbf{x}_{i,j}$ can be identified as class k, since $\mathrm{AD}_k > \mathrm{AD}_{k'} > \mathrm{AD}_{k''}$. The procedures of ADRC is summarized as follows.

---

**Algorithm ADRC**

---

**Input:** Available HSI $\mathbf{X}$, dictionary $\mathbf{D}$, and a testing pixel $\mathbf{x}_{i,j}$.
**Step 1:** Compute $\hat{\boldsymbol{\alpha}}_{i,j}$ according to (17.13).
**Step 2:** Calculate the AD of each class based on $\hat{\boldsymbol{\alpha}}_{i,j}$ according to (17.14).
**Step 3:** Identify the class with the maximum AD.

$$\kappa = \max_k \{\mathrm{AD}_k \mid k \in [1, K]\}$$

**Output:** class($\mathbf{x}_{i,j}$) = $\kappa$.

---

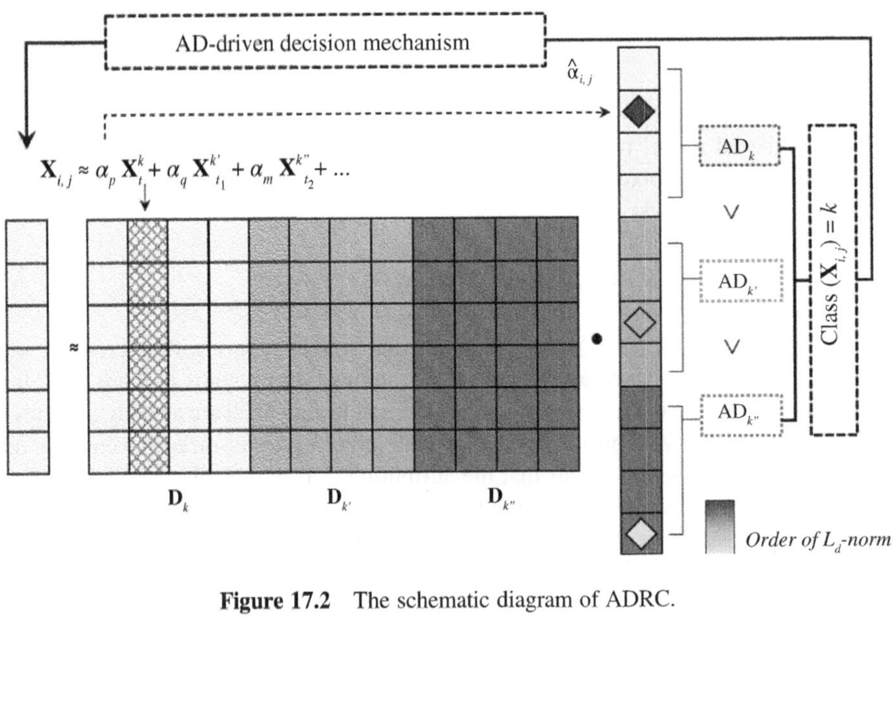


**Figure 17.2** The schematic diagram of ADRC.

## 17.3.2 Approach in the Spectral-Spatial Domain

1. *Joint processing framework based on group sparsity (LSDA-GSRC)*

LSDA-GSRC integrating LSDA with GSRC is proposed for hyperspectral imagery classification. Firstly, LSDA–GSRC adopts LSDA to project the input data to a lower-dimensional subspace that preserves the local neighborhood information and the discriminant information. Then, the projected data can be linearly represented by group-structured training samples to classify the testing set by measuring the class approximation error. In a word, LSDA–GSRC takes advantage of both the local manifold structure and group SR for better characterization and classification for hyperspectral data set. LSDA–GSRC is a two-step process:

1. Local manifold subspace-based projection by LSDA

Let $\mathbf{X} = \{\mathbf{x}_1, \mathbf{x}_2, \ldots \mathbf{x}_n\}$ be a HSI, where $\mathbf{x}_i = [x_{i1}, x_{i2}, \ldots, x_{id}]^T$ denotes a spectral vector associated with an image pixel $i \in S$. $n$ and $d$ represent the total number of pixels and spectral bands, respectively. The principle of a subspace-based projection is to find a transformation matrix $\mathbf{P} = \{\mathbf{p}_1, \mathbf{p}_2, \ldots, \mathbf{p}_m\}$ that maps $\mathbf{X}$ to $\mathbf{Z} = \{\mathbf{z}_1, \mathbf{z}_2, \ldots \mathbf{z}_n\} \in \mathbb{R}^{m \times n}$ $(m \ll d)$ via $\mathbf{Z} = \mathbf{P}^T\mathbf{X}$. Under assumption that $\mathbf{x}_i \in \mathcal{M}$ ($\mathcal{M}$ denotes the underlying manifold), one can build a graph $G$ by searching its $t$ nearest neighbors (i.e. $N(\mathbf{x}_i) = \{\mathbf{x}_i^1, \ldots, \mathbf{x}_i^t\}$) and putting an edge between $\mathbf{x}_i$ and $N(\mathbf{x}_i)$. In order to develop both geometrical and discriminant structure of data manifold, LSDA is adopted to construct within-class graph $G_w$ and between-class graph $G_b$ by splitting $N(\mathbf{x}_i)$ into two subsets, $N_w(\mathbf{x}_i)$ and $N_b(\mathbf{x}_i)$, which can be expressed as

$$\begin{cases} N_w(\mathbf{x}_i) = \{x_i^j | l(x_i^j) = l(x_i), \quad 1 \le j \le t\} \\ N_b(\mathbf{x}_i) = \{x_i^j | l(x_i^j) \ne l(x_i), \quad 1 \le j \le t\} \end{cases} \tag{17.15}$$

where $l(\mathbf{x}_i)$ is the class label of $\mathbf{x}_i$. Therefore, the weight matrices $\mathbf{W}_w$ and $\mathbf{W}_b$ of $G_w$ and $G_b$ can be defined as

$$\begin{aligned} \mathbf{W}_{w,ij} &= \begin{cases} 1, & \text{if } \mathbf{x}_i \in N_w(\mathbf{x}_j) \text{ or } \mathbf{x}_j \in N_w(\mathbf{x}_i) \\ 0, & \text{otherwise} \end{cases} \\ \mathbf{W}_{b,ij} &= \begin{cases} 1, & \text{if } \mathbf{x}_i \in N_b(\mathbf{x}_j) \text{ or } \mathbf{x}_j \in N_b(\mathbf{x}_i) \\ 0, & \text{otherwise} \end{cases} \end{aligned} \tag{17.16}$$

To satisfy the criterion that connected pointes in $G_w$ are close enough and the ones in $G_b$ stay as distant as possible, the following two objective functions are defined for $\mathbf{Z}$:

$$\min \sum_{ij} (\mathbf{z}_i - \mathbf{z}_j)^2 \mathbf{W}_{w,ij} \tag{17.17}$$

$$\max \sum_{ij} (\mathbf{z}_i - \mathbf{z}_j)^2 \mathbf{W}_{b,ij} \tag{17.18}$$

Since $\mathbf{Z} = \mathbf{P}^T\mathbf{X}$, the objective function (17.17) can be reduced to $(\mathbf{P}^T\mathbf{X}\mathbf{D}_w\mathbf{X}^T\mathbf{P} - \mathbf{P}^T\mathbf{X}\mathbf{W}_w\mathbf{X}^T\mathbf{P})$, where $\mathbf{D}_w$ is a diagonal matrix with $D_{w,\,ii} = \sum_j \mathbf{W}_{w,\,ij}$. Similarly, the objective function (17.18) can be reduced to $\mathbf{P}^T\mathbf{X}\mathbf{L}_b\mathbf{X}^T\mathbf{P}$, where $\mathbf{L}_b = \mathbf{D}_b - \mathbf{W}_b$ is the Laplacian matrix of $G_b$, and $\mathbf{D}_b$ is a diagonal matrix with $D_{b,ii} = \sum_j \mathbf{W}_{b,ij}$. Considering the influence of $D_{w,ii}$ to $\mathbf{x}_i$, it is suggested to impose a constraint $\mathbf{Z}^T\mathbf{D}_w\mathbf{Z} = 1$ to the objective function. Then the entire optimization problem can be further represented as

$$\arg\max_{\mathbf{P}} \mathbf{P}^T\mathbf{X}[\alpha\mathbf{L}_b + (1-\alpha)\mathbf{W}_w]\mathbf{X}^T\mathbf{P} \quad \text{s.t.} \quad \mathbf{P}^T\mathbf{X}\mathbf{D}_w\mathbf{X}^T\mathbf{P} = 1 \tag{17.19}$$

where $\alpha \in [0, 1]$ is a suitable constant. The solution to (17.19) can be found by solving the following generalized eigenvalue problem:

$$\mathbf{X}[\alpha\mathbf{L}_b + (1-\alpha)\mathbf{W}_w]\mathbf{X}^T\mathbf{P} = \lambda\mathbf{X}\mathbf{D}_w\mathbf{X}^T\mathbf{P} \tag{17.20}$$

Let $\mathbf{P}$ be the solution to (17.20) with the decreasing eigenvalues and the result of LSDA can be obtained by $\mathbf{Z} = \mathbf{P}^T\mathbf{X}$ with $\mathbf{P} = \{\mathbf{p}_1, \mathbf{p}_2, \ldots, \mathbf{p}_m\}$.

2. Classification with reduced data by GSRC

The projected testing set is linearly represented by group-structured training samples via GSRC. Therefore, the final objective function of LSDA-GSRC can be represented as

$$\hat{\mathbf{A}}^L = \min_{\mathbf{A}^L} \frac{1}{2}\|\phi(\mathbf{Y}) - \phi(\mathbf{D})\mathbf{A}^L\|_F^2 + \lambda\sum_{g\in G}\omega_g\left\|\mathbf{A}_g^L\right\|_2 \tag{17.21}$$

where $\phi(\mathbf{Y})$ and $\phi(\mathbf{D})$ are projected by LSDA with $\mathbf{Y}$ and $\mathbf{D}$, $\mathbf{A}^L$ is a group sparse matrix, and $\mathbf{A}_g^L$ refers to the coefficients of each group corresponding to $\phi(\mathbf{D})$. Finally, the class label of the center testing pixel $\mathbf{y}$ in $\mathbf{Y}$ is determined by the minimum total residual error:

$$\text{class}(\mathbf{y}) = \arg\min_k \left\|\phi(\mathbf{Y}) - \phi(\mathbf{D})\delta_k\left(\hat{\mathbf{A}}^L\right)\right\|_F^2 \tag{17.22}$$

where $\delta_k\left(\hat{\mathbf{A}}^L\right)$ is the indicator operation that can zero out all the elements in $\hat{\mathbf{A}}^L$ that do not belong to class $k$. The implementation of LSDA-GSRC is as follows:

---

**Algorithm LSDA-GSRC**

---

**Input:** Available training data $\mathbf{D} = \{\mathbf{d}_i\}_{i=1}^n$ and a testing set in a spatial neighborhood $\mathbf{Y} = \{\mathbf{y}_j\}_{j=1}^t$.
**Step 1:** Normalize $\mathbf{D}$ and $\mathbf{Y}$ to have unit $l_2$-norm.
**Step 2:** Obtain $\phi(\mathbf{D})$ and $\phi(\mathbf{Y})$ by LSDA using $\mathbf{D}$ and $\mathbf{Y}$.
**Step 3:** Obtain weight matrix $\hat{\mathbf{A}}^L$ in GSRC according to (17.21).
**Step 4:** Compute the residual according to (17.22).
**Step 5:** Identify the class label of the center pixel $\mathbf{y}$ in $\mathbf{Y}$.
**Output:** class($\mathbf{y}$).

---

2. *Post-processing framework based on neighborhood activity analysis (NADRC)*

ADRC changes the perspective of analysis and decision mechanism of the SR-based models, while the acquisition of coefficient still depends on the mutual response of spectral feature. In this case, the spectral information of the testing pixel has an important influence on the determination of its class, such as the same object with different spectrums, existed in HSI may produce spectral variation [47, 48]. Based on the aforementioned AD-driven model, a novel spectral-spatial framework is established to rectify the misclassification under the framework that only uses in the spectral information called neighborhood ADRC (NADRC). The basic assumption is that adjacent pixels in spatial position have a high probability to belonging to the same class [49].

Specifically, Figure 17.3 illustrates the schematic diagram of the proposed NADRC. Following some basic conditions in ADRC, H represents the image of AD, where the dimensional information of each pixel is updated by its class-dependent AD. For a testing pixel $\mathbf{x}_{i,j}$, a $\widehat{w} \times \widehat{w}$ pixel-sized neighborhood centered on its position is extracted from H and denoted as $\mathrm{H}_{i,j}$. In the case shown in Figure 17.3, it is very likely to lead to a misclassification via the AD-based decision, if there is a spectral variation of $\mathbf{x}_{i,j}$ resulting in an inaccurate response of $\hat{\boldsymbol{\alpha}}_{i,j}$. But at the same time, it can be noticed that most of its surrounding active atoms mainly concentrated in another class, which indicates that $\mathbf{x}_{i,j}$ has a high probability to locate in a homogeneous area. Therefore, it is available to take advantage of this spatial coherence to adjust the determination.

In this context, another concept is defined toward the image of AD called neighborhood AD (NAD). The NAD evaluates the contribution of different class-dependent sub-dictionaries to the representation of the testing neighborhood. The NAD of a certain class is calculated by the sum of AD as follows:

$$\mathrm{NAD}_k = \sum_{v=1}^{\widehat{w}^2} \mathrm{AD}_k^{(v)}, \tag{17.23}$$

where $v$ denotes the index of vector in the neighborhood. In this case of Figure 17.3, it is obvious that $\mathrm{NAD}_k > \mathrm{NAD}_{k'} > \mathrm{NAD}_{k''}$, in spite of $\mathrm{AD}_k < \mathrm{AD}_{k'} < \mathrm{AD}_{k''}$, such that the class label of $\mathbf{x}_{i,j}$ can be corrected as class $k$. The pseudo code of NADRC is summarized as follows:

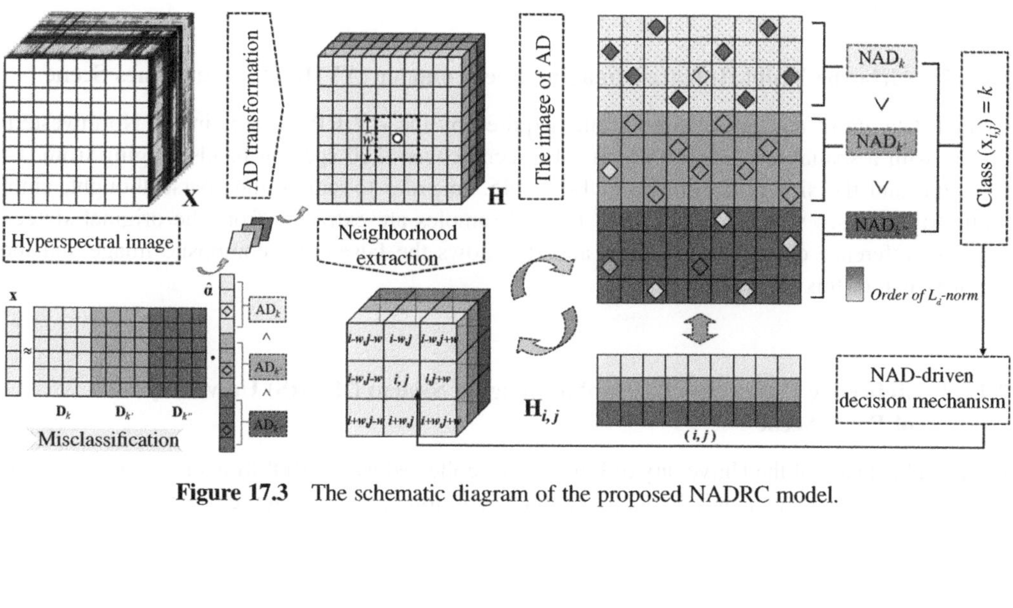


**Figure 17.3** The schematic diagram of the proposed NADRC model.

**Algorithm NADRC**

**Input:** Available HSI **X**, dictionary **D**, a testing pixel $\mathbf{x}_{i,j}$, and window size $\widehat{w}$.
**Step 1:** Obtain the class-dependent AD of each pixel in **X** according to (17.14).
**Step 2:** Form the activity image H based on Step 1.
**Step 3:** Extract $\widehat{w} \times \widehat{w}$ pixel-sized neighborhood $\mathrm{H}_{i,j}$ centered on the location of current testing pixel $\mathbf{x}_{i,j}$.
**Step 4:** Calculate the NAD of each class from $\mathrm{H}_{i,j}$ according to (17.23).
**Step 5:** Identify the class with the maximum NAD.

$$\kappa = \max_{k} \{\mathrm{NAD}_k \mid k \in [1, K]\}$$

**Output:** $\mathrm{class}(\mathbf{x}_{i,j}) = \kappa$.

## 17.4 Experimental Results and Analysis

In this section, we first provide brief descriptions of the two data sets used in the experiment. The detailed description of the applied data sets is given in Section 17.4.1. The parameter tuning related to different algorithms is presented in Section 17.4.2. We verify the performance of different algorithms in the spectral domain and the spectral-spatial domain. Then, we analyzed the experimental results and performance assessments in Section 17.4.3.

### *17.4.1 Experimental Data*

Two data sets presented in this section are Airborne Visible/Infrared Imaging Spectrometer (AVIRIS) Indian Pines Scene and Reflective Optics Spectrographic Imaging System (ROSIS) University of Pavia Scene, which can be generally used for the evaluation of the classifier.

#### 17.4.1.1 Airborne Visible/Infrared Imaging Spectrometer (AVIRIS) Indian Pines Scene

The first data are of the Indian Pines scene acquired by the AVIRIS sensors in the Northwestern Indiana, with a spatial resolution of 20 m. The scene covers 220 spectral bands ranging from 0.4 to 2.5 μm, and the size of the image is 145 × 145. In order to satisfy the sparse thought, eight ground-truth classes with a total of 8624 labeled samples are extracted from the original sixteen categories reference data. Figure 17.4(a) and (b) shows the false-color composite image and the reference map of this scene, respectively.

#### 17.4.1.2 Reflective Optics Spectrographic Imaging System (ROSIS) University of Pavia Scene

The second data are of the University of Pavia scene collected by the ROSIS over a downtown area near the University of Pavia in Italy, with a spatial resolution of 1.3 m. After removing 12 bands with

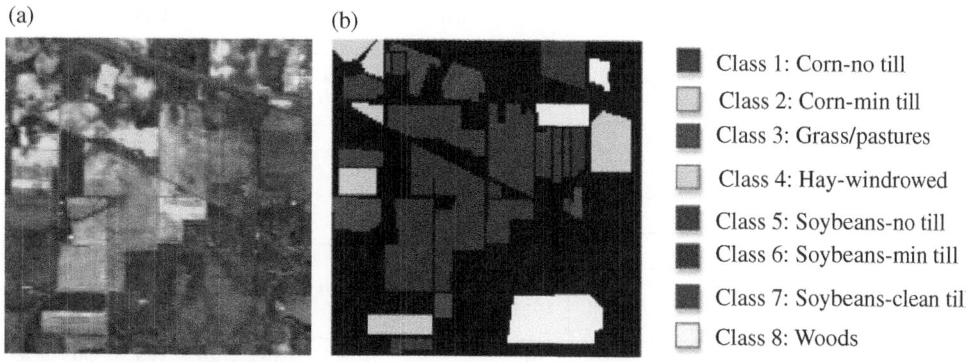


**Figure 17.4** The Airborne Visible/Infrared Imaging Spectrometer (AVIRIS) Indian Pines scene: (a) false-color composite image, (b) reference map.

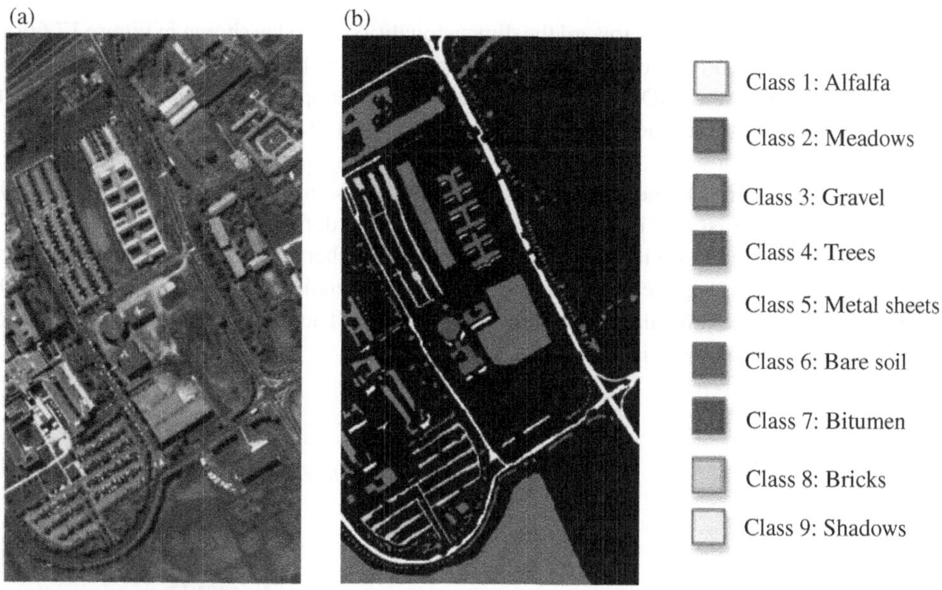


**Figure 17.5** The Reflective Optics Spectrographic Imaging System (ROSIS) University of Pavia scene: (a) false-color composite image, (b) reference map.

high noise and water absorption, the scene has 103 spectral bands ranging from 0.43 to 0.86 μm, with $610 \times 340$ pixels. Nine ground-truth classes with a total of 42 776 labeled samples are contained in the reference data. Figure 17.5(a) and (b) shows the false-color composite image and the reference map of this scene, respectively.

### *17.4.2 Parameter Tuning*

The regularization parameter $\lambda$ in all the representation-based models is basically set to the values ranging from $10^{-3}$ to $10^{-1}$. For the dimensionality reduced parameter $m$ in LPP, we report the

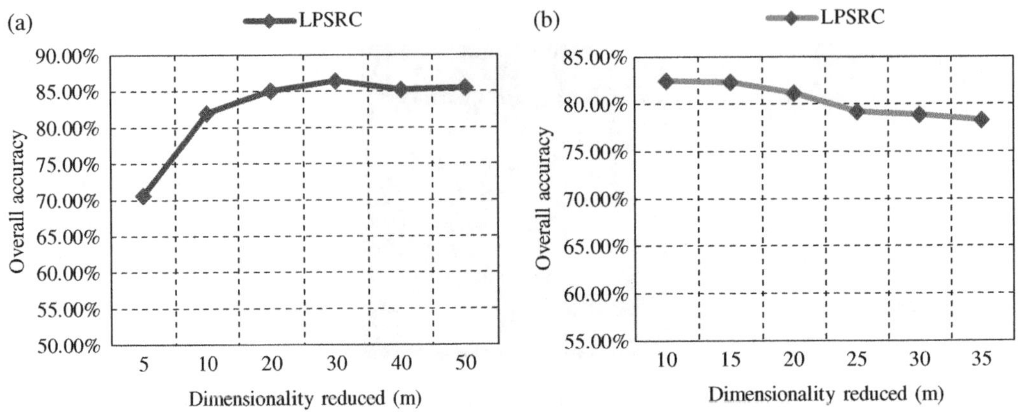


**Figure 17.6** Dimensionality reduced parameter ($m$) versus OA for (a) Indian Pines data set using LPSRC, (b) University of Pavia data set using LPSRC.

sensitivity of the LPSRC algorithm toward the dimensionality reduced $m$ shown in Figure 17.6(a) and (b). At last, we set $m = 30$ for the Indian Pines data set and $m = 10$ for the University of Pavia data set. And, for the LSDA-GSRC algorithm, we also adopted $m = 30$ and $m = 10$ for the two data sets, respectively. In addition, the spatial neighborhood size in LSDA-GSRC is generally adopted as $t = 9$.

The other two important parameters are investigated in details with respect to ADRC and NADRC in our experiment. The first one is the type of metric norm $l_d$ defined in the AD of NADRC. The relationship between $d$ and OA with respect to the different number of labeled samples $n$ is illustrated in Figure 17.7. As can be seen, the sensitivity of ADRC to $d$ is relatively low. Moreover, the NADRC generally obtained excellent results under both of the settings. Therefore, both $l_1$-norm and $l_2$-norm can be applied for measurement in this case.

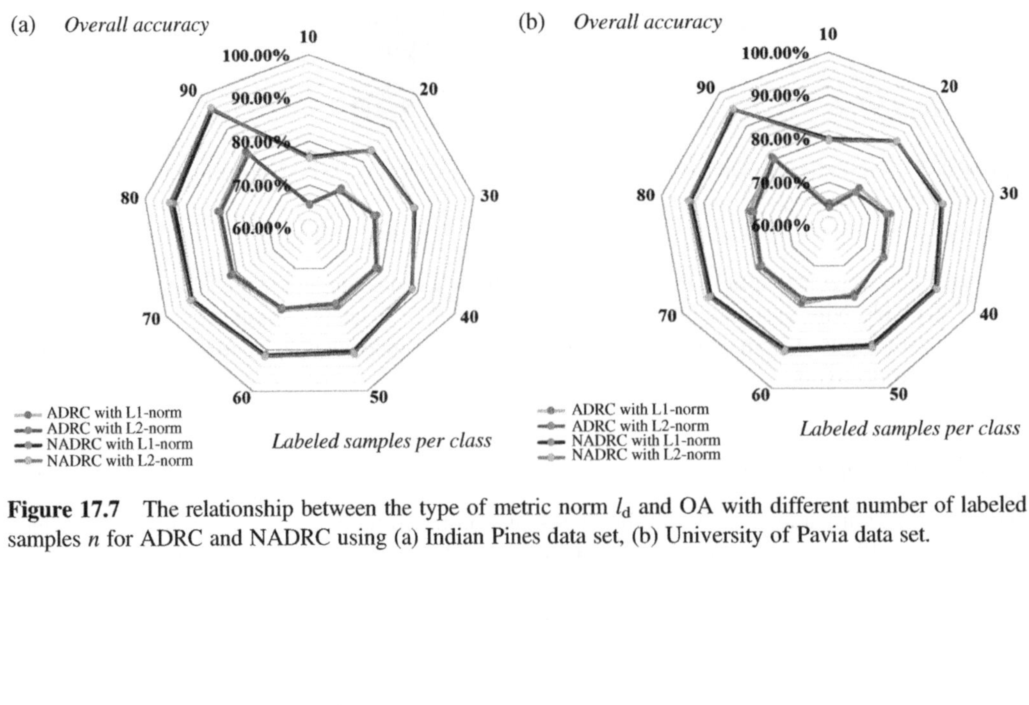


**Figure 17.7** The relationship between the type of metric norm $l_d$ and OA with different number of labeled samples $n$ for ADRC and NADRC using (a) Indian Pines data set, (b) University of Pavia data set.

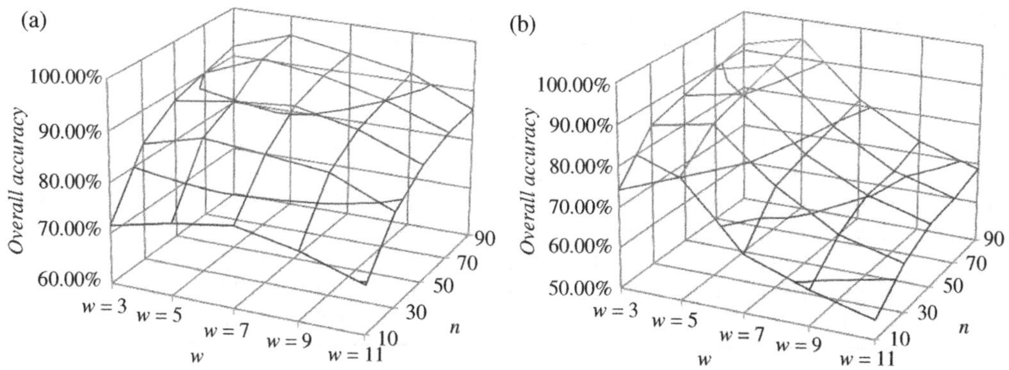


**Figure 17.8** The relationship between the size of neighborhood $\widehat{w}$ and OA with different number of labeled samples $n$ for NADRC using (a) Indian Pines data set, (b) University of Pavia data set.

The second parameter is the size of neighborhood $\widehat{w}$ used in NADRC. With the consideration of differences in the distribution of ground objects, the optimal condition of spatial consistency in different scenes may not be the same. Therefore, the relationship between $\widehat{w}$ and OA with respect to $n$ toward three data sets is evaluated and reported in Figure 17.8. As can be seen, the overall accuracies mostly reach peak at first and then decreasing as a whole, with the increase of $\widehat{w}$. It indicates that different scenes generally have optimal $\widehat{w}$ with regard to the spatial coherence, and over-correction will happen and interfere the results with the exceeded size. Therefore, $\widehat{w}$ is set to 5 for the two data sets. In addition, the parameters that control the amount of penalty and the spread of the RBF kernel are chosen by cross-validation.

## *17.4.3 Results and Analysis*

In this section, we selected different numbers of training samples for the two data sets to test and verify the proposed LPSRC, ADRC, LSDA-GSRC, and NADRC models. In order to satisfy the sparse thought in the SR-based model, the size of the training set starts from 30 training samples per class and increases to 90 training samples per class. For comparative purpose, the SVM and the classic SR-based models, including SRC and CRC, are first considered in the spectral domain. A post-processing model for spatial coherence, i.e. SVM-MRF and the JR-based models, including GSRC and JSRC, are further considered in the spectral-spatial domain. The overall classification accuracy (OA) and category classification accuracy (CA) are used as the main evaluation indexes to analyze the effectiveness and stability of the four algorithms proposed in this chapter. Tables 17.1 and 17.2 show the overall classification accuracies acquired by different methods with different number of labeled samples. As illustrated in Tables 17.3 and 17.4, we randomly selected 90 and 90 labeled samples per class with a total of 720 and 810 samples to construct a dictionary and the training model, the corresponding classification maps are presented in Figures 17.9 and 17.10. Besides, the selected training samples constitute the approximately 8.35 and 1.89% of the labeled samples in the reference map, and the other remained samples are used in validation. From these results, the following conclusions can be drawn:

The classification results demonstrate that the overall accuracy has a positive relationship with the number of labeled samples, and each model has the same law, that is the overall accuracy is increased by the number of labeled samples. Besides, this phenomenon only be satisfied under a certain number

**Table 17.1** OA (in percent) for the Indian Pines data set using different size of training samples

| Samples (per class) | SVM | CRC | SRC | ADRC | LPSRC | SVM-MRF | JSRC | GSRC | LSDA-GSRC | NADRC |
|---|---|---|---|---|---|---|---|---|---|---|
| 240 (30) | 58.08 | 74.32 | 75.95 | 75.29 | 79.20 | 78.44 | 80.11 | 83.49 | 85.34 | 85.49 |
| 320 (40) | 61.55 | 75.98 | 77.96 | 78.36 | 80.82 | 82.09 | 83.61 | 86.49 | 87.36 | 88.49 |
| 400 (50) | 61.90 | 76.79 | 79.91 | 79.36 | 83.24 | 82.74 | 85.35 | 88.87 | 90.42 | 90.87 |
| 480 (60) | 64.91 | 78.16 | 80.67 | 80.48 | 84.65 | 84.99 | 86.69 | 89.52 | 92.37 | 92.52 |
| 560 (70) | 66.20 | 78.34 | 81.72 | 82.14 | 85.20 | 87.33 | 88.62 | 91.03 | 93.20 | 94.03 |
| 640 (80) | 67.51 | 79.12 | 82.32 | 82.52 | 86.36 | 87.84 | 89.47 | 91.13 | 94.89 | 95.13 |
| 720 (90) | 68.54 | 81.04 | 83.34 | 83.38 | 86.39 | 88.68 | 90.54 | 93.08 | 94.99 | 96.30 |

**Table 17.2** OA (in percent) for the University of Pavia data set using different size of training samples

| Samples (per class) | SVM | CRC | SRC | ADRC | LPSRC | SVM-MRF | JSRC | GSRC | LSDA-GSRC | NADRC |
|---|---|---|---|---|---|---|---|---|---|---|
| 270 (30) | 72.31 | 70.22 | 74.90 | 74.86 | 75.58 | 82.54 | 79.89 | 82.30 | 85.01 | 87.62 |
| 360 (40) | 75.64 | 72.28 | 75.09 | 75.44 | 77.54 | 85.24 | 81.88 | 84.49 | 86.94 | 89.66 |
| 450 (50) | 78.35 | 73.47 | 76.88 | 77.73 | 80.10 | 87.96 | 82.08 | 85.67 | 87.24 | 90.11 |
| 540 (60) | 79.80 | 74.26 | 79.21 | 78.38 | 80.87 | 88.67 | 83.71 | 86.17 | 88.95 | 91.01 |
| 630 (70) | 81.94 | 75.45 | 78.97 | 78.72 | 81.76 | 90.72 | 84.54 | 87.66 | 90.99 | 92.86 |
| 720 (80) | 82.30 | 76.85 | 79.95 | 79.53 | 81.78 | 90.50 | 85.11 | 88.56 | 91.24 | 93.50 |
| 810 (90) | 83.14 | 77.53 | 80.46 | 80.81 | 82.42 | 90.09 | 87.11 | 91.31 | 93.36 | 95.09 |

**Table 17.3** OA and CA (in percent) for the Indian Pines data set (90 training samples per class)

| Class | SVM | CRC | SRC | ADRC | LPSRC | SVM-MRF | JSRC | GSRC | LSDA-GSRC | NADRC |
|---|---|---|---|---|---|---|---|---|---|---|
| Corn-no till | 54.75 | 85.07 | 78.54 | 77.63 | 82.05 | 74.36 | 86.21 | 90.93 | 89.60 | 93.29 |
| Corn-min till | 58.29 | 73.26 | 81.88 | 83.44 | 90.05 | 89.69 | 94.39 | 96.66 | 97.80 | 98.68 |
| Grass | 90.86 | 95.77 | 97.56 | 97.06 | 97.79 | 97.89 | 98.23 | 99.16 | 100.00 | 98.39 |
| Hay-wind-rowed | 98.82 | 99.59 | 99.93 | 99.91 | 100.00 | 99.45 | 99.88 | 99.96 | 100.00 | 100.00 |
| Soybeans-no till | 71.25 | 61.16 | 81.79 | 81.99 | 85.02 | 88.26 | 94.95 | 95.27 | 99.55 | 90.08 |
| Soybeans-min till | 56.46 | 72.61 | 70.64 | 70.70 | 70.66 | 86.06 | 80.44 | 88.34 | 86.44 | 95.75 |
| soybeans-clean till | 69.16 | 84.85 | 91.43 | 91.82 | 92.35 | 98.99 | 96.03 | 98.50 | 99.88 | 99.84 |
| Woods | 94.60 | 97.99 | 99.47 | 99.54 | 99.69 | 98.65 | 99.61 | 99.98 | 99.77 | 100.00 |
| OA | 68.54 | 81.04 | 83.33 | 83.38 | 85.20 | 88.68 | 90.54 | 93.08 | 94.69 | 96.30 |

**Table 17.4** OA and CA (in percent) for the University of Pavia data set (90 training samples per class)

| Class | SVM | CRC | SRC | ADRC | LPSRC | SVM-MRF | JSRC | GSRC | LSDA-GSRC | NADRC |
|---|---|---|---|---|---|---|---|---|---|---|
| Alfalfa | 77.25 | 81.48 | 76.62 | 76.07 | 80.98 | 85.21 | 65.74 | 68.66 | 84.98 | 97.74 |
| Shadows | 80.47 | 77.35 | 78.82 | 79.80 | 54.45 | 88.42 | 88.16 | 90.63 | 90.45 | 91.06 |
| Meadows | 80.47 | 56.07 | 79.96 | 78.34 | 86.40 | 89.11 | 89.81 | 94.76 | 94.23 | 94.42 |
| Bricks | 93.81 | 92.27 | 94.45 | 95.07 | 96.57 | 95.99 | 96.60 | 90.99 | 93.44 | 98.47 |
| Gravel | 99.39 | 99.93 | 99.85 | 99.72 | 99.93 | 99.67 | 100.00 | 100.00 | 100.00 | 100.00 |
| Tree | 86.24 | 75.90 | 79.19 | 79.34 | 97.68 | 95.50 | 78.30 | 85.01 | 89.68 | 99.30 |
| Bare soil | 92.89 | 62.03 | 89.83 | 89.80 | 82.13 | 95.73 | 98.71 | 99.52 | 99.37 | 99.17 |
| Bitumen | 82.49 | 69.31 | 70.97 | 71.39 | 84.61 | 92.51 | 84.27 | 90.04 | 94.27 | 97.04 |
| Metal sheet | 99.85 | 83.74 | 98.54 | 98.18 | 91.43 | 99.87 | 97.63 | 96.63 | 99.43 | 99.08 |
| OA | 83.14 | 77.53 | 80.46 | 80.81 | 84.42 | 90.09 | 87.11 | 91.31 | 93.28 | 95.12 |

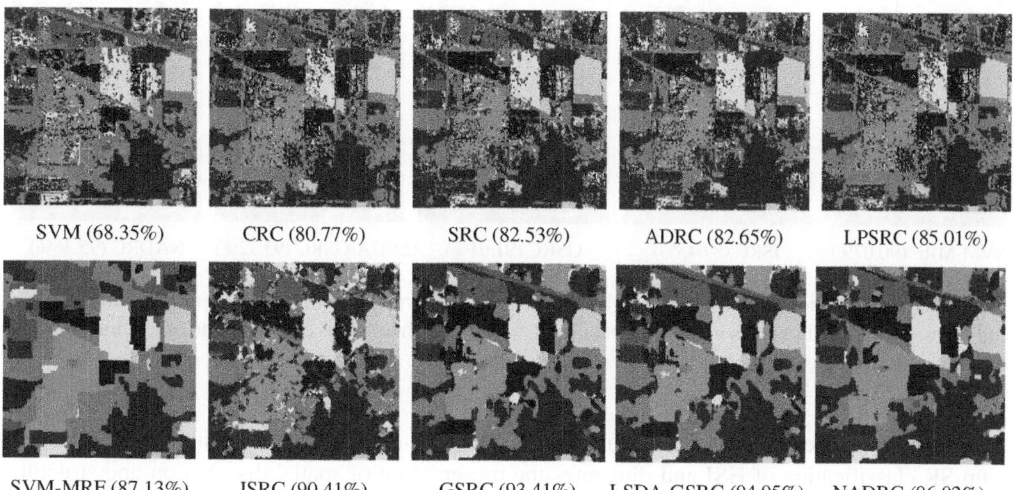


**Figure 17.9** Classification maps obtained by the different tested methods for the Indian Pines data set. In all cases, 720 labeled samples in total (90 per class) were used. These maps correspond to one of the 20 Monte Carlo runs conducted in each case.

of the labeled samples, the growth trend would be stopped when the labeled samples reach a certain number.

Compared with the SVM, SR-based model generally provides rather plausible results, which proves that the SRC is suitable for the HSI classification tasks. Moreover, the accuracy of CRC is relatively low than that of SRC, which indicates the $l_1$-norm is more suitable for HSI. Compared with SR-based models, ADRC provides similar accuracies as SRC, which underlines a simplified procedure without calculating residual in ADRC.

For the classifier based on dimensionality reduction, LPSRC is better than the original SRC model, which proves the spectral feature optimization based on popular learning is effective for SR classification of HSI. Similarity, compared with GSRC, LSDA-GSRC has achieved better classification

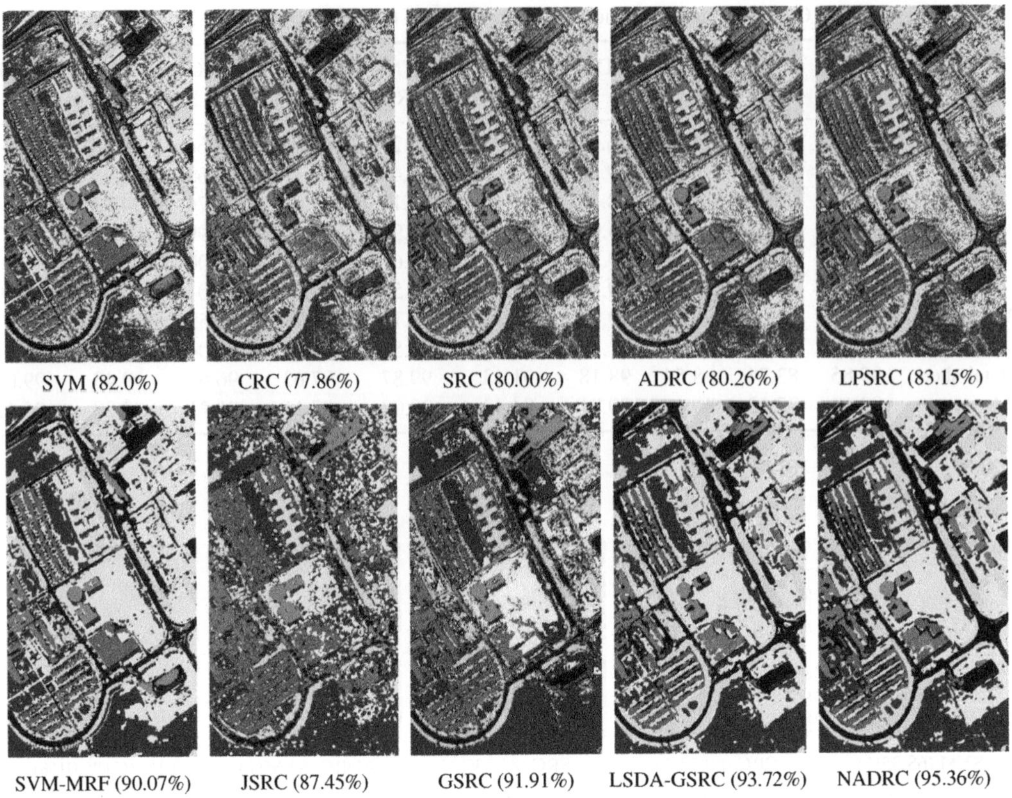


**Figure 17.10** Classification maps obtained by the different tested methods for the University of Pavia data set. In all cases, 810 labeled samples in total (90 per class) were used. These maps correspond to one of the 20 Monte Carlo runs conducted in each case.

results, which verifies the spectral feature optimization based on popular learning is effective for group SR classification of HSI and illustrates the preservation of spatial distribution and structural discrimination information is important to the coordinated representation classification of spectral image spatial and spectral information.

For the spectral-spatial domain: The improvement of SVM-MRF, JSRC, NADRC toward their original counterparts, i.e. SVM, SRC, and ADRC, proved the effectiveness of exploiting spatial information. SVM-MRF model outperforms the SVM classifier with two data sets, which verifies the improvement brought by the integration of spatial coherence. This can also be confirmed by the increasing overall accuracies of JSRC toward SRC in terms of the Indian Pines and the University of Pavia scene. Compared with ADRC model, NADRC also brings stable and significant improvements in the two data sets, which not only supports that the spatial extension from AD to NAD-driven framework is effective but also confirmed that the composed overall constraint and decision mechanism is reliable for the HSIC task, with respect to the innovation exploration of the sparse coefficient.

As the classification results shown in Figures 17.9 and 17.10, in the Indian Pines data set, when 90 training samples of each class are used, the proposed LPSRC obtains the accuracy with an OA of 85.01%, 2.48% higher than SRC. The classification accuracy of the LSDA-GSRC model is 94.95%, which is 1.54 and 12.42% higher than GSRC and SRC model, respectively. And the classification accuracy of the NADRC model is 96.02%, which is 13.37 and 13.49% higher than ADRC and SRC model, respectively. In the University of Pavia data set, when 90 training samples of each class are used, the proposed LPSRC obtains the accuracy with an OA of 83.15%, 3.15% higher than SRC. The classification accuracy of the LSDA-GSRC model is 93.72%, which is 1.81 and 13.72% higher than GSRC and SRC model, respectively. And the classification accuracy of the NADRC model is 95.36%, which is 15.1, and 15.36% higher than ADRC and SRC model, respectively. Based on the above experimental results, it can be proved that the proposed LPSRC, LSDA-GSRC, ADRC, and NADRC models are effective and stable for HSIC.

## 17.5 Conclusion

From the perspective of potential sparse characteristic of HSI, this chapter conducts studies and proposes four new classification methods for the in-depth coordinate of spatial and spectral information, called LPSRC, LSDA-GSRC, ADRC, and NADRC. The LPSRC method combines popular LPP learning algorithm and the SR model, effectively solve the problem of band redundancy under limited conditions. The LSDA-GSRC model combines the popular LSDA with the GSRC model. Applying LSDA to project the input data to a local manifold subspace and representing the projected testing set by GSR for classification. The ADRC is an equivalent and simplified model to the SRC model, which uses the AD-driven decision mechanism. Deeply understand and explore the sparse coefficient for concise and simplified mechanism. The NADRC model put forward an NAD of the matrix-formed coefficient, which takes advantage of spatial coherence in terms of the overall activity of each class based on the extension of AD. This chapter uses two sets of real hyperspectral data sets to analyze and verify the effectiveness and stability of the proposed methods by designing comparative experiment under different condition.

## Acknowledgments

This work was supported by the China Postdoctoral Science Foundation under Grant 2022T150080 and 2020M680925, the Open Research Fund of Key Laboratory of Digital Earth Science, Institute of Remote Sensing and Digital Earth, Chinese Academy of Sciences under Grant 2020LDE001, the Fundamental Research Funds for Central Universities under Grant 3132022249, and the National Nature Science Foundation of China under Grant 42101350 and 61922013.

## References

1. Plaza, A., Benediktsson, J.A., Boardman, J.W. et al. (2009). Recent advances in techniques for hyperspectral image processing. *Remote Sens. Environ.* 113: S110–S122.
2. Fauvel, M., Tarabalka, Y., Benediktsson, J.A. et al. (2013). Advances in spectral-spatial classification of hyperspectral images. *Proc. IEEE* 101 (3): 652–675.
3. Bioucas-Dias, J.M., Plaza, A., Dobigeon, N. et al. (2012). Hyperspectral unmixing overview: geometrical, statistical, and sparse regression-based approaches. *IEEE J. Sel. Top. Appl. Earth Obs. Remote Sens.* 5 (2): 354–379.
4. Su, H., Yong, B., and Du, Q. (2016). Hyperspectral band selection using improved firefly algorithm. *IEEE Geosci. Remote Sens. Lett.* 13 (1): 68–72.
5. Li, J., Bioucas-Dias, J., and Plaza, A. (2012). Spectral–spatial hyperspectral image segmentation using subspace multinomial logistic regression and Markov random fields. *IEEE Trans. Geosci. Remote Sens.* 50 (3): 809–823.
6. Yu, H., Gao, L., Liao, W. et al. (2017). Multiscale superpixel-level subspace-based support vector machines for hyperspectral image classification. *IEEE Geosci. Remote Sens. Lett.* 14 (11): 2142–2146.

7. Du, P., Bai, X., Tan, K. et al. (2020). Advances of four machine learning methods for spatial data handling: a review. *J. Geovis. Spat. Anal.* 4 (13): 1–25.
8. Gao, L., Yao, D., Li, Q. et al. (2017). A new low-rank representation based hyperspectral image denoising method for mineral mapping. *Remote Sens.* 9 (11): 1145.
9. Zhuang, L. and Bioucas-Dias, J.M. (2018). Fast hyperspectral image denoising and inpainting based on low-rank and sparse representations. *IEEE J. Sel. Top. Appl. Earth Obs. Remote Sens.* 11 (3): 730–742.
10. Tang, Y.Y., Yuan, H., and Li, L. (2014). Manifold-based sparse representation for hyperspectral image classification. *IEEE Trans. Geosci. Remote Sens.* 52 (12): 7606–7618.
11. Li, W., Liu, J., and Du, Q. (2016). Sparse and low rank graph-based discriminant analysis for hyperspectral image classification. *IEEE Trans. Geosci. Remote Sens.* 54 (7): 4094–4105.
12. Yu, H., Gao, L., Liao, W. et al. (2020). Global spatial and local spectral similarity-based manifold learning group sparse representation for hyperspectral imagery classification. *IEEE Trans. Geosci. Remote Sens.* 58 (5): 3043–3056.
13. Li, W., Du, Q., and Zhang, B. (2015). Combined sparse and collaborative representation for hyperspectral target detection. *Pattern Recognition* 48 (12): 3904–3916.
14. Li, W. and Du, Q. (2015). Collaborative representation for hyperspectral anomaly detection. *IEEE Transactions on Geoscience and Remote Sensing* 53 (3): 1463–1474.
15. Yang, A.Y., Sastry, S.S., Ganesh, A. et al. (2010). Fast ℓ 1-minimization algorithms and an application in robust face recognition: a review. 17th IEEE International Conference on Image Processing (ICIP), Hong Kong, China (December 2010), 1849–1852.
16. Chen, Y., Nasrabadi, N.M., and Tran, T.D. (2013). Hyperspectral image classification via kernel sparse representation. *IEEE Trans. Geosci. Remote Sens.* 51 (1): 217–231.
17. Li, W., Du, Q., Zhang, F., and Hu, W. (2016). Hyperspectral image classification by fusing collaborative and sparse representations. *IEEE J. Sel. Top. Appl. Earth Obs. Remote Sens.* 9 (9): 4178–4187.
18. Jia, S., Deng, B., Zhu, J. et al. (2017). Superpixel-based multitask learning framework for hyperspectral image classification. *IEEE Trans. Geosci. Remote Sens.* 55: 2575–2588.
19. Niu, Y. and Wang, B. (2016). Hyperspectral anomaly detection based on low-rank representation and learned dictionary. *Remote Sens.* 8: 289.
20. Gao, L., Yu, H., Zhang, B., and Li, Q. (2016). Locality-preserving sparse representation-based classification in hyperspectral imagery. *J. Appl. Remote Sens.* 10: 042004.
21. Roweis, S.T. and Saul, L.K. (2000). Nonlinear dimensionality reduction by locally linear embedding. *Science* 290 (5500): 2323–2326.
22. Belkin, M. and Niyogi, P. (2001). Laplacian eigenmaps and spectral techniques for embedding and clustering. *NIPS* 14: 585–591.
23. Niyogi, X. (2004). Locality preserving projections. *Neural Information Processing Systems* 16: 153.
24. Tenenbaum, J.B., De Silva, V., and Langford, J.C. (2000). A global geometric framework for nonlinear dimensionality reduction. *Science* 290 (5500): 2319–2323.
25. Jia, S., Xie, Y., Tang, G., and Zhu, J. (2016). Spatial-spectral-combined sparse representation-based classification for hyperspectral imagery. *Soft Comput.* 20 (12): 4659–4668.
26. Zou, J., Li, W., Chen, C., and Du, Q. (2016). Scene classification using local and global features with collaborative representation fusion. *Information Sciences* 348: 209–226.
27. Yu, H., Gao, L., Liao, W., and Zhang, B. (2018). Group sparse representation based on nonlocal spatial and local spectral similarity for hyperspectral imagery classification. *Sensors* 18 (6): 1695.
28. Gao, S., Tsang, I., and Chia, L. (2013). Laplacian sparse coding, hypergraph Laplacian sparse coding, and applications. *IEEE Trans. Pattern Anal. Mach. Intell.* 35 (1): 92–104.
29. Liu, G., Lin, Z., Yan, S. et al. (2013). Robust recovery of subspace structures by low-rank representation. *IEEE Trans. Pattern Anal. Mach. Intell.* 35 (1): 171–184.
30. Chen, Y., Nasrabadi, N.M., and Tran, T.D. (2011). Hyperspectral image classification using dictionary-based sparse representation. *IEEE Trans. Geosci. Remote Sens.* 49 (10): 3973–3985.

31. Li, J., Zhang, H., Huang, Y., and Zhang, L. (2014). Hyperspectral image classification by nonlocal joint collaborative representation with a locally adaptive dictionary. *IEEE Trans. Geosci. Remote Sens.* 52 (6): 3707–3719.
32. Li, J., Zhang, H., and Zhang, L. (2014). Column-generation kernel nonlocal joint collaborative representation for hyperspectral image classification. *ISPRS Journal of Photogrammetry and Remote Sensing* 94: 25–36.
33. Sun, X., Qu, Q., Nasrabadi, N.M. et al. (2013). Structured priors for sparse-representation-based hyperspectral image classification. *IEEE Geosci. Remote Sens. Lett.* 11 (7): 1235–1239.
34. Li, W. and Du, Q. (2015). A survey on representation-based classification and detection in hyperspectral remote sensing imagery. *Pattern Recognition Letters* 83 (2): 115–123.
35. Zhang, L., Yang, M., and Feng, X. (2011). Sparse representation or collaborative representation: which helps face recognition? IEEE International Conference on Computer Vision, ICCV 2011, Barcelona, Spain (November 6–13).
36. Liu, J., Wu, Z., Li, J. et al. (2016). Probabilistic-kernel collaborative representation for spatial-spectral hyperspectral image classification. *IEEE Trans. Geosci. Remote Sens.* 54 (4): 2371–2384.
37. Liu, J., Wu, Z., Xiao, L., and Yan, H. (2020). Learning multiple parameters for kernel collaborative representation classification. *IEEE Tran. Neur. Net. Lear.* 2020. https://doi.org/10.1109/TNNLS.2019.2962878.
38. Su, H., Yu, Y., Du, Q., and Du, P. (2020). Ensemble learning for hyperspectral image classification using tangent collaborative representation. *IEEE Trans. Geosci. Remote Sens.* 58 (6): 3778–3790.
39. Su, H., Zhao, B., Du, Q., and Du, P. (2019). Kernel collaborative representation with local correlation features for hyperspectral image classification. *IEEE Trans. Geosci. Remote Sens.* 57 (2): 1230–1241.
40. Su, H., Zhao, B., Du, Q. et al. (2018). Multifeature dictionary learning for collaborative representation classification of hyperspectral imagery. *IEEE Trans. Geosci. Remote Sens.* 56 (4): 2467–2484.
41. Zhang, X., Xu, C., Li, M., and Sun, X. (2015). Sparse and low-rank coupling image segmentation model via nonconvex regularization. *Int. J. Pattern Recognit. Artif. Intell.* 29 (2): 1555004.
42. Zhang, L., Yang, M., and Feng, X. (2011). Sparse representation or collaborative representation: which helps face recognition? IEEE International Conference on Computer Vision, ICCV 2011, Barcelona, Spain (Novemebr 2011), 471–478.
43. Tropp, J.A. and Wright, S.J. (2010). Computational methods for sparse solution of linear inverse problems. *Proc. IEEE* 98 (6): 948–958.
44. Sprechmann, P., Ramirez, I., Sapiro, G., and Eldar, Y. (2011). C-HiLasso: a collaborative hierarchical sparse modeling framework. *IEEE Trans. Signal Processing* 59 (9): 4183–4198.
45. He, X., Yan, S., Hu, Y. et al. (2005). Face recognition using Laplacianfaces. *IEEE T. Pattern Anal.* 27 (3): 328–340.
46. Cai, D., He, X., and Han, J. (2005). Document clustering using locality preserving indexing. *IEEE Trans. Knowl. Data Eng.* 17 (12): 1624–1637.
47. Keshava, N. and Mustard, J.F. (2002). Spectral unmixing. *IEEE Signal Process. Mag.* 19 (1): 44–57.
48. Quintano, C., Fernández-Manso, A., Shimabukuro, Y., and Pereira, G. (2012). Spectral unmixing. *Int. J. Remote Sens.* 33 (17): 5307–5340.
49. He, L., Li, J., Liu, C., and Li, S. (2018). Recent advances on spectral-spatial hyperspectral image classification: an overview and new guidelines. *IEEE Trans. Geosci. Remote Sens.* 56 (3): 1579–1597.

# 18

# Collaborative Classification Based on Hyperspectral Images

Junping Zhang[1], Xiaochen Lu[2], and Tong Li[3]
[1] *School of Electronics and Information Engineering, Harbin Institute of Technology, Harbin, China*
[2] *School of Information Science and Technology, Donghua University, Shanghai, China*
[3] *Satellite Application Division, Shanghai Institute of Satellite Engineering, Shanghai, China*

## 18.1 Introduction

Over the past decades, land-cover classification has become a main topic in hyperspectral (HS) remote-sensing applications. HS image enables a detailed separation of similar ground materials due to its high spectral resolution. However, the low spatial resolution, mixed pixels, and the "the same object but different spectrum" and "the same spectrum but different objects" phenomena bring challenging problems to analysis of HS data. With the availability of multiple-source image data from operational earth observation satellites, the demand for a higher classification accuracy of remotely sensed images has encouraged an increasing usage of distinct characteristic information collected from different sources. Combining these multiple-source data is believed to offer strengthened capabilities for land-cover classification than the single HS image type [1, 2].

In remote-sensing area, panchromatic (PAN) image is possibly one of the most widely used data sources, because of its very high spatial resolution and relatively low-cost accessibility. And, PAN image usually presents much more details of the ground objects than HS image [3]. The spatial details reflect the contextual relations that can be used to describe a pixel more accurately. That's where PAN's strength lies. Generally, PAN image covers the visible and near infrared wavelength ranges with only a single spectral band. Thus, it does not provide much spectral information [4].

Compared with PAN, multispectral (MS) image has multiple bands and covers the visible-near infrared, and sometimes even short wave or middle wave infrared wavelength ranges. Thus, it involves much more spectral information than PAN image and is usually employed for land-cover classification and target recognition in the past decades. Conversely, it has lower spatial resolution than PAN image. By dividing the wavelength range into more spectral bands, HS image presents far more spectral features than usual MS image [5]. Hence, HS image can achieve much higher classification accuracies and is able to recognize more categories [6].

*Advances in Hyperspectral Image Processing Techniques*, First Edition. Edited by Chein-I Chang.

At the end of the infrared range, it generally refers to that thermal infrared (TIR) image covering the long-wave infrared (LWIR) wavelength ranges, which is different from the reflective imaging mode, collects the thermal radiations of land-covers rather than the reflection of sunlight, thus providing all-day and all-weather imaging capabilities. The thermal radiations contain temperature and emissivity information, which makes it possible to detect thermal sources. Likewise, by dividing the LWIR wavelength ranges into tens of spectral bands, LWIR HS images contain much more TIR spectral information that can be used to distinguish the different components of minerals and atmosphere [7]. Nevertheless, the noise and blurred edges of infrared image severely limit its application in remote-sensing image classifications.

Considering the different characteristics of these image types, the comprehensive utilization of these multisource remote-sensing images attracts more and more attention of researchers. Therefore, this chapter focuses on the collaborative classification of HS-centered multisource images, including HS image, high-resolution PAN, LWIR HS images, etc. In the following subsections, we will first review the current existing works of collaborative processing of multisource images and present the main problems and challenges in multisource image collaborative processing. Then, we will take two state-of-the-art methods for example to discuss how to effectively explore the advantages of HS and other type of images.

### *18.1.1 Hyperspectral and Panchromatic Images*

HS image has high spectral resolution and presents plentiful spectral details. However, owing to the constraint of sensors, HS image usually has low spatial resolution, which limits its application in some circumstances. PAN image, on the contrary, although has insufficient spectral information, its high spatial quality attracts wide interests in target detection and land-cover classification tasks. Obviously, it is of great significance to combine the two different kinds of data to obtain new data with both ideal spatial and spectral information.

For the collaborative classification of HS and high-resolution PAN images, existing works can be summarized to three basic techniques: (1) classification based on pixel-level fusion; (2) classification based on multiple-feature extraction and combination [1, 2]; and (3) classification based on decision-level fusion, also called multi-classifier fusion.

Classification based on pixel-level fusion aims at producing a new HS image that has the same spatial resolution with the PAN image, meanwhile preserves the spectral information of the original HS image [8, 9], and then performing conventional classification [10, 11] on the new HS image to generate a high-resolution thematic map. Due to the similar mechanism with MS image pan-sharpening, MS pan-sharpening algorithms can be employed for the fusion of HS and PAN images in a manner. Aside from the pan-sharpening algorithms, some researchers also address this problem from the perspective of spectral unmixing and matrix factorization. Such algorithms include spectral constraint nonnegative matrix factorization (NMF), sparse constraint NMF. Experiments and analyses have demonstrated the effectiveness of these methods in the aspect of improving the spatial quality of HS image.

However, several critical issues still should be noted. First, HS data usually have hundreds of spectral bands, which contain much more detailed spectral information than MS data [12–19]. Thus, conventional pan-sharpening algorithms may present severe spectral distortion. Second, HS and MS/PAN data are acquired by different platforms in general. The differences between the spectral responses and resolutions, atmospheric and sensor conditions severely affect the data fusion performance. Last but not least, unlike MS data, HS data are normally acquired in a much wider spectral range than the PAN data. The difficulty that arises consists in defining fusion models that yield good results in the part of the HS spectral range that is not covered by the PAN data.

For these reasons, more researches have been conducted on the feature- and decision-level fusions for HS and PAN images. In such schemes, feature extraction is first conducted on both HS and PAN images separately [20]. And then, the extracted features are processed by different orders. The quality of features impacts the classification results to a great extent so that most researches focus on the feature extraction methods. Due to the high spatial resolution, spatial features of PAN images include the texture features described by Gabor filters, gray-level co-occurrence matrix, etc., shape features described by pixel shape index, length and width indices, etc., edge features described by high-pass filters [21], etc. Each of them can be integrated with the spectral features of HS image, thereby improving the classification results through feature- or decision-level fusion algorithms [22–25]. In the past decade, the object-oriented classification methods attract considerable concerns because of its wonderful performance in eliminating the noise and speckles on the thematic maps. Based on image segmentation techniques [26], object-oriented methods divide PAN images into thousands of continuous and non-overlap regions. And classification, thus, will be carried on object level rather than pixel level, which ensures the continuity of land-covers. Inspired by image segmentation, some other works focus on investigating collaborative spectral-spatial feature classification for HS image to address the lack of PAN images [27–31].

In summary, the main idea of collaborative classification for HS and PAN images is to fully explore the abundant spectral information of HS image, as well as the high-resolution property of PAN images, so as to overcome the defects of either single image type.

### *18.1.2 Hyperspectral and Multispectral Images*

In many earth observation applications, a potential improvement of the spatial resolution of HS images is required. However, due to the deficiency of spectral information of high-resolution PAN image, pixel-level fusion of HS and PAN images, i.e. HS image sharpening by PAN image, always leads to a large spectral distortion of the enhanced HS image. Therefore, the fusion of HS and MS data has a possibility to produce a new data that contributes to the accurate identification and classification of an area observed at a fine spatial and spectral resolution. For this reason, fusion of HS data with MS data, known as hyper-sharpening, received a considerable interest in recent years.

The earliest related algorithm dates back to the usage of wavelet transform. Its performance highly depended on the spectral resampling method. Working in the wavelet domain, the maximum a posteriori (MAP) estimation method shows noise resistance property [32]. Furthermore, incorporating principal component analysis into the method of MAP can greatly reduce the computational cost and show an improvement in noise tolerance [33]. Apart from those MAP-based methods, some researchers focus on the generalization of MS pan-sharpening algorithms to solve HS and MS fusion issues and make substantial successes [34]. Some representative methods include sparse spatio-spectral representation method, coupled sparse tensor factorization-based fusion method, deep convolutional neural networks [35–37], and "HySure" method, referring to HS super-resolution [38, 39], which formulates data fusion as a convex optimization problem and uses a form of vector total variation-based regularization.

In the past several years, another category referring to the matrix factorization-based techniques attracted increasing attentions, because of their easy implementation, high efficiency, and fewer parameters. In such techniques, the HS and MS images are decomposed into some basis and coefficient matrices. The fused HS data is subsequently obtained by multiplying the enhanced basis and coefficient matrices. A representative method is the coupled NFM method [40], which alternately unmixes the low-resolution HS data and high-resolution MS data. The algorithm yields higher image quality over conventional methods although it is time-consuming due to the large number of NMF steps. Other similar algorithms include the dictionary-pair learning method [41], the spectral-constrained NMF method [42], and joint-criterion NMF-based (JNMF) hyper-sharpening algorithm [43].

Currently, owing to the insufficiency of data sources and imaging devices, most of the researches use synthesized data to conduct experiments and test algorithms. Massive works show that for synthetic data, existing algorithms have demonstrated their feasibility and considerable potentials for hyper-sharpening tasks [44, 45]. Nonetheless, it is worth to note that for real remote-sensing data, the difference between different remote-sensing systems is much more complicated, even if they cover the same spectral range. The different spectral resolutions and the potential radiance difference caused by different illumination, atmospheric, or sensor conditions, etc., make it difficult to describe the spectral response relations accurately. Hence, the hyper-sharpening task still has much potential to be further investigated.

### *18.1.3 LWIR Hyperspectral and Visible Images*

LWIR (also called thermal infrared, TIR) wavelength general refers to a spectral range of 8–14 μm. Many ground objects that contain oxide, sulfate, silicate, etc., have apparent difference of emissive features in the LWIR wavelength range, thus creating a unique "spectral signature" of a material, which can be used to distinguish from each other.

Unlike the visible imagers, LWIR sensors collect the thermal radiation of ground objects, which is determined by land surface temperature and emissivity, rather than the reflected sunlight, thus offering a day/night capability to sense the relatively unique material signature. LWIR HS imaging makes use of this for the purpose of classification or other applications. LWIR HS imaging is an all-weather, full-area, noncontact, and nondestructive technology. Hence, it has been successfully applied in the indoor and outdoor thermographic inspection of the defects of solid materials and building structures in the engineering scale. Besides, TIR sensors are more robust to poor lighting or lighting changes and also to dust and fog. Up until now, LWIR HS data could be acquired by several sensors [46], such as the Interferometer Atmospheric Sounding Instrument (IASI) on the METOP-A satellite, the Cross-track Infrared Sounder on the NPOESS system, and so on. Images from thermal sensors are generally characterized by anomalous pixels. Some of them appear as dark or saturated pixels, while others show intermittent response variations, thereby leading to significant noise and low definition. The severity of anomalous behavior strongly depends on the sensor or environmental conditions and the acquisition modality [47].

Compared with the TIR data, the visible image collects the reflected sunlight within the visible wavelength ranges (0.4–0.7 μm), thus providing an accordant imagery with human vision. Visual color image usually presents much more spatial details and clarity. Even small ground objects such as houses, single trees, and shrubs can be observed. Nevertheless, the less spectral information leads to a lack of the ability to discriminate some ground objects that have similar colors. Besides, the visual data usually have a much higher spatial resolution than LWIR HS data, thereby probably yielding a multi-scale fusion problem.

Apart from the traditional fusion algorithms designed for general HS images, an effective collaborative processing strategy for infrared HS and visible images is the decision-level-based fusion, which employs the state-of-the-art classifiers to acquire the predicted maps for LWIR HS and visible images individually, and then merge the predicted labels or probabilities of both images, according to a proper criterion. Such strategies highly depend to the performance of base classifiers and the fusion criteria. An alternative method for better mining the characteristics of both images is to adopt a hierarchical fusion strategy. In such strategies, the infrared HS and visible images are firstly classified to identify the grass, trees, and man-made objects. Then, the emissivity features of infrared HS image are used to discriminate the different materials of man-made objects, such as asphalt, concrete, granite, and so on. Additionally, a post-processing is usually employed to fully utilize the advantages of high-resolution characteristics of the visible image. Anyway, the main idea of collaborative processing of LWIR HS and visible images is to fully explore the respective features of both images to pursue the better land-cover classification results [48].

## 18.2 Problems and Challenges in Multisource Image Collaborative Processing

In multisource remote-sensing images, different spectral ranges and resolutions usually reflect the various attributes and features about the observed objects from different aspects, thus, taking full use of the complementarity and difference among these varied images will maximize the advantages. Consequently, to fully explore the potential information of multisource images, it is necessary to develop appropriate collaborative processing techniques for various scenarios [49–54]. Nonetheless, there are several critical problems and challenges that we have to face.

One thing that should be noted firstly is the impact of probable differences of imaging conditions. Due to the limitation of current technology, multisource images are mostly acquired by different platforms, e.g. airborne and spaceborne. Different acquisition times or illumination, atmospheric conditions, even the sensors of different performance may cause serious spectral distortion in fusion results, especially for pixel-level-based fusion methods. This problem might be overcome in future, thanks to the fast development of sensor technology. But it should be still carefully considered at the moment.

Second, in HS centered multisource image fusion, the problems of HS image still exist. The small- or unbalanced-sample problem [55], nonlinear feature, "the same object but different spectrum," and "the same spectrum but different objects" phenomena, etc., severely puzzle the further application of multisource images all the time [56].

Last but not least, due to the large data amount, multisource images, however, always suffer from high feature redundancy and even, the curse of dimensionality, which has plagued researchers for a long time. In order to cope with this problem, varieties of feature selection and feature extraction methods have been proposed for the sake of dimension reduction. Unfortunately, few feature selection and extraction methods show commendable robustness for different types of multisource images [57–60].

Therefore, a comprehensive exploration of collaborative classification of multisource images is of great significance in enhancing the ability of data utilization and information acquisition, as well as the extension of the remote-sensing application.

## 18.3 Collaborative Classification of Hyperspectral and High-Resolution Panchromatic Images

### *18.3.1 Methodology*

PAN images, generally, have higher spatial resolution with more details of ground objects, compared with HS images. Combining spatial features of PAN images is believed to offer strengthened capabilities for land-cover classification. Here, we introduce a self-learning approach based on image segmentation and semi-supervised learning for collaborative classification of HS and PAN images [1–2]. The self-learning algorithm is comprised of two steps. In the first step, the learning algorithm itself predicts the candidates by utilizing the output of a spectral-based classifier and the spatial-spectral information of labeled samples through image segmentation technique. The second step is an active learning (AL)-like algorithm, i.e. the learning algorithm automatically selects the samples with the highest uncertainty from the candidate set. In this way, the learning

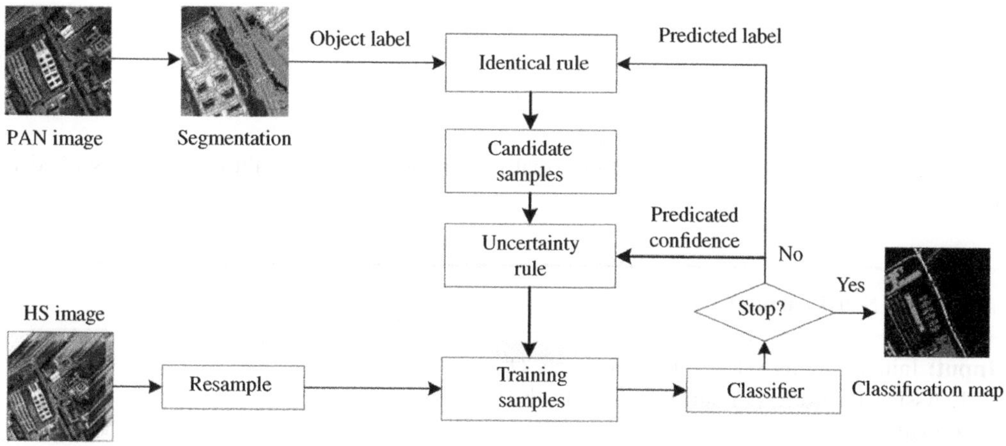


**Figure 18.1** Main procedures of the proposed self-learning-based synergetic classification approach.

machine can select the most useful samples as new training samples automatically and get their class labels by itself at the same time. As a result, no extra cost is required for labeling the selected samples, which realizes the goal of "self-learning." The scheme of self-learning strategy can be described by the following flowchart (Figure 18.1).

The self-learning strategy is based on the assumption that under the optimal segmentation scale, all pixels belonging to a same object have an identical class label. Thus, for a labeled sample, the pixels that share the same object can be labeled with high confidence as belonging to the same class with this labeled sample. Here we call it the object label. The self-learning strategy considers the consistency between the object labels and predicted labels of the classifier, thereby determining which pixels ought to be new training samples in the learning procedure. In this way, this algorithm integrates the spatial information and spectral features of labeled samples to construct new candidate set without human interaction.

Suppose $X^L = \{(x_1^L, y_1), (x_2^L, y_2), \ldots, (x_n^L, y_n)\}, y_k \in \{1, 2, \ldots, C\}$ denotes a set of $n$ labeled samples, in which every labeled sample $x_k^L, k \in \{1, 2, \ldots, n\}$ belongs to an individual object. $C$ is the number of classes. $O_k^U$ denotes the set of unlabeled samples inside the object where $x_k^L$ locates. $O^U = \bigcup_{k=1}^{n} O_k^U$ is the set of all unlabeled samples locating in those objects. Therefore, for each sample $x_i^U$ in $O^U$, we have

$$\text{If } \forall x_i^U \in O_k^U, \quad \text{then } x_i^U :\rightarrow y_k \tag{18.1}$$

in which $y_k$ is the label of $k$th object. Suppose $x_i^U$ is predicted by the classifier $x_i^U : \rightarrow \tilde{y}_i$, which has been trained by $X^L$. $\tilde{y}$ denotes the predicted labels. Then,

$$\text{If } \tilde{y}_i = y_k, \quad \text{then } (x_i^U, y_k) \rightarrow X^C \tag{18.2}$$

where $X^C$ is the candidate set $X^C = \{x_i^U \mid \tilde{y}_i = y_k\}$.

In other words, for each iteration of self-learning procedure, the candidate set is updated by selecting the samples that have the identical predicted labels with object labels. After each iteration, the training set and unlabeled set will also be updated,

$$
\begin{aligned}
X^L &= X^L \cup X^I \\
O^U &= O^U \setminus X^I \\
X^C &= \emptyset
\end{aligned} \tag{18.3}
$$

in which $X^I \subseteq X^C$ denotes the selected informative samples. The algorithm can be described as follows:

---

**Algorithm Self-Learning**

---

**Input:** Initial training set $X^L$, unlabeled sets $O_k^U$;

1. Assign the corresponding labels $y_k$ to the samples of $O_k^U$;

**Repeat**

2. Train the SVM model using $X^L$;
3. Classify the unlabeled set $O^U$ to obtain the predicted labels using the trained SVM model;
4. Construct the candidate set $X^C$ according to (2) and (3) and optimize the candidate set $X^C$;
5. Select the informative samples $X^I$ from $X^C$ by applying an uncertainty-rule;
6. Update training set and candidate set according to (4) and (6) and remove $X^I$ from $O^U$ according to (5);

**Until:** A stopping criterion is satisfied;

**Output:** SVM classifier retrained by the final training set $X^L$.

---

Figure 18.2 shows the main idea of constructing the candidate set. Figure 18.2(a) represents three objects belonging to different classes, which is obtained by the image segmentation method. Figure 18.2(b) shows the confidence of the spectral-based classifier, e.g. the predicted probabilities of support vector machine (SVM). Then the candidate set can be constructed as shown in Figure 18.2(c), in which the color cells comprise the candidate set. And the new training samples in Figure 18.2(d) can be selected through the uncertainty rule. In some cases, there exist a few of mislabeled samples, which are probably caused by the misclassification or suboptimal segmentation. A simple approach to reduce these samples is to assign a confidence threshold for the predicted probabilities. Obviously, the segmentation scale is quite crucial to the final result, and an over-segmentation would be preferred since it is necessary to ensure that the given samples with different class labels should not locate in a single object.

In order to select the informative samples, different AL methods can be adopted in this self-learning strategy, e.g. margin sampling, breaking ties (BT), modified BTs (MBT), and modified margin sampling (MMS). As an SVM-specific AL strategy, margin sampling heuristic strategy samples the candidates lying within the margin by computing their distance to the separating hyperplane. BT strategy aims at finding the samples with minimum difference between two most possible classes. MBT technique has the ability to promote diversity in the selection process. MMS combines the idea of MBT and margin sampling strategies and aims at dividing the candidate set into several groups according to their hypothetical labels and then applying margin sampling to select the uncertain samples that are closest to the separating hyperplane.

Furthermore, in order to attain a steady result and enhance the classification result with fewer iterations, we can imagine that incorporating the diversity of samples could find more precise decision rules by using as few samples as possible. The main idea of integrating diversity into AL is to select a batch of

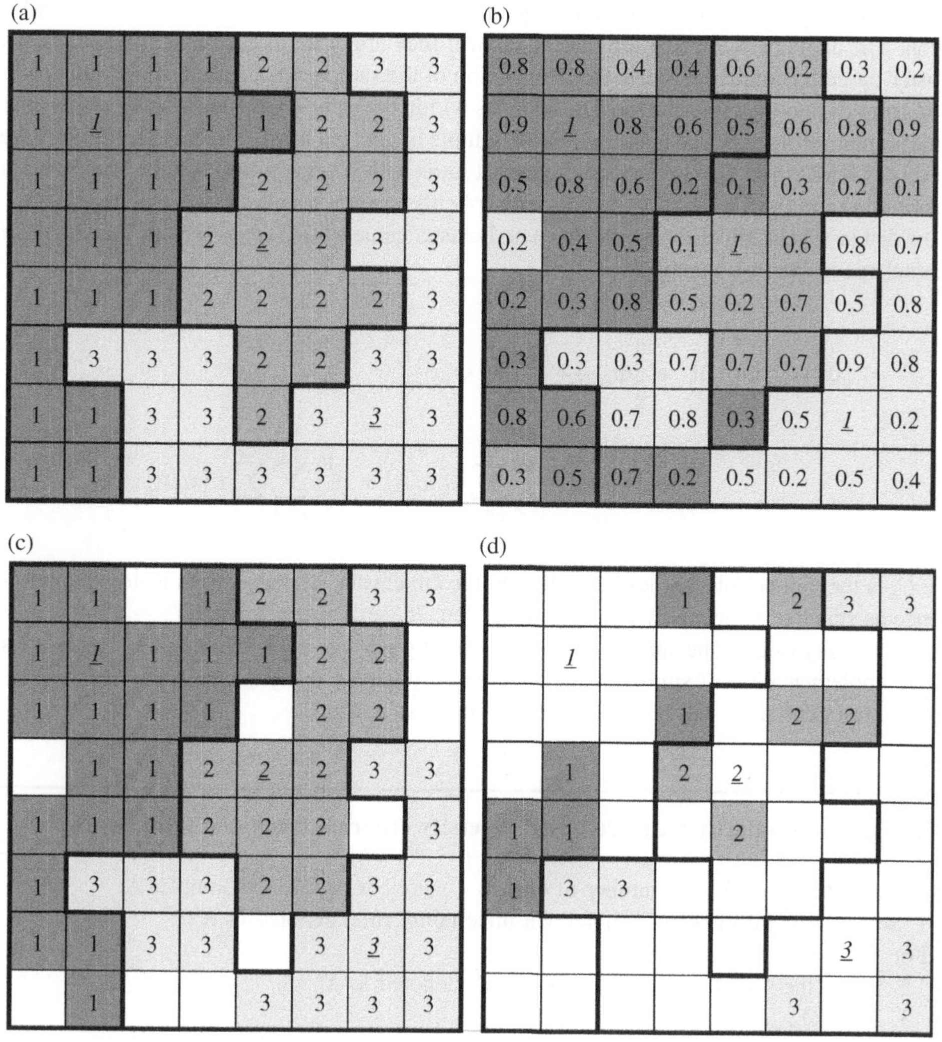

**Figure 18.2** The main idea of self-learning. (The pixels with underlined italic labels denote the labeled samples.) (a) The objects where labeled samples locate, (b) the estimated confidence of classifier, (c) candidates, (d) selected uncertain samples.

unknown samples that have low confidence (i.e., the most uncertain ones) and are diverse from each other simultaneously. There are lots of works focusing on measuring the diversity of samples, which can be divided into two categories, i.e. similarity measure-based criterion and cluster-based criterion.

Typical similarity measure-based diversity criterions include spatial Euclidean distance, spectral angle mapping (SAM), kernel cosine angle (KCA), etc. For instance, spatial Euclidean distance computes the Euclidean distance between the two-dimensional (2-D) coordinates of two candidates in the spatial domain of the image. KCA is a similarity measure in the spectral domain. Unlike SAM, KCA is defined in the kernel space,

$$\mathrm{KCA} = \frac{\Phi(x_i)\cdot\Phi(x_j)}{\left\|\Phi(x_i)\right\|\cdot\left\|\Phi(x_j)\right\|} = \frac{k(x_i, x_j)}{\sqrt{k(x_i, x_i)\cdot k(x_i, x_i)}} \tag{18.4}$$

where $x_i$ and $x_j$ represent two samples, $\Phi(\cdot)$ is a nonlinear mapping function, and $k(\cdot,\cdot)$ is the kernel function. The angle between two samples is small if they are close to each other and vice versa.

Cluster-based techniques evaluate the distribution of the samples in a feature space and group the similar samples into the same cluster. The samples within the same cluster are usually correlated and provide similar information, so a representative sample is selected from each cluster. Kernel $k$-means clustering (KKM) works also in the kernel space. It starts with several initial clusters. Since the cluster centers in the kernel space cannot be expressed explicitly, several pseudocenters will be selected, and then the distance between each sample $\Phi(x_i)$ and cluster center $\Phi(\mu_\upsilon)$ ($\mu_\upsilon$ is the $\upsilon$th pseudocenter) in the kernel space can be computed.

$$\begin{aligned} \|\Phi(x_i)-\Phi(\mu_k)\|^2 &= \left\|\Phi(x_i)-\frac{1}{C_k}\sum_{j=1}^{C_k}\Phi(x_j)\right\|^2 \\ &= k(x_i,x_i)-\frac{2}{C_k}\sum_{j=1}^{C_k}k(x_i,x_j)+\frac{1}{C_k^2}\sum_{j=1}^{C_k}\sum_{l=1}^{C_k}k(x_j,x_l) \end{aligned} \tag{18.5}$$

where $C_k$ is the number of samples belonging to the $k$th cluster. The algorithm is iterated until convergence as standard $k$-means clustering.

Either diversity rule can be applied in the self-learning strategy after obtaining the candidate sample. The implementation of similarity measure-based diversity criteria, e.g. spatial Euclidean distance, KCA, and as such, can be generalized as follows.

---

**Algorithm of similarity measure-based diversity criterion combined self-learning**

---

**Input**: Candidate set $X^{\text{Cand}}$, number of classes $C$, number of selected samples $N$;

1. Select most informative samples via uncertainty rule, denoted as $X^I$,

**For** $c$ = 1-$C$

2. Select the samples of $c$th class from $X^I$, denoted as $X_c^I$;
3. $X_c^L = \emptyset$;
4. If the number of samples in $X_c^I$ is less than $N/C$, then put them into $X_c^L$,

**Otherwise**:

5. Pick out the most uncertain sample from $X_c^I$, and put it into $X_c^L$;

**Repeat**

6. For each sample $\{x_i | x_i \in X_c^I \cap x_i \notin X_c^L\}$, compute the mean value of distance with the samples in $X_c^L$;
7. Pick out the sample that has minimum mean value and put it into $X_c^L$;

**Until**: $X_c^L$ has $N/C$ samples;

**End for**

**Output**: Totally $N$ samples included in $X^L = \cap_{c=1}^{C} X_c^L$.

---

whereas the cluster-based diversity criterion can be implemented via the following algorithm,

**Algorithm of Cluster-Based Diversity Criterion Combined Self-Learning**

**Input**: Candidate set $X^{\text{Cand}}$, number of classes $C$, number of selected samples $N$;
1. Select most informative samples via uncertainty rule, denoted as $X^I$,
**For** $c = 1$ to $C$
2. Select the samples of $c$ th class from $X^I$, denoted as $X_c^I$;
3. $X_c^L = \emptyset$;
4. If the number of samples in $X_c^I$ is less than $N/C$, then put them into $X_c^L$,
**Otherwise**:
**Repeat**
5. Apply kernel k-means clustering to $X_c^I$, the cluster number is $N/C$;
6. Select the most uncertain sample of each cluster, and put it into $X_c^L$;
**Until**: $X_c^L$ has $N/C$ samples;
**End for**
**Output**: Totally $N$ samples included in $X^L = \cap_{c=1}^{C} X_c^L$.

Figure 18.3 shows how the diversity criteria work by taking SVM for example. Figure 18.3(a) shows the primary hyperplane obtained by the training set (the labeled samples). Figure 18.3(b) shows the retrained hyperplane without considering the diversity rule. Two candidates per class are selected by the uncertainty rule. It can be seen that the AL heuristics only concentrate on the samples lying on the boundaries of different classes. Obviously, it is unsatisfactory as some samples have been misclassified. Figure 18.3(c) shows the retrained hyperplane with similarity measures. The first sample is selected according to its confidence, while the subsequent ones are selected according to their distance to the previous ones. Figure 18.3(d) shows the retrained hyperplane with sample clustering. The selected samples are picked from each cluster. By contrast, the diversity-based ALs not only focuses on the uncertainty but also considers the similarity between the selected samples. Therefore, the selected samples scatter around the cluster and can overcome the problem of bias sampling.

The self-learning procedure is iterated until an automatic or manual stop criterion is met, for instance, a maximal iteration is met, or the decision boundary remains unchanged. Generally, the within-class variance keeps invariant when the decision boundary is almost not changed, which can be used to define the stop criterion of self-learning procedure.

### *18.3.2 Experiments*

In this subsection, we present the experimental results performed on four data sets, which have been widely used in ground object classification.

1. Data set of University of Pavia. It was collected over the University of Pavia, Italy, by the Reflective Optics System Imaging Spectrometer (ROSIS) system. It consists of 103 spectral bands after removing the noisy bands, and 610 × 340 pixels for each band with a pixel resolution

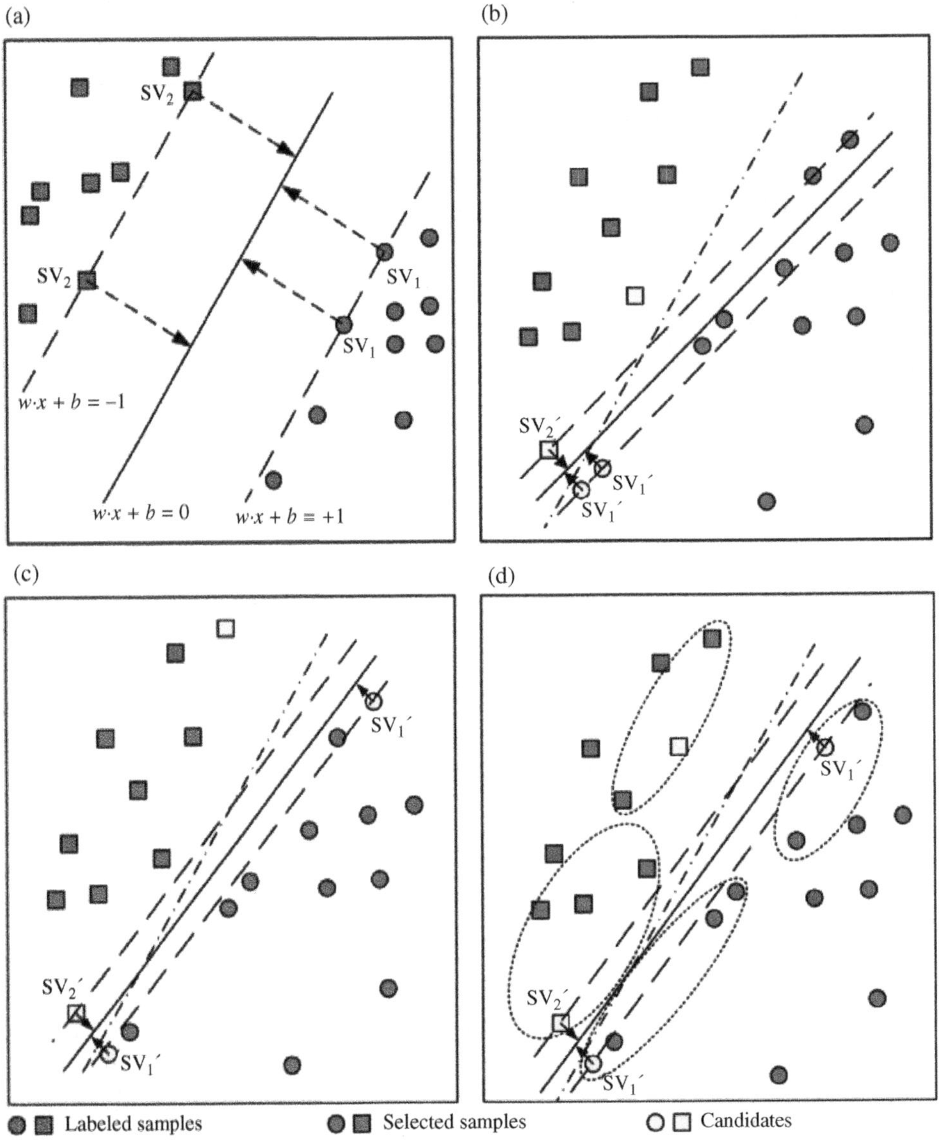


**Figure 18.3** The impact of sample diversity (The dot-dash lines in (b–d) are the hyperplane in (a).). (a) Initial training samples and classification boundary, (b) The selected samples by AL and the retrained classification boundary after one iteration, (c) The selected samples by similarity measurement-based diversity and the retrained classification boundary after one iteration, (d) The selected samples by cluster-based diversity and the retrained classification boundary after one iteration.

of 1.3 m. We compute the average value for band 1–65 to simulate the high-resolution PAN image, and down-sample the HS image by a factor of 2 to obtain the low-resolution HS image with a pixel resolution of 2.6 m. The ground truth image has the same resolution as the original HS image with nine classes.

2. Data set of San Diego. The second data set is a low-altitude AVIRIS HS image of a portion of the North Island of the U.S. Naval Air Station in San Diego, CA, USA. This HS image consists of 126 bands of size 400 × 400 pixels with a spatial resolution of 3.5 m per pixel. We use the average of bands 6–36 to synthesize a PAN image. Then the HS image is subsampled to a lower scale by a factor of 4 (i.e. a resolution of 14 m). The ground truth image has the same resolution as the original HS image with eight classes inside.
3. Data set of Indian Pines. This image is a well-known scene taken in 1992 by the AVIRIS sensor over the Indian Pines region in Northwestern Indiana. It has 144 × 144 pixels and 200 spectral bands with a pixel resolution of 20 m. Nine classes including different categories of crops have been labeled in the ground truth image. Likewise, we simulate the low-resolution HS and high-resolution PAN images by spatial and spectral degradation of the original HS image.
4. Data set of the University of Houston. The last data set includes an HS image provided by the 2013 IEEE GRSS Data Fusion Contest and a co-registered PAN image collected by the WorldView-II satellite. The HS image is collected over the University of Houston campus and the neighboring urban area and consists of 144 bands with a spatial resolution of 2.5 m. For the sake of avoiding the registration error caused by different platforms, a subset of size 640 × 320 is used and subsampled by a factor of 4. The PAN image has a spatial resolution of 0.5 m and is subsampled to the same scale as the original HS image. A ground truth image including 12 classes is available, which also has the same size as the original HS image.

All images have been radiometrically calibrated (shown in Figure 18.4).

To validate the effectiveness of the self-learning strategy, in this sub-section, it is compared with two other semi-supervised learning methods by using SVM with a Gaussian kernel as the classifier. The first one is called neighbor-based semi-supervised self-learning, abbreviated to NBSL, and the second one is called spatial neighborhood information-semi-supervised learning, abbreviated to SNI-SSL. Both methods use the neighborhood information of labeled samples in HS image alone. For convenience, the method introduced in this book is abbreviated to SBSL hereinafter. And the diversity-combined strategies are called SBSL-SPA, SBSL-KCA, and SBSL-KKM, denoting for the spatial Euclidean distance-based, kernel cosine angle-based, and kernel $k$-means clustering-based SBSL, respectively. In all cases, we conduct 10 independent Monte Carlo runs with respect to the labeled training set from the ground truth images. All the graphics display the average values of 10 experiments. The maximum iteration is 20. The segmentation map of the PAN image is obtained by using the full-$\lambda$ schedule.

The classification results are shown in Figure 18.5–18.8. The first point on each curve represents the OA obtained by using the SVM algorithm alone. Apart from the advantage of exploiting unlabeled samples as is shown by all of these SL approaches, the OAs increase as the number of training samples increases and incline to converge after a small number of iterations. This deserves special attention since these samples are generated in a relatively lower-cost, easier, and, more importantly, self-learning fashion, i.e. no extra cost of human labeling or determination is required during the learning procedure.

By comparing the six strategies in these figures, we can see that the SBSL approach is superior to NBSL and SNI-SSL in most cases. This is because compared with NBSL, SBSL utilizes the spatial information provided by the corresponding PAN image and aims at selecting those most informative samples in a single object (usually having a lower spectral similarity with the known sample), while NBSL selects the uncertain samples locating in the neighborhood of the labeled samples (generally having higher spectral similarity). And in most cases, the diversity-based SL approaches can attain higher accuracies. KKM-based SBSL seems to be much more stable than the others.

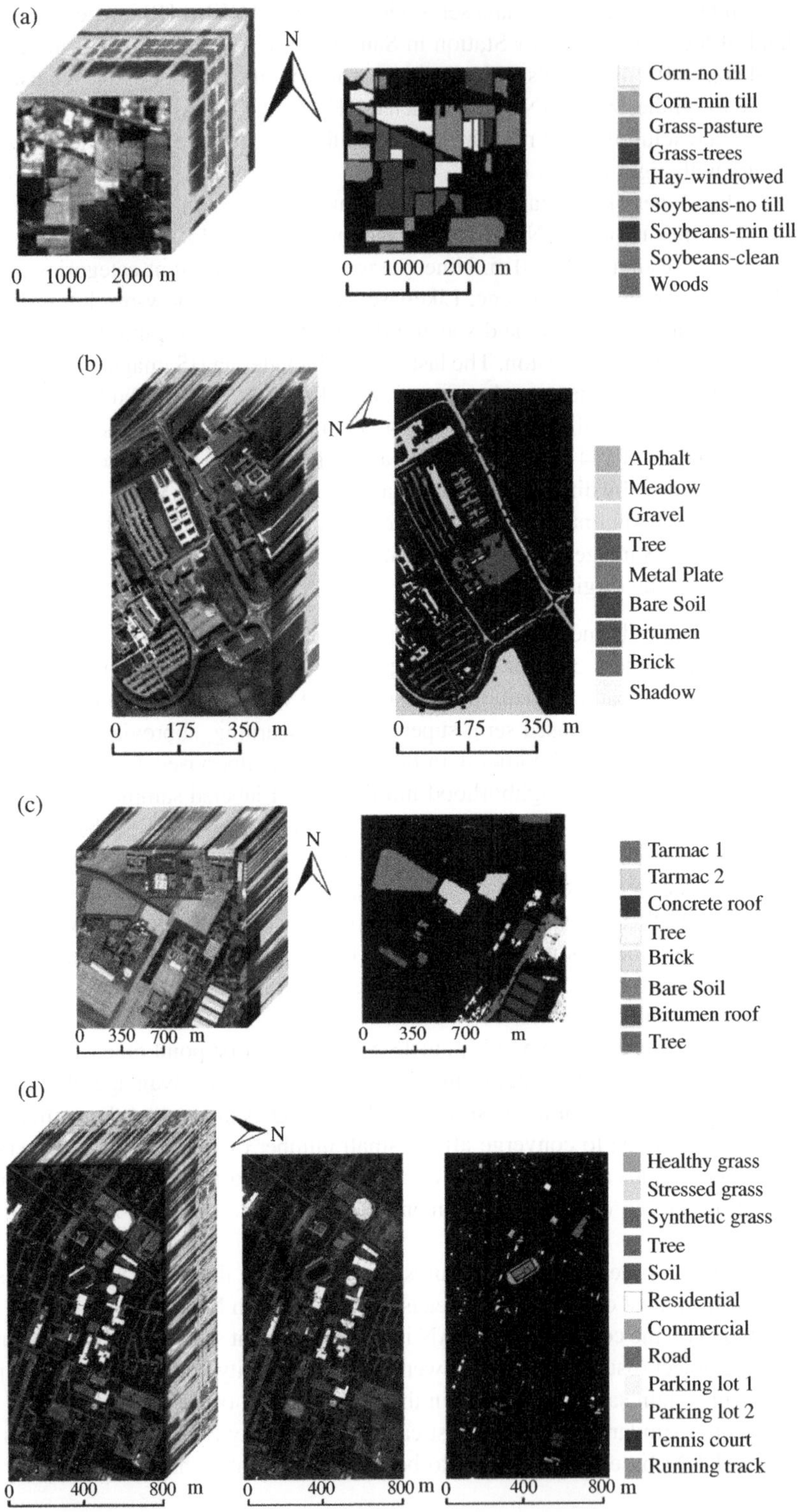


**Figure 18.4** Experimental data sets. (a) Data set of Indian Pines, (b) data set of University of Pavia, (c) data set of San Diego, (d) data set of University of Houston.

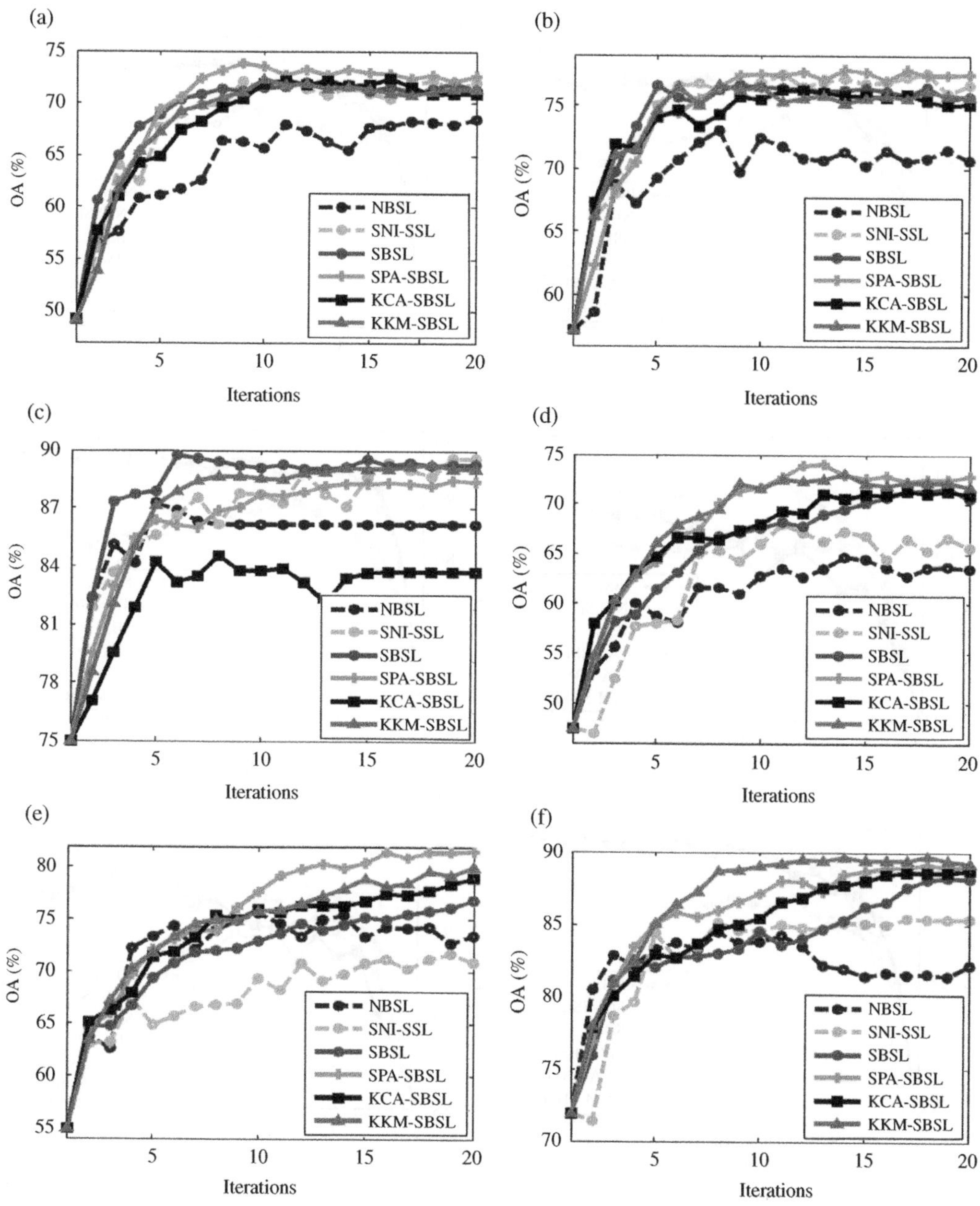


**Figure 18.5** Overall accuracies obtained for the data set of University of Pavia. (The numbers of initial training samples are 3, 5, and 10 per class, respectively, and the selected samples during each iteration are 45, 45, and 90, respectively.) (a) 3 samples per class (MBT), (b) 5 samples per class (MBT), (c) 10 samples per class (MBT), (d) 3 samples per class (MMS), (e) 5 samples per class (MMS), (f) 10 samples per class (MMS).

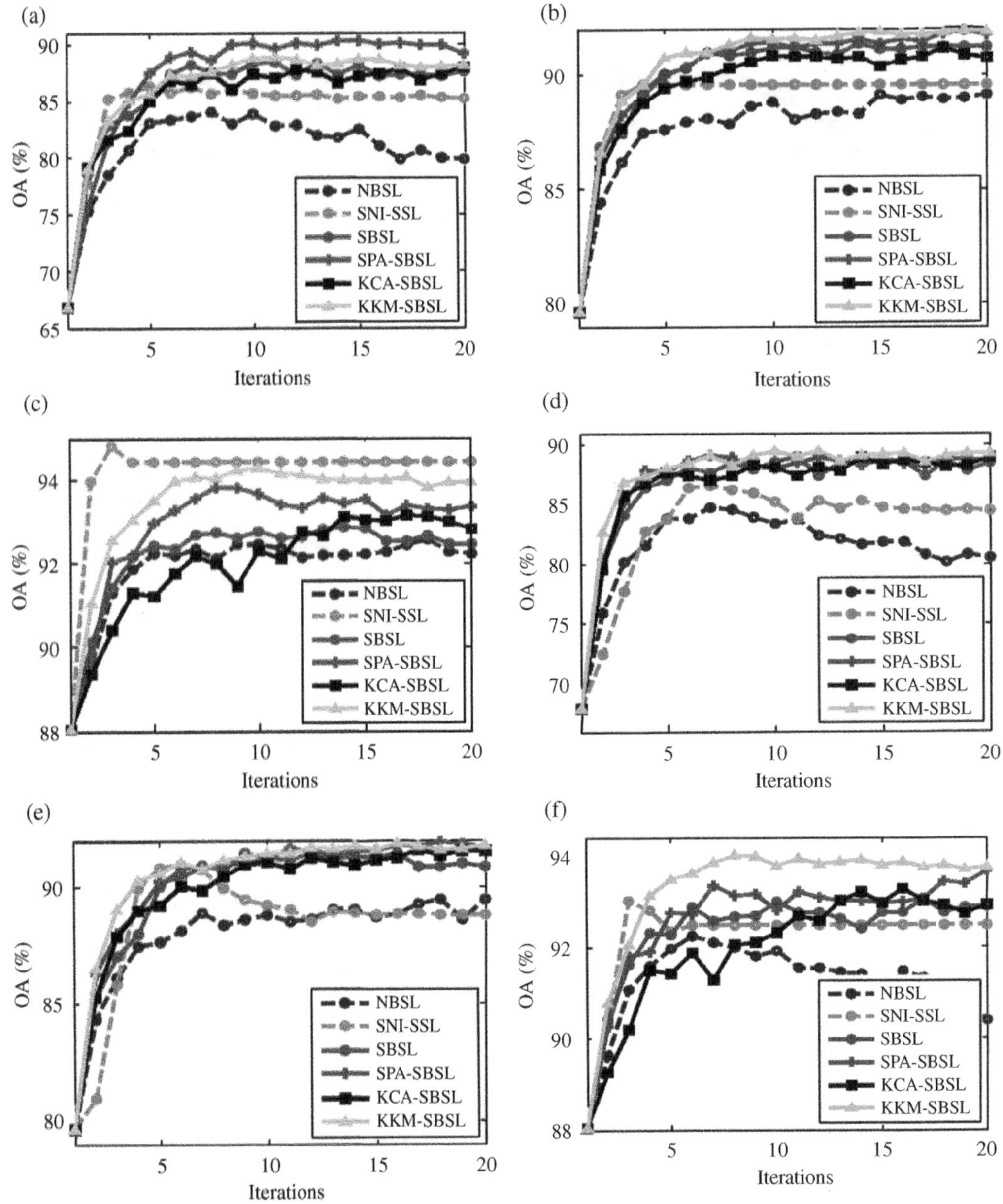


**Figure 18.6** Overall accuracies obtained for the data set of San Diego. (The numbers of initial training samples are 3, 5, and 10 per class, respectively, and the selected samples during each iteration are 40, 40 and 64, respectively.) (a) 3 samples per class (MBT), (b) 5 samples per class (MBT), (c) 10 samples per class (MBT), (d) 3 samples per class (MMS), (e) 5 samples per class (MMS), (f) 10 samples per class (MMS).

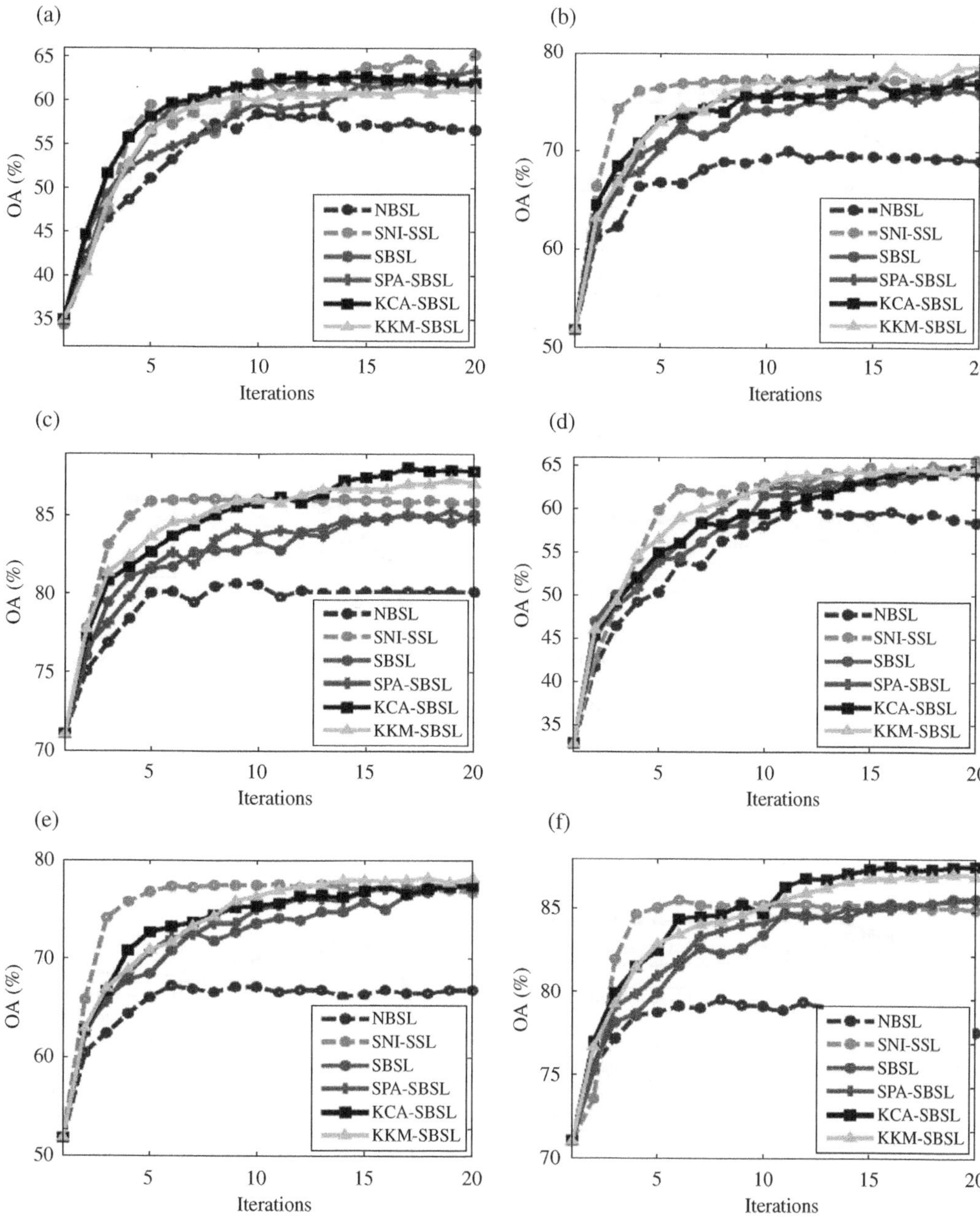


**Figure 18.7** Overall accuracies obtained for the data set of Indian Pines. (The numbers of initial training samples are 3, 5, and 10 per class respectively, and the selected samples during each iteration are 27, 27, and 36, respectively.) (a) 3 samples per class (MBT), (b) 5 samples per class (MBT), (c) 10 samples per class (MBT), (d) 3 samples per class (MMS), (e) 5 samples per class (MMS), (f) 10 samples per class (MMS).

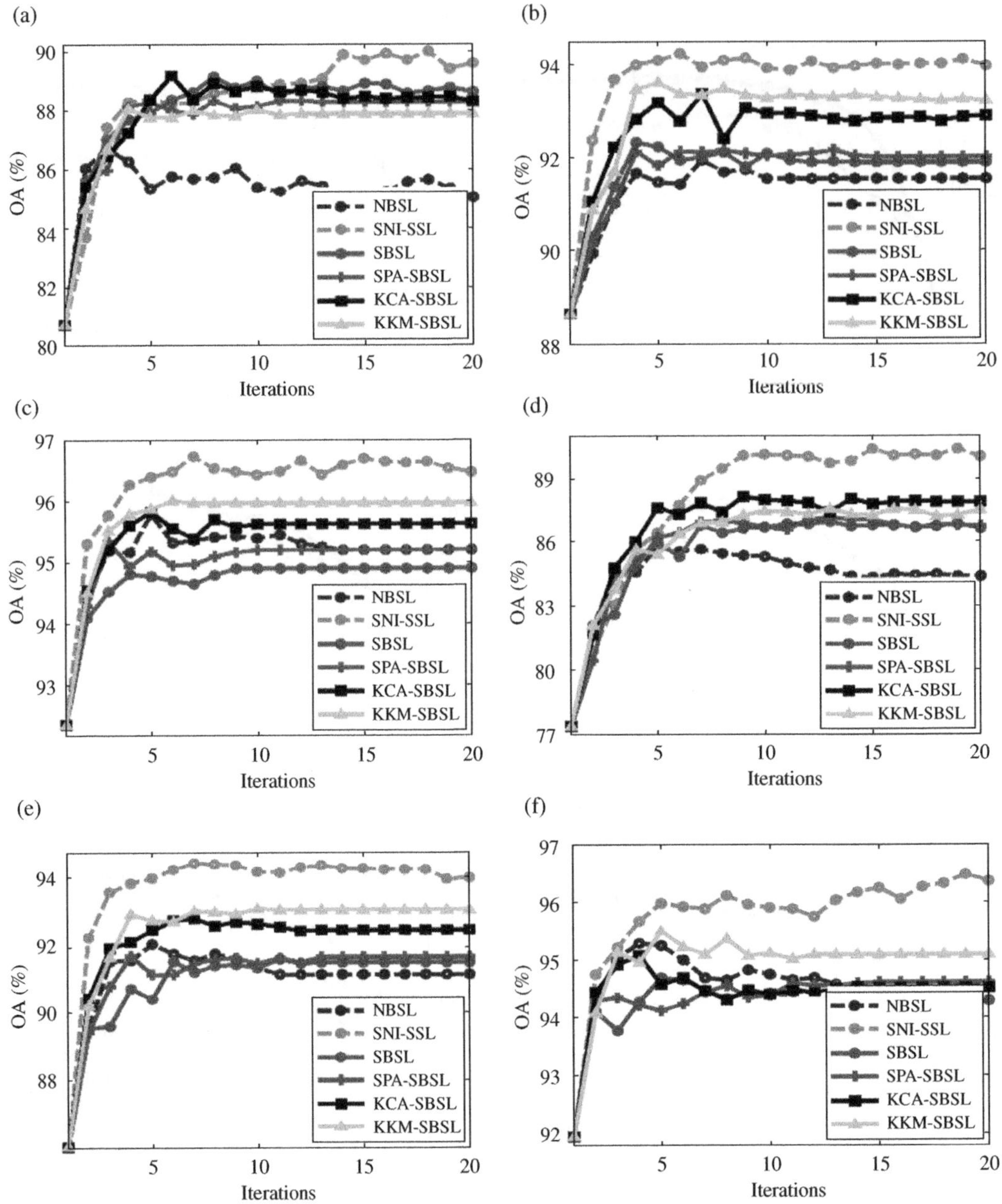


**Figure 18.8** Overall accuracies obtained for the data set of University of Houston. (The numbers of initial training samples are 3, 5, and 10 per class, respectively, and the selected samples during each iteration are 24, 36, and 60, respectively.) (a) 3 samples per class (MBT), (b) 5 samples per class (MBT), (c) 10 samples per class (MBT), (d) 3 samples per class (MMS), (e) 5 samples per class (MMS), (f) 10 samples per class (MMS).

Furthermore, as a matter of fact, due to the complex and dense distribution of ground objects for urban areas, the segments on PAN images usually turn to be small and scattered. In such cases, the SL approaches seem to converge in much fewer iterations; for instance, KKM and KCA converge within only five iterations. This is quite meaningful since it takes a smaller computation cost.

Additionally, it should be noted that the segmentation scale of the PAN image has a rather great impact on the classification results. As is presented above, an improper segmentation scale will sometimes lead to failure for the self-learning procedure. To better understand this point, we conduct a comparison on the University of Pavia Dataset. The segmentation parameters contain a scale level and a merge level, which decide the edge intensity and merge cost.

The comparisons are conducted under different amount of training samples and AL approaches. Figure 18.9(a)–(c) show the segmentation results under different scale and merge levels: (a) scale level = 20, merge level = 60; (b) scale level = 40, merge level = 90; (c) scale level = 60, merge level = 90. Figure 18.9(a) represents a typical case of over-segmentation, in which several objects have been segmented into much more undesired small regions. In this case, the amount of unlabeled samples is limited since only part of the pixels near the labeled samples can be considered as candidates. While Figure 18.9(c) displays a typical case of under-segmentation, in which segments representing different land-covers are incorrectly merged together. Then in such case, the classification result may be unstable

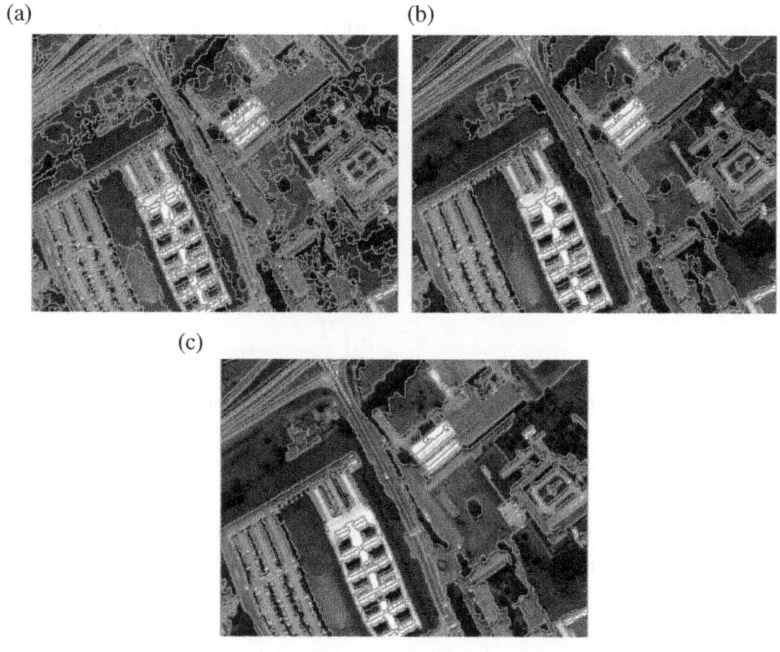


**Figure 18.9** The impact of segmentation scale on overall classification accuracies obtained on the University of Pavia data set. The second row represents the overall accuracies obtained by different sampling strategies (i.e. margin sampling, breaking ties, and modified breaking ties) under 5 labeled samples per class with 90 selected samples per iteration. And the bottom row represents the results under 10 labeled samples per class with 180 selected samples per iteration. (a) Level 1, (b) level 2, (c) level 3, (d) 5 labeled samples, margin sampling, (e) 5 labeled samples, breaking ties, (f) 5 labeled samples, MBT, (g) 10 labeled samples, margin sampling, (h) 10 labeled samples, breaking ties, (i) 10 labeled samples, MBT.

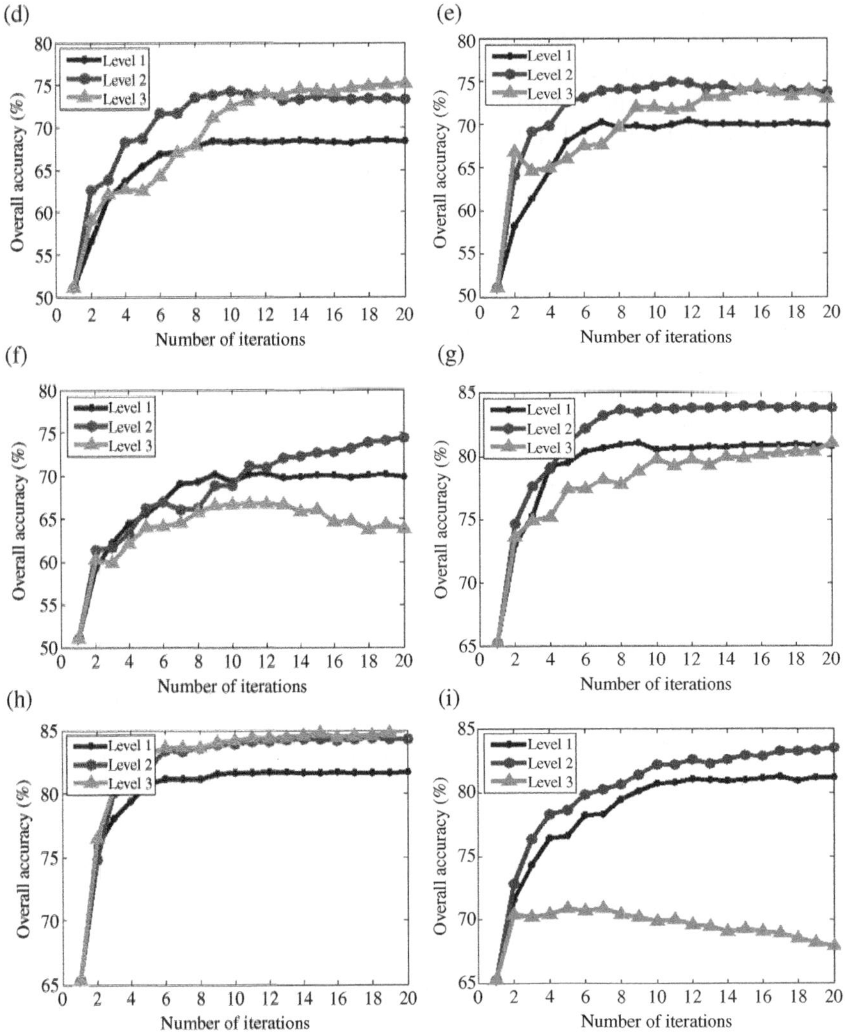


**Figure 18.9** (*Continued*).

due to the possibly mislabeled samples. Figure 18.9(b) displays the segmented image under relatively appropriate parameters, where most segment boundaries in the upper right corner are coherent to real image edges.

Figure 18.9(d)–(i) display part of the results. The horizontal axis represents the number of iterations, which also represents the number of unlabeled samples selected during the learning procedure. And the vertical axis represents the overall accuracy (OA) values. The first point on each curve represents the overall accuracy obtained by using SVM algorithm alone, which is generally affected by the labeled samples (50.08 and 65.29% corresponding to 5 and 10 labeled samples per class). From these figures, it can be seen that the proposed self-learning algorithm can incorporate unlabeled samples in a few objects into the training set, thereby significantly improving the classification result, for instance, nearly 75%/80% after 10 iterations when using 5/10 labeled samples per class, respectively.

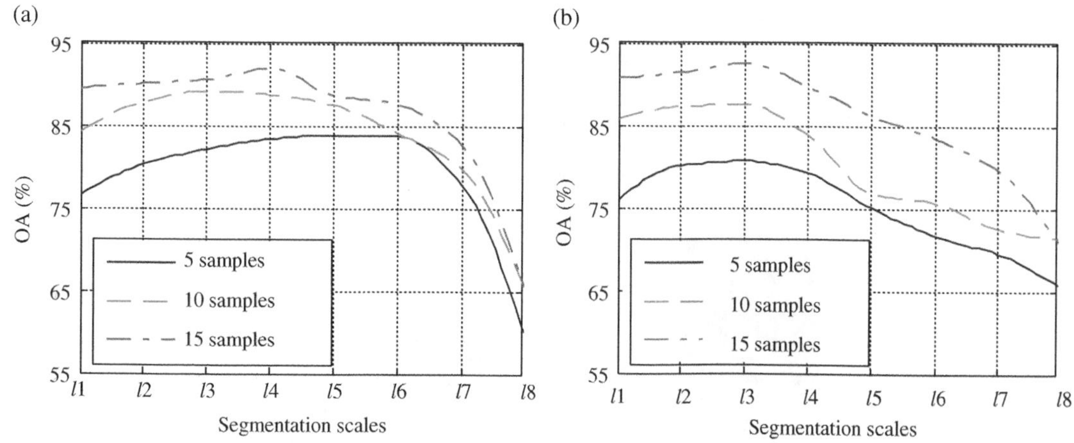


**Figure 18.10** Impact of different segmentation levels on the classification results with different sampling strategies. (a) MBT, (b) MMS.

However, these figures also reveal the impact of segmentation results. We can see that in most cases, level 2 (i.e. when the segmentation levels are selected appropriately) performs better than the other two levels, since it has relatively stable improvement of overall accuracy. Level 1 has lower accuracies compared with level 2 in these figures, which is mainly because of the insufficient unlabeled samples and relatively less information of those samples under the condition of over-segmentation. On the other side, level 3 has quite unsteady results that sometimes it has excellent improvement, e.g. Figure 18.9(h), while sometimes it seems to be meaningless, e.g. Figure 18.9(i). This is possibly because the mislabeled samples in different objects will decrease the performance of self-learning in the case of under-segmentation.

A more intuitive comparison of overall accuracies (OAs) with respect to segmentation levels has been given in Figure 18.10, in which the horizontal axis represents eight segmentation levels from finer scales to coarser scales, and the vertical axis shows the final overall accuracies after the learning processes. The results reveal that the classification results will depend on the segmentation results to a great extent. Obviously, under-segmentation sometimes may lead to the involving of mislabeled samples, which possibly causes the deterioration of the classifier. Therefore, in order to avoid this problem, we prefer to select over-segmentation rather than under-segmentation.

## 18.4 Collaborative Classification of Infrared Hyperspectral and Visible Images

### *18.4.1 Methodology*

In this subsection, we elaborate a decision level-based collaborative classification method for infrared HS and visible (VIS) color images [7]. The main framework is shown in Figure 18.11. First, preprocesses have been conducted on LWIR HS and VIS images. Then for LWIR HS image, we present a semi-supervised feature extraction method called semi-supervised local discriminant analysis ($S^2$LDA) for sparse representation classification. For VIS image, texture features and spectral features have been combined for classification. The second stage is to merge the results of both classifiers to generate a final thematic map.

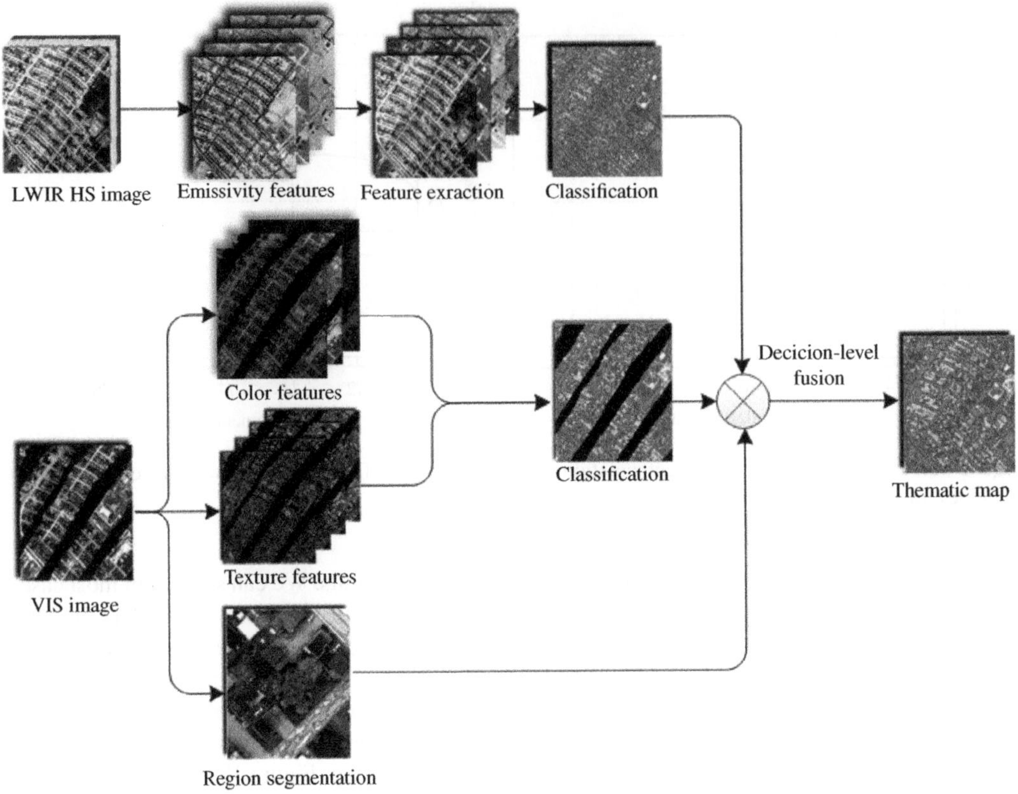


**Figure 18.11** The main procedures of collaborative classification method for LWIR HS and visible images.

Unlike conventional HS image with wavelength ranging in 0.4–2.5 μm, LWIR HS reflects both the emissivity features and temperatures of land-covers. It should be noted that only the emissivity features are determined by the materials of ground objects. For instance, Figure 18.12 shows the emissivity spectra of some typical materials, from which we can see that the different materials usually have unique emissivity features that can be used to distinguish them easily.

Therefore, it is necessary to extract the emissivity data from the detected radiance data first. This procedure is called temperature emissivity separation (TES). For LWIR HS data, TES method is usually underdetermined. In order to solve this underdetermined problem, several TES algorithms have been proposed. Here, we introduce the normalized emissivity method (NEM).

For the radiance data of LWIR image, we have

$$L(\lambda, T) = \tau(\lambda)\left[\varepsilon B(\lambda, T) + (1-\varepsilon)L_{\downarrow}(\lambda)\right] + L_{\uparrow}(\lambda) \tag{18.6}$$

where $L_{\downarrow}$, $L_{\uparrow}$, and $\tau$ denote for downward atmospheric radiance, upward atmospheric radiance, and atmospheric transmittance of LWIR bands, respectively. Suppose for the LWIR HS image, there are some objects that have a pre-defined maximum emissivity $\varepsilon_{\max}$, thus, according to the Planck's law and Kirchhoff's law,

$$T_{\lambda} = \frac{c_2}{\lambda \ln\left(\frac{\tau\varepsilon_{\max} c_1}{\lambda^5\pi\left[L(\lambda, T) - L_{\uparrow} - \tau(1-\varepsilon_{\max})L_{\downarrow}\right]} + 1\right)} \tag{18.7}$$

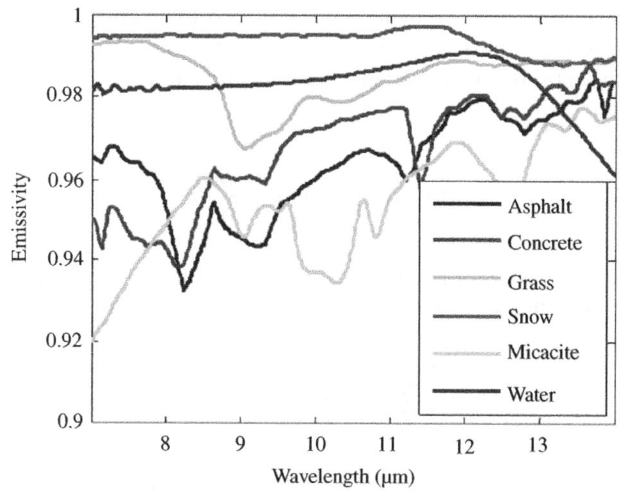


**Figure 18.12** Emissivities of different land-covers.

where $c_1 = 3.7418 \times 10^{-16}\ \mathrm{W \cdot m^2}$, $c_2 = 14\,388 \times 10^{-6}\ \mathrm{m \cdot K}$. By setting $T = \max(T_\lambda)$, we have,

$$\varepsilon(\lambda) = \frac{L(\lambda,T) - L_\uparrow - \tau L_\downarrow}{\tau\left[B(\lambda,T) - L_\downarrow\right]} = \frac{L(\lambda,T) - L_\uparrow - \tau L_\downarrow}{\tau c_1 / \left[\pi\lambda^5\left(e^{c_2/\lambda T} - 1\right)\right] - \tau L_\downarrow} \tag{18.8}$$

Generally, by omitting the influence of atmosphere, Eq. (4-2) can be written as

$$T_\lambda = \frac{c_2}{\lambda \ln\left(\frac{\varepsilon_{\max} c_1}{\lambda^5 \pi L(\lambda)} + 1\right)} \tag{18.9}$$

And then,

$$\varepsilon(\lambda) = \frac{L(\lambda)}{B(\lambda,T)} = \frac{\pi L(\lambda)\lambda^5\left(e^{c_2/\lambda T} - 1\right)}{c_1} \tag{18.10}$$

Once the emissivity data of land-covers are obtained, they can be used for object recognition and classification instead of the original HS data. Nevertheless, since the emissivity data have same dimensions with the original HS data, a feature extraction procedure should be conducted for dimension reduction.

Semi-supervised local discriminant analysis is a semi-supervised feature extraction (FE) method, which combines the advantages of the supervised method, local fisher discriminant analysis (LFDA), and the unsupervised method, neighborhood preserving embedding (NPE). By making full use of the available samples, the proposed method attempts to discover the global discriminative structure of the data, meanwhile preserving the local neighbor spatial information.

First let $X = \{(x_i, y_i)\}_{i=1}^{n},\ y_i \in \{1, 2, \ldots, C\}$ denote the labeled samples, in which $C$ is the number of classes, and $z_i = W^{\mathrm{T}} \cdot x_i$ denotes the projected sample in the low-dimensional feature space using projection matrix $W$.

Based on the classical supervised FE method, local discriminant analysis (LDA), LFDA modifies the expression of the between-class covariance and the within-class covariance, but still seeks the directions of which the ratio of two covariance matrices is maximized. The optimization problem of LFDA is as follows:

$$w_{\text{LFDA}} = \arg\max_{w} \frac{w^{\text{T}} S_b w}{w^{\text{T}} S_w w} \tag{18.11}$$

where $w$ is the transformation vector and $w^T$ denotes the transpose of $w$. $S_b$ and $S_w$ are the between-class covariance and the within-class covariance matrices, respectively, which can be expressed as

$$\begin{aligned} S_b &= \frac{1}{2}\sum_{i,j=1}^{n} V_{b(i,j)}\left(x_i - x_j\right)\left(x_i - x_j\right)^T \\ S_w &= \frac{1}{2}\sum_{i,j=1}^{n} V_{w(i,j)}\left(x_i - x_j\right)\left(x_i - x_j\right)^T \end{aligned} \tag{18.12}$$

$$\begin{aligned} V_{b(i,j)} &= \begin{cases} A_{i,j}(1/n - 1/n_l) & y_i = y_j = l \\ 1/n & y_i \neq y_j \end{cases} \\ V_{w(i,j)} &= \begin{cases} A_{i,j}/n_l & y_i = y_j = l \\ 0 & y_i \neq y_j \end{cases} \end{aligned} \tag{18.13}$$

in which $n_l$ is the number of samples in the $l$th class. $A_{i,\,j}$ denotes the similarity of the sample pair $x_i$ and $x_j$,

$$A_{i,j} = e^{-\left\|x_i - x_j\right\|^2/t} \tag{18.14}$$

The solution to the optimization problem is equivalent to solving the following generalized eigenvalue problem

$$S_b w = \lambda S_w w \tag{18.15}$$

LFDA combines the ideas of LDA and locality preserving projections (LPP). It can preserve the local discriminative information of data effectively. However, LFDA only utilizes the discriminative information and neighbor structure information of labeled samples, while loses the structure information of unknown samples.

NPE is a non-supervised FE method, which seeks a projection that preserves neighboring data structure in the low-dimensional feature space. Given a data set $X$, NPE first constructs an adjacency graph on the data set as locally linear embedding (LLE) does. It can characterize the local geometry of these patches by linear coefficients that reconstruct each data point from its neighbors. Reconstruction errors are measured by the cost function,

$$\min \sum_i \left\| x_i - \sum_j Q_{ij} x_j \right\|, \quad \text{s.t.} \sum_j Q_{ij} = 1, j = 1, 2, \ldots, k \tag{18.16}$$

where $Q$ is the reconstruction weights matrix. NPE supposes that if the data point $x_i$ in space $R^D$ can be reconstructed by $Q$, then the corresponding point $y_i$ in the mapping space $R^d$ can also be reconstructed by $Q$. Therefore, the projection matrix $P$ can be obtained by solving the following minimization problem:

$$\min \sum_i \left\| z_i - \sum_j Q_{ij} z_j \right\|, \quad \text{s.t. } z_i^T z_i = I \tag{18.17}$$

The projection matrix of NPE can be optimized as follows:

$$w_{\text{NPE}} = \arg\max_w \frac{w^T X X^T w}{w^T X M X^T w} \tag{18.18}$$

where $M = (I - Q)^T (I - Q)$, $I$ is an identity matrix, and $X$ is the original data.

The solution is equivalent to solving the following generalized eigenvalue problem:

$$XX^T w = \lambda XMX^T w \tag{18.19}$$

Contrary to LFDA, NPE considers the structure information of unlabeled samples while it loses the discriminative information of labeled information.

By combining LFDA and NPE, $S^2$LDA can make full use of both labeled samples and unlabeled samples in a novel and efficient way. Suppose the training data set $X = [X_{\text{labeled}}, X_{\text{unlabeled}}]$, and the labeled and unlabeled set can be expressed as

$$\begin{aligned} X_{\text{labeled}} &= \{(x_i, y_i)\}_{i=1}^{n}, y_i \in \{1, 2, \ldots, C\} \\ X_{\text{unlabeled}} &= \{x_i\}_{i=n+1}^{N} \end{aligned} \tag{18.20}$$

where $C$ is the number of classes. Without loss of generality, we normalize the original data and assume that the labeled samples are ordered according to their labels, with data matrix of $l$th class $X^{(l)} = \left[x_1^{(l)}, x_2^{(l)}, \ldots, x_{n_k}^{(l)}\right]$, where $x_i^{(l)}$ is the $i$th sample in $l$ th class. Then we can rewrite the expression of the between-class covariance and the within-class covariance in formula (18.12) as follows:

$$S_m = S_b + S_w = X_{\text{labeled}} L_m X_{\text{labeled}}^T \tag{18.21}$$

where $L_m = D_m - V_m$, $V_m = V_b + V_w$.

$$S_w = X_{\text{labeled}} L_w X_{\text{labeled}}^T \tag{18.22}$$

where $L_w = D_w - V_w$. $D_m$ and $D_w$ are diagonal matrices, in which the diagonal elements are the sum of corresponding lines in matrices $V_m$ and $V_w$, respectively:

$$D_m = \sum_{j=1}^{n} V_{m(i,j)}, \quad D_w = \sum_{j=1}^{n} V_{w(i,j)} \tag{18.23}$$

Therefore, if we define $S_b = X_{\text{labeled}} L_b X_{\text{labeled}}^T$, $L_b = L_m - L_w$, and the problem (18.19) can be rewritten as follows:

$$XL_bX^Tw = \lambda XL_wX^Tw \tag{18.24}$$

from which we can see that the eigenvalue problem of LFDA has a similar form with NPE. Then we define the following matrices:

$$\begin{aligned} P &= \begin{bmatrix} (L_b)_{n\times n} & 0 \\ 0 & 0_{(N-n)\times(N-n)} \end{bmatrix} \quad Q = \begin{bmatrix} (L_w)_{n\times n} & 0 \\ 0 & 0_{(N-n)\times(N-n)} \end{bmatrix} \\ I &= \begin{bmatrix} 0_{n\times n} & 0 \\ 0 & I_{(N-n)\times(N-n)} \end{bmatrix} \quad M = \begin{bmatrix} 0_{n\times n} & 0 \\ 0 & M_{(N-n)\times(N-n)} \end{bmatrix} \end{aligned} \tag{18.25}$$

Now we can reformulate the generalized eigenvalue problems of LFDA and NPE as

$$\begin{aligned} XPX^Tw &= \lambda XQX^Tw \\ XIX^Tw &= \lambda XMX^Tw \end{aligned} \tag{18.26}$$

The main idea of our approach is to combine the two formulas in (18.26) together. Therefore, we get the generalized eigenvalue problem of the proposed S$^2$LDA as

$$X(P+I)X^Tw = \lambda X(Q+M)X^Tw \tag{18.27}$$

And the covariance matrices $S_b$ and $S_w$ in (18.5) can be fixed as follows:

$$\begin{aligned} S^- &= X(P+I)X^T \\ &= [X_{\text{labeled}}, X_{\text{unlabeled}}] \begin{bmatrix} (L_b)_{n\times n} & 0 \\ 0 & I_{(N-n)\times(N-n)} \end{bmatrix} [X_{\text{labeled}}, X_{\text{unlabeled}}]^T \end{aligned} \tag{18.28}$$

$$\begin{aligned} S_- &= X(Q+M)X^T \\ &= [X_{\text{labeled}}, X_{\text{unlabeled}}] \begin{bmatrix} (L_w)_{n\times n} & 0 \\ 0 & M_{(N-n)\times(N-n)} \end{bmatrix} [X_{\text{labeled}}, X_{\text{unlabeled}}]^T \end{aligned} \tag{18.29}$$

Matrices $S^-$ and $S_-$ are both symmetric and positive semi-definite. By solving the eigenvalue problem above, we can get the projection matrix. Feature extracted by the S$^2$LDA method can preserve both the class discriminative and local neighbor spatial information. The extracted features are used to be classified by sparse representation (SR) classification method. Sparse representation classification model aims at solving the following optimization problem,

$$\hat{\alpha} = \arg\min_{\alpha} \frac{1}{2}\|x - D\alpha\|_2^2, \quad \text{s.t. } \|\alpha\|_0 \le K \tag{18.30}$$

where $\alpha$ is the coefficient to be solved, $D$ denotes the over-complete dictionary composed of training samples. $x$ is the unknown sample. $K$ is the predefined sparse degree. The problem can be solved by orthogonal matching pursuit (OMP) method. And the class of $x$ is determined by minimizing the reconstruction error,

$$\text{Class}(x) = \arg\min_{i=1,2,\ldots,c} \|x - D_i\hat{\alpha}_i\|_2 \tag{18.31}$$

For visible color image, the spectral features include the R/G/B values of each pixel, which can be arranged by

$$F_{\text{spectral}} = \left[x_i^r, x_i^g, x_i^b\right]^T \tag{18.32}$$

Nevertheless, because of the "the same spectrum but different objects" problem, the spatial texture features have been extracted to combine with the spectral features. Texture features of high-resolution remote-sensing image often play an important role in distinguishing those ground objects with similar spectral features. Numerous methods have been proposed to extract the texture features from image regions. The gray-level co-occurrence matrix (GLCM) provides a measure of the probability of occurrence of two gray levels separated by a given distance in a given direction. A 3-D Gabor filter can capture the specific orientation, scale, and wavelength-dependent properties of HS image and provide an efficient means of texture features representation. The conventional 2-D Gabor wavelet has already been proved superior for representing the texture features in the natural scene and aerial photographs. Hence, we apply a 2-D Gabor filter on the intensity image, which is derived from the RGB bands.

$$G_{s,d}(x,y) = \frac{\|k\|}{\delta^2} \cdot e^{-\frac{\|k\|^2 \cdot (x^2+y^2)}{2\delta^2}} \cdot \left( e^{ik\cdot\begin{pmatrix} x \\ y \end{pmatrix}} - e^{-\frac{\delta^2}{2}} \right) \tag{18.33}$$

where

$$k = \frac{\pi}{2^{s+1}} \cdot e^{i\cdot(\pi d/8)}, \quad \delta = 2\pi \tag{18.34}$$

The Gabor filter function can be expressed as (18.33). The parameters $s = [0, 1, 2, 3]$ and $d = [0, 1, 2, 3, 4, 5, 6, 7]$ determine the four scales and eight directions of the Gabor function, respectively. The Gabor filter is applied on intensity component,

$$\begin{aligned} F_{s,d} &= G_{s,d} * I \\ I &= (R + G + B)/3 \end{aligned} \tag{18.35}$$

Thus, the texture features of each pixel can be described as follows:

$$F_{\text{texture}} = \left[x_i^{0,0}, x_i^{1,2}, \ldots, x_i^{3,7}\right]^T \tag{18.36}$$

in which $x_i^{s,d}$ is the texture feature of $i$th sample on $s$th scale and $d$th direction.

Finally, the color and texture features of the VIS image are combined and classified by the joint sparse representation (JSR) model. Suppose there are totally $S$-type features, the JSR classification method can be described as follows:

$$\hat{A}^s = \arg\min_{\alpha^s} \frac{1}{2}\sum_{s=1}^{S} \|x^s - D^s\alpha^s\|_2^2, \quad \text{s.t. } \|A\|_{\text{row},0} \leq K \tag{18.37}$$

where $A = [\alpha^1, \alpha^2, \ldots, \alpha^S]$ is the row-sparse coefficient matrix to be solved, $D^s$ is the dictionary corresponding to the $s$th feature type. $K$ is the predefined sparse degree. Once the coefficient matrix $\hat{A}$ is solved, the class of sample $x$ can be determined also by minimizing the reconstruction error,

$$\text{Class}(x) = \arg\min_{i=1,2,\ldots,c} \sum_{s=1}^{S} \|x^s - D_i^s\hat{\alpha}_i^s\|_2 \tag{18.38}$$

The JSR model can be solved by joint OMP method. Figure 18.13 shows a summary of JSR algorithm.

To numerically integrate the JSR classification results of the extracted features for LWIR HS and VIS images, multi-classifier fusion should be taken on the decision level. Generally, decision-level fusion can be defined as the process of integrating information from individual data sources to output a more credible result after each data source has undergone a preliminary classification. Yet varieties of decision-level fusion techniques have been proposed, e.g. majority voting, behavior knowledge space, Bayesian decision theory, and D–S evidence theory, fuzzy integral. Opinion pool (OP) is also named as consensus rule, which has several forms, e.g. linear opinion pools (LIP) and logarithmic opinion pools (LGP). D–S theory and fuzzy Integral have the ability to represent and synthesize uncertain information, since they can combine objective evidence for a hypothesis with the system's expectation of the importance of that evidence to the hypothesis.

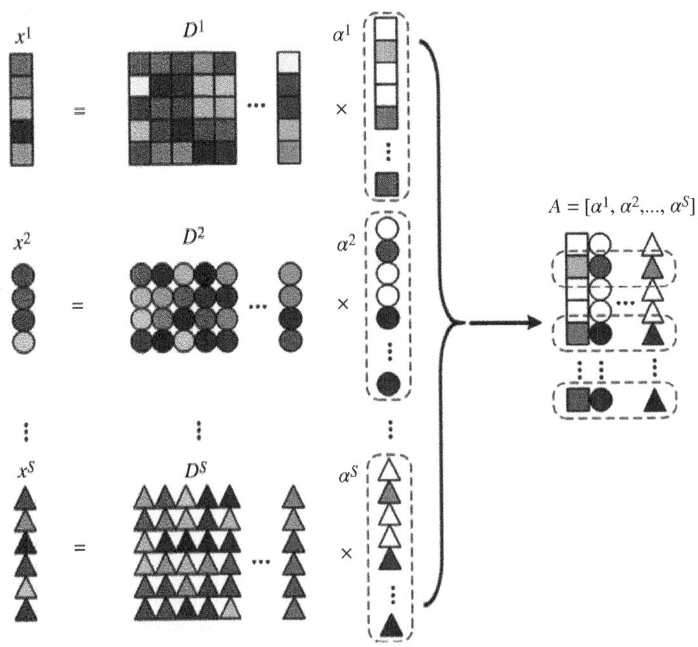


**Figure 18.13** Joint sparse representation model for multi-feature classification.

High-resolution images usually present more ground details, thereby making the extraction of finer detailed spatial information possible. However, the problem of excessive heterogeneity occurs that the ground objects of same category may reflect different spectral features. This noisy phenomenon appears as many isolated pixels and a lack of spatial coherency (speckles and holes in classified areas), which makes it not suitable to extract the thematic map from HR image only by the traditional pixel-based classification methods. Image segmentation makes use of both spatial information and spectral features, and seeks to divide an image into spatially continuous disjoint and un-overlapping homogeneous regions. Therefore, the salt-and-pepper noise or spots usually caused by "the same object with different spectrum" or "the same spectrum with different objects" have been eliminated.

During the past few decades, various segmentation techniques have been proposed, e.g. optimal edge detector, watershed algorithm, and region-growing. In this paper, a segmentation method, called full $\lambda$-schedule (FLS) algorithm, based on edge detection and region merging has been conducted on VIS image to acquire an optimal segmentation result. This segmentation technique comprises two steps. First, boundary detection and region segmentation have been applied to generate a preliminary segmentation result. The optimal segmentation scale can be selected manually. However, image segmentation based on edge detection always leads to over-segmentation. Therefore, the second step that refers to region merging is indispensable. By defining a merging cost $t_{i,j}$ for each pair of adjacent regions $(r_i, r_j)$, the FLS algorithm has been conducted to merge those similar neighbor regions:

$$t_{i,j} = \frac{(|r_i| \cdot |r_j|)/(|r_i| + |r_j|) \cdot \|U_i - U_j\|^2}{\text{length}(\partial(r_i, r_j))} \tag{18.39}$$

where $|r_i|$ is the area of region $r$. $\|U_i - U_j\|^2$ denotes the Euler distance between the average spectral vectors of two regions, and length$(\partial(r_i, r_j))$ is the length of common boundary. The merging level can be selected manually to obtain a relatively better result.

As a result of segmentation, the entire image consists of thousands of regions. Each region represents a single ground object that consists of pixels with similar spectral and spatial features. Therefore, in this book, we present a context-based fusion strategy (CBP). This strategy is based on the assumption that the probability of each object can be represented by those pixels that comprise this object. An effective approach is to add the probability of every pixel together. Then the conventional LIP and LGP can be modified as

$$\begin{aligned} \omega &= \arg \max_{j \in \{1,2,\ldots,c\}} \left( \sum_i \alpha_L \cdot p^L(\omega_j|x_i) + \sum_i \alpha_V \cdot p^V(\omega_j|x_i) \right) \\ \omega &= \arg \max_{j \in \{1,2,\ldots,c\}} \ln \left( \left[ \sum_i p^L(\omega_j|x_i) \right]^{\alpha^L} \cdot \left[ \sum_i p^V(\omega_j|x_i) \right]^{\alpha^V} \right), \end{aligned} \tag{18.40}$$

in which $p(\omega_j)$ is the posterior probabilities of a single object belonging to class $\omega_j$. $c$ is the number of classes. $p^L$, $p^V$ denote the probabilities of the two classifiers. The softmax model is applied for mapping the reconstruction error to the predicted probabilities of JSR classifier in this book. $x_i$ is the $i$th pixel locating in the current object. A normalization procedure should be performed on $\sum_i p(\omega_j \mid x_i)$.

In this way, pixels inside an object have the same probabilities as $p(\omega_j)$. Besides, the accuracies of each class derived from the confusion matrices for valid sample set are utilized to determine the weights $\alpha^L$, $\alpha^V$.

### 18.4.2 Experiments

The study data set is shown in Figure 18.14, which is accessible by 2014 IEEE Geoscience and Remote Sensing Committee (GRSS) Data Fusion Contest (DFC). The data set involves two image sources acquired at different spectral ranges and spatial resolutions: a coarser-resolution LWIR (TIR) HS data set and fine-resolution data acquired in the visible (VIS) wavelength range.

The two data sources cover an urban area near Thetford Mines in Québec, Canada. The thermal HS image was acquired using the "Hyper-Cam," an airborne LWIR HS imager based on a Fourier-transform spectrometer (FTS). It consists of 84 spectral bands in the 7.8–11.5 μm region with a pixel resolution of about 1 m. It has been calibrated to at-sensor spectral radiance units. A digital color camera, mounted on the same airborne platform, was also used to acquire visible imagery data that consisted of multiple color sub-images associated with distinct flight-lines. The visible image consists of three bands (R/G/B) with a pixel resolution of about 0.2 m. It is composed of a series of spatially disjoint sub-images separated by missing data. The sizes of two images are $874 \times 751$ and $4386 \times 3769$ pixels, respectively. The two airborne data sources were collected simultaneously and georeferenced. A ground truth map is provided along with the data subset. This map has the same spatial resolution and is registered with the airborne color data subset.

The TES algorithm is first carried out to attain the emissivity of ground objects for the LWIR HS image. Since the data used in this chapter was acquired by airborne platform, the influence of atmosphere is omitted. Thus, in this chapter, the atmospheric radiance $L_\downarrow$ and $L_\uparrow$ are equal to 0, and atmospheric transmittance $\tau$ is set to 1. Figure 18.15 shows the TES results. The temperature retrieval result shows that for most ground objects, the surface temperature ranges from 15 to 25. This estimation is quite close to the actual temperature in this area. Therefore, in the following experiments, we use the derived emissivity features for classification instead of the original radiance data.

Feature extraction methods, aim at reducing the data dimensions while keep the classification accuracy. We would like to show the different performance of several common HS feature extraction methods, e.g. principal component analysis (PCA), semi-supervised discriminant analysis (SDA), semi-supervised LDA (SELD), semi-supervised LFDA (SELF), $S^2$LDA, with regard to the HS image classification accuracy. First, we down-sample the ground truth image to the size of LWIR HS image by nearest sampling, and randomly select some samples as the atoms of SR classifier. The classifications are conducted under different sample numbers and feature numbers. The average accuracies of five independent runs are shown in Figure 18.16. Figure 18.16(a) shows the results

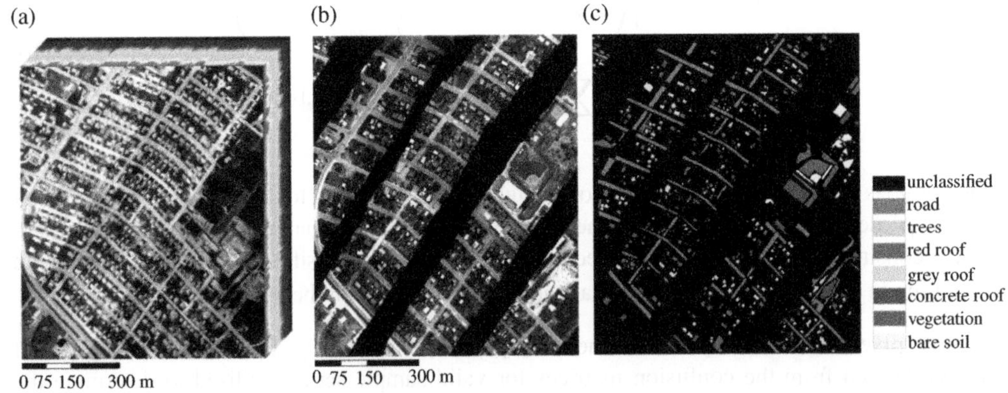


**Figure 18.14** The GRSS DFC data set. (a) LWIR HS image, (b) VIS image, (c) ground truth map.

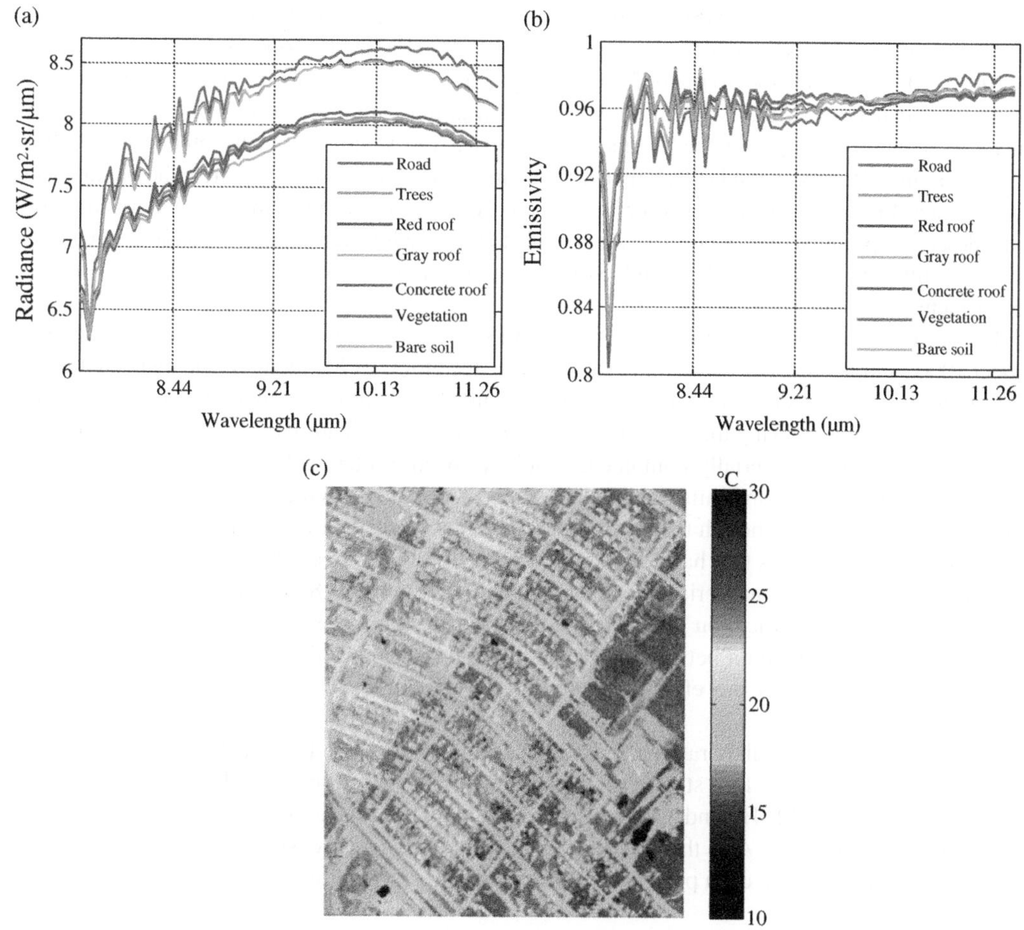


**Figure 18.15** Radiance of Hyper-Cam LWIR HS data and temperature-emissivity separation result. (a) Radiance, (b) emissivity features, (c) temperature retrieval result.

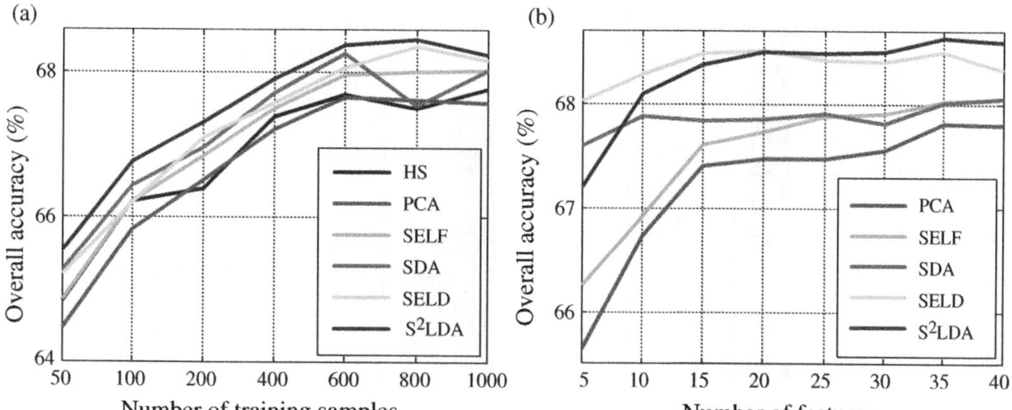


**Figure 18.16** Comparison of FE methods under different numbers of extracted features and training samples. (a) Using first 30 features, (b) selecting 1000 samples per class.

when 50, 100, 200, 400, 600, 800, and 1000 samples per class are selected randomly with 30 major feature components preserved after feature extraction, while Figure 18.16(b) shows the results when 5, 10, 15, 20, 25, 30, 35, and 40 major feature components are preserved with 1000 samples per class selected. It can be seen that with the increment of training samples, the OA of each FE method increases and mostly surpasses the original HS features. $S^2$LDA outperforms the others on the whole.

From Figure 18.16b, it can be seen that with the increment of extracted features, the overall accuracy of each FE method increases at first and scarcely varies when more than 20 features are used. This deserves special attention because we are able to use much fewer features to achieve a relatively stable result. Obviously, $S^2$LDA and SELD methods have more acceptable results in these experiments.

The second part includes the classification of VIS image. In this experiment, we also randomly select 1000 samples per class as training samples, i.e. the atoms of SR method, and analyze the classification results of spectral and spatial features. Figure 18.17 shows the overall accuracies of VIS image classification by using the color features (SPE) and texture features (TEX). The color and texture features can also be serially combined, which has been written as SC.

The figure shows that combining the spatial features has greatly enhanced the classification accuracies for VIS image, even though the result of texture features seems to be rather poor. This is due to some misclassified objects that have similar spectral features, e.g. trees and vegetation. JSR method outperforms the conventional serially combining method. Figure 18.18 shows the confusion matrices of spectral feature classification and JSR methods, where we can see that by combining texture features, the misclassification between trees and vegetation, roofs and roads, has dropped to a large extent. This demonstrates the effectiveness of the collaborative classification of spatial-spectral features.

Last, we present the collaborative classification results of LWIR HS and VIS images based on decision-level fusion. The first step of the collaborative classification method is to obtain the classification results of LWIR HS and VIS images individually. For this reason, we also select 1000 sample per class for training. And the LWIR HS image is classified by SR classifier after extracting 30 features of $S^2$LDA. For each pixel, we obtain a predicted probability vector for the seven classes.

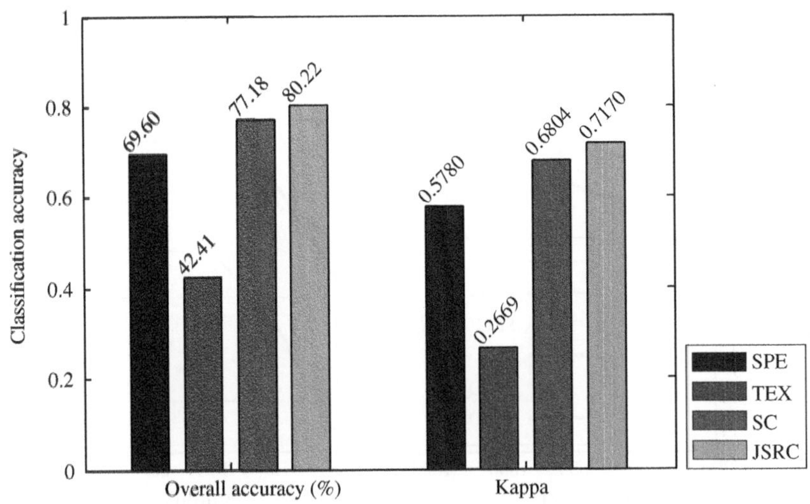


**Figure 18.17** Classification overall accuracies and Kappa coefficients.

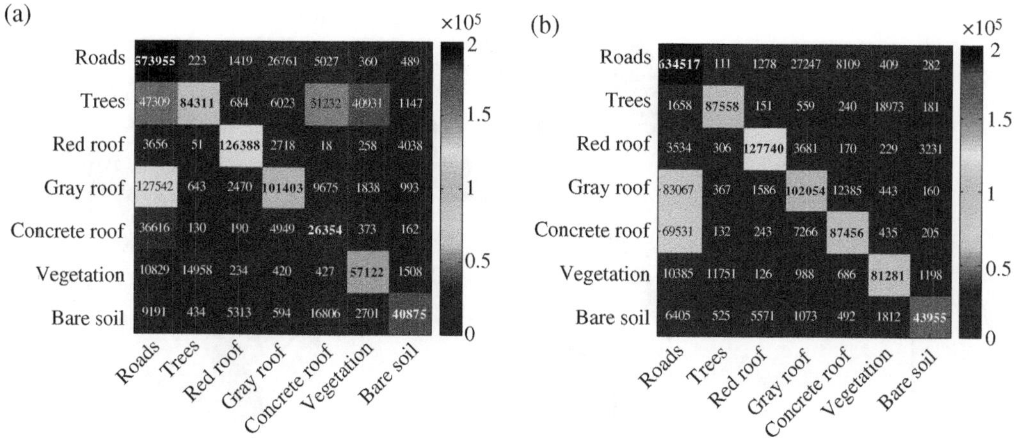


**Figure 18.18** Confusion matrices of classifications using (a) spectral features, (b) spatial-spectral features-based JSR classification method.

Then, we up-sample the probability map to the size of ground truth image by bilinear interpolation. Finally, the probability map is fused with that of the VIS image by several decision-level fusion strategies, e.g. LIP, LGP, D–S Evidence theory, fuzzy integral (FUZ), as well as the presented context-based LIP (CLIP) and context-based LGP (CLGP).

Figure 18.19 shows the Kappa coefficients of different methods, from which it can be seen that the presented CLIP and CLGP have much higher accuracies compared with the conventional decision fusion methods. More importantly, conventional pixel-based classification methods always lead to the noisy phenomenon in the classification maps, e.g. Figure 18.20(a) and (b). This problem also remains for pixel-based decision fusion methods, e.g. Figure 18.20(d). It sometimes does not conform

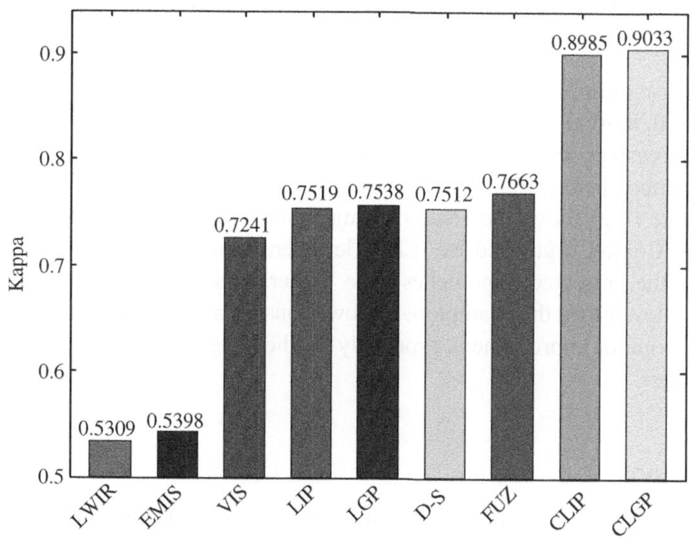


**Figure 18.19** Kappa coefficient of the synergetic classification for infrared HS and visible MS images. (a) Classification of LWIR, (b) classification of VIS, (c) segmentation of VIS, (d) LGP classification result, (e) CLIP classification result, (f ) CLGP classification result.

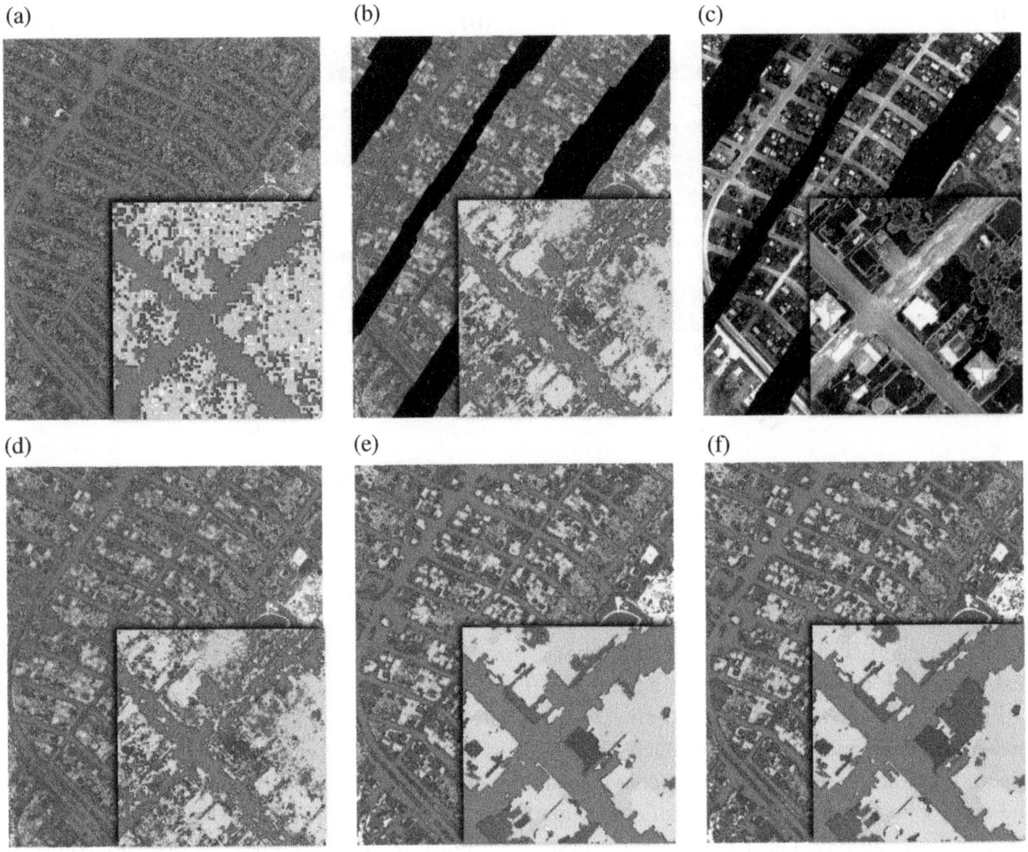


**Figure 18.20** Comparison of fusion results. (a) Classification of LWIR, (b) classification of VIS, (c) segmentation of VIS, (d) LGP classification result, (e) CLIP classification result, (f) CLGP classification result.

to the reality, because most observed ground objects are spatially continuous. By using the context-based fusion method, most of the speckles have been removed and the trees are holonomic. The contours of most objects can be seen clearly, e.g. Figure 18.20(e) and (f). Hence, the classification map is more helpful for human perception.

In addition, Figure 18.21 shows the accuracy comparison with some other methods presented in the 2014 Data Fusion Contest organized by IEEE Geoscience and Remote Sensing Committee. The result depicts that the presented approaches have higher accuracies than most of the participated teams. Though it is lower than the champion, it shows considerable potential because of its flexibility and the plenty of room for improvement, especially for the images with different imaging modalities or wavelength ranges.

## 18.5 Conclusions

HS image enables a detailed separation of similar ground materials. But the high dimensionality, limited, or unbalanced training samples bring a series of problems for the subsequent processing and applications. Therefore, a higher classification accuracy of HS image has encouraged an increasing usage of multisource images, e.g. high-resolution PAN image, MS image, and infrared image. However, the collaborative processing of multisource images also suffered from some new problems like

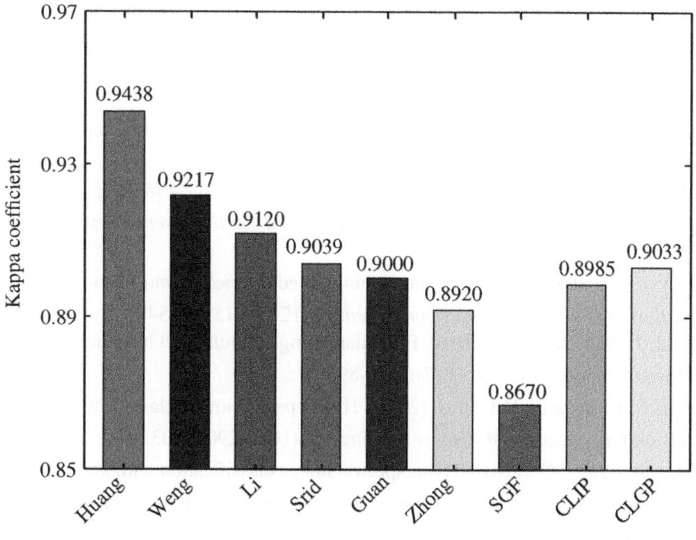


**Figure 18.21** Comparison with some methods from the IEEE GRSS Data Fusion Contest.

different imaging conditions, different spectral ranges, data redundancy, and even the curse of dimensionality. Hence, his chapter focuses on the HS-centered multisource image collaborative classification methods and deeply analyzes the probable problems and challenges in multisource image synergetic processing. In order to elaborate the main idea of how to fully explore the complementary information of multisource images, we take two collaborative classification methods for example. In the first situation, a self-learning-based classification algorithm is presented for the HS and high-resolution PAN images to effectively solve the small-sample problem. By combining the semi-supervised classification and image segmentation techniques, the self-learning algorithm utilizes the plentiful spectral information of HS image and the spatial information of PAN image to produce more available training samples automatically, thereby achieving much better results compared with supervised classification algorithm. In the second situation, a decision-level fusion method is presented for LWIR HS and VIS images aiming at producing a high spatial resolution classification map. After extracting the spectral features of the LWIR HS images and the spatial-spectral features of the VIS image separately, the context-based decision fusion method is conducted to merge the classification results of both images. The method has been demonstrated to be a promising approach compared with the other state-of-the-art methods on the data set provided by the 2014 IEEE GRSS Data Fusion Contest. We have to point out that the benefit of collaborative processing of multisource images should only originate from the comprehensive exploration of the complementary and divergent information of those image types.

## References

1. Lu, X., Zhang, J., Li, T., and Zhang, Y. (2016). A novel synergetic classification approach for hyperspectral and panchromatic images based on self-learning. *IEEE Transactions on Geoscience and Remote Sensing* 54 (8): 4917–4928.
2. Lu, X., Zhang, J., Li, T., and Zhang, Y. (2016). Incorporating diversity into self-learning for synergetic classification of hyperspectral and panchromatic images. *Remote Sensing* 8 (10): 804.
3. Zhang, J., Li, T., Lu, X., and Cheng, Z. (2016). Semantic classification of high-resolution remote-sensing images based on mid-level features. *IEEE Journal of Selected Topics in Applied Earth Observations and Remote Sensing* 9 (6): 2342–2353.

4. Li, T., Zhang, J., Lu, X., and Zhang, Y. (2017). SDBD: a hierarchical region-of-interest detection approach in large-scale remote sensing image. *IEEE Geoscience and Remote Sensing Letters* 14 (5): 699–703.
5. Li, T., Zhang, J., and Zhang, Y. (2014). Classification of hyperspectral image based on deep belief networks. *IEEE International Conference on Image Processing* 5132–5136.
6. Lu, X., Zhang, J., Li, T., and Zhang, Y. (2017). Hyperspectral image classification based on semi-supervised rotation forest. *Remote Sensing* 9 (9): 924.
7. Lu, X., Zhang, J., Li, T., and Zhang, G. (2015). Synergetic classification of long-wave infrared hyperspectral and visible images. *IEEE Journal of Selected Topics in Applied Earth Observations and Remote Sensing* 8 (7): 3546–3557.
8. Lu, X., Zhang, J., Yu, X. et al. (2019). Hyper-sharpening based on spectral modulation. *IEEE Journal of Selected Topics in Applied Earth Observations and Remote Sensing* 12 (5): 1534–1548.
9. Lu, X., Zhang, J., Li, T., and Zhang, Y. (2016). Pan-Sharpening by multilevel inter-band structure modeling. *IEEE Geoscience and Remote Sensing Letters* 13 (7): 892–896.
10. Porta, C.J.D., Bekit, A.A., Bernard, H.L. et al. (2019). Hyperspectral image classification via compressive sensing. *IEEE Transactions on Geoscience and Remote Sensing.* 57 (10): 8290–8303.
11. Liu, L., Huang, W., and Wang, C. (2018). Hyperspectral image classification with kernel-based least-squares support vector machines in sum space. *IEEE Journal of Selected Topics in Applied Earth Observations and Remote Sensing* 11 (4): 1144–1157.
12. Chen, Y., Wang, Y., Gu, Y. et al. (2019). Deep learning ensemble for hyperspectral image classification. *IEEE Journal of Selected Topics in Applied Earth Observations and Remote Sensing* 12 (6): 1882–1897.
13. Yu, C., Zhao, M., Song, M. et al. (2019). Hyperspectral image classification method based on CNN architecture embedding with hashing semantic feature. *IEEE Journal of Selected Topics in Applied Earth Observations and Remote Sensing* 12 (6): 1866–1881.
14. Paoletti, M.E., Haut, J.M., Fernandez-Beltran, R. et al. (2019). Capsule networks for hyperspectral image classification. *IEEE Transactions on Geoscience and Remote Sensing* 57 (4): 2145–2160.
15. Le J., Yazdanpanah A.P., Regentova E. et al. (2015). A deep belief network for classifying remotely-sensed hyperspectral data. International Symposium on Visual Computing, Las Vegas (14–16 December 2015), 682–692.
16. Wang, L., Zhang, J., Liu, P. et al. (2017). Spectral-spatial multi-feature-based deep learning for hyperspectral remote sensing image classification. *Soft Computing* 21 (1): 213–221.
17. Tao, C., Pan, H., Li, Y. et al. (2015). Unsupervised spectral spatial feature learning with stacked sparse autoencoder for hyperspectral imagery classification. *IEEE Geoscience and Remote Sensing Letters* 12 (12): 2438–2442.
18. Ma, X., Wang, H., and Wang, J. (2016). Semisupervised classification for hyperspectral image based on multi-decision labeling and deep feature learning. *ISPRS Journal of Photogrammetry and Remote Sensing* 120: 99–107.
19. Lv, F., Han, M., and Qiu, T. (2017). Remote sensing image classification based on ensemble extreme learning machine with stacked autoencoder. *IEEE Access* 5: 9021–9031.
20. He, N., Paoletti, M.E., Haut, J.M. et al. (2019). Feature extraction with multiscale covariance maps for hyperspectral image classification. *IEEE Transactions on Geoscience and Remote Sensing* 57 (2): 755–769.
21. Pan, B., Shi, Z., and Xu, X. (2017). Hierarchical guidance filtering-based ensemble classification for hyperspectral images. *IEEE Transactions on Geoscience and Remote Sensing* 55 (7): 4177–4189.
22. Xia, J., Ghamisi, P., Yokoya, N. et al. (2018). Random forest ensembles and extended multiextinction profiles for hyperspectral image classification. *IEEE Transactions on Geoscience and Remote Sensing* 56 (1): 202–216.
23. Xia, J., Yokoya, N., and Iwasaki, A. (2017). Hyperspectral image classification with canonical correlation forests. *IEEE Transactions on Geoscience and Remote Sensing* 55 (1): 421–431.
24. Feng, W., Huang, W., and Bao, W. (2019). Imbalanced hyperspectral image classification with an adaptive ensemble method based on SMOTE and rotation forest with differentiated sampling rates. *IEEE Geoscience and Remote Sensing Letters* 16 (12): 1879–1883.
25. (a) Xia, J., Falco, N., and Benediktsson, J.A. (2017). Hyperspectral image classification with rotation random forest via KPCA. *IEEE Journal of Selected Topics in Applied Earth Observations and Remote Sensing* 10 (4): 1601–1609; (b)Wang, J., Jiao, L., Liu, H. et al. (2015). Hyperspectral image classification by spatial-spectral derivative-aided kernel joint sparse representation. *IEEE Journal of Selected Topics in Applied Earth Observations and Remote Sensing* 8 (6): 2485–2500.

26. Mei, S., Ji, J., Geng, Y. et al. (2019). Unsupervised spatial–spectral feature learning by 3D Convolutional autoencoder for hyperspectral classification. *IEEE Transactions on Geoscience and Remote Sensing* 57 (9): 6808–6820.
27. Chen, J., Xia, J., Du, P. et al. (2016). Combining rotation forest and multiscale segmentation for the classification of hyperspectral data. *IEEE Journal of Selected Topics in Applied Earth Observations and Remote Sensing* 9 (9): 4060–4072.
28. Santara, A., Mani, K., Hatwar, P. et al. (2017). BASS net: Band-adaptive spectral-spatial feature learning neural network for hyperspectral image classification. *IEEE Transactions on Geoscience and Remote Sensing* 55 (9): 5293–5301.
29. Li, Y., Zhang, H.K., and Shen, Q. (2017). Spectral-spatial classification of hyperspectral imagery with 3D convolutional neural network. *Remote Sensing* 9 (1): 67.
30. Merentitis, A. and Debes, C. (2015). Many hands make light work-on ensemble learning techniques for data fusion in remote sensing. *IEEE Geoscience and Remote Sensing Magazine* 3 (3): 86–99.
31. Huang, X. and Zhang, L. (2013). An SVM ensemble approach combining spectral, structural, and semantic features for the classification of high-resolution remotely sensed imagery. *IEEE Transactions on Geoscience and Remote Sensing* 51 (1): 257–272.
32. Hardie, R.C., Eismann, M.T., and Wilson, G.L. (2004). MAP estimation for hyperspectral image resolution enhancement using an auxiliary sensor. *IEEE Transactions on Image Processing* 13 (9): 1174–1184.
33. Zhang, Y., Backer, B., and Scheunders, P. (2009). Noised-resistant wavelet-based Bayesian fusion of multispectral and hyperspectral images. *IEEE Transactions on Geoscience and Remote Sensing* 47 (11): 3834–3843.
34. Chen, Z., Pu, H., Wang, B. et al. (2014). Fusion of hyperspectral and multispectral images: a novel framework based on generalization of pan-sharpening methods. *IEEE Geoscience and Remote Sensing Letters* 11 (8): 1418–1422.
35. Yang, J., Zhao, Y., and Chan, J. (2018). Hyperspectral and multispectral image fusion via deep two-branches convolutional neural network. *Remote Sensing* 10 (5): 800.
36. Palsson, F., Sveinsson, J.R., and Ulfarsson, M.O. (2017). Multispectral and hyperspectral image fusion using a 3-D-convolutional neural network. *IEEE Geoscience and Remote Sensing Letters.* 14 (5): 639–643.
37. Zhou, F., Hang, R., Liu, Q. et al. (2019). Pyramid fully convolutional network for hyperspectral and multispectral image fusion. *IEEE Journal of Selected Topics in Applied Earth Observations and Remote Sensing.* 12 (5): 1549–1558.
38. Simões, M., Bioucas-Dias, J., and Almeida, L.B. (2015). A convex formulation for hyperspectral image superresolution via subspacebased regularization. *IEEE Transactions on Geoscience and Remote Sensing.* 53 (6): 3373–3388.
39. Simões, M., Bioucas-Dias, J., and Almeida, L.B. (2014). Hyperspectral Image superresolution: an edge-preserving convex formulation. *International Conference on Image Processing* 4166–4170.
40. Yokoya, N., Yairi, T., and Iwasaki, A. (2012). Coupled nonnegative matrix factorization unmixing for hyperspectral and multispectral data fusion. *IEEE Transactions on Geoscience and Remote Sensing* 50 (2): 528–537.
41. Song, H., Huang, B., Zhang, K. et al. (2014). Spatio-spectral fusion of satellite images based on dictionary-pair learning. *Information Fusion* 18: 148–160.
42. An, Z. and Shi, Z. (2014). Hyperspectral image fusion by multiplication of spectral constraint and NMF. *Optik* 125: 3150–3158.
43. Karoui, M.S., Deville, Y., Benhalouche, F.Z. et al. (2016). Hypersharpening by joint-criterion nonnegative matrix factorization. *IEEE Transactions on Geoscience and Remote Sensing* 55 (3): 1660–1670.
44. Zhang, K., Wang, M., and Yang, S. (2017). Multispectral and hyperspectral image fusion based on group spectral embedding and low-rank factorization. *IEEE Transactions on Geoscience and Remote Sensing* 55 (3): 1363–1371.
45. Yokoya, N., Grohnfeldt, C., and Chanussot, J. (2017). Hyperspectral and multispectral data fusion: a comparative review of the recent literature. *IEEE Geoscience and Remote Sensing Magazine* 5 (2): 29–56.

46. Santini, F., Palombo, A., Dekker, R.J., and Pignatti, S. (2014). Advanced anomalous pixel correction algorithms for hyperspectral thermal infrared data: the TASI-600 case study. *IEEE Journal of Selected Topics in Applied Earth Observations and Remote Sensing* 7 (6): 1939–1404.
47. Wang, N., Wu, H., Li, C., and Li, Z. (2011). Temperature and emissivity retrievals from hyperspectral thermal infrared data using linear spectral emissivity constraint. *IEEE Transactions on Geoscience and Remote Sensing* 49 (4): 1291–1303.
48. Liao, W., Huang, X., Coillie, F.V. et al. (2015). Processing of multiresolution thermal hyperspectral and digital color data: outcome of the 2014 IEEE GRSS Data Fusion Contest. *IEEE Journal of Selected Topics in Applied Earth Observations and Remote Sensing* 8 (6): 2984–2996.
49. Thoonen, G., Mahmood, Z., and Peeters, S. (2012). Multisource classification of color and hyperspectral images using color attribute profiles and composite decision fusion. *IEEE Journal of Selected Topics in Applied Earth Observations and Remote Sensing* 5 (2): 510–521.
50. Mahmood, Z., Akhter, M.A., Thoonen, G. et al. (2013). Contextual subpixel mapping of hyperspectral images making use of a high resolution color image. *IEEE Journal of Selected Topics in Applied Earth Observations and Remote Sensing* 6 (2): 779–791.
51. Goodenough, D.G., Dyk, A., Niemann, K.O. et al. (2013). Processing hyperion and ALI for forest classification. *IEEE Transactions on Geoscience and Remote Sensing* 41 (6): 1321–1331.
52. Stavrakoudis, D.G., Dragozi, E., Gitas, I.Z. et al. (2014). Decision fusion based on hyperspectral and multispectral satellite imagery for accurate forest species mapping. *Remote Sensing* 6 (8): 6897–6928.
53. Kattenborna, T., Maacka, J., Faßnachtb, F. et al. (2015). Mapping forest biomass from space-fusion of hyperspectral EO1-hyperion data and Tandem-X and Worldview-2 canopy height models. *International Journal of Applied Earth Observation and Geo information* 35: 359–367.
54. Delalieux, S., Zarco-Tejada, P.J., Tits, L. et al. (2014). Unmixing-based fusion of hyperspatial and hyperspectral airborne imagery for early detection of vegetation stress. *IEEE Journal of Selected Topics in Applied Earth Observations and Remote Sensing* 7 (6): 2571–2582.
55. Ma, X., Wang, H., Geng, J. et al. (2016). Hyperspectral image classification with small training set by deep network and relative distance prior. *IEEE International Geoscience and Remote Sensing Symposium* 3282–3285.
56. Merentitis, A., Debes, C., and Heremans, R. (2014). Ensemble learning in hyperspectral image classification: toward selecting a favorable bias-variance tradeoff. *IEEE Journal of Selected Topics in Applied Earth Observations and Remote Sensing* 7 (4): 1089–1102.
57. Kaufman, J.R., Eismann, M.T., and Celenk, M. (2015). Assessment of spatial-spectral feature-level fusion for hyperspectral target detection. *IEEE Journal of Selected Topics in Applied Earth Observations and Remote Sensing* 8 (6): 1–11.
58. Chang, N., Vannah, B., and Yang, Y.J. (2014). Comparative sensor fusion between hyperspectral and multispectral satellite sensors for monitoring microcystin distribution in Lake Erie. *IEEE Journal of Selected Topics in Applied Earth Observations and Remote Sensing* 7 (6): 2426–2442.
59. Hu J., Mou L., Schmitt A. et al. FusioNet: a two-stream convolutional neural network for urban scene classification using PolSAR and hyperspectral data. Joint Urban Remote Sensing Event, Dubai (6–8 March 2017), 1–4.
60. Cao, Q., Zhong, Y., Ma, A. et al. (2018). Urban land use/land cover classification based on feature fusion fusing hyperspectral image and LiDar data. *IEEE International Geoscience and Remote Sensing Symposium* 8869–8872.

# 19

# Class Feature-Weighted Hyperspectral Image Classification

Shengwei Zhong[1], Jiaojiao Li[2], Xiaodi Shang[3], Shuhan Chen[4], and Chein-I Chang[5]
[1]*School of Computer Science and Engineering, Nanjing University of Science and Technology, Nanjing, China*
[2]*State Key Laboratory of Integrated Services Networks, School of Telecommunications Engineering, Xidian University, Xian, China*
[3]*Center for Hyperspectral Imaging in Remote Sensing (CHIRS), School of Information and Technology, Dalian Maritime University, Dalian, China*
[4]*Department of Electrical Engineering, Zhejiang University, Hangzhou, China*
[5]*Remote Sensing Signal and Image Processing Laboratory, Department of Computer Science and Electrical Engineering, University of Marylandw Baltimore County, Baltimore, MD, USA*

This chapter studies a recently developed class feature (CF)-weighted hyperspectral image classification (CFW-HSIC), which extracts features from classes that can be used to weight significance of each of classes for classification. Specifically, it can be used to address imbalanced class and background issues. Two types of class features (CFs) are of particular interest, intra-class features (Intra-CFs) and inter-class features (Inter-CFs) to capture class intra-pixel and inter-pixel information, respectively. These features are then used to calculate the CF probability for each of classes as its class significance for classification. Such obtained CF-calculated probabilities not only automatically allocate an appropriate training sample size for each of classes for classification, but also provide class weights to generalize commonly used overall accuracy (OA) and average accuracy (AA) to CF-weighted classification measures, CF-OA and CF-AA, respectively. Accordingly, the conventional HSIC can be further extended to CFW-HSIC, which can effectively deal with classification of imbalanced classes as well as class variability (CV)/class separability (CS).

## 19.1 Introduction

Hyperspectral image classification (HSIC) is a fundamental task for image analysts who interpret image scenes according to different applications such as agriculture, environmental monitoring, forestry, and natural resources. Two major issues arising in HSIC have been overlooked over the past

*Advances in Hyperspectral Image Processing Techniques*, First Edition. Edited by Chein-I Chang.

years. One is the imbalanced class issue. A general approach, called sample ratio (SR) which has been widely used for training sample selection, is based on the proportion of class size. In this case, a large class generally has a large number of training samples, while a small class has a small number of training samples. Unfortunately, SR may lead to a situation where a small class may end up with very few or even no training samples at all that can be selected for this class. Moreover, SR may also result in a case that the classification rate from a large class may eventually dominate the accuracy rate of a small class. This brings up a challenging issue regarding how to select appropriate numbers of training samples for individual classes with various sizes, specifically small classes. Another issue is the background (BKG) effect. In HSIC, the BKG has never become an issue since classes of interest are all provided by ground truth and all BKG data samples have been excluded from classification. However, in reality, BKG data samples are simply unlabeled data samples which are not known by ground truth but may be very important and crucial to certain applications. For example, for the Purdue Indiana Indian Pines test site, the classes of interest are crops which are important to agriculture and everything other than crops is considered as an uninterested object and categorized into BKG. However, for military applications, the railroads and highways in the BKG are of particular interest for transportation. Besides, some of BKG data samples may actually belong to certain known classes but cannot be identified by ground truth. As a result, their effects do have a significant impact on classification and cannot be underestimated. Unfortunately, the commonly used average accuracy (AA) and overall accuracy (OA) are generally calculated by the ground-truth-provided data samples and do not include such BKG data samples for classification. So, it gives rise to an approach which must be able to factor class characteristics into classification.

Recently, Zhong *et al.* developed a class feature (CF) analysis-based approach to address the abovementioned issues by developing CF descriptors that can capture class information for classification [1]. Since a spectral CF is defined as a spectral property of a class that exhibits its particular characteristics distinct from that of other classes, two types of CF descriptors, intra-class features (Intra-CFs) and inter-class features (Inter-CFs), were custom-designed in [1] for this purpose with an Intra-CF and an Inter-CF corresponding to variability within a class, class variability (CV) and class separability (CS) between classes, respectively. For each CF descriptor, we can further calculate CF probabilities of each of classes where the CF probability of each class can be used as a quantitative indicator to allocate its training sample size as well as to weight its significance in calculating OA. By virtue of these CF probabilities, a new CF-based classification criterion can be derived, called CF-OA which includes the commonly used AA and OA as its special cases and the conventional HSIC can be further generalized to CF-weighted HSIC (CFW-HSIC). It turns out that CFW-HSIC can resolve both imbalanced class and BKG issues. To fully explore the utility of CFW-HSIC, we refer details to [1].

## 19.2 Class Feature Descriptors

Assume that there are $M$ classes of interest to be classified, $\{C_m\}_{m=1}^{M}$. Let $\boldsymbol{\mu}_m$ be the sample mean of the $m$th class, $C_m$. Two types of CF descriptors are derived as follows.

### 19.2.1 Intra-Class Feature Descriptors for Class Variability

Intra-CF descriptors calculate sample statistics within a class, i.e. CV. One simple Intra-CF criterion is to measure how tightly data samples are clustered, that is, the more samples are clustered in a class, the denser the class is, the better structure the class is. With this interpretation, an Intra-CF can be characterized by class density (CD) which calculates the ratio of class mass $||\boldsymbol{\mu}||^2 = \boldsymbol{\mu}^T\boldsymbol{\mu}$ to class area measured by $\sigma^2$ by

$$\vartheta_{\text{CD}}^{\text{Intra-CF}}(C_m) = \frac{\boldsymbol{\mu}_m^T\boldsymbol{\mu}_m}{\sigma_m^2} \tag{19.1}$$

where $\sigma_m^2$ is the variance of the $m$th class, $C_m$, defined by

$$\sigma_m^2 = \frac{1}{|C_m|}\sum_{\mathbf{r}\in C_m}(\mathbf{r}-\boldsymbol{\mu}_m)^T(\mathbf{r}-\boldsymbol{\mu}_m) \tag{19.2}$$

with $|C_m|$ defined as the number of elements in $C_m$.

So, to characterize CD in (19.1), two features are needed, class mean, $\boldsymbol{\mu}$, which can be considered as "mass" of a class and class variance, $\sigma^2$, which can be considered as "momentum" of a class. If a class is significant, its class mean, i.e. mass, $\boldsymbol{\mu}$, should be sufficiently large, while its class variance $\sigma^2$, i.e. momentum, should be also relatively small so that the data samples in the class will be clustered together. This indicates that the class is highly dense. That is, the higher density is the class, the more clustered are the data samples in the class, thus the more significant is the class. On the other hand, CD can be also considered as signal-to-noise ratio (SNR) generally used in communications and signal processing, that is, the ratio of the class mass considered as signal energy specified by $||\boldsymbol{\mu}||^2$ to the class variance considered as noise energy $\sigma^2$.

Now, let $n_m$ be the number of data samples in class $C_m$ and $N$ be the total number of data samples to be used for classification. Then the SR can be defined by

$$\vartheta_{\text{SR}}^{\text{Intra-CF}}(C_m) = \frac{n_m}{N} \tag{19.3}$$

which can be also considered as an Intra-CF criterion. However, SR in (19.3) only calculates the size of each class related to the entire data set size, $N$, which is fixed but it does not measure the variance within the class as CD does in (19.1). Therefore, CD is a better Intra-CF criterion than SR in the sense that CD includes class mean as class mass and also takes care of data sample variations within the class by variance $\sigma^2$ as opposed to SR which only measures the ratio of a class to the data set size.

An interesting remark on CD and SR in regard to how to deal with imbalanced class is worthwhile. Since a small class generally has a higher value of CD, CD can be used to take care of smaller classes. By contrast, SR intends to put more weights on larger classes and less weights on smaller classes. So, it places more focus on larger classes in classification by sacrificing smaller classes. Consequently, the issue of imbalanced classes can be effectively resolved by CD but not SR.

### *19.2.2 Inter-Class Feature Descriptors for Class Separability*

Unlike CD which only focuses on the class itself with no regard to other classes, Inter-CFs measure the degree that a class interacts with other classes, i.e. CS. A criterion, to be called class FR (CFR), can be derived from the commonly used Fisher's ratio (FR) in Fisher's linear discriminant analysis [2] and defined by

$$\vartheta_{\text{CFR}}^{\text{Inter-CF}}(C_i) = \frac{m(\boldsymbol{\mu}_i, \boldsymbol{\mu}_{i^*})}{\sigma_i^2 + \sigma_{i^*}^2} \tag{19.4}$$

where $m(\boldsymbol{\mu}_i, \boldsymbol{\mu}_{i^*})$ is the minimal distance from the class mean, $\boldsymbol{\mu}_i$, to all other class means, $\{\boldsymbol{\mu}_j\}_{j=1, j\neq i}^{M}$ by finding

$$i^* = \arg\left\{\min_{1 \leq j\neq i \leq M} m(\boldsymbol{\mu}_i, \boldsymbol{\mu}_j)\right\} \tag{19.5}$$

and $\sigma_{i^*}^2$ is the variance of class $C_{i^*}$ defined by

$$\sigma_{i^*}^2 = \frac{1}{|C_{i^*}|}\sum_{\mathbf{r}\in C_{i^*}} (\mathbf{r} - \boldsymbol{\mu}_{i^*})^T(\mathbf{r} - \boldsymbol{\mu}_{i^*}). \tag{19.6}$$

The numerator of (19.4) measures CS as opposed to the denominator of (19.4) to measure CV by $\sigma_i^2 + \sigma_{i^*}^2$. More specifically, CFR measures the ratio of CS specified by (19.5) to CV specified by (19.6).

### 19.2.3 Total Class Features

By combining $\vartheta_{\text{CD}}^{\text{Intra-CF}}(C_i)$ and $\vartheta_{\text{CFR}}^{\text{Inter-CF}}(C_i)$, we can further define a new CF criterion, called total class feature (TCF), defined by multiplying their values as

$$\vartheta_{\times}^{\text{TCF}}(C_m) = \vartheta_{\text{CD}}^{\text{Intra-CF}}(C_m) \times \vartheta_{\text{CFR}}^{\text{Inter-CF}}(C_m). \tag{19.7}$$

This criterion is designed to take care of CD and CS for compromise.

### 19.2.4 Calculation of CF Probabilities

Now we can use (19.1), (19.3), (19.4), and (19.7) to define various types of CF probabilities, denoted by $p_m^{\text{CF}} = p^{\text{CF}}(C_m)$ for the $m$th class, $C_m$ with CF specified by CD, SR, CFR, and TCF as follows:

$$p_m^{\text{CD}} = p^{\text{CD}}(C_m) = \frac{\vartheta_{\text{CD}}^{\text{Intra-CF}}(C_m)}{\sum_{j=1}^{M}\vartheta_{\text{CD}}^{\text{Intra-CF}}(C_j)} \tag{19.8}$$

$$p_m^{\text{SR}} = p^{\text{SR}}(C_m) = \vartheta_{\text{SR}}^{\text{Intra-CF}}(C_m) = \frac{n_m}{N} \tag{19.9}$$

$$p_m^{\text{CFR}} = p^{\text{CFR}}(C_m) = \frac{\vartheta_{\text{CFR}}^{\text{Inter-CF}}(C_m)}{\sum_{j=1}^{M}\vartheta_{\text{CFR}}^{\text{Inter-CF}}(C_j)} \tag{19.10}$$

$$p_m^{\text{TCF}} = p^{\text{TCF}}(C_m) = \frac{\vartheta_{\times}^{\text{TCF}}(C_m)}{\sum_{j=1}^{M}\vartheta_{\times}^{\text{TCF}}(C_j)}. \tag{19.11}$$

## 19.3 Allocation of Class Training Sample Size

A commonly used approach to selecting training samples is randomly sampling, particularly used for cross validation. In particular, it randomly selects training samples according to a certain percentile such as 5, 10%, etc., of each class. In this case, it assumes that the importance of all classes is equally likely, which is exactly the case defined by $P_{AA}$. However, there are two major problems with this approach. One is determination of how much percentile needs to be used for training sample selection. The other is the same chosen percentile applied to each of classes regardless of their importance. So, for a class with a small number of data samples, it encounters an issue that there will be no or very few training samples can be selected if the percentile is too small. By contrast, for a class with a large number of data samples, it may result in over-sampling with too many training samples to be selected than what it needs. It seems that no work is reported in the literature regarding how to design a criterion to select individual size of training samples for each class, namely, $n_m^{\text{training}}$ for class $C_m$. The approach presented in this section is primarily devoted to resolving this imbalanced class issue. It takes advantage of CF probabilities in Section 19.3 to develop an automatic and effective means of allocating $n_m^{\text{training}}$ for a given total number of training data samples, $N^{\text{training}}$.

For a given $N^{\text{training}}$, instead of using a pre-chosen percentile, we allocate the number of training samples for each class constrained by $N^{\text{training}}$ in accordance with its CF probability. More specifically, for a class, $C_m$, the size of training samples, $n_m^{\text{training}}$, to be allocated for $C_m$ is defined by

$$n_m^{\text{training,CF}} = \lceil N \times p^{\text{CF}}(C_m) \rceil \tag{19.12}$$

where $p^{\text{CF}}(C_m)$ is the CF probability of $C_m$. For example, if CFR in (19.4) is used, $p^{\text{CFR}}(C_m)$ in (19.10) is then used to calculate (19.12).

As a result of (19.12), the total number of training samples allocated for all classes may not be exactly equal to the pre-chosen number of training samples, $N^{\text{training}}$, but can be approximated by

$$\left| \sum_{m=1}^{M} n_m^{\text{training,CF}} - N^{\text{training}} \right| \le M. \tag{19.13}$$

However, on some occasions,

$$n_m^{\text{training,CF}} \ge \left(\frac{n_m}{2}\right). \tag{19.14}$$

In this case, we can set $n_m^{\text{training,CF}}$ to the upper bound, $n_m^{\text{UB}} = (n_m/2)$. On the other hand, if

$$n_m^{\text{training,CF}} < \left(\frac{n_m}{100}\right), \tag{19.15}$$

then we set $n_m^{\text{training}}$ to the lower bound, $n_m^{\text{LB}} = (n_m/100)$. Nevertheless, for each class, these upper bound $n_m^{\text{UB}}$ and lower bound $n_m^{\text{LB}}$ can be chosen to adapt different needs. The details of implementing an algorithm for selecting CF-allocated training sample class size can be summarized as follows.

**CF-Allocated Training Sample Class Size Selection Algorithm**

1. Initial Conditions:
   (a) Specify the total number of training samples required to be selected, $N^{\text{training}}$.
   (b) Select a measure from (19.1), (19.3), (19.4), and (19.7).
   (c) Determine the lower bounds, $\{n_m^{\text{LB}}\}_{m=1}^{M}$, and upper bounds, $\{n_m^{\text{UB}}\}_{m=1}^{M}$, used in (19.13) and (19.14).
   (d) $\overline{N}^{(0)} = N^{\text{training}}$ and $k = 0$.
2. For a given CF, calculate CF probabilities via (19.8)-(19.11).
3. At the $k$th iteration, for each class in $\{C_m\}_{m=1}^{M}$, use $\overline{N}^{(k)}$ to calculate (19.12) to find $n_m^{\text{training,CF}(k)}$.
4. If $n_m^{\text{training,CF}(k)} \geq n_m^{\text{UB}}$ in (19.14), let $\overline{n}_m^{(k)} = n_m^{\text{UB}}$ and $\tilde{n}_m^{(k)} = n_m^{\text{UB}} - n_m^{\text{training,CF}(k)}$ which will be negative.
5. If $n_m^{\text{training,CF}(k)} < n_m^{\text{LB}}$ in (19.15), let $\overline{n}_m^{(k)} = n_m^{\text{LB}}$ and $\tilde{n}_m^{(k)} = n_m^{\text{LB}} - n_m^{\text{training,CF}(k)}$ which will be positive.
6. Calculate $\overline{N}^{(k)} = \sum_{m=1}^{M} \overline{n}_m^{(k)}$ and $\tilde{N}^{(k)} = \sum_{m=1}^{M} \tilde{n}_m^{(k)}$.
7. If $k = 0$, go to step 9. Otherwise, continue.
8. Check if

$$\left| \overline{N}^{(k)} - N^{\text{training}} \right| \leq M. \tag{19.16}$$

   If yes, the algorithm is terminated. Otherwise, let $\overline{N}^{(k+1)} = \overline{N}^{(k)} + \tilde{N}^{(k)}$.
9. Let $k \leftarrow k+1$ and go to step 3.

Interestingly, according to [3], the training sample size selection used by edge preserving filtering (EPF)-based methods developed in [4] can be also considered as a special case of the above CF-allocated sample size selection algorithm. That is, for a given $N^{\text{training}}$ obtained from a given percentile (%), we determine

$$n_{\text{average}} = \left( \frac{N^{\text{training}}}{M} \right). \tag{19.17}$$

If $(n_m/2) \geq n_{\text{average}}$, then $n_i^{\text{training}} = n_{\text{average}}$. In this case, $n_m^{\text{training}}$ is over-sampling for large classes. Otherwise, it is under-sampling. We let $n_m^{\text{training}} = (n_m/2)$ for small classes. So, in terms of the above algorithm, $n_m^{\text{UB}}$ in (19.14) is set to $n_m^{\text{UB}} = n_{\text{average}}$ and $n_m^{\text{LB}} = 0$ in (19.15).

Although some works such as [5, 6] also looked into the issue in training samples for classification, their focuses were placed on sampling strategy of selecting training samples not on determination of training sample size to be used for each class which is exactly the goal that CF probabilities would like to accomplish.

## 19.4 CFW-HSIC

The CFW-HSIC presented in this section extends the traditional HSIC by including the CF-Allocated Training Sample Class Size Selection Algorithm developed in Section 19.3 for each class, $n_m^{\text{training,CF}}$, in HSIC so that a CF-weighted OA (CF-OA), $P_{\text{CF-OA}}$, can be used to address accuracy of small classes effectively.

**CFW-HSIC**

1. Select a CF descriptor.
2. Find CF probabilities for each class.
3. According to calculated CF probabilities, a CF-Allocated Training Sample Class Size Selection Algorithm is used to find $n_m^{\text{training,CF}}$ for each class.
4. Use CF probabilities for each class to calculate CF-OA. It should be noted that when CF is specified by SR, $P_{\text{CF-OA}}$ is reduced to $P_{\text{OA}}$.

By concluding this section, five remarks on CF-allocated training sample class size used in CFW-HSIC are worth being mentioned in the following:

1. It should be noted that CF is not designed according to classification difficulty. That is, a class with high CF does not necessarily imply that it is easy to be classified and vice versa. On the contrary, depending upon the image data to be processed, a class with a high CF probability may be actually difficult to be classified. Nevertheless, CF is indeed very closely related to and associated with class size. Because of that, the use of CF can prevent smaller classes from being overwhelmed and compromised by large classes as demonstrated in the experiments conducted in Section 19.5. Furthermore, in many practical applications, smaller classes generally provide crucial information for data analysts. Unlike SR which weighs more heavily on larger classes than on smaller classes in classification, CF can take care of classification of imbalanced classes by including their significance measured by Intra-CFs, Inter-CFs, or TCFs.
2. As pointed out in the introduction and Section 19.2, Intra-CFs only measure CV and have no interaction with other classes. By contrast, Inter-CFs not only measure CS from other classes but also CV, both of which are closely related to classification. TCFs try to take advantage of both Intra-CFs and Inter-class CFs. As a consequence, TCFs are expected to be right in between. The experiments performed in Section 19.5 indeed demonstrate the characteristics of these three CF criteria.
3. For a given fixed number of training samples, $N^{\text{training}}$, it is important to allocate an appropriate class size of training samples for each of classes with constraints imposed by $N^{\text{training}}$. It is true that a class with a high CD or a high CFR usually requires fewer samples to train and vice versa. In other words, adding more training samples for a large class with a low CF in CD and CFR to improve classification seems a right way to do. However, with $N^{\text{training}}$ being fixed, this may not be an effective way in an overall performance. This is because a larger class may require more training samples to improve its accuracy by the same level than that required for a smaller class

with a higher CF probability. As an example, if we are given $N$ dollars to purchase different items (i.e. each item corresponds to one class) weighted by their needs (the need is weighted by CF), what is the best and effective way to allocate this money to achieve our best need. Apparently, spending more money on items which we need (i.e. classes with high CF) is better than those items which are less needed (i.e. classes with low CF) using a fixed budget of $N$ dollars (i.e. a given fixed number of training samples). Using the same token, when it comes to class size allocation of training sample among different classes, CF provides a guideline for such allocation. As will be demonstrated in the following experiments, using CF can indeed increase classification accuracy with a significant amount up to 8% depending upon the classifiers to be used for classification. This was also shown in [4] where the numbers of training samples selected for smaller classes in the Purdue data (classes 9, 7, 1, and 16) with less than 100 data samples were set to 50% of the entire class size. Unfortunately, such selection was empirical according to (19.17) and no explanation was given in [4]. Interestingly, CF provides such justification as will be shown in the Purdue data experiments.

4. Furthermore, the CF-allocated training sample class size selection algorithm can be considered as an adaptive algorithm to balance CF of each class by adjusting $n_m^{\mathrm{UB}}$ used in (19.14) for smaller classes and $n_m^{\mathrm{LB}}$ used in (19.15) for larger classes. Since the total number of training samples to be used is fixed at $N^{\mathrm{training}}$, the class size of training samples allocated by (19.14) and (19.15) are actually against each other. For example, if we would like to impose less significance on smaller classes, the $n_m/2$ in (19.14) can be reduced to $n_m/3$, while increasing $n_m/100$ in (19.15) to $n_m/50$ to emphasize larger classes. This also illustrates the selection rule used in [4] where in [4], $n_m^{\mathrm{UB}} = n_{\mathrm{average}}$ and $n_m^{\mathrm{LB}} = 0$ in (19.14) and (19.15).

## 19.5 Experiments

The experiments conducted in this section constitute a very important part of the work presented in this chapter since the experimental results provide many interesting and intriguing findings that have never been explored in the past and reported in the literature. Also, CF is completely determined by data itself and not by a classifier. So, the conclusions drawn from the conducted experiments can be also applied to any classifier.

In the following experiments, three classifiers were selected to represent different types of classifiers which can take advantages of CF to improve classification. One is the commonly used spectral classifier, support vector machine (SVM), using RBF kernels with the same parameters used in [4]. A second one is recently developed spectral-spatial (SS) classifiers, the EPF-based methods in [4] where a comprehensive comparative analysis was performed among most existing SS techniques and the four EPF-based techniques, EPF-B-c, EPF-G-c, EPF-B-g, and EPF-G-g, were shown to be best classification techniques where "B" and "G" are used to specify bilateral filter and guided filter, respectively, and "g" and "c" indicate that the first principal component and color composite of three principal components are used as reference images [4]. According to this study, the EPF-G-g seemed to perform better in overall performance among the four EPF-based methods. In this case, the EPF-G-g was selected for our experiments as a representative to perform classification in comparison with CFW-HSIC using Intra-CFs, Inter-CFs, and Total CFs as criteria. Nevertheless, the other three EPF-based methods can be also used for experiments with similar conclusions. A third one is an iterative version of EPF-based methods recently developed in [7] which improved EPF-based methods using feedback loops. For fair comparison, the iterative EPF-G-g (IEPF-G-g) which was developed in [7] as the counterpart of EPF-G-g was also used for experiments.

### 19.5.1 *Purdue University's Indiana Indian Pines*

The first image to be used for experiments is a well-known airborne visible infrared imaging spectrometer (AVIRIS) image scene, Purdue Indiana Indian Pine test site shown in Figure 19.1(a), its ground truth in Figure 19.1(b), and aerial view in Figure 19.1(c). Table 19.1 also tabulates all the specific types of 16 classes. It has the size of 145 × 145 pixel vectors taken from an area of mixed agriculture and forestry in Northwestern Indiana, USA, with details of band and wavelength given in the caption.

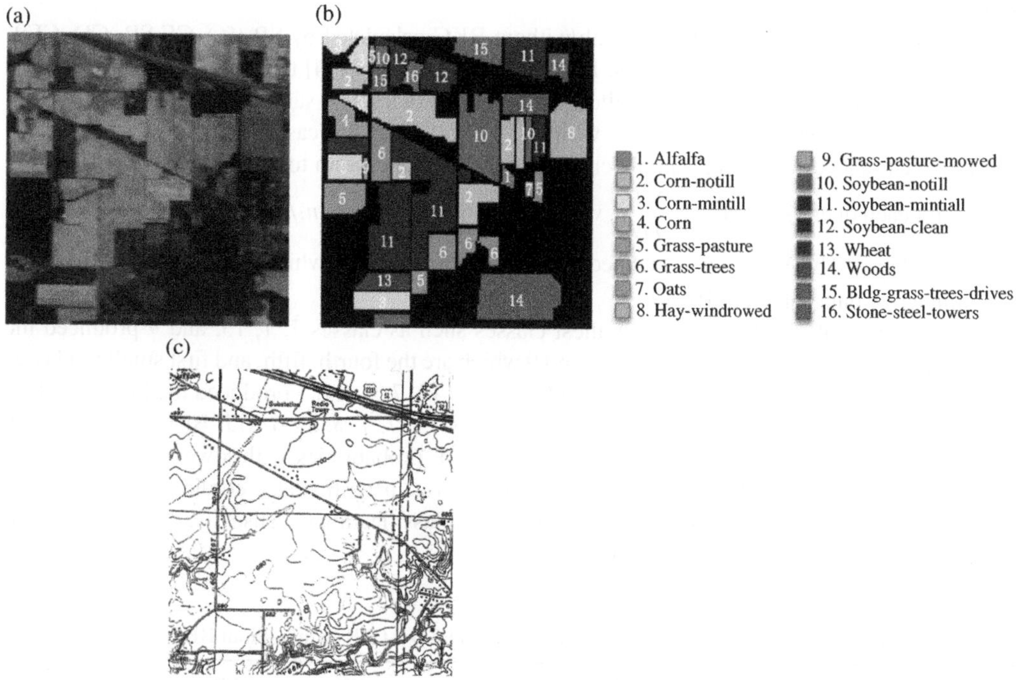


**Figure 19.1.** AVIRIS image scene: Purdue Indiana Indian Pines test site: (a) band 186 (2162.56 nm), (b) ground-truth map and classes by different degrees of gray intensities, and (c) a USGS Quadrangle map of the test site, where its color maps can be found in Ref. [8], figure 10/with permission of IEEE.

**Table 19.1** Class labels of Purdue Indiana Indian Pines with number of data samples in each class

| | | | | | |
|---|---|---|---|---|---|
| Class 1 (46) | Alfalfa | Class 7 (28) | Grass/pasture-mowed | Class 13 (205) | Wheat |
| Class 2 (1428) | Corn-notill | Class 8 (478) | Hay-windrowed | Class 14 (1265) | Woods |
| Class 3 (830) | Corn-min | Class 9 (20) | Oats | Class 15 (386) | Bldg-grass green-drives |
| Class 4 (237) | Corn | Class 10 (972) | Soybeans-notill | Class 16 (93) | Stone-steel towers |
| Class 5 (483) | Grass/pasture | Class 11 (2455) | Soybeans-min | Class 17 (10249) | BKG |
| Class 6 (730) | Grass/trees | Class 12 (593) | Soybeans-clean | | |

The data set is available at website https://purr.purdue.edu/publications/1947/serve/1?el=1. It was recorded in June 1992 with 220 bands including water absorption bands (bands 104–108, 150–163, 220). This scene is a very interesting data set because it is made up of various sizes of imbalanced classes. For example, there are 4 classes with data samples less than 100 (class 9 with 20 data samples, class 7 with 28 data samples, class 1 with 46 data samples, and class 16 with 93 data samples) and 3 classes with data samples greater than 1000 (class 11 with 2455 data samples, class 2 with 1428 data samples, and class 14 with 1265 data samples). So, this scene can be used to exhibit and explore the full utility of CF in classification and to prevent smallest classes being compromised by large classes.

Table 19.2 tabulates CF probabilities with/without BKG calculated by SR (%), CF-SR, CD, CFR, and CD × CFR for Purdue's Indian Pines. It should be noted that in [4] there are no CF probabilities calculated for EPF-G-g and IEPF-G-g to allocate the number of training samples for each class. Instead, these numbers were determined by (19.17) with $N^{\text{training}} = 1025$. In this case, the same $N^{\text{training}}$ was also used in the CF-allocated training sample class size selection algorithm to allocate the size of training samples for each of classes, $\left\{n_i^{\text{training}}\right\}_{i=1}^{M}$, with lower bounds, $\{n_i^{\text{LB}} = n_i/100\}_{i=1}^{M}$ and upper bounds, $\{n_i^{\text{UB}} = n_i/2\}_{i=1}^{M}$. The $N^{\text{training}}$ is calculated by the percentile of 10% which is the ratio of 1025 to the total number of data samples, 10249.

As shown in Table 19.2, the three smallest classes such as classes 7, 1, 13, and 9 produced the highest CD. Comparing to CD, class 16, 13, and 9 which are the fourth, fifth, and first smallest classes produced three highest CFR values. For Total CFs, class 13, 7, 9, and 1 produced the highest values without BKG while the highest TCFs occurred in class 13, 7, 1, and 16 with BKG. Additionally, Table 19.2 also shows that BKG generally had lower CF probabilities with lowest CD values and the fourth lowest CD × CFR values. The fact that BKG was insignificant according to CF provided the reason of why BKG has not been a major concern in classification because discarding BKG from classification may not be as serious as it is.

**Table 19.2** CF probabilities calculated by SR, CD, CFR, and CD × CFR with/without BKG for Purdue's data

| Class | CF-SR | CD | CFR | CD × CFR | SR (10%) (BKG) | CD (BKG) | CFR (BKG) | CD × CFR (BKG) |
|---|---|---|---|---|---|---|---|---|
| 1 | 0.0045 | 0.2061 | 0.0686 | 0.1712 | 0.0022 | 0.2047 | 0.0806 | 0.2107 |
| 2 | 0.1393 | 0.0139 | 0.0016 | 0.0003 | 0.0679 | 0.0138 | 0.0019 | 0.0003 |
| 3 | 0.0810 | 0.0187 | 0.0016 | 0.0004 | 0.0395 | 0.0186 | 0.0019 | 0.0005 |
| 4 | 0.0231 | 0.0079 | 0.0173 | 0.0017 | 0.0113 | 0.0079 | 0.0203 | 0.0020 |
| 5 | 0.0471 | 0.0096 | 0.0322 | 0.0037 | 0.0230 | 0.0095 | 0.0379 | 0.0046 |
| 6 | 0.0712 | 0.0296 | 0.0527 | 0.0189 | 0.0347 | 0.0294 | 0.0062 | 0.0023 |
| 7 | 0.0027 | 0.2168 | 0.0814 | 0.2138 | 0.0013 | 0.2152 | 0.0957 | 0.2631 |
| 8 | 0.0466 | 0.0765 | 0.0686 | 0.0635 | 0.0227 | 0.0759 | 0.0806 | 0.0781 |
| 9 | 0.0020 | 0.1318 | 0.1140 | 0.1820 | 0.0010 | 0.1308 | 0.0082 | 0.0137 |
| 10 | 0.0948 | 0.0261 | 0.0017 | 0.0005 | 0.0462 | 0.0259 | 0.0020 | 0.0007 |
| 11 | 0.2395 | 0.0222 | 0.0019 | 0.0005 | 0.1168 | 0.0221 | 0.0022 | 0.0006 |
| 12 | 0.0579 | 0.0083 | 0.0015 | 0.0001 | 0.0282 | 0.0082 | 0.0017 | 0.0002 |
| 13 | 0.0200 | 0.1764 | 0.1140 | 0.2437 | 0.0098 | 0.1751 | 0.1340 | 0.2999 |
| 14 | 0.1234 | 0.0247 | 0.0322 | 0.0096 | 0.0602 | 0.0245 | 0.0379 | 0.0119 |
| 15 | 0.0377 | 0.0132 | 0.0220 | 0.0035 | 0.0184 | 0.0131 | 0.0259 | 0.0043 |
| 16 | 0.0091 | 0.0184 | 0.3886 | 0.0865 | 0.0044 | 0.0182 | 0.4567 | 0.1065 |
| BKG | N/A | N/A | N/A | N/A | 0.5125 | 0.0070 | 0.0062 | 0.0006 |

If we further compare the training sample sizes selected for the four smallest classes, i.e. classes 9, 7, 1, and 16 by EPF in [4] to the sizes selected by the three CF measures, CD, CFR, and CD × CFR, we immediately discovered that both sets were nearly identical except class 1 with 25 training samples selected by EPF which went beyond 50% of the total data samples, 46 and with 23 training samples allocated by CF, which was exactly 50% set by the upper bound, $n_1^{\mathrm{UB}} = (46/2) = 23$. This showed that CF-allocated training sample sizes can well explain and justify the rationale of the training sample sizes selected by EPF for smaller classes to ensure that the classification accuracies of these classes would not be compromised by larger classes. Using the CF probabilities in Table 19.2 and (19.12)–(19.15), we can find the number of training samples allocated for each class and tabulated in Table 19.3.

A noted above, finding $\left\{n_i^{\mathrm{training}}\right\}_{i=1}^{M}$ for CF is necessary. In order to satisfy the constraints specified by (19.12)–(19.15), the total number of training samples, 1025 required in [4], the actual number of data sample, $\overline{N}$, used to achieve (19.12)–(19.13) is generally not equal to exactly 1025 but approximates $\sum_{i=1}^{M} n_i^{\mathrm{training}} \approx 1025$. Nevertheless, the total number of training samples used for all the three classifiers using CF was very close to 1025. As also noted in [4], the selection of the training samples by EPF/IEPF-based methods was not based on SR but rather empirical. For example, in order to take care of smaller classes, EPF/IEPF-based methods determined the numbers of training samples selected for the 4 smallest classes, classes 9, 7, 1, and 16 were 50% of the entire data samples in each of these 4 classes. These numbers are exactly the same numbers of training samples allocated by CD, CR, and CD × CFR for these four classes. This indicated that EPF/IEPF-based methods realized the issue of training sample size for smaller classes but unfortunately, they did not provide a guide for how the numbers of these training samples, i.e. the training sample sizes were determined for these classes. The proposed CF probabilities do offer specific rules of how a training sample size can be determined by (19.12)–(19.15).

**Table 19.3** Number of training samples allocated for each of classes for Purdue's data

| | | Number of training samples allocated in each of classes | | | | | |
|---|---|---|---|---|---|---|---|
| Class | Number of samples in classes | EPF-G-g | SR (10%) | CF-SR | CD | CFR | CD × CFR |
| 1 | 46 | 25 | 5 | 23 | 23 | 23 | 23 |
| 2 | 1428 | 83 | 143 | 87 | 46 | 15 | 15 |
| 3 | 830 | 78 | 84 | 48 | 62 | 9 | 9 |
| 4 | 237 | 68 | 24 | 34 | 26 | 58 | 23 |
| 5 | 483 | 79 | 49 | 63 | 32 | 107 | 51 |
| 6 | 730 | 78 | 74 | 92 | 98 | 175 | 256 |
| 7 | 28 | 14 | 3 | 14 | 14 | 14 | 14 |
| 8 | 478 | 66 | 48 | 239 | 239 | 228 | 239 |
| 9 | 20 | 10 | 3 | 10 | 10 | 10 | 10 |
| 10 | 972 | 81 | 98 | 77 | 86 | 10 | 10 |
| 11 | 2455 | 99 | 246 | 31 | 73 | 25 | 25 |
| 12 | 593 | 73 | 60 | 46 | 28 | 6 | 6 |
| 13 | 205 | 70 | 21 | 23 | 103 | 103 | 103 |
| 14 | 1265 | 90 | 127 | 27 | 82 | 107 | 131 |
| 15 | 386 | 65 | 39 | 169 | 44 | 73 | 48 |
| 16 | 93 | 46 | 10 | 44 | 47 | 47 | 47 |
| Total | 10249 | 1025 | 1034 | 1027 | 1013 | 1010 | 1010 |

Using the number of training samples provided in Table 19.3 for each class, Tables 19.4 and 19.5 tabulate $P_A(C_i)$, $P_{AA}$, $P_{OA}$, and $P_{CF\text{-}OA}$ produced by, SVM, EPF-G-g, and IEPF-G-g, respectively, where N/A indicates "not applicable." Since EPF/IEPF did not use CF to weigh classes, their $P_{OA}$ was calculated in the same way that was done in [4] and their results are tabulated in an extra row, i.e. the second last row in Table 19.5. Here, a note on $P_{AA}$ is worthwhile. $P_{AA}$ in (19.15) is calculated based on an assumption that the significance of all classes is equally likely, that is, $p_i^{CF} = (1/M)$ for $1 \le i \le M$. It was shown in the illustrative examples in [8] that it was not an appropriate criterion to measure classification performance. So, in the following experiments, the results on $P_{AA}$ will not be discussed for comparison. Nevertheless, their results are included for reference.

According to the results in Tables 19.4 and 19.5, how much classification performance can be improved was closely related to how good the used classifier is. For example, SVM only performed spectral classification and its classification performance was the worst among the three classifiers. In this case, SVM using CF improved $P_{OA}$ by 10–13% over SVM without using CF. On the other hand, EPF-G-g includes an edge-preserving filter which captures spatial information. Its classification outperformed SVM by nearly 10–12%. However, if EPF-G-g used CF to allocate the number of training samples and weigh classification accuracy for each of classes, $P_{CF\text{-}OA}$ improved $P_{OA}$ by 3–4.5%. Now, if IEPF-G-g was used by including feedback loops, its $P_{OA}$ was improved by 3.3% over $P_{OA}$ produced by EPF-G-g. But if IEPF-G-g further used CF to allocate the number of training samples and weigh classification accuracy for each of classes, its $P_{CF\text{-}OA}$ could be even improved by 1.5% over $P_{OA}$ that was produced by IEPF without using CF. Given the fact that the $P_{OA}$ produced by IEPF was 0.9819 which is already extremely high, 1.5% improvement offered by CF is indeed very difficult and significant. As for four smallest classes with the number of data samples less than 100, CF took very care of them by keeping their accuracy rates nearly one, specifically, IEPF. There is an interesting observation on accuracy of the smallest class, class 9, with using SR as CF for EPF and IEPF in Table 19.7. Since there are only 20 data samples in class 9, two training samples are

**Table 19.4** $P_A(C_i)$, $P_{AA}$, and $P_{CF\text{-}OA}$ produced by SVM for Purdue's data without BKG

| | SR (10%) | CF-SR | CD | CFR | CD × CFR |
|---|---|---|---|---|---|
| Class | $P_A(C_i)$ | $P_A(C_i)$ | $P_A(C_i)$ | $P_A(C_i)$ | $P_A(C_i)$ |
| 1 | 0.6522 | 0.9565 | 0.9783 | 0.9130 | 0.9130 |
| 2 | 0.8179 | 0.7962 | 0.6772 | 0.4895 | 0.5763 |
| 3 | 0.6819 | 0.6843 | 0.7566 | 0.2699 | 0.4289 |
| 4 | 0.7848 | 0.8945 | 0.7764 | 0.9072 | 0.9494 |
| 5 | 0.9296 | 0.9441 | 0.8903 | 0.9689 | 0.9275 |
| 6 | 0.9712 | 0.9534 | 0.9699 | 0.9890 | 0.9986 |
| 7 | 0.4643 | 0.9286 | 0.9643 | 1.0000 | 1.0000 |
| 8 | 0.9854 | 0.9958 | 0.9958 | 0.9958 | 0.9979 |
| 9 | 0.6000 | 0.9000 | 1.0000 | 0.9000 | 0.9500 |
| 10 | 0.7881 | 0.8940 | 0.8652 | 0.5885 | 0.5967 |
| 11 | 0.8477 | 0.4395 | 0.6949 | 0.5210 | 0.5817 |
| 12 | 0.8010 | 0.8364 | 0.6998 | 0.2901 | 0.2479 |
| 13 | 0.9805 | 0.9756 | 0.9902 | 1.0000 | 1.0000 |
| 14 | 0.9478 | 0.7383 | 0.9217 | 0.8775 | 0.9391 |
| 15 | 0.7047 | 0.9534 | 0.7021 | 0.8083 | 0.7358 |
| 16 | 0.8602 | 1.0000 | 1.0000 | 1.0000 | 1.0000 |
| $P_{AA}$ | 0.8011 | 0.8682 | 0.8677 | 0.7824 | 0.8027 |
| $P_{CF\text{-}OA}$ | 0.8467 | 0.7486 | 0.9544 | 0.9663 | 0.9731 |

**Table 19.5** $P_A(C_i)$, $P_{AA}$, $P_{OA}$, and $P_{CF\text{-}OA}$ produced by EPF-G-g and IEPF-G-g for Purdue's data without BKG

| | EPF/IEPF-G-g | SR (10%) | CF-SR | CD | CFR | CD × CFR |
|---|---|---|---|---|---|---|
| Class | $P_A(C_i)$ | $P_A(C_i)$ | $P_A(C_i)$ | $P_A(C_i)$ | $P_A(C_i)$ | $P_A(C_i)$ |
| 1 | 0.9783/0.9783 | 0.8913/0.9783 | 0.9783 /0.9783 | 1.0000/0.9783 | 1.0000/0.9783 | 1.0000/1.0000 |
| 2 | 0.8894/0.9531 | 0.9384/0.9923 | 0.8866/0.9811 | 0.8158/0.907 | 0.6793/0.8081 | 0.5798/0.8746 |
| 3 | 0.9060/0.9976 | 0.8807/0.9554 | 0.8916/0.9530 | 0.9205/0.9831 | 0.5747/0/7916 | 0.5482/0/7241 |
| 4 | 1.0000/1.0000 | 0.9958/0.9958 | 1.0000/1.0000 | 1.0000/1.0000 | 1.0000/1.0000 | 1.0000/0.9873 |
| 5 | 0.9545/1.0000 | 0.9752/0.9752 | 0.9627/0.9710 | 0.8903/0.8944 | 0.9814/1.0000 | 0.9524/1.0000 |
| 6 | 1.0000/0.9959 | 1.0000/0/9986 | 0.9945/1.0000 | 1.000/0.9986 | 1.0000/0.9986 | 1.0000/0.9986 |
| 7 | 0.9643/0.9643 | 0.2857/0.9643 | 0.9643/0.9286 | 0.9643/1.0000 | 0.9643/0.9643 | 0.9643/1.0000 |
| 8 | 1.0000/1.0000 | 1.0000/1.0000 | 1.0000/1.0000 | 1.0000/1.0000 | 1.0000/1.0000 | 1.0000/1.0000 |
| 9 | 1.0000/1.0000 | 0.6000 /0.6000 | 0.8000/1.0000 | 1.0000/1.0000 | 0.9500/1.0000 | 1.0000/1.0000 |
| 10 | 0.9794/0.9588 | 0.8971/0.9609 | 0.9691/0.9743 | 0.9877/0.9784 | 0.6975/0.8076 | 0.6214/0.7243 |
| 11 | 0.9145/0.9788 | 0.9857/0.9906 | 0.5874/0.7409 | 0.8680/0.9430 | 0.8823/0.8827 | 0.8933/0.7890 |
| 12 | 0.9949/0.9848 | 0.9966/0.9983 | 0.9966/0.9966 | 0.9427/0.9882 | 0.0152/0.6206 | 0.3541/0.5363 |
| 13 | 1.0000/1.0000 | 0.9902/0.9902 | 0.9902/0.9902 | 1.0000/1.0000 | 0.9951/1.0000 | 1.0000/0.9951 |
| 14 | 0.9905/0.9984 | 0.9897/0.9945 | 0.7747/0.9043 | 0.9407/0.9992 | 0.9186/0.9992 | 0.9960/0.9945 |
| 15 | 0.9430/0.9767 | 0.9689/0.9974 | 1.0000/1.0000 | 0.9974/1.0000 | 1.0000/1.0000 | 0.8782/1.0000 |
| 16 | 1.0000/1.0000 | 1.0000/0.9892 | 1.0000/0.9892 | 1.0000/1.0000 | 1.0000/1.0000 | 1.0000/0.9892 |

randomly selected. In this case, EPF-G-g using CF = SR (2%) yielded 5% accuracy compared to 100% accuracy produced by EPF-G-g using 10 randomly selected training samples. But the $P_{SR\text{-}OA} = 0.9564$ produced by EPF-G-g using CF = SR (2%) was higher than $P_{OA} = 0.9486$ EPF-G-g without using CF. This showed that a higher $P_{OA}$ can be achieved by sacrificing accuracy of small classes. However, such incident can be avoided and corrected if feedback loops of spatial information are included in IEPF-G-g as shown in Table 19.5 with 100% accuracy for both cases. The above experiments clearly demonstrated the advantages and benefits provided by CF.

### *19.5.2 University of Pavia*

A second hyperspectral image data shown in Figure 19.2(a) was collected and recorded by the ROSIS-03 satellite sensor over an urban area surrounding the University of Pavia, Italy. It is of size 610 × 340 × 115 with a spatial resolution of 1.3-m/pixel and a spectral coverage ranging from 0.43 to 0.86 μm with spectral resolution of 4 nm (12 most noisy channels were removed before experiments). Nine classes of interest are considered for this image. Figure 19.2(b) provides its ground-truth map along with color class labels in Figure 19.2(c). Table 19.6 also tabulates the number of data samples in parentheses collected for each class.

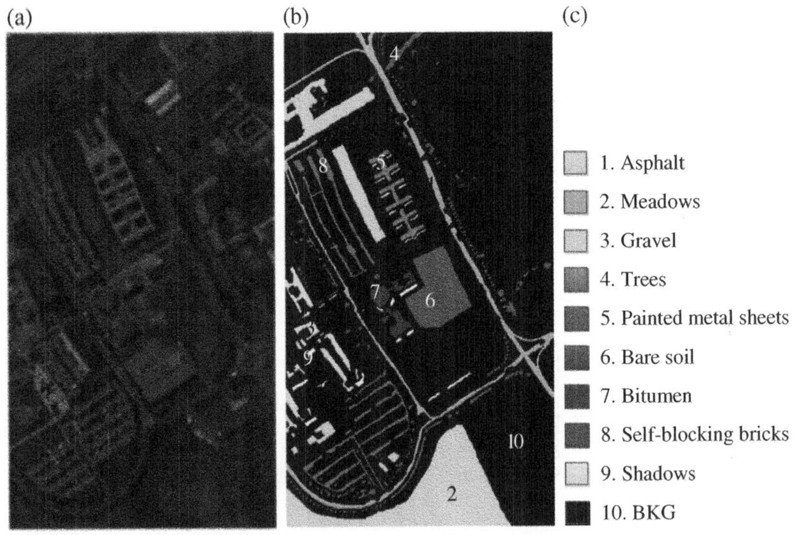


**Figure 19.2.** Ground truth of University of Pavia scene with nine classes: (a) University of Pavia scene, (b) ground-truth map, and (c) classes by different degrees of gray intensities, where its color maps can be found in Ref. [8], figure 12/with permission of IEEE.

**Table 19.6** Class labels of University of Pavia with number of data samples in each class

| Class 1 (6631) | Asphalt | Class 5 (1345) | Painted metal sheets | Class 9 (947) | Shadows |
|---|---|---|---|---|---|
| Class 2 (18649) | Meadows | Class 6 (5029) | Bare soil | Class 10 (164624) | BKG |
| Class 3 (2099) | Gravel | Class 7 (1330) | Bitumen | | |
| Class 4 (3064) | Trees | Class 8 (3682) | Self-blocking bricks | | |

The University of Pavia scene has a data structure similar to that of Salinas where all classes are also relatively large with no particularly small classes except three classes (i.e. class 9 with 947 data samples, class 7 with 1330 data samples, and class 5 with 1345 data samples) relatively small compared to three largest classes (class 2 with 18649 data samples, class 1 with 6631 data samples, and class 6 with 5029 data samples). In this case, Kang et al. [4] selected 285 training samples for each class. So, in the following experiments, the same total number of training samples, $N = 2565 = 285 \times 9$, was used by CF-allocated training sample class size-finding algorithm to allocate size of the training samples for each of classes, $\left\{n_i^{\text{training}}\right\}_{i=1}^{M}$, with lower bounds, $\{n_i^{\text{LB}} = n_i/100\}_{i=1}^{M}$ and upper bounds, $\{n_i^{\text{UB}} = n_i/2\}_{i=1}^{M}$. Specifically, the percentile used by SR was 6% calculated based on 2565. In this case, the percentile used by SVM for selection of training samples was set to 6%. Table 19.7 tabulates CF probabilities with/without BKG calculated by SR, CD, CFR, and CD × CFR for this scene. Table 19.8 tabulates the number of training samples allocated by CF probabilities in Table 19.7.

Using the number of training samples provided in Table 19.8 for each class, Table 19.9 tabulates $P_A(C_i)$, $P_{AA}$, $P_{OA}$, and $P_{CF\text{-}OA}$ produced by EPF-G-g and IEPF-G-g, respectively. Once again, since

**Table 19.7** CF probabilities calculated by SR, CD, CFR, and CD × CFR for University of Pavia

| Class | CF-SR | CD | CFR | CD×CFR | CF-SR (BKG) | CD (BKG) | CFR (BKG) | CD × CFR (BKG) |
|---|---|---|---|---|---|---|---|---|
| 1 | 0.1550 | 0.0462 | 0.0044 | 0.0087 | 0.0320 | 0.0457 | 0.0143 | 0.0164 |
| 2 | 0.4360 | 0.0409 | 0.0403 | 0.0713 | 0.0899 | 0.0405 | 0.0675 | 0.0682 |
| 3 | 0.0491 | 0.1503 | 0.0016 | 0.0101 | 0.0101 | 0.1488 | 0.0051 | 0.0190 |
| 4 | 0.0716 | 0.0529 | 0.0967 | 0.2215 | 0.0148 | 0.0524 | 0.3188 | 0.4170 |
| 5 | 0.0314 | 0.0381 | 0.0589 | 0.0970 | 0.0065 | 0.0377 | 0.1940 | 0.1825 |
| 6 | 0.1176 | 0.0263 | 0.0403 | 0.0458 | 0.0242 | 0.0260 | 0.0047 | 0.0030 |
| 7 | 0.0311 | 0.3644 | 0.0044 | 0.0686 | 0.0064 | 0.3608 | 0.0143 | 0.1292 |
| 8 | 0.0861 | 0.2669 | 0.0016 | 0.0180 | 0.0178 | 0.2642 | 0.0051 | 0.0338 |
| 9 | 0.0221 | 0.0141 | 0.7520 | 0.4591 | 0.0046 | 0.0140 | 0.3715 | 0.1296 |
| BKG | N/A | N/A | N/A | N/A | 0.7938 | 0.0098 | 0.0047 | 0.0012 |

**Table 19.8** Numbers of training samples allocated for each of classes for University of Pavia

| | | Number of training samples allocated in each of classes | | | | | |
|---|---|---|---|---|---|---|---|
| Class | Number of samples in classes | EPF-G-g | SR (2%) | CF-SR | CD | CFR | CD×CFR |
| 1 | 6631 | 285 | 398 | 398 | 133 | 67 | 67 |
| 2 | 18649 | 285 | 1119 | 1119 | 187 | 328 | 270 |
| 3 | 2099 | 285 | 126 | 126 | 431 | 21 | 39 |
| 4 | 3064 | 285 | 184 | 184 | 152 | 787 | 839 |
| 5 | 1345 | 285 | 81 | 81 | 110 | 479 | 367 |
| 6 | 5029 | 285 | 302 | 302 | 76 | 328 | 174 |
| 7 | 1330 | 285 | 80 | 80 | 665 | 36 | 260 |
| 8 | 3682 | 285 | 221 | 221 | 766 | 37 | 68 |
| 9 | 947 | 285 | 57 | 57 | 41 | 474 | 474 |
| Total | 42776 | 2565 | 2568 | 2568 | 2561 | 2557 | 2558 |

*Source:* Based on Ref. [4].

**Table 19.9** $P_A(C_i)$, $P_{PR}(C_i)$, $P_{PR}(C_i)$, $P_{OA}$, $P_{CF\text{-}OA}$, and $P_{PR}$ produced by EPF-G-g and IEPF-G-g for University of Pavia

| | EPF/IEPF-G-g | SR (2%) | CF-SR | CD | CFR | CD × CFR |
|---|---|---|---|---|---|---|
| Class | $P_A(C_i)$ | $P_A(C_i)$ | $P_A(C_i)$ | $P_A(C_i)$ | $P_A(C_i)$ | $P_A(C_i)$ |
| 1 | 0.9646/0.9894 | 0.9858/0.9973 | 0.9851/0.9973 | 0.8815/0.9887 | 0.9739/0.9827 | 0.9332/0.9759 |
| 2 | 0.9853/0.9884 | 1.0000/0.9996 | 0.9991/1.0000 | 0.9976/0.9495 | 0.9849/0.9915 | 0.9841/0.9932 |
| 3 | 0.9095/0.9462 | 0.8766/0.9929 | 0.8113/0.9285 | 0.8823/0.9995 | 0.7527/0.9295 | 0.7732/0.9819 |
| 4 | 0.9484/0.9426 | 0.8877/0.9912 | 0.9112/0.9543 | 0.9471/0.9951 | 0.9654/0.9977 | 0.9648/0.9997 |
| 5 | 1.0000/1.0000 | 0.9993/0.9933 | 1.0000/0.9985 | 1.0000/1.0000 | 1.0000/0.9993 | 0.9993/1.0000 |
| 6 | 1.0000/1.0000 | 0.9847/1.0000 | 0.9905/1.0000 | 0.9887/0.9897 | 1.0000/1.0000 | 0.9895/1.0000 |
| 7 | 0.9962/0.9932 | 0.9669/1.0000 | 0.9654/1.0000 | 1.0000/1.0000 | 0.9902/1.0000 | 1.0000/1.0000 |
| 8 | 0.9764/0.9853 | 0.9870/0.9984 | 0.9897 /0.9973 | 0.9981/1.0000 | 0.9783/0.9957 | 0.9924/0.9764 |
| 9 | 1.0000/1.0000 | 1.0000/1.0000 | 1.0000/0.9989 | 1.0000/0.9989 | 1.0000/1.0000 | 1.0000/1.0000 |
| $P_{AA}$ | 0.9756/0.9828 | 0.9653/0.9976 | 0.9614/0.9861 | 0.9661/0.9913 | 0.9606/0.9885 | 0.9596/0.9919 |
| $P_{OA}$ | 0.9147/0.9851 | N/A | N/A | N/A | N/A | N/A |
| $P_{CF\text{-}OA}$ | N/A | 0.9797/0.9982 | 0.9786/0.9925 | 0.9731/0.9968 | 0.9955/0.9992 | 0.9875/0.9986 |

SVM cannot compete with EPF/IEPF methods, their results for University of Pavia are not included here either. Because EPF/IEPF did not use CF to weigh classes, $P_{OA}$ was calculated according to [4] and their results are tabulated in the second last row in Tables 19.8. As shown in Table 19.9, its $P_{CF\text{-}OA}$ produced by EPF-G-g using CD and CFR or CD $\times$CFR also improved significantly over its $P_{OA} = 0.9147$ by 6.5% up to 7.3%, respectively. Interestingly, when EPF was replaced by IEPF as a classifier, its $P_{OA}$ suddenly jumped to 0.9851, which is already very high and nearly perfect with only 1.5% away from 100%. In this case, improving its rate will be considered extremely difficult. However, CF-weighted IEPF actually did it by pulling its $P_{OA}$ from 0.9851 up to 0.9966 by 1% and to 0.9992 by 1.5% in Table 19.9.

A few interesting findings can be observed from the above experimental results.

1. As demonstrated by all the three image experiments, CFW-HSIC performed better than HSIC without using CF no matter what a classifier is used. This is a tremendous advantage for any classifier to be used for any image scene in classification.
2. CF seems particularly useful for a classifier which is not effective such as SVM.
3. In HSIC, there is always a great challenge in dealing with classification of imbalanced classes [9]. Unlike [9] which expanded smaller classes and reduced large classes by orthogonal subspace projection, we resolve this issue by introducing CF to weigh each of classes according to its significance without changing data samples. In particular, CF can retain the accuracy of smaller classes so that these classes will not be compromised by large classes. On the other hand, the classification produced by large classes will not over-dominate the entire classification accuracy if CF is appropriately imposed.
4. One of the most important novel ideas in this chapter is to use CF probabilities to allocate a desired size of training samples used for each class. An automatic algorithm for selecting CF-allocated training sample class size is developed to address this issue. This algorithm is adaptive and can be adjusted to vary with the total number of training samples, $N^{\text{training}}$, to be used for classification. The selection training sample class size used in [4] can be considered as a special case of the algorithm.
5. As also shown in the above experiments, IEPF-G-g always performed very well and its produced $P_{OA}$ was also very high. Generally speaking, it is very difficult to increase such high $P_{OA}$ even with slight improvement. CFW-HSIC successfully improved OA to some extent from 1% over $P_{OA} = 0.98$ to 8% over $P_{OA} = 0.91$.

## 19.6 Novelties

Several novelties derived from CFW-HSIC are briefly described as follows:

1. The first novelty is to design criteria to capture CFs, referred to as Intra-CF, Inter-CF, and TCF. The Intra-CF can be used to characterize properties within a class, i.e. CV, while Inter-CF can be used to characterize properties between classes, i.e. CS. The TCS combines an intra-CF and an inter-CF into a single CF. By virtue of these three CFs, we can address the issue of imbalanced classes. Although FR has been used in pattern classification, it is indeed used as a criterion to design and develop "classifiers" not "class information" as we propose in the chapter. Specifically, the criteria of CD and CFR defined for classes are absolutely new. FR is calculated for pair-wise classes not separability of a class different from all other classes as CFR is designed, whereas CD is defined to measure the density of a class which has also never been used for classification. Technically speaking, they are developed from different design rationales. The concepts of intra-class variability and inter-class separability are very new. To our best knowledge, they cannot be found in the existing literature. Similarly, it is also true for TCF which combines CD and CFR.

2. A second novelty is to use the obtained CFs to calculate CF-probabilities to prioritize each class in terms of class significance. In other words, the class can be prioritized by their associated CF probabilities according to its class significance. The higher the CF probability calculated for a class is, the more significant the class is.
3. A third novelty is to use CF probabilities to allocate the number of training samples for each of classes. Generally, the selection of training samples is carried out by a certain percentile (%) of ground-truth samples for each class. As a result, the total number of training samples varies with the chosen percentile (%). Instead of predetermining percentile (%), it assumes that the total number of training samples, $N^{\text{training}}$, is given to be used for classification. Then, CF probabilities are calculated to allocate the number of training samples among $N^{\text{training}}$ training samples for each class. More specifically, an algorithm can be designed by using CF probabilities to determine the number of training samples required for each class not by its class size as commonly done by percentile (%) or SR but rather by its CF probability.
4. A fourth novelty is to use CF probabilities assigned to each of classes as weights to extend HSIC to CFW-HSIC. The higher the CF probability of a class is, the more weight the class is assigned and thus, the more significant the class is. So, when it comes to calculating OA, CF probabilities can be factored in OA to avoid small classes being compromised or overwhelmed by large classes. As will be demonstrated in the experiments, such CF-CF-OA indeed improves OA significantly. Interestingly, SR can be also used to calculate its associated CF, denoted by CF-SR which determines the number of training samples for each of classes, $\left\{n_i^{\text{training,SR}}\right\}_{i=1}^{M}$. This is quite different from the traditional SR, referred to as SR (%) in the sense that SR (%) selects % of class data samples as training samples.
5. The fifth novelty is a generation of $P_{AA}$ and $P_{OA}$ to $P_{CF\text{-}OA}$. Over the past years, $P_{AA}$ has not received as much as attention as $P_{OA}$ has. This is unfortunate since $P_{AA}$ can be as important as $P_{OA}$. This is because $P_{AA}$ considers that all classes are equally significant in which case $P_{AA}$ can avoid the accuracy of small classes being overwhelmed or dominated by the accuracy of large classes as what $P_{OA}$ does. Most importantly, $P_{CF\text{-}OA}$ can be considered as a generalized version of $P_{AA}$ and $P_{OA}$ by including CF as a measure of class significance with CF = $1/M$ to derive $P_{AA}$ and CF = SR to derive $P_{OA}$.
6. The last and sixth novelty is to use CF probabilities to address the BKG issue. Since the BKG class generally has a large number of widespread data samples, its CF probability is usually very small. As a result, including CF probabilities as weighting factors in classification can reduce the effect of BKG on classification as illustrated by experiments. This may explain why the BKG issue is not of major concern and excluded from HSIC.

Finally, in order to further justify the motivation of developing CFs for classification, the following simple experiments were performed for the Purdue data in Figure 19.1 using only 5% of data samples (i.e. 512 data samples) as training samples. Using SR, CD, CFR, and CD × CFR as CFs, the CF-allocated training sample size selection algorithm was implemented to determine the number of training samples needed to be selected for each of 16 classes, $\left\{n_i^{\text{training,CF}}\right\}_{i=1}^{16}$, tabulated in Table 19.10, where SR is the traditional method to select 5% out of data samples in each of classes and CF-SR is to use SR as CF to calculate SR probabilities to determine $\left\{n_i^{\text{training,SR}}\right\}_{i=1}^{16}$. As a result, class 1 can now have only two data samples and both class 7 and class 9 can have only one data sample to be used as training samples. Tables 19.11 tabulates the $P_A(C_i)$, $P_{AA}$, $P_{OA}$, and $P_{CF\text{-}OA}$ produced by EPF-G-g and IEPF-G-g, respectively, where the accuracies of classes 1, 7, and 9 using SR (5%) are all zeros with very low $P_{AA}$, but $P_{OA}$ was still very high, 0.9086. However, if we factor in the CF-SR probabilities in

**Table 19.10** Numbers of training samples allocated for each of classes for Purdue's data using 5% of data samples as training samples

| Class | Number of samples in classes | Number of training samples allocated in each of classes | | | | | |
|---|---|---|---|---|---|---|---|
| | | EPF-G-g | SR (5%) | CF-SR | CD | CFR | CD × CFR |
| 1 | 46 | 23 | 2 | 22 | 23 | 23 | 23 |
| 2 | 1428 | 41 | 71 | 41 | 41 | 15 | 20 |
| 3 | 830 | 36 | 41 | 22 | 44 | 12 | 14 |
| 4 | 237 | 33 | 12 | 17 | 25 | 66 | 40 |
| 5 | 483 | 35 | 24 | 30 | 50 | 35 | 45 |
| 6 | 730 | 33 | 36 | 43 | 28 | 95 | 66 |
| 7 | 28 | 14 | 1 | 14 | 14 | 8 | 11 |
| 8 | 478 | 33 | 24 | 118 | 18 | 6 | 9 |
| 9 | 20 | 10 | 1 | 10 | 10 | 10 | 10 |
| 10 | 972 | 35 | 49 | 36 | 16 | 21 | 14 |
| 11 | 2455 | 46 | 123 | 27 | 51 | 25 | 35 |
| 12 | 593 | 33 | 30 | 22 | 80 | 47 | 96 |
| 13 | 205 | 33 | 10 | 11 | 70 | 95 | 103 |
| 14 | 1265 | 40 | 63 | 15 | 22 | 38 | 20 |
| 15 | 386 | 34 | 19 | 76 | 25 | 6 | 8 |
| 16 | 93 | 33 | 5 | 21 | 7 | 22 | 6 |
| Total | 10249 | 512 | 511 | 525 | 524 | 524 | 520 |

selection of training samples for each of classes as shown in Table 19.11, the accuracies of classes 1, 7, and 9 were not only very high nearly 100%, but also both $P_{AA}$ and $P_{OA}$ were very high as well. These simple experiments provide hard evidence that how the accuracies of small classes are sacrificed and compromised by that of large classes when the total number of training samples is very limited to practical constraints. This demonstrates the use of CF is crucial to small classes and the concept of CF can indeed address such imbalanced class issue. It is worth noting that the training sample size of each class used by EPF-G-g in Table 19.8 was not SR but rather determined by the training sample average, $n^{\text{average}} = (N/M) = (512/16) = 32$, in (19.17), which can be also considered as another type of CF.

As a concluding remark, we should point it out that the goal of this chapter is not to develop a new HSIC method but rather to introduce the new concept of CFs into HSIC deal with issues of imbalanced classes and BKG. CF probabilities can be used to select the training sample size of each of classes as well as generalize OA to CF-OA. Using CF-OA, any HSIC method can be further extended to CFW-HSIC. In other words, CFW-HSIC is determined by CFs based on data statistics. It is independent of a specific HSIC method. This implies that CFW-HSIC can be also applied to any HSIC method. In our experiments, the SS classification methods such as EPF-based methods were particularly selected to justify the utility of the proposed CF-based approach to HSIC. There are two reasons that EPF-based methods in [4] were selected for experiments. One is that EPF-based methods have been shown to perform better than many existing HSIC methods. The other is that their codes are available at the authors' website [3] so that readers can repeat what we did in our experiments for their comparison. Of course, CFW-HSIC can be also applied to any other HSIC method in a similar manner such as deep learning networks [10–12], and convolutional neural networks [13–19] for the same purpose. Accordingly, for each specific HSIC method, CFW-HSIC can improve the HSIC method by including CFs in classification.

**Table 19.11** $P_A(C_i)$, $P_{AA}$, $P_{OA}$, and $P_{CF\text{-}OA}$ produced by EPF-G-g and IEPF-G-g for Purdue's data without BKG

| | EPF/IEPF-G-g | SR (5%) | CF-SR | CD | CFR | CD × CFR |
|---|---|---|---|---|---|---|
| Class | $P_A(C_i)$ | $P_A(C_i)$ | $P_A(C_i)$ | $P_A(C_i)$ | $P_A(C_i)$ | $P_A(C_i)$ |
| 1 | 1.0000/0.9783 | 0/0.9783 | 0.9783/0.9783 | 1.0000/0.9783 | 1.0000/1.0000 | 0.9783/0.9783 |
| 2 | 0.8550/0.8536 | 0.9209/0.9846 | 0.8522/0.8845 | 0.8648/0.8880 | 0.4433/0.8277 | 0.5735/0.6814 |
| 3 | 0.6193/0.9337 | 0.6470/0.9590 | 0.8651/0.9675 | 0.8988/0.9289 | 0.3759/0.7084 | 0.1771/0.7000 |
| 4 | 1.0000/1.0000 | 0.9451/1.0000 | 1.0000/1.0000 | 1.0000/1.0000 | 1.0000/1.0000 | 1.0000/1.0000 |
| 5 | 0.9151/0.9317 | 0.9130/0.9482 | 0.9275/0.9213 | 0.9896/0.9917 | 0.9234/0.9710 | 0.9627/0.9834 |
| 6 | 0.9740/0.9781 | 0.9904/0.9986 | 0.9918/0.9986 | 0.9986/0.9986 | 1.0000/1.0000 | 0.9918/1.0000 |
| 7 | 0.9643/0.9643 | 0/0 | 0.9643/0.9286 | 0.9643/0.9286 | 0.9643/0.9643 | 0.9643/0.9286 |
| 8 | 1.0000/1.0000 | 1.0000/1.0000 | 1.0000/1.0000 | 0.9937/0.99958 | 0.7803/1.0000 | 0.8933/0.9812 |
| 9 | 1.0000/1.0000 | 0/0 | 0.9500/1.0000 | 1.0000/1.0000 | 1.0000/1.0000 | 1.0000/1.0000 |
| 10 | 0.8549/0.8508 | 0.9496/0.9825 | 0.9136/0.9198 | 0.6337/0.7500 | 0.7397/0.8220 | 0.6193/0.7325 |
| 11 | 0.8676/0.8909 | 0.9625/0.9564 | 0.8318/0.9312 | 0.8114/0.9063 | 0.7853/0.8318 | 0.8310/0.8896 |
| 12 | 0.9578/0.9983 | 0.9477/0.99983 | 0.7066/0.9022 | 0.9966/0.99983 | 0.9410/0.99949 | 0.9933/0.9983 |
| 13 | 0.9951/0.9902 | 0.9902/0.9902 | 0.9902/0.9902 | 0.9902/0.9902 | 1.0000/1.0000 | 0.9951/0.9902 |
| 14 | 0.9834/1.0000 | 0.9858/0.9747 | 0.7028/0.9170 | 0.7889/0.7984 | 0.9929/0.9968 | 0.9874/0.9992 |
| 15 | 0.9922/0.9896 | 0.5285/0.8575 | 1.0000/0.9170 | 0.8860/0.7984 | 0.0389/0.9968 | 0.0285/0.9992 |
| 16 | 0.9892/0.9892 | 0.9892/0.9677 | 0.9892/0.9892 | 0.9892/0.9892 | 0.9892/0.9892 | 0.9892/0.9892 |
| $P_{AA}$ | 0.9355/0.99593 | 0.7356/0.8498 | 0.9165/0.958 | 0.9254/0.9415 | 0.8109/0.9189 | 0.8115/0.9176 |
| $P_{OA}$ | 0.8926/0.9282 | 0.9086/0.9660 | N/A | N/A | N/A | N/A |
| $P_{CF\text{-}OA}$ | N/A | N/A | 0.8617/0.9369 | 0.9661/0.9675 | 0.9337/0.9689 | 0.9572/0.9747 |

## 19.7 Conclusion

This chapter introduces three types of CF, Intra-CF, Inter-CF and TCF, to account for significance of classes of interest so as to address issues of imbalanced classes and BKG. The CFs derived from these criteria can be used to calculate CF probabilities to prioritize significance of each of classes. These CF-calculated probabilities are then used to determine the size of training samples for each of classes and also used as weights to calculate CF-OA which can be considered as a generalization of the commonly used classification measures, AA and OA. Including CF-calculated probabilities into HSIC derives CFW-HSIC which can be considered as an extension to the conventional HSIC. Specifically, when CF is specified by SR, the SR-weighted HSIC is reduced to commonly used HSIC. On the other hand, when $P_{AA}$ is used, it assumes that CF is weighted equally for all the classes. Using the idea of CF to weigh different classes in classification, CFW-HSIC has significant advantages over HSIC. Most importantly, in order to avoid small classes being compromised by large classes, Intra-CFs can be used to retain their crucial information in classification. In order for a classifier to be effective, Inter-CFs can be used to increase the classification accuracy of a classifier. Additionally, TCFs are further used to take care of Intra-CFs and Inter-CFs in terms of CV and CS. Another key novel idea resulting from CF is development of an adaptive algorithm for users to automatically determine how many training samples should be allocated to each class with a given fixed total number of training samples, $N^{\text{training}}$, that users can afford in real applications. This CF-allocated training sample class size selection algorithm can adapt to balancing the classification between small and large classes. To our best knowledge, no such algorithm has been reported in the literature. Finally, we conduct a comprehensive analysis and study via extensive experiments to demonstrate that the CF-OA, $P_{CF\text{-}OA}$, can truly reflect the significance of each class which plays in classification, while the commonly used $P_{AA}$ and $P_{OA}$ cannot. Finally, we hope that the work presented in this chapter will lead to several new directions to HSIC. For example, CF-calculated probabilities can be used to prioritize all classes of interest to be processed progressively, one class at a time as progressive HSIC. Such an idea can be also used to expand and navigate one-class classification [20–26] in a progressive manner as progressive one-class classification. A third one is that CF-calculated probabilities can provide class information using information theory to help select variable band numbers for each of classes for HSIC [27]. A fourth one is that CF-calculated probabilities can be also used to develop a progressive band processing approach to HSIC [28].

## References

1. Zhong, S., Chang, C.-I, Li, J. et al. (2019). Class Feature Weighted Hyperspectral Image Classification. *IEEE Journal of Selected Topics in Applied Earth Observations and Remote Sensing* 12 (12): 4728–4745.
2. Bishop, C.M. (2006). Machine Learning and Pattern Recognition. Springer.
3. Xudong Kang's Homepage [Online]. http://xudongkang.weebly.com.
4. Kang, X., Li, S., and Benediktsson, J.A. (2014). Spectral-spatial hyperspectral image classification with edge-preserving filtrering. *IEEE Transaction on Geoscience and Remote Sensing* 52 (5): 2666–2677.
5. Millard, K. and Richardson, M. (2015). On the importance of training data sample selection in random forest image classification: a case study in peatland ecosystem mapping. *Remote Sensing* 7 (7): 8489–8515.
6. Liang, J., Zhou, J., Qian, Y. et al. (2017). On the sampling strategy for evaluation of spectral-spatial methods in hyperspectral image classification. *IEEE Transactions on Geoscience and Remote Sensing* 55 (2): 862–880.
7. Zhong, S., Chang, C.-I, and Zhang, Y. (2019). An iterative edge preserving filtering for hyperspectral image classification. *IEEE Geoscience and Remote Sensing Letters* 16 (1): 90–94.
8. Chang, C.-I (2019). Statistical detection theory approach to hyperspectral image classification. *IEEE Transactions on Geoscience and Remote Sensing* 57 (4): 2057–2074.

9. Li, J.J., Du, Q., Li, Y., and Li, W. (2018). Hyperspectral image classification with imbalanced data based on orthogonal complement subspace projection. *IEEE Transactions on Geoscience and Remote Sensing* 56 (7): 3838–3851.
10. Lu, X., Chen, Y., and Li, X. (2018). Hierarchical recurrent neural hashing for image retrieval with hierarchical convolutional features. *IEEE Transactions on Image Processing* 27 (1): 106–120.
11. Li, S., Song, W., Fang, L. et al. (2019). Deep learning for hyperspectral image classification: an overview. *IEEE Transactions on Geoscience and Remote Sensing* 57 (4): 6690–6709.
12. Fang, L., Liu, G., Li, S. et al. (2019). Hyperspectral image classification with squeeze multibias network. *IEEE Transactions on Geoscience and Remote Sensing* 57 (3): 1291–1301.
13. Li, Y., Zhang, H., and Shen, Q. (2017). Spectral–spatial classification of hyperspectral imagery with 3D convolutional neural network. *Remote Sensing* 9 (1): 67.
14. Zhou, X., Liu, N., Tang, F. et al. (2019). A deep manifold learning approach for spatial-spectral classification with limited labeled training samples. *Neurocomputing* 331 (28): 138–149.
15. Han, M., Cong, R., Li, H. et al. (2018). Joint spatial-spectral hyperspectral image classification based on convolutional neural network. *Pattern Recognition Letters* 130: 38–45.
16. Chen, Y., Jiang, H., Li, C. et al. (2016). Deep feature extraction and classification of hyperspectral images based on convolutional neural networks. *IEEE Transactions on Geoscience and Remote Sensing* 54 (10): 6232–6251.
17. Zhao, W. and Du, S. (2016). Spectral–spatial feature extraction for hyperspectral image classification: a dimension reduction and deep learning approach. *IEEE Transactions on Geoscience and Remote Sensing* 54 (8): 4544–4554.
18. Zhong, Z., Li, J., Luo, Z., and Chapman, M. (2018). Spectral–spatial residual network for hyperspectral image classification: a 3-D deep learning framework. *IEEE Transactions on Geoscience and Remote Sensing* 56 (2): 847–858.
19. Ghamisi, P., Yokoya, N., Li, J. et al. (2017). Advances in hyperspectral image and signal processing: a comprehensive overview of the state of the art. *IEEE Geoscience and Remote Sensing Magazine* 5 (4): 37–78.
20. Sanchez-Hernandez, C., Boyd, D.S., and Foody, G.M. (2007). One-class classification for mapping a specific land-cover class: SVDD classification of fenland. *IEEE Transactions on Geoscience and Remote Sensing* 45 (4): 1061–1073.
21. Munoz-Mari, J., Bovolo, F., Gomez-Chova, L. et al. (2010). Semisupervised one-class support vector machines for classification of remote sensing data. *IEEE Transactions on Geoscience and Remote Sensing* 48 (8): 3188–3197.
22. Li, W., Guo, Q., and Elkan, C. (2011). A positive and unlabeled learning algorithm for one-class classification of remote-sensing data. *IEEE Transactions on Geoscience and Remote Sensing* 49 (2): 717–725.
23. Li, W. and Guo, Q. (2014). A new accuracy assessment method for one-class remote sensing classification. *IEEE Transactions on Geoscience and Remote Sensing* 52 (8): 4621–4632.
24. Mack, B. and Waske, B. (2017). In-depth comparisons of maxEnt, biased SVM and one-class SVM for one-class classification of remote sensing data. *Remote Sensing Letters* 8 (3): 290–299.
25. Hossain, M.A., Jia, X., and Benediktsson, J.A. (2017). One-class oriented feature selection and classification of heterogeneous remote sensing images. *IEEE Journal of Selected Topics in Applied Earth Observation and Remote Sensing* 9 (4): 1606–1612.
26. Deng, X., Li, W., Liu, X. et al. (2018). One-class remote sensing classification: one-class vs. binary classifiers. *International Journal of Remote Sensing* 39 (6): 1890–1910.
27. Song, M., Shang, X., Wang, Y. et al. (2019). Class information-based band selection for hyperspectral image classification. *IEEE Transactions on Geoscience and Remote Sensing* 57 (11): 8394–8416.
28. Song, M., Yu, C., Xie, H., and Chang, C.-I (2020). Progressive band selection processing of hyperspectral image classification. *IEEE Geoscience and Remote Sensing Letters* 17 (10): 1762–1766.

# 20

# Target Detection Approaches to Hyperspectral Image Classification

Chein-I Chang[1,2], Bai Xue[1], and Chunyan Yu[2]
[1]*Department of Computer Science and Electrical Engineering, University of Maryland Baltimore County, Baltimore, MD, USA*
[2]*Center for Hyperspectral Imaging in Remote Sensing (CHIRS), Information and Technology College, Dalian Maritime University, Dalian, China*

## 20.1 Introduction

Generally speaking, detection theory is a statistical binary decision theory which formulates a detection problem as a binary hypothesis testing problem via two hypotheses to describe two different scenarios specified by "*null hypothesis*" under a probability distribution, $p_0$, and "*alternative hypothesis*" $H_1$, under a probability distribution, $p_1$. So, for each data sample $\mathbf{r}$, a detector $\delta(\mathbf{r})$ is actually a binary decision maker that finds a threshold value, $\tau$, to determine which hypothesis is true for a given data sample $\mathbf{r}$. As a result of using $\tau$ as a decision threshold value, there are four decisions, two correct decisions with a decision on $H_0$ when $H_0$ is true, known as true negative (TN) and a decision on $H_1$ when $H_1$ is true known as true positive (TP), also referred to as detection probability, $P_D$, and two wrong decisions with a decision on $H_1$ when $H_0$ is true, known as type I error, also referred to as false alarm probability, $P_F$, and a decision on $H_0$ when $H_1$ is true, known as type II error, also referred to as miss probability, $P_M$. Accordingly, a detector $\delta(\mathbf{r})$ is in fact determined by two key elements, its detector statistics $\Lambda(\mathbf{r})$ and the threshold value $\tau$. Assume that the prior probabilities of $H_0$ and $H_1$ are given by $P(H_0)$ and $P(H_1)$, respectively, and also let $c_{ij}$ be the cost incurred by a decision saying $H_i$ when $H_j$ is true. Then, the threshold $\tau$ is generally given by

$$\tau = \frac{(c_{10} - c_{00})P(H_0)}{(c_{01} - c_{11})P(H_1)}. \tag{20.1}$$

Depending upon the available knowledge of $P(H_j)$ and $c_{ij}$, three general approaches are developed to find an optimal detector. When the prior knowledge of $P(H_j)$ and $c_{ij}$ is completely known, it is called Bayesian approach and its derived detector is referred to as Bayes detector. When only prior

*Advances in Hyperspectral Image Processing Techniques*, First Edition. Edited by Chein-I Chang.

knowledge of $c_{ij}$ is available but the prior knowledge of $P(H_j)$ is unknown, in this case, $\tau$ is a function of $P(H_0)$. The resulting detector is to find the least favorable probability distribution of $P(H_0)$, to be called minimax detector. When the prior knowledge of $P(H_j)$ and $c_{ij}$ is completely unknown, in this case, we cannot determine $\tau$ by (20.1). Under this situation, a well-known approach is Neyman–Pearson (NP) detection theory which maximizes $P_D$ subject to $P_F \leq \beta$. Using the constraint $P_F = \beta$, we can go back to determine the value of $\tau$. The resulting detector is called NP detector. Unlike the previous Bayes and minimax approaches whose detection performance is evaluated by (20.1), NP detector cannot take advantage of (20.1) and must rely on the inter-relationship between $P_D$ and $P_F$. This leads to the development of receiver operating characteristic (ROC) analysis which plots the curve of $P_D$ versus $P_F$ to evaluate the detection performance of a detector. Interestingly, regardless of which approach is taken, all their resulting detector statistics turn out to be the same likelihood ratio test (LRT) given by $\Lambda(\mathbf{r}) = p_0(\mathbf{r})/p_1(\mathbf{r})$ with three different ways to determine the threshold $\tau$. Consequently, the design of an optimal detector leads to finding an optimal threshold value $\tau^*$. For those who are interested in the above derivations, we refer the details to [1–3] and [4, Chapter 2].

On the other hand, classification is a class membership labeling process which assigns a data sample to its particular class. It is generally performed by preselecting a set of training samples to train a classifier and then evaluate its classification performance on test data samples. Two commonly used criteria to measure classification performance are average accuracy (AA) and overall accuracy (OA). However, there are two major issues in using these two measures. One is the imbalance class issue. While AA considers all classes are equally likely, OA is generally not affected much by the classification accuracy of smaller classes, specifically, the accuracy of very small classes which can be very likely overwhelmed or sacrificed by the accuracy of large classes. The other is background (BKG) issue. Since AA and OA are calculated by the ground truth, BKG data samples are removed from classification. However, such BKG data samples are simply unlabeled data samples and generally have significant impact on labeled data samples in classification. Discarding BKG data samples is not realistic in practical applications [4]. As a matter of fact, a classifier may have a very high AA and OA but may end up with very poor classification rate when BKG data samples are factored into classification. Unfortunately, over the past years, these two issues have been overlooked and have not received much attention [5–34].

Detection and classification are well-established theories and generally are considered as two different separate areas. The first work to make a link between these two was reported by Chang [35] who interpreted a multiple-class classification as a multiple hypothesis testing problem. In particular, the two issues encountered in classification can be actually resolved by detection theory. Specifically, it considers a CI as signal specified by $H_1$ and all other classes and BKG as noise specified by $H_0$. In this case, BKG and imbalanced class issues can be solved by signal detection in noise. Moreover, a new detection measure, called precision rate, can be also introduced in classification to take care of BKG issue into classification accuracy.

## 20.2 Signal Detection Theory

A signal detection problem is generally described by a binary hypothesis testing problem as follows [1]:

$$\begin{aligned} &H_0: \quad \text{noise}: Y \approx p_0(\mathbf{r}) \\ &\text{versus} \\ &H_1: \quad \text{signal + noise}: Y \approx p_1(\mathbf{r}) \end{aligned} \tag{20.2}$$

where $Y$ is an observable random variable with observations specified by $\mathbf{r}$. The hypotheses $H_0$ and $H_1$ are called "*null hypothesis*" which represents signal absence, i.e. noise only and "*alternative hypothesis*" which represents signal presence, respectively, each of which is governed by probability density functions, $p_0(\mathbf{r})$ and $p_1(\mathbf{r})$. Assume that the prior probabilities of $H_0$ and $H_1$ are given by $P(H_0)$ and $P(H_1)$, respectively. Also, let $c_{ij}$ be the cost incurred by a decision saying $H_i$ when $H_j$ is true. Then, the solution to (20.2) is the Bayes detector $\delta^{\text{Bayes}}(\mathbf{r})$ which is a LRT, $\Lambda(\mathbf{r})$, given by

$$\delta^{\text{Bayes}}(\mathbf{r}) = \begin{cases} 1; & \text{if } \Lambda(\mathbf{r}) \geq \tau \\ 0; & \text{if } \Lambda(\mathbf{r}) < \tau \end{cases} \tag{20.3}$$

where $\Lambda(\mathbf{r}) = p_0(\mathbf{r})/p_1(\mathbf{r})$, the threshold $\tau$ is given by

$$\tau = \frac{P(H_0)(c_{10} - c_{00})}{P(H_1)(c_{01} - c_{11})}. \tag{20.4}$$

Using (20.3), the $P_D$ and $P_F$ can be calculated by

$$P_D = \int_{\Lambda(\mathbf{r}) \geq \tau} p_1(\mathbf{r})d\mathbf{r} \tag{20.5}$$

$$P_F = \int_{\Lambda(\mathbf{r}) \geq \tau} p_0(\mathbf{r})d\mathbf{r} \tag{20.6}$$

with $P_{TN} = 1\text{-}P_F$, referred to as true no-signal (TN) probability corresponding to signal absence and $P_M = 1\text{-}P_D$ referred to as miss probability corresponding to no signal detection when there is a signal present. Since finding probabilities in (20.5) and (20.6) is not realistic and may not be applicable to real-world classification problems, a general approach is to calculate $P_D$, $P_F$, $P_{TN}$, and $P_M$ directly from the data samples. According to [35], we define

- $n_{\text{signal}}$ = the number of signal samples.
- $n_D$ = the number of detected signal samples.
- $n_{\text{no-signal}}$ = the number of no signal samples.
- $n_{TN}$ = number of no-signal samples detected as no signal samples.
- $n_F$ = the number of no-signal samples detected as signal samples.
- $n_M$ = the number of signal samples detected as no-signal samples, i.e. signal samples missed by detection.
- $N$ = total number of data samples = $n_{\text{signal}} + n_{\text{no-signal}}$.

So, we can have

$$n_{\text{signal}} = n_D + n_M \tag{20.7}$$

$$n_{\text{no-signal}} = n_{TN} + n_F \tag{20.8}$$

$$N = n_{\text{signal}} + n_{\text{no-signal}} \tag{20.9}$$

$$P(H_0) = p(\text{no-signal}) = \frac{n_{\text{no-signal}}}{N} \tag{20.10}$$

$$P(H_1) = p(\text{signal}) = \frac{n_{\text{signal}}}{N}. \tag{20.11}$$

Using the above definitions, we can further define the following various signal detection measures that can be applied to pattern classification.

- Signal detection probability, also noted as recall rate,

$$\mathrm{P_D} = \mathrm{P_{recall}} = \frac{n_\mathrm{D}}{n_\mathrm{signal}}. \tag{20.12}$$

- Signal predictive value (SPV), also denoted by precision rate

$$\mathrm{P_{SPV}} = \mathrm{P_{Precision}} = \mathrm{P_{PR}} = \frac{n_\mathrm{D}}{n_\mathrm{D} + n_\mathrm{F}}. \tag{20.13}$$

- No-signal predictive value (NPV)/rate (NPR)

$$\mathrm{P_{NPV}} = \frac{n_\mathrm{TN}}{n_\mathrm{TN} + n_\mathrm{M}}. \tag{20.14}$$

- Overall accuracy (OA)

$$\begin{aligned}\mathrm{P_{OA}} &= \frac{n_\mathrm{D} + n_\mathrm{TN}}{n_\mathrm{signal} + n_\mathrm{no\text{-}signal}} \\ &= \mathrm{P}(H_1)\frac{n_\mathrm{D}}{n_\mathrm{signal}} + \mathrm{P}(H_0)\frac{n_\mathrm{TN}}{n_\mathrm{no\text{-}signal}} = \frac{n_\mathrm{signal}}{N}\mathrm{P_D} + \frac{n_\mathrm{TN}}{N}\mathrm{P_{TN}} \\ &= \frac{n_\mathrm{signal}}{N}\frac{n_\mathrm{D}}{n_\mathrm{signal}} + \frac{n_\mathrm{no\text{-}signal}}{N}\frac{n_\mathrm{TN}}{n_\mathrm{no\text{-}signal}} = \frac{n_\mathrm{D} + n_\mathrm{TN}}{N}.\end{aligned} \tag{20.15}$$

Figure 20.1 describes the relationships among all definitions defined above.

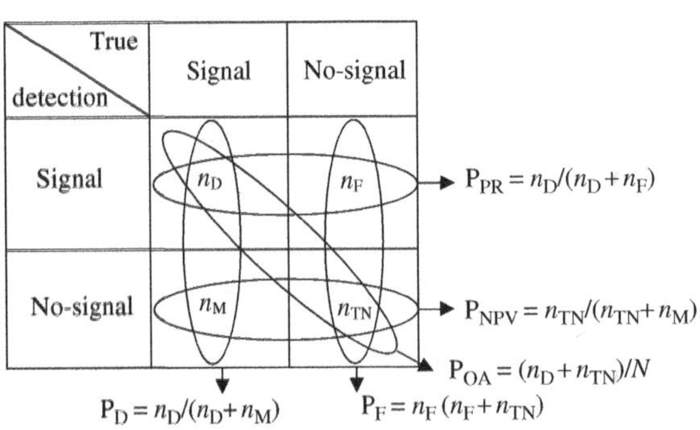


**Figure 20.1** Graph relationships among definitions (20.12)–(20.15).

## 20.3 Binary Classification Theory

This section translates the signal detection problem specified by the hypotheses $H_0$ and $H_1$ in (20.7) to a binary classification problem where two hypotheses $H_1$ and $H_2$ represent two classes, class of no interest (CNI) corresponding to no signal (i.e. signal absence) and CI corresponding to signal presence, respectively. Then, the hypothesis testing problem in (20.7) can be re-expressed as a two-class classification problem by

$$\begin{aligned} H_1 : &\quad \text{class of no interet} \approx p_1(\mathbf{r}) \\ \text{versus} & \\ H_2 : &\quad \text{class of interest} \approx p_2(\mathbf{r}). \end{aligned} \tag{20.16}$$

Since finding probability mass function for $p_j(y)$ in (20.16) is not feasible in real-world classification problems, an alternative approach to (20.16) is to construct a binary classification confusion matrix based on four decisions in Figure 20.1 where $P_D$, $P_F$, $P_M$, and $P_{TN}$ are reinterpreted as classification of interest (CI), false classification (FC) detection, misclassification (MS), and classification of CNI, respectively. Table 20.1 describes a CI-versus-CNI classification confusion matrix.

According to Table 20.1 and Figure 20.1, several binary classification criteria can be defined in the following:

$n_{CI}$ = the number of samples in CI.
$n_{CIC}$ = the number of samples in CI classified into CI.
$n_{CNI}$ = the number of samples in CNI.
$n_{CNIC}$ = number of samples in CNI classified into CNI.
$n_{FC}$ = the number of samples in CNI classified into CI.
$n_{MC}$ = the number of samples in CI classified into CNI.
$N$ = total number of data samples = $n_{CI} + n_{CNI}$.

$$P_{CI} = \frac{n_{CIC}}{n_{CIC} + n_{MC}} \tag{20.17}$$

$$P_{FC} = \frac{n_{FC}}{n_{CNI}} = \frac{n_{FC}}{n_{FC} + n_{CNIC}} \tag{20.18}$$

$$P_{PR} = \frac{n_{CIC}}{n_{CIC} + n_{FC}} \tag{20.19}$$

$$P_{OA} = \frac{n_{CIC} + n_{CNIC}}{N}. \tag{20.20}$$

Using (20.17)–(20.20), Figure 20.2 also shows a two-class confusion matrix corresponding to Table 20.2.

**Table 20.1** Confusion matrix for binary classification

| Classification \ True | Class of interest | Class of no interest |
|---|---|---|
| Class of interest | CI (classification of interest) | FC (false classification) |
| Class of no interest | MC (misclassification) | CNI (classification of no interest) |

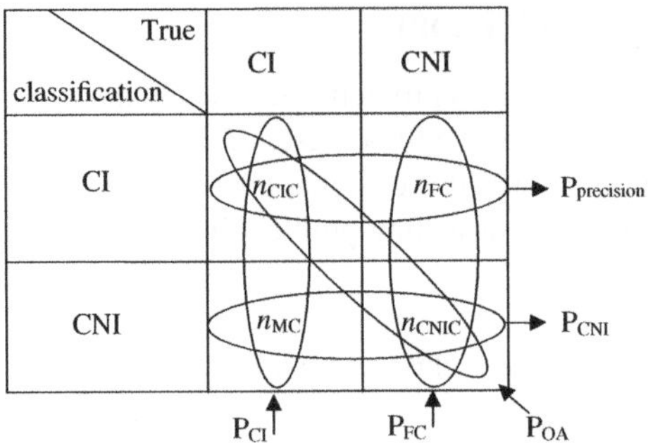


**Figure 20.2** Two-class confusion matrix.

**Table 20.2** One-to-one correspondence between Bayes detection theory and Bayes $M$-class classification theory

| Bayesian detection theory | Bayesian classification theory |
|---|---|
| Signal and no signal | $M$ classes: $\{C_i\}_{i=1}^{M}$ classification |
| Binary hypothesis testing problem (binary hypotheses: $H_1$, $H_0$) | $M$-hypothesis testing problem ($M$ hypotheses: $\{H_i\}_{i=1}^{M}$) |
| Parameter set $\Lambda = \{1, 2\}$ | Parameter set $\Lambda = \{1, 2, \ldots, M\}$ |
| Observation space: $\mathbf{r} \in \Gamma$ | Data space: data samples, $\mathbf{r} \in \{\mathbf{r}_i\}_{i=1}^{N}$ |
| Detector: $\delta(\mathbf{r}) \geq \tau \Rightarrow \mathbf{r} \in H_1$; $\delta(\mathbf{r}) < \tau \Rightarrow \mathbf{r} \in H_0$ | A set of $M(M-1)/2$ pair-wise detectors, $\{\delta_j(\mathbf{r})\}_{j=1}^{M(M-1)/2}$ Or $M-1$ between class hyperplanes or boundary curves $\{\delta_j(\mathbf{r}) = \tau_j\}_{j=1}^{M-1}$ |
| Prior probabilities, $P(H_0)$ and $P(H_1)$ | Class prior probabilities: $\{p(C_i)\}_{i=1}^{M}$ |
| ROC analysis: detector statistics: $P_D$, $P_F$, $P_M$, $P_{TN}$ | $M$-class confusion matrix |
| $P_D = \dfrac{n_D}{n_S}$ | $P_A(C_i) = P(\hat{C}_i \mid C_i) = \dfrac{n_{ii}}{\sum_{j=1}^{M} n_{ji}}$ |
| Cost of detection: $P(H_1)P_D + P(H_0)P_{TN}$ | Classification measure: $P_{OA} = \sum_{i=1}^{M} p(C_i)P_A(C_i)$ |
| A posteriori detector: | A posteriori classifier: |
| $\hat{\delta}_{MAP}(\mathbf{r}) = \arg\{\max_{j=0,1}\{\hat{\delta}_j(\mathbf{r})\}\}$ | $\hat{\delta}_{MAP}(\mathbf{r}) = \arg\{\max_{j\in\{1,2,\ldots,M\}}\{\hat{\delta}_j(\mathbf{r})\}\}$ |
| $P_{SPV} = \dfrac{n_D}{n_D + n_F}$ | $P_{PR}(\hat{C}_j) = \sum_{i=1}^{M} p(C_i)P(C_i \mid \hat{C}_j)$ |

It is worth noting that there is a significant difference between the detection in Figure 20.1 and the binary classification in Figure 20.2. A detector performs signal detection finds an appropriate threshold $\tau$ which corresponds to finding boundaries between signals and noise in signal detection. By contrast, a binary classifier performs classification by assigning each data sample to one of two classes. It does not find boundaries as a detector does. In light of image processing, a detector is designed to find edges in the data, whereas a binary classifier segments the image data into foreground and BKG classes, each of which represents one single class.

Table 20.2 describes one-to-one correspondence between Bayes detection theory and Bayes $M$-class classification theory.

## 20.4 Multiple Hypotheses Testing Theory

Theoretically speaking, a binary hypothesis testing problem (20.7) can be simply extended to a multiple-hypotheses testing problem by expanding two hypotheses to $M$ hypotheses.

### *20.4.1 Multi-Target Detection Problems*

Assume that there are $M$ signals and each signal is specified by one hypothesis as follows:

$$\begin{aligned} H_0 &: \quad s_1 \approx p_1(\mathbf{r}) \\ &\vdots \\ H_M &: \quad s_M \approx p_M(\mathbf{r}) \end{aligned} \tag{20.21}$$

where $p_j(\mathbf{r}) = p(\mathbf{r} \mid H_j)$ is the probability density function used to describe the $j$th hypothesis, $H_j$. In general, there are two ways to perform an $M$-hypotheses testing problem [36, pp. 182–184].

#### 20.4.1.1 One-Against-All (Winner-Take-All) Approach

One common approach uses the one-against-all strategy, i.e. $C_j$ versus $\{C_i\}_{i=1,i\neq j}^{M}$, which finds the hypothesis with the largest a priori probability $p_{j^*}(\mathbf{r})$ satisfying

$$j^*(\mathbf{r}) = \arg\{\max_{1 \le j \le M} p_j(\mathbf{r})\} = \arg\{\max_{1 \le j \le M} p(\mathbf{r} \mid H_j)\} \tag{20.22}$$

in which case $y$ is assigned to the $j^*$th signal. This is generally called generalized maximum likelihood detector (GMLD). Since (20.22) selects the class which yields the maximal prior probability $p(\mathbf{r} \mid H_j)$ over all hypotheses, $\{H_j\}_{j=1}^{M}$, it is also referred to as one-against-all classification. In a neural network, such class membership assignment is also known as winner-take-all classification.

#### 20.4.1.2 One-Against-One Approach

An alternative approach makes use of one-to-one strategy which runs through all possible pairs of binary hypotheses out of $M$ hypotheses, i.e. $C_j$ versus $C_{i,\, i\neq j}$. Then, the signal detection is performed by a majority vote.

More specifically, let $J = C_2^M = \frac{M(M-1)}{2}$ be the number of pairs of binary hypotheses derived from $M$ hypotheses and $n_j(\mathbf{r})$ be the number of binary hypothesis tests that declare the $y$ observation to be the $j$th signal among $J$ pair-wise binary hypothesis testing problems. Then, the observation $y$ will be declared to be the $j^*$th signal if it satisfies

$$j^*(\mathbf{r}) = \arg\{\max_{1 \le j \le J} n_j(\mathbf{r})\}. \tag{20.23}$$

#### 20.4.1.3 Multi-Class Classification Problems

Following an approach similar to translation of a binary detection problem to a binary classification problem in Section 20.3 we can also translate a multi-target detection problem to a multi-class classification problem by reformulating (20.24) as

$$\begin{aligned} H_1 &: \quad \text{class } 1 \approx p_1(\mathbf{r}) \\ &\ \vdots \\ H_M &: \quad \text{class } M \approx p_M(\mathbf{r}) \end{aligned} \tag{20.24}$$

where $p_j(\mathbf{r}) = p(\mathbf{r} \mid H_j)$ is the probability density function of the observation $y$ under hypothesis $H_j$.

Analogous to the binary classification problem in (20.16), the $M$-class classification problem specified by (20.24) will not be addressed by finding probability mass functions for $p_j(\mathbf{r})$ for $1 \le j \le M$ but rather an $M$-class confusion matrix constructed in Figure 20.3 which is modified from Figure 20.2.

Once (20.24) is formulated, how to select an effective measure is very crucial to validate classification performance. Based on Figure 20.3, similar classification measures obtained for binary classification can be also extended and derived as follows:

- $M$ = the number of classes.
- $n_{ij}$ = the number of samples in the $j$th class to be classified into the $i$th class.
- $n_{jj}$ = the number of samples in the $j$th class correctly classified into the $j$th class.
- $n_j$ = the number of data samples in the $j$th class, i.e. $n_j = \sum_{i=1}^{M} n_{ij}$.
- $\hat{n}_{ij}$ = the number of data samples classified in the $i$th class, which are supposed to in the $j$th class, $C_j$.
- $\hat{n}_{ij} = n_{ij}$ with $n_{ij}$ interpreted from a priori point of view and $\hat{n}_{ij}$ interpreted from a posteriori point of view.
- $\hat{n}_i$ = the number of data samples classified in the $i$th class, i.e. $\hat{n}_i = \sum_{j=1}^{M} \hat{n}_{ij}$.

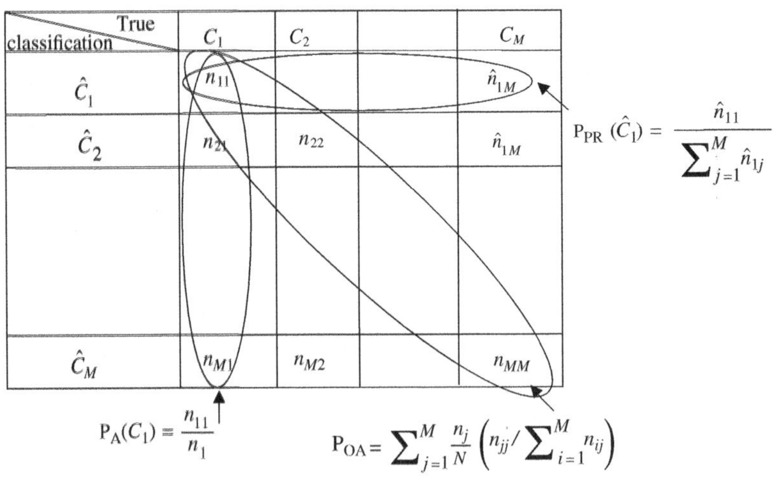


**Figure 20.3** $M$-class confusion matrix.

- $\tilde{n}_j = \sum_{i=1,i\neq j}^{M} n_{ij}$ = the number of samples in the $i$th class with $i \neq j$ falsely classified into the $j$th class.
- $N$ = total number of data samples, $N = \sum_{j=1}^{M} n_j$.
- $C_j$ = set of samples in the $j$th class.
- $\tilde{C}_j$ = set of data samples from classes $\{C_i\}_{i=1,i\neq j}^{M}$ classified into $C_j$,

$$p(C_j) = \frac{n_j}{N}.$$

In the following, we define several measures that can be used for classification.

- Accuracy

$$\mathrm{P_A}(C_j) = \text{accuracy of the } j\text{th class} = \mathrm{P}(C_{jj} \mid C_j) = \frac{n_{jj}}{n_j}. \tag{20.25}$$

- Misclassification rate

$$\begin{aligned} p\left(\tilde{C}_j | C_j\right) &= \text{misclassification rate of the } j\text{th class, } C_j \\ &= \sum_{i=1}^{M} p(C_{ij}|C_j) = \frac{\sum_{i=1,i\neq j}^{M} n_{ij}}{\sum_{i=1}^{M} n_{ij}} = 1 - p(C_{jj}|C_j) = 1 - \mathrm{P_A}(C_j). \end{aligned} \tag{20.26}$$

$$\mathrm{P}\left(\tilde{C}\right) = = \sum_{j=1}^{M} p(C_j) p\left(\tilde{C}_j \mid C_j\right) = \text{misclassification rate.} \tag{20.27}$$

- Precision rate (PR)

$$\mathrm{P_{PR}}(\hat{C}_i) = p(\hat{C}_i \mid \text{classification}) = \frac{\hat{n}_{ii}}{\sum_{j=1}^{M} \hat{n}_{ij}}. \tag{20.28}$$

$$\mathrm{P_{PR}} = \sum_{i=1}^{M} p(\hat{C}_i) p_{\mathrm{PR}}(\hat{C}_i) = \frac{1}{N} \sum_{i=1}^{M} \hat{n}_i p_{\mathrm{PR}}(\hat{C}_i). \tag{20.29}$$

- Overall accuracy (OA)

$$\begin{aligned} \mathrm{P_{OA}} &= p_{\text{overall-accuracy}}\left(\{C_j\}_{j=1}^{M}\right) \\ &= \sum_{j=1}^{M} p(\mathrm{C}_j) p(\mathrm{C}_{jj}|\mathrm{C}_j) = \frac{1}{N} \sum_{j=1}^{M} n_j \left(\frac{n_{jj}}{\sum_{i=1}^{M} n_{ij}}\right). \end{aligned} \tag{20.30}$$

- Average accuracy (AA)

$$\mathrm{P_{AA}} = \frac{1}{M} \sum_{j=1}^{M} \mathrm{P_A}(C_j) = \frac{1}{M} \sum_{i=1}^{M} \frac{n_{ii}}{n_i}. \tag{20.31}$$

## 20.5 Iterative Constrained Energy Minimization

In order to cope with the BKG issue, it has been shown that the constrained energy minimization (CEM) developed in [37–40] and linearly constrained minimum variance (LCMV) [41, 42] were the two hyperspectral imaging detection techniques that are able to suppress BKG effectively without prior knowledge.

In HSI classification, one of most widely used classifier is pure-pixel-based support vector machine (SVM). When it performs $M$-class classification, it adopts two approaches, one-against-all/winner-take-all and one-against-one as discussed in Section 20.4 where one-against-one is generally preferred [36]. This section presents a third approach which converts an $M$-class classification to a standard signal detection problem. It is one against rest, i.e. $C_j$ versus $C_j^c$ where $C_j^c$ is the set containing all data samples including BKG pixels other than the data samples in $C_j$. The idea is to consider one CI, $C_j$, as the alternative hypothesis $H_1$, while the rest of classes including BKG grouped to a single class, $C_j$ versus $C_j^c$ as the null hypothesis $H_0$. The classification is then performed by binary classification as discussed in Section 20.3. Such an approach is referred to as one-against-rest as opposed to one-against-all/winner-take-all and one-against-one carried out by most classification methods. A key difference among them is that such detection approach requires $M$ binary classifications, one for each class compared to $M$–1 binary classifications required by one-against-all/winner-take-all and $M(M-1)/2$ binary classifications required by one-against-one.

More specifically, supposed that the $j$th class, $C_j$ is a CI to be classified. The multi-target detection specified by (20.24) can be reinterpreted as a binary detection problem by specifying hypothesis $H_j$ as the CI, $C_j$ to be classified and the rest of classes $\{C_i\}_{i=1,i\neq j}^{M}$ to be considered as CNI, then (20.23) can be translated to a binary classification problem as follows:

$$\begin{aligned} &H_0: \quad \text{class of no interest} = \{C_i\}_{i=1,i\neq j}^{M} \approx p_j^c(\mathbf{r}) \\ &\text{versus} \\ &H_j: \quad \text{class of interet} \approx p_j(\mathbf{r}) \end{aligned} \tag{20.32}$$

where $\mathbf{r}$ is a data sample vector corresponding to $y$ in (20.25).

In order for (20.32) to perform $M$-class classification, (20.32) must be carried out $M$ times to classify $M$ classes, one hypothesis to specify a particular class. This is different from the one-against-all/winner-take-all approach in Section 20.4.1.1 which requires comparisons of $M-1$ hypotheses and the one-against-one approach in Section 20.4.1.2 which requires $\binom{M}{2} = (M(M-1)/2)$ binary hypothesis tests. But as usual, finding probability mass functions of $p_j^c(\mathbf{r})$ and $p_j(\mathbf{r})$ in (20.32) is not practical in many applications. In addition, the hypothesis $H_0$ in (20.32) does not include the BKG class. Interestingly, both these two issues can be resolved by a well-known hyperspectral subpixel target detection technique, called CEM [37–40] which performs detection of a target of interest while suppressing BKG without knowing BKG knowledge. In this case, CEM further converts the binary classification problem in (20.32) to performing target detection using the class signature obtained from the CI specified by $H_1$ as the desired target signature without performing NP detection. The great benefit derived from CEM is that it takes care of BKG issue by suppressing unknown BKG compared to (20.32) which simply ignores the BKG class as most works on hyperspectral image classification (HSIC) [5–34] are reported in the literature. Unfortunately, when CEM was developed in [38], it was specifically designed for subpixel target detection and did not take into account inter-pixel spatial information. So, directly applying CEM to classification is not applicable. To further resolve

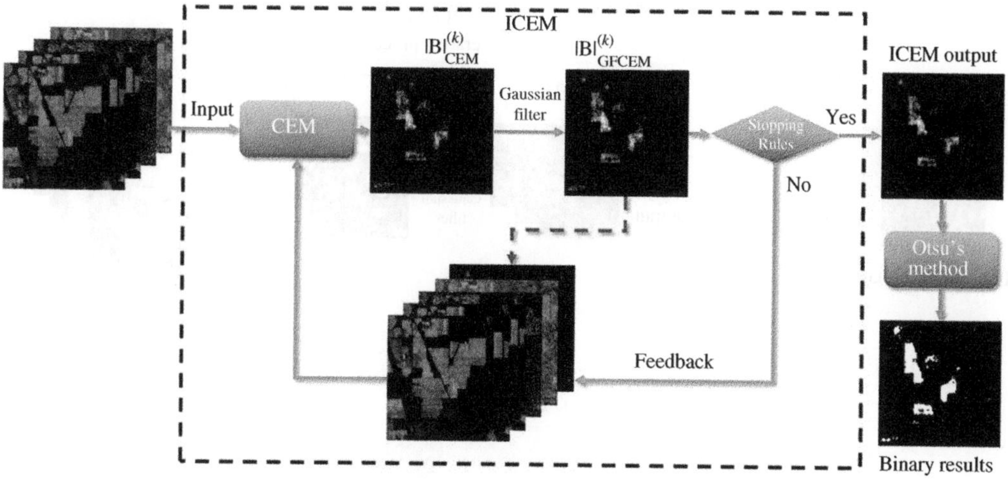


**Figure 20.4** A graphic implementation of ICEM.

this dilemma, a recent approach was developed to extend CEM to iterative CEM (ICEM) [43] by including Gaussian filters to capture spatial information directly from each of $M$ CEM-detected classification maps and feeding back these $M$-obtained spatial images to be combined with the currently being processed data cube to create a new image cube in an iterative manner. In this case, the newly formed image cube after each iteration contains more spatial information than the one obtained in the previous iteration. That is, the more feedback, the more spatial information captured. As a consequence, ICEM has both spectral and spatial information to perform classification well. For a quick reference, a graphic diagram of implementing ICEM is shown in Figure 20.4 with details referred to [43] where the block diagram in the dark gray dashed line shows how ICEM is implemented.

## 20.6 Iterative Linearly Constrained Minimum Variance

Since CEM only performs one class at a time, it requires CEM being performed $M$ times to be able to classify $M$ classes. Thus, it cannot classify $M$ classes simultaneously in one-shot operation. In order for CEM to perform as a classifier, it was further extended to LCMV which performs $M$-class classification all together at once [39, 42]. Such classification is referred to as all-against-all classification which indicates that $M$ classes are classified simultaneously, $\{C_j\}_{j=1}^{M}$ versus $\{C_i\}_{j=1}^{M}$.

Like CEM, LCMV is also a pixel-based technique without accounting for inter-pixel spatial information. To address this issue, LCMV is further extended to iterative LCMV (ILCMV) for classification [44, 45] in a similar manner that ICEM was developed. However, there are two key differences. One is how BKG is suppressed. In ICEM, when the classification of a class, $C_i$, is performed, BKG includes all other classes $\{C_j\}_{j=1,j\neq i}^{M}$. As a consequence, the BKG suppression will also suppress all classes $\{C_j\}_{j=1,j\neq i}^{M}$ other than the class, $C_i$, currently being processed. As for ILCMV all classes of interest, $\{C_j\}_{j=1}^{M}$ will be classified simultaneously at once. In this case, the BKG suppression only takes effect on all data samples not in $\{C_j\}_{j=1}^{M}$ including BKG. The other is that ICEM acts as a subpixel target detector but performs as a binary classifier. By contrast, ILCMV is a multiple-target

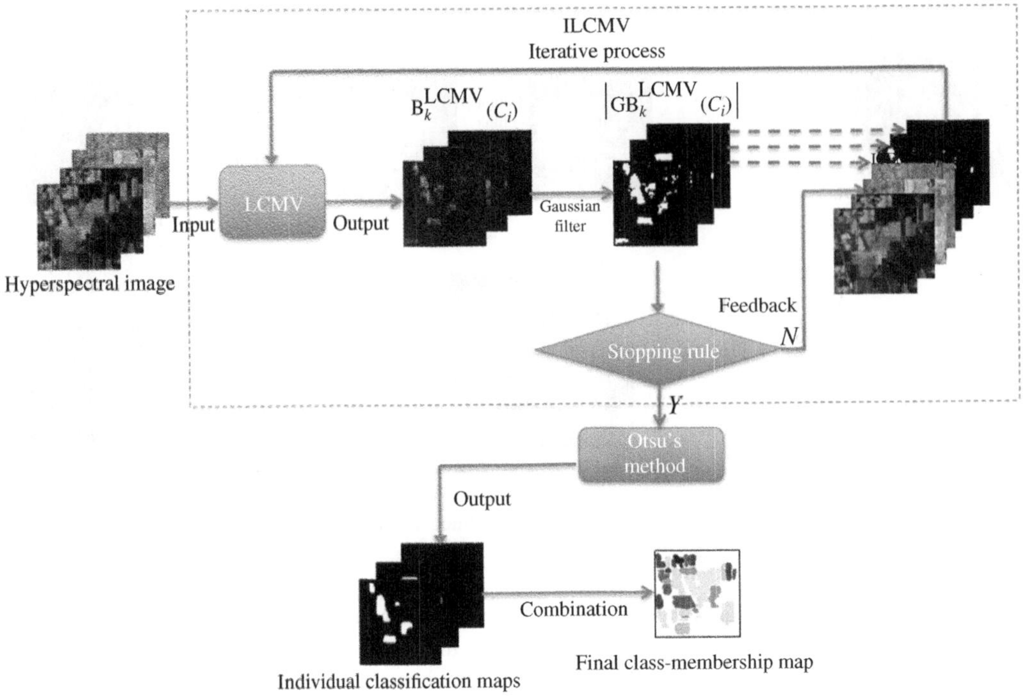


**Figure 20.5** A graphic implementation of ILCMV with the block diagram in the gray dotted line to show the implementation of ILCMV.

mixed pixel classifier but can be used to perform multi-class classification by including feedback loops as shown in Figure 20.5 where the block diagram in the gray dotted line is included to show the implementation of ILCMV. For more details, we refer to [45].

Table 20.3 lists how many passes are required by SVM, CEM, and LCMV to classify $M$ classes where each pass indicates that a classier is performed once. For example, there are two ways for SVM to perform an $M$-class classification. One is one-against-one which requires $M(M\text{-}1)/2$ passes to yield a majority vote. The other is one-against-all/winner-take-all which requires $M$-1 comparison by (20.32). As for CEM, it requires $M$ passes, each of which is to classify data samples into one particular class. The most efficient way is LCMV which classifies all the $M$ classes simultaneously in one pass.

As we can see from Table 20.3, the subpixel detection-based CEM and the mixed pixel classification-based LCMV are very different from traditional classification techniques because they implement different classification strategies of how to classify multiple classes in different passes.

**Table 20.3** Comparisons in passes required by SVM, CEM, and LCMV

| SVM | | CEM | LCMV |
|---|---|---|---|
| One-against-one $C_j$ versus $C_{i,\ i\neq j}$ | One-against-all $C_j$ versus $\{C_i\}_{i=1,i\neq j}^{M}$ | One-against-rest $C_j$ versus $\cup_{i=1,i\neq j}^{M} C_i$ | All-against-all $\{C_j\}_{j=1}^{M}$ versus $\{C_j\}_{j=1}^{M}$ |
| $M(M\text{-}1)/2$ passes | $M$-1 passes | $M$ passes | One pass |

## 20.7 How to Convert Detection to Classification

A salient difference between target detection and class classification lies on the fact that detection is performed on only targets of interest while classification is performed by assigning a class membership to each of data samples. As a result, spatial correlation among data samples in the same class provides crucial information compared to target detection whose spatial correlation may not be needed such as subpixel targets. This also explains the reason that a direct extension of detection to classification is generally inappropriate and ineffective. Accordingly, the issue of how to capture spatial information to be included in detection is very challenging. As shown in [37, 39], CEM and LCMV only deal with precise targets and are not robust to target variations. So, they cannot be directly applied to classification. This is because the data samples in the same class may have a wide range of spectral variations which cause a breakdown of CEM and LCMV. The first work of extending CEM to classification was reported in [43] which developed ICEM to capture and feed spatial information obtained from the CEM-detection maps back to the input in an iterative manner. In this case, ICEM converts target detection to binary classification by iteratively including feedbacks of spatial information captured from CEM-detection maps so that ICEM can perform $M$-class classification problems by binary classification in $M$ passes via constraining one class at a time per pass. Similarly, ILCMV also extends multiple-target detection to multiple-class classification by iteratively feeding back of spatial information captured from LCMV multiple-target detection maps in one single pass. Figure 20.6 illustrates how the idea of detection is extended to classification.

As noted above, when CEM and ILCMV perform classification, the targets to be constrained cannot cover all the data sample vectors in classes. In this case, the class sample mean for each class is used to represent its target class and constrained by ICEM and ILCMV. Additionally, the iterative process implemented by ICEM and ILCMV also plays a key role in taking care of such spatial variations caused by data samples in a class through iterations.

## 20.8 Real Image Experiments

Two popular real hyperspectral images, Purdue University's Indiana Indian Pines and University of Pavia, which are available on website http://www.ehu.eus/ccwintco/index.php?title=Hyperspectral_Remote_Sensing_Scenes, were used for experiments.

### *20.8.1 Purdue University's Indiana Indian Pines*

The first real image to be used for experiments is a well-known Airborne Visible/Infrared Imaging Spectrometer (AVIRIS) image scene, the Purdue University Indiana Indian Pines test site shown in Figure 20.7(a), its ground-truth map in Figure 20.7(b) along with different classes highlighted by

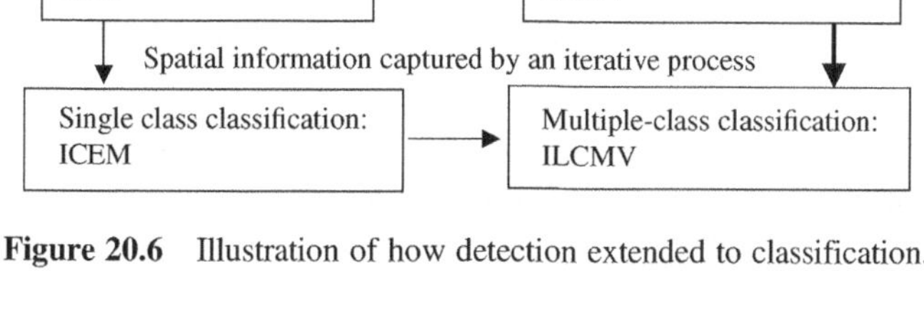


**Figure 20.6** Illustration of how detection extended to classification.

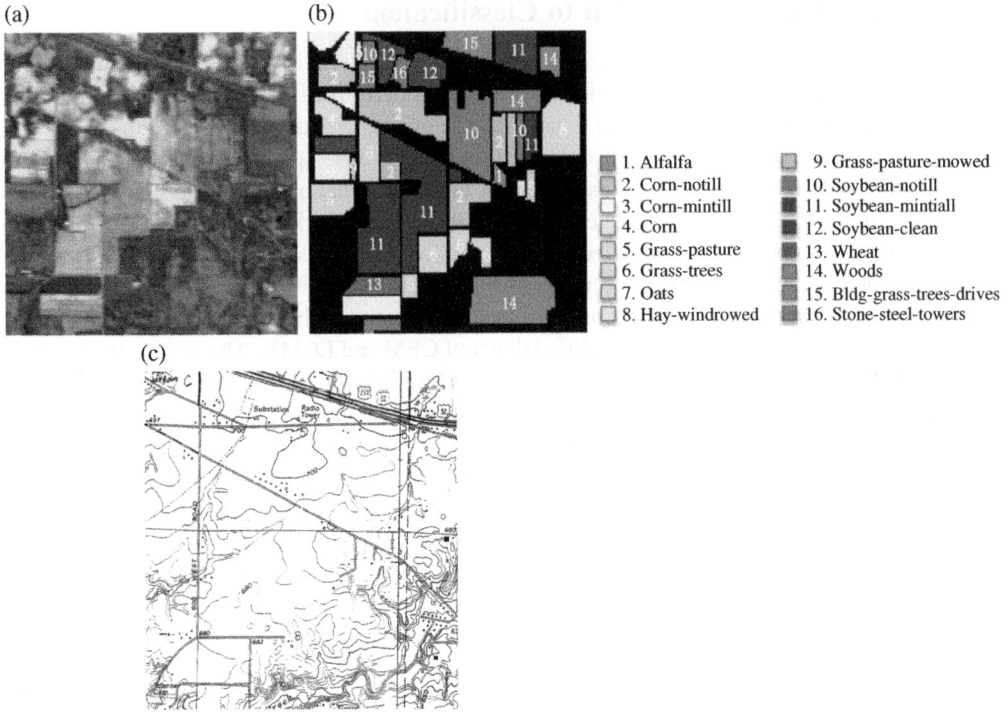


**Figure 20.7** AVIRIS image scene: Purdue Indiana Indian Pines test site; (a) band 186 (2162.56 nm), (b) ground-truth map and classes by different degrees of gray intensities, and (c) A USGS Quadrangle map of the test site, where its color maps can be found in Ref. [35], figure 10/with permission of IEEE.

various colors and an aerial view in Figure 20.7(c). Table 20.4 also tabulates all the specific types of 16 classes. It has the size of 145 × 145 pixel vectors taken from an area of mixed agriculture and forestry in Northwestern Indiana, USA, with details of band and wavelength given in the caption. The data set is available at website https://purr.purdue.edu/publications/1947/serve/1?el=1. It was recorded in June 1992 with 220 bands including water absorption bands (bands 104–108, 150–163, 220).

**Table 20.4** Class labels of Purdue Indiana Indian Pines with number of data samples in each class

| | | | | | |
|---|---|---|---|---|---|
| Class 1 (46) | Alfalfa | Class 7 (28) | Grass/ pasture-mowed | Class 13 (205) | Wheat |
| Class 2 (1428) | Corn-notill | Class 8 (478) | Hay-windrowed | Class 14 (1265) | Woods |
| Class 3 (830) | Corn-min | Class 9 (20) | Oats | Class 15 (386) | Bldg-grass green-drives |
| Class 4 (237) | Corn | Class 10 (972) | Soybeans-notill | Class 16 (93) | Stone-steel towers |
| Class 5 (483) | Grass/pas-ture | Class 11 (2455) | Soybeans-min | Class 17 (10249) | BKG |
| Class 6 (730) | Grass/trees | Class 12 (593) | Soybeans-clean | | |

The Purdue University's Indian Pines scene is a very interesting data set because it contains 16 classes with various imbalanced classes. For example, there are 4 classes with data samples less than 100 (class 9 with 20 data samples, class 7 with 28 data samples, class 1 with 46 data samples, and class 16 with 93 data samples) and 3 classes with data samples greater than 1000 (class 11 with 2455 data samples, class 2 with 1428 data samples, and class 14 with 1265 data samples). This scene provides an excellent data set to demonstrate imbalanced classes and BKG issues.

### *20.8.2 University of Pavia*

A second hyperspectral image data shown in Figure 20.8(a) was collected and recorded by the ROSIS-03 satellite sensor over an urban area surrounding the University of Pavia, Italy. It is of size $610 \times 340 \times 115$ with a spatial resolution of 1.3-meter per pixel and a spectral coverage ranging from 0.43 to 0.86 μm with spectral resolution of 4 nm (12 most noisy channels were removed before experiments). Nine classes of interest are considered for this image. Figure 20.8(b) provides its ground-truth map along with color class labels in Figure 20.8(c).

Since the main goal of this section is to demonstrate the concepts presented in this paper, the classifiers to be used for experiments are selected based on the following reasons. For pure-pixel classification, SVM-based approaches are most preferred. In a recent work [12], an edge preserving filter (EPF)-based approach was developed where the four EPF-based techniques, EPF-B-c, EPF-G-c, EPF-B-g, and EPF-G-g, were proposed with "B" and "G" used to specify bilateral filter and guided filter, respectively, and "g" and "c" indicate that the first principal component and color composite of three principal components are used as reference images [12]. According to a comprehensive comparative analysis conducted in [12], it was shown that these four classifiers perform well among most recently developed spectral-spatial best classification techniques.

As for mixed pixel classification, CEM and LCMV recently developed in [43–45] can be considered as a good alternative to SVM-based approaches. Therefore, in the following experiments, three types of classifiers, pure-pixel SVM-based EPF methods, subpixel detection-based ICEM, and the

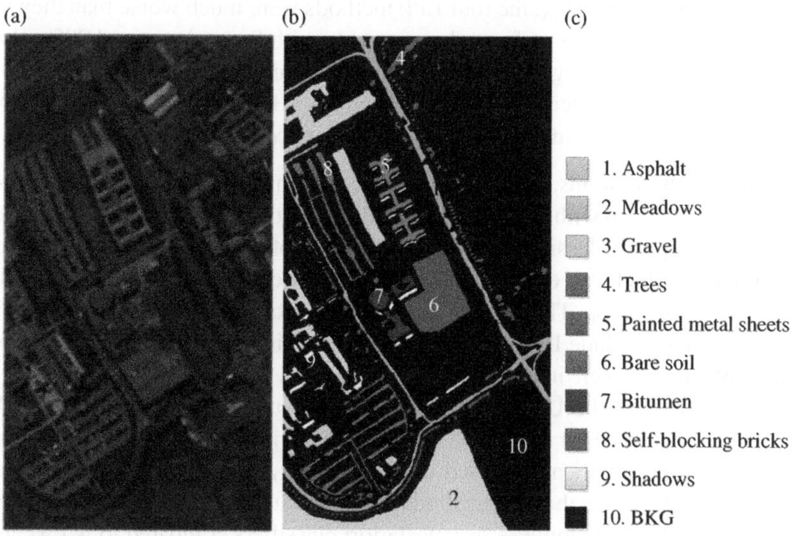


**Figure 20.8** Ground truth of University of Pavia scene with nine classes: (a) University of Pavia scene, (b) ground-truth map, and (c) classes by different degrees of gray intensities, where its color maps can be found in Ref. [35], figure 12/with permission of IEEE.

mixed pixel classification-based ILCMV, were selected for a comparative study where the used parameters were selected to be the same as those used in [12], [43, 45]. In all experiments, the BKG class was excluded for classification but BKG pixels were used to calculate $P_{PR}$ for evaluation of a posteriori classification. It should be noted that the ICEM implemented in this section was different from the ICEM in [43] with two aspects. One is that for fairness of comparison, no band selection and nonlinear expansion were performed prior to classification. Another is that the desired signatures used for class sample means in the following experiments by ICEM were fixed and provided by ground-truth locations, while the desired signatures used for class sample means by the ICEM in [43] were updated and varied after each iteration. Also, the ILCMV implemented here is a special case of the iterative target-constrained interference-minimized filter (ITCIMF) develop in [45] with the undesired signature set **U** set to an empty set, i.e. $\mathbf{U} = \emptyset$. Since, it was shown that ILCMV could do better without **U**. For more details, we refer to [45].

Tables 20.5–20.6 tabulate the classification results in terms of $P_A$ in (28), $P_{AA}$ in (33), $P_{OA}$ in (32), and $P_{PR}$ in (46) where the best results of $P_A(C_i)$, $P_{PR}(C_i)$, $P_{AA}$, $P_{OA}$, and $P_{PR}$ are boldfaced.

According to Tables 20.5–20.6, the three image data sets exhibit quite different image characteristics through three types of classifiers, four EPF-based methods (EPF-B-c, EPF-G-c, EPF-B-g, and EPF-G-g), ICEM, and ILCMV.

1. Purdue's Indian Pines:
   - $P_{AA}$ and $P_{OA}$: All the three types of classifiers performed very well with ICEM being the best which has better rates than four EPF methods by 2–3% for $P_{AA}$ and 6% for $P_{OA}$. Similarly, the second best was ILCMV which also has better rates than four EPF methods by 1.5% for $P_{AA}$ and 3.5–4% for $P_{OA}$.
   - $P_{PR}$: The four EPF methods worked poorly with $P_{PR}$ around 44.5%, while ICEM was the best with $P_{PR}$ = 94.23% followed by ILCMV with $P_{PR}$ = 92.64%.

- $P_{AA}$ and $P_{OA}$: The four EPF methods did very well even better than it did for the other two image data sets with EPF-G-c being the best among all the four EPF-based methods. Interestingly, ICEM and ILCMV were not comparable to the EPF methods where ILCMV was the worst with its $P_{AA}$ and $P_{OA}$ lower than those produced by the four EPF methods by 12–20%.
- $P_{PR}$: In contrast to $P_{AA}$ and $P_{OA}$, the four EPF methods were much worse than they were for the other two image data sets and performed very poorly with $P_{PR}$ only around 20% which was the worst among all the three image data sets. On the other hand, both ICEM and ILCMV did worse than other two image sets in terms of $P_{AA}$, $P_{OA}$, and $P_{PR}$. However, ICEM and ILCMV did much better than all the four EPF methods did by at least twice or three times $P_{PR}$.

Based on the above observations, it is very obvious to note that the four pure pixel SVM-based EPF methods performed really well with very high $P_{AA}$ and $P_{OA}$ but with very low $P_{PR}$. This indicates that EPF methods work as a priori classifiers because BKG was not included in classification due to their use of the ground truth which can exclude BKG from classification. On the other hand, ICEM generally did best in the sense of $P_{PR}$. This implies that ICEM works well as a posteriori classifier because its classified data samples include BKG pixels as a posteriori information for classification. ILCMV was somewhat right in between in terms of $P_{AA}$, $P_{OA}$, and $P_{PR}$. The most intriguing finding is Table 20.6 which shows that the University of Pavia seems more interesting than the Purdue data because it offers hard evidence on difference between a priori classification and a posteriori classification. That is, a priori classification based on $P_{AA}$ and $P_{OA}$ is quite different from a posteriori classification based on PR where the EPF methods were shown to be effective a priori classifiers with very high $P_{AA}$ and $P_{OA}$ but very low $P_{PR}$ as a posteriori classifiers compared to ICEM and ILCMV,

**Table 20.5** $P_A$, $P_{OA}$, $P_{AA}$, and $P_{PR}$ calculated by four EPF-based methods, ICEM, and ILCMV for Purdue's Indian Pines

| | EPF-B-g | | EPF-B-c | | EPF-G-g | | EPF-G-c | | ICEM | | ILCMV | |
|---|---|---|---|---|---|---|---|---|---|---|---|---|
| Class | $P_A(C_i)$ | $P_{PR}(C_i)$ | $P_A(C_i)$ | $P_{PR}(C_i)$ | $P_A(C_i)$ | $P_{PR}(C_i)$ | $P_A(C_i)$ | $P_{PR}(C_i)$ | $P_A(C_i)$ | $P_{PR}(C_i)$ | $P_A(C_i)$ | $P_{PR}(C_i)$ |
| 1 | 0.9957 | 0.3901 | 0.9783 | 0.4012 | 0.9978 | 0.3882 | **1.0000** | 0.4240 | 0.9565 | 0.7333 | 0.9565 | **0.7719** |
| 2 | 0.8288 | 0.6681 | 0.8303 | 0.6558 | 0.8268 | 0.6556 | 0.8183 | 0.6518 | **0.9734** | **0.9398** | 0.9601 | 0.9196 |
| 3 | 0.8375 | 0.6842 | 0.8401 | 0.6767 | 0.8361 | 0.6725 | 0.8286 | 0.6701 | **0.9807** | **0.9346** | 0.9699 | 0.8974 |
| 4 | 0.9975 | 0.4190 | **0.9983** | 0.4460 | **0.9983** | 0.4117 | **0.9983** | 0.4105 | 0.9831 | 0.9628 | 0.9873 | **0.9286** |
| 5 | 0.9509 | 0.2946 | 0.9484 | 0.3032 | **0.9549** | 0.2981 | 0.9507 | 0.3110 | 0.9441 | **0.9383** | 0.8944 | 0.9290 |
| 6 | **0.9978** | 0.3474 | 0.9971 | 0.3487 | 0.9925 | 0.3353 | 0.9959 | 0.3514 | 0.9836 | 0.8831 | 0.9712 | **0.8997** |
| 7 | 0.9750 | 0.6000 | 0.9643 | 0.6189 | 0.9786 | 0.6182 | 0.9750 | 0.6502 | **1.0000** | 0.6512 | **1.0000** | **0.7568** |
| 8 | **1.0000** | 0.6664 | **1.0000** | 0.6561 | **1.0000** | 0.6818 | **1.0000** | 0.6798 | 0.9895 | **0.9834** | 0.9874 | 0.9574 |
| 9 | 0.9200 | 0.3985 | 0.8900 | 0.4788 | 0.9650 | 0.5173 | 0.8700 | **0.6899** | **1.0000** | 0.5405 | **1.0000** | 0.5263 |
| 10 | 0.9105 | 0.5878 | 0.9105 | 0.5845 | 0.9060 | 0.5888 | 0.9011 | 0.5718 | **0.9527** | **0.9344** | 0.9393 | 0.9185 |
| 11 | 0.8802 | 0.8001 | 0.8796 | 0.7910 | 0.8691 | 0.7950 | 0.8718 | 0.7904 | **0.9744** | **0.9565** | 0.9470 | 0.9417 |
| 12 | 0.9523 | 0.5672 | 0.9546 | 0.5641 | 0.9528 | 0.5470 | 0.9671 | 0.5293 | **0.9663** | **0.9409** | 0.9545 | 0.8943 |
| 13 | 0.9951 | 0.6619 | 0.9951 | 0.6702 | **0.9956** | 0.6396 | 0.9980 | 0.5449 | 0.9756 | **0.9662** | 0.9854 | 0.9352 |
| 14 | 0.9720 | 0.3014 | 0.9730 | 0.2976 | 0.9686 | 0.3029 | 0.9736 | 0.3006 | **0.9747** | **0.9770** | 0.9352 | 0.9689 |
| 15 | 0.9655 | 0.1253 | 0.9617 | 0.1264 | 0.9420 | 0.1249 | 0.9642 | 0.1296 | **0.9767** | **0.9642** | 0.9067 | 0.9669 |
| 16 | **1.0000** | 0.3658 | **1.0000** | 0.3652 | **1.0000** | 0.3631 | **1.0000** | 0.3494 | 0.9785 | **0.8667** | 0.9892 | 0.8598 |
| $P_{AA}$ | 0.9487 | | 0.9451 | | 0.9490 | | 0.9445 | | **0.9756** | | 0.9615 | |
| $P_{OA}$ | 0.9154 | | 0.9155 | | 0.9106 | | 0.9112 | | **0.9726** | | 0.9509 | |
| $P_{PR}$ | 0.4462 | | 0.4463 | | 0.4439 | | 0.4442 | | **0.9423** | | 0.9264 | |

**Table 20.6** $P_A$, $P_{OA}$, $P_{AA}$, and $P_{PR}$ calculated by four EPF-based methods, ICEM, and ILCMV for University of Pavia

| | EPF-B-g | | EPF-B-c | | EPF-G-g | | EPF-G-c | | ICEM | | ILCMV | |
|---|---|---|---|---|---|---|---|---|---|---|---|---|
| Class | $P_A(C_i)$ | $P_{PR}(C_i)$ | $P_A(C_i)$ | $P_{PR}(C_i)$ | $P_A(C_i)$ | $P_{PR}(C_i)$ | $P_A(C_i)$ | $P_{PR}(C_i)$ | $P_A(C_i)$ | $P_{PR}(C_i)$ | $P_A(C_i)$ | $P_{PR}(C_i)$ |
| 1 | 0.9811 | 0.1743 | **0.9815** | 0.1732 | 0.9755 | 0.1769 | 0.9797 | 0.1772 | 0.7659 | **0.4660** | 0.7656 | 0.3293 |
| 2 | 0.9882 | 0.3644 | 0.9876 | 0.3642 | 0.9882 | 0.3690 | **0.9889** | 0.3735 | 0.8470 | **0.9046** | 0.8168 | 0.7865 |
| 3 | 0.9432 | 0.3238 | **0.9441** | 0.3268 | 0.9428 | 0.3330 | 0.9436 | 0.3338 | 0.9066 | **0.5828** | 0.4574 | 0.4126 |
| 4 | 0.9563 | 0.0963 | 0.9513 | 0.0958 | 0.9502 | 0.0975 | **0.9852** | 0.1025 | 0.6711 | **0.3121** | 0.7997 | 0.2693 |
| 5 | 0.9995 | 0.3924 | 0.9995 | 0.3994 | **1.0000** | 0.3811 | 0.9988 | 0.3950 | 0.9978 | 0.6546 | 0.9844 | **0.6925** |
| 6 | 0.9980 | 0.1046 | 0.9979 | 0.1051 | 0.9980 | 0.1047 | **0.9987** | 0.1043 | 0.9336 | **0.9800** | 0.7112 | 0.7432 |
| **7** | **0.9995** | 0.3781 | 0.9993 | 0.3722 | 0.9980 | 0.3868 | 0.9986 | 0.3970 | 0.9654 | **0.7892** | 0.8564 | 0.7624 |
| 8 | 0.9757 | 0.2016 | 0.9746 | 0.2028 | 0.9763 | 0.2035 | **0.9834** | 0.2038 | 0.8541 | **0.4650** | 0.7917 | 0.3019 |
| **9** | **1.0000** | 0.0996 | **1.0000** | 0.0977 | **1.0000** | 0.0806 | **1.0000** | 0.0752 | 0.8638 | 0.3405 | 0.7856 | **0.3483** |
| $P_{AA}$ | 0.9824 | | 0.9817 | | 0.9810 | | **0.9863** | | 0.8673 | | 0.7743 | |
| $P_{OA}$ | 0.9836 | | 0.9830 | | 0.9824 | | **0.9865** | | 0.8444 | | 0.7763 | |
| $P_{PR}$ | 0.2029 | | 0.2027 | | 0.2026 | | 0.2035 | | **0.6470** | | 0.5037 | |

both of which were reasonably good a priori classifiers but very effective a posteriori classifiers with high $P_{PR}$ compared to a priori classifiers.

As a concluding remark, the experiments conducted in this section suggest that to effectively perform HSI classification, only using $P_{AA}$ and $P_{OA}$ to measure classification performance is not sufficient since it merely performs a priori classification. It must be coupled with $P_{PR}$ to deal with a posteriori classification which can evaluate the effectiveness of a classifier.

## 20.9 Conclusion

This chapter develops statistical detection approach to HSIC which is based on hyperspectral subpixel single target detector, CEM, and subpixel multiple-target detector, LCMV. It is quite different from conventional SVM-based spectral-spatial approaches reported in the literature where it replaces the pure-pixel-based SVM with the mixed-pixel-based CEM/LCMV. In order to capture spatial information, a set of Gaussian filters are implemented to capture spatial information from CEM/LCMV-detected maps which is fed back iteratively to create new hyperspectral data cubes to be reprocessed by CEM next-round iteration so as to improve classification performance. Such an iterative CEM/LCMV (ICEM/ILCMV) can be considered as a spectral-spatial mixed filter, that is, the more iterations carried out by ICEM/ILCMV, the more spatial information captured. As a result of ICEM, there are many unique features can be derived as follows:

1. ICEM/ILCMV can work as a subpixel classifier by replacing the commonly used SVM as mixed pixel classification technique.
2. The real-valued CEM/LCMV-detection maps can be thresholded such as Otsu's method to produce binary maps to perform classification.
3. Using a one-against-rest strategy for a binary hypothesis testing problem converts a multi-classification confusion matrix to a binary confusion matrix. As a result, three new performance measures, detection rate, false alarm rate, precision rate, are introduced in addition to the commonly used OA and accuracy.
4. The idea that takes advantage of iterative feedbacks of Gaussian-filtered CEM/LCMV detection maps to capture spatial information can be also applied to many other pixel-based spectral filters and opens up a new direction for HSIC. Several investigations along this line are currently undertaken.

## References


1. Poor, H.V. (1994). *An Introduction to Signal Detection and Estimation*. New York: Springer.
2. Chang, C.-I (2022). Hyperspectral target detection: hypothesis testing, signal-to-noise ratio and spectral angle theories. *IEEE Transactions on Geoscience and Remote Sensing* 60: 1–23.
3. Chang, C.-I (2022). Hyperspectral anomaly detection theory: a dual theory of hyperspectral target detection. *IEEE Transactions on Geoscience and Remote Sensing* 60: 1–20.
4. Chang, C.-I (2013). *Hyperspectral Data Processing: Algorithm Design and Analysis*. John Wiley & Sons.
5. Melgani, F. and Bruzzone, L. (2004). Classification of hyperspectral remote sensing images with support vector machines. *IEEE Transactions on Geoscience and Remote Sensing* 42 (8): 1778–1790.
6. Benediktsson, J.A., Palmason, J.A., and Sveinsson, J.R. (2005). Classification of hyperspectral data from urban areas based on extended morphological profiles. *IEEE Transactions on Geoscience and Remote Sensing* 43 (3): 480–491.
7. Vamps-Valls, G. and Bruzzone, L. (2005). Kernel-based methods for hyperspectral image classification. *IEEE Transactions on Geoscience and Remote Sensing* 43 (6): 1351–1362.

8. Tarabalka, Y., Benediktsson, J.A., and Chanussot, J. (2009). Spectral-spatial classification of hyperspectral imagery based on partitional clustering techniques. *IEEE Transactions on Geoscience and Remote Sensing* 47 (8): 2973–2987.
9. Tarabalka, Y., Fauvel, M., Chanussot, J., and Benediktsson, J.A. (2010). SVM- and MRF-based method for accurate classification of hyperspectral images. *IEEE Geoscience and Remote Sensing Letters* 9 (4): 736–740.
10. Li, J., Bioucas-Dias, J.M., and Plaza, A. (2010). Semisupervised hyperspectral image segmentation using multinominal logistic regression model with active learning. *IEEE Transactions on Geoscience and Remote Sensing* 48 (11): 4085–4098.
11. Zhang, B., Li, S., Jia, X. et al. (2011). Adaptive hyperspectral Markov random field approach for classification of hyperspectral imagery. *IEEE Geoscience and Remote Sensing Letters* 8 (11): 4085–4098.
12. Li, J., Bioucas-Dias, J.M., and Plaza, A. (2011). Hyperspectral image segmentation using a new Bayesion approach with active learning. *IEEE Transactions on Geoscience and Remote Sensing* 49 (10): 3947–3960.
13. Chen, Y., Nasabardi, N., and Tran, T.D. (2011). Hyperspectral image segmentation using dictionary-based sparse representation. *IEEE Transactions on Geoscience and Remote Sensing* 49 (10): 3973–3985.
14. Tarabalka, Y., Fauvel, M., Chanussot, J., and Benediktsson, J.A. (2011). Segmentation and classification of hyperspectral images using minimum spanning forest grown from automatically selected makers. *IEEE Transactions on Systems, Man, and Cybernetics-Part B: Cybernetics* 40 (5): 1267–1279.
15. Fauvel, M., Tarabalka, Y., Benediktsson, J.A. et al. (2013). Advances in spectral-spatial classification of hyperspectral images. *Proceedings of IEEE* 101 (3): 652–675.
16. Kang, X., Li, S., and Benediktsson, J.A. (2014). Spectral-spatial hyperspectral image classification with edge-preserving filtrering. *IEEE Transactions on Geoscience and Remote Sensing* 52 (5): 2666–2677.
17. Guo, X., Huang, X., Zhang, L. et al. (2016). Support tensor machines for classification of hyperspectral remote sesning imagery. *IEEE Transactions on Geoscience and Remote Sensing* 54 (6): 3248–3264.
18. Fu, W., Li, S., Fang, L. et al. (2016). Hyperspectral image classification via shape-adaptive joint sparse representation. *IEEE Journal of Selected Topics in Applied Earth Observations and Remote Sensing* 9 (2): 556–567.
19. Kang, X., Li, S., Fang, L. et al. (2015). Extended random walker-based classification of hyperspectral images. *IEEE Transactions on Geoscience and Remote Sensing* 53 (1): 144–153.
20. Lu, T., Li, S., Fang, L. et al. (2016). Set-to-set distance-based spectral spatial classification of hyperspectral images. *IEEE Transactions on Geoscience and Remote Sensing* 54 (12): 7122–7134.
21. Lu, T., Li, S., Fang, L. et al. (2017). From subpixel to superpixel: a novel fusion framework for hyperspectrakl image classification. *IEEE Transactions on Geoscience and Remote Sensing* 55 (8): 4398–4411.
22. Guo, X., Huang, X., Zhang, L. et al. (2016). Support tensor machines for classification of hyperspectral remote sesning imagery. *IEEE Transactions on Geoscience and Remote Sensing* 54 (6): 3248–3264.
23. Ma, X., Wang, H., and Wang, J. (2016). Semisupervised classification of hyperspectral image based on multi-level labeling and deep feature learning. *JSPRS Journal Photogrammetry and Remote Sensing* 120: 99–107.
24. Camps-Valls, G., Gomez-Chova, L., Munoz-Mari, J. et al. (2006). CKs for hyperspectral image classification. *IEEE Geoscience and Remote Sensing Letters* 3 (1): 93–97.
25. Gurram, P. and Kwon, H. (2013). Contextual SVM using Hilbert space embedding for hyperspectral classification. *IEEE Geoscience and Remote Sensing Letters* 10 (5): 1031–1035.
26. Peng, J., Zhou, Y., and Chen, C. (2015). Region-kernel-based supprt vector machines for hyperspectral image classification. *IEEE Transactions on Geoscience and Remote Sensing* 53 (9): 4810–4824.
27. Sun, S., Zhong, P., Xiao, H., and Wang, R. (2015). Active learning with Gaussian process classifier for hyperspectral image classification. *IEEE Transactions on Geoscience and Remote Sensing* 53 (4): 1746–1760.
28. Pullanagari, R.R., Kereszturi, G., Yule, I.J., and Ghamisi, P. (2017). Assessing the performance of multiple spectral-spatial features of a hyperspectral image for classification of urban land cover classes using support vector machines and artificial neural network. *Journal of Applied Remote Sensing* 11 (2): 026009-1–026009-21.
29. Lu, X., Yuan, Y., and Zheng, X. (2017). Joint dictionary learning for multispectral change detection. *IEEE Transactions on Cybernetics* 47 (4): 884–897.

30. Lu, X., Zheng, X., and Yuan, Y. (2017). Remote sensing scene classification by unsupervised representation learning. *IEEE Transactions on Geoscience and Remote Sensing* 55 (9): 5148–5157.
31. Hossain, M.A., Jia, X., and Benediktsson, J.A. (2016). One-class oriented feature selection and classification of heterogeneous remote sensing images. *IEEE Journal of Selected Topics in Applied Earth Observations and Remote Sensing* 9 (4): 1606–1612.
32. Lu, X., Chen, Y., and Li, X. (2018). Hierarchical recurrent neural hashing for image retrieval with hierarchical convolutional features. *IEEE Transactions on Image Processing* 27 (1): 106–120.
33. Lu, X., Zhang, W., and Li, X. (2018). A hybrid sparsity and distance-based discrimination detector for hyperspectral images. *IEEE Transactions on Geoscience and Remote Sensing* 56 (3): 1704–1717.
34. Lu, X., Wang, B., Li, X., and Zheng, X. (2017). Exploring models and data for remote sensing image caption generation. *IEEE Transactions on Geoscience and Remote Sensing* 55 (9): 5148–5157.
35. Chang, C.-I (2019). A statistical detection theory approach to hyperspectral image classification. *IEEE Transactions on Geoscience and Remote Sensing* 57 (4): 2057–2074.
36. Bishop, C.H. (2006). *Pattern Recognition and Machine Learning*. New York: Springer.
37. Chang, C.-I (2016). *Real Time Progressive Hyperspectral Image Processing: Endmember Finding and Anomaly Detection*. New York: Springer.
38. Harsanyi, J.C. (1993). Detection and classification of subpixel spectral signatures in hyperspectral image sequences. Department of Electrical Engineering. University of Maryland, Baltimore County, MD.
39. Chang, C.-I (2003). *Hyperspectral Imaging: Techniques for Spectral Detection and Classification*. New York, NY: Kluwer Academic/Plenum Publishers.
40. Chang, C.-I (2013). *Hyperspectral Data Processing: Algorithm Design and Analysis*. Wiley.
41. Frost, O.L. III (1972). An algorithm for linearly constrained adaptive array processing. *Proceedings of the IEEE* 60: 926–935.
42. Chang, C.-I (2002). Target signature-constrained mixed pixel classification for hyperspectral imagery. *IEEE Transactions on Geoscience and Remote Sensing* 40 (2): 1065–1081.
43. Xue, B., Yu, C., Wang, Y. et al. (2017). A subpixel target approach to hyperspectral image classification. *IEEE Transactions on Geoscience and Remote Sensing* 55 (9): 5093–5114.
44. Yu, C., Xue, B., Wang, Y. et al., (2017). Multi-Class constrained background suppression approach to hyperspectral image classification. *2017 IEEE/GRSS International Geoscience and Remote Sensing Symposium* (IGARSS), (23–28 July 2017), Fort Worth, Texas, USA.
45. Yu, C., Xue, B., Song, M. et al. (2018). Iterative target–constrained interference-minimized classifier for hyperspectral image classification. *IEEE Journal of Selected Topics in Applied Earth Observation and Remote Sensing* 11 (4): 1095–1117.

# Index



*Advances in Hyperspectral Image Processing Techniques*, First Edition. Edited by Chein-I Chang.

Printed and bound by CPI Group (UK) Ltd, Croydon, CR0 4YY
17/09/2026
14964324-0005